Lecture Notes in Networks and Systems 629

The series "Lecture Notes in Networks and Systems" publishes the latest developments in Networks and Systems—quickly, informally and with high quality. Original research reported in proceedings and post-proceedings represents the core of LNNS.

Volumes published in LNNS embrace all aspects and subfields of, as well as new challenges in, Networks and Systems.

The series contains proceedings and edited volumes in systems and networks, spanning the areas of Cyber-Physical Systems, Autonomous Systems, Sensor Networks, Control Systems, Energy Systems, Automotive Systems, Biological Systems, Vehicular Networking and Connected Vehicles, Aerospace Systems, Automation, Manufacturing, Smart Grids, Nonlinear Systems, Power Systems, Robotics, Social Systems, Economic Systems and other. Of particular value to both the contributors and the readership are the short publication timeframe and the world-wide distribution and exposure which enable both a wide and rapid dissemination of research output.

The series covers the theory, applications, and perspectives on the state of the art and future developments relevant to systems and networks, decision making, control, complex processes and related areas, as embedded in the fields of interdisciplinary and applied sciences, engineering, computer science, physics, economics, social, and life sciences, as well as the paradigms and methodologies behind them.

Indexed by SCOPUS, INSPEC, WTI Frankfurt eG, zbMATH, SCImago.

All books published in the series are submitted for consideration in Web of Science.

For proposals from Asia please contact Aninda Bose (aninda.bose@springer.com).

Mohamed Ben Ahmed ·
Anouar Abdelhakim Boudhir ·
Domingos Santos · Rogerio Dionisio ·
Nabil Benaya
Editors

Innovations in Smart Cities Applications Volume 6

The Proceedings of the 7th International Conference on Smart City Applications

Springer

Editors
Mohamed Ben Ahmed
Faculty of Science and Technology,
Computer Engineering Department
Abdelmalek Essaadi University
Tangier, Morocco

Anouar Abdelhakim Boudhir
Faculté des sciences et techniques de Tanger
Université Abdelmalek Essaâdi
Tangier, Morocco

Domingos Santos
Interdisciplinary Centre of Social Science
Polytechnic Institute of Castelo Branco
Castelo Branco, Portugal

Rogerio Dionisio
DiSAC-Digital Services, Applications
and Content
Polytechnic Institute of Castelo Branco
Castelo Branco, Portugal

Nabil Benaya
FST Al-Hoceima, Physics Department
Abdelmalek Essaadi University
Al-Hoceima, Morocco

ISSN 2367-3370 ISSN 2367-3389 (electronic)
Lecture Notes in Networks and Systems
ISBN 978-3-031-26851-9 ISBN 978-3-031-26852-6 (eBook)
https://doi.org/10.1007/978-3-031-26852-6

This Springer imprint is published by the registered company Springer Nature Switzerland AG
The registered company address is: Gewerbestrasse 11, 6330 Cham, Switzerland

Committees

Conference Chair

Rogério Dionisio	Polytechnic Institute Castelo Branco, Portugal

Conference Co-chairs

Mohamed Ben Ahmed	FST, Tangier, UAE University, Morocco
Anouar Boudhir Abdelhakim	FST, Tangier, UAE University, Morocco
İsmail Rakıp Karaş	Karabuk University, Turkey
Bernadetta Kwintiana Ane	University of Stuttgart, Germany

Conference Steering Committee

Rogério Dionisio	Polytechnic Institute Castelo Branco, Portugal
Domingos Santos	Polytechnic Institute Castelo Branco, Portugal
İsmail Rakıp Karaş	Karabuk University, Turkey
Alias Abdul Rahman	Universiti Teknologi Malaysia
Mohamed Wahbi	EHTP Casablanca, Morocco
Mohammed Bouhorma	FST, Tangier UAE University, Morocco
Chaker El Amrani	FST, Tangier UAE University, Morocco
Bernard Dousset	UPS, Toulouse, France
Rachid Saadane	EHTP Casablanca, Morocco
Ali Youness	FS, Tetouan, Morocco

Local Organizing Committee

Domingos Santos	Polytechnic Institute Castelo Branco, Portugal
Rogério Dionisio	Polytechnic Institute Castelo Branco, Portugal
Fernando Ribeiro	Polytechnic Institute Castelo Branco, Portugal
José Metrôlho	Polytechnic Institute Castelo Branco, Portugal
Ana Vaz Ferreira	Polytechnic Institute Castelo Branco, Portugal
João Neves	Polytechnic Institute Castelo Branco, Portugal
Cristina Calmeiro	Polytechnic Institute Castelo Branco, Portugal
Rogério Ribeiro	Polytechnic Institute Castelo Branco, Portugal

Nuno Caseiro	Polytechnic Institute Castelo Branco, Portugal
Rita Tavares	Polytechnic Institute Castelo Branco, Portugal

Technical Program Committee

Abdel-Badeeh M. Salem	Ain Shams University, Egypt
Abdullah Elen	Bandirma Onyedi Eylül University, Türkiye
Abdullah Emin Akay	Bursa Technical University, Türkiye
Abdurrahman Eymen	Erciyes University, Türkiye
Accorsi, Riccardo	Bologna University, Italy
Adib Habbal	Karabuk University, Türkiye
Adnan Alajeeli	Karabuk University, Türkiye
Aftab Ahmed Khan	Karakoram International University, Pakistan
Ahmad S. Almogren	King Saud University, Saudi Arabia
Ahmed Kadhim Hussein	Babylon University, Iraq
Alabdulkarim Lamya	King Saud University, Saudi Arabia
Alghamdi Jarallah	Prince Sultan University, Saudi Arabia
Ali Jamali	Universiti Teknologi Malaysia
Ali Jamoos	Al-Quds University, Palestine
Alias Abdul Rahman	Universiti Teknologi Malaysia
Aliihsan Sekertekin	Cukurova University
Ana Paula Silva	Polytechnic Institute of Castelo Branco, Portugal
Ana Ferreira	Polytechnic Institute Castelo Branco, Portugal
Anabtawi Mahasen	Al-Quds University, Palestine
Anton Yudhana	Universitas Ahmad Dahlan, Indonesia
Arlindo Silva	Polytechnic Institute of Castelo Branco, Portugal
Arif Çağdaş Aydinoglu	Gebze Technical University, Türkiye
Arioua Mounir	UAE, Morocco
Arturs Aboltins	Technical University of Riga, Latvia
Assaghir Zainab	Lebanese University, Lebanon
Aydın Üstün	Kocaeli University, Türkiye
Bahadır Ergun	Gebze Technical University, Türkiye
Barış Kazar	Oracle, USA
Bataev Vladimir	Zaz Ventures, Switzerland
Behnam Atazadeh	University of Melbourne, Australia
Bessai-Mechmach Fatma Zohra	CERIST, Algeria
Beyza Yaman	Dublin City University, Ireland
Biswajeet Pradhan	University of Technology Sydney, Australia
Berk Anbaroğlu	Hacettepe University, Türkiye
Burhan Selcuk	Karabuk University, Türkiye
Bulent Bayram	Yildiz Technical University, Türkiye

Carlos Cambra	Universidad de Burgos, Spain
Caner Ozcan	Karabuk University, Türkiye
Caner Güney	Istanbul Technical University, Türkiye
Cláudio Barradas	Polytechnic Institute of Santarém, Portugal
Cristina Calmeiro	Polytechnic Institute Castelo Branco, Portugal
Cumhur Şahin	Gebze Technical University, Türkiye
Daniel Ferreira de Azevedo	Polytechnic Institute of Viseu, Portugal
Damir Žarko	Zagreb University, Croatia
Darko Stefanovic	University of Novi Sad, Serbia
Dominique Groux	UPJV, France
Edward Duncan	The University of Mines & Technology, Ghana
Eehab Hamzi Hijazi	An-Najah University, Palestine
Eftal Şehirli	Karabuk University, Türkiye
El Mhouti Abderrahim	FS, Tetouan UAE University, Morocco
El Hebeary Mohamed Rashad	Cairo University, Egypt
Elif Sertel	Istanbul Technical University, Türkiye
Emre Yücer	Karabuk University, Türkiye
Emrullah Sonuç	Karabuk University, Türkiye
Enrique Arias	Castilla-La Mancha University, Spain
Fatmagül Kılıç Gül	Yıldız Technical University, Türkiye
Fátima Domingues	Instituto de Telecomunicações, Portugal
Faiza DIB	FSTH, UAE University, Morocco
Fernando Velez	University of Beira Interior, Portugal
Fernando Reinaldo	Polytechnic Institute Castelo Branco, Portugal
Ferhat Atasoy	Karabuk University, Türkiye
Filip Biljecki	National University of Singapore
Florent Bruguier	University of Montpellier, France
Füsun Balık Şanlı	Yıldız Technical University, Türkiye
Francesc Anton Castro	Technical University of Denmark
Ghulam Ali Mallah	Shah Abdullatif University, Pakistan
Habibullah Abbasi	University of Sindh, Pakistan
Helder Fontes	INESC-TEC, Porto, Portugal
Hakan Kutucu	Karabuk University, Türkiye
Hanane Reddad	USMS University, Morocco
Hande Demirel	İstanbul Technical University, Türkiye
Hugo Marques	Polytechnic Institute Castelo Branco, Portugal
Huseyin Bayraktar	General Directorate of GIS, Türkiye
Hüseyin Pehlivan	Gebze Technical University, Türkiye
Huseyin Topan	Bulent Ecevit University, Türkiye
Huseyin Zahit Selvi	Konya Necmettin Erbakan University
İbrahim Baz	İstanbul Ticaret University, Turkiye
İlhami Muharrem Orak	Karabuk University, Türkiye

Ilker Türker	Karabuk University, Türkiye
Indubhushan Patnaikuni	RMIT - Royal Melbourne Institute of Technology, Australia
İsa Avcı	Karabuk University
Ismail Rakip Karas	Karabuk University, Türkiye
Ismail Büyüksalih	Bimtaş A.Ş., Türkiye
Ivin Amri Musliman	Universiti Teknologi Malaysia
J. Amudhavel	VIT Bhopal University, Madhya Pradesh, India
Jaime Lioret Mauri	Polytechnic University of Valencia, Spain
José Javier Berrocal Olmeda	Universidad de Extremadura, Spain
João Neves	Polytechnic Institute Castelo Branco, Portugal
João Lourenço Marques	Universidade de Aveiro, Portugal
José Benjamim Fonseca	University of Trás-os-Montes and Alto Douro, Portugal
José Cabral	University of Minho, Portugal
José Metrôlho	Polytechnic Institute Castelo Branco, Portugal
Jus Kocijan	Nova Gorica University, Slovenia
Kadir Ulutaş	Karabuk University
Kasım Ozacar	Karabuk University
Khoudeir Majdi	IUT, Poitiers University, France
Labib Arafeh	Al-Quds University, Palestine
Loncaric Sven	Zagreb University, Croatia
Lotfi Elaachak	FSTT, UAE, Morocco
Luis Farinha	Polytechnic Institute Castelo Branco, Portugal
Luís Quinta-Nova	Polytechnic Institute Castelo Branco, Portugal
Mademlis Christos	Aristotle University of Thessaloniki, Greece
Maria Joao Simões	Universidade da Beira Interior, Portugal
Mehmet Akbaba	Karabuk University, Türkiye
Mete Celik	Erciyes University, Türkiye
Miranda Serge	Nice University, France
Mónica Costa	Polytechnic Institute of Castelo Branco, Portugal
Mohamed El Ghami	University of Bergen, Norway
Mohammad Sharifikia	Tarbiat Modares University, Iran
Mousannif Hajar	Cadi Ayyad University, Morocco
Mufit Cetin	Yalova University, Türkiye
Muhamad Uznir Ujang	Universiti Teknologi Malaysia
Mike Horhammer	Oracle, USA
Muhammad Imzan Hassan	Universiti Teknologi Malaysia
Muhammed Kamil Turan	Karabuk University, Türkiye
Murat Yakar	Mersin University, Türkiye
Murat Lüy	Kırıkkale University, Türkiye
Mustafa Akgul	Istanbul University, Türkiye

Keynotes

Responsible and Resilient Smart Cities; The Model Proposed for African Smart Cities: Case Study of Cape Verde

Loide Monteiro

CEO of Smart City Foundation, Cape Verde

Abstract. The sustainable development challenges for small archipelagic cities are exponentially bigger when compared to the large metropolises on the continent. One of the biggest problems of these small cities face is related to disorderly urban expansion. From night to day, new peripheral neighborhoods are borne, bringing social and urban problems of an unplanned expansion. Hence, it is essential that these cities leverage their progress on sustainable pillars, creating smart solutions tailored to their challenges.

The Extended Information Systems Success Measurement Model

Teodora Lolic

University of Novi Sad, Serbia

Abstract. Technological innovations such as the introduction or upgrade of information systems in work environments are, globally, one of the most common changes people face daily in their workplaces. Organizational management is consequently seeking to observe and track the return of the investment; therefore, it is necessary to enable the stakeholders to adequately measure the success of information systems. Information systems success has been an emerging research topic for the past decade and still is. Many theoretical models and frameworks have been developed with the aim of giving the most accurate measures for IS success. However, according to the current state of the art, no model for measuring the success of information systems observing the factors that emphasize the importance of people's reactions to changes in the work environment has been found.

How Can Sensing and Communication Make Cities Smart?

Susana Sargento

University of Aveiro, Portugal

Biography: Susana Sargento is a Full Professor in the University of Aveiro and a senior researcher in the Institute of Telecommunications, where she is leading the Network Architectures and Protocols (NAP) group. She received her PhD in 2003 in Electrical Engineering in the University of Aveiro, being a visiting student at Rice University in 2000 and 2001. She joined the Department of Computer Science of the University of Porto between 2002 and 2004, and she was a Guest Faculty of the Department of Electrical and Computer Engineering from Carnegie Mellon University, USA, in August 2008, where she performed Faculty Exchange in 2010/2011. She was the founder of Veniam, which builds a seamless low-cost vehicle-based internet infrastructure, and she was the winner of the 2016 EU Prize for Women Innovators.

2050 Now – Creating the Future Sustainable Cities Today

Andy Van Den Dobbelsteen

TU Delft, The Netherlands

Biography: Andy van den Dobbelsteen is full professor of Climate Design & Sustainability and Sustainability Coordinator of TU Delft. Next to this he is Principal Investigator with the AMS Institute in Amsterdam. He publishes a lot and delivers lectures nationally and internationally. At TU Delft Andy teaches students how to design sustainable cities, buildings and technology. He was faculty advisor to three successful TU Delft Solar Decathlon teams and is responsible for the free online course Zero-Energy Design, winner of the edX Online Prize 2020. His short film Energy Slaves won the Dutch Oscar for commissioned films. Together with his team, Andy has led and conducted many national and international research projects around energy transition, climate adaptation, circularity and other sustainability themes.

For his work in the domain of sustainability, Andy became Knight in the Order of the Dutch Lion. In 2019, he received the Academic Society Award from the Dutch Royal Institute of Engineers for the way he communicates scientific knowledge to the general public.

Smart City as a Distributed Platform: Towards a System for Citizen-Oriented Management

Juan Corchado

University of Salamanca, Spain

Abstract. This keynote will present success stories regarding especially smart cities. All these fields require the development of interactive, reliable and secure systems which we are capable of building thanks to current advances. Deepint.net, a tool developed by DCSc and BISITE, will be presented. Several use cases of intelligent systems will be presented, and it will be analyzed how the different processes have been optimized by means of tools that facilitate decision making.

Contents

Solving the Dynamic Ambulance Relocation and Dispatching Problem Using a Novel Metaheuristic

Maryam Alami Chentoufi(✉), Rachid Ellaia, and Mohamed Lamrabet

Laboratory of Study and Research in Applied Mathematics (LERMA), Engineering for Smart and Sustainable Systems Research Center (E3S), Mohammadia School of Engineers, Mohammed V university in Rabat, Ibn Sina Avenue, BP. 765, Agdal, Rabat, Morocco
Maryam.lami12@gmail.com

Abstract. Proper deployment and a good management of emergency vehicles leads to shorter response times to an emergency call and thus reduces suffering, mortality and adverse health consequences for patients. In this context, we proposed a dynamic management of emergency vehicles in Morocco through the SOS accident system. Indeed, decision-making in these emergency situations remains a difficult task, given the number of variables that come into play on the one hand and the dynamism and unpredictability of some of them on the other. The general field in which the present project falls is in Vehicle Routing Problem (VRP), which involves building routes at minimum cost so that vehicles can visit exactly once each customer who has been geographically distributed. We are going to be more particularly interested in a management of ambulances in real time since it will be done on the basis of information which will become accessible over time, which leads to the resolution of the problem of the dynamic elaboration of vehicle routes with time window (DVRPTW). To carry out the current project, we first developed a mathematical model of the response time of emergency vehicles, and calculated the travel time of the vehicles. Then, we used a novel metaheuristic based on Passing Vehicle Search and Tabu search algorithm for the optimization part, which allowed us to take into account most of the constraints that appear in the practice and which constrain emergency vehicles during their daily missions.

Keywords: Vehicle routing problem · Real-time emergency response · Vehicle assignment · Vehicle allocation · PVS algorithm · Tabu search · Emergency vehicles management

1 Introduction

With the construction of urbanization and the increasing traffic vehicles, the frequency and impact of traffic accidents are intensifying. The research on emergency rescue is getting more and more attention. When traffic accident occurs, rescue timeliness is the key to emergency rescue. Reasonable arrangement of emergency vehicle path can avoid congestion and shorten the travel time, so that the accident loss can be reduced.

M. Ben Ahmed et al. (Eds.): SCA 2022, LNNS 629, pp. 1–23, 2023.
https://doi.org/10.1007/978-3-031-26852-6_1

Emergency systems operate in highly dynamic environments and are subject to fairly frequent disruptions. These disruptions are due to the dynamic arrival of service requests throughout the day. This project is therefore part of a context where we are seeking to define a dynamic distribution strategy for emergency vehicles to optimize response time in a changing environment while making the most of the information and technologies available, more specifically telecommunications tools and distributed architectures. The vehicle distribution strategy must adapt to the variations observed in the environment and ensure the best possible compromise between the quality of the solution provided and the computing time required. We present in the second section of the state of art the situation studied in a specific context relating to road safety in Morocco which has been discussed in several scientific articles whether at the level of analysis of statistical data of road traffic mortality in Morocco [1], or at the level of study on the evolution of road safety in Morocco [2] and the quality of its management system [3]. In Sect. 3, we modeled the emergency vehicle response time, which is the period between receiving the call and arriving at the incident scene of an ambulance. Next, we will find the optimal path between the accident site and each of the emergency vehicles available, while calculating the travel time generated by our model for all vehicles. We then present a novel metaheuristic based on Passing Vehicle search and Tabu search algorithm to identify the emergency vehicle that will have the shortest travel time. Section 4 will be devoted to the discussion of the various results obtained.

2 State of Art and Related Works

2.1 The Vehicle Routing Problem

The vehicle routing problem is a combinatorial optimization and operations research problem; it falls under the category of transportation problems which is based on the fact that a fleet of vehicles must be routed to visit a set of customers at a minimum cost, subject to vehicle constraint and driving time constraint. In the case of medical emergency management, this problem includes dynamic elements, and unknowns that will be managed by graphical information systems (GIS) and global positioning systems (GPS), and consequently manage vehicle routes.

To solve Dynamic VRPs, several works based on metaheuristics have been proposed over the last decade, including:

- Among the most cited articles in the DYNAMIC VRPS literature, the study by Bent and al [4], which presents a multiple plan approach (MPA) and another multiple scenario approach (MSA).
- Novoa and al [5] adopted a dynamic approach based on the rollout algorithm [6] to solve static VRP with stochastic demands. In this problem the customer demands follow a known probabilistic distribution and the exact customer demand is only revealed at the time of delivery.
- The work done by Potvin and Benyahia, where they associated each vehicle with a vector of attribute values reflecting the effect of inserting new service requests into its current route, while using the genetic algorithm (GA) [7]. On a similar problem Cheung and al [8] also applied their own implementation of the genetic algorithm.

This time the problem is characterized by tough windows of time when mail collection and delivery must take place. Always with the genetic algorithm, but this time on another version of the Dynamic VRPs, Hemert and Poutre [9] introduced the notion of successful regions, which are poles, where there are more likely potential customers. The problem addressed this time is a problem of collecting loads from customers and delivering to a single central depot.
- The article of Azi and al [10] whose idea is to set the routes, from the start, of the vehicles from the depot and the decision to be taken for dynamic requests only concerns their acceptances or rejection according to their proximity to the vehicles. Routes already planned. This work was based on the Neighborhood Search (NS) algorithm.
- In their article, Montemmani and al [11] have broken down the planning period into sub-periods during which a re-optimization takes place which takes into account new dynamic queries. The static problem is solved using the ant colony system.

In this context, several projects have been launched to create tools to optimize, monitor and manage emergency interventions. Among these projects, we can cite the MERCURE project, which is based on a monitoring system that optimizes resources, travel times and takes care of all the constraints relating to the medical field such as the level of emergency, pathology, medical skills, localization..., which allows to quickly and dynamically solve the problem of management of emergency vehicles with a high level of precision, and this thanks to the most recent models and algorithms seated of emergency vehicles with a high level of precision, and this thanks to the most recent models and algorithms seated of emergency vehicles such as the global location coverage model (LSCM), which aims to minimize the number of ambulances needed to cover all demand points. As well as the Maximum Coverage Locating Problem (MCLP) model, which aims to maximize population coverage subject to the limited availability of ambulances. However, to get closer and closer to reality, several dynamic models are starting to emerge to periodically update ambulances, which do not depend only on sophisticated system technologies, but also on the availability of search heuristics. Fast and accurate, allowing variations in travel time during the day to be included in the shortest path calculations.

As part of our project, we will build on the projects and models launched previously, to establish our own model that shapes the vehicle dispatch system, considering communications from the transportation network, and the vehicle dispatch center, which minimizes response times for emergency vehicles, so that they interact more quickly with their environment, by using the hybridization of the algorithms we have just mentioned.

2.2 Moroccan Road Safety Policy

The road safety policy in Morocco has seen an important turning point in the management of the problem of traffic accidents. In order to create a general mobilization and federate all efforts around the subject, the Government has put in place an Integrated National Security Strategy covering the period 2017–2026. It is part of a long-term vision aimed at the development of responsible behavior and the emergence of safer roads, the objective being to reduce the number of deaths due to accidents by 50 pc by 2026, or less than 1,900 dead and not more than 3,000 in 2020.

Morocco has adopted a new road safety strategy for 2026 which sets an ambitious and quantifiable target for reducing road fatalities by half compared to its current level, as a solution of this issue.

In this context, the idea of the proposed project focuses on four axes:

- Help provided to victims of traffic accidents through the SOS accident application.
- Management of accidents on the road.
- Assessment and decision support in terms of accident risks.
- Post accident care.

2.2.1 Help Provided to Victims of Traffic Accidents through the SOS Accident Application

First, we are going to set up an application (Android and iOS) that would allow Internet users to be quickly put in touch with the nearest security post to report an emergency situation, especially in the event of an accident (Fig. 1). Indeed, the SOS accident application would put the driver in touch with the medical regulation center, by also providing the geolocation data of the telephone (Fig. 2) and the data entered during the creation of his account on the application, in order to allow rapid and appropriate emergency response. Overview of the use of the application:

After installing the application, the user is prompted to create an account and provide the following information (Fig. 3):

- Last name *;
- First name *;
- Address *;
- CIN *;
- Phone number *;
- Blood groups, Rhesus; (Possibility to import the receipt).
- Allergy: Yes / No (If Yes, give the user the possibility to import the proof).
- History of illness; Yes / No (If Yes give the user the possibility to import the medical file).
- etc…

This falls within the framework of the ambition: "Assistance provided to victims of traffic accidents". Indeed, the effectiveness of road accident rescue services is often a question of time. People who receive prompt and adequate treatment are more likely to survive. In order to improve the quality of assistance (support on the scene and during the transfer of victims).

2.2.2 Back Office Application Dedicated to Emergency Personnel

Another back office application dedicated to emergency personnel is offered as part of this project, it allows:

Management of internet users' requests through (Fig. 4):

- The consultation on the map of the geographical position of ambulances and that communicated by the requesters for help (from the SOS accident application).
- Communication of the command to the ambulance and transfer of the GPS position to reach the victim and take charge.
- Entering requests received by phone.
- Search (by Application No., by region, by CIN).
- The update…,
- etc.

The follow-up of emergency requests through:

- The implementation of indicators and dashboards for piloting and monitoring accidents.
- The creation of a repository of data for the management of indicators related to road accidents.

Principle of vehicle geolocation: The geolocation of vehicles makes it possible to know at any time, in near real time, where the vehicles equipped with geolocation devices are located.

The GPS geolocation system connects to the satellite positioning network to know its location. The data is then transmitted to the driver and / or to the operator of the medical regulation center, generally by mobile cellular networks. Operation of the emergency vehicle geolocation system Emergency vehicle geolocation systems (ambulances) work with two key elements:

- GPS trackers connected to the satellite positioning network and installed on all vehicles in the fleet, allowing emergency vehicles to be geolocated,
- Fleet management software, connected to the trackers, generally by the mobile cellular network, which indicates the positioning of vehicles equipped with these trackers.

Overview of the use of the application:

- At the level of the SOS accident application, the user connects to the application and makes a request for the nearest ambulance to join him immediately,
- At the Back office level, the operator consults the geographical position of the ambulances and that of the user on the map to find the nearest available vehicle,
- The operator communicates the command to the ambulance and transfers the route to reach the victim and take charge of him to his vehicle's GPS.

Post-accident management: Among the objectives of this project is to address the priority of emergency vehicles in a traffic light intersection. In order to facilitate traffic in the declared area (optimization of red lights) The emergency vehicle detection procedure is based on GPS technology.

2.2.3 Statistical Data Analysis

The application allows the display of statistics relating to road safety (Fig. 5):

- Category of users most exposed to the risk: vulnerable users (killed, seriously injured, and slightly injured.)
- Category of users most exposed to the risk: vulnerable users (Trucks, Buses, Cars, Pedestrians, Passenger cars, Cyclists and mopeds).
- Vehicles involved by category, Vehicles involved by age category.
- Geographic variation of traffic accidents.
- Variation by month of traffic accidents.
- When several accidents are reported in the same accident accumulation zone, the latter is declared a black zone: therefore, a group of experts (commission) should travel to study the level of safety in this conflict point.

The application allows the display of statistics relating to road safety (Fig. 5):

- Category of users most exposed to the risk: vulnerable users (killed, seriously injured, slightly injured.)
- Category of users most exposed to the risk: vulnerable users (Trucks, Buses, Cars, Pedestrians, Passenger cars, Cyclists and mopeds).
- Vehicles involved by category, Vehicles involved by age category.
- Geographic variation of traffic accidents.
- Variation by month of traffic accidents.
- When several accidents are reported in the same accident accumulation zone, the latter is declared a black zone: therefore, a group of experts (commission) should travel to study the level of safety in this conflict point.

2.2.4 Assessment and Decision Support in Terms of Accident Risks

Analyzes will be carried out by the commission in order to establish a local diagnosis of the problem of accidents in an accident accumulation zone (Fig. 6).

2.2.5 Mobile App Preview

Fig. 1. Authenticating users with Fingerprint (TouchID)

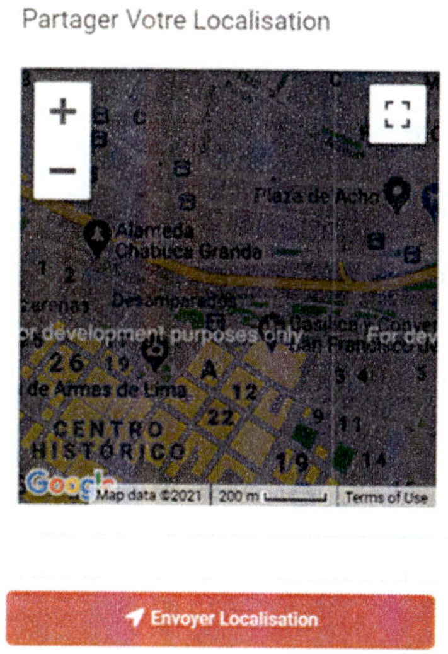

Fig. 2. Sharing the real-time user location

2.2.6 Web App Preview

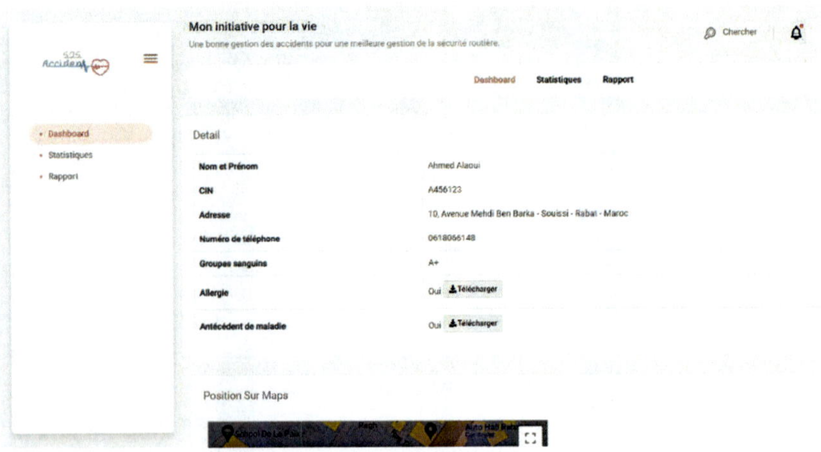

Fig. 3. Patient information web form

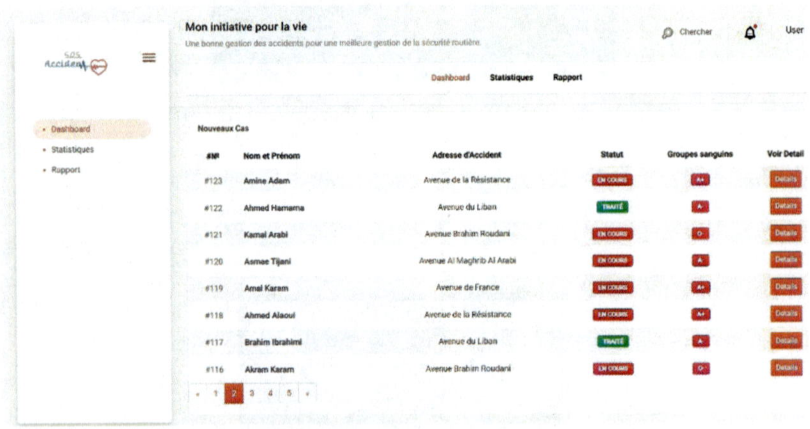

Fig. 4. The dashboard offered by the proposed web application

Fig. 5. Benchmarking safety data

Fig. 6. Accident accumulation zone

3 Dynamic Management of Emergency Vehicles

3.1 The Response Time Model of the Emergency Vehicle

Our goal is to model the response time of the emergency vehicle. This time corresponds to the period between the reception of the call and the arrival at the scene of the incident of an ambulance. Therefore, we will first conduct an approximate approach to the speed of the vehicle, which is based on the different forces that apply to it. In this approach, we will rely on a study of the physical dynamics of the vehicle which will provide us with a model of the travel time where the emergency vehicle is motorized (we consider for example that the driving wheels are the front wheels), and which takes into account the influence of gravity, air, vehicle engine and ground reaction.

3.1.1 Expression of Travel Time

After any calculation done, we end up with the following differential equation in speed:

$$\left(M_T a + 4\frac{J}{a}\right)\frac{dv}{dt} + h.av^2 = 2\Gamma - M_T.g(bsin(\alpha) + acos(\alpha)) \quad (1)$$

It is indeed a homogeneous equation in terms of dimensions. A simple resolution of this differential equation leads us to the following relation which links the distance travelled D by the emergency vehicle and the travel time T:

$$T = \frac{1}{\lambda\beta} log(e^{\lambda^2}\frac{D}{2}.\left(1 + \sqrt{1 + 2.e^{-2\lambda^2 D}}\right) \quad (2)$$

With: $\lambda^2 = \frac{h.a}{M_T a + \frac{4J}{a}}$ and $\beta^2 = \frac{2\Gamma - M_T.g(bsin(\alpha)+acos(\alpha))}{M_T a + \frac{4J}{a}}$.
Γ: Torque monitor.
M_T: Total mass of the vehicle.
α: Angle of inclination of the road relative to the horizontal.
a: Wheel radius.
b: Distance between the line of action of $(\overrightarrow{R_N})$ and the vertical passing through the center C.
J: Moment of inertia of a wheel with respect to its axis.
x_0: Abscissa of the vehicle departure.
h: Coefficient of proportionality relative to air resistance.

3.2 Travel Time Calculation

The first step is to find the optimal path between the place of the accident and each of the emergency vehicles available, we implemented in python a function which allows calculating the distance of the shortest route between the destination and the "origin". And then, we defined a function shortest distance, which will calculate the distance of the shortest path between the place of the accident and the place of the vehicle. Next, we will use this function to calculate the estimated travel time for each emergency vehicle. To do this, we defined a travel time function which will return a list of the travel times of all the vehicles. This function will be modified later to take into account the influence of traffic in our model. Now, we just calculate the travel time generated by our preliminary model already established.

3.3 Optimization

3.3.1 Travel Time Calculation

The emergency services have to respect the two major constraints:

They are highly dynamic: most or all requests are unknown at the beginning and we have no information about their arrival time;

The response time must be very low because lives can be in danger.

To answer these criteria, we are going to extend the well-known VRP (Vehicle Routing Problem) to Dynamic Vehicle Rounting Problem with Time Window (DVPRTW).

3.3.2 VRP

The static VRP is defined on an oriented graph G=(V, A) in which G=(V, A) in which $V=v_0, v_1, \ldots, v_N$ is the set of vertices and where vertex v0 is a depot at which are based m identical vehicles of capacity Q, while the remaining N vertices represent customers.

The minimization of the total route length can be defined by the following optimization problem:

$$Min \sum_{k=1}^{m} \sum_{(i,j)\in A} C_{ij} X_{ijk} C_{ij} \tag{3}$$

is the travel cost on the edge (v_i,v_j) under constraints (respect of vehicle capacity, assignment of each customer to a unique vehicle, respect of the number of vehicles available, Continuity of the road "a vehicle entering the road must exit").

is the travel cost on the edge (v_i, v_j) under constraints (respect of vehicle capacity, assignment of each customer to a unique vehicle, respect of the number of vehicles available, Continuity of the road "a vehicle entering the road must exit").

3.3.3 DVRPTW

In the static VRP, vehicles must be routed to visit a set of customers at minimum cost, and we suppose that all orders for all customers are known in advance. However, when all or part of the data remains unknown and is dynamically revealed during service, the planned routes can be predefined based on the new information required. By taking into account these variations observed in the environment (accidents, congestion, arrival of a new urgent call, etc.) as well as the quality-time issue (managing the limited fleet of emergency vehicles, satisfactory coverage of the territory served), we are then faced with a dynamic VRP.

To switch from a VRP to a DVRPTW, we add two constraints:

It is very likely that new patient requests and cancelled requests may occur all the time, even after the optimization process has started. The dynamism consists in receiving several requests during the evolution of the simulation and it proves to be so much useful to reduce the costs in vehicle routing problems. The date when the request i arrives is noted gi as the generation date of request i:

The time windows constraint which consists in having 2 time limits associated with each requesti: [ai, bi]. The vehicle must start the customer service before bi, but if any of them arrives at customer i before ai; it must wait: So the smaller the time window of are quest is the harder will it be to find a good insertion place in a vehicle road.

3.3.4 A Novel Metaheuristic for VRP and VRPTW Problem

Our resolution of both VRP and VRPTW is based on hybridization between the Passing Vehicle Search algorithm and the Tabu search algorithm.

Passing vehicle search algorithm (PVS), considers the mathematics of the passage of a vehicle on a two-lane highway [12]. Similar to other metaheuristic methods, PVS is a population-based method that requires starting an initial set of solutions and then seeking the optimal solution by following the mathematical characteristics of vehicles that pass on a two-lane highway.

Building upon some of his previous work, Fred Glover proposed in 1986 a new approach, which he called tabu Search, to allow LS methods to overcome local optima. (In fact, many elements of this first TS proposal, and some elements of later TS elaborations, were introduced in Glover, 1977 [13], including short term memory to prevent the reversal of recent moves, and longer term frequency memory to reinforce attractive components.).

The Proposed Algorithm:

Step 1: Define the parameters such as population size (PS), stopping criteria (Number Of Generations (NOG), Maximum function evaluations (FE), and error) number of design variables (DV), bounds on design variables ((LB),(UB)).

Define the function for optimization:

Minimize f(x)

subject to $X=\{X_1, X_2, \ldots, X_{DV}\}$, where $LB_i \leq X_i \leq UB_i$

Step 2: Generate an initial set of solutions.

Step 3: Set generation counter t = 1

Step 4: Initialize the tabu list and all the individuals

Step 5: Select any three set of parameters from the given solution space and out of which one solution represents the Current/Back Vehicle (BV) and the other two solutions are selected randomly and represent the Front Vehicle (FV) and the oncoming Vehicle (OV).

Step 6: Evaluate the Distance (D) and Velocities (V) for the three set of parameters (vehicles).

The distances between the vehicles are calculated

based on the fitness value. Arrange the population in ascending order and calculate the distance using Eq. 4. From the expressions, it can be observed that the value of the distance is normalized to 1. The velocity corresponding to the vehicle is calculated using Eq. 5

$$D_k = \frac{N_k}{PS} \quad (4)$$

$$V_k = R_k(1 - D_k) \quad (5)$$

where k=(1,2,3) and R_k is a random number ∈(0,1)

Step 7: Depending upon the velocity and distance of the set of parameters, update the solution space and if the solution is better than previous solution, accept the new solution (new parameters).

Step 8: Fill in the Tabu list with the new solution
Step 9 : If (Tabu list.size > max tabu size)

Tabu list.remove First()

End if

Step 8: Repeat the procedure until the termination criteria is satisfied, else go to Step 2

End

4 Results

In this section, we are going to visualize the results of the implementation of both VRP and VPRTW codes under python. As mentioned previously, we've adopted in the resolution hybridization between Taboo Search (TS) and Passing Vehicle Search (PVS).

4.1 VRP Results

We are going to represent two samples from the Data that we have worked on (Table 1 and Table 2):

The first table contains 11 nodes while that the second contains 22 nodes. For each node, is attributed a coordinate and a section randomly. The code under python uses Tabu moves, which are improving due to Passing Vehicle Search algorithm over the generations. We have executed this code on 20 generations and each generation is composed of 20 individuals. In addition, we have fixed the probability that an individual Route will mutate to a rate of 40%. For the first sample, we have chosen 4 tracks and 3 tracks for the second. Therefore, the program will provide a resolution with 3 different paths.

Table 1. 1st sample of VRP database.

Node	*Coordinate*	*Section*
1	87.951292	2.658162
2	33.466597	66.682943
3	91.778314	53.807184
4	20.526749	47.633290
5	9.006012	81.185339
6	20.032350	2.761925
7	77.181310	31.922361
8	41.059603	32.578509
9	18.692587	97.015290
10	51.658681	33.808405
11	44.563128	47.541734

Table 2. 2nd sample of VRP database.

Node	*Coordinate*	*Section*
1	37.806330	50.599689
2	9.961241	20.337535
3	28.186895	70.415357
4	62.129582	6.183050
5	50.376904	42.796106
6	71.285134	43.671987
7	34.156316	49.113437
8	85.201575	71.837519
9	27.466659	1.394696
10	97.985778	44.746239
11	40.730003	98.400830
12	73.799860	61.076693
13	85.076449	17.029328
14	16.052736	11.899167
15	20.160527	67.238380
16	22.730186	99.725333
17	77.196570	88.503677
18	18.494217	31.971191
19	72.743919	16.071047
20	4.153569	41.981262
21	79.027680	95.034639
22	14.145329	40.690329

4.1.1 Sample 1

We represent in the tables below the overall best path length, Run time and Current path length for each generation of each track for the first sample (Table 3, Table 4, Table 5, Table 6):

Table 3. Data of first track of sample 1

Generation	*Overall best path length*	*Run time (s)*	*Current path length*
1	157.070	0.080	157.070
2	157.070	0.090	157.070
3	157.070	0.101	157.070
4	157.070	0.111	157.070
5	157.070	0.122	157.070
6	157.070	0.134	157.070
7	157.070	0.145	157.070
8	157.070	0.155	157.070
9	157.070	0.166	157.070
10	157.070	0.177	157.070
11	157.070	0.187	157.070
12	157.070	0.196	157.070
13	157.070	0.206	157.070
14	157.070	0.219	157.070
15	157.070	0.230	157.070
16	157.070	0.239	157.070
17	157.070	0.252	157.070
18	157.070	0.262	157.070
19	157.070	0.271	157.070
20	157.070	0.278	157.070

- At generation 1, we obtain the optimal route with First best path length: 157.070.
- At generation 20, we obtain the optimal route with Final best path length: 157.070.

⇨ **No improvement over generations**

Table 4. Data of second track of sample 1

Generation	*Overall best path length*	*Run time (s)*	*Current path length*
1	48.281	0.005	48.281
2	48.281	0.013	48.281
3	48.281	0.023	48.281
4	48.281	0.033	48.281
5	48.281	0.042	48.281
6	48.281	0.051	48.281
7	48.281	0.060	48.281
8	48.281	0.070	48.281
9	48.281	0.079	48.281
10	48.281	0.088	48.281
11	48.281	0.096	48.281
12	48.281	0.104	48.281
13	48.281	0.114	48.281
14	48.281	0.125	48.281
15	48.281	0.133	48.281
16	48.281	0.141	48.281
17	48.281	0.151	48.281
18	48.281	0.159	48.281
19	48.281	0.168	48.281
20	48.281	0.173	48.281

- At generation 1, we obtain the optimal route with First best path length: 48.281.
- At generation 20, we obtain the optimal route with Final best path length: 48.281.

⇨ **No improvement over generations**

The program notice the total length of vehicle paths, which is equal to 479.235 and Total Runtime: 1.899. Here is a visualization of the Complete Vehicle Paths (Fig. 7):

4.1.2 Sample 2

Same as the first sample, we represent the informations's tracks in the tables below (Table 7):

Commentary: For all the tracks, the total path length is improved from the first generation to the last. This is due to the hybridization with Passing Vehicle Search algorith, which have such an effect to improve the path length at each generation. The program notice the total length of vehicle paths, which is equal to 712.441 and Total Runtime: 2.629. Here is a visualization of the Complete Vehicle Paths (Fig. 8):

Table 5. Data of second track of sample 2

Generation	*Overall best path length*	*Run time (s)*	*Current path length*
1	236.172	0.013	236.172
2	236.172	0.030	236.172
3	236.172	0.046	236.172
4	236.172	0.063	236.172
5	236.172	0.087	236.172
6	236.172	0.105	236.172
7	236.172	0.121	236.172
8	236.172	0.138	236.172
9	236.172	0.157	236.172
10	228.781	0.012	228.781
11	228.781	0.032	228.781
12	228.781	0.051	228.781
13	226.916	0.014	226.916
14	226.916	0.028	226.916
15	226.916	0.043	226.916
16	226.916	0.056	226.916
17	226.916	0.071	226.916
18	226.916	0.086	226.916
19	226.916	0.100	226.916
20	226.916	0.109	226.916

- At generation 1, we obtain the optimal route with First best path length: 147.332.
- At generation 20, we obtain the optimal route with Final best path length: 147.332.

⇨ **No improvement over generations**

4.2 VRPTW Results

For the Database in this section, we have worked with a JSON (Java Script Object Notation) files. We executed the Python code on 27 samples.

Each JSON file contains 100 customers and a depart that represents the depot from where the vehicles are dispatched to different localizations, and each one have 5 variables: “coordinates”, ”demand”, “due_time”, “ready_time” and “service_time”, the matrix of distances, the max vehicle number and the vehicle capacity.

To solve the VRPTW, we used another code always by hybridizing PVS and TS under python. But this time, the approach that we have adopted to represent the results is quite different from the previous one (VRP). Here, the code execution specifies for each generation the number of evaluated individuals unlike the VRP code where the number of individuals is fixed before the execution. Moreover, this approach is based generally on analyzing the fitness function by identifying it minimum, maximum, average and standard deviation.

Table 6. Data of fourth track of sample 1

Generation	Overall best path length	Run time (s)	Current path length
1	126.551	0.033	126.551
2	126.551	0.039	126.551
3	126.551	0.044	126.551
4	126.551	0.050	126.551
5	126.551	0.055	126.551
6	126.551	0.060	126.551
7	126.551	0.066	126.551
8	126.551	0.071	126.551
9	126.551	0.077	126.551
10	126.551	0.082	126.551
11	126.551	0.088	126.551
12	126.551	0.094	126.551
13	126.551	0.099	126.551
14	126.551	0.104	126.551
15	126.551	0.112	126.551
16	126.551	0.117	126.551
17	126.551	0.122	126.551
18	126.551	0.127	126.551
19	126.551	0.133	126.551
20	126.551	0.135	126.551

- At generation 1, we obtain the optimal route with First best path length: 126.551.
- At generation 20, we obtain the optimal route with Final best path length: 126.551.

⇨ **No improvement over generations**

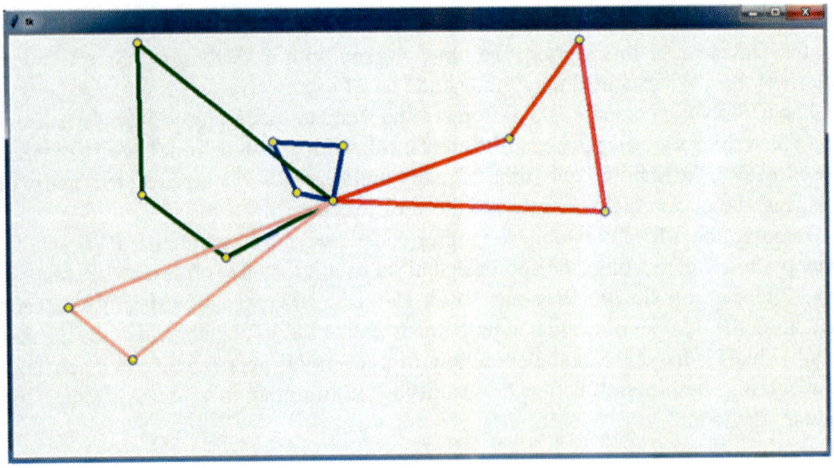

Fig. 7. Visualization of the complete vehicle paths

Table 7. Data of first track of sample 2

Generation	*Overall best path length*	*Run time (s)*	*Current path length*
1	255.988	0.0350	255.988
2	246.781	0.0240	246.781
3	225.566	0.018	225.566
4	225.566	0.044	225.566
5	225.566	0.062	225.566
6	225.566	0.084	225.566
7	225.566	0.103	225.566
8	225.566	0.121	225.566
9	225.566	0.139	225.566
10	225.566	0.157	225.566
11	225.566	0.180	225.566
12	225.566	0.214	225.566
13	225.566	0.245	225.566
14	225.566	0.270	225.566
15	225.566	0.297	225.566
16	225.566	0.323	225.566
17	225.566	0.348	225.566
18	225.566	0.381	225.566
19	225.566	0.406	225.566
20	225.566	0.423	225.566

- At generation 1, we obtain the optimal route with First best path length: 255.988.
- At generation 20, we obtain the optimal route with Final best path length: 225,566.

The execution under python of 3 Data files of the 27 samples that we worked on, gives the results below (Table 10):

The test is executed on 300 generations and for each generation the number of evaluated individuals is marked. We find also the different variables that can describe the fitness function. As we can see, the min_fitness, avg_fitness and std_fitness is changing increasley while max_fitness is stabilized on a constant value 1.939079e?05 (Table 11).

Table 8. Data of second track of sample 2

Generation	*Overall best path length*	*Run time (s)*	*Current path length*
1	269.567	0.016	255.988
2	264.340	0.016	264.340
3	264.340	0.034	264.340
4	264.340	0.051	264.340
5	260.729	0.015	260.729
6	260.729	0.033	260.729
7	253.138	0.012	253.138
8	253.138	0.029	253.138
9	253.138	0.049	253.138
10	246.679	0.016	246.679
11	246.679	0.035	246.679
12	246.679	0.060	246.679
13	246.679	0.079	246.679
14	246.679	0.100	246.679
15	246.679	0.123	246.679
16	246.679	0.139	246.679
17	234.760	0.015	234.760
18	234.760	0.030	246.679
19	234.760	0.048	246.679
20	234.760	0.058	234.760

The test is executed on 100 generation. As we can see, the min_fitness and avg_fitness is changing increasley and std_fitness decreasley while max_fitness is stabilized on a constant value 0.000126 (Table 12).

The test is executed on 300 generation. As we can see, the min_fitness and avg_fitness is changing increasley and std_fitness decreasley while max_fitness is stabilized on a constant value 1.527980e?05.

Table 9. Data of third track of sample 2

Generation	*Overall best path length*	*Run time (s)*	*Current path length*
1	236.172	0.013	236.172
2	236.172	0.030	236.172
3	236.172	0.046	236.172
4	236.172	0.063	236.172
5	236.172	0.087	236.172
6	236.172	0.105	236.172
7	236.172	0.121	236.172
8	236.172	0.138	236.172
9	236.172	0.157	236.172
10	228.781	0.012	228.781
11	228.781	0.032	228.781
12	228.781	0.051	228.781
13	226.916	0.014	226.916
14	226.916	0.028	226.916
15	226.916	0.043	226.916
16	226.916	0.056	226.916
17	226.916	0.071	226.916
18	226.916	0.086	226.916
19	226.916	0.100	226.916
20	226.916	0.109	226.916

- At generation 1, we obtain the optimal route with First best path length: 236.172.
- At generation 20, we obtain the optimal route with Final best path length: 226.916.

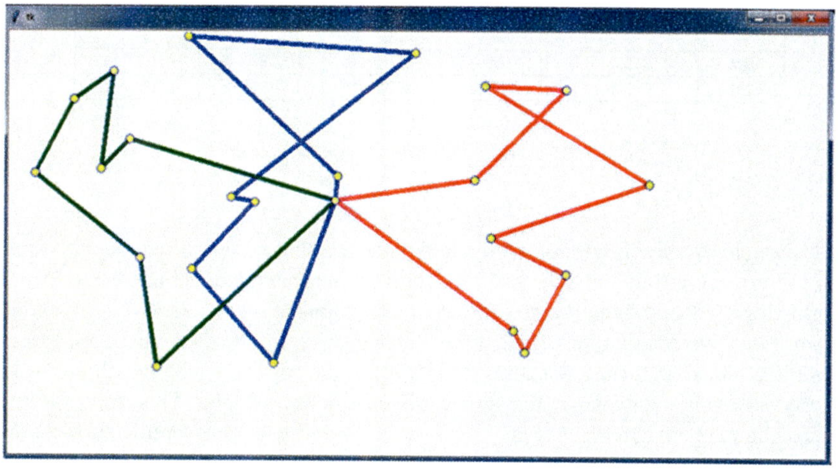

Fig. 8. Visualization of the complete vehicle paths

Table 10. VRPTW results for C204

Generation	Evaluated_individuals	Min_fitness	Max_fitness	Avg_fitness	Std_fitness
0	349	1,279579e-05	1.939079e-05	1.568041e-05	1.272774e-06
1	343	1.279030e-05	1.939079e-05	1.577911e-05	1.334861e-06
2	332	1.279030e-05	1.939079e-05	1.580546e-05	1.279526e-06
...					
99	346	1.876928e-05	1.939079e-05	1.930955e-05	1.796519e-07
...					
298	352	1.939079e-05	1.939079e-05	1.939079e-05	2.020941e-07
299	347	1.939079e-05	1.939079e-05	1.939079e-05	2.020941e-07

Table 11. VRPTW results for R101

Generation	Evaluated_individuals	Min_fitness	Max_fitne ss	Avg_fitn ess	Std_fitness
0	73	9.7633459e-05	0.000126	0.000110	6.610346e-06
1	68	0.000101	0.000126	0.000111	6.795503e-06
2	76	0.000101	0.000126	0.000112	5.656645e-06
...					
99	66	0.000126	0.000126	0.000126	5.7152149e-12

Table 12. VRPTW results for "customized data"

Generation	Evaluated_individuals	Min_fitness	Max_fitness	Avg_fitness	Std_fitness
0	349	1,164306e-05	1.527980e-05	1.304262e-05	5.847981-07
1	343	1. 164306e-05	1.527980e-05	1.306696e-05	6.105491-07
2	332	1. 164306e-05	1.527980e-05	1.305746e-05	5.926039-07
...					
99	346	1.433146e-05	1.527980e-05	1.526794e-05	1.053627e-07
...					
298	352	1.527980e-05	1.527980e-05	1. 527980e-05	1.623772e-12
299	347	1.527980e-05	1.527980e-05	1.527980e-05	1.623772e-12

5 Conclusion

In this article, we proposed the first dynamic management system of emergency vehicles in Morocco throug the SOS accident, then we presented a novel metaheuristic by hybridizing PVS and Tabu search algorithm for optimization problems in a dynamic environment. We dealt more specifically with the problem of the dynamic development of vehicle routes with time windows (VRPTW) at the service of the health sector in Morocco, namely the dynamic management of emergency vehicles. Therefore, we first proposed a mathematical model of the response time of emergency vehicles. We proceeded in a second part to the optimization process. Finally, we presented the results obtained by our approach, where we dealt with the problem of vehicle routing (VRP) with and without a time window. These encouraging results have enabled us to judge the technique used is an efficient method for the resolution of this type of dynamic problem.

Acknowledgement. – Special thanks to Mrs. Zineb el Mahboub et Mrs. Iqueram Boulif for their contribution in this projet, which is a part of the Road Safety Project (CNRST – Ministry of equipment, transport, logistics and water) 2021/2024.

– NARSA 2021 – Mon initiative pour la vie.

Conflict of Interest. The authors declare that they have no conflict of interest.

References

1. Mbarek, A., Jiber, M., Yahyaouy, A., Sabri, A.: Road traffic mortality in Morocco: analysis of statistical data. In: 2020 International Conference on Intelligent Systems and Computer Vision (ISCV), pp. 1–7. Fez, Morocco (2020)
2. El Gameh, M., et al.: Quantitative analysis and study on the evolution of road safety in Morocco. Int. J. Res. Stud. Sci., Eng. Technol. **1**, 210–215 (2014)
3. Ennajih, D., Laaraifi, A., Elgameh, M., Sallik, A., Chaouch, A., Echchelh, A.: The quality management system on road safety in Morocco. Int. J. Res. Stud., Sci. Eng. Technol. **5**, 1–6 (2018)
4. Bent, R.W., Van Hentenryck, P.: Scenario-based planning for partially dynamic vehicle routing with stochastic customers. Oper. Res. **52**, 977–987 (2004)
5. Novoa, C., Storer, R.: An approximate dynamic programming approach for the vehicle routing problem with stochastic demands. Eur. J. Oper. Res. **196**(2), 509–515 (2009)
6. Bertsekas, D.P., Tsitsiklis, J.N., Wu, C.: Rollout algorithms for combinatorial. J. Heuristics **3**(3), 245–262 (1997)
7. Benyahia, I., Potvin, J.-Y.: Decision support for vehicle dispatching using genetic programming. IEEE Trans. Syst., Man, Cybern. – Part A: Syst. Humans **28**(3), 306–314 (1998)
8. Cheung, B.K.S., Choy, K.L., Li, C.-L., Shi, W., Tang, J.: Dynamic routing model and solution methods for fleet management with mobile technologies. Int. J. Prod. Econ. **113**(2), 694–705 (2008)
9. Orozco, J.-A., Barcel.: Reactive and proactive routing strategies with real-time traffic information. Procedia – Soc. Behav. Sci. **39**, 633–648 (2012). https://doi.org/10.1016/j.sbspro.2012.03.136
10. Azi, N., Gendreau, M., Potvin, J.-Y.: A dynamic vehicle routing problem with multiple delivery routes. Ann. Oper Res **199**(1), 103–112 (2011)
11. Crainic, T.G., Gendreau, M., Potvin, J.-Y.: Intelligent freight-transportation systems: Assessment and the contribution of operations research. Transport. Res. Part C: Emerg. Technol. **17**(6), 541–557 (2009). https://doi.org/10.1016/j.trc.2008.07.002
12. Savsani, P., Savsani, V.: PVS: A novel meta-heuristic algorithm. Appl. Math. Mode. **40**, 3951–3978 (2016)
13. Glover, F., Laguna, M.: Tabu Search, pp. 858–859. Kluwer Academic Publishers (1997)

Environmental and Climate Risk Management in Public Procurement: A Proposed Decision Support Tool

Tarik El Haddadi[1](✉), Mohamed Ben Ahmed[2], Taoufik Mourabit[1], Oumaima El Haddadi[3], and Ahmad El Allaoui[3]

[1] Natural Risks Research Team, Faculty of Science and Technology, Abdelmalek Essaadi University, Tangier, Morocco
elhaddadi.tarik@gmail.com

[2] Computer Science Team, Faculty of Science and Technology, Abdelmalek Essaadi University, Tangier, Morocco

[3] Data Science and Competitive Intelligence Team (DSCI), ENSAH, Abdelmalek Essaadi University, Tetouan, Morocco

Abstract. Sustainable public procurement (SPP) is a powerful lever for the public sector to respond to ecological and climate challenges. Admittedly, such an approach appears to be costly and complicated to put in place. Indeed, the integration of sustainable development requirements into public procurement brings significant changes in the consumption habits of decision-makers and public procurement practitioners in Morocco. In order to support public purchasers in their SPP approach, so-called decision-making tools appear to be an essential element. The consideration of environmental and climate issues in public procurement depends on a multitude of different elements influencing the decision-making process as all public decisions must be in common accord with the public interest. Therefore, the understanding of the decision-making process and the availability of decision-support tools are of utmost importance. In this context, this research paper presents our research work and we hereby propose decision support tools for Moroccan public procurement practitioners. Given the complexity of the public sector and the interdisciplinary nature of the concept of sustainability, our approach is inspired by the techniques of Multiple Criteria Decision Making (MCDM). Thus, we chose the AHP method for the development of the tool. And for the deployment, we used qualitative and quantitative data for the parameterisation of the tool, in order to make decision about the ecological purchase of computers taking into account the climatic risks. The tool provides a ranking of the best possible choices as an output.

Keywords: Sustainable public procurement (SPP) · Decision-making · Climate issues · MCDM · AHP

M. Ben Ahmed et al. (Eds.): SCA 2022, LNNS 629, pp. 24–35, 2023.
https://doi.org/10.1007/978-3-031-26852-6_2

1 Introduction

Public procurement is the largest single market in the world in terms of volume [1, 2]. This renders public procurement is a powerful tool for the public sector to meet the different needs of public services and also to ensure the implementation of different political, social and environmental policies [3, 4]. Accordingly, purchasers need to reconsider their approach to accommodate SPP to meet environmental and climate challenges [5]. Consequently, thinking has focused on optimising public procurement so that it can consider sustainable development (SD) considerations. The integration of SD requirements in public procurement brings significant changes to production and consumption habits of different economic agents.

Considering that environmental and climate issues in public procurement requires a so-called sustainable decision [6], the latter depends on a multitude of different elements influencing the decision-making process. Like all public decisions, the choice must be made between individual interests specific to the organisation and collective interests. In this perspective, each choice must in principle have consequences that are favourable to environmental and climate issues [7]. Therefore, the understanding of the decision-making process and the availability of decision sup-port tools are essential. In this context, we have developed in the following the different elements of decision making as well as our proposal for a decision-making model for public purchasers in Morocco.

2 Multi-criteria Decision Making and Sustainability

2.1 Multi-criteria Decision

Multidimensionality is an intrinsic fact of the concept of sustainability. The indicators and criteria of sustainability are multidisciplinary and are the subject of a multitude of studies from a range of disciplines. As a result, multi-criteria approaches, and techniques are used to study the phenomenon of sustainability. The discipline has subsequently become widely studied via multi-criteria decision support techniques such as MCDM (Multiple Criteria Decision Making) belonging to a branch of study known as operations research. This aims to find the best solutions among others to solve a given problem and in this case to the infiltration of sustainability and particularly the aspect related to the effects of climate change in the public procurement process. This facilitates operational or strategic decision-making in a heterogeneous and/or contradictory context [8]. It is in this context that Janeiro [9] believes that the issue of sustainability falls within the scope of analysis of MCDM. This is confirmed by the work of Diaz-Balteiro [10] in their analysis of 268 articles. In this framework they demonstrated that the most used MCDM techniques for sustainability assessment are the Analytic Hierarchy Process (AHP) and the Weighted Average Method (WAM).

The first method is simple (AHP), systematic and flexible, and it explains its frequent use by researchers from several disciplines to compare objectives or alternatives [11, 12]. The second method (WAM), also known as the geometric weighted average method, is a mathematical calculation and represents to some extent practical simplifications of the approaches Multi Attribute Utility Theory (MAUT) and Multi Attribute

Value Analysis (MAVT). In this perspective, several approaches have been interested in studying the correlations between the different criteria/indicators in order to be able to propose classifications of sustainability. It should also be noted that a high percentage of MCDM techniques has been hybridised with the SWOT technique -Strengths Weaknesses Opportunities Threats- allowing for the consideration of stakeholder preferences in relation to the initially suggested indicators. As sustainability is a multi-disciplinary field, many works approach sustainability analysis via MCDM techniques inspired mainly by algorithmic and mathematical methods of multi-criteria analysis.

2.2 Green/sustainable Public Procurement and Decision Making

The decision on green or sustainable public procurement depends on a variety of conflicting criteria such as cost/price, quality, ecological/carbon footprint, environmental impact, social impact and regulation. These different criteria easily come into conflict. This makes the structuring of the issue central to the decision-making process. For this reason, the approaches and methods developed in the context of decision support provide solutions for structuring and planning issues involving several criteria in an attempt to provide practitioners and decision-makers with a clearer vision for the choice of the most optimal solutions (alternatives). Understandably, each alternative must be accompanied by its performance, allowing the classification/sorting of the different alternatives/variants. Several approaches are used, including: AHP, ELECTRE, PROMETHEE, TOPSIS, MAVT and MAUT among others to address sustainability issues. Most of these methods have been the subject of numerous scientific studies [10, 13–16]. Decision-making in green public procurement is so complicated, due to the financial, regulatory and technical constraints that govern the public procurement process. This explains the scarcity of tools that help practitioners and decision makers to formulate sustainable criteria in public procurement [14, 17–19]. Incidentally, there are some works proposing tools in the field of APD such as [14, 20].

In order to contribute to sustainable/ecological decision making in this multi-criteria context, we proposed an approach and a tool inspired by the work cited above, which was the subject of the article [21]. The latter is an analysis of the state inventory of SPP in Morocco, which used data analysis methods for automatic data retrieval from knowledge bases.

3 Methodology

In order to use SPP as a tool to reduce the negative impacts of ICT on the environment in general and particularly those related to climate change risks, the method must facilitate the adoption of SPP in order to achieve the desired goal [17, 22]. To achieve this facilitation objective, the tool/method must be derived based on a participatory approach between the different stakeholders (e.g., public decision-makers, environmental experts, public purchasers, suppliers…). Indeed, if public procurement practitioners participate in the development of the different tools, the understanding of the issue will be better, and the results are more likely to be accepted [23]. In a participatory approach we have tried to propose a method for sustainable decision making in the context of IT equipment procurement. Our approach is inspired by multi-criteria decision-making assessments of life cycle environmental impacts in accordance with the recommendations of authors such as [18, 24, 25], as well as the result of a detailed literature review and a field investigation.

3.1 The AHP Model of Decision-Making

The choice of the AHP is based on our literature review which showed that AHP is one of the most used methods in sustainable decision making, allowing the selection of the most advantageous alternatives via ranking of the performance of different solutions [26]. Typically, the data fed into the AHP are qualitative and quantitative data, resulting from interviews/questionnaires with experts [27, 28]. The AHP is widely used by several disciplines either exclusively or in combination with other methods. Generally, the scales adopted range from 1 to 9, known as Saaty scales [29]. Our approach is adapted to the method presented by Saaty [29] and the model is parameterised with the criteria and sub-criteria mentioned in Table 1. The objective is to make decision on the ecological purchase of computers taking into account the climate risks.

The choice of weights for each criterion and sub-criteria is based on the literature review, the expertise of our research team and based on a sample of public procurement practitioners. The weights must be adapted to each case and result from a sort of compromise between the various stakeholders. The criteria are defined according to the literature review, the research team's expertise in environmental analysis, in addition to the participation of public procurement practitioners.

3.2 The Classification of Alternatives

The choice of criteria is at the centre of the evaluation process, but the quantification of indicators is so complicated that it requires "scientific" expertise in addition to being considerably time-consuming and costly. Since the target of our decision-making model (public purchasers) is not in principle specialists in environmental issues, our choice of indicators is geared towards efficient and easy-to-implement means of verification. Therefore, TYPE 1 or ISO 14024 labels are the most suitable to indicate the environmental performance of a product or service.

The use of these labels is also justified by the absence of indexes at the national level (Morocco). As the index of durability and repeatability which specifies the level of durability of a given piece of equipment on standard scales and the restricted number of manufacturers which brings indications relative to the carbon footprint of their product, we leaned towards the consideration of ecolabels such as TCO and EPEAT and Energy Star as reliable indicators.

In addition to environmental performance, our evaluation model introduced three other aspects. The first is related to the basic technical aspects, necessary to satisfy the technical needs defined by the purchasing departments. The second part is the economic aspect, mainly related to cost while the third is the legal aspect to provide reflections on the conformity of the possible choices with the principles governing public procurement.

Table 1. Decision-making in the green procurement of computers taking into account climate risks.

Criteria	Sub-criteria	Description/objectives	Indicator/points
Technical performance (C1) * 40%	Technical requirements (C1) * 40%	Presence of the minimum technical requirements	**Satisfied:** meets the minimum requirement **Not satisfied**: does not meet the minimal requirements **Very satisfied**: offers better technical performance than requested
Environmental performance (C2) * 40%	Energy efficiency (C21) * 10%	Reduction of greenhouse gas emissions from energy consumption	**Satisfied**: ENERGYSTAR, TCO or EPEAT certification **Not satisfied**: absence of labels
	Reduce the depletion of natural resources (energy, water, metals, etc.) (C22) * 30%	Any exploitation of resources is by nature energy consuming and a major emitter of greenhouse gases. In this context, it is necessary to extend the life of equipment	**Satisfied**: Type I or ISO 14024 eco-labels such as TCO or EPEAT **Not satisfied**: absence of labels
	Waste management (C23) * 10%	End-of-life management via recyclability and reuse. As well as the requirement for reusable and/or recyclable packaging	**Satisfied**: Type I or ISO 14024 eco-labels such as TCO or EPEAT **Not satisfied**: absence of labels
	Environmental training (C24) * 10%	The obligation to produce a training manual on green management of computers must be provided. Of which energy management functions must be present on the hardware itself	**Satisfied**: a clear description of how energy management works **Not satisfied**: lack of information

(*continued*)

Table 1. *(continued)*

Criteria	Sub-criteria	Description/objectives	Indicator/points
	Other Impacts (C25) * 10%	Is the alternative better or worse in terms of the various environmental impacts of ICT?	**Satisfied**: Type I or ISO 14024 eco-labels such as TCO or EPEAT **Not satisfied**: absence of labels
Economic performance (C3) * 10%	Profitability (C31) * 50%	What are the overall costs? Acquisition and operating costs	**Overall cost/acquisition price**
	Concurrence (C32) * 50%	Are there alternatives on the national and international market?	**The number of national alternatives / the number of international alternatives**
Legislative support (C4) * 10%	Regulation (C4) * 100%	Does the acquisition of the alternative comply with the principles governing public control?	**Satisfied**: Compliant **Not satisfied**: Not in conformity

* **Weight.**

Finally, to speed up the evaluation and comparison process, we have introduced into the tool elements based on data analysis methods for automatic data extraction from knowledge bases. In particular, data related to ENERGYSTAR, TCO or EPEAT labelling. The extracted data as well as the data obtained from the evaluation of other indicators are processed in the Python ecosystem for analysis via a multi-criteria decision-making module for the classification of alternatives.

4 Demonstration and Application

4.1 Demonstration

The decision-making tool [30] is open at the level of analysis on the AHP methodology. Based on the structuring of the problem specified in Table 1 (objectives, criteria, sub-criteria), a hierarchical decision structure on four levels (Objectif, Criteria, Sub-Criteria and Alternative) was defined as shown in Fig. 1.

The AHP method comprises four steps. The first step consists of defining the criteria comparison matrix via the Saaty table [31] and based on Table 2.

To ensure the hierarchical structuring of the criteria with the alternatives, the other three steps continue to compare the criteria and the alternatives as follows:

- Step 1: Calculate the eigenvector or the relative weights and λ_{max} for each matrix of order n

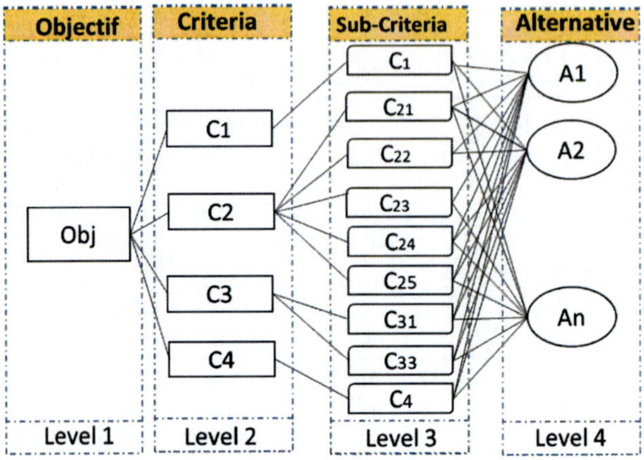

Fig. 1. The hierarchical decision structures

Table 2. Average random index (RI) based on Matrix

S. No Size of Matrix (x)	1	2	3	4	5	6	7	8	9	10
Random Consistency Index (RI)	0	0	0.52	0.89	1.11	1.25	1.35	1.40	1.45	1.49

- Step 2: Compute the consistency index for each matrix of order n by the formula: $CI = (\lambda_{max} - n)/(n - 1)$.
- Step 3: The consistency ratio is then calculated using the formulae: $CR = CI/RI$.

So, first we apply the method on the criteria, and we obtain the weights of each criterion, as presented in Table 3.

Table 3. Criteria pairwise comparison.

	Technical performance	Environmental performance	Economic performance	Legislative support	Weight
Technical performance	1	1	5	7	0.43673124
Environmental performance	1	1	5	5	0.40149695
Economic performance	1/5	1/5	1	3	0.10567994
Legislative support	1/7	1/5	1/3	1	0.05609187

$$\lambda_{max} = 4{,}12,\ CI = \frac{\lambda_{max} - n}{n - 1} = 0{,}04,\ CR = \frac{CI}{RI} = 0.0445 < 0{,}10(ACCEPTABLE) \quad (1)$$

The method must then be applied to the sub-criteria to obtain the local weights of each sub-criterion in relation to the criterion. The objective is to be able to multiply each local weight by the global weight of the criterion as shown in Table 4 and 5.

Table 4. Pairwise comparison sub cretaria of performance environnementale.

	C21	C22	C23	C24	C25	Local weight	Global Weight
C21	1	3	3	7	7	0.503	0.202
C22	1/3	1	1	3	3	0.185	0.074
C23	1/3	1	1	3	3	0.185	0.074
C24	1/7	1/5	1/3	1	2	0.067	0.026
C25	1/7	1/5	1/3	1/2	2	0.059	0.023

$$\lambda_{max} = 5,\ 123,\ CI = \frac{\lambda_{max} - n}{n - 1} = 0{,}031,\ CI/RI = 0.027 < 0{,}10(ACCEPTABLE) \quad (2)$$

Table 5. Pairwise comparison sub-criteria of performance economic.

	Profitability	Concurrence	Local weight	Global weight
Profitability	1	3	0,66	0,06974876
Concurrence	1/3	1	0,34	0,03593118

Finally, and after calculating all the matrices and that all the racial consistencies are acceptable, we obtain the desired weight vectors which will determine the best possible choice of alternative which in our case is the best computer (Table 6).

4.2 Application

To test the tool at hand, we took as an example an administration X which has defined as a purchase need, and a computer with the following technical performances: an intel core i5 processor of 5th generation with a storage capacity superior to one terra-byte and a RAM capacity of 8 Go. Our tool processes the data stored in its (knowledge) database and proposes a list of computers with equivalent or better technical performance than those required. The list of computers can also be compiled from suppliers' proposals in the context of public procurement processes.

The afore-mentioned lists will be automatically analysed for environmental and technical aspects thanks to the database enriched with directories relating to the "Energy Star", "TCO" and "EPEAT" labels, whereas for the other economic and legal aspects the data are filled in manually after the deliberation of the committee designated for the evaluation of alternatives. The tool processes and analyses the different data and produces as output a table (Table 6) detailing the different ratings and weightings, resulting in the classification of the different alternatives. Thus, according to the said classification, the best alternative to be considered is alternative 5.

Table 6. Final classification of alternatives.

Criteria	Technical performance	Environmental performance					Economic performance		Legislative support		
Weight Criteria	0.43673124	0.40149695					0.105		0.056		
Sub-Criteria	C1	C21	C22	C23	C24	C25	C31	C32	C4		
Weight Sub-Criteria	0.43673124	0.202	0.074	0.074	0.026	0.023	0.069	0.035	0.056	$\vec{C} \times \vec{A} = \sum_{i-1,j-1}^{n,m} c_i \times a_j^*$	Rank
Alternative 1	0.2	0.6	0.6	0.6	0.6	0.6	0.2	1	1	0.43280352	7
Alternative 2	0.6	0.4	0.4	0.4	0.4	0	0.4	1	1	0.5316016	3
Alternative 3	0.2	0.6	0.6	0.4	0.2	0.2	0.2	0.6	1	0.42531269	8
Alternative 4	0.4	0.8	0.8	0.4	0.2	0.4	0.8	1	1	0.58644258	2
Alternative 5	0.8	0.6	0.6	0.6	0.6	0.1	0.6	0.6	1	0.6965981	1
Alternative 6	0.8	0	0	0	0	0	0.4	0.4	1	0.44698499	6
Alternative 7	0.8	0.2	0.2	0.2	0	0	0.2	0.8	1	0.51728636	4
Alternative 8	0.2	0.8	0.8	0.6	0.4	0.2	0.6	0.8	1	0.49379411	5

* **ci**: weight of sub-criteria number i, **aj**: the alternative number j, **n** = 9 and **m** = 8.

5 Discussion and Conclusion

5.1 Discussion

The use of models inspired by multi-criteria decision-making methods are widely recommended in such studies as quantitative data are difficult to measure. In our case, the choice depends on several parameters. Firstly, data related to the environmental and technical performance of the computers to be purchased including economic and regulatory data, hence the need to use qualitative indicators generally expressing the opinions of experts in each field [32]. This qualitative data poses a problem when assessing uncertainties, which has led researchers such as Lindfors and J. Ammenberg [14] to reflect on the issue by proposing qualitative assessments of uncertainty based on the level of consensus between the parties involved. It should be pointed out in this respect that any qualitative judgement emanating from a discussion and exchange of data is necessarily non-deterministic, allowing the introduction of the human dimension into the ecological assessment process [33]. Indeed, the integration of the opinions of experts and stakeholders runs the risk of having subjective opinions oriented towards one's own interests (bias), hence the importance of weighting the different opinions in order to counteract this problem of bias [34]. The use of algorithms has been shown to be useful in decision-making situations, but in the case of rarities or lack of knowledge the gap must be closed or resolved by assumptions or compromises resulting from discussion between the different stakeholders. Infinite quantitative and qualitative measures must be integrated into mathematical approaches that allow meaningful results to be obtained and provide the possibility of classification and differentiation between different alternatives.

5.2 Conclusion

The method we propose is oriented towards the purchase of IT equipment, but it can be adapted to other types of purchases such as works or services. It is the environmental objectives that guide the structuring of the steps, as the criteria and sub-criteria are flexible for each objective. The method is open to the integration of large sources of knowledge (multitude of indicators), which saves a lot of time during the analysis, so the method is well oriented towards decision making in a multi-criteria context such as green public procurement. Although the results are not of optimal precision, this type of method allows classification and choice among a multitude of alternatives, thus offering a more informed and quality decision.

References

1. World Bank Group: Benchmarking-Public-Procurement-2017-Assessing-Public-Procurement-Regulatory-Systems-in-180-Economies.pdf (2016)
2. Darnall, N., Stritch, J.M., Hsueh, L., Bretschneider, S.: A framework for understanding sustainable public purchasing. Acad. Manag. Proc. **2018**(1), 15677 (2018). https://doi.org/10.5465/AMBPP.2018.15677abstract
3. Kheifets, B.A., Chernova, V.Y.: Public procurement as an instrument for implementing economic policy (experience of EU countries). RUDN J. Econ. **28**(3), 568–584 (2020). https://doi.org/10.22363/2313-2329-2020-28-3-568-584
4. Malolitneva, V., Dzhabrailov, R.: Strategic public procurement: facilitating sustainable development in Ukraine. Eur. J. Sustain. Dev. **8**(2), 91 (2019). https://doi.org/10.14207/ejsd.2019.v8n2p91
5. Nina, N., Hans, Q., Streng, J., van Dijk, L.: Public procurement as a strategic instrument to meet sustainable policy goals: the experience of Rotterdam. Transp. Res. Procedia **46**, 285–292 (2020). https://doi.org/10.1016/j.trpro.2020.03.192

6. Biberos-Bendezú, K., Cárdenas, Ú., Kahhat, R., Vázquez-Rowe, I.: Introducing environmental decision-making criteria to foster green public procurement in Peru. Integr. Environ. Assess. Manag. **18**, 1206–1220 (2021). https://doi.org/10.1002/ieam.4488
7. Pouikli, K.: Towards mandatory Green Public Procurement (GPP) requirements under the EU Green Deal: reconsidering the role of public procurement as an environmental policy tool. ERA Forum **21**(4), 699–721 (2020). https://doi.org/10.1007/s12027-020-00635-5
8. Kumar, A., et al.: A review of multi criteria decision making (MCDM) towards sustainable renewable energy development. Renew. Sustain. Energ. Rev. **69**, 596–609 (2017). https://doi.org/10.1016/j.rser.2016.11.191
9. Janeiro, L., Patel, Martin K.: Choosing sustainable technologies. Implications of the underlying sustainability paradigm in the decision-making process. J. Cleaner Prod. **105**, 438–446 (2015). https://doi.org/10.1016/j.jclepro.2014.01.029
10. Diaz-Balteiro, L., González-Pachón, J., Romero, C.: Measuring systems sustainability with multi-criteria methods: a critical review. Eur. J. Oper. Res. **258**(2), 607–616 (2017). https://doi.org/10.1016/j.ejor.2016.08.075
11. Guliyeva, A.E., Lis, M.: Sustainability management of organic food organizations: a case study of Azerbaijan. Sustainability **12**(12), 5057 (2020). https://doi.org/10.3390/su12125057
12. Spanidis, P.-M., Roumpos, C., Pavloudakis, F.: A fuzzy-AHP methodology for planning the risk management of natural hazards in surface mining projects. Sustainability **13**(4), 2369 (2021). https://doi.org/10.3390/su13042369
13. Lindfors, A., Feiz, R., Eklund, M., Ammenberg, J.: Assessing the potential, performance and feasibility of urban solutions: methodological considerations and learnings from biogas solutions. Sustainability **11**(14), 3756 (2019). https://doi.org/10.3390/su11143756
14. Lindfors, A., Ammenberg, J.: Using national environmental objectives in green public procurement: Method development and application on transport procurement in Sweden. J. Cleaner Prod. **280**, 124821 (2021). https://doi.org/10.1016/j.jclepro.2020.124821
15. Wang, J.-J., Jing, Y.-Y., Zhang, C.-F., Zhao, J.-H.: Review on multi-criteria decision analysis aid in sustainable energy decision-making. Renew. Sustain. Energ. Rev. **13**(9), 2263–2278 (2009). https://doi.org/10.1016/j.rser.2009.06.021
16. Cinelli, M., Coles, S.R., Kirwan, K.: Analysis of the potentials of multi criteria decision analysis methods to conduct sustainability assessment. Ecol. Indic. **46**, 138–148 (2014). https://doi.org/10.1016/j.ecolind.2014.06.011
17. Bernal, R., San-Jose, L., Retolaza, J.L.: Improvement actions for a more social and sustainable public procurement: a Delphi analysis. Sustainability **11**(15), 4069 (2019). https://doi.org/10.3390/su11154069
18. Cheng, W., Appolloni, A., D'Amato, A., Zhu, Q.: Green public procurement, missing concepts and future trends – a critical review. J. Cleaner Prod. **176**, 770–784 (2018). https://doi.org/10.1016/j.jclepro.2017.12.027
19. Liu, J., Xue, J., Yang, L., Shi, B.: Enhancing green public procurement practices in local governments: Chinese evidence based on a new research framework. J. Cleaner Prod. **211**, 842–854 (2019). https://doi.org/10.1016/j.jclepro.2018.11.151
20. Vidal, R., Sánchez-Pantoja, N.: Method based on life cycle assessment and TOPSIS to integrate environmental award criteria into green public procurement. Sustainable Cities Soc. **44**, 465–474 (2019). https://doi.org/10.1016/j.scs.2018.10.011
21. Haddadi, T.E., Mourabit, T., Haddadi, A.E.: Sustainable public procurement in morocco: An investigative survey regarding tender preparation. Sustainable Prod. Consumption **26**, 33–43 (2021). https://doi.org/10.1016/j.spc.2020.09.002
22. Sönnichsen, S.D., Clement, J.: Review of green and sustainable public procurement: Towards circular public procurement. J. Cleaner Prod. **245**, 118901 (2020). https://doi.org/10.1016/j.jclepro.2019.118901

23. Lang, D.J., et al.: Transdisciplinary research in sustainability science: practice, principles, and challenges. Sustain. Sci. **7**(S1), 25–43 (2012). https://doi.org/10.1007/s11625-011-0149-x
24. Alhola, K., Nissinen, A.: Integrating cleantech into innovative public procurement process – evidence and success factors. J. Public Procurement **18**(4), 336–354 (2018). https://doi.org/10.1108/JOPP-11-2018-020
25. Testa, F., Annunziata, E., Iraldo, F., Frey, M.: Drawbacks and opportunities of green public procurement: an effective tool for sustainable production. J. Cleaner Prod. **112**, 1893–1900 (2016). https://doi.org/10.1016/j.jclepro.2014.09.092
26. Velasquez, M., Hester, P.T.: An Analysis of multi-criteria decision making methods. Int. J. Oper. Res. **10**(2), 56–66 (2013)
27. Chavosh Nejad, M., Mansour, S., Karamipour, A.: An AHP-based multi-criteria model for assessment of the social sustainability of technology management process: a case study in banking industry. Technol. Soc. **65**, 101602 (2021). https://doi.org/10.1016/j.techsoc.2021.101602
28. Fagundes, M.V.C., Hellingrath, B., Freires, F.G.M.: Supplier selection risk: a new computer-based decision-making system with fuzzy extended AHP. Logistics **5**(1), 13 (2021). https://doi.org/10.3390/logistics5010013
29. Saaty, T.L.: Decision making with the analytic hierarchy process. Int. J. Serv. Sci. **1**(1), 83 (2008). https://doi.org/10.1504/IJSSCI.2008.017590
30. El Haddadi, T., Ben Ahmed, M., Mourabit, T.: Establishment of a watch platform of public sustainable purchase in Morocco. In: Ben Ahmed, M., Boudhir, A.A., Karaş, İR., Jain, V., Mellouli, S. (eds.) SCA 2021. LNNS, vol. 393, pp. 435–444. Springer, Cham (2022). https://doi.org/10.1007/978-3-030-94191-8_35
31. Saaty, R.W.: The analytic hierarchy process—what it is and how it is used. Math. Model. **9**(3–5), 161–176 (1987). https://doi.org/10.1016/0270-0255(87)90473-8
32. Mendoza Jiménez, J., Hernández López, M., Franco Escobar, S.E.: Sustainable public procurement: from law to practice. Sustainability **11**(22), 6388 (2019). https://doi.org/10.3390/su11226388
33. Martinez-Alier, J., Munda, G., O'Neill, J.: Weak comparability of values as a foundation for ecological economics. Ecol. Econ. **26**(3), 277–286 (1998). https://doi.org/10.1016/S0921-8009(97)00120-1
34. Ernst, A., Biß, K.H., Shamon, H., Schumann, D., Heinrichs, H.U.: Benefits and challenges of participatory methods in qualitative energy scenario development. Technol. Forecast. Soc. Change **127**, 245–257 (2018). https://doi.org/10.1016/j.techfore.2017.09.026

Evaluating Dimensionality Reduction Approaches on Erstwhile Hyperion and Newly Launched PRISMA Datasets

Kavach Mishra(✉) and Rahul Dev Garg

Geomatics Engineering Group, Civil Engineering Department, Indian Institute of Technology Roorkee, Roorkee, Uttarakhand, India
{kmishra,rdgarg}@ce.iitr.ac.in

Abstract. Hyperspectral images have rich feature information, plagued by redundancy and noise. Dimensionality reduction approaches can aid in extracting meaningful information from such voluminous data. Three such approaches have been tested on Hyperion and PRISMA data of Ahmedabad, India. The eigenvalues and visual inspection of the components chosen from the scree plots, show that the kernel principal component analysis (k-PCA) outperforms both standardized and unstandardized principal component analyses (PCA) and minimum noise transform (MNF) in both datasets. Stacked k-PCA components 3, 2 and 1 in Hyperion and 4, 2 and 1 in PRISMA differentiate natural and built-up land cover classes, thereby choosing them for further image processing tasks.

Keywords: Hyperion · PRISMA · PCA · k-PCA · MNF

1 Introduction

Earth observation applications involving peak accuracies, like biophysical parameter retrieval [1], mineral characterization [2], and material distinction [3], require hyperspectral sensors.

A hyperspectral sensor is a passive imaging sensor with a bandwidth of 1–10 nm and can view a target in more than 200 channels, mostly in the visible and near-infrared (VNIR) and short-wave infrared (SWIR) wavelengths [4]. Each pixel in a hyperspectral image has a continuous spectrum instead of a discrete one for a multispectral pixel. Hughes' phenomenon [5] can affect this rich spectral content which is voluminous and has signal noise and feature repetition.

Multispectral data pre-processing routines handle hyperspectral data inadequately, necessitating new procedure development or re-imagining existing algorithms for such data. For instance, minimum noise fraction (MNF) [6] was developed from principal component analysis (PCA) [7] for transforming the hyperspectral data into an orthogonal subspace and generating noise-free, uncorrelated components having maximal information.

PCA and MNF consider a linear data distribution within the image, true if homogeneity is present in the scene being imaged. However, non-linear spectral reduction

M. Ben Ahmed et al. (Eds.): SCA 2022, LNNS 629, pp. 36–45, 2023.
https://doi.org/10.1007/978-3-031-26852-6_3

techniques like kernel-PCA (k-PCA) [8–11] must be utilized for scenes wherein heterogeneity is evident, like an urban area. k-PCA uses a kernel function to define the subspace, and a kernel matrix's eigenvectors are calculated.

Here, linear (PCA, MNF) and non-linear (k-PCA) spectral reduction approaches have been evaluated on two real-world hyperspectral datasets acquired over Ahmedabad, India. Visual inspection of the generated components, their eigenvalues and scree plots help to identify the best approach and the number of components to be adopted for further processing.

2 Related Works

In [12], the advantage of MNF over PCA in eliminating noise and producing noise-free uncorrelated components is highlighted. MNF outperforms PCA in generating components with lower eigenvalues from AVIRIS-NG data of Kalaburgi, India [13]. [14] conclude that narrowband ratios derived from PCA components are the most robust in the lithological mapping of Rajasthan, India, from new spaceborne hyperspectral datasets. A singular value decomposition (SVD) strategy is compared with an eigenvalue decomposition (EVD) strategy for performing PCA of PRecursore IperSpettrale della Missione Applicativa (PRISMA) [15] and lunar hyperspectral data in [16]. At the same time, PCA and MNF outperform independent component analysis (ICA) [17] in identifying various minerals on the AVIRIS-NG image of Jahazpur, India [18]. In [9], k-PCA performs better than PCA in extracting urban information using support vector machines and extended morphological profiles.

3 Study Area

The study area is a major part of western Ahmedabad, comprising educational institutes, large commercial establishments and housing societies with wide roads in a gridded manner. The minor part in eastern Ahmedabad has defunct cloth mills' lands and the crowded streets of the city's oldest neighbourhoods.

4 Datasets

Images from the Hyperion [19] and PRISMA [15] have been used here. A comparative assessment of their technical parameters is presented in [2]. The Hyperion image acquired on November 4, 2002, is a terrain-corrected level-1 file. The PRISMA image taken on October 5, 2020, is both terrain and atmospherically corrected level-2 file.

5 Methodology

Figure 1 shows the overall methodology, and the following sub-sections describe the major steps.

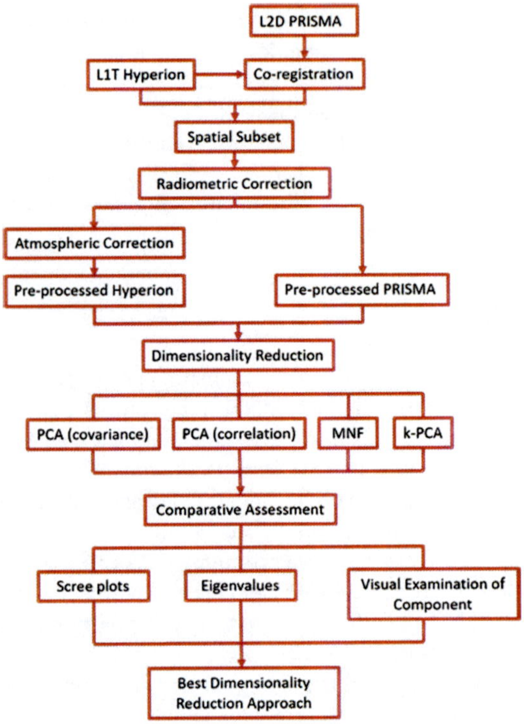

Fig. 1. Overall methodology

5.1 Hyperspectral Image Pre-processing

Pre-processing of the Hyperion data involves spatial subsetting to yield the area of interest in square format, radiometric correction (manual inspection of every band for noise or no information and along-track de-striping), atmospheric correction using Fast Line-of-sight Atmospheric Analysis of Hypercubes (FLAASH) [20] and scaling of surface reflectance.

For PRISMA data, pre-processing includes stacking VNIR and SWIR data cubes in the Environment for Visualizing Images (ENVI) standard format using "prismaread" [21], co-registration with the Hyperion data, spatial subsetting to give the study area and radiometric correction.

5.2 Dimensionality Reduction

The processed Hyperion and PRISMA datasets were subjected to PCA using correlation and covariance matrices.

Let $\mathbf{Z}$ be an $M \times N$ matrix such that $N < M$ and every data point $\mathbf{z}_i$ is of size $N \times 1$, where M = number of dimensions, N = number of observations and $i = 1, 2, \ldots, n$. The pseudocode of PCA is specified in Table 1 below.

Table 1. Pseudocode of PCA

Algorithm: PCA (covariance or correlation)
1: Compute $\underline{\mathbf{z}} = \frac{1}{n}\sum_{i=1}^{n} \mathbf{z}_i$
2: Subtract $\underline{\mathbf{z}}$ from each $\mathbf{z}_i$ in $\mathbf{Z}$ to obtain $\mathbf{Z}_c$
3: Compute covariance matrix $\mathbf{P} = \mathbf{Z}_c\mathbf{Z}_c^T$
or
4: Determine $\mathbf{Z}_c^* = \mathbf{Z}_c\mathbf{D}^{-1}$, where $\mathbf{D}$ is a diagonal scaling matrix
5: Compute correlation matrix $\mathbf{R} = \mathbf{Z}_c^*\mathbf{Z}_c^{*T}$
6: Eigenvectors and eigenvalues of $\mathbf{P}$ or $\mathbf{R}$, i.e., $\{\lambda_i, \mathbf{u}_i\}_{i=1,\ldots,M}$ provided $\lambda_1 \geq \lambda_2 \geq \ldots \geq \lambda_M$
7: Return $\{\lambda_i, \mathbf{u}_i\}_{i=1,\ldots,k}$, i.e., top principal components (PCs)

The datasets are also subjected to MNF, which performs noise removal and data reduction using two cascaded PCA operations. After the first transformation, the noise variance is 1, and the bands are uncorrelated. The second transformation applied to the first transformation's output yields normalized original bands possessing greater signal-to-noise ratios (SNRs) [6]. MNF is unaffected if scales vary in bands.

For k-PCA, the performance depends on the selected kernel function. Here, a Gaussian kernel function [22] has been chosen, and the number of samples has been varied from 112 to 2000 to minimize the γ parameter and the computation time for both the datasets. In Hyperion's case, the minimum γ and computation time was 0.0099 and 14.9 s, respectively, when 250 samples were chosen for clustering. In PRISMA's case, the minimum γ and computation time was 0.01219 and 0.256 s, respectively, when 118 samples were chosen.

6 Results and Discussion

Tables 2 and 3 show the eigenvalues for the first 10 components generated from Hyperion and PRISMA datasets, respectively. Regardless of the method, the eigenvalues are higher for Hyperion than PRISMA as the processed Hyperion data has more bands (128) than the processed PRISMA data (112). For both datasets, k-PCA components' eigenvalues are the lowest, suggesting their adoption for further processing. For more clarity, the scree plots were generated, as shown in Fig. 4 and analyzed. As a thumb rule, only those components from each method were taken for visual examination whose eigenvalues fell before the break in slope of the scree plot. So, in Hyperion's case, 5 components were selected for PCA (covariance), PCA (correlation), and k-PCA each, while 6 were selected for MNF. In PRISMA's case, 7 components were selected for PCA (correlation) and MNF each, while 6 were selected for PCA (covariance) and k-PCA each. Figures 2 and 3 show these components in Hyperion and PRISMA cases, respectively, along with false colour composites (FCCs) of both datasets.

Table 2. Eigenvalues for Hyperion data

Component No.	PCA (covariance)	PCA (correlation)	MNF	k-PCA
1	0.317519	113.170812	88.776927	0.006151
2	0.021621	10.697869	25.380355	0.000434
3	0.003006	2.566022	12.502330	0.000111
4	0.000773	0.319485	9.661760	0.000057
5	0.000299	0.173107	7.219272	0.000011
6	0.000242	0.093799	6.654141	0.000006
7	0.000224	0.070772	6.072628	0.000005
8	0.000168	0.066590	5.698659	0.000004
9	0.000149	0.059068	5.344767	0.000003
10	0.000139	0.053953	4.799366	0.000003

Table 3. Eigenvalues for PRISMA data

Component No.	PCA (covariance)	PCA (correlation)	MNF	k-PCA
1	0.189049	89.273329	30.880417	0.004469
2	0.037245	16.761151	22.418770	0.000905
3	0.006225	4.662466	12.391390	0.000150
4	0.000712	0.370488	8.188408	0.000076
5	0.000315	0.274257	6.264823	0.000018
6	0.000189	0.090200	5.754276	0.000008
7	0.000132	0.071407	5.436626	0.000005
8	0.000054	0.058741	5.027927	0.000004
9	0.000050	0.055443	4.628532	0.000003
10	0.000049	0.047528	4.562713	0.000001

Figure 2 shows that river water is highlighted in PC1 of PCA (correlation), PCA (covariance) and k-PCA, followed by the arterial roads. River water and built-up area are equally prominent in PC2 of PCA (correlation), leading to mixing between them. In PC3 of PCA (correlation), mixing between vegetation and river water is evident. Noise is also evident from PC3 onwards. In PCA (covariance), vegetated areas are visible clearly in PC2. PC3, PC4 and PC5 of PCA (covariance) are noisier than PCA (correlation). Noise is evident in every MNF component. In MNF1, the soil is highlighted, while built-up and vegetated areas in the eastern portion are highlighted in MNF2. Vegetation and soil are highlighted together in MNF3, indicating that these cannot be easily segregated. Vegetated areas are highlighted in PC2 of k-PCA, while arterial roads and built-up

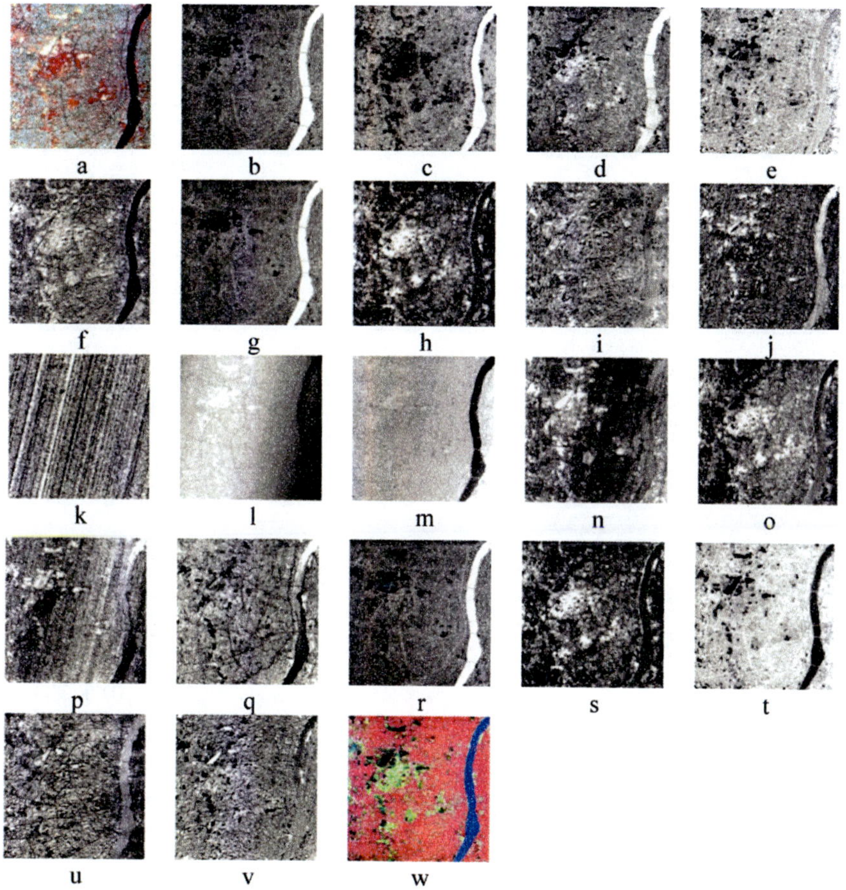

Fig. 2. Results of dimensionality reduction approaches for Hyperion data. a. Hyperion FCC (R:40, G: 18, B:8). b – f shows PC1, PC2, PC3, PC4 and PC5 of PCA (correlation). g – k shows PC1, PC2, PC3, PC4 and PC5 of PCA (covariance). l – q shows MNF1, MNF2, MNF3, MNF4, MNF5, and MNF6 of MNF. r - v shows PC1, PC2, PC3, PC4 and PC5 of k-PCA. w. FCC of k-PCA components (R: PC3, G: PC2; B: PC1) showing distinct natural and built-up land cover classes

regions are separately visible in PC3 of k-PCA. PC4 and PC5 of k-PCA are dominated by noise.

It is evident from visual inspection that the major artificial and natural surfaces in the form of water, vegetation, soil, roads and built-up can be distinguished in the FCC of k-PCA components: PC3, PC2 and PC1 in the red (R), green (G) and blue (B) channels respectively (Fig. 2w).

Fig. 3. Results of dimensionality reduction approaches for PRISMA data. a. PRISMA FCC (R:37, G: 24, B: 15). b – h shows the first to the seventh component of PCA (correlation). i – n shows the first to the sixth component of PCA (covariance). o – u shows the first to the seventh component of MNF. v – aa shows the first to the sixth component of k-PCA. ab. FCC of k-PCA components (R: PC4, G: PC2; B: PC1) showing distinct natural and built-up land cover classes

Figure 3 highlights water in PC5 of PCA (correlation), while the rest of the image is noisy. Bare soil and vegetation regions are highlighted clearly in PC1 and PC2 of PCA (correlation). Urban areas are not highlighted in any component. The noise content increases steadily from PC3 onwards. Soil and built-up regions experience mixing in PC1

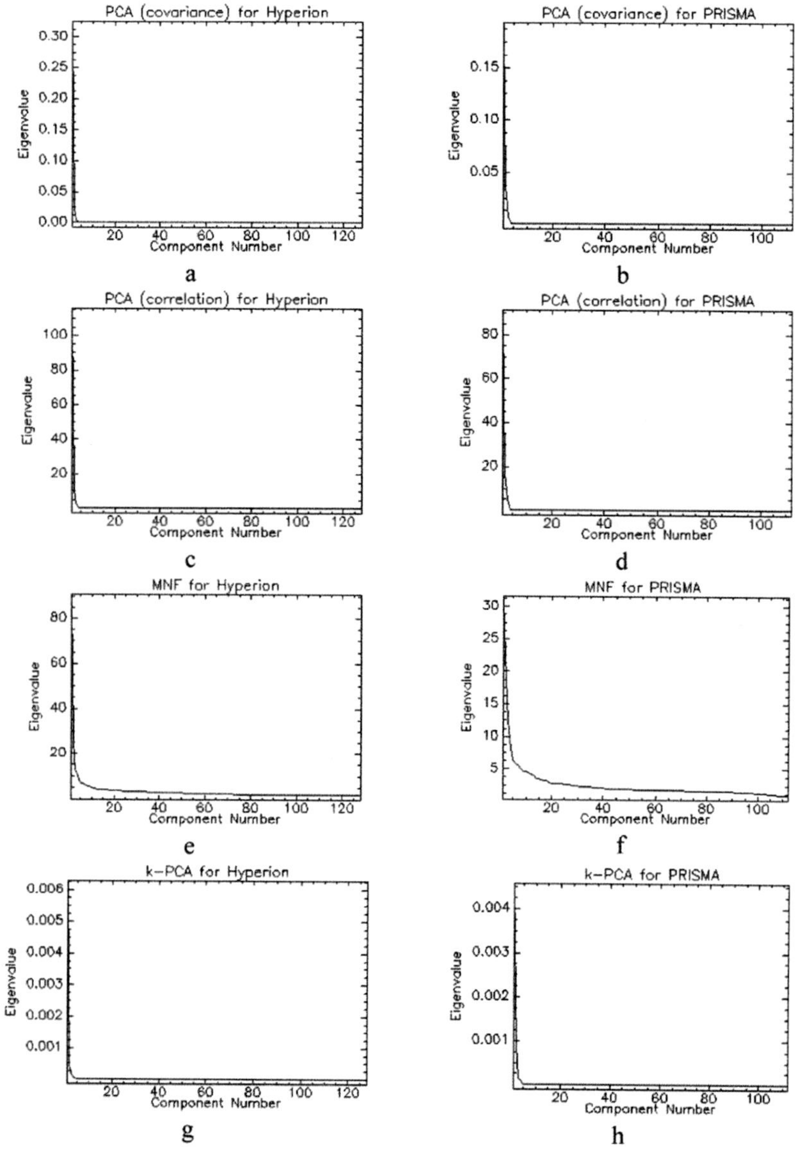

Fig. 4. Scree plots. a, c, e and g are scree plots of PCA (covariance), PCA (correlation), MNF and k-PCA for Hyperion data. b, d, f and h are scree plots of PCA (covariance), PCA (correlation), MNF and k-PCA for PRISMA data

of PCA (covariance), which can be differentiated upon designating a threshold value. Vegetation appears more pronounced in PC2. The remaining components are dominated by noise and a mix-up of features. MNF1 and MNF2 highlight the water areas, while

MNF3 and MNF4 highlight vegetation and soil, respectively. Built-up is highlighted in MNF5, along with some soil and river water regions. MNF6 and MNF7 contain noise. Examining FCCs of MNF components (R:5, G:4, B:3 and R:4, G:3, B:2) shows that built-up areas are intimately mixed with vegetation and soil, making their extraction as a separate class almost impossible. The background noise mixed with each land cover feature can be easily classified as one of these classes if these components are selected for information extraction. Among k-PCA components, PC1, PC2 and PC4 highlight vegetation, soil, water, built-up and roads distinctly, which is also evident in Fig. 3ab.

7 Conclusion

PCA (covariance and correlation), MNF and k-PCA were executed on two same-season Hyperion and PRISMA images of Ahmedabad, India. Each method's eigenvalues were higher for Hyperion than PRISMA owing to Hyperion having more bands than PRISMA. k-PCA had the lowest eigenvalues in both cases and outperformed both PCA and MNF. FCC from k-PCA components, i.e., (R: PC3, G: PC2, B: PC1) in Hyperion's case and (R: PC4, G: PC2, B: PC1) in PRISMA's case distinguish the natural and artificial land covers clearly. A comparison with ICA and partial informational correlation (PIC) [23] is ongoing. Future research also includes testing these approaches on the super-resolved outputs generated from Hyperion and PRISMA data.

Acknowledgements. The authors are grateful to the United States Geological Survey (USGS) and the Italian Space Agency (ASI) for providing Hyperion and PRISMA datasets free of charge.

References

1. Darvishzadeh, R., Skidmore, A., Atzberger, C., van Wieren, S.: Estimation of vegetation LAI from hyperspectral reflectance data: effects of soil type and plant architecture. Int. J. Appl. Earth Obs. Geoinf. **10**(3), 358–373 (2008)
2. Tripathi, P., Garg, R.D.: First impressions from the PRISMA hyperspectral mission. Curr. Sci. **119**(8), 1267–1281 (2020)
3. Mishra, K., Siddiqui, A., Kumar, V.: A comparative assessment of the efficacy of super resolved airborne hyperspectral outputs in urban material and land cover information extraction. Int. Arch. Photogramm. Remote. Sens. Spat. Inf. Sci. **42**(5), 653–658 (2018)
4. Goetz, A.F., Vane, G., Solomon, J.E., Rock, B.N.: Imaging spectrometry for Earth remote sensing. Science **228**(4704), 1147–1153 (1985)
5. Hughes, G.: On the mean accuracy of statistical pattern recognizers. IEEE Trans. Inf. Theory **14**(1), 55–63 (1968)
6. Green, A.A., Berman, M., Switzer, P., Craig, M.D.: A transformation for ordering multispectral data in terms of image quality with implications for noise removal. IEEE Trans. Geosci. Remote Sens. **26**(1), 65–74 (1988)
7. Jolliffe, I.T., Cadima, J.: Principal component analysis: a review and recent developments. Philos. Trans. R. Soc. A: Math., Phys. Eng. Sci. **374**(2065), 20150202 (2016)
8. Scholkopf, B., Smola, A.J., Muller, K.R.: Nonlinear component analysis as a kernel eigenvalue problem. Neural Comput. **10**(5), 1299–1319 (1998)

9. Fauvel, M., Chanussot, J., Benediktsson, J.A.: Kernel principal component analysis for the classification of hyperspectral remote sensing data over urban areas. EURASIP J. Adv. Signal Process/ **2009**(1), 1–14 (2009). https://doi.org/10.1155/2009/783194
10. van der Maaten, L.J.P, Postma, E.O., van den Herik, H.J.: Dimensionality reduction: a comparative review. Tilburg University Technical Report, TiCC-TR 2009-005 (2009)
11. Tipping, M.E.: Sparse kernel principal component analysis. In: Advances in Neural Information Processing Systems, vol. 13, pp. 633–639. The MIT Press, Cambridge MA USA (2000)
12. Siddiqui, A., Chauhan, P., Kumar, V., Jain, G., Deshmukh, A., Kumar, P.: Characterization of urban materials in AVIRIS-NG data using a mixture tuned matched filtering (MTMF) approach. Geocarto Int. **37**(1), 332–347 (2022)
13. Priyadarshini, K.N., Sivashankari, V., Shekhar, S., Balasubramani, K.: Comparison and evaluation of dimensionality reduction techniques for hyperspectral data analysis. Proceedings **24**, 6 (2019). https://doi.org/10.3390/IECG2019-06209
14. Tripathi, P., Garg, R.D.: Feature extraction of DESIS and PRISMA hyperspectral remote sensing datasets for geological applications. Int. Arch. Photogramm. Remote. Sens. Spat. Inf. Sci. **44**, 169–173 (2021)
15. Pignatti, S., et al.: Environmental products overview of the Italian hyperspectral prisma mission: the SAP4PRISMA project. In: 2015 IEEE International Geoscience and Remote Sensing Symposium (IGARSS), pp. 3997–4000. IEEE (2015)
16. Tripathi, P., Garg, R.D.: Comparative analysis of singular value decomposition and eigen value decomposition based principal component analysis for Earth and lunar hyperspectral image. In: 2021 11th Workshop on Hyperspectral Imaging and Signal Processing: Evolution in Remote Sensing (WHISPERS), pp. 1–5. IEEE (2021)
17. Hyvärinen, A., Oja, E.: Independent component analysis: algorithms and applications. Neural Netw. **13**(4–5), 411–430 (2000)
18. Tripathi, P., Garg, R.D.: Data dimensionality reduction based geological interpretation of aviris-ng hyperspectral data. In: Buckley, S., et al. (eds.) OSA Optical Sensors and Sensing Congress 2021 (AIS, FTS, HISE, SENSORS, ES), OSA Technical Digest, paper HTh2B.4. Optica Publishing Group (2021)
19. Datt, B., McVicar, T.R., Van Niel, T.G., Jupp, D.L., Pearlman, J.S.: Pre-processing EO-1 Hyperion hyperspectral data to support the application of agricultural indexes. IEEE Trans. Geosci. Remote Sens. **41**(6), 1246–1259 (2003)
20. Adler-Golden, S. M., et al.: Atmospheric correction for short-wave spectral imagery based on MODTRAN4. In: SPIE, Imaging Spectrometry V, vol. 3753, pp. 61–69 (1999)
21. Busetto, L., Ranghetti, L.: prismaread: A tool for facilitating access and analysis of PRISMA L1/L2 hyperspectral imagery. https://github.com/IREA-CNR-MI/prismaread. Last accessed 30 Jun 2020
22. Shawe-Taylor, J., Christianini, N.: Kernel Methods for Pattern Analysis. Cambridge University Press, Cambridge UK (2004)
23. Paul, S., Kumar, D.N.: Partial informational correlation-based band selection for hyperspectral image classification. J. Appl. Remote Sens. **13**(4), 046505 (2019)

Check for updates

Intelligent Multi-sensor Mobile System

Hind Mestouri[1](✉) and Saida Bahsine[2]

[1] Laboratory of Mathematical Computer and Communication System UCA, Safi, Morocco
h.mestouri@uca.ma

[2] Laboratory of Fundamental and Applied Physics UCA, Safi, Morocco

Abstract. This paper proposes an intelligent mobile detector system, using multiple sensors, for applications exploring inaccessible environments, controlled remotely via Wifi, using the Arduino board. The system features a gas sensor for leak detection (such as smoke and butane), and sends an alert or turns on a fan. A temperature sensor is used, in case of fire, to send an alert or trigger a water system. The system will be mobile to exploit places inaccessible by humans, and sends the collected results via Wifi to a computer or Smartphone. An initial prototype is established; different detection results are obtained and discussed.

Keywords: Gas sensor · Temperature sensor · Arduino board · Arduino Sorf-ward · Fritzing · Vertuino · WIFI · DC motor

1 Introduction

Connected objects, internet of things, are a recent efficient way to collect data and even to process them in some cases. The explosion of this technology is due to the miniaturization of electronic components, but also to their low cost and eases of implementation. A connected object is equipped with sensors that communicate the measured value via a network interface, either locally or via the Internet, in order to observe the measurement or to process it and then give the object the signal to act in accordance with the various actuators that can be made available to it [1].

In the case where the human presence is not possible, for example in the case of natural disasters, a mobile robot equipped with several sensors, can explore a non-accessible environment, collect different types of information and send them remotely. The main objective of this work is to realize an intelligent system able to control and analyze data sent by a robot through a programmable board. It is a small robot controlled manually via Wi-Fi. Its main mission is to explore places where humans cannot access by measuring the temperature and concentration of smoke. This robot has as a base a programmable card, from which we will specify the tasks that the robot must do (forward, backward….) using the ESP8266 model (Wi-Fi model), and thanks to an Android platform, we can visualize the different data collected by the robot (the temperature value and the concentration of smoke).

In this article we start by presenting the characteristics of the different components used, and the software used. A simulation model and a first prototype are presented, we have performed several measurements and tests, and we will present the different results and discussions.

M. Ben Ahmed et al. (Eds.): SCA 2022, LNNS 629, pp. 46–57, 2023.
https://doi.org/10.1007/978-3-031-26852-6_4

2 Components and Software

In this part, we will present the components and the different software used for the realization of our intelligent multi-sensor system [2].

2.1 Components

Arduino UNO. Arduino is an open-source project that provides a platform for prototyping interactive objects, consisting of an electronic board and a programming environment. The electronic board is not equipped with any type of connector other than the USB port, so it uses a different module (Wifi, Bluetooth, memory card reader, etc.) which makes it cheap, extremely adaptable to different projects and economical from an energy point of view [3] (see Fig. 1).

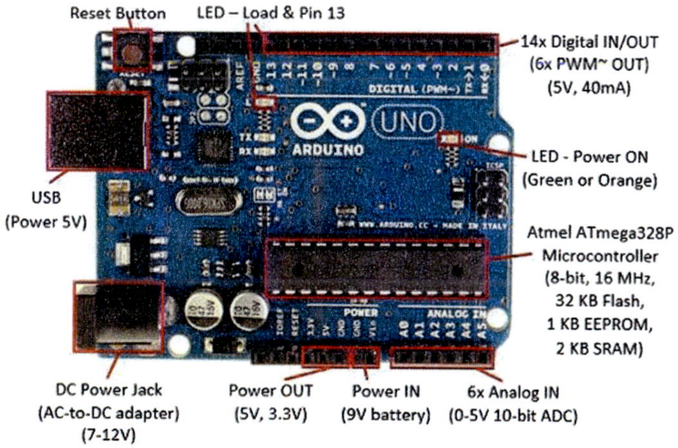

Fig. 1. Arduino UNO

There are different types of Arduino boards of different sizes and with different features. For our project we chose the Arduino UNO board, one of the most standard boards and one that meets our needs perfectly.

The Arduino UNO board is based on an ATMega328 microcontroller clocked at 16 MHz. It is the most economical microcontroller board of Arduino. Connectors located on the outer edges of the board allow plugging a series of complementary modules.

Temperature Sensor LM35. The LM35 is an integrated circuit, calibrated at the factory, or use as a precision temperature sensor (see Fig. 2).

Specifications [2, 4]:

- Typical consumption 60 μA.
- Accuracy: ±0.75 °C (typical).

- Calibrated directly in degrees Celsius.
- Probe gain (output): 10 mV/°C.
- Supply voltage: +4 to +30 V (+20 V recommended).

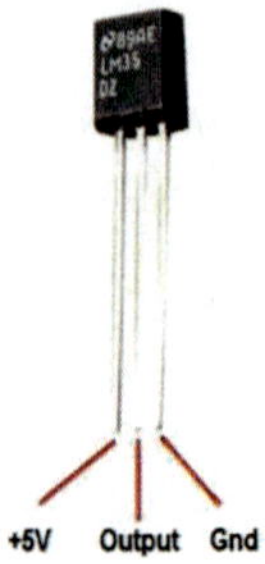

Fig. 2. Temperature sensor LM35

Gas Sensor MQ-2. Gas Sensor MQ-2 (Fig. 3) is a sensor with an analog output (A0), which signals the presence of smoke by raising the output voltage. The sensor delivers a voltage proportional to the gas concentration. When the gas concentration is higher, the output voltage is higher too.

In our case, the reading of the reference voltage on the Arduino is in reference condition, output A0 returns a value around 200 mV and this value rises quickly in the presence of gas.

Specifications [2, 4]:

- Vcc 5 V input;
- GND input;
- Analog output A0;
- Digital output D0;
- Detects: smoke, propane, butane, methane, alcohol, hydrogen, LPG.

Fig. 3. MQ-2 gas sensor

DC Moto. A DC motor (Fig. 4) is an electromechanical converter that allows the bi-directional conversion of energy between an electrical installation with direct current and a mechanical device; depending on the energy source. In simple terms, this means that a DC motor can convert electricity into mechanical energy. DC motors have the particularity of being able to operate in both directions, depending on the way in which the current is applied to it.

Fig. 4. DC motor

Breadboard. The breadboard (Fig. 5) is a solderless device for making brief models with equipment and for testing circuit designs. Most electronic sections of electronic circuits can be interconnected by installing their wires or terminals in the openings and, a little later, establishing a relationship through wires, if necessary. The test plate consists of metal strips placed underneath the plate and associated with the openings located at the most noticeable location on the plate. The metal strips are arranged in a fan-like pattern as shown below. Note that the top and base openings are connected evenly and separated in the middle in an upward direction [5].

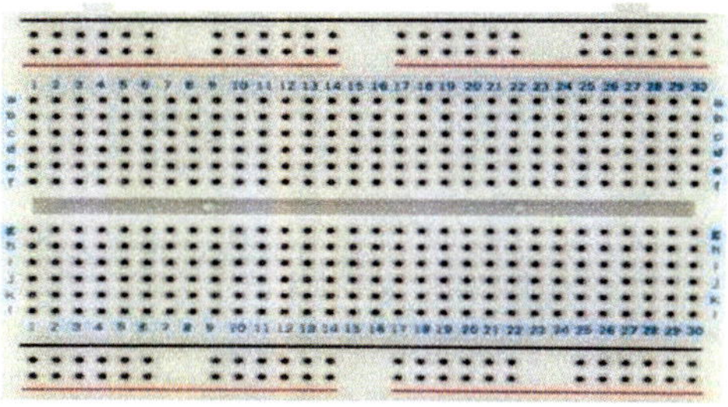

Fig. 5. Breadboard

Servomotor. Servomotors are actuators. Widely used in model making and industry, their main characteristic is their "torque", i.e. the rotational force they can exert. The more torque a servomotor has, the more it will be able to operate heavy "members" such as moving an arm that carries a load (see Fig. 6).

Fig. 6. Servomotor

ESP8266 WiFi. The ESP8266 WiFi (see Fig. 7) module is a standalone network with a built-in TCP/IP protocol stack that can give a microcontroller access to your Wi-Fi network. The ESP8266 is capable of hosting an application or offloading all Wi-Fi network functions from another application processor. Each ESP8266 module is preprogrammed with AT control configuration firmware, which means you can simply plug it into your Arduino device and get as much WiFi-capability as the WiFi Shield offerings (and that's just out of the box)! The ESP8266 module is an extremely cost-effective board with a huge and still growing community.

The supply voltage for these modules must be 3.3 V, but the RX pin can only take a maximum voltage of 3.3 V. You will need a voltage divider bridge to bring back the 5 V voltage delivered by the Arduino board so as not to damage the RX pin of the ESP8266 module. The ESP8266 pins have 6 pins, only 4 of which are commonly used. In the order:

- Rx / Tx (serial link);
- Reset which must be at HIGH (+3,3 V) to make the module work, LOW makes a reset of the module;
- CH_PD which must be at HIGH (+3,3 V) to make the module work or LOW to put it in standby;
- 2 GPIO pins.

Fig. 7. Wifi ESP8266

Integrated Circuit L293D. The L293D (see Fig. 8) is a power bridge composed of several transistors and relays that enables the rotation of a motor. The L293D is a double

H-bridge, which means that it is possible to use it to drive four separate motors (in one direction only) thanks to its 4 channels. It is also possible to make up two H-bridges to drive two separate motors, in both directions and independently of each other. It is important to note that the L293D can deliver a maximum of 600 mA, so please choose your motors accordingly. The technical characteristics of the L293D component are:

- Number of H-bridges: 2;
- Max Current Continuous: 600 mA (x2);
- Max peak current <2 ms: 1200 mA;
- VS Max Motor Supply: 36 V;
- VSS Max Logic Supply: 7 V;
- Number of pins: 16 DIP;
- Voltage loss: 1.3 V.

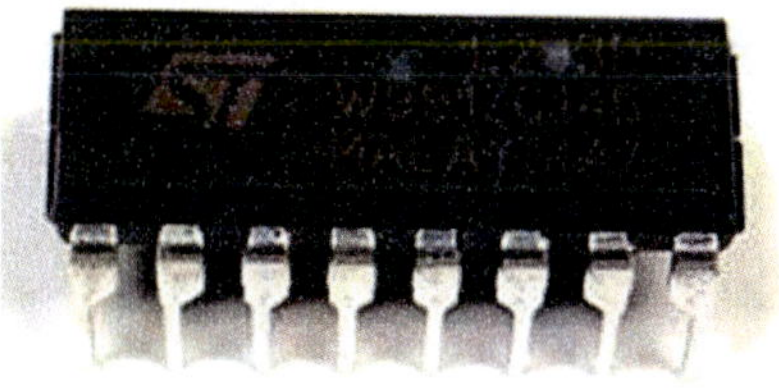

Fig. 8. Integrated circuit L293D

2.2 Softwares

Arduino Software. The Arduino software is a free hardware board with a microcontroller on it that can be programmed to analyze and produce electrical signals in order to perform a variety of tasks such as controlling a robot.

Fritzing Software. The Fritzing software is an open-source hardware initiative that makes electronics accessible as a creative material for everyone. It offers a software tool, a community site, and services in the spirit of Arduino and processing, fostering a creative ecosystem that allows users to document their prototypes, share them with others, teach electronics in a classroom, and layout and build [6].

Virtuino Application. Virtuino is an Android application designed to control the robot remotely through a dynamic interface, this interface allows visualizing everything that happens at the robot, also to control its motors to make it move forward, backward, right, and left [7].

3 Development of a Simulation Model and Prototype

3.1 Simulation Model

For this study we have established a simulation model of our system, thanks to the software Fritzing, knowing that we have the different characteristics of each component that we will use (Fig. 9).

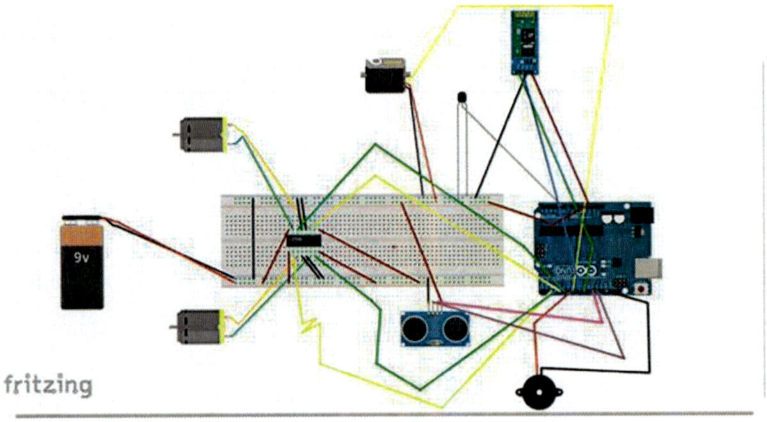

Fig. 9. Simulation of the global connection diagram

3.2 Creating a Prototype

We started with the assembly and connection of the components of our system using the connection cables and the frame, which is a plate on which all the components of the robot are placed and which allows it to be held in balance. As shown in Fig. 10.

The presented assembly allows measuring continuously the temperature, it is possible to visualize the detected values during a time interval that will be determined in the test.

This visualization is done at the level of the Arduino IDE. It also allows to detect the presence of gas, thanks to a gas sensor, which measures the concentration of smoke. A bell will react according to the value detected by the sensor.

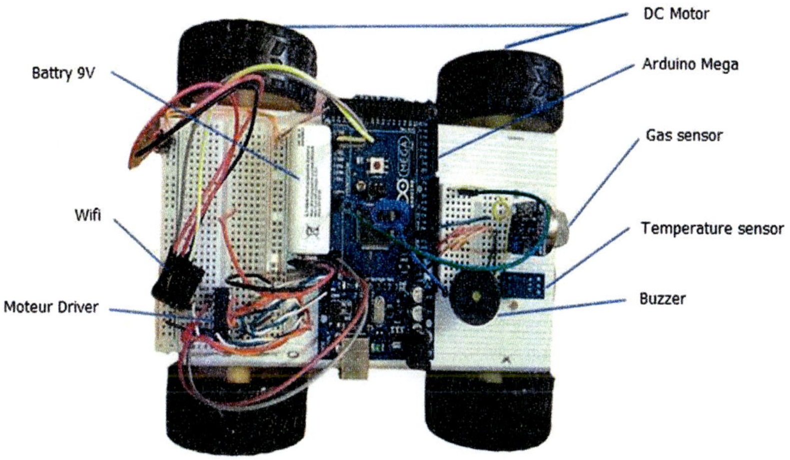

Fig. 10. Assembly of the prototype system

3.3 Block Diagram

In order to understand how the program works, and to make it more readable, we preferred to put it in the form of a flowchart (see Fig. 11).

We had to create the Arduino program to read the detected data and generate the appropriate outputs. Once the quantities are detected, it reads them and displays them in IDE Arduino, as it triggers the alarm for the detection of gas and displays the results of the detected temperature, while turning the servomotor to explore all the surroundings of the system.

4 Results and Discussions

4.1 Temperature Measurements

We first measured the initial value of the temperature in the absence of flame, the maximum value that the sensor can detect is 100 °C. In order not to damage it, we measure the maximum distance corresponding to this value, we find 5 cm, we make vary this distance until 45 cm by a step of 5 cm and we measure. The calibration is repeated several times, varying the intensity of the fire. Normally the temperature value detected is displayed in mV, and then converted into degrees Celsius, but we have modified the code to display it in Celsius directly for better readability and ease of measurement. The results are shown in Fig. 12.

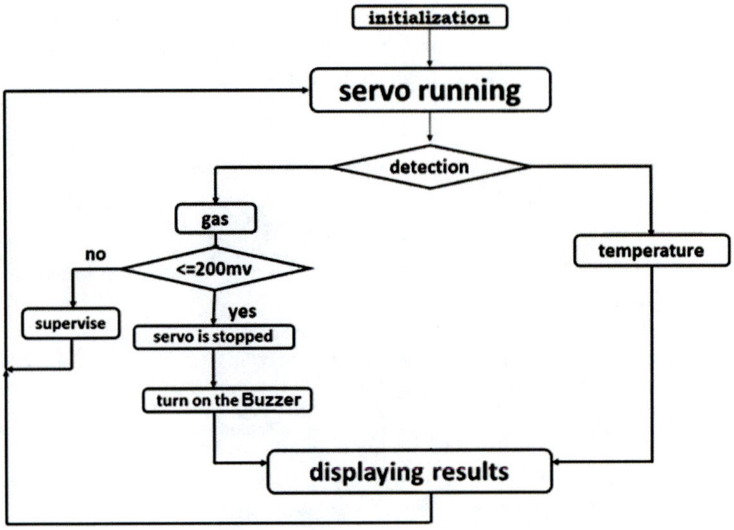

Fig. 11. Flowchart of the system

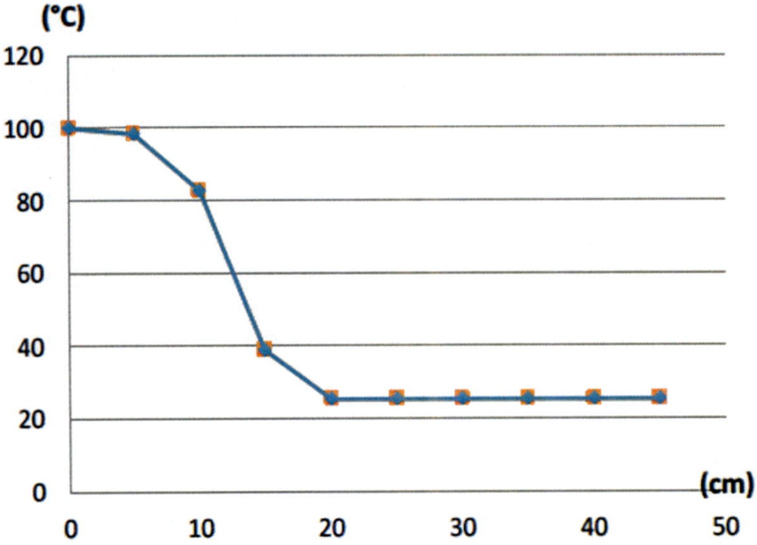

Fig. 12. Graph showing the calibration of the temperature sensor

We note that the variation of the measured value and the distance are inversely proportional. Between 5 cm and 15 cm and between 25 cm and 45 cm the measured value is almost stable, it undergoes a small variation. Between 15 cm and 25 cm the relationship between the distance and the measured value is approximately linear. The temperature measurements start to take the same value of 25.2 °C from appropriate fire intensity.

For this sensor we have done, however, an angular calibration. As before, we measure the initial value that the sensor detects in the absence of flame, then we measure the distance corresponding to the maximum value of the sensor at a well defined angle, the angular interval varies from 0° to +40° by a step of 5°. We repeat the calibration several times by varying the intensity of the fire.

We found that the sensor calibration measurements begin to take on the same 252 mV value at the appropriate fire intensity. The maximum value is at the angle Ø° to the sensor, which means that the sensor reaches its maximum response when the flame source is in front of it.

4.2 Gas Detection

Similarly for the gas sensor we measure the initial value captured by the sensor in the absence of smoke, then we measure the distance corresponding to the maximum value of the sensor at a well defined angle. We note that this sensor has a very wide angular range which does not help us to direct us to the source of gas with precision, also the vapor state of gas forces us to vary the distance of the source of gas vertically with respect to the sensor. We note that the higher the gas concentration, the higher the output voltage. Also the alarm is triggered in the presence of gas, we can also connect an alarm.

4.3 Driving the Robot

First of all, we launch the Virtuino application on the Smartphone, we connect to the server thanks to its IP address contained in the virtuino library. Our control interface consists of four buttons: up (forward), down (reverse), left (turn left) and right (turn right). The concentration of smoke and the degree of temperature captured by the robot are displayed in a new interface of the application. Figure 13 shows an example of the Virtuino application display [7, 8].

4.4 Discussions

The system is able to detect different signals at the same time, and we can combine the different results detected, temperature and gas, to shout an intelligent system, which will analyze the results, for example if the temperature is high and there is no presence of gas, in this case, it will not send an alert of presence of fire, it will just display the value of temperature detected. On the other hand, in the case of detection of high temperature or presence of smoke, it will trigger the alarm. The small robot is controlled manually via WiFi. Its main mission is to explore places where humans cannot access by measuring the temperature and concentration of smoke. This robot has as a base a programmable card, from which we have specified the tasks that the robot must perform (forward, backward...), using the ESP8266 model (Wi-Fi model), and thanks to an Android platform, we could visualize the different data captured by the robot (temperature and gas values).

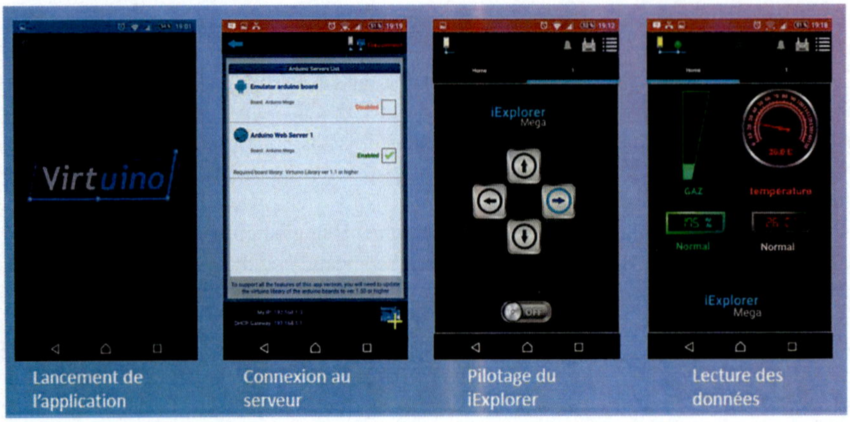

Fig. 13. The different steps to control the robot

5 Conclusion

In this work we were able to create a kind of monitoring prototype. We managed to make a first version of the system that performs its main functions, temperature and gas detection. The second step was to add to our system a Wifi ESP8266 model, which has the function of sending the data detected by the sensors using Wifi to an Android platform installed on a computer or Smartphone, this platform will allow to visualize the results, giving it an IP address that exists in the same network as the address of the wifi card. Moreover, the mobility has made the system more usable, for example in rescue and surveillance situations, in case of disasters, where the access for a human being contains risk, this mobility is controlled and manipulated by the same platform, adding a DC motor, manipulated remotely.

References

1. Gannavaram, T.K.V., Kandhikonda, U.M., Bejgam, R., Keshipeddi, S.B., Sunkari, S.: A brief review on Internet of Things (IoT). In: 2021 International Conference on Computer Communication and Informatics (ICCCI), pp. 1-6. Coimbatore, India (2021). https://doi.org/10.1109/ICCCI50826.2021.9451163
2. https://learn.adafruit.com/dht
3. https://www.arduino.cc
4. Karvinen, T., et al.: Les capteurs pour Arduino et Raspberry Pi. Tutoriels et projets. Du- nod, 304 p. (2014)
5. Gannavaram, T.K.V., Bejgam, R., Keshipeddi, S.B., Sunkari, S., Aluvala, V.K.: Conversion of sound energy into electrical energy in highly populated areas. In: 2021 6th International Conference on Communication and Electronics Systems (ICCES), pp. 32–36 (2021). https://doi.org/10.1109/ICCES51350.2021.9489219
6. https://fritzing.org
7. https://virtuino.com/

8. Bejgam, R., Gannavaram, T.K.V.: Integrating machine to machine communication (M2M) and MQTT protocol techniques for conversion of water motor pump into a smart system. In: 2021 Second International Conference on Electronics and Sustainable Communication Systems (ICESC), pp. 982–987. Coimbatore, India (2021). https://doi.org/10.1109/ICESC51422.2021.9532848

Integrated System for Pressure Ulcers Monitoring and Prevention

Luis Fonseca[1], Fernando Reinaldo[1,2(✉)], José Metrôlho[1,2], Filipe Fidalgo[1,2], Rogério Dionísio[1,2], Arlindo Silva[1,2], Osvaldo Santos[1,2], and Mohammad Amini[3]

[1] Polytechnic Institute of Castelo Branco, Castelo Branco, Portugal
fribeiro@ipcb.pt
[2] DiSAC – Digital Services, Applications and Content, Castelo Branco, Portugal
[3] SENSOMATT, Lda, Portugal

Abstract. Pressure ulcers are a critical issue for patients and healthcare professionals, requiring their frequent monitoring, with a consequent impact on healthcare costs. This problem has been gaining attention and approaches have been proposed, using sensor-based systems, to facilitate this monitoring and help health caregivers to achieve greater effectiveness in the treatment of this type of ulcer. In this paper, the architecture, and the prototype of a new system for pressure ulcer monitoring and prevention are presented. It considers information related to both intrinsic and extrinsic predisposing factors and it addresses the components of data acquisition, data analysis, and production of complementary support to well-informed clinical decision-making. The system includes a pressure ulcer management portal and a mobile application, that allows caregivers to manage clinical information about pressure ulcers of the patients and uses data acquired from a pressure sensor sheet under the mattress to provide useful information for monitoring the patients. Considering the situation of each patient, the system will produce indicators/alerts to healthcare professionals, simultaneously improving pressure-ulcer patient care quality and safety and minimizing the burnout in healthcare professionals.

Keywords: E-health · Pressure ulcers prevention · Sensor-based systems · Smart healthcare

1 Introduction

Quality of healthcare is an important goal for patients, healthcare professionals, and the healthcare system. However, close monitoring is often still essential to implement specific prevention measures and ensure patients' quality of life. It is a healthcare intensive task and many technological approaches have been proposed to help both improve the outcome for patients and alleviate the burnout risk of healthcare professionals.

Particularly, in patients at risk of developing Pressure Ulcers (PU), or patients with PU, intrinsic data about the patient (e.g., limited mobility, poor nutrition, comorbidities, aging skin) [1–4] and/or extrinsic data (e.g., pressure from hard surfaces, shearing

M. Ben Ahmed et al. (Eds.): SCA 2022, LNNS 629, pp. 58–68, 2023.
https://doi.org/10.1007/978-3-031-26852-6_5

from involuntary muscle movements, excessive moisture) are very important to provide the healthcare provider with additional information which can facilitate the definition of appropriate monitoring schedule and treatment. Thus, close monitoring is essential to ensure their quality of life and that their condition does not become more serious. The improvement of these services can be supported by systems that can monitor the status of patients in real-time and provide information for acting according to individual diagnoses. These solutions help in monitoring and treatment, generating alerts and recommendations that are relevant to support healthcare professionals' decisions/actions, reducing their burden, contributing to an improvement in the patient's quality of life, and a decrease in health service costs [5]. In this sense, the use of IoT technologies associated with ML technologies creates new opportunities to develop a large range of solutions.

In this paper, we present a system, denominated SensoMatt, for pressure ulcer prevention cantered around body pressure data acquired from a sensor sheet placed under the bedridden patient. The system produces indicators/alerts to healthcare professionals, simultaneously minimizing the burnout in healthcare professionals and improving pressure-ulcer patient care quality and safety.

The remainder of this paper will be as follows. Section 2 presents the goal and main requirements of this proposal. Section 3 provides an overview of related works. Section 4 provides an overview of the proposed solution, including the proposed architecture technologies and main functionalities. Finally, Sect. 5 presents the conclusion and directions for future work.

2 Goal and Requirements

The main goal of this project is to conceptualize, model, and develop a system that could be suitable for supporting healthcare professionals' decisions when taking care of pressure ulcer patients. The research and development are focused on the establishment of indicators based on two ways of acquiring data: a sensor sheet (placed in the bed of the patients) and data observed and inserted by healthcare professionals. Data will be stored resulting in intelligence reports and alarms for healthcare professionals for preventing and lowering the risk of pressure wounds developing further, so that actions can be adjusted according to the generated alarms.

Based on the demands and needs for this specific target of application, for this system the functional requirements that were identified are:

- The system should allow registering of users and editing of their personal data;
- The system should allow the creation and validation of indicators KPI (Key Performance Indicators);
- The system should allow the building of dashboards with direct and/or parameterized consultation, with the option to produce reports and the respective consultation;
- The system should generate alerts and allow users to consult them.

The non-functional requirements are:

- Portability - Limitations regarding hardware and software on which the system will be implemented and on the ease of migration;
- Security – blocking unauthorized access;
- Usability – Ease and accessibility in using the system.

The developed functionalities are displayed in the use case diagram of Fig. 1:

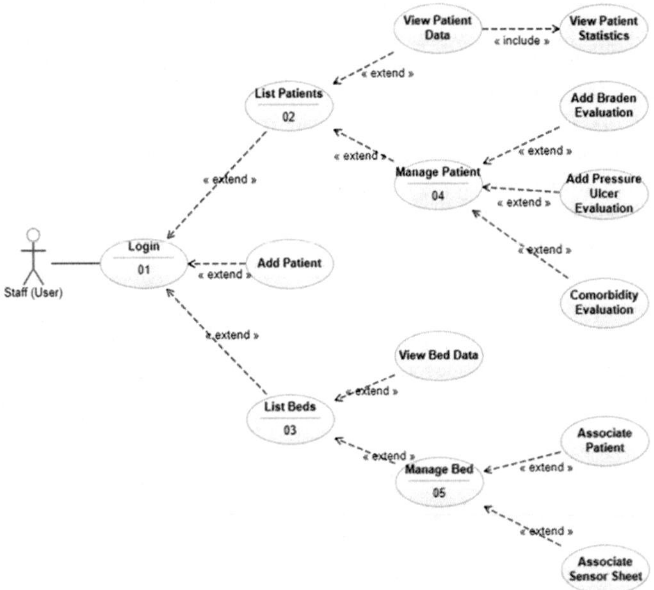

Fig. 1. Use case diagram.

The staff role represents every user with access to the platform, this access must be given by the system administrator. The Staff User is required to login information to have access to the platform, once entered, the staff user will be redirected to whichever page they were trying to access. The system should allow the Staff User to see a list of patients. The patient list view can be searched for patient information (name, age, birthdate). It should also allow the Staff User to create new patients in the system or select a specific patient from the list and manage his medical record. The patient's clinical information should include information on risk factors for pressure ulcers (intrinsic and extrinsic) as well as other relevant Inputs for pressure injury predictive algorithms. This includes providing information about the actions that are taken on that patient's condition but also information about the patient's condition and pressure ulcers. Staff Users should be able to monitor and visualize the patient's condition at each moment, as well as analyse statistics about the patient's pressure map in the most critical parts of the patient, over time. Whenever necessary, the user must be allowed to enter information about any actions he has taken that change the patient's conditions. Finally, the system should allow the staff user to see a list of beds. The beds list view can be searched by location or bed identification. It should also allow the staff user to insert new beds in the system or select a specific bed from the list and manage its features and status. It must also be allowed to assign a patient with a bed. The implementation of these requirements will be described in Sect. 4.

3 Related Work

In the last decades, there have been several research works to propose technical approaches to gather sensorial information deemed useful for pressure ulcer prevention. The fundamental goal of these approaches is to acquire the information needed to classify the lying position of the patient using some sort of pressure sensor usually placed below the patient's sheets (e.g., [6, 7]). Furthermore, other different approaches based or complemented by several types of inertial sensors [8–10], video images [11, 12], and others [13], also get satisfactory results. The fact that pressure sensors do not need to be attached to the patient is relatively flexible in the information they provide, are easily installed, and can be used successfully with different algorithms for lying position classification, seems to have increased, over the years, of the number of approaches based on this type of sensor. The information acquired from the sensors is processed in varying ways according to the sensor type. Mostly it included simple processing steps of the numerical raw data obtained from the sensors, but, mainly in the case of the pressure sensors, some effort is made by various authors to obtain higher-level features from the original data matrix (e.g., [6, 7]).

The algorithms can be used in pressure ulcer prevention approaches in the following three stages:

- preprocessing of data: In this stage, the processing is usually relatively simple, but in some works, more sophisticated algorithms were used (e.g. [14]).
- lying position classification: this is a very important factor for pressure ulcer prevention, and it can be implemented using many combinations of different sensors and algorithms. Some form of pressure sensor matrix is possible, and it is the most common sensor used. Regarding the classification algorithms, neural networks (including deep networks) (e.g. [6–8, 15]), support vector machines [9, 13], principal component analysis [13], and tree or rule-based systems [8, 16] are some of the usual approaches. Neural network-based approaches are the most popular, followed by support vector machines.
- decision towards ulcer prevention: considering, as input, data about the lying position and other sources (patient history, etc.) generate outputs to help caregivers to decide what actions to apply to the patient.

However, most research works stop at lying position classification. The ones that go forward tend to contribute to pressure ulcer prevention in the following ways:

i. Monitoring [14–17] – data gathered from the sensors can be analysed by caregivers. This can include pressure maps or other values (e.g., time since the last change in position, physiological data, etc.). These values can be displayed, usually in graphical form or using mobile apps.

ii. Notifications [9–11, 14–19] – Raising of alarms when the patient is resting in the same position for longer than a specified amount of time. This reduces the risk of the patient resting for extended periods in the same position (the most common cause of pressure ulcers) and can save the caregiver's time if the patient changes position spontaneously. Other alarms can be raised if the patient moves too much – when restlessness is a risk – or gets up from bed (maybe falling).
iii. Personalization e.g. [9, 14] – allow that data to be visualized as is relevant for that particular patient. Information regarding the patient's medical history can be inputted into the application, to be displayed or used in some decision-making processes.
iv. Actuation [16, 17] – these are not common and tend to be very expensive. As an example, a few beds have pressure actuators that can control pressure in specific areas. Some of these actuators can be controlled remotely by caregivers.
v. Prediction [20] – offer some kind of prediction of pressure ulcer occurrence based on diverse sources of information, both sensor-based and obtained in other ways, such as patient history or physiological data.

Good results can be achieved in lying position detection using a variety of techniques. However, these results are yet rarely tested in real conditions and are measured in some form of a simulated environment. Several approaches raise alarms that are helpful to the caregivers and health professionals, namely as a warning system for the position changing cycle and as an alarm for dangerous situations such as restlessness or getting up of bed (or falling). So, there are several successful approaches to the lying position detection problem with clear practical implications, yet there are less successful cases of approaches able to use this information in conjunction with other data sources to effectively predict pressure ulcer development risk.

4 A System for Pressure Ulcers Monitoring and Prevention

In the context of what was mentioned in the previous section, the architecture and prototype of a new approach that is being developed to fill some of the identified gaps, and contribute to a more complete decision support system, are presented below. This system is also based on data from a pressure map, constructed from the acquisition of values using a pressure sensor sheet, and personalized data from the respective patient being monitored.

4.1 Architecture and Technologies

This section contains information regarding the overall implemented architecture of the system, split into two sections, one for the sensor sheet data and one for the clinical data.

The sensor sheet data section portrays the system capable of extracting data from the sensor sheet and, using artificial intelligence algorithms, the creation of alarms to be used by the applications used by the staff. The clinical data section is the section responsible for clinical data that will also be used in the algorithms, according to each patient's physical and medical characteristics the algorithm will have different outcomes.

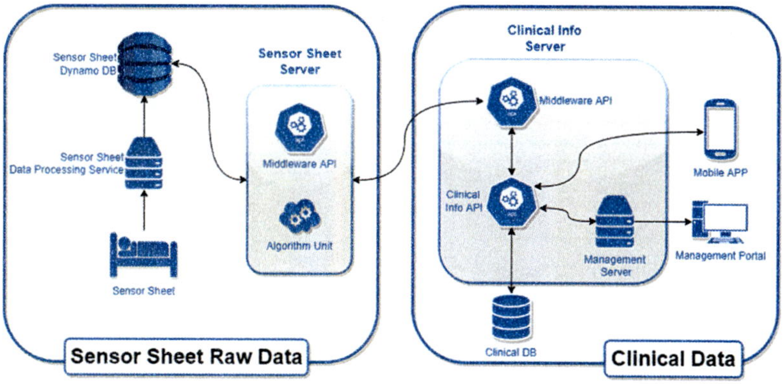

Fig. 2. Architecture diagram.

The architecture on the left side of the Fig. 2 contains the sensor sheet, which gathers the data of the bedded patient, a data processing service that receives the data from the sensor sheet and cleans it so only the required data is stored in a database. The two other elements on the sensor sheet raw data section are responsible for passing the raw data through the algorithm unit so that body positions and body parts that are under pressure can be identified. This information can be translated into alarms to be used in the management portal and application. Alarms will be transmitted through an API that ensures communication between the different sections.

The right side of the diagram contains the architecture for the clinical data section of the architecture which accounts for the data acquired through the usage of both the management portal and the mobile application, the database (Clinical DB) is where the information regarding the patients, beds, and users is stored, this can be accessed through the backend, encapsulated inside a server where the clinical info API is responsible for managing data on the database, this API is the communication bridge between the management server/mobile application and the clinical database, the server also contains the Middleware API responsible for communicating with the left section of the architecture for the algorithmic processes regarding the patient's health condition. The front end of the architecture is hosted on a React server for the management portal which can be accessed from any browser with internet access.

4.2 Prototype

Most healthcare systems already have an application to facilitate patient data management, mostly generalized so they can have as much information as possible. For the SensoMatt system, there was a simplistic approach in which only the data considered relevant to pressure wounds (intrinsic/extrinsic factors) is managed. This way the responsible medical staff can view and edit information related to this subject more quickly and easily.

A pressure sensor sheet under the mattress provides useful information about the patient's body parts under pressure. Placing the pressure sensors underneath the mattress

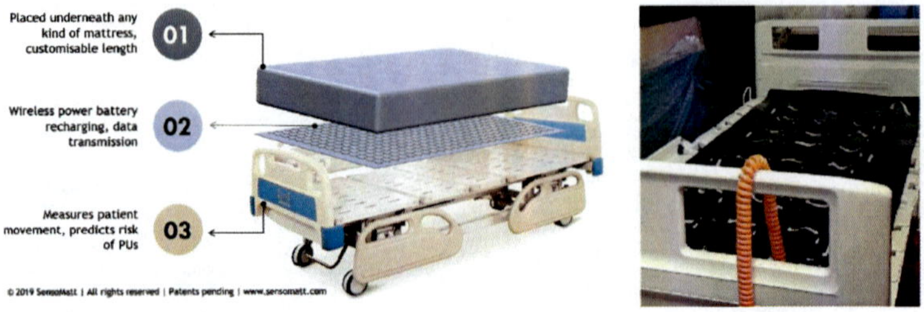

Fig. 3. Left: Sensor sheet in the bed. Right: Sensor sheet.

is a less intrusive solution for the patient. This way there are no changes in temperature or friction, it does however pose a challenge when trying to accurately measure the pressure and identify the patient's body position when lying down. The sensor sheet layout shown in Fig. 3 (left), presents the 3 corresponding intervening factors for its functioning. First the bed mattress, for an all-around solution the sensor sheet should be put below the mattress and be able to correctly measure the pressure distribution despite the differences in mattresses. Secondly, the sensor sheet measures patient movement and pressure and sends the measured data to the Sensor Sheet Data Processing service. Thirdly, the data obtained by the sensor sheet is registered, which allows producing a pressure distribution map of the body and identifying the posture of the bedridden patient, using Artificial Intelligence algorithms. Based on this information, it is possible to identify the parts of the body which are under pressure, and this information is used to predict the risk of pressure ulcers. The sensor sheet is presented in Fig. 3 (right). In its current version, it contains an array of 4×10 sensors, and a sheet sensor with a higher number of sensors is being developed.

The information is later processed using technologies such as cloud computing, mobile computing, and artificial intelligence to provide patients and clinical staff with personalized technology for the early detection and prevention of pressure ulcers.

Considering each patient's situation, together with the clinical information, the system will produce indicators/alerts to healthcare professionals, simultaneously improving the quality and safety of pressure ulcer patient care and minimizing the burnout on healthcare professionals. In the prototype's current stage, alerts are based on rules that consider the patient's existing ulcers, the patient's body position and for how long the same position is kept. The platform includes a web portal and a mobile application for managing information on pressure ulcers using information obtained by the sensor sheet. The management portal was designed to convene a faster way of evaluating the patient's state concerning pressure ulcers, the patient is added into the system with the filling of a simple form that includes some medical parameters like height and weight as well as some other possibly related health issues. The patient can then be associated with a bed in the system and be monitored by a sensor sheet when available. After admission, the medical staff can perform multiple evaluations, be it relevant medical information like comorbidities or Braden[1] scale evaluations that measure the risk of pressure ulcers or

[1] https://www.bradenscale.com/.

evaluate the patient's afflicting pressure ulcers so they can be recorded and monitored throughout the bedding period.

In Fig. 4 there is a portion of the pressure ulcer evaluation screen in which the clinical staff can evaluate the state of the patient's ulcers, in here the staff can select a body part from a body image mapped with the multiple body parts, after selecting a body part they are prompted to input the wound's width, depth, and pressure ulcer stage, this can be done for multiple body parts and once the user is finished they can complete the evaluation by clicking the "Evaluation Complete" button.

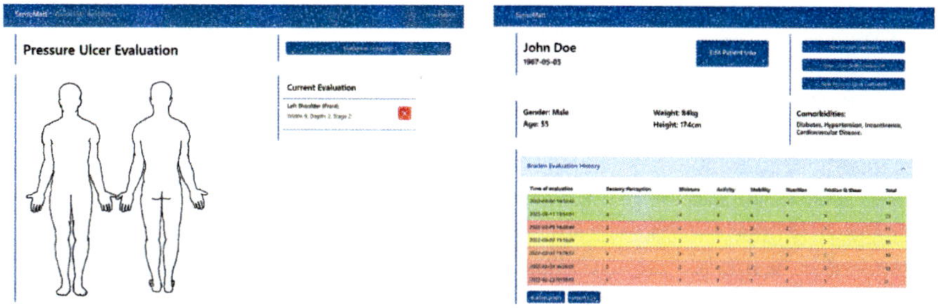

Fig. 4. Left: Pressure Ulcer Evaluation Screen. Right: Braden evaluation table.

The evaluations are then stored and can be viewed in a table, displaying all the previous evaluations along with the parameters for each body part's ulcer, for easier assessment of the evolution of the patient's wounds each body part can be selected from the table so the staff can view a graphical representation of it.

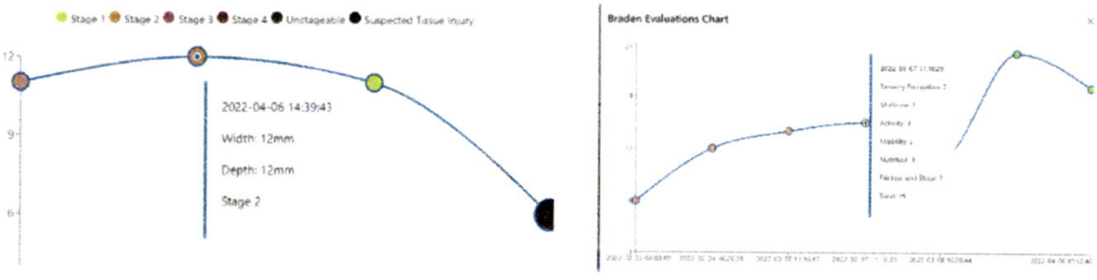

Fig. 5. Left: Pressure Ulcer graphical representation. Right: Braden evaluation chart.

Figure 5 shows the graphical representation of a simulated pressure ulcer and its evolution over 4 different evaluations, each dot represents an ulcer, the Y-axis represents the depth of the wound, the size of the dot represents the width, and the colour represents the corresponding stage. The dots can also be hovered for a more detailed report on the evaluation of a certain ulcer at a certain time. Apart from this graphical representation the system also has a graphical representation for the Braden scale evaluations and all other data collected from the medical staff interactions with the platform.

The mobile app has a more simplistic approach to the functions included in the platform considering screen size and other conditioning factors. As such the main goal of the app is for healthcare professionals to have remote access to the patient's information and most importantly getting the same alerts the platform gets but in the shape of notifications for quicker access.

5 Conclusion and Future Work

In this paper, a system has been presented to support caregivers of patients with pressure ulcers. The architecture and the prototype of the new system for pressure ulcer monitoring and prevention were presented in their present state. This system is based on pressure data collection, using a pressure sensor sheet, and lying position determination. An information system to explore the acquired data, including back-office storage of relevant information and a mobile application for healthcare professionals are still under development, with specific care being taken to ensure that privacy, security, and other specific regulations in the healthcare field are fully met. Considering the situation of each patient, the system will produce indicators/alerts to healthcare professionals, simultaneously improving pressure-ulcer patient care quality and safety and minimizing the burnout in healthcare professionals. Work is also being done with healthcare professional partners to adjust different aspects of the prototype system, that will influence the requirements for the mobile application being developed, the definition of the data that is both useful to store in the information system, and the several aspects of the visualization, alarm, and predictive components that these professionals consider more helpful in managing patients with pressure ulcers.

Thus, future work will evolve in two directions. On the one hand, a sensor sheet is being developed with a greater number of sensors to obtain more accurate pressure map images. It will also be necessary to obtain pressure map images from a larger sample of patients. These will be important to train the body position detection algorithms and thus improve their accuracy. On the other hand, predictive algorithms will be studied and implemented to make recommendations to health professionals. The results will be compared with other similar works and evaluated by potential users.

Acknowledgements. This research has received funding from the P2020 and EU FEDER Programme [PO CENTRO 17/SI/2019] under grant agreement No. 70107 [SENSOMATT].

References

1. Bluestein, D., Javaheri, A.: Pressure ulcers: prevention, evaluation, and management. Am. Fam. Physician. **78**, 1186–1194 (2008)
2. Garcia, A.D., Thomas, D.R.: Assessment and management of chronic pressure ulcers in the elderly. Med. Clin. North Am. **90**, 925–944 (2006). https://doi.org/10.1016/j.mcna.2006.05.018
3. National Pressure Ulcer Advisory Panel, European Pressure Ulcer Advisory Panel and Pan Pacific Pressure Injury Alliance: Pressure injury alliance. prevention and treatment of pressure ulcers: quick reference guide. In: Emily, H. (ed.) Cambridge Media, Osborne Park, Australia (2014)

4. National Institute for Health and Care Excellence: Pressure ulcers. Quality standard [QS89] (2015). https://www.nice.org.uk/guidance/qs89. Last accessed 12 Jul 2021
5. Ribeiro, F., Fidalgo, F., Silva, A., Metrôlho, J., Santos, O., Dionisio, R.: Literature review of machine-learning algorithms for pressure ulcer prevention: challenges and opportunities. Informatics **8**, 76 (2021). https://doi.org/10.3390/informatics8040076
6. Rodríguez, A.P., Gil, D., Nugent, C., Quero, J.M.: In-bed posture classification from pressure mat sensors for the prevention of pressure ulcers using convolutional neural networks. In: Rojas, I., Valenzuela, O., Rojas, F., Herrera, L.J., Ortuño, F. (eds.) IWBBIO 2020. LNCS, vol. 12108, pp. 338–349. Springer, Cham (2020). https://doi.org/10.1007/978-3-030-45385-5_30
7. Matar, G., Lina, J.-M., Kaddoum, G.: Artificial neural network for in-bed posture classification using bed-sheet pressure sensors. IEEE J. Biomed. Health Inform. **24**(1), 101–110 (2020). https://doi.org/10.1109/JBHI.2019.2899070
8. Alinia, P., Samadani, A., Milosevic, M., Ghasemzadeh, H., Parvaneh, S.: Pervasive lying posture tracking. Sensors **20**, 5953 (2020). https://doi.org/10.3390/s20205953
9. Bernal Monroy, E., Polo Rodríguez, A., Estevez, M., Medina, J.: Fuzzy Monitoring of in-bed postural changes for the prevention of pressure ulcers using inertial sensors attached to clothing. J. Biomed. Inform. **107**, 103476 (2020). https://doi.org/10.1016/j.jbi.2020.103476
10. Minteer, D.M., et al.: Pressure ulcer monitoring platform—a prospective, human subject clinical study to validate patient repositioning monitoring device to prevent pressure ulcers. Adv. Wound Care **9**(1), 28–33 (2020). https://doi.org/10.1089/wound.2018.0934
11. Qidwai, U., Al-Sulaiti, S., Ahmed, G., Hegazy, A., Ilyas, S.K.: Intelligent integrated instrumentation platform for monitoring long-term bedridden patients. In: 2016 IEEE EMBS Conference on Biomedical Engineering and Sciences (IECBES), pp. 561–564 (2016). https://doi.org/10.1109/IECBES.2016.7843512
12. Bennett, S.L., Goubran, R., Rockwood, K., Knoefel, F.: Monitoring the relief of pressure points for pressure ulcer prevention: a subject dependent approach. In: 2013 IEEE International Symposium on Medical Measurements and Applications (MeMeA), pp. 135–138 (2013). https://doi.org/10.1109/MeMeA.2013.6549722
13. Lee, H.J., Hwang, S.H., Lee, S.M., Lim, Y.G., Park, K.S.: Estimation of body postures on bed using unconstrained ECG measurements. IEEE J. Biomed. Heal. Informatics. **17**, 985–993 (2013). https://doi.org/10.1109/JBHI.2013.2252911
14. Hsiao, R.-S., Mi, Z., Yang, B.-R., Kau, L.-J., Bitew, M.A., Li, T.-Y.: Body posture recognition and turning recording system for the care of bed bound patients. Technol. Heal. Care. **24**, S307–S312 (2016). https://doi.org/10.3233/THC-151088
15. Pongthanisorn, G., Viriyavit, W., Prakayapan, T., Deepaisam, S., Somlertlamvanich, V.: ECS: elderly care system for fall and bedsore prevention using non-constraint sensor. In: 2020 International Electronics Symposium (IES), pp. 340–344 (2020). https://doi.org/10.1109/IES50839.2020.9231781
16. Abdelmoghith, A., Shaaban, R., Alsheghri, Z., Ismail, L.: IoT-based healthcare monitoring system: bedsores prevention. In: 2020 Fourth World Conference on Smart Trends in Systems, Security and Sustainability (WorldS4), pp. 64–69 (2020). https://doi.org/10.1109/WorldS450073.2020.9210319
17. Barroso, I., et al.: Assisted caretaking system for geriatric home care. In: Stephanidis, C., Antona, M., Gao, Q., Zhou, J. (eds.) HCII 2020. LNCS, vol. 12426, pp. 439–449. Springer, Cham (2020). https://doi.org/10.1007/978-3-030-60149-2_34
18. Lee, C.-N., Yang, S.-C., Li, C.-K., Liu, M.-Z., Kuo, P.-C.: Alarm system for bed exit and prolonged bed rest. In: 2018 International Conference on Machine Learning and Cybernetics (ICMLC), pp. 439–443 (2018). https://doi.org/10.1109/ICMLC.2018.8527027

19. Liu, J., Chen, X., Chen, S., Liu, X., Wang, Y., Chen, L.: TagSheet: sleeping posture recognition with an unobtrusive passive tag matrix. In: IEEE INFOCOM 2019 - IEEE Conference on Computer Communications, pp. 874–882 (2019). https://doi.org/10.1109/INFOCOM.2019.8737599
20. Hong, Y.-S.: Smart care beds for elderly patients with impaired mobility. Wirel. Commun. Mob. Comput. **2018**, 1780904 (2018). https://doi.org/10.1155/2018/1780904

Smart Rust - Use of Drones to Combat Yellow Wheat Rust

Rui Alves[1(✉)], Paulo Matos[2,3], João Ascensão[1], Diogo Camelo[1], and Fernanda Pança[1]

[1] Instituto Politécnico de Bragança, Bragança, Portugal
{rui.alves,pmatos}@ipb.pt,
{a34505,a36739,a41425}@alunos.ipb.pt
[2] Research Centre in Digitalization and Intelligent Robotics (CeDRI), Instituto Politécnico de Bragança, Campus de Santa Apolónia, 5300-253 Bragança, Portugal
[3] Laboratório para a Sustentabilidade e Tecnologia em Regiões de Montanha (SusTEC), Instituto Politécnico de Bragança, Campus de Santa Apolónia, 5300-253 Bragança, Portugal

Abstract. In recent times, the combination of IoT or Cloud Computing concepts with agriculture has grown dramatically, in part due to rapid environmental changes such as extreme weather conditions, soil deterioration and lack of soil moisture. With the use of these modern concepts in agriculture, systems are built to effectively manage the field of cultivation, where with the help of sensors, it's possible to monitor light, air humidity, temperature, soil humidity or with the use of drones, the application of pesticides in pest control. As agriculture provides, among other things, food to ensure human nutrition, there must be a good management and care of crops. The Mediterranean basin, including Portugal, is an area highly vulnerable to climate change, suffers from the fungus that produces the yellow wheat rust, the evolution of which has led to breeds adapted to higher temperatures. This disease affects wheat and other cereals in various countries and threatens the future and stability of cereal crops, which, if not stopped in good time, can cause considerable social and economic damage. Due to this problem this paper presents a solution for the monitoring of yellow rust in wheat, using drones, sensors, and artificial intelligence processes, with the main objective of reducing the impact of this fungus on production.

Keywords: Drone · IoT · Deep learning · Wheat · Yellow rust

1 Introduction

The current food production is enough for about 7 billion people [10], but almost one billion are hungry or malnourished due to poverty and unequal distribution.

This work has been supported by FCT - Fundação para a Ciência e Tecnologia within the Project Scope: UIDB/05757/2020.

M. Ben Ahmed et al. (Eds.): SCA 2022, LNNS 629, pp. 69–78, 2023.
https://doi.org/10.1007/978-3-031-26852-6_6

Thus, with the expected increase in population [4] for the next decades, the best forecasts point to the need to increase agricultural production by 60% to 100% [2] in order to guarantee food for the entire population, including the percentage that lives in a state of extreme malnutrition.

Among the various crops, resulting of good adaptability to environmental conditions [6], ease of storage, the possibility of integration in the most varied culinary dishes among many other reasons, cereal production has an important role in feeding the world's population. Food and Agriculture Organization (FAO) reports an annual increase of 0.9% [3] in global cereal production in 2021, largely attributed to increased maize production. The same FAO estimates that the use of cereals will increase in 2021/22, by 1.1%, as a result of expansions in food consumption, feed use and other applications. However, world estimates of cereal production and use at the end of the seasons in 2022 are above the initial expected levels, but remain below the record levels achieved in 2018/19, making the world cereal trade below the record level of 2020/21, mainly due to an expected fall in the corn trade and reflecting the impact of the disturbances caused by the war in Ukraine.

In many other areas of activity, climate change has a considerable impact on the growth and development of agricultural production [1]. In this way, regions at high risk of climate change (e.g. the entire Mediterranean basin) begin to suffer the impact of pests and plant diseases. A good example is the fungus that produces yellow wheat rust, whose evolution has made it adapt to higher temperatures.

Using drones, sensors, and artificial intelligence processes, this paper presents a solution for monitoring yellow rust in wheat, with the main objective of reducing the impact of this fungus on production. The rest of the paper is organized in: Sect. 2 where a short description of the fungus that causes yellow rust is presented and how it can be fought; Sect. 3 illustrates the details of the proposed solution for monitoring this fungus; Sect. 4 describes the conclusions of the work and goals for future work.

2 Yellow Rust Fungus

The fungus *Puccinia striiformis f. sp. tritici* [16] is responsible for yellow wheat rust [11]. The origin of the name is the result of the yellow color (see image Fig. 1) of the plant after being infected, something that results from the distribution of the urea of the fungus in the leaves. The appearance of this disease occurs mainly at the beginning of the growing period, since conditions such as high altitude, high humidity, rainfall and mild temperatures between 7 and 15 °C are present [4].

Yellowish-green spots on the leaf of the infected plant are usually the first symptoms of the disease; they become yellowish after the establishment of the pathogen. When the disease progresses, the initial lesions form pustules on the underside of the leaf called urea. When the infection reaches higher levels, the plant suffers from early defoliation.

Fig. 1. Rust contaminated wheat leaf.

Being considered a pathogen, the fungus that causes this disease, which does not only affect wheat, if not treated in a timely manner, can affect cereal production. Most of the existing control measures accentuate the use of pesticides, but when the disease is detected at an advanced stage, these measures become ineffective.

Another possible combat approach is the use of preventive measures such as: cultivation of stable and disease-resistant varieties; search in the fields for plants with suspected contamination, in order to prevent contamination from spreading to healthy plants, However, in large areas it is totally impossible to carry out this type of searches manually, which raises the need of technical solutions as the ones proposed in this paper,

3 Proposed Solution

In the Fig. 2 is represented the architecture proposed for the mitigation of the problem described in Sect. 2. This solution consists of 3 components:

- drone.
- mobile application.
- intelligent system.

The general functioning of this solution accentuates, in capturing photos of the wheat plant. The farmer, using the drone, analyzes the land where he wants to monitor the presence of yellow rust. During this process, the drone will capture photos of the plants, sending them to the intelligent system, along with the GPS coordinates. In case of a positive analysis, the server sends to the mobile application the coordinates of the plant with suspected yellow rust for a second analysis or application of a possible treatment. In the Sects. 3.1, 3.2, 3.3 the technical details of the 3 main components of the solution will be discussed.

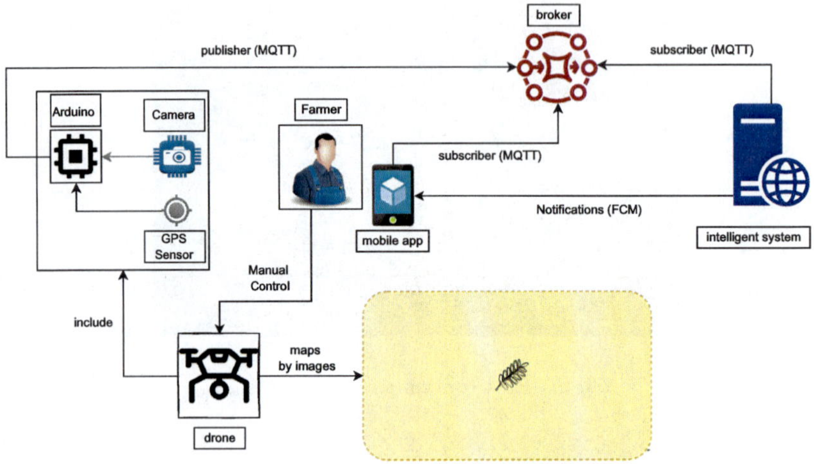

Fig. 2. Overview of the proposed architecture.

3.1 Drone

The solution produced is not dependent on a specific model of drone, it should only meet 2 essential requirements: operate at altitudes above 5 m; the range of the remote control should be sufficient to monitor the selected terrain. Its operation is based on two modes: capture of images of wheat plants; verification of plants identified by the intelligent system as suspected of yellow rust. At this stage of the work, both processes are done manually, that is, it is the farmer who controls the drone. For the operation of these two modes of operation was added to the drone, as illustrated in the figure 3.2, an Arduino MKR NB 1500, a high-resolution camera and a GPS sensor.

When the first mode of operation is activated the drone traverses the selected terrain capturing the images of the plants. The capture of the photos, in the current version, is initiated at the moment when the drone reaches 1.50 m altitude and is performed from 10 s to 10 s. Sequentially it is published in a Broken, using LTE-M, the string in Base64 of the captured image as well as the longitude and latitude of the place where the photo was captured. This information is then collected by the intelligent system as will be explained in Sect. 3.3.

In context where these lands are located in isolated areas, the use of tools such as Smart Rust is compromised due to the weak support for the Internet [14].

It was, therefore, necessary to add mechanisms to the presented solution in this paper that make it possible to overcome the difficulties identified. To overcome the lack of Internet support in some regions, communication between medication organizers and institutional servers is carried out using NB-IoT or LTE-M.

These technologies are currently part of the LTE network and any place on the planet with network coverage, is fertile ground for NB-IoT and LTE-M that significantly improve IoT devices energy consumption. In cases where LTE-M doesn't exist, NB-IoT can be used to replace it. It has the same mechanisms and works in a similar way.

Finally, the second mode of operation consists of checking the positive cases of yellow rust detected by the intelligent system. When the farmer receives in the mobile application the coordinates of a plant with possible contamination, the second mode of operation of the drone allows a closer analysis of the suspected plant. The information received in the mobile application allows the farmer to drive a drone to the coordinates received. The accuracy of this routing is guaranteed because the arduino chip, present in the drone, publishes in Broker, every 5 s, the GPS coordinates in which it is located, enabling, as described in Sect. 3.2 draw in the mobile application the path made by the drone and the distance it is from the point to be analyzed. When the drone hits the plant with suspected contamination, it captures a new set of photos again. It then sends it to the server where it will be analyzed in detail by the farmers.

3.2 Mobile Application

This application has as main functions: the registration and management of drones; the listing of missions (field flights) and their information; the map overview from the drone current position to a certain position inside a land where yellow rust was found; controlling drone mode of operation. In Fig. 3 it is possible to see the layout built for the mobile application.

To add/register a drone in the application, the user has a view for that, where it only has to insert it's name. The user can also view a list existent of missions (land trips), Fig. 3 left side, which are inserted automatically from the intelligent system, with latitude and longitude coordinates where yellow rust was found, if any, as well as the information about the wheat, (if it has a yellow rust problem or not).

To allow the drone to do certain operations, the user can send commands to it via application. To do this, the user have to select a land trip, the associated drone, and one of the two operation modes. If the user want's to observe in a visual way the current position of the drone until it gets to the land coordinates, a satellite view of this was implemented, as visible in Fig. 3 right side. Both positions, the current one from the drone and the one where the yellow rust was found (land coordinates), are inserted into the app from the intelligent system. In the case of the second operation mode, the drone position will be updated each 5 s in the satellite view.

All communications from the intelligent system with the mobile app, are done via Firebase Cloud Messaging (FCM), which is a cross-platform messaging solution that allows send messages at no cost.

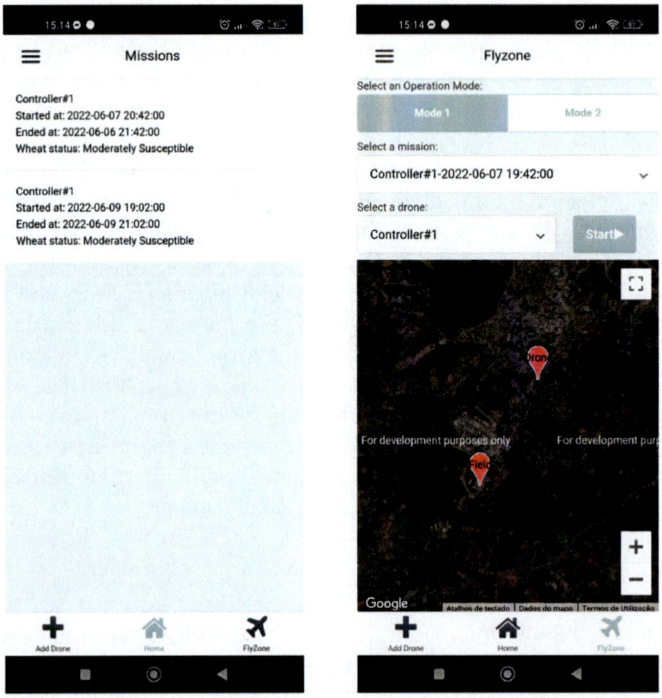

Fig. 3. Layout of the developed application.

3.3 Intelligent System

In this paper, the designed intelligent system is used to identify the level of yellow rust in wheat leaves, which will make a significant difference in the life cycle of these plants.

The designed intelligent system is composed by three components:

- MQTT client;
- wheat leaves detector;
- classifier of yellow rust levels in wheat leaves;

The MQTT Client is responsible for receiving the pictures taken by the drone and send via the MQTT Broker. The wheat leaves detector is responsible to detect the wheat leaves on the pictures captured by the drone and extract them to separated pictures, which have just the wheat leaf displayed (Fig. 4). The yellow rust levels classifier in wheat leaves is responsible to identify the level of yellow rust in the pictures extracted by the detector.

The detector and the classifier of the intelligent system are both deep learning models that are based on Convolutional Neural Networks (CNNs), which are a well-established Deep Learning Algorithm applied to Computer Vision.

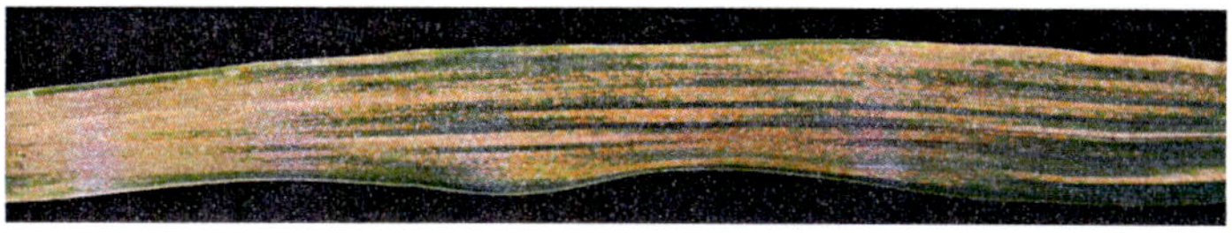

Fig. 4. Example of a detected wheat leaf.

The MobileNetV2 were the base model used to develop these models. The MobileNetV2 is a convolutional neural network architecture that pursue to perform well on mobile devices, being as computational light as possible [9,12,15].

In the Fig. 5 can be observed a flowchart where is present the structure of the operation of the designed intelligent system.

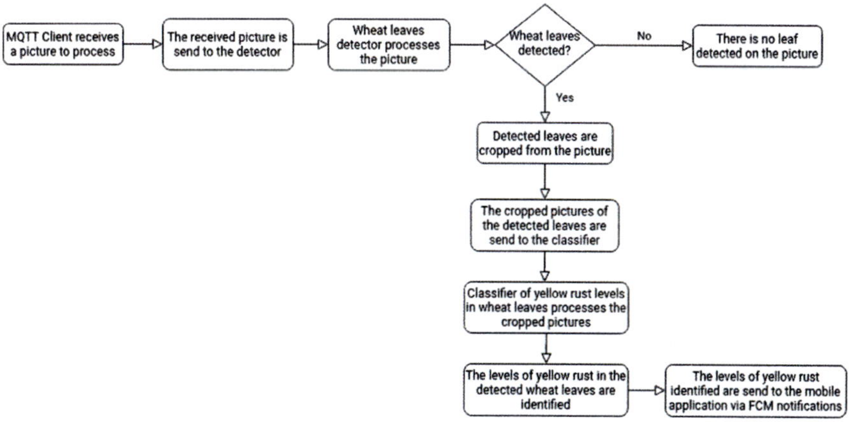

Fig. 5. Structure of the operation of the designed intelligent system.

The MQTT Client of the designed intelligent system is a process that is always running and it's responsible to receive the pictures send by the drone via the MQTT Broker and sending those pictures to the detector. The wheat leaves detector after receiving those pictures, it's going to pre-process them, where their shape are resized to match the inputs of the detector model and every pixel are normalized. After the pre-processing, the pictures are given to the detector which is going to detect the wheat leaves on those pictures, and scrop the detected wheat leaves to separated images. Then, those separated pictures of the detected wheat leaves are given to the yellow rust levels classifier, which will identify the level of yellow rust in the detected leaves. The levels of yellow rust are [8]:

- No disease (0: No signs of infection);
- Resistant (R: Minor signs of infection);

- Moderately Resistant (MR: Small and medium signs of infection);
- Moderately Resistant-Moderately Susceptible (MRMS);
- Moderately Susceptible (MS: Medium signs of infection);
- Susceptible (S: Major signs of infection).

With the levels of yellow rust in the detected wheat leaves identified, these are sent to the Mobile Application via FCM notifications like mentioned before. In the mobile application, the farmer can then observe the detected wheat leaves and their level of yellow rust.

4 Conclusion

As described throughout the work, yellow rust on wheat has a considerable impact on the development of wheat production. Although the solution is focused on wheat, this disease is common to a set of plants which increases the need for solutions for its control and monitoring. However, for lack of other solutions, the fight against these diseases is done using pesticides, which sometimes becomes ineffective because the detection is made advanced phase of the disease. On the other hand, the negative impact on the environment makes the use of pesticides an impractical solution. Thus, alternative disease control and monitoring solutions, as illustrated in this paper, are necessary so that the detection and control of yellow rust is possible without endangering the environment.

The solution presented is another solution that makes use of intelligent techniques to reduce the complexity of some of the processes in the agricultural sector. However, although intelligent solutions are often used in various areas of activity they often become highly complex solutions [5,13] sometimes requiring additional efforts so that they can be used as in the academic solutions that exist in the lithium area for the detection of yellow rust, which use photos [7,17] or analysis of airborne particles to detect the presence of yellow rust. In addition to this being a solution with low implementation/acquisition costs, the prototype presented in this article, tries to remove some of this complexity, facilitating ease of use so that the solution can be used by as many farmers as possible with as little effort as possible.

The current state of the work is dependent on: the field fly tests and training of the intelligent system since all other components exist and developed or implemented. Conceptually, the presented solution provides an efficient mechanism in fighting yellow wheat rust. Despite the advanced development, which allowed to validate the robustness of some of the concepts of architecture, the solution is still far from being usable by a farmer. In this way, the following points were left for future work:

- The big challenge for the future is to put autonomous driving in the 2 modes of drone operation.
- Conduct a study to quantify the success rate of the tool in detecting yellow rust.

- Analyze the feasibility of using the tool for detection of yellow rust in other crops.
- Optimize the sending of GPS coordinates by the drone in the second mode of operation.
- Analyze other deep learning algorithms applied on computer vision and check their effect on the intelligent system.
- Change the system to allow the drone in operation mode 2 to be able to apply other methods of controlling yellow rust.

References

1. Climate impacts on agriculture and food supply. https://climatechange.chicago.gov/climate-impacts/climate-impacts-agriculture-and-food-supply. Accessed 15 June 2022
2. Feeding the world sustainably. https://www.un.org/en/chronicle/article/feeding-world-sustainably. Accessed 15 June 2022
3. World cereal production, utilization, stocks, and trade all likely to contract in 2022/23. https://www.fao.org/worldfoodsituation/csdb/en/. Accessed 15 June 2022
4. Berners-Lee, M., Kennelly, C., Watson, R., Hewitt, C.N.: Current global food production is sufficient to meet human nutritional needs in 2050 provided there is radical societal adaptation. Elementa: Sci. Anthropocene **6**, June 2018. https://doi.org/10.1525/elementa.310,52
5. Camelo, D.M.A., Santos, J.C.C., Martins, M.P.G., Gouveia, P.D.F.: Modelling academic dropout in computer engineering using artificial neural networks. In: Rocha, Á., Adeli, H., Dzemyda, G., Moreira, F., Ramalho Correia, A.M. (eds.) WorldCIST 2021. AISC, vol. 1366, pp. 141–150. Springer, Cham (2021). https://doi.org/10.1007/978-3-030-72651-5_14
6. Chloupek, O., Hrstkova, P.: Adaptation of crops to environment. Züchter Genet. Breed. Res. **111**(7), 1316–1321 (2005)
7. Guo, A., et al.: Wheat yellow rust detection using uav-based hyperspectral technology. Remote Sensing **13**(1) (2021). https://www.mdpi.com/2072-4292/13/1/123
8. Hayit, T., Erbay, H., Varçın, F., Hayit, F., Akci, N.: Determination of the severity level of yellow rust disease in wheat by using convolutional neural networks. J. Plant Pathol. **103**(3), 923–934 (2021). https://doi.org/10.1007/s42161-021-00886-2
9. Howard, A.G., et al.: Mobilenets: efficient convolutional neural networks for mobile vision applications. arXiv preprint arXiv:1704.04861 (2017)
10. Kc, K.B., et al.: When too much isn't enough: does current food production meet global nutritional needs? PLoS ONE **13**(10), e0205683 (2018)
11. Ma, J., Huang, X., Wang, X., Chen, X., Qu, Z., Huang, L., Kang, Z.: Identification of expressed genes during compatible interaction between stripe rust (puccinia striiformis) and wheat using a cdna library. BMC Genomics **10**, 586 (2009)
12. PapersWithCode: Mobilenetv2 explained. https://paperswithcode.com/method/mobilenetv2. Accessed 25 June 2021
13. Pinto, A., Correia, A., Alves, R., Matos, P., Ascensão, J., Camelo, D.: ehealthcare-a medication monitoring approach for the elderly people. In: International Conference on Wireless Mobile Communication and Healthcare. pp. 221–234. Springer (2022). https://doi.org/10.1007/978-3-031-06368-8_15

14. Ruiz-Martínez, I., Esparcia, J.: Internet access in rural areas: Brake or stimulus as post-covid-19 opportunity? Sustainability **12**(22) (2020). https://www.mdpi.com/2071-1050/12/22/9619
15. Sandler, M., Howard, A., Zhu, M., Zhmoginov, A., Chen, L.C.: Mobilenetv 2: inverted residuals and linear bottlenecks. In: Proceedings of the IEEE Conference on Computer Vision and Pattern Recognition, pp. 4510–4520 (2018)
16. Zemeca, L., Bankina, B.: Characterization of yellow rust (puccinia striiformis westend.): review, pp. 69–76, December 2018
17. Zhang, T., Yang, Z., Xu, Z., Li, J.: Wheat yellow rust severity detection by efficient df-unet and uav multispectral imagery. IEEE Sens. J. **22**(9), 9057–9068 (2022)

Industrial Digitalization Solutions for Precision Forestry Towards Forestry 4.0

Pedro M. B. Torres[1,2](✉), Geoffrey Spencer[1], Luís Neto[2], Gil Gonçalves[2], and Rogério Dionísio[1,3,4]

[1] Instituto Politécnico de Castelo Branco, 6000-084 Castelo Branco, Portugal
pedrotorres@ipcb.pt

[2] SYSTEC - Research Center for Systems and Technologies, 4200-465 Porto, Portugal

[3] DiSAC - R&D Unit for Digital Services, Applications and Content, 6000-767 Castelo Branco, Portugal

[4] INESC TEC - Institute for Systems and Computer Engineering, Technology and Science, 4200-465 Porto, Portugal

Abstract. This paper presents machine digitalization solutions with particular application in forest machines, such as harvesters and wood processing machines. In line with all the requirements of Industry 4.0, this type of machines also needs digitization to align with the concept defined as Forestry 4.0, where we think of a smarter forest in which all stakeholders, humans, forest producers, machines and factories communicate. For machine manufacturers is a step that must be taken to modernize machines, enabling remote access services for maintenance, productivity monitoring, and management of forest operations. It consists of developing cyber-physical systems around the machines with digital twins that allow the simulation and identification of faults that may occur. A solution is presented to enable CAN Bus communication between the controller, operator joysticks, and sensors/actuators, as well as a Digital Twin solution to emulate machine operations.

Keywords: Industry 4.0 · Forestry 4.0 · CAN bus · Digital twin · Cyber-physical systems

1 Introduction

Forestry 4.0 [1], or intelligent precision forest, is a new paradigm based on Industry 4.0 [2] and processes digitization, which will have an increasing contribution to the development of forest machines. Some of the main international machinery manufacturers already incorporate in their equipment various types of sensors that allow them to collect relevant information about the forest (e.g. forest

M. Ben Ahmed et al. (Eds.): SCA 2022, LNNS 629, pp. 79–86, 2023.
https://doi.org/10.1007/978-3-031-26852-6_7

density), about the performance of the machine (e.g. the position of the machine, percentage of time in operation) and on the progress of forestry operations (e.g. type of wood cut and estimated volume). The trend is for there to be an increasing volume of high-precision and low-cost information. As the development of new sensors for equipment advances, it is also necessary to progress in the technologies for communicating the data obtained to central systems, preferably "in real time". Central systems will necessarily have to store and process large volumes of information, the so-called Big Data. Here too, there are challenges related to interoperability or the Internet of things (IoT), which are related to the integration, compatibility and exchange of information collected from different systems and/or different equipment. Information systems, too, undergo a transformation process in order to be able to make use of the large volumes of information and provide useful interfaces to monitor operations and to support planning and decision-making in a multi-user logic. The ultimate goal is to have "reactive" systems that allow adapting planning to what is actually happening on the ground. Manufacturers of forestry implements and tools must align the development of their products with the Forest 4.0 concept, considering more modular, intelligent and interoperable products. In Portugal, CUTPLANT SOLUTIONS, S.A essentially develops forestry machinery, with the VICORT brand, from which the processing heads and forestry fellers stand out, already well established in the Iberian market and with some projection in international markets, such as Brazil and Indonesia. Processing heads and forestry harvesters are typically attached to caterpillars machines that were originally built for public works and construction, and are physically and functionally adapted to become machines that operate in the forest for felling and processing trees (precise felling, pruning , peeling and stacking). The machine undergoes several changes, with different forest protections, lighting and cooling systems being placed on it, in addition to undergoing changes at the hydraulic and electronic level to be able to operate the processing heads or other implements that are coupled to them in a more efficient way. Forest machines operate in very hostile environments, often requiring maintenance interventions. The causes of machine inoperability are diverse, caused either by bad operation, bad maintenance, accidents or due to the violent wear and tear to which they are subjected on a daily basis. Typically, forest machine manufacturers export all over the world, making it necessary to have a team capable of being ready to assist with all maintenance requests received, in order to keep the customer satisfied and with full machine availability. However, an assistance team capable of being ready in any region of the world, requires the allocation of many technical resources (human resources, vehicles, tools, etc.), which, due to the costs involved, is currently only achievable by multinational companies. In addition to the adverse context of the forest, the causes that give rise to the need for assistance to the machine in the field are related to the bad handling of the machine and the failure to carry out preventive maintenance by the purchaser of the equipment. Thus, it is identified the need for sensors on board these machines that allow, in addition to assisting the operator in the correct operation of the machine, also alert for preventive maintenance. To make it viable for small and medium-sized companies to export machines and reduce assistance costs, it is necessary to achieve more robust electromechanical solutions in the forestry context, more intelligent solutions that can self-diagnose problems and assist in preventive maintenance. It also helps to operate the machine, making it more precise and adequate. Solutions that include virtual tools for training operators and resellers. Thus, the SMARTCUT consortium [3] identifies the following problems in processing heads and forestry harvesters:

- High assistance costs for forest machines;
- Low interoperability with existing machines and forest 4.0 decision support systems;
- Electromechanical systems with low robustness for the forest context;
- Systems without the ability to store the history of operation and predictive capacity for preventive maintenance; and
- Closed training simulators with limited functionality.

The maintenance costs of forest machines are one of the most relevant elements in forestry operations. In most situations, the highest cost does not come from the repair itself, but from the machine's downtime and the impact that this stop may have on the service chain, leading to the stop of downstream equipment due to lack of supply. The identification of the nature of the faults and the speed of response of the technical assistance is a critical success factor. This issue is amplified when the company selling forest machines intends to operate in foreign markets and its framework is that of an SME, in which human, technical and financial resources are more limited, being dependent on local representatives with the same problems, but who also have to be technically trained in order to convey confidence to potential customers. The project focuses on the design and development of a solution for sensing forest equipment based on CAN technology, which allows the execution of telediagnosis, telemaintenance and forest monitoring tasks by the company producing the equipment and/or its representatives, in each Country of operation. It also focuses on the design and development of an innovative tool for authoring virtual training scenarios that will take advantage of a physical simulator/emulator to recreate, via virtual reality, the operation and failure conditions of the equipment and, simultaneously, the urographic and physical features found in the terrain. This tool will make it possible to train and/or improve the operational efficiency of forestry equipment operators and maintenance technicians, contributing to the reduction of breakdowns, faster repairs, contributing to the optimization of production. This authoring tool will grant autonomy to the company as it will not only allow the management and maintenance of all existing training scenarios but also the expeditious creation of new training scenarios based on immersive virtual reality. All this will allow the company to optimize the operation training and tele-assistance process and contribute to an increase in its competitiveness.

The remainder of this paper is organized as follows: Sect. 2 describes the state of the art and related works. Section 3 describes how the Digital Twin concept can be applied on forestry machines to overcome some of the identified issues. Section 4 presents the experimental results achieved and, finally, Sect. 5 presents the conclusions and future work.

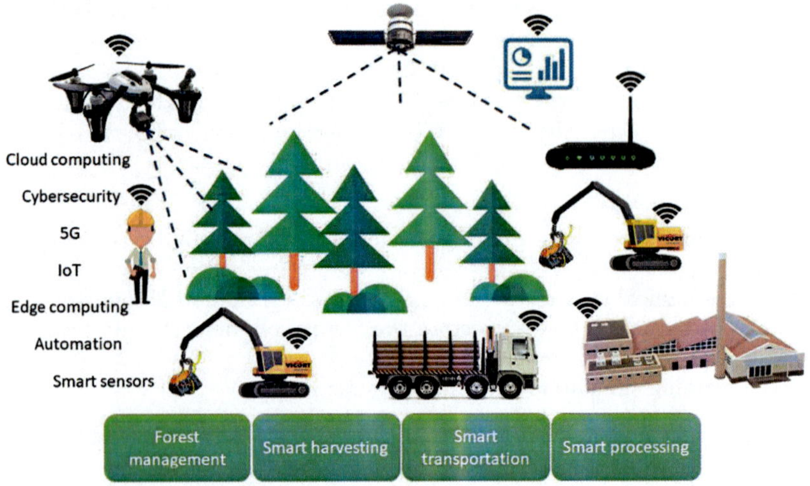

Fig. 1. Forestry 4.0 concept.

2 State of the Art

With the boom of the 4th industrial revolution and the growth of technology, nowadays any sector talks about digitalization and the applicability of Industry 4.0 concepts, whether in cities (smart cities), in transports (smart mobility), in health (smart health), in our homes (smart homes), in agriculture (smart agriculture) and in practically everything around us. It's the age of connectivity. In the forest, the concept is also already adopted by forest and machinery producers, and by all entities on the sector. Nowadays work up with a vision and a need to digitize processes and products, to increase productivity and gains in each process that is operated. In this line, there are many works focused on the digitalization of the forest, Forestry 4.0. In [4], a framework is presented to provide a holistic view of Forestry 4.0 from a forestry supply chain perspective. The digital technologies relevant to each of the business activities in the chain, the network infrastructure, the system intelligence, and the digital ecosystem of the collaborative forest supply chain are presented. In [5], the authors systematically reviews and analyses the state-of-the-art technologies applied in the forest supply chain and reports on the current and potential impacts of technologies on the transformation of the forest supply chain towards the"Forest Industry 4.0". In [6] the authors exploring the possibilities for technological interventions in forests to drastically improve their ecosystem. Intelligent systems for sensing, monitoring, and methods for analysis to be used in applications such as forest fire incidents, illegal logging of trees, poaching, etc. are discussed. As illustrated in Fig. 1, nowadays connectivity and cooperation between all agents is important for a smarter management of processes and natural resources, where there is knowledge of the status of the entire production chain, from the forest to the final consumer. All this is only possible due to the existence of intelligent sensors, inside the forest and the machines, the automation of processes, the edge computing that can bring the works of artificial intelligence closer to the products, the concept of all connected devices provided by IoT, the importance of low latency and high bitrates provided by 5G, cybersecurity to ensure infrastructure resilience and integrity, cloud computing among other emerging technologies.

3 Digital Twins

The term Digital Twin is used to describe models that accurately represent a physical asset. In the context of industry, the Reference Model for Industry 4.0 adopts this concept, which is coined Component I4.0 [7]. These components can encapsulate all aspects about an asset that can be represented digitally (e.g. software, communication, physical models), and make them available through an I4.0 standardized interface. This way, the plant is "aware" of its assets, and these can collaborate towards common goals. This opens up possibilities such as advanced simulation techniques, development of predictive models, reconfiguration in response to changes and plug&play of new components.

Component based philosophy is well established in the software engineering domain [8]. For industrial applications, the standard IEC 61499 defines a software component model [9] that can be used to build control and monitoring applications. The building block of this standard is the Function Block (FB). A FB encapsulates a unit of software logic, exposing its services through a set of standard interfaces. A parallel can be established between the I4.0 Component and IEC 61499. Several FB's can be used and interconnected to represent different aspects of an asset; this composition of FB's, in its turn, can represent the Component I4.0 (Digital Twin) of the asset itself. The following paragraphs of this section detail how this logic is being applied to machinery in the SMART-CUT project.

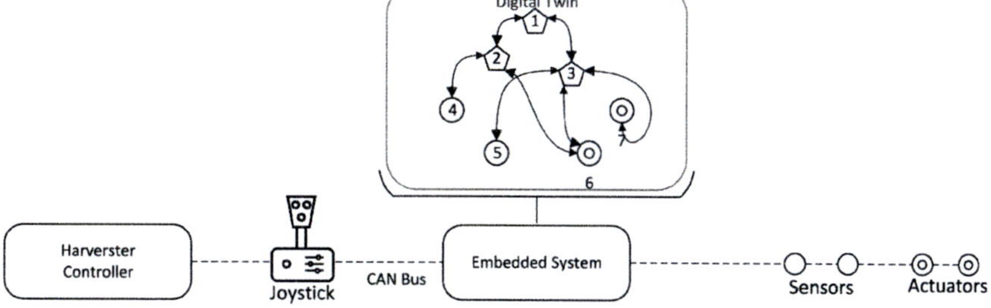

Fig. 2. Digital Twin applied to forestry machines.

By employing the IEC 61499 standard, FB's can be used to represent subsystems of a machine. Figure 2 illustrates an abstract composition of FB's, using different shapes, enumerated from 1 to 7. Some of these shapes represent sensors (4,5) and actuators (6,7), which are physically connected to the machine CAN bus. These specific FB's, can encapsulate the software necessary to drive these sensors and actuators, or can simply represent these sub-systems, holding information such as make and model, as well as runtime data gathered through the CAN bus. Other FB's in the composition (1,2,3) may embed and apply data processing techniques, or embed software to send the gathered data through a gateway, using HTTP services. The Digital Twin block consists in the composition of FB's that represent a specific forestry machine, with a specific configuration of sub-systems and software, at a specific time in the machine life cycle. In a technical perspective, the Digital Twin block (Fig. 2) is also a runtime environment responsible for hosting and executing the FB's software.

This component approach allows for a machine to be updated with new sensors, actuators and software. The Digital Twin of the machine can be extended as long as the required software is abstracted by a FB. Several instances of the machine Digital Twin can be created and maintained to respond to specific problems. The designs can be reused to create the Digital Twin's of new machines, saving development costs. This approach relies on the effort that was made to digitalize human interfaces (joysticks, buttons and consoles) as well as the digitalization of sensor and actuator interfaces through CAN bus. The next chapters detail the communication and digitalization developments on these machines.

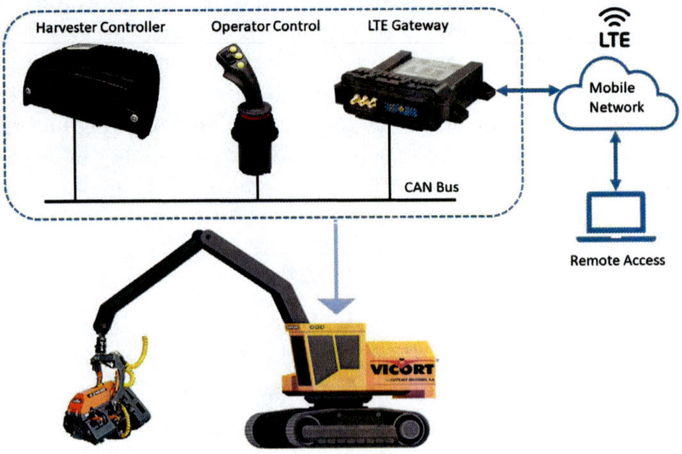

Fig. 3. Harvesters digitization overview.

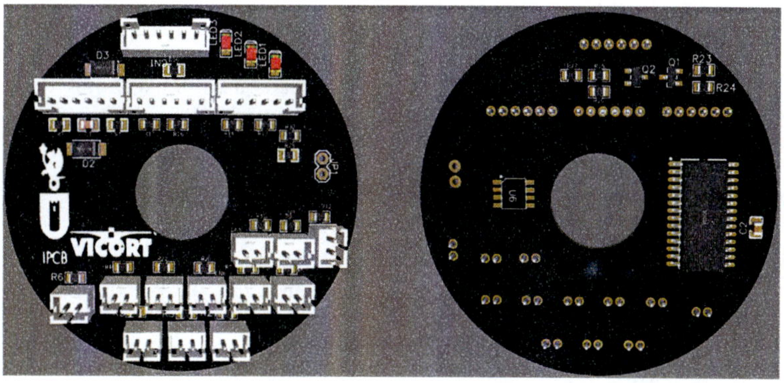

Fig. 4. PCB for Joystick interfaces. a) top layer view b) bottom layer view.

4 Communications

This section presents some of the experimental results achieved and realized namely in establishing CAN Bus communications at the level of operator controls (Joysticks) and other machine components (sensors/actuators). To better understand the digitization architecture of the harvesters, Fig. 3 shows an overview of the configuration implemented in the VICORT machines. A CAN Bus communication system is implemented between the different machine resources for standardizing communication, but also aligning the VICORT machines with what exists in competitors. Remote access through the CAN Bus/LTE gateway allows leveraging the development of telemaintenance and remote monitoring solutions. Figure 4 shows one of the PCB boards developed for joysticks that allows all joystick operations to communicate with the main controller over CAN Bus. These joysticks are being tested in real scenarios on harvesters for validation.

5 Conclusions

The paper presents a solution for digitizing Harverters based on establishing a standard communication between all machine assets via the CAN Bus protocol. An LTE/CAN Bus gateway was implemented to enable remote access of the machines for telemaintenance and remote monitoring tasks. A Digital Twin was also implemented, essential to emulate the functionality of the Harverters, allowing fault simulation and operation training for the operators. As future work, other hardware modules will be developed for specific CAN Bus communications for sensors and actuators, as well as the implementation of remote monitoring systems involving dashboards, database, and interaction with virtual reality solutions for immersive simulation environments. LoRa (Long Range) technology based on LoRaWAN (Long Range Wide Area Network) protocol will also be considered in future developments in order to enable communication between different machines in the forest.

Acknowledgement. This research was supported by the project "SMARTCUT - Diagnóstico e Manutenção Remota e Simuladores para Formação de Operação e Manutenção de Máquinas Florestais", financed by Portuguese funds through the program COMPETE 2020, under the grant number POCI-01-0247-FEDER-048183.

References

1. Müller, F., Jaeger, D., Hanewinkel, M.: Digitization in wood supply - a review on how Industry 4.0 will change the forest value chain. Comput. Electron. Agri. **162**, 206–218 (2019)
2. Xu, L.D., Xu, E.L., Li, L.: Industry 4.0: state of the art and future trends. Int. J. Prod. Res. **56**(8), 2941–2962 (2018). https://doi.org/10.1080/00207543.2018.1444806
3. SMARTCUT project, Available online. https://www.vicort.com/projetos-cofinancia/dos-ue/smartcut. (Accessed 21 June 2022)
4. Feng, Y., Audy, J. F.: Forestry 4.0: a framework for the forest supply chain toward industry 4.0. Gestao e Producao, **27**(4), e5677 (2021). https://doi.org/10.1590/0104-530X5677-20
5. He, Z., Turner, P.A.: Systematic review on technologies and industry 4.0 in the forest supply chain: a framework identifying challenges and opportunities. Logistics **5**, 88 (2021). https://doi.org/10.3390/logistics5040088
6. Singh, R., Gehlot, A., Akram, S.V., Thakur, A.K., Buddhi, D., Das, P.K.: Forest 4.0: digitalization of forest using the Internet of Things (IoT). J. King Saud Univ. - Comput. Inf. Sci. **34**, 5587–5601 (2021)
7. Wagner, C., et al.: The role of the Industry 4.0 asset administration shell and the digital twin during the life cycle of a plant. In: 2017 22nd IEEE International Conference on Emerging Technologies and Factory Automation (ETFA), pp. 1–8 (2017)
8. Vale, T., et al.: Twenty-eight years of component-based software engineering. J. Syst. Softw. **111**, 128–148 (2016). Elsevier
9. Vyatkin, V.: The IEC 61499 standard and its semantics. In: IEEE Industrial Electronics Magazine, vol. 3, no. 4, pp. 40–48 (2009). IEEE (2009)

Sentiment Analysis in Drug Reviews Based on Improved Pre-trained Word Embeddings

Nouhaila Bensalah[1(✉)], Habib Ayad[1], Abdellah Adib[1], and Abdelhamid Ibn el farouk[2]

[1] Team Data Science & Artificial Intelligence, University of Hassan II Casablanca, Casablanca 20000, Morocco
nouhaila.bensalah@etu.fstm.ac.ma , abdellah.adib@fstm.ac.ma

[2] Teaching, Languages and Cultures Laboratory Mohammedia, Mohammedia, Morocco

Abstract. Sentiment analysis (SA) is the process of ranking documents based on their polarity (positive, negative, or neutral). It has a wide range of applications, including opinion mining, health informatics, and recommender systems. In the pharmaceutical industry, automatic analysis of people's online opinions can be used to evaluate a large number of user opinions and extract useful information regarding drug efficacy and risk factors, which can be used to improve pharmacovigilance systems. Approaches for sentiment analysis have evolved over time from simple rules to powerful machine learning techniques such as deep learning, which has emerged as an emerging technology in several natural language processing (NLP) applications. Sentiment analysis is aware of this accomplishment, and various deep learning-based systems have subsequently demonstrated their effectiveness over previous approaches, achieving peak scores on typical sentiment analysis datasets. As transformers have revolutionized the field of NLP, achieving peak scores for different NLP tasks, we propose to combine a pre-trained model called BERT (Bidirectional Encoder Representations from Transformers) with static embedding model, which has not been studied before. The obtained vectors were used as input features for a Bidirectional Gated Recurrent Unit (BiGRU). The experimental results reveal that the proposed model outperforms the state of the art in sentiment analysis of drug reviews.

Keywords: Sentiment classification · Drugs reviews · Contextual embedding models · Static embedding models

1 Introduction

With the growth of Web 2.0 platforms, there is a massive volume of user-generated information, commonly known as social media. As a result, many researchers were devoted the past few years to developing powerful algorithms for sentiment analysis of user-generated data [6]. Sentiment analysis refers to a sort of subjectivity analysis which examines sentiment in a particular textual

M. Ben Ahmed et al. (Eds.): SCA 2022, LNNS 629, pp. 87–96, 2023.
https://doi.org/10.1007/978-3-031-26852-6_8

unit in order to understand the polarity (i.e., positive, negative, or neutral) of people's attitudes regarding various aspects of a subject. Sentiment analysis techniques have evolved over the years from simple rules to machine learning (ML), deep learning (DL) and transfer learning approaches, which have emerged as an emerging strategy in many automatic language processing tasks [3,11]. Deep Learning techniques have surpassed other standard ML techniques for sentiment analysis in recent years. These algorithms are able to detect feelings or attitudes from textual data, without the need for any feature engineering. There are several deep learning-based approaches that can be used for sentiment analysis, such as Recurrent Neural Networks (RNNs) and Convolutional Neural Networks (CNNs). Compared to machine learning models, they give significantly better results. Nevertheless, with machine learning and data analysis techniques, the data set must come from the same domain as the one used for training. Therefore, each time the domain changes, a new model must be developed. Many attempts have been made in the field of NLP; several techniques have been developed, including rule-based approaches, ML, DL, and transfer learning. Yet, most research in the field of Sentiment Analysis has focused on broad domains (such as electronic reviews, movie reviews, and restaurant reviews), with relatively limited research in the health and medical areas [20]. In this work, we combine the static embedding model with pre-trained BERT language model [8] for Sentiment Analysis using a BiGRU-based Deep learning model. Briefly, to encode linguistic information, the pre-trained BERT language model parameters were intensively trained on massive amounts of unannotated dataset. Recent literature reviews have revealed that when using BERT, outstanding results were achieved but with a very high training time. However, one of the drawbacks of contextual models, such as BERT is that they require massive computational resources for training in order to generate representations using the Sentiment Analysis specific dataset. On the other hand, static embeddings have significant advantages over the contextual ones generated using pretrained language models in terms of speed and computational resources required for training. Furthermore, static embedding models may handle domain-specific datasets having distinct features from those learnt by BERT during training. Furthermore, they obtained satisfactory results while needing less training time. In this paper, static word embeddings is combined with contextualized embeddings which can optimize SA effectiveness in drug comments. Additionally, a BiGRU layer was incorporated on the top of the resulting features for further training to enhance the final context vectors. The model showed great results with a small training dataset due to the fact that the proposed SA model leverages the advantages of the pre-trained BERT model and static embedding models.

The structure of this paper is as follows. After discussing previous works Sect. 2, we present the used dataset and describe the model proposed in this word Sect. 3. Then, we evaluate the model and discuss the findings Sect. 4. Lastly, we provide a conclusion based on the experiments Sect. 5.

2 Related Works

While sentiment analysis has been widely used in various application areas, it has received less attention in the pharmaceutical industry Earlier efforts in Sentiment Analysis of Drug reviews relied heavily on rules [13] and sentiment lexicons (such as SentiWordNet 3.0 [2]) to identify the overall polarity (positive, neutral or negative) of a particular drug review. However, lexicon-based techniques struggle with phonetic and typographical misspellings. Therefore, current research has focused on ML and DL techniques to address these challenges. In [1], Abidin et al. used six distinct sets of n-gram characteristics to classify tweets. According to their study, the k-Nearest Neighbors (k-NN) approach outperforms the NB algorithm by 81.2% in terms of F1-score. Gräßer et al. [9] collected a dataset of drug reviews from the Drugs.com website, which offers drug information for patients and health professionals. Each drug review has a rating between 0 and 9, indicating the patient's degree of satisfaction with the drug. The reviews were classified into three groups based on their scores: positive (rating ≥ 7), negative (rating ≤ 4) and neutral (rating in [3,6]). The authors classified the drug reviews based on Logistic Regression (LR), reaching an accuracy of 92.24%. Lately, DL models have yielded impressive scores for Sentiment Analysis in the health sector. Yadav et al. [19] evaluate the results of a CNN vs. standard ML algorithms, namely Support Vector Machines (SVM), Random Forest, and a multilayer perceptron (MLP), for aspect-based sentiment analysis of terms such as "drug" and "medical condition". As compared to the standard algorithms, the CNN model achieved substantially greater scores. Several techniques towards obtaining pre-training contextual representations have been emerged in recent years, including Semi-supervised Sequence Learning [7], Embeddings from Language Models (ELMo) [15] and Universal Language Model Fine-tuning (ULMFit) [10]. Since last year, various solutions for the Sentiment Analysis task have been established adopting language models. These efforts achieved cutting-edge results [12,16,17], but have rarely focused on the Sentiment Analysis of drug reviews. Biseda and Mo [5] assessed the efficiency of Bidirectional Encoder Representations from Transformers (BERT) across multiple tasks, such as Sentiment Analysis of drug reviews. They used the dataset generated by Gräßer et al. [9]. The authors classified the drug reviews into three categories: highly negative (rating ≤ 3), highly positive (rating ≥ 8) and neutral (rating in [4,7]). Their experiments yielded 90.6% accuracy for the Sentiment Analysis task. In this paper, we propose an enhanced word embedding model by integrating contextual and static embeddings. Our suggested embedding model is based on contextual models that have shown strong potential in a variety of NLP tasks, as well as static embedding models.

3 The Proposed Approach

Given a text sentence $S = \{w_1, w_2, ..., w_t, \ ..., w_n\}$, where w_i refers the i-th word in the sentence, we combine the BERT derived embeddings at word level with the static word vectors. In this paper, we use a CNN architecture (CNN-sentence) to get the sentence vector (the context vector) from the contextual embeddings. The CNN architecture consists of 64 filters with windows of 2, 3, and 5 words. To avoid the problems associated with gradient vanishing, we used a linear Rectification unit (ReLU) [14] as the activation function. To provide the most significant features, a max-pooling layer is used. The extracted characteristics are then combined. Finally, a global max-pooling layer is applied on the resulting features, see Fig. 1. In addition, we adopt BiGRU architecture to capture the local features using the static embeddings.

Next, the Multiply layer is used to combine the static vectors generated using the BiGRU model and the context vector obtained using CNN. Specifically, we multiply each static vector h_i with the context vector p. The proposed unified model is able of selecting the significant features from the static embeddings along with the generated context vector from BERT model.

Four convolutional layers are applied on the resulting embeddings. The size of the windows for the four convolutionallayers are 4, 5, 6 and 7, respectively. Max pooling is added after each convolution with a pooling length of 3. A dropout of 30% is added at the end of each of the convolutional layers. It is used to randomly drop units from the Neural Network during training and hence avoid overfitting. Finally, a nonlinear activation func-tion is applied to introduce non-linearity into the output of the Convolutionaland the Max pooling layers. In this work, we have adopted the Rectified LinearUnit (ReLU) as an activation function. The last block is composed of a classification layer using softmax activation function. It is often used at the neural network's final layer for Multi-class classification to generate the probability distribution over sentiment labels.

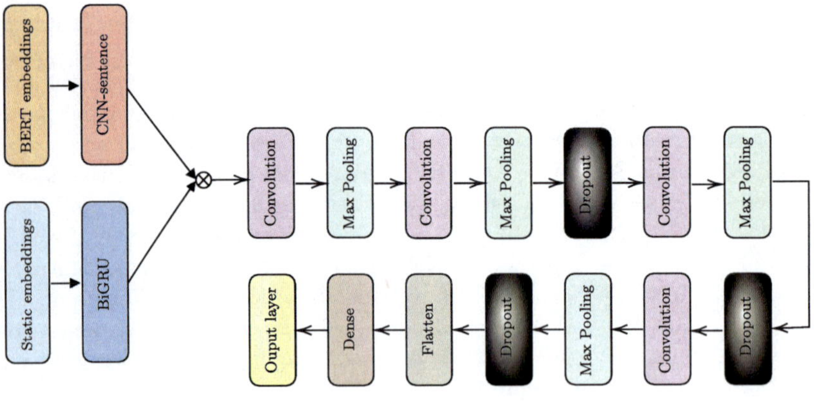

Fig. 1. The proposed approach

4 Experimental Settings and Results

4.1 Dataset

We use the dataset proposed in [9], which is a collection of drug reviews gathered from the website Drugs.com. Each drug review includes a score ranging from 0 to 9, indicating the patient's level of satisfaction with the drug. There are a total of 215,063 drug reviews in the corpus. Furthermore, the developers of this corpus classified the drug reviews into three levels of polarity depending on the rating: negative (class 0, rating ≤ 4), neutral (class 1; rating in [3,6]) and positive (class 2; rating ≥ 7).

4.2 Experimental Settings

We should first preprocess the drug reviews in order to represent them using the GloVe word embedding model. Data preparation is a key stage in which extraneous information is removed in order to speed up the learning process of classification models and boost accuracy. In this work, the NLP toolbox in Python was employed to preprocess the data. The text is first changed to lower case, links, punctuation and HTML (HyperText Markup Language) tags are removed. In this study, we use a 75:25 split ratio where 75% of data are used for the models' training while 25% of data are taken for models' testing. In this paper, we address the SA task using 3-classes proposed in [9], and also using 10 classes. As previously stated, each drug comment is assigned an integer score (rating) ranging from 0 to 9, indicating overall patient satisfaction. As a result, we consider these ratings to be the classes for our problem. For 10-class dataset, Fig. 2 illustrates the distribution of various drug reviews based on their class (degree of satisfaction). There is a significant imbalanced distribution of the classes, with those having higher polarized ratings predominating (such as 1, 8, 9 and 10).

Fig. 2. Class distribution in training and test datasets (10 classes).

The 3-class dataset also has an unbalanced distribution of the classes (see Fig. 3). This figure indicates that the distribution of the three classes in test and training sets is similar. The reviews with positive polarity (class 1) are around 66%, while the reviews with negative polarity (class -1) represent approximately 25% and only 9% for the reviews with neutral polarity (class 0).

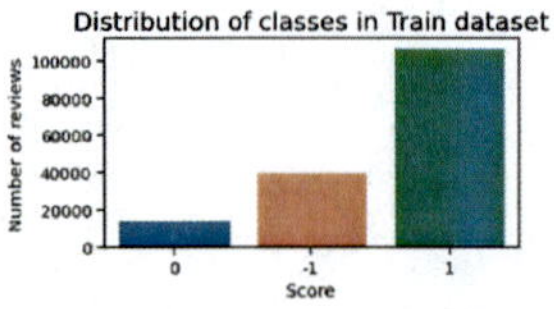

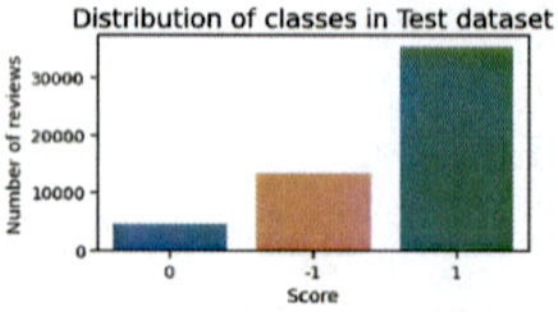

Fig. 3. Class distribution in training and test datasets (3 classes).

Before splitting the data, data is shuffled to decrease variance and assure model generalizability. Shuffling also facilitates the training data to be more representative of the general data distribution and helps to minimize model overfitting. Because the drug reviews varied in length, it was required to pad and truncate the texts to equalize their lengths. According to the cumulative distribution function throughout the duration of reviews, nearly all processed texts are less than or equal to 100 tokens in length. As a result, we limited the length of processed reviews to 250 tokens. For static embeddings, the nltk.tokenizer module [4] is devoted to the task of tokenizing, or dividing a text into its constituent parts. For BERT embeddings, we adopt the WordPiece tokenizer by Wu et al. [18], which breaks down any word that is not found in the vocabulary into sub-words to overcome the Out Of Vocabulary (OOV) problem.

4.3 Results

To assess our systems, we have used the F1-score which is the most commonly used for text classification. The micro-averaged and macro-averaged have been carried out for F1-score. In the macro-average, we show the mean of the values collected individually for each class. In the micro-average, the average metric is computed by adding the contributions of each class.

Results for 10 Classes. In order to assess the efficiency of the suggested system in analyzing the sentiment of drug reviews, we show the performance of the proposed model applied only on the static word embeddings. Moreover, we show the results obtained using BERT embeddings alone. Based on Tables 1, 2 and 3, the results show that our approach achieves better results than the other considered systems As previously mentioned, there is a very unbalanced distribution of classes, with a predominance of the most polarized ratings. This created a challenge when trying to categorize the poorly represented reviews. The results show that regardless of the model used, the performance of each class is dependent on the number of instances of each class in the training set. As a result, classes with a larger number of training instances generally perform better than those with a smaller number of instances in the training set.

Table 1. Results of the proposed approach

Label	F1-score
1	0.78
2	0.60
3	0.59
4	0.56
5	0.57
6	0.55
7	0.55
8	0.58
9	0.64
10	0.78
Micro avg.	0.68
Macro avg.	0.62

Table 2. Results of the proposed approach applied on GloVe embeddings

Label	F1-score
1	0.74
2	0.56
3	0.47
4	0.47
5	0.48
6	0.43
7	0.46
8	0.53
9	0.56
10	0.76
Micro avg.	0.63
Macro avg.	0.57

Table 3. Results of the proposed approach applied on BERT embeddings

Label	F1-score
1	0.77
2	0.53
3	0.50
4	0.50
5	0.51
6	0.46
7	0.49
8	0.56
9	0.59
10	0.79
Micro avg.	0.65
Macro avg.	0.59

4.4 Results for 3 Classes

Tables 4, 5 and 6 show the results using the dataset with three polarities using the proposed model, the proposed model applied only on the static word embeddings and the proposed model applied on BERT embeddings alone. As observed using the dataset with ten classes, the proposed model provides significantly better results than the other considered systems. These results can be explained by the fact that our approach leverages the GloVe model which extracts features associated with the language in which most drug reviews are written. In addition, the proposed model takes advantage of the BERT model, which has achieved outstanding results on a variety of NLP-related tasks in different languages.

Table 4. Results of the proposed approach

Label	F1-score
0	0.66
1	0.95
–1	0.88
Micro avg.	0.92
Macro avg.	0.83

Table 5. Results of the proposed approach applied on GloVe embeddings

Label	F1-score
0	0.65
1	0.94
–1	0.85
Micro avg.	0.89
Macro avg.	0.80

Table 6. Results of the proposed approach applied on BERT embeddings

Label	F1-score
0	0.65
1	0.94
–1	0.87
Micro avg.	0.91
Macro avg.	0.82

5 Conclusion

Since language representation models such as BERT are increasingly used for classification problems, we proposed a new method to combine static word embeddings with BERT-based language models. Intuitively, the reason why our hybrid strategy outperforms the BERT-based model using a domain-specific dataset (drug reviews) is due to the fact that BERT was trained on Wikipedia only. An alternative strategy would be to train language models on this dataset. While such a solution is likely to lead to an improvement in overall performance, it is not always feasible in practice. Updating the language models is extremely expensive. On the other hand, static word vectors play a crucial role in handling emerging keywords such as popular hashtags and recent topics. Based on the experimental results, the integration of static word vectors is a promising strategy for improving language models.

Funding. This work was supported by the Ministry of Higher Education, Scientific Research and Innovation, the Digital Development Agency (DDA) and the CNRST of Morocco (Alkhawarizmi/2020/01).

References

1. Abidin, T.F., Hasanuddin, M., Mutiawani, V.: N-grams based features for Indonesian tweets classification problems. In: 2017 International Conference on Electrical Engineering and Informatics (ICELTICs) (2017), pp. 307–310
2. Baccianella, S., Esuli, A., Sebastiani, F.: SentiWordNet 3.0: an enhanced lexical resource for sentiment analysis and opinion mining. In: Proceedings of the International Conference on Language Resources and Evaluation, LREC (2010)
3. Bensoltane, R., Zaki, T.: Towards Arabic aspect-based sentiment analysis: a transfer learning-based approach. Soc. Netw. Anal. Min. **12**(1), 1–16 (2022)
4. Bird, S., Loper, E.: NLTK: the natural language toolkit. Association for Computational Linguistics
5. Biseda, B., Mo, K.: Enhancing pharmacovigilance with drug reviews and social media. CoRR abs/2004.08731 (2020)
6. Cardie, C.: Sentiment analysis and opinion mining. Comput. Linguistics **40**(2), 511–513 (2014)
7. Dai, A.M., Le, Q.V.: Semi-supervised sequence learning. In: Advances in Neural Information Processing Systems, vol. 28 (2015). Curran Associates Inc
8. Devlin, J., Chang, M., Lee, K., Toutanova, K.: BERT: pre-training of deep bidirectional transformers for language understanding. In: Proceedings of the 2019 Conference of the North American Chapter of the Association for Computational Linguistics: Human Language Technologies, NAACL-HLT, pp. 4171–4186 (2019)
9. Gräßer, F., Kallumadi, S., Malberg, H., Zaunseder, S.: Aspect-based sentiment analysis of drug reviews applying cross-domain and cross-data learning. In: Proceedings of the 2018 International Conference on Digital Health, DH, pp. 121–125 (2018)
10. Howard, J., Ruder, S.: Universal language model fine-tuning for text classification. In: Proceedings of the 56th Annual Meeting of the Association for Computational Linguistics, ACL, pp. 328–339 (2018)

11. Hutto, C., Gilbert, E.: VADER: a parsimonious rule-based model for sentiment analysis of social media text. In: Proceedings of the International AAAI Conference on Web and Social Media **8**, 216–225 (2014)
12. Li, X., et al.: Enhancing BERT representation with context-aware embedding for aspect-based sentiment analysis. IEEE Access **8**, 46868–46876 (2020)
13. Na, J., Kyaing, W.Y.M., Khoo, C.S.G., Foo, S., Chang, Y., Theng, Y.L.: Sentiment classification of drug reviews using a rule-based linguistic approach. In: The Outreach of Digital Libraries: A Globalized Resource Network - 14th International Conference on Asia-Pacific Digital Libraries, ICADL, vol. 7634, pp. 189–198 (2012)
14. Nair, V., Hinton, G.E.: Rectified linear units improve restricted Boltzmann machines. In Proceedings of the 27th International Conference on Machine Learning (ICML-10), pp. 807–814 (2010)
15. Peters, M.E., Neumann, M., Iyyer, M., Gardner, M., Clark, C., Lee, K., Zettlemoyer, L. Deep Contextualized Word Representations. In Proceedings of the 2018 Conference of the North American Chapter of the Association for Computational Linguistics: Human Language Technologies, NAACL-HLT (2018), pp. 2227–2237
16. Sun, C., Huang, L., Qiu, X.: Utilizing BERT for aspect-based sentiment analysis via constructing auxiliary sentence. In: Proceedings of the 2019 Conference of the North American Chapter of the Association for Computational Linguistics: Human Language Technologies, pp. 380–385. NAACL-HLT (2019)
17. Sweidan, A.H., El-Bendary, N., Al-Feel, H.: Sentence-level aspect-based sentiment analysis for classifying adverse drug reactions (ADRs) using hybrid ontology-XLNet transfer learning. IEEE Access **9**, 90828–90846 (2021)
18. Wu, Y., et al.: Google's neural machine translation system: bridging the gap between human and machine translation. CoRR abs/1609.08144 (2016)
19. Yadav, S., Ekbal, A., Saha, S., Bhattacharyya, P.: Medical sentiment analysis using social media: towards building a patient assisted system. In: Proceedings of the Eleventh International Conference on Language Resources and Evaluation, LREC (2018)
20. Zafra, S.M.J., Valdivia, M.T.M., Molina-González, M.D., López, L.A.U.: How do we talk about doctors and drugs? sentiment analysis in forums expressing opinions for medical domain. Artif. Intell. Med. **93**, 50–57 (2019)

Combined Feature Selection and Rule Extraction for Credit Applicant Classification

Siham Akil[1](✉), Sara Sekkate[2], and Abdellah Adib[1]

[1] Team: Data Science and Artificial Intelligence, Laboratory of Mathematics, Computer Science and Applications (LMCSA), Faculty of Sciences and Technologies, Mohammedia, Morocco
siham.ak2@gmail.com , abdellah.adib@fstm.ac.ma

[2] Higher National School of Arts and Crafts of Casablanca Hassan II University of Casablanca, Casablanca, Morocco

Abstract. A sensitive area such as credit risk assessment has always been a high priority and quite difficult for financial institutions to make financial decisions. In order to have a more relevant and accurate classification model, it is necessary for models to be interpreted and understood by financial professionals. The interaction and interpretation of the model becomes more limited as the number of features increases. In order to reduce the number of non significant features, feature selection must be implemented on the original data. The purpose of this study is not only to find a set of logical rules in an easily interpretable form for classification but also to select a set of informative features that can solve the classification problem. To this end, we present a combination of feature selection and rule extraction techniques. The experiment were carried out with the Australian Credit Dataset from the UCI Machine Learning Repository consisting of 13 attributes and a decision variable. The experimental results show that the used techniques are efficient in terms of interpretability, comprehensibility and accuracy.

Keywords: Credit scoring · Credit risk assessment · Rule extraction · Feature selection · Decision rules

1 Introduction

Almost all financial institutions – banks in particular – use credit scoring models to assess credit risk. These models employ a probabilistic measure of credit risk, often known as a risk rating or chance of default. The purpose of credit scoring is to ascertain how different borrower characteristics affect their likelihood of default, by using statistical methods that are calculated through software and

M. Ben Ahmed et al. (Eds.): SCA 2022, LNNS 629, pp. 97–104, 2023.
https://doi.org/10.1007/978-3-031-26852-6_9

historical data, typically the performance history of the borrowers or the history of the loans made. Additionally, they generate "scores," which are notes that quantify the default risk of present or prospective borrowers, allowing for the division of these distinct borrowers into risk classes: accepted or refused.

Support vector machines (SVM) [1], Random Forest (RF) [2], neural networks (NN) [3,4], and other machine learning techniques have been applied and have proven to be quite useful for credit risk assessment employing credit scoring models. However, some methodological approaches must be used in data processing to provide a performing model. As long as the number of features increases, the number of necessary calculations rises as well, residing that the interpretation and interactivity of the model is reduced. The solution is to implement two main techniques that solve these problems consisting of feature selection and rule extraction.

A combination of these two processes can be considered as a solution to these dimensionality and interpretability problems. Feature selection is the process of selecting the most consistent, non-redundant and narrowing down the set of features to those that are most relevant to the machine learning model [5]. It provides a simpler model, shorter training time, reduced variance thereby increasing accuracy and avoiding the curse of high dimensionality [6]. On the other hand, rule extraction is a technique that solves the black box problem of classification algorithms, their presumed complexity and the difficulty of fully understanding their underlying logic [7]. Hence, the need to present them in a more observable and understandable way, particularly in cases where the classification process is important, such as in sensitive areas like credit evaluation [8]. It consists in extracting simple, logical and understandable rules that can explain the behavior of machine learning models by revealing the internal knowledge of the trained models [9].

In this study, in order to ensure both a good interpretation and higher accuracy of credit scoring models and to improve the accuracy of the model, we focus on feature selection methods to retain only relevant features in credit evaluation and subsequently use rule extraction methods to extract logical rules and present them in a way that humans can easily interpret them. The rest of the paper is organized as follows: Sect. 2 summarizes the related work on credit scoring. The experimental setup and data description is given in Sect. 3. Results are reported and discussed in Sect. 4. Section 5 concludes the paper.

2 Related Works

Considerably, many studies have been conducted on feature selection for financial decision support, with the aim to reduce the original number of features and improve model performances [5]. Cheng-Lung Huang et al. [1] constructed three hybrid SVM-based approaches: SVM using grid search to optimize model parameters, SVM using grid search and F-score to select relevant features and SVM using Genetic Algorithm (GA) in order to both optimize and select relevant features. The authors in [2] used two Feature Selection (FS) techniques, namely Random Forest (RF) and F-score, to select the most relevant features by combining them with the standard SVM, and subsequently compare them. Kozodoi, Nikita, et al. [10], designed a FS framework for credit scoring using GA NSGA-II which comprises three main steps: Fast non-dominated sorting, diversity preservation and population update. Yue Lie et al. [11] investigated existing GA approaches for FS, then they created and developed a method based on the usage of subpopulations with various types of data for fitness assessment. In order to deal with the problem of unbalanced data distribution, the authors of [12] aimed to evaluate the instances in terms of the entropy of their features. The authors in [13], examined multiple FS techniques such as Chi-Square, Information-Gain and Gain-Ratio and evaluated five Machine Learning (ML) classifiers such as Bayesian, Naïve Bayes, SVM, Decision Tree (C5.0) and RF with the aim to find the appropriate input predictors to build credit scoring models. For the credit scoring task, in conjunction with five ML classifiers, Akil et al. [5] have demonstrated the significance of filter and embedding feature selection strategies. These techniques were discovered to be faster, less to compute, and very efficient in producing an optimal subset that improved model performance. An effective credit scoring model requires a good trade-off between model accuracy and comprehensibility [14]. Indeed, classification algorithms are black boxes that do not offer good comprehensibility and interpretability [9]. In this context, various studies have been carried out on rule extraction for credit scoring problems [9,14]. In their study [15], Hruschka and Ebecken proposed a GA method for extracting rules from neural networks based on clustering the activation values of hidden units. To accurately categorize samples from a given data set, the authors first created a back-propagation network and trained it. This neural network model was then used to extract classification rules using the created Genetic Clustering Algorithm (CGA). Bazan et al. [16] aimed to compare the performance of a classifier based on computing all minimally consistent decision rules with a method based on two-layer learning. In the first learning layer, the collection of classifiers is derived from a portion of the original training dataset. Then, in the second layer, they guide the classifier using the model extracted from the created classifier based on the performance of the remaining training data. Hayashi et al. [17] considered a continuous Re-RX rule extraction method, in which an input unit was designed for each continuous attribute in the dataset and the discrete attributes were converted into a binary input string, with the aim of generating decision trees using J48graft that achieves both higher accuracy and good model interpretability. Sagi and Rokach [8] explored an approach where a decision forest is transformed into an interpretable decision tree. The methodology adopted consists of creating a set of rule conjunctions that represent the original decision forest; then they are hierarchically organized to build a new decision tree in order to find a compromise between accuracy and interpretability. Many authors have conducted extensive research to create robust credit scoring models using feature selection and other rule extraction techniques, but they have not yet been collectively studied in conjunction. The above discussion shows that the search for an interpretable and robust credit scoring model remains a potentially important research area.

3 Background

3.1 Feature Selection

Feature selection is a crucial aspect in data preprocessing, useful on a variety of fronts such as curse dimensionality and reduction in training time. It aims at reducing the number of original features to an optimal subset, which can be used to provide equal or better results than the original set. For this study, we seek to explore the most relevant features to characterize the reliability of credit applicants. Below are the feature selection methods used in this work:

1- Analysis of variance: is a statistical test that measures the interdependence of two or more variables that differ significantly from each other, by calculating the F-test, which is the ratio of between-class variability to within-class variability that enables to determine the importance of a feature in the discriminant analysis. We compute the F-test value of a given feature as follows:

$$F = \frac{\sum_{j=1}^{J} N_j(\bar{x_J} - \bar{x})/(J-1)}{\delta^2} \tag{1}$$

where J is the number of classes, N_j is the number of instances in jth class, $\bar{x_J}$ is the mean of instances X in class j, $\bar{x_J}$ indicates the mean value for all instances, and δ^2 is the pool variance.

2- Kendall Rank Correlation : named after Maurice Kendall [18], who developed it in 1938, is a statistic measure that provides the ordinal connection between two quantities based on the τ coefficient calculated as follows [19]:

$$\tau = \frac{n_c - n_d}{\binom{n}{2}} \tag{2}$$

where n_c and n_d are the number of concordant and discordant pairs, respectively and $\binom{n}{k} = \frac{n(n-1)}{2}$ is the binomial coefficient for the number of items from n items. Kendall's τ coefficient of correlation varies between 1 and -1, it will take a value of 1 (high) when observations have an identical or similar rank, and a value of -1 (low) when the observations have a dissimilar rank between two features.

3- Random Forest: is a supervised ML algorithm, composed of many decision trees. In fact, each tree in the random forest can determine the importance of a feature based on its ability to increase node purity. The higher the increment in node purity, the higher the importance of the feature. This process is performed for each tree, averaged across all trees, and then normalized to 1. Therefore, the sum of the importance scores calculated by a random forest is equal to 1.

3- Gradient Boosting Decision Tree (GBDT): Feature selection using a gradient boosting decision tree falls into the category of embedded methods. After the construction of the boosted trees, an importance score is assigned to each feature. This score is calculated for each boosted decision tree based on the extent to which each feature separation point improves the performance measure, weighted by the number of observations for which the node is responsible [20]. The more an attribute is employed to form key decisions with decision trees, the upper is its relative importance.

3.2 Rule Extraction

1- Decision Tree: is a widely used data mining technique that is transparent and creates a set of production rules for decision making. It provides decision rules by extracting IF-THEN rules that contain potential information from the data. The IF-THEN rules make it easier to understand how the sample propagates through the tree during prediction. A rule is created for each path from the root to the leaf node to extract the rule from the decision tree. Each split criterion along the specified path is combined with the AND operator to form the antecedent (IF part) of the rule. The leaf node contains the class predictions that form the consequent part (THEN part) [21]. A decision tree can be transformed into an IF-THEN classification rule by tracing the path from the root node to each leaf node in the tree [22].

2- Random Forest: is a tree-based learning model composed of many decision trees [23]. When building individual trees, an attempt is made to use feature clustering and randomness to create an uncorrelated forest of trees where the predictions are more accurate than the predictions of the individual tree. Each node in each tree of a random forest can be transformed into an elementary rule. It leads to extract a large collection of rules from a set of trees that build and integrate multiple decision trees during the training phase, based on how often they appear [24]. Finally, the most common rules that represent robust and powerful data models are combined to form predictions.

4 Experimental Results and Discussion

In this work, Australian Credit Dataset, which is one of the most common datasets in the credit scoring field has been used to implement a combination of FS and rule extraction. The dataset has been obtained from the UCI Machine Learning Repository [25]. However, for confidentiality reasons, the names and values of all features are replaced with symbolic data. This includes 690 loan applicants with 14 features, including 383 creditworthy customers (Class 1) and 307 defaulting customers (Class 0). In this study, we randomly split this dataset into a 70% for training set and a 30% for testing set.

To evaluate our model, we use a very popular evaluation metric in ML to assess credit scoring models, that is, accuracy, which generally describes the performance of the model in all classes. Otherwise, the accuracy is the fraction of predictions that the model gets right.

Table 1. Reduced number of features.

Feature selection technique	Kendall Rank Correlation	ANOVA	RF	GBDT
Number of retained features	6	4	6	6

Table 1 shows the number of features held by each FS technique. Employing Kendall's rank correlation, ANOVA, RF, and GBDT, they have reduced the original number of the feature set by 7, 9, 7, and 7, respectively. Thus, only relevant features are selected and can be used for rule extraction.

From Table 2, we notice that the performance obtained using the feature selection techniques was almost similar and ranged from 77% to 88%. The best result was obtained using GBDT as a FS technique and Decision Tree (DT) as a classifier also devoted to rule extraction, as the decision tree algorithm produces accurate and interpretable models, it is also fast, both at the time of construction and application.

Table 2. Empirical results of our experiments.

Feature Selection technique	Rule Extraction method	Accuracy (%)	
		Train	Test
Kendall Rank Correlation	Decision Tree	85.3002	86.4734
	Random Forest	79.7101	77.2946
ANOVA	Decision Tree	86.9565	86.4734
	Random Forest	85.5072	86.9565
RF	Decision Tree	85.0931	86.4734
	Random Forest	85.7142	84.0579
GBDT	Decision Tree	86.3354	88.4057
	Random Forest	84.4720	85.9903

Table 3 shows that we obtained a total number of rules 7, 8, 7 and 8 respectively using DT and 24, 22, 23 and 24 using RF for Kendall Rank Correlation, ANOVA, RF and GBDT respectively. Subsequently, we eliminated all the insignificant rules that have an importance of 0, therefore we have an efficient, transparent and convincing system for financial professionals with fewer decision rules.

Interpretability of machine learning algorithms is necessary especially when the intended application involves important decisions. We argue that simplicity, stability and predictability are the minimum requirements for an interpretable model.

Table 3. Number of total and filtered rules.

Feature Selection technique	Rule Extraction method	Total number of rules	Number of filtered rules
Kendall Rank Correlation	DT	7	2
	RF	24	4
ANOVA	DT	8	3
	RF	22	6
RF	DT	7	2
	RF	23	6
GBDT	DT	8	2
	RF	24	4

5 Conclusion

In this paper, we examined four FS techniques in conjunction with two rule extraction methods for credit risk assessment using the Australian Credit dataset from the UCI ML repository. It was revealed that the combination of these techniques was a very important approach for credit scoring residing in a reduced number of features and fewer decision rules while maintaining a promising classification performance and provided a solution to the trade-off between model interpretability and accuracy. These results encourage us to pursue further research in this direction and in particular for the rule extraction techniques that deserve more attention.

Future studies could explore this approach further by using other FS and rule extraction techniques to improve the performance and comprehensibility of the models. We also aim to test our approach on other databases.

Acknowledgements. This work was supported by the Ministry of Higher Education, Scientific Research and Innovation, the Digital Development Agency (DDA) and the CNRST of Morocco (Alkhawarizmi/2020/01).

References

1. Huang, C.L., Chen, M.C., Wang, C.J.: Credit scoring with a data mining approach based on support vector machines. Expert Syst. Appl. **33**(4), 847–856 (2007)
2. Shi, J., Zhang, S.Y., Qiu, L.M.: Credit scoring by feature-weighted support vector machines. J. Zhejiang Univ. Sci. C **14**(3), 197–204 (2013)
3. Ha, V.S., Nguyen, H.N.: Credit scoring with a feature selection approach based deep learning. MATEC Web Conf. **54**, 1–5 (2016)
4. Munkhdalai, L., Namsrai, O.E., Ryu, K.H.: Credit scoring with deep learning. In: 4th International Conference on Information, System and Convergence Applications, pp.1–5 (2018)

5. Siham, A., Sara, S., Abdellah, A.: Feature selection based on machine learning for credit scoring : an evaluation of filter and embedded methods. In: 2021 International Conference on Innovations in Intelligent SysTems and Applications, INISTA 2021 - Proceedings (2021)
6. Rtayli, N., Enneya, N.: Selection features and support vector machine for credit card risk identification. Procedia Manufact. **46**, 941–948 (2020)
7. Friedman, J.H., Popescu, B.E.: Predictive learning via rule ensembles. Annal. Appl. Stat. **2**(3), 916–954 (2008)
8. Sagi, O., Rokach, L.: Explainable decision forest: transforming a decision forest into an interpretable tree. Inf. Fusion **61**, 124–138 (2020)
9. Hayashi, Y.: Application of a rule extraction algorithm family based on the Re-RX algorithm to financial credit risk assessment from a Pareto optimal perspective. Oper. Res. Perspectives **3**, 32–42 (2016)
10. Kozodoi, N., Lessmann, S., Papakonstantinou, K., Gatsoulis, Y., Baesens, B.: A multi-objective approach for profit-driven feature selection in credit scoring. Decis. Support Syst. **120**(March), 106–117 (2019)
11. Liu, Y., Ghandar, A., Theodoropoulos, G.: island model genetic algorithm for feature selection in non-traditional credit risk evaluation, pp. 2771–2778 (2019)
12. Carta, S., Ferreira, A., Recupero, D.R., Saia, M., Saia, R.: A combined entropy-based approach for a proactive credit scoring. Eng. Appl. Artif. Intell. **87**, 103292 (2020)
13. Trivedi, S.K.: A study on credit scoring modeling with different feature selection and machine learning approaches. Technol. Soc. **63**, 101413 (2020)
14. Mashayekhi, M., Gras, R.: Rule extraction from decision trees ensembles: new algorithms based on heuristic search and sparse group lasso methods. Int. J. Inf. Technol. Decision Making **16**(06), 1707–1727 (2017)
15. Hruschka, E.R., Ebecken, N.F.: Applying a clustering genetic algorithm for extracting rules from a supervised neural network. Proceed. Int. Joint Conf. Neural Netw. **3**(2), 407–412 (2000)
16. Bazan, J.G.: Classifiers Based on Two-Layered Learning. In: Tsumoto, S., Słowiński, R., Komorowski, J., Grzymała-Busse, J.W. (eds.) RSCTC 2004. LNCS (LNAI), vol. 3066, pp. 356–361. Springer, Heidelberg (2004). https://doi.org/10.1007/978-3-540-25929-9_42
17. Hayashi, Y., Oishi, T.: High accuracy-priority rule extraction for reconciling accuracy and interpretability in credit scoring. N. Gener. Comput. **36**(4), 393–418 (2018). https://doi.org/10.1007/s00354-018-0043-5
18. Kendall, M.G.: A new measure of rank correlation. Biometrika **30**(1/2), 81–93 (1938)
19. Kendall, M.G.: Rank correlation methods (1948)
20. Tian, Z., Xiao, J., Feng, H., Wei, Y.: Credit risk assessment based on gradient boosting decision tree. Procedia Comput. Sci. **174**, 150–160 (2020)
21. Griselda, L., Joaquín, A., et al.: Using decision trees to extract decision rules from police reports on road accidents. Procedia. Soc. Behav. Sci. **53**, 106–114 (2012)
22. Vasilev, N., Mincheva, Z., Nikolov, V.: Decision tree extraction using trained neural network. SMARTGREENS 2020 - Proceedings of the 9th International Conference on Smart Cities and Green ICT Systems, pp. 194–200 (2020)
23. Mashayekhi, M., Gras, R.: Rule extraction from random forest: the RF+HC methods. In: Barbosa, D., Milios, E. (eds.) CANADIAN AI 2015. LNCS (LNAI), vol. 9091, pp. 223–237. Springer, Cham (2015). https://doi.org/10.1007/978-3-319-18356-5_20
24. Bénard, C.: SIRUS interpretable RF, vol. 130 (2021)
25. Dua, D., Graff, C.: UCI machine learning repository (2017)

Low-Cost Global Navigation Satellite System (Low-Cost GNSS) for Mobile Geographic Information System (GIS)

Muhammad Ali Sammuneh(✉), Rani El Meouche(✉), Mojtaba Eslahi, and Elham Farazdaghi

Institut de Recherche en Constructibilité (IRC), ESTP Paris, 94230 Cachan, France
{msammuneh,relmeouche,meslahi,efarazdaghi}@estp-paris.eu

Abstract. Geolocation by GNSS (Global Navigation Satellite System) is developing regularly with new infrastructures, and new equipment such as correction networks, in particular, Centipede (free), low-cost receivers, and antennas. In addition, the evolution of GIS (Geographic Information Systems) has led to the creation of so-called mobile GIS applications for smartphones and tablets. This allows the collection, storage, and visualization of geographic information via these applications. All of these less expensive means of acquiring geographic information have led to a study on the investment of data from low-cost GNSS in mobile GIS. For this, we have conducted GNSS observation surveys with low-cost antennas based on the Centipede network to determine their precisions in RTK (Real Time Kinematic). The SW-Maps mobile GIS application served as the base application for this research. At the level of mobile GIS applications, SW-Maps offers great advantages in terms of data collection, storage, and visualization. The storage is done in several formats (.shp, .csv, .xls, .kml…), it is then possible to verify the consistency and reliability of the data. Therefore, the data collected in this study with SW-Maps are consistent with reality. This led to their use of ArcGIS Pro and the creation of a StoryMap.

Keywords: GNSS low cost · Centipede project · Real-time kinematic positioning (RTK) · Mobile GIS

1 Introduction

According to the French Society of Photogrammetry and Remote Sensing [23], "a GIS is a computer system allowing, from various sources, to gather and organize, manage, analyze and combine, elaborate and present geographically localized information, contributing to particular to the management of space." By this definition, we notice that the fundamental elements of the functioning of a GIS are geographical information. Thus, with the emergence of GNSS spatial positioning techniques and their continuous development, access to geographic data has been a major accomplishment [11]. Nowadays, precision ranging from centimeters to millimeters in differential positioning is approached, due to high precision equipment including the receivers and antennas.

M. Ben Ahmed et al. (Eds.): SCA 2022, LNNS 629, pp. 105–117, 2023.
https://doi.org/10.1007/978-3-031-26852-6_10

These "high precision" antennas and receivers are not accessible to everyone because of their relatively high cost. Therefore, for reasons of cost optimization, some actors such as private companies, farmers, and individuals have initiated numerous studies to test the accuracy of positioning with low-cost antennas. This is one of the main reasons for launching the Centipede project, an open access RTK (Real Time Kinematic) base network which we will discuss further in this project.

At the GIS level, the concept of mobile GIS has emerged. It consists of installing one or more GIS applications (SW-Maps, Collector, Survey123, Locus, Qfield, Input...) on a smartphone or a tablet in order to collect a multitude of geolocalized information that is directly stored there via an internal connection. With such flexibility in data acquisition, procedures are needed to ensure firstly the choice of the best available applications for referencing spatial information and secondly the consistency and reliability of data integration. We have here two concepts (mobile GIS and low-cost GNSS) which have some economic advantages and are therefore accessible to a large number of users. Therefore, it would be important to study the use of low-cost GNSS data in mobile GIS applications. This paper tries to address the issue of investing low-cost GNSS data in mobile GIS applications and find out the appropriate procedures for ensuring the consistency and reliability of the data for integration into a GIS mobile application, as well as choosing the most suitable GIS applications to integrate and use the low-cost GNSS data reliably and efficiently.

The main objective of this project is to investigate the use of low-cost GNSS for the implementation of a mobile GIS at the level of accuracy of low-cost receivers through:

- A comparative analysis between the uncalibrated IP66 and the calibrated IP67 antennas in terms of accuracy.
- RTK survey using the Centipede network and investment of the data in carefully selected mobile GIS applications.
- Providing a procedure to ensure consistency and reliability of the data.

1.1 Low-Cost GNSS

A GNSS, or Global Navigation Satellite System, is a generic name for a group of artificial satellites composed mainly of constellations of the American (GPS), the Russian (GLONASS), the European (Galileo), and the Chinese (Beidou) that send position and timing data from their high orbits.

Using a GNSS receiver integrated with a smartphone gives the absolute position (standalone) with around ten meters of accuracy [14]. The idea is to connect the smartphone with a GNSS receiver able to receive corrections of its position in real time via another GNSS station. On other hand, we want that these corrections will be free and accessible.

The Centipede GNSS network responds to our exigence because Centipede RTK is a collaborative network of GNSS bases open and available to anyone in the coverage area. Public institutes, individuals, and private actors such as farmers or other public partners extend the network [24]. Typical nominal accuracy for RTK systems is 1 cm horizontally and 2 cm vertically [16].

GNSS gives the position of an element everywhere and in real time. They are the most accurate systems used by geolocation tools [25]. The number of studies on the accuracy of low-cost GNSS receivers and dual-frequency receivers is very few, while comprehensive research on low-cost single-frequency receivers has been done so far because they have been in use for a long time [7].

Bellon et al. [2] have attempted real-time monitoring of rapid deformation using low-cost GNSS receivers. They investigated the performance of low-cost GNSS to see if this type of sensor can be used for real-time monitoring of landslides. The advantages of topographic techniques for landslide monitoring, in particular high precision GNSS and their high cost, are presented. In this study, a U-blox receiver (350 euros) and a Garmin antenna (40 euros) were used. According to the authors, a level of accuracy of 1 cm can be obtained in real time with low-cost instruments for a few hundred Euros.

The accuracy of low-cost GNSS receivers has been confirmed by several verifiers. Cina et Piras called low-cost receivers mass-market GNSS receivers [3]. They concluded that this kind of receiver could be used in static post-processing mode to reach millimeter accuracy for monitoring or surveying. Five years later Hodgson [6] evaluate a low-cost receiver/antenna in RTK FIX mode showing accuracies of 3.7 cm for the horizontal component and 4.2 cm for the height. The ZED-F9P receiver is tested in various terrain situations by Janos and Kuras in 2021 [8] and found that open sky condition is necessary to obtain precision measurements. Hamza et al. [5] compare geodetic and low-cost receivers with calibrated and uncalibrated antennas in absolute and relative modes. Their test condition is short baselines under an open-sky. All kinds of antennas detected spatial displacements smaller than 1 cm in relative positioning mode and 2 cm in absolute mode. Tuini et al. (2022) crustal deformation study uses the term "cost-effective device" for low cost [20]. They compared a low-cost receiver (u-Blox ZED F9P) with two geodetic devices for the same geodetic antenna. They used static post processing mode to calculate position components based on GNSS permanent network stations. The differences between u-Blox ZED F9P and the geodetic receivers is 1–2 mm in horizontal components and 5–8 mm in vertical one. The question of the reliability of the u-blox ZED-F9P RTK engine is tested by Sanna et al. (2022) and it is more efficient than RTKLib [15].

1.2 Mobile GIS

Mobile GIS provides a bridge between the desktop and the terrain, enabling capturing, storing, updating, manipulating, analyzing and displaying geospatial data and information in real time in the field. This can happen by integrating different technologies and devices.

Mobile GIS provides a bridge between the desktop and the terrain, enabling capturing, storing, updating, manipulating, analyzing and displaying geospatial data and information in real time in the field. This can happen by integrating different technologies and devices and open access databases that support spatial analysis and field surveys (Jolma et al. [9]; Steiniger and Hay [18]; Singh et al. [17]; Lindgren [10]; Fan et al. [4]; Yu et al. [21]; Nowak et al. [12]).

Nowak et al. [12] collected state-of-the-art on mobile GIS applications for environmental field surveys. In this review, the authors discuss the capabilities and availability of mobile GIS applications for on-site environmental data collection. A special aspect of these applications is the fact that they not only provide geographic positions but also allow filling in attributes that will facilitate further processing for GIS analysis. They described applications such as Mapit Spatial, QField, Open Data Kit (ODK), SW Maps, Locus GIS, Global Mapper Mobile, Mobile Data Collection, and Collector for ArcGIS, etc. They focused on the use of Collector for ArcGIS for the description of environmental phenomena through a practical case. Finally, the authors state that mobile devices (smartphones, tablets…) and mapping applications are a solution to improve the workflow of geographic data collection. They also explain the limited number of scientific publications on the use of the described mobile applications [12].

As it is mentioned mobile GIS has various applications. Mbock et al. [13] used the mobile GIS and vertex tool in forest management. They have used an ESRI mobile GIS application (ArcPad) and a tool called Vertex used in forestry in order to establish a geo-referenced inventory of trees and to estimate their heights. The GIS application helped to identify, characterize and geo-reference the tree species surveyed. This allowed for the mapping of the distribution of tree species.

Mobile technology offers many advantages. Among them, we can mention the possibilities provided by mobile GIS applications to improve maps by marking points of interest, roads, house numbers and the advantages of downloading photos for better orientation. Barliba et al. [1] describe the main advantages of the devices (including the mobile phones and tablets equipped with a GPS receiver) due to the fact that they are convenient for any mobile user, they are easy to use and have a low purchase cost compared to professional tools. They also present the advantages and disadvantages of using Android and iPad or iPhone smartphones and tablets. The applications "HERE Map Creator" and "Mapillary" were used respectively to map and update terrain data, and then create a Street View map. Later, the results obtained from "HERE Map Creator" were compared also with "Open Street Map". According to the authors, mobile GIS allows for real-time updates and can involve more people working simultaneously. However, the existence of satellite images that are not regularly updated and often poorly geo-referenced can mislead users.

Thirion E. [19] utilise SIG pour représenter la distribution spatiale de composés chimiques ou organiques selon le type de compartiments d'un organisme. He presents another aspect of applying GIS, by using the typical GIS tools (which generally have the function of collecting, structuring, analyzing and retrieving data) to represent the spatial distribution of carbon and nitrogen compounds within a tree species (beech). This research aims to determine the physiological mechanisms and morphological modifications induced by strong constraints (defoliation or drought) at different dates. To achieve this aim, one thousand 10-year-old beech trees planted in a plant breeding facility were kept under a roof in order to avoid rainfall. The researchers analyzed the impact of these constraints on the carbon and nitrogen metabolisms of the trees. After a phase of quantification of nitrogen compounds and carbohydrate reserves in the different compartments of the trees, the researchers decided to use GIS to represent them in the compartments of the aboveground and belowground structure of a tree. The use of a GIS vector layer, cut into distinct objects in a discrete space, of the polygon type was chosen over a raster layer that represents the space in a continuous regular mesh. This option is justified by the fact that the vector layer is simpler to implement and it allows, through the visual appearance of the vector layer, to precisely delineate the compartments studied.

The methodology and the study area are presented in the next section. In Sect. 3, we have discussed the temporal evolution of the accepted corrections. Section 4 deals with issues related to antenna studies. Integration of the project with mobile GIS and its applications are presented in Sect. 5. The discussions and conclusions are given in Sect. 6.

2 Methodology

Our work, compared to the previous studies, is to test the ZED-F9P receiver and IP66 antenna for topographic surveying using a simple android GIS application in a real-time kinematic mode based on Centipede Network Stations (see Fig. 1). Our strategy is to:

- Compare the solutions of calibrated and uncalibrated low-cost antenna
- Compare the Centipede station's solution results
- Compare GIS solutions

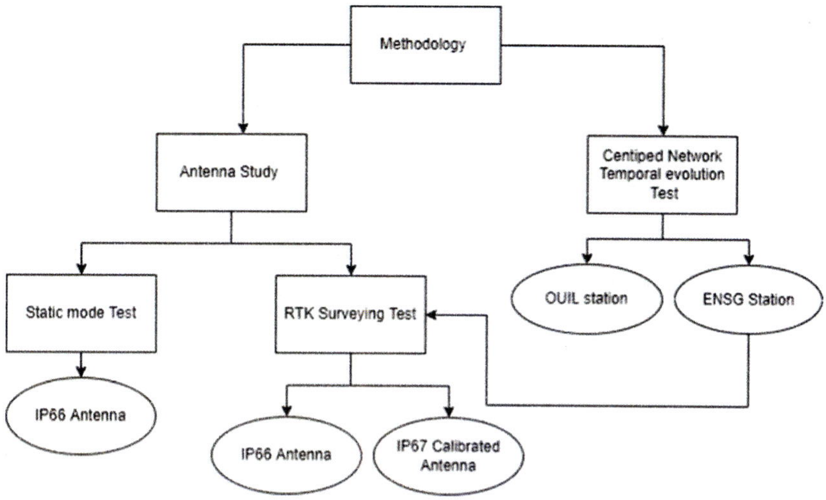

Fig. 1. Reference points and the study area – Cachan campus of ESTP Paris

For the study, we have low-cost GNSS receivers (ZED-F9P) with two Survey multi-band external GNSS antenna including the calibrated IP67 and no calibrated IP66 [26]. We use high accuracy Leica GS16 receiver/antenna. The aim is to determine the RTK position accuracy of the F9P/IP66 compared to the F9P/IP67 and Leica GS16. The goal is to test the possibility of using no calibrated antenna instead of high precision or calibrated antenna for surveying work. To do this, the RTK surveys of the same area will be performed with a GS18 and then with F9P/IP66 and F9P/IP67 low-cost receivers and antennas. The SW Maps – GIS and Data Collector smartphone application is used to collect, store and visualize the readings for the low-cost antennas whereas the Centipede network points (see below) are used as a canvas. For the Leica antenna, the commercial RTK base network (ORPHEON) is considered for the survey [22].

The study takes place on Cachan campus of ESTP Paris (*École spéciale des travaux publics, du bâtiment et de l'industrie).* The school already has a network of known points in planimetric (RGF 93), altimetric (IGN 69) coordinates, and open areas (low-rise buildings) offer good conditions for GNSS observations.

3 Temporal Evolution of Centipede Received Corrections

The Centipede is free of charge which is a good point economically speaking, but there is no guarantee concerning the well-being of stations. Centipede station has a correction range of 30 km. For this test, we identified two stations of this network ENSG and OUIL of respective baselines (19.6 km and 18.1 km) from the study area (see Fig. 2). It was, therefore, necessary to verify if these stations are trustworthy.

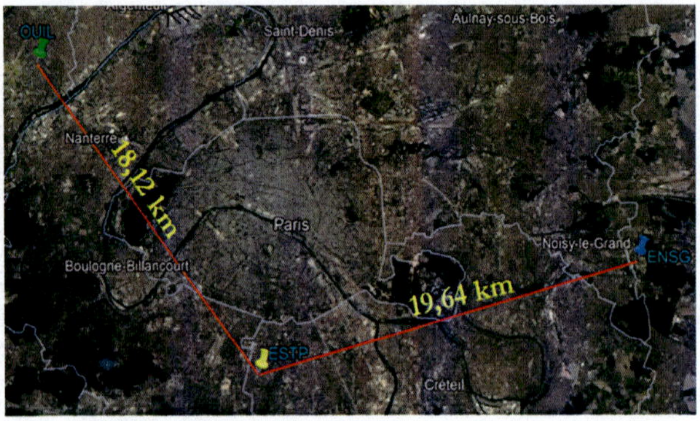

Fig. 2. ENSG and OUIL stations of Centipede

Two references known coordinates REF1 and REF2 are used. Thirty minutes RTK survey with an acquisition rate of 5 s on our two references by receiving the corrections from Centipede network RTK for both base points OUIL and ENSG. Simultaneously, two F9b/IP67, one on REF1 and the other on REF2 receive respectively corrections from the OUIL and ENSG bases of the network. The u-center GNSS evolution software for Windows is used at this stage where the RTK FIXED mode has to be effective during the survey, which means that the fixation of the ambiguities is applied and that the coordinates received are corrected by Centipede base points (see Fig. 3).

Once the projected coordinates were obtained, deviations were determined between each of them and the true coordinates of REF1 and REF2. These differences corroborate the behavior of the corrections transmitted by the two bases to our references with the time (see Fig. 4).

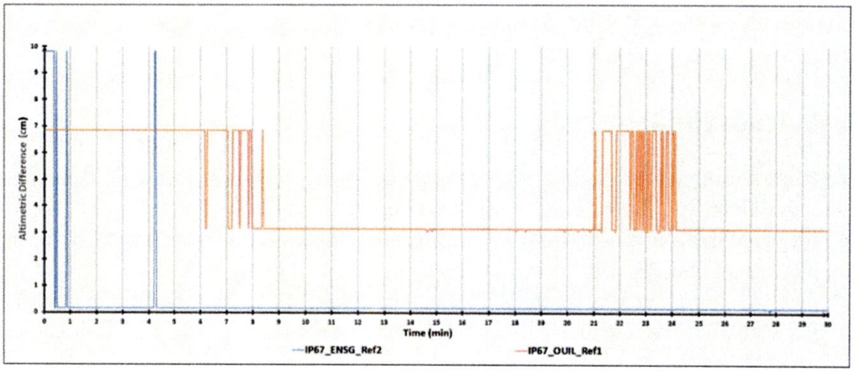

Fig. 3. Temporal evolutions of the altimetric coordinate difference of references Ref1 and Ref2 obtained by observations in RTK mode (30 min) on the permanent OUIL and ENSG stations of the free Centipede network.

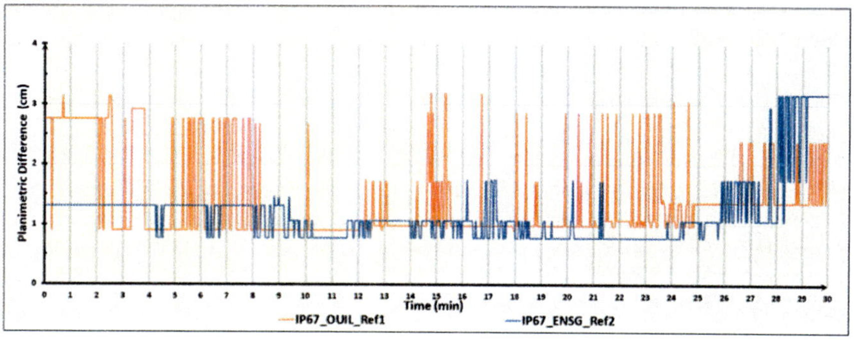

Fig. 4. Temporal evolutions of the planimetric coordinate difference of references Ref1 and Ref2 obtained by observations in RTK mode (30 min) on the permanent OUIL and ENSG stations of the free Centipede network

The graphs show that the corrections from the ENSG station are much more trustworthy than OUIL. We see that the results of the OUIL station have an altimetric difference of 7 cm for the first six minutes and then it goes down to reach 3 cm and then there is a perturbation between 21 and 24 min. The amplitude of perturbation is 4-cm. Concerning the ENSG station the altimetric difference is well than 1 cm; nevertheless, we have three 10 cm difference perturbation values for three observations. We can consider these three perturbations as anomalies statistically speaking.

The planimetric difference of ENSG station is close to 1 cm level for the first 26 min, to reach 3 cm max at the end. The perturbation amplitude is about 1 cm. On the other hand, the OUIL station has a planimetric difference level range between 1 and 3 cm all the observation time.

This translates well to the fact that for works of this type within the ESTP, we advise using rather the ENSG than OUIL even if in reality both provide corrections of the same network. We encourage more reliable studies of Centipede network stations if we want to use them in topographic works.

4 Antenna Studies

The purpose of this study is to examine the accuracy of a low-cost receiver/antenna in static and kinematic mode. For the static mode, we use, the ZED F9P receiver, IP66 antenna, and GNSS u-center software to collect the observations (rate 30 s) for 1h30 in ubx file format and RTKlib to convert the observation file into RINEX format and to calculate the coordinates based on RGP baseline stations. To have a convertible ubx file it is necessary to make sure that in u-center software view/message view/RXM (Receiver Manager) the Almanac is loaded and RAWX (Multi GNSS Raw Measurement Data) is enabled.

We compare the known coordinates of REF1 with the calculated static solution via Online GNSS Network calculations, based on Bernese GNSS Software [27]. The solution has an estimated internal accuracy (millimeters): [SN: 4.6 SE: 3.0 SH: 14.4]. The difference between the coordinates is 5 mm in the horizontal component and 21 mm in the height component. These results are for long distance baselines ranging between 4 km and 30 km with 1h30-observation time. Hence, we have an estimation of the accuracy under these conditions and we will examine the IP67 antenna in the future study.

For the RTK mode, the practical use of the ZED F9P receiver is with an IP66 antenna or IP67 calibrated one for topographic survey works. Therefore, we use RTK Fix mode based on a permanent GNSS station Centipede network and via the SW Map application, the same feature's points of an area (already surveyed using LEICA GS16) are surveyed by two surveyors successively point by point to have the same conditions. The first surveyor has a ZED F9P receiver, IP66 antenna, and SW-map android application to collect the RTK-Fixed observation. The second one uses IP67 calibrated antenna instead of IP66. We compare the differences between LEICA set points and the IP66 and IP67 to examine the difference between these two kinds of the antenna in surveying works. Finally, we have a set of 34 points. The deviation is the difference of the coordinate in horizontal and vertical components between these sets. We calculate the deviations between the set of points from IP66 to IP67 in the first step and GS16 Leica in the second step.

Figure 5 compare the low-cost GNSS F9P/IP66 based on centipede corrections to the Leica G16-Orpheon using an ENSG station. The accuracy class for both planimetric and altimetric is mainly about 4 cm. This value is much better than those imposed by legislation in several application areas (Agriculture, Cadastral Surveying > (1/5000), Aerial Photogrammetry etc. and therefore for GIS). This validates the possibility to use low cost GNSS instead of high-precision antennas for these areas.

The comparison of the low-cost GNSS F9P/IP66 and F9P/IP67 based on centipede corrections is provided using the ENSG station (see Fig. 6). Here the accuracy of the survey is such that:

- The difference is about 2 cm. The absolute difference between the accuracies is 1 cm, which justifies the possibility of using one of the two antenna types.

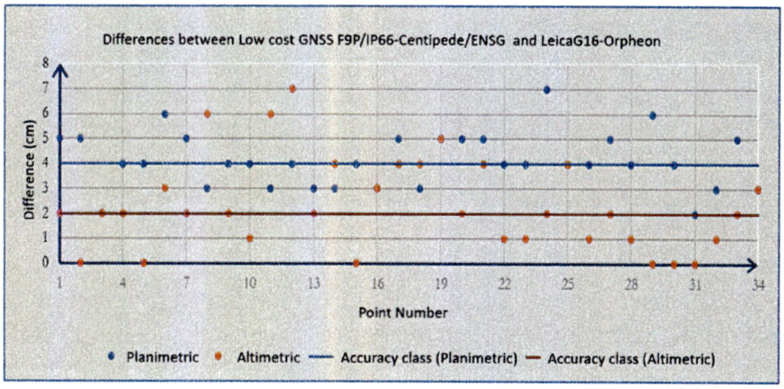

Fig. 5. Differences between the low-cost GNSS F9P/IP66 based on centipede corrections to the Leica G16-Orpheon using ENSG station.

- This means that we can use an IP66 antenna instead of calibrated IP67 since the accuracy (Max = 3 cm) is satisfactory for many applications.

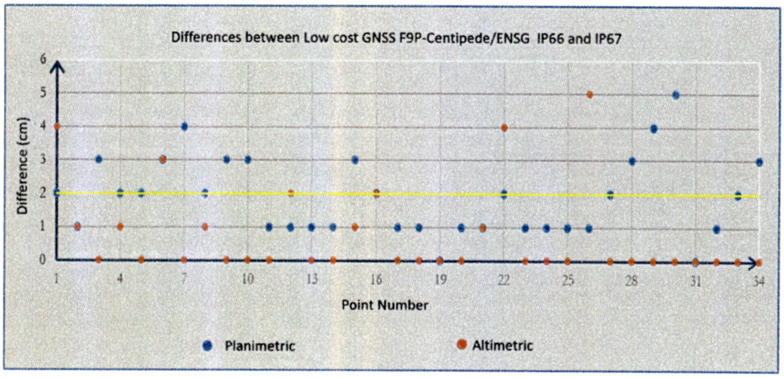

Fig. 6. Differences between the low-cost GNSS F9P/IP66 and F9P/IP67 based on centipede corrections using ENSG stations.

5 Integration to GIS

This is an important step in this study, in which we reviewed previous efforts, including the search tools available today, to evaluate applications for collecting, storing, and visualizing GNSS data that are compatible with smartphones. It is important to keep in mind that the GNSS data we have been talking about is not just ordinary data, rather it is data from low cost GNSS hardware.

In this perspective of developing low-cost mobile GIS, it was therefore imperative to reference-free and/or open source applications as a priority and to add those that, despite their paying services, met our criteria.

It is necessary to remind that until now, the objective of our work is to be able to elaborate mobile GIS from data coming from low cost GNSS hardware. This is why in addition to our two low cost antennas (IP66/IP67: about 300€) coupled with the Ublox receivers (70€) which are themselves low cost; we had to work with the Centipede network of free access RTK bases. Hence, the need to focus on free applications independently of the operating system (Android/IOS) including other features specific to smartphones.

As a result of our research, applications were found that allow some to connect directly to the Centipede caster via an NTRIP Client and therefore receive real-time corrections from the network (SW_Maps, RTKgps+) and others connecting via the same process with the difference that here the NTRIP client service is dissociated from the GIS data collection applications (Input, Qfield, mobile mapping applications), but which require Lefebure. NTRIP client to connect them to the Centipede caster Lefebure allows connecting SW_Maps to Centipede). Among them, the one that was mainly used for our low cost antenna detail surveys is SW_Maps. The Input and Lefebure applications were used for connection tests.

The data that has been collected from SW-Maps Pro were processed on Arcgis Pro. This leads to the realization of StoryMaps through a map published online (see Fig. 7). Table 1 summarizes some mobile GIS applications.

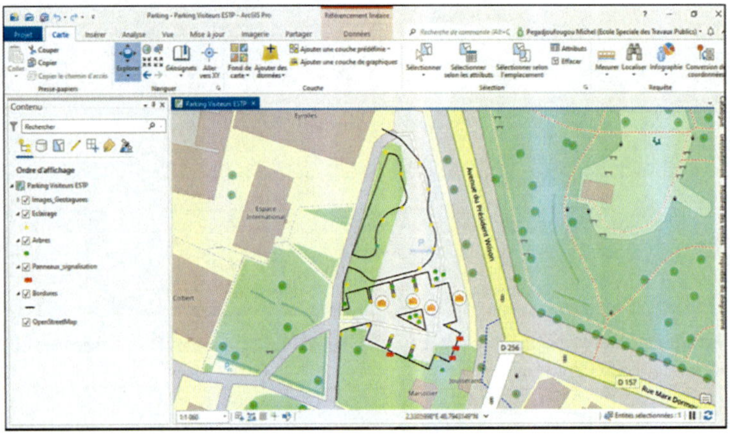

Fig. 7. Visualization on Arcgis Pro

Table 1. Comparison of the mobile GIS applications.

App name	Interface	Operating system	GNSS external	Open source	Supporting formats	Connection cable	Manipulation
SW-Map		Android (Play store)	Yes	Yes	.shp / .csv / .xls / .json / .kml	USB B & C	Simple to use
Input		Android (Play store) / IOS (App store)	Yes	No (39€/month)	.shp / .csv / .xls / .kml / .dxf	USB B & C / Lightening	Simple to use
Ofield	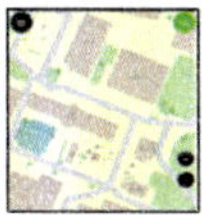	Android (Play store)	Yes	Yes	.shp / .csv / .xls / .json / .dxf	USB B & C	Simple to use
Global Mapper Pro		Android (Play store) / IOS (App store)	Yes	No (45€/year)	.shp / .csv / .png / .dxf / .txt / .tif	USB B & C / Lightening	Simple to use

6 Conclusion

In this research, we have conducted GNSS observation surveys with low cost antennas based on the Centipede network to determine their precisions in RTK. The SW-Maps mobile GIS application served as the base application for this research.

The results of the real-time observations (RTK) carried out in this study have confirmed on the one hand the possibility of using an uncalibrated antenna (IP66) and on the other hand the use of this antenna instead of high precision receivers.

Indeed the average accuracies through the ENSG and OUIL points in planimetry (±2.5 cm) and altimetry (±2.5 cm) of the uncalibrated antenna compared to the calibrated antenna justify this first possibility.

Knowing that topographic works are subject to the decree of September 16, 2003, on accuracy classes, we obtained average accuracy classes of ±3 cm in planimetry than in altimetry. The accuracy classes obtained correspond to those used for agricultural work, cadastral work, photography, for 1/500 surveys, and therefore usable for GIS applications.

At the level of mobile GIS applications, SW-Maps offers great advantages in terms of data collection, storage and visualization. The storage is done in several formats (.shp, .csv, .xls, .kml…), it is then possible to verify the consistency and reliability of the data. Therefore, the data collected in this study with SW-Maps are consistent with reality. This led to their use of Arcgis Pro and the creation of a StoryMap.

During this project, some lines of thought could not be deepened due to time constraints. We suggest the advanced exploitation of the applications: Qfield, Input, and

Global Mapper Pro which are all functional with Android and IOS. Moreover, the installation of a Centipede point at ESTP Paris (Cachan campus) and studies on the simultaneous use of ENSG and OUIL points would be an advantage for future studies.

Acknowledgement. This research was partially supported by SNCF reseau, we thank Jean Christophe Michelin. We would also thank the group of students from ESTP, H.Toure, M.TAMBA, M.KONE and D.MASSOUO who participated in the project.

References

1. Bârliba, F.C., Gheorghescu, I.C., Moscovici, A.M.: The advantages of using mobile GIS technology. Res. J. Agric. Sci. **49**, 20–28 (2017)
2. Bellone, T., Dabove, P., Manzino, A.M., Taglioretti, C.: Real-time monitoring for fast deformations using GNSS low-cost receivers. Geomat. Nat. Hazards Risk **7**, 458–470 (2016). https://doi.org/10.1080/19475705.2014.966867
3. Cina, A., Piras, M.: Performance of low-cost GNSS receiver for landslides monitoring: test and results. Geomat. Nat. Hazards Risk **6**, 497–514 (2015). https://doi.org/10.1080/19475705.2014.889046
4. Fan, G., Chen, F., Li, Y., Liu, B., Fan, X.: Development and testing of a new ground measurement tool to assist in forest GIS surveys. Forests **10**, 643 (2019). https://doi.org/10.3390/f10080643
5. Hamza, V., Stopar, B., Ambrožič, T., Sterle, O.: Performance evaluation of low-cost multifrequency GNSS receivers and antennas for displacement detection. Appl. Sci. **11**, 6666 (2021). https://doi.org/10.3390/app11146666
6. Hodgson, M.E.: On the accuracy of low-cost dual-frequency GNSS network receivers and reference data. GIScience Remote Sens. **57**, 907–923 (2020). https://doi.org/10.1080/15481603.2020.1822588
7. Jackson, J., Saborio, R., Anas Ghazanfar, S., Gebre-Egziabher, D., Davis, B.: Evaluation of Low-Cost, Centimeter-Level Accuracy OEM GNSS Receivers. University of Minnesota, Department of Aerospace Engineering and Mechanics (2018)
8. Janos, D., Kuras, P.: Evaluation of low-cost GNSS receiver under demanding conditions in RTK network mode. Sensors **21**, 5552 (2021). https://doi.org/10.3390/s21165552
9. Jolma, A., et al.: Chapter ten free and open source geospatial tools for environmental modelling and management. In: Environmental Modelling, Software and Decision Support, pp. 163–180. Elsevier (2008). https://doi.org/10.1016/S1574-101X(08)00610-8
10. Lindgren, A.: Implementation of a Volunteered Geographic Information (VGI) Mobile Application for Plant Inventory. A graduate project for the degree of Master of Science (2017)
11. Manzini, N.: Surveillance de santé structurale des ouvrages d'art incluant les systèmes de positionnement par satellites (2020)
12. Nowak, M.M., Dzióbۘ, K., Ludwisiak, Ł, Chmiel, J.: Mobile GIS applications for environmental field surveys: a state of the art. Glob. Ecol. Conserv. **23**, e01089 (2020). https://doi.org/10.1016/j.gecco.2020.e01089
13. Philippes, M., Gilles Yvans, A.: Utilisation des SIG mobiles et l'outil vertex dans la gestion forestière : une application dans la forêt de l'école nationale des eaux et forêts de Mbalmayo, vol. 3, 133–147 (2018)
14. Saka, M.H., Kavzoglu, T., Ozsamli, C., Alkan, R.M.: Sub-metre accuracy for stand alone GPS positioning in hydrographic surveying. J. Navig. **57**, 135–144 (2004). https://doi.org/10.1017/S0373463303002571

15. Sanna, G., Pisanu, T., Garau, S.: Behavior of low-cost receivers in base-rover configuration with geodetic-grade antennas. Sensors **22**, 2779 (2022). https://doi.org/10.3390/s22072779
16. Seeber, G.: Satellite Geodesy. Walter de Gruyter (2003)
17. Singh, P.S., Chutia, D., Sudhakar, S.: Development of a web based GIS application for spatial natural resources information system using effective open source software and standards. J. Geogr. Inf. Syst. **4**, 261–266 (2012). https://doi.org/10.4236/jgis.2012.43031
18. Steiniger, S., Hay, G.J.: Free and open source geographic information tools for landscape ecology. Ecol. Inf. **4**, 183–195 (2009). https://doi.org/10.1016/j.ecoinf.2009.07.004
19. Thirion, E.: Utilisation d'un Système d'Information Géographique pour représenter la distribution spatiale de composés chimiques ou organiques selon le type de compartiments d'un organisme. Exemple de la répartition de la concentration de glucides non structuraux dans un arbre. Cah Tech INRA, pp. 1–11 (2021)
20. Tunini, L., Zuliani, D., Magrin, A.: Applicability of cost-effective GNSS sensors for crustal deformation studies. Sensors **22**, 350 (2022). https://doi.org/10.3390/s22010350
21. Yu, H., Liu, X., Kong, B., Li, R., Wang, G.: Landscape ecology development supported by geospatial technologies: a review. Ecol. Inf. **51**, 185–192 (2019). https://doi.org/10.1016/j.ecoinf.2019.03.006
22. Orpheon-network, the first 100% full GNSS network in France (2019). https://reseau-orpheon.fr/en/. Accessed 12 Sep 2022
23. S.F.P.T. (French Society of Photogrammetry and Remote Sensing) homepage. http://www.sfpt.fr/
24. Centipede, Collaborative, Open and Open Source Real Time Kinematic. Centipede RTK. https://docs.centipede.fr/. Accessed 12 Sep 2022
25. IGN: produire et diffuser les données géographiques et forestières en France – Portail IGN – IGN. https://www.ign.fr/. Accessed 12 Sep 2022
26. Survey GNSS Multiband Antenna (IP66). ArduSimple. https://www.ardusimple.com/product/survey-gnss-multiband-antenna/. Accessed 12 Sep 2022
27. Calculs GNSS Réseau en ligne|RGP. https://rgp.ign.fr/SERVICES/calcul_online.php. Accessed 8 Sep 2022

Machine Learning Algorithms and the Prediction of the Direction of Movement of Moroccan Agribusiness Stock

Ismail Ouaadi[1(✉)] and Aomar Ibourk[2]

[1] Management Sciences, Faculty of Legal, Economic and Social Sciences Agdal, Mohammed V University of Rabat, Rabat, Morocco
ismail.ouaadi@gmail.com

[2] Microeconometrics, LARESSGD University of Marrakech, Marrakech, Morocco

Abstract. In this paper, we attempt to introduce machine learning algorithms that can be used as a way to address climate change issues and their impact on the stock market, and through this, to provide same new useful insights that allow to predict direction of movement of stock market. The most well-known type of machine learning algorithms are: gradient boosting, k-nearest neighbours, k-means, random forest, and artificial neural network, which allow for the development and use of algorithms and statistical models to analyze and draw conclusions and patterns from data. In this study, we introduce these algorithms to show the performance of the developed models in predicting the stock direction of movement of agri-food companies, which are listed on the Casablanca Stock Exchange, based on climate change indexes in the case of Morocco. The results showed that these models produces forecasts with a different level of accuracy, depending on the method of the climate change indexes computation.

Keywords: Stock market · Direction of movement prediction · Drought indexes · Machine learning

1 Introduction

Given that no agricultural policy has been able to achieve its goals of ensuring the population's food security or helping to advance economic growth, several government efforts demonstrate how dependent the Moroccan economy has been on rainfall. This failure was probably caused by either bad policies or a flawed underlying decision-making process. The significant connection between finance industries, particularly stock market trading, and climate change that has been the subject of numerous research merits note here. Additionally, the digital age allows for the collection and storage of enormous amounts of data, which in turn presents certain challenges for decision-makers, and required skills to handle this data according to the phases of processing, analysis, and decision-making. Additionally, the development of potent artificial intelligence-related algorithms

M. Ben Ahmed et al. (Eds.): SCA 2022, LNNS 629, pp. 118–129, 2023.
https://doi.org/10.1007/978-3-031-26852-6_11

has encouraged additional and more trustworthy study. These algorithms have proven to be quite successful in forecasting and making predictions in a number of sectors.

Our effort aims to introduce artificial intelligence approaches in order to suggest ways to deal with climate change variables and to predict the direction of movement of corporate stock. In general, artificial intelligence refers to a computer's capacity to carry out operations that resemble human intellect. It includes a group of methods that precisely accomplish this goal using machine learning methods. Some of these sophisticated and intricate techniques are classified in three categories of machine learning techniques: supervised, unsupervised and reinforcement learning algorithms. These machine learning techniques allow the creation and application of algorithms to examine and derive conclusions and patterns from data. Here, we use five technique, namely: gradient boosting, k-nearest neighbours, k-means, random forest, and artificial neural network, to demonstrate how well these models perform in forecasting the movement of stock market for all Casablanca Stock Exchange agro-food companies, as a function of climate change variables in the case of Morocco.

This introduction describes the organization of our work as follows: The use of machine learning techniques in direction of movement of stock market prediction is covered in the next section, which also provides some theoretical background. The Sect. 3, which is devoted to techniques, contains a description of the data we used and how our model was created. We present and discuss our findings in Sect. 4, which is followed by a summary of our paper, a limit, and future work in the Sect. 5.

2 Literature Review

The importance of stock market have been getting intention either in financial business or in scientific community. Consequently, it constitutes a crucial motivator behind the recent advance in financial engineering and computation. Given that machine learning encompasses powerful computational algorithms, it got more attention and always apply to financial matters. Therefore, it shows more success in stock market area, revealed by many works, which we can cite as an example, the work of Patel et al., (2015) [8], in which authors try to predict future values of stock market index. To build an hybrid models for prediction, they used the followed machine learning algorithms: Artificial Neural Networks (ANN), Support Vector Regression (SVR) and Random Forest. Based on 10 years of historical data of Indian stock market, authors have selected two indices: CNX Nifty and S&P Bombay Stock Exchange. They made their experiments for 1–10, 15 and 30 days. In intention to evaluate the accuracy of these models, authors computed four accuracy measures: Mean Absolute Percentage Error (MAPE), Mean Absolute Error (MAE), Relative Root Mean Squared Error (rRMSE), and Mean Squared Error (MSE). Given that the existing literature, with the same purpose, have exposed the use of single stage prediction, this study shows the performance of the two-stage approach, which was revealed by the results found with the experiments done in this research.

While, Huang et al., (2005) in [3] seek to find out the performance of SVM and its variants models in predicting financial movement direction, by forecasting the weekly movement direction of NIKKEI 225 index. Doing so, the authors compare SVM accuracy with Linear Discriminant Analysis, Quadratic Analysis, and Elman Back-propagation Neural Networks. Meanwhile, they found that SVM and models that combine SVM with other classifiers methods outperform the remain classification methods, this conclusion is given as result of many experiments of these models that take as features macroeconomic variables such as: term structure of interest rates (TS), short-term interest rate (ST), long-term interest rate (LT), consumer price index (CPI), industrial production (IP), government consumption (GC), private consumption (PC), gross national product (GNP) and gross domestic product (GDP). Where the predicted output is the weekly changes of the NIKKEI 225 index, which is determined as a direction function that has two values: 1 if the given index at time t is larger than that at time $t-1$, and -1 otherwise. Noticed that data used to complete this study are gathered from Yahoo and Pacific Exchange Rate Service belonging to January 1, 1990 until December 31, 2002.

In the same perspective, the work of [7] attempts to predict the direction of movement of stock and stock price index in a context of Indian stock market. The authors used four machine learning models, namely: ANN, SVM, Random Forest (RF) and Naive-bayes. To compare the accuracy of the models between them, authors have used two approaches as input (computation of technical parameters and representing technical parameters as trend deterministic data), and doing models' evaluation according to two metrics (accuracy and F-measures), based on 10 years of historical data from 2003 to 2012. As results, they found that, in the first approach, RF shows better results than the other prediction models, while in the second approach the performance of all models has increased.

Also, many other works have performed in other research fields like education, where authors try to predict the student grade retention using six machine learning algorithms [5], and future teacher pedagogical effectiveness [4]. In our work we have used some machine learning techniques such as Gradient Boosting, K-Nearest Neighbour, Neural Networks and K-means clustering which is an unsupervised learning algorithm, to predict the direction movement of agro-business companies in the case of Morocco stock market.

3 Methodology

The implementation of machine learning algorithms has notably increased these days, due to the success of these techniques in many research fields. To build a machine learning model that is able to perform predictions properly and accurately, a set of steps should be achieved and followed carefully. According to many works [2,5], we can present the most important steps related to this topic: *i*) data preprocessing, which consists of cleaning and handling data; *ii*) model training and validating and *iii*) accuracy evaluation. Following this approach, this section is organized as follow: Firstly, we give a brief description of the dataset and the used features. Secondly, we present the proposed approach for data cleaning and preprocessing. Finally, we explain the predictive models chosen to be executed.

3.1 Data Description and Preprocessing

In this study, we attempted to predict stock market monthly direction of movement of stock related to the companies listed on the Casablanca Stock Exchange and operating in the agri-food industry, based on some drought indexes. The monthly stock data are gathered from the Moroccan stock market called Casablanca Stock Exchange (available at https://www.casablanca-bourse.com). Where precipitation and temperature data are gathered from the World Bank Group, Climate Change Knowledge Portal (CCKP Database, available at https://climateknowledgeportal.worldbank.org/download-data), data provided in this database are measured monthly, given that temperature is recorded by mean, min and max. Our work covers the period from 1998 until 2019. These data are required for computing monthly direction of movement of stock and drought indexes, hereafter we give a description.

As target, the movement of the stock values of the involved companies (ΔMF) are computed, based on the rate of change between the current month and the previous month, as follow (Eq. 1):

$$\Delta MF(n) = \frac{MF(n) - MF(n-1)}{MF(n-1)} \tag{1}$$

where:

MF(n) = MF of the current month
MF(n − 1) = MF of the previous month

Where the predicted output is this monthly movement of the agri-industries, which is determined as a direction function that has two values: 1 if the given movement at time t is greater than 0, and −1 otherwise.

We selected the Standardized Precipitation Index (*SPI index*) as a characteristic because it enables the description of meteorological drought over a variety of periods. The calculation is as follows (Eq. 2):

$$SPI = \frac{Pt_i - Pt_m}{\sigma} \tag{2}$$

where:

Pt_i = precipitation per month of the year;
Pt_m = average precipitation of the series on the time scale considered;
σ = standard deviation of the series on the considered time scale.

The Standardized Precipitation-Evaporation Index (*SPEI index*), which is based on potential evapotranspiration (*PET*) and precipitation (*Pt*), is calculated at the end. As cited in [6], to compute the indexes, the SPEI R library [1,10] is called. This allows us to utilize two computational methods: Thornthwaite method and Hargreaves method. Table 1 gives a summary of the descriptive statistics of the data obtained following the above computations of the features and the target variable.

Table 1. Descriptive statistics of the data.

	Month	PET_harg	PET_thornth	SPI	SPEI_thotnth	SPEI_harg	var_dire
count	262	262	262	262	262	262	262
mean		118.802	77.286	0.010	0.010	0.011	1
std		51.475	52.398	0.971	0.979	0.973	
min	1	48.730	12.358	−3.172	−2.605	−2.395	
25		64.510	30.269	−0.590	−0.672	−0.704	
50		116.603	65.320	−0.049	0.028	−0.012	
75%		171.342	114.340	0.662	0.612	0.712	
max	12	210.362	183.667	2.476	2.122	2.194	

var_dire: Stock market shares direction of movement of agribusiness companies; PET_harg: PET index computed based on Hargreaves method; PET_thornth: PET index computed based on Thornthwaite method; SPI: Standardized precipitation index; SPEI_thotnth: SPEI index computed based on Thornthwaite method; SPEI_harg: SPEI index computed based on Hargreaves method.

Given that the first two months of 1998 were excluded because they were utilized to determine the SPEI index for the subsequent month (here March 1998), our dataset consists of 262 entries for each variable. Additionally, this dataset contains six variables, five of which are intended to be features (SPI, SPEI & PET Thornthwaite, and SPEI & PET Hargreaves Indexes), with Variation serving as the predicted variable. According to the equation above, these properties are measured using the SPEI R package and exported to a csv file for further processing.

Figure 1 illustrates the distribution of all variables (SPI, SPEI & PET Thornthwaite, SPEI & PET Hargreaves Indexes and stock market fluctuation). The analysis of the sub-plots allows us to determine the distribution of these variables. Here, we will normalize this data before we used it to perform our models. Figures 2 and 3 present the correlation between the target and features, indeed it shows the absence of any linear relationship among the input and the outputs of the model, as we can observe in the matrix correlation which displays coefficients near to 0.

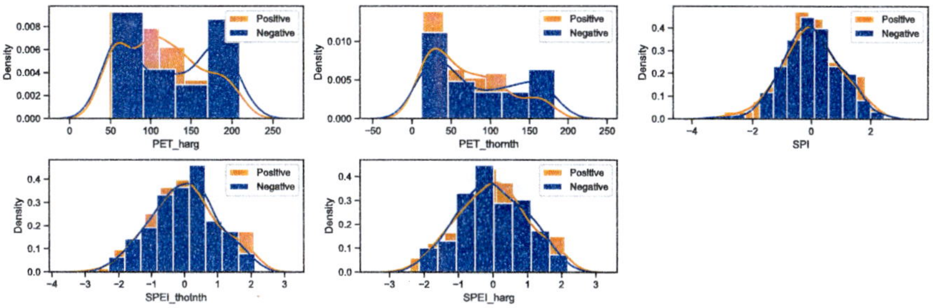

Fig. 1. Data analysis plot.

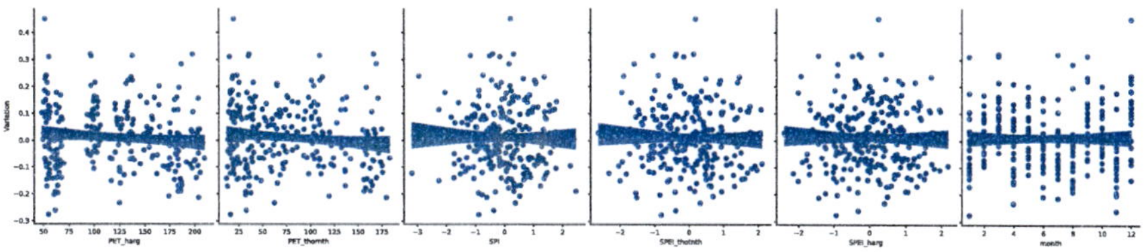

Fig. 2. Correlation analysis chart between the direction of movement of agri-food stocks and other characteristics.

3.2 Models

For the aim of classification in this study, we have selected five machine learning algorithms: gradient boosting, k-nearest neighbours, k-means, random forest, and artificial neural network. These algorithms are used to divide the target variable (direction of movement) into two categories, where a positive change indicates an increase in the company's share value (var_dire = 1) and a negative change indicates a reduction in the company's share value (var_dire = −1).

Before we use our data into our models, this dataset should be preprocessed firstly, this step can enhances the performance of the models, and should be taken carefully as mentioned in [5]. The training step of machine learning algorithm is the most sensitive one for data cleaning, for this reason all criteria related to this purpose are satisfied. Some technical aspects of the used algorithms are listed below.

Qualified as the simplest machine learning algorithm, K-Nearest Neighbours (KNN) can be implemented easily. Thus, The KNN is used in many studies, it allows the classification of a given data point based on its k-nearest neighborhoods. The distance between the two data points is measured using the Manhattan distance (Eq. 3) or Euclidean distance (Eq. 4).

$$d(x, y) = \|x - y\| = \sum_{i=1}^{n} |x - y| \tag{3}$$

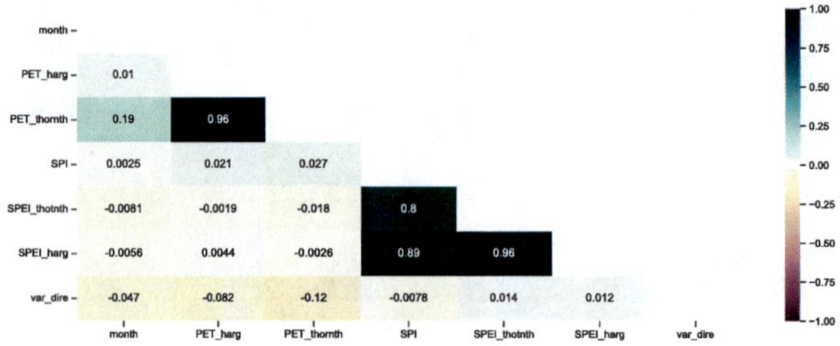

Fig. 3. Features correlation matrix.

$$d(x,y) = \|x-y\| = \sqrt{\sum_{i=0}^{n} (x_0 - y_0)^2} \tag{4}$$

where (x,y) are vectors.

Secondly, the gradient boosting (GBT)is an ensemble method, like decision tree and random forest, it combines multiple decision trees to create a powerful model. The GBT algorithm can be used for either classification or regression problems. In our case it used for the classification issue, and it allows to understand the importance of each predictor. It is more sensible to the parameter setting than other ensemble method, but can perform better if the parameter is set correctly.

The last supervised learning algorithm used in our work is Artificial Neural Network (ANN), it is considered as one of the famous technique in machine learning, and it's the one that put artificial intelligence back in the spotlight after a long period of oblivion [9]. Recent developments in machine learning of ANN have enabled the emergence of a revolutionary kind of machine learning techniques, namely, Deep Learning which deploys a complex model of neural networks. In this study, we employ a relatively simple methods of ANN, called multilayer perceptron (MLP), it viewed as a generalization of linear models that perform multiple stages of processing to come to a decision.

As we have introduced earlier, we chose K-Means Clustering (k-means) as an unsupervised learning algorithm. K-means is one of the commonly used clustering algorithms. Its approach is simple, it tries to find cluster centers that are representative of certain region of data. To do so, the algorithm performs and alternates between two steps: assigning each data point to the closest cluster center, and in a next step it set each cluster center as the mean of the data point that are assigned to it. When there is no longer changes in the assignment of instances to clusters the algorithm is braked and it declares the completion of the task.

4 Experiments and Results

This section is reserved to describe the experiment methodology, to report the findings of the training and the test of these algorithms, and to discuss the results lastly.

4.1 Experiment

Fitting Models: To fit our model, we have used Scikit-learn from python packages. Once the dataset is loaded, it is split into training, test and validation dataset. The first figure (Fig. 4) shows model fitted with all features (SPI Index, SPEI and PET Indexes) to predict monthly direction of movement of agro-food companies stock. Similarly, the second figure (Fig. 6) shows model fitted with SPI Index, PET and SPEI Thornthwaite Index as features for predicting the same monthly direction of movement. Finally, Fig. 5 demonstrates the model fitted given the same indexes computed by the Hargreaves method.

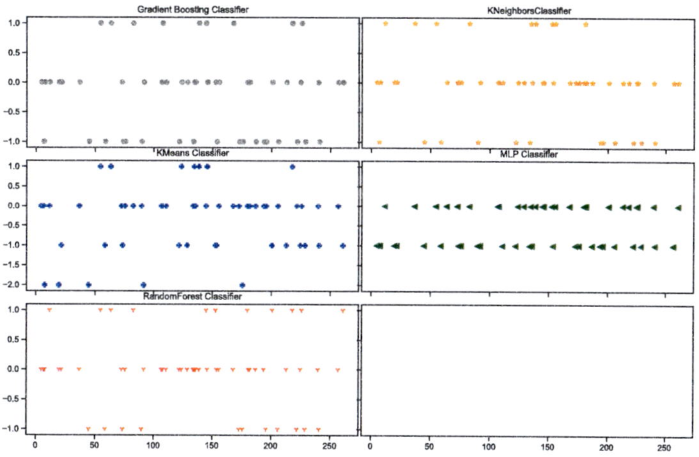

Fig. 4. Model fitted with the overall features.

These three models, Figs. 4, 5 and 6, show the difference between the real direction of movement and the predicted one. Here, we can observe that each model can predict the same output differently, given that the 0 value means that the model get the right answer, other value are significance of the bad prediction.

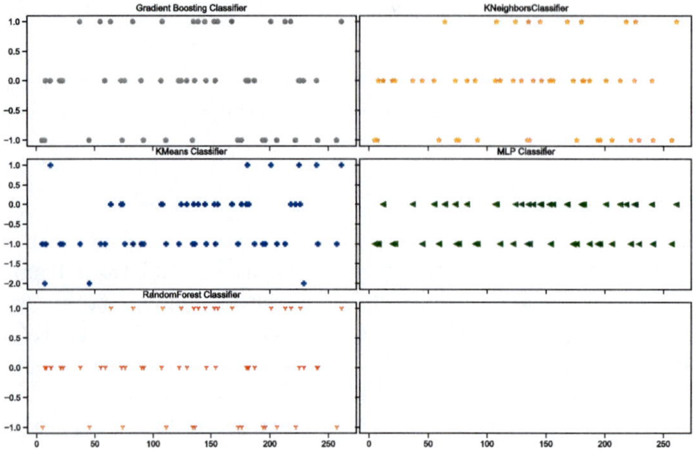

Fig. 5. Model fitted with SPEI index according to Hargreaves method.

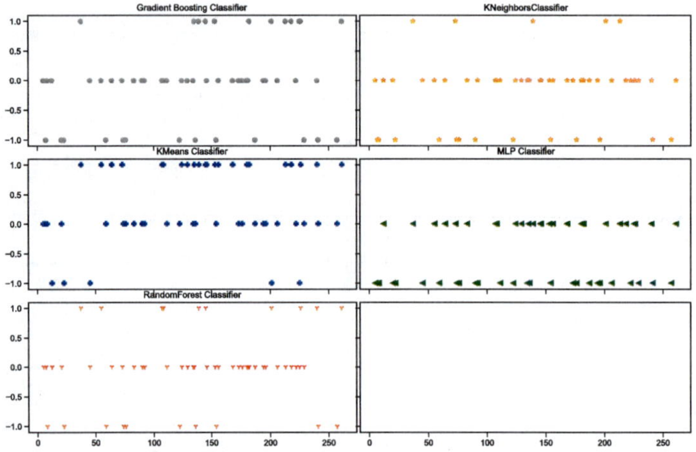

Fig. 6. Model fitted with SPEI index according to Thornthwaite method.

Forecasting and Simulation: Figures 7, 8 and 9 illustrate the performance metrics results of the predictions made by these five models. According to these charts, we can judge the success of these models, where the value of each metric indicates the performance of a given model. In other words, the model can predict the outputs by a probability of this value. To perform a rigorous accuracy assessment of these models, the classification accuracy, recall, precision, F-Measure, macro and weighted average metrics are computed as cited by [4].

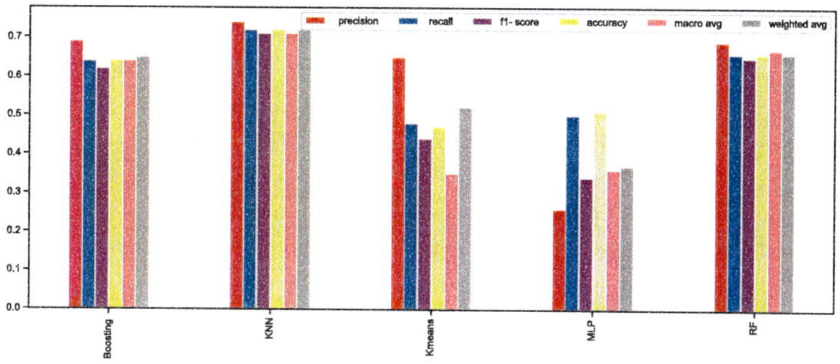

Fig. 7. Model with overall features prediction results.

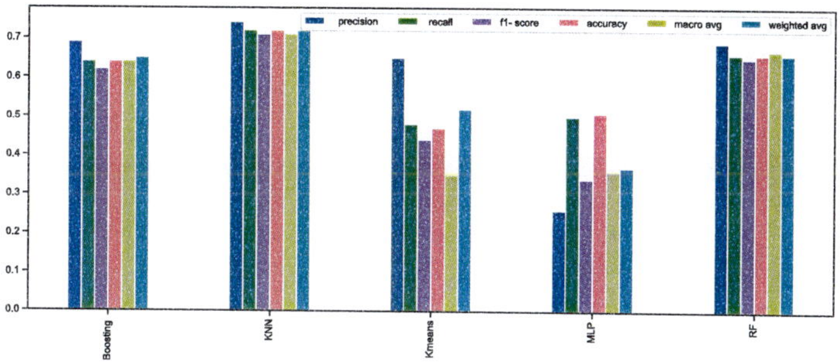

Fig. 8. Model with Thornthwaite method prediction results.

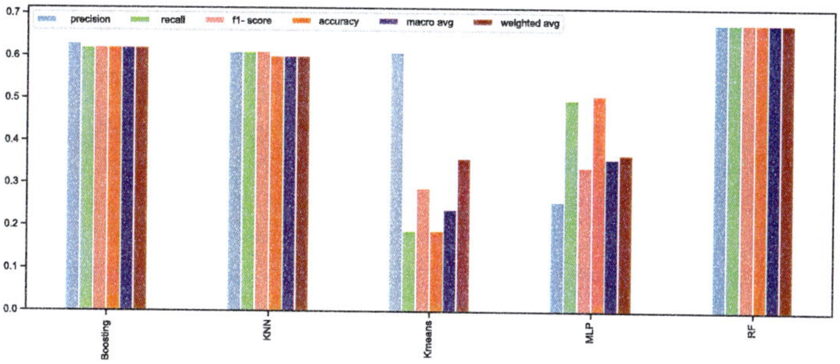

Fig. 9. Model with Hargreaves method prediction results.

4.2 Results and Discussion

Based on the results shown in Figs. 7, 8 and 9, using the machine learning algorithms the direction of movement of agro-food industry in Casablanca Stock Exchange can be predictable by the drought indexes, and maybe it can be predicted also by other climate change indexes. It can be inferred that some indexes perform better than other as features. From these figures, we can say that the models with SPEI index computed according to Thornthwaite method and those that used the overall features have the highest values of performance metrics than the Hargreaves method of calculating of the same index. Comparing the kind of machine learning, we can deduce that KNN and RF models perform very well that the others. Where in the Hargreaves method the GB and RF modes get the highest scores.

Our results are consistent with those found in the literature, which means that machine learning algorithms provide powerful tools to make predictions than classical tools. Most of studies have compared their findings with some benchmark models of forecasting using other method like Moving Average and its variants or other kind of machine and deep learning algorithms. This strategy will be covered in further research to better understanding the relationship among stock market direction of movement and climate change or environmental indexes in case of Morocco.

Finally, we have implemented models with a limited number of parameters, but same kind of models that use more parameters can provide better results. In addition, we have used many accuracy metrics to evaluate the performance of our models, which can present contradictions of the performance evaluation of the same models. It is, however, possible to use same relevant metrics to evaluate the accuracy and robustness of machine learning models.

5 Conclusion

The results of this study support the use of machine learning methods for financial data forecasting. Given the method of index computation, the GB, KNN, RF, K-means, and ANN approaches used in this study, which form the basis of the developed models, produce forecasts with varying degrees of accuracy.

Given that the traditional econometric models focus on the explanations of the relationships, machine learning models focus on predictions. Precisely, ANN models which is considered as a black boxes, this means that it isn't a good choice for determining the impact of independent variables on the dependent variable or a suitable method for analyzing the causal relationship, unlike the ensemble learning algorithms or even the KNN and K-means which can show the importance of each features. However, as we can find in multiple previous studies and the results of this study, machine learning models often show high forecasting accuracy.

In our coming works, we will try to model the direction of movement of stock market with different environmental indexes as new features and by introducing new techniques of machine learning, in intention to get a powerful models that can provide more accurate and robust results.

References

1. Beguería, S., Vicente-Serrano, S.M., Reig, F., Latorre, B.: Standardized precipitation evapotranspiration index (SPEI) revisited: parameter fitting, evapotranspiration models, tools, datasets and drought monitoring. Int. J. Climatol. **34**(10), 3001–3023 (2014)
2. Chong, E., Han, C., Park, F.C.: Deep learning networks for stock market analysis and prediction: methodology, data representations, and case studies. Expert Syst. Appl. **83**, 187–205 (2017)
3. Huang, W., Nakamori, Y., Wang, S.Y.: Forecasting stock market movement direction with support vector machine. Comput. Oper. Res. **32**(10), 2513–2522 (2005)
4. Ibourk, A., Hnini, K., Ouaadi, I.: Analysis of the pedagogical effectiveness of teacher qualification cycle in Morocco: a machine learning model approach. In: Studies in Computational Intelligence. Springer (in press)
5. Ibourk, A., Ouaadi, I.: An exploration of student grade retention prediction using machine learning algorithms. In: Fakir, M., Baslam, M., El Ayachi, R. (eds.) International Conference on Business Intelligence, CBI 2022. LNBIP, vol. 449, pp. 94–106. Springer, Cham (2022). https://doi.org/10.1007/978-3-031-06458-6_8
6. Ouaadi, I., Ibourk, A.: The contribution of deep learning models: application of LSTM to predict the Moroccan GDP growth using drought index. In: Studies in Computational Intelligence. Springer (in press)
7. Patel, J., Shah, S., Thakkar, P., Kotecha, K.: Predicting stock and stock price index movement using trend deterministic data preparation and machine learning techniques. Expert Syst. Appl. **42**(1), 259–268 (2015)
8. Patel, J., Shah, S., Thakkar, P., Kotecha, K.: Predicting stock market index using fusion of machine learning techniques. Expert Syst. Appl. **42**(4), 2162–2172 (2015)
9. Russell, S., Norvig, P.: Artificial Intelligence: A Modern Approach, Global Edition, 4th Foundations, vol. 19, p. 23. Pearson Education, Hoboken (2021)
10. Vicente-Serrano, S.M., Beguería, S., López-Moreno, J.I.: A multiscalar drought index sensitive to global warming: the standardized precipitation evapotranspiration index. J. Clim. **23**(7), 1696–1718 (2010)

Using Practical Activities for Robotics E-Learning: Case of Study on Web-Based Robotics Remote Labs

Meriem L. Aarizou(✉) and Nasr Eddine Berrached

Intelligent Systems Research Laboratory, Université des Sciences et de la Technologie d'Oran, Oran, Algeria
meriem.aarizou@univ-usto.dz

Abstract. Teaching robotics to engineering student requires doing many labs' practical activities (PAs). This type of activity helps students to put robotic concepts and theory into practice in real environments. In an E-Learning environment, Virtual and Remote Laboratories (VRL), particularly web-based laboratories, are used in PAs to remotely teach practical concepts of robotics. But how can a remote PA in robotics that uses VRLs be as effective as a conventional PA for students? And what are the characteristics that a robotic VRL platform should have to fulfill the requirements of an academic training? In this article, we are interested in identifying technical requirements that are related to implementing a web-based robotic VRL. We also focus on the educational aspect of a web-based robotic VRL in a remote practical activity. Furthermore, we study, from an instructor point of view, whether or not a remote PA can be as efficient as a conventional PA in classrooms and laboratories. At the end of the study, we were able to select five main categories of criteria with which we were able to evaluate the state of the art of E-Learning robotics.

Keywords: Requirements · Robotics · Practical activities · Remote-practical activities · E-learning · Distance learning · Virtual laboratories · Remote laboratories · Virtual and remote laboratories

1 Introduction

Remote access laboratories are considered as didactic tools for "*Experimental remote training*". They bring together physical and virtual resources that, we believe, are important for practical remote sessions. These laboratories are divided into two types: Virtual Laboratories (VL) which are software simulations that give graphical representation of a laboratory; and Remote Laboratories (RL) which are a set of real devices and physical instruments located on

M. Ben Ahmed et al. (Eds.): SCA 2022, LNNS 629, pp. 130–144, 2023.
https://doi.org/10.1007/978-3-031-26852-6_12

a remote site. Virtual and Remote Laboratories (VRL) are the combination of both laboratories. They are generally used when instructors have to teach some concepts in emulated environments before testing them on real systems.

From a technical point of view, Web-based VRLs are particularly useful for remote teaching, as they:

- Can be available to a large number of students,
- Can be accessed from a web page, which does not require installing additional software,
- Can be used from a web browser, and thus from any device like computers, smartphones or tablets,
- Are easy to integrate into an e-learning program.

On the other hand, if web-based VRLs are used to teach robotics, they will bring to institutions, instructors and students many benefits such as:

1. Providing a deep understanding of basic concepts of robotics in virtual environments, and the possibility to execute robotic commands on real systems.
2. Providing low-cost solutions to universities and institutions by sharing expensive robotic resources to many other institutions.
3. Reducing the number of sessions in laboratories and classrooms and
4. Allowing a large number of students to attend practical sessions.

Despite the number of merits attributed to using VRLs in education, some researchers have doubts about their effectiveness comparing to conventional local laboratories. Ma and Nickerson [1] have compared physical/simulated laboratories and remote laboratories from 60 research articles. Several opinions have raised about the pedagogical value of VRLs: for example, authors in [2] consider that, despite the fact that RLs provide experiments from real instruments, their manipulations are not as realistic as in real laboratories. On the other hand, other studies, such as [3] and [4] show that learners are motivated and ready to use remote laboratories, as they consider that using RLs can be very useful for their education and more efficient than using conventional laboratories. Finally, researchers in [5] and [6] believe that there is no difference between using conventional PAs and remote PAs. Ma and Nickerson believe that the reason behind this difference of opinion is the lack of standards by which we can evaluate the effectiveness of VRLs in remote teaching.

In this article we focus on defining a set of criteria with which we evaluate the effectiveness of remote PAs via VRLs in teaching robotics. We limited our study to web-based VRLs applications and we only selected academic and published papers that studied the pedagogical aspect of web-based VRLs. We mainly focus on answering the following couple of questions:

1. What are the main requirements for building pedagogical platforms that can be used for PAs in robotics?
2. What are the specifications for remote PAs so they can be as effective as conventional hands-on robotics sessions?

In our study, we first determined the main goals of teaching robotics in-situ, by selecting the pedagogical requirements of a conventional robotics PA. In the second section, we discussed how we can guarantee those requirements in a remote environment, and what are the limitations that can be caused if we try to apply the requirements on a remote environment. The set of criteria for remote PAs in robotics is identified in the third section. The last section consists on evaluating the existing pedagogical platforms for teaching robotics.

2 Robotics PA Analysis

In the first step of our study, we set up a questionnaire form that we distributed to a number of Algerian instructors, researchers, PhD students and teachers, from the Electronics and Automatics departments. From both departments 10 teachers: three professors, three assistant teachers and four part-time teachers, have agreed to give us their answers. All teachers had experience in teaching robotics at the university and had conducted scientific researches in the Robotics field. The questionnaire can be accessed from [9] and can be used by other researchers under CC BY 4.0 License.

In this study, we wanted to identify four types of information:

(i) General information about the activity and the learners;
(ii) Information about the pedagogical content of the activity;
(iii) Information about the tools and materials that are generally used during the training sessions; and
(iv) Information about the administrative side of the training session.

First of all, it is important to note that our study is based on academic programs of robotics that are intended to students in Automatics or Electronics. Therefore, the set of criteria that we will elaborate in the following, is related to academic training rather than professional training. Academic training in robotics usually requires technical and theory prerequisites, as well as mathematics and algebra background. Academic programs are also annual or semi-annual, with weekly sessions and are intended to university/college students.

All the instructors who filled our questionnaire declared that the best way for students to work during practical sessions is in teams. Students must be allowed to work in teams of three to five to carry out their activities. This can help in solving the problem of lack of equipment, because several students will be able to do their work on a single device. On the other hand, because of the lack of robot devices and equipment, some instructors had to use simulation software to do the PA.

Instructors evaluate their students according to the work they carry out during their activities and the reports they deliver at the end of the session. However, this evaluation can be tedious and tiresome to teachers if the number of students in class is high, which is, most of the time, more than 20 students under the supervision of a single instructor. In addition, the lack of equipment is the most recurrent problems during training sessions.

Typically for any robotics PA, instructors must teach theory to students, either at the beginning of the practical session, or during another course session. PAs have to be programmed to allow student to design their robotic systems (prototypes can also be made if necessary) and to make algorithms and programs controlling real robots. The PA must not be restricted to one particular type of software or hardware.

Furthermore, we deduced from the answers we had that it is important for students to have access to suitable training benches and to real robotic devices. While virtual systems can be suitable to pedagogical activities, they can't replace robotic devices during the training. Virtual systems can rather be used alongside real robots if there is not enough real devices or if the training requires a more theoretical understanding. Instructors believe that simulations do not teach technical skills to students.

Finally, instructors who answered our questionnaire consider that the ideal number of students per class would not exceed ten. Although the time duration of the sessions can vary according to the activity, instructors believe that a period from 1 h 30 min to 2 h would be sufficient and optimal for a practical session.

3 Limitations of Remote PAs in Robotics

In this section, we will discuss if the requirements we selected from our analysis can be carried out in a remote distributed environment such in E-Learning programs.

As a matter of fact, we received some mixed opinions about how feasible distance teaching robotics is. While some instructors believe that it will be particularly useful as it will bring many institutions together, help to share many experiences and increase the number of teaching hours. Others sees that the main problem would be Internet speed as it would certainly affects the PA experience. Moreover, teachers who were not in favor of using remote PA, consider it anti-educational since students will not be able to handle real robot. According to them, it will decrease the quality of experience during practical sessions. These diverged opinions however are perfectly in line with Ma and Nickerson's study [1]. We believe that a remote PA in robotics can be as effective as a typical PA in robotics, if we are able to guarantee the same requirements and objectives of a robotic PA, in a remote environment.

On the other hand, we agree that remote training may face communication issues, as it would be difficult for teachers to remotely follow their students' progress during the practical sessions. In addition, students will not be able to request assistance from their teachers during the activity nor to work together. Therefor, it is important to develop an E-learning platform that could overcome these limitations by providing real time communication and monitoring.

Doing experiments on remote devices imply that neither students nor trainers can see their manipulations on the remote robot. Therefore, it is important to implement interactive manipulation platforms that provide real time visual feedback from the remote site. Many technologies can be used for that purpose such as: Virtual reality techniques, 3D modelization and video streaming. On the other hand, using those tools will affect the Internet speed and increase time delay during manipulation.

4 Remote Practical Activities Requirements in Robotics

Based on the study we carried out, we were able to define a set of criteria which will help us evaluating VRLs platforms for teaching robotics. These criteria will also able useful for building platform for remote robotics PAs. We classified our criteria into 05 categories. We excluded technical aspects from our study such as transmission protocol and software architecture, and only considered aspects which have an impact on the pedagogical values of the remote teaching.

4.1 General Requirements

The first category includes elements that describe the general aspect of the application. These criteria will define the type of laboratory used in the robotics training, the number and the type of robotic devices that are used during sessions and whether they are intended for general public or a limited audience.

By combining the criteria C1.3, C1.4 and C1.5 (see Table 1) we can estimate if a session can contain a large number of students and whether the number of benches is sufficient for all students.

Table 1. General requirements.

Code	Criteria	Values
C1.1	Type of laboratory	RL, VL or VRL
C1.2	Type of device	Robot arm, mobile robot, ...
C1.3	Number of devices	1, 2, ...
C1.4	Access mode	Open/restricted
C1.5	User mode	Single-user/multi-user

4.2 Requirements for Distance Learning

The criteria defined in this section describe the educational aspects of sessions. These requirements mainly identify the communication mode between instructors and students and the type of training. Criteria C2.1 and C2.2 (see Table 2) specify the type of activity and if it requires theoretical courses. The criteria from C2.3 to C2.7 describe if there is real-time interactions between students and teachers, if teachers can supervise their students and if it is possible to them to ask questions and give answers. Criteria C2.8 and C2.9 are about sharing documents and other course materials between trainers and learners through the platform.

Table 2. E-learning requirements.

Code	Criteria	Values
C2.1	Type of training	Further training, academic training, ...
C2.2	E-learning integration	Yes/No
C2.3	Real-time communication	Instructor-to-instructor, Instructor-to-student
C2.4	Real-time monitoring	Yes/No
C2.5	Students evaluation	Yes/No
C2.6	Questions/Answers	Yes/No
C2.7	Team work	Yes/No
C2.8	Documentation	Yes/No
C2.9	Reports delivery	Yes/No

4.3 Requirements for the Pedagogical Content

The criteria defined in this category describe the program covered during PAs. These elements may vary from one activity to another and depend on the criterion C3.1 (Table 3). PA platforms should allow creating robot models, guide students to make prototypes if necessary, implement programs and test them remotely on robots. These elements are described below by criteria C3.2 to C3.5.

Table 3. Pedagogical requirements.

Code	Criteria	Values
C3.1	Concepts	Modeling, commands, ...
C3.2	Modelization	Yes/No
C3.3	Prototyping	Yes/No
C3.4	Distance programming	Yes/No
C3.5	Tests on remote robots	Yes/No

4.4 Requirements for Ergonomic and Client Interface

This category defines the way that students and teachers can interact with the remote system. Students and instructors must have a detailed view of the remote site, and the platform should visualize all the information about the remote site and devices such as: speeds and accelerations, SLAM data, navigation map, positions and orientations of each element, etc. Users should be able as well to change some of these parameters in order to have more interactivity with the remote site. These elements are described by criteria C4.1, C4.2 and C4.3 of Table 4.

Table 4. Interface requirements.

Code	Criteria	Values
C4.1	Feedback	Image, Video, 2D/3D, ...
C4.2	Display of parameters	Yes/No
C4.3	Interactivity	Yes/No

4.5 Requirements for Administrative Management

This category describes the criteria that will help administrators to manage the course. By combining criteria C5.1 to C5.4 in Table 5 we could identify the access mode to the training session and the period during which the practical activities are open.

Table 5. Administrative management requirements.

Code	Criteria	Values
C5.1	Online inscription	Yes/No
C5.2	Schedule access	Yes/No
C5.3	Preset hourly volume	Yes/No

5 Related Work on Remote PA in Robotics

In our study on remote PAs in teaching robotics, we could only find a few scientific papers which evaluate the effectiveness of remote PAs in education. Less work has been found on remote PAs in robotics: Of the 60 articles selected by Ma and Nickerson in [1], none of them were related to robotics education. And from 82 research papers on VLs and RLs in education [7] and 20 papers on VLs in science and technology education [8], only nine articles were dealing with the pedagogical aspect of remote PA in robotics. Among these articles, we have excluded applications that do not correspond to the web-based VRL architecture and MOOC and licensed applications. We selected research papers that clearly indicate the educational program in their study. All of these articles refer only to students' opinions the end of the training to evaluate the effectiveness of their tools. The period of our bibliographic research is from 2003 to 2018.

5.1 RoboLab

The "RoboLab" platform [12] is a VRL developed at the University of Alicante for teaching robotics on manipulator arms. RoboLab has been integrated into the educational programs of the University of Alicante since 1999–2000. It is used to teach "Robots and sensorial systems" in three different engineering courses: "Computer Engineering", "Computer Systems" and "Management Computing". The educational teams developed six PAs in order to:

1. Teach and introduce the robotics basics: robot components, coordinate system, etc.
2. Study the kinematics of the Scorbot ER-IX robot arm (05 Dof robot with a grip)
3. Develop computer vision and image processing algorithms to detect and locate objects on an image.
4. Use direct and inverse kinematic models to pick the detected objects
5. Implement visual techniques to move objects using the robot in a virtual environment.
6. Write control programs to move objects using the manipulator arm.

During the 2001–2002 academic year, a questionnaire was given to 50 students from different specialties, after each activity, to evaluate the effectiveness of RoboLab. The results showed that despite the increasing number of students interested in RoboLab, the majority (about 60% of students) prefer real experiments and activities in real laboratories, as they consider that teachers' supervision and guidance as well as team work are essential for the learning process.

5.2 AutomatL@abs

AutomatL@abs [13] is a collaborative project between seven universities in Spain. It aims to share the resources of several academic institutions and to group their laboratory systems (VRL, VL and RL) within the same platform. AutomatL@abs is deployed so that each university can teach its own pedagogical program in automation and robotics, using laboratories that are located in other universities. In this project, all VRLs have been integrated in a LMS tool, called eMersion [14]. The authors focused on providing the following pedagogical and administrative features, as they consider them essential for remote PAs in robotics:

1. Documentation: to enable students to carry out their activities.
2. Common workspace: where students and their teachers can interact and communicate synchronously or asynchronously.
3. Working groups: where students interact with each other during the activity.
4. Schedules: in order to organize students access to VRLs.
5. The sharing of students' experimental results.
6. Evaluation platform: so teachers can evaluate their students before and during their practical activities.

Among the AutomatL@abs laboratories are the UNED's automatic control VRLs using DC Motors, heat flow systems and three-tank systems; and the Robotics VRL of the University of Alicante which is a new implementation of RoboLab [12], and uses a 05 DoF manipulator arm. These labs have three visual feedback modes:

1. The virtual mode where only a three-dimensional representation displays the manipulation results of the remote robot;
2. The video mode which provides a video streaming of the remote system;
3. And the augmented reality mode where a three-dimensional representation of the robot is overlaid on video stream with automatic perspective correction. This feature provides better information about the operation of the real robot than the virtual representation, and encourages students to more interact with the remote system.

The authors had evaluated this project during the 2008–2009 academic year, proposing a questionnaire to 120 students from the seven universities. The results published in their article show that 88% of the students were satisfied with their pedagogical experience with AutomatL@bs' VRLs, and 67% found that this mode of teaching was very useful. comparing to the usual teaching. Nevertheless, 25% of the students were indifferent to this mode of teaching and found it indistinguishable from classical teaching, and 9% were dissatisfied with their teaching experience, because they feel that replacing a PA with a remote PA will not allow students to interact directly with real robots. In addition, some students had opinions on the technical quality of the VRLs, about 20% were neutral or not in favor to the use of these tools because of the quality of their Internet network speed.

5.3 RoboUALab

RoboUALab [15] was initially developed to address the problem of students having lack of interest during an AutomatL@ab activity. In this work, researchers had thought of using "blended-learning" methodology which combines face-to-face work with distance learning in PAs. Therefore, students must use the VRL for an in-site PA under the supervision of the teacher, then complete the rest of the experiments remotely. This new implementation of robotics VRLs has been integrated into AutomatL@bs, and is also maintained by the University of Alicante.

The RoboUALab VL implements a number of necessary features for teaching robotics and automation:

- Kinematics: where students can manipulate the robot by specifying the values of the joints and the robot's effector.
- Trajectory generation: where students can perform movements record the trajectory.
- Environment modeling: where users enter virtual objects into the robot workspace for pick and place operations.
- Dynamics: allowing students to change the dynamic parameters of the robot (such as mass and inertia) and observe the result on the virtual robot.
- Programming: Users can program routines on the robot in simulation.

The RoboUALab RL allows users to execute commands already tested on the VL. Other devices can also be remotely controlled such as the camera and a Programmable Logic Controller for the lighting of the remote site.

To evaluate the effectiveness of this approach, the authors had also proposed a questionnaire to 50 students who had undergone this training. 82% of the students were in favor of using RoboUALab while 12% were neutral. The authors had also wanted to compare this approach and the VRLs used to traditional teaching methods. 82% think that this methodology is better while the remaining 18% do not find any difference. In general, when compared with the results of the assessments made in [13,14], the results show that the majority of students are in favor of using VRLs in their remote PAs in robotics. Furthermore, results have shown that students attach more importance to documentation and support provided by teachers.

5.4 VR RoboticLab

Tzafestas et al. [15], researchers from the National Polytechnic University of Athens, had developed a VRL to study the effect of remote PAs on education compared to the use of conventional PAs. The platform's interface was designed to teach how to program a robot manipulator arm in three control modes:

- Direct remote operation command: where each command is immediately transferred to the remote robot, while ensuring a 2D visualization of the feedback.
- Indirect control: where commands are generated via a control panel, tested then executed locally a virtual 3D robot on the web interface. They are afterwards sent to be executed on the remote robot.
- Manual control: this control mode offers the same functionalities as the previous mode except that the user uses a virtual "teach pendant" on his web page to send commands.

This study was carried out on 60 students who were enrolled into the "robot kinematics and control" course. Students were divided into three main groups. Each group was divided into six teams of three to five students. Each team had different sessions with the same experiments to perform. At the end of one session all students perform an evaluation test on the real robot. During this test, the examiner takes notes about the different skills acquired by students and the possible errors they might do. Team work and the time required for students to complete the test were also noted. The results of this test allowed the researchers to assess student performance and the level of difficulties encountered during the tests.

5.5 IVL

Noguez and Sucar [16] bring a new concept to VLs in robotics by integrating intelligent tutoring. Intelligent Tutoring System (ITS) is a software tool that allows students to be assisted in e-learning. It provides students with customized instructions and remarks each step of the training process, without any intervention from the instructor. This type of system is generally built from several cognitive models.

In this work, the authors used a VL to accompany the students during the "mobile robotics" course, which was taught to 3rd and 4th year engineer students, at the Institute of Technology and Higher Studies in Monterrey, Mexico. The pedagogical goal of this training is that students should be able to build mobile robots by themselves. The first basic notions such as mechanical design and configuration are put into practice using the VL, while advanced notions such as artificial intelligence are treated in a classrooms.

The VL system was developed for the mechanical and electronic configuration, and the kinematic modelization of a mobile robot.

Four types of PAs were done using this platform: the first PA consists on teaching the mechanical aspects of a mobile robot, the second and third PAs were meant to explore the fundamental properties of Infrared sensors, and the fourth PA allows students to learn how to control and command the mobile robot. The role of the smart tutor is to supervise the experiments of each student by providing assistance at each step.

5.6 RFCP

RFCP [16,17] is a platform developed by National University of Distance Education (UNED) and the University of Complutense of Madrid, for teaching mobile robots control. It is used "Systems and Control Engineering" master students, and has two types of laboratories: an VL called RFC_SIM and an VRL called RFC_EXP.

RFC_SIM is used to test training control and obstacles avoidance algorithms on virtual Moway-type mobile robots. This tool offers students the possibility to define the number of robots to be used and their physical properties, the number of moving obstacles, their size, their speed, as well as the obstacle avoidance algorithm. Students can also choose the type of training (free, circle or aligned training) and change the positions of robots and obstacles during the experiment.

The RFC_EXP laboratory is used to test training control strategies on real robots remotely. For this purpose, the current positions of the robots are calculated in real time from the captured video images. Students can interact with the remote system through RFC SIM and see the results on the video return.

Four sessions of remote PAs are offered to students. For each session the student must download the documents in which the activity is explained, follow the instructions on RFC_SIM, then move to RFC_EXP for the real experiment.

6 Analysis

In this section we will evaluate the projects we mentioned above, according to the criteria we identified in previous sections. The analysis is provided in Table 6. The acronym N/A (Not Applicable) is used when the criterion is not clearly specified in the article.

First of all, it is important to mention that none of the projects met all of the criteria. Nevertheless, all these projects are maintained by academic institutions (schools and universities) and are used for formal academic training. Therefore, access to these laboratories is restricted to students enrolled in these courses. Most of these projects deal with issues related to communication between students, real-time monitoring, and the availability of online documentation. The use of LMS tools would solve many of these problems, because they allow synchronous communication and interaction between students and teachers, good administrative management (access, schedule, etc.), and remote monitoring (MCQs, tests, reporting, etc.). Nevertheless, according to Table 6, some projects integrating LMS platforms do not take sufficient advantage of those tools. Other projects, not using LMS, implemented other tools to provide these features, such as using ITS.

The economic and organizational benefits of remote PAs are not explored in the projects we studied. As a mater of fact, the robots used in the different platforms are cheap or student-built robots, and using schedules to organize platform access was not mentioned in most of the projects. Moreover, the use of resources during an activity is only done by one student at a time, and collaborative group work has not been provided by any of the platforms (except for the platforms that are used in conjunction with a conventional PA), which implies that the problem of resource management was not taking under consideration. The notions taught in all the remote PAs projects are basic robotics notions: modeling, control, manipulation, etc. No work considered teaching advanced robotics such as autonomous navigation or SLAM location. In addition, the evaluation of these laboratory tools has been done based on the answers and students' impressions of their experiences. We consider that the best way to evaluate the effect of remote PAs is to consider student' results in exams and long-term projects. Finally, visual feedback is important for real-time manipulation during a hands-on activity. The most commonly used technology is the 3D representation of the robot in real time. We consider that the implementation of this technology is the best way to display the remote site as it provides concise information about remote sites and increases the interactivity of the student.

Table 6. Evaluation of state of the art.

Project	VR robotic Lab	IVL	RoboLab	AutomatL@bs	RoboUALab	RFCP
C1.1	VRL	VL	VRL	VRL	VRL	VL and RL
C1.2	SCARA robot arm	Mobile robot Mobile robot	05 DoF Robot arm	05 DoF Robot arm	05 DoF Robot arm	Mobile Robot
C1.3	1	1	1	N/A	1	Many
C1.4	Restricted	Restricted	Restricted	Restricted	Restricted	Restricted
C1.5	Single-user	Single-user	Single-user	Single-user	Single-user	Single-user
C2.1	Engineering	Engineering	Engineering	N/A	Engineering	Master
C2.2	No	No	No	Yes	Yes	Yes
C2.3	N/A	No	No	Yes	Yes	No
C2.4	Yes	Yes(ITS)	No	Yes	Yes	No
C2.5	Yes	Yes(ITS)	N/A	Yes	Yes	N/A
C2.6	N/A	No	N/A	No	No	No
C2.7	No	No	No	Yes	Yes	No
C2.8	Yes	N/A	Yes	Yes	Yes	Yes
C2.9	N/A	N/A	N/A	Yes	Yes	Yes
C3.1	Kinematics and Control	Mechanics, Sensors, Kinematics and Differentials	Robotics, Kinematic, Objects Detection and Visual servoing	N/A	Kinematics, Path Generation Manipulation and Programming	Formation control and Obstacle Avoidance
C3.2	No	Yes	No	N/A	No	No
C3.3	No	Yes	No	N/A	No	No
C3.4	No	No	Yes	N/A	Yes	Yes
C3.5	Yes	No	Yes	N/A	Yes	Yes
C4.1	Video and 2D/3D	3D	Video and 2D/3D	Video, 2D/3D and Augmented reality	Video, 2D/3D and Augmented reality	Video and 2D
C4.2	No	Yes	Yes	Yes	Yes	Yes
C4.3	Yes	No	Yes	Yes	Yes	Yes
C5.1	Yes	N/A	Yes	Yes	Yes	Yes
C5.2	N/A	N/A	N/A	Yes	Yes	N/A
C5.3	Yes	Yes	Yes	Yes	Yes	N/A

7 Conclusion and Future Work

In this article we conducted a study in order to standardize and put requirements on practical robotics e-learning. We have set up a questionnaire in order to build a set of conditions and elements that real instructors, who have taught robotics in university and institutions, consider important according to their pedagogical and technical experience. We then consider putting those conditions and elements in a e-learning environment and study the possibility of creating a remote practical activity which assure the same pedagogical objectives of robotics instructors. Our study was formalized into 5 types of pedagogical requirements which were used to study the efficiency of educational platforms for teaching robotics. We limited our study on web-based robotic VRLs applications, which are published and described in scientific and research papers, and which indicate clearly their educational program. Those works have clearly addressed the pedagogical aspect of using robotic VRL in e-Learning. By using the requirements we selected we could have a global and general evaluation of the pedagogical aspect of those applications. As future work, we will use the requirements we selected to deploy our remote Teaching platform in robotics. Instructors who have participated to the survey will use the new tested system and a deep evaluation on students can be performed. A software module will be added to the VRL application that reduces and optimize bandwidth consumption during the remote PAs, so users can have a better experience on performing remote manipulation on real robots, even with a limited bandwidth network.

References

1. Ma, J., Nickerson, J.V.: Hands-on, simulated, and remote laboratories: a comparative literature review. ACM Comput. Surv. (CSUR) **38** (2006)
2. Zorica, N., Jan, M., Andrew, N.: Remote laboratories versus virtual and real laboratories. In: 33rd Frontiers in Education Conference (FIE2003), vol. 1 (2003)
3. Cooper, M., Donnelly, A., Ferreira, J.M.M.: Remote controlled experiments for teaching over the internet: a comparison of approaches developed in the pearl project. In: Proceedings of the 19th Annual Conference of the Australian Society of Computers in Learning in Tertiary Education (2002)
4. Scanlon, E., Colwell, C., Cooper, M., Di Paolo, T.: Remote experiments, re-versioning and re-thinking science learning. Comput. Educ. **43** (2004)
5. Ogot, M., Elliott, G., Glumac, N.: An assessment of in-person and remotely operated laboratories. J. Eng. Educ. **92** (2003)
6. Sonnenwald, D.H., Whitton, M.C., Maglaughlin, K.L.: Evaluating a scientific collaboratory: results of a controlled experiment. ACM Trans. Comput.-Hum. Interact. (TOCHI) **10** (2003)
7. Heradio, R., de la Torre, L., Galan, D., Cabrerizo, F.J., Herrera-Viedma, E., Dormido, S.: Virtual and remote labs in education: a bibliometric analysis. Comput. Educ. **98** (2016)
8. Potkonjak, V., et al.: Virtual laboratories for education in science, technology, and engineering: a review. Comput. Educ. **95** (2016)
9. Aarizou, M.L.: From practical training to remote practical training in robotics: opportunities and challenges (2019). https://forms.gle/oYAB5dbcdeK6LrQQ7
10. Tzafestas, C.S., Palaiologou, N., Alifragis, M.: Virtual and remote robotic laboratory: comparative experimental evaluation. IEEE Trans. Educ. **49** (2006)
11. Noguez, J., Sucar, L.: Intelligent virtual laboratory and project-oriented learning for teaching mobile robotics. Int. J. Eng. Educ. **22** (2006)
12. Torres, F., Candelas-Herıas, F.A., Mendez, S.T.P., Pomares, J., Gil, P., Zamora, F.G.O.: Experiences with virtual environment and remote laboratory for teaching and learning robotics at the University of Alicante (2006)
13. Vargas, H., Sanchez, J., Jara, C.A., Candelas, F.A., Torres, F., Dormido, S.: A network of automatic control web-based laboratories. IEEE Trans. Learn. Technol. **4** (2011)
14. Gillet, D., Ngoc, A.V.N., Rekik, Y.: Collaborative web-based experimentation in flexible engineering education. IEEE Trans. Educ. **48** (2005)
15. Jara, C.A., Candelas, F.A., Puente, S.T., Torres, F.: Hands-on experiences of undergraduate students in automatics and robotics using a virtual and remote laboratory. Comput. Educ. **57** (2011)

16. Fabregas, E., Farias, G., Dormido-Canto, S., Guinaldo, M., Sanchez, J., Bencomo, S.D.: Platform for teaching mobile robotics. J. Intell. Robot. Syst. **81** (2016)
17. Galan,D., et al.: Online virtual control laboratory of mobile robots. In: IFAC-PapersOnLine, vol. 51 (2018)

Smart Home and Machine Learning as a Sustainable Healthcare Solution: Review and Perspectives

Eloutouate Lamiae[1](✉), Gibet Tani Hicham[2], Elouaai Fatiha[1], and Bouhorma Mohammed[1]

[1] FSTT, Abdelmalek Essaadi University, Tetouan, Morocco
lamiae.elo@gmail.com
[2] FPL, Abdelmalek Essaadi University, Tetouan, Morocco

Abstract. Over the past two years, the world has witnessed one of the worst pandemics due to the outbreak of coronavirus (covid19), which has infected hundreds of millions and claimed the lives of millions across the globe. If we have learned anything from this pandemic, it is that the actual healthcare systems are unreliable under situations of enormous pressure. Accordingly, the present investigation tackles smart healthcare paradigm as a solution to transform the classical healthcare model into a sustainable one. Therefore, this paper reviews the most advances on remote healthcare monitoring technologies and introduces a novel smart home architecture combined with cloud computing and machine learning to create a sustainable solution for healthcare. Furthermore, a case study of a patient with heart disease is suggested to highlight the importance of using machine learning to automate medical monitoring at home. Additionally, an investigation of human behavior using neural network transformers is suggested as a perspective of the research in hand to examine patients' activities at home using surveillance camera thus constructing a resilient remote healthcare model.

Keywords: Smart healthcare · Smart home · Cloud computing · Machine learning

1 Introduction

Until recently, countries and governments around the world have established healthcare systems and regulations that guarantee their citizens safety anytime and anywhere. However, with the first inflation of the covid19 pandemic [1] cases, back on 2020, everyone came to the realization that more is required to be done in order to sustain the healthcare system and make it more robust in case of large-scale emergencies [2]. On the one hand, millions of people lost their life and many more others still suffering from long term conditions caused by covid19 [3]. Subsequently, the world can't afford another pandemic [4], especially with the same way of interventions on healthcare and economics. On the other hand, advanced computer science technologies such as sensors, wearables, Internet of Things (IoT), cloud computing, etc. have proven their reliability in many sectors,

M. Ben Ahmed et al. (Eds.): SCA 2022, LNNS 629, pp. 145–155, 2023.
https://doi.org/10.1007/978-3-031-26852-6_13

including healthcare. In light of these developments, radical measures need to be taken to preserve the life of citizens no matter their health conditions or requirements for medical attention without the need of overwhelming hospitals nor intensive care units [5]. Taking this into account, smart home technologies combined with the previous mentioned innovations can present a key solution for healthcare systems in a pandemic case scenario or worst. Accordingly, this paper will start from reviewing the recent advances on smart home and smart healthcare, then briefly highlights the main technologies involved in smart healthcare and demonstrates a proposed smart home architecture using machine learning to establish a sustainable healthcare paradigm capable of overcoming any crisis. Furthermore, this paper introduces neural network transformers as a key mechanism to scale smart home automation to a new level, where the user behavior can be observed and intelligent decisions regarding his health condition can be made automatically to simplify his life.

The smart healthcare innovative model has found it beginnings since 2009 by IBM in their new born proposed concept of "Smart Planet" [6]. Correspondingly, smart healthcare by definition [7] is a novel health service system that employ sensors, wearable and IoT to respond to medical situation remotely and in an intelligent fashion.

By investigating literature, one can clearly perceive the various architectural and technological innovation proposed as an approach to modernize healthcare and make it more resilient. Whereas for example, authors on the [8] paper, proposed an architecture that combines Blockchain, IoT, Fog Computing and Artificial Intelligence to monitor a patient health condition remotely and updates the physicians on real time which can then accelerate the diagnosis and plan medical interventions. While the authors proposal was innovative, they created a misperception by employing the Artificial Intelligence concept as a synonym for machine learning, considering that they have used supervised learning algorithms (machine learning type) to train their proposed intelligent model. The authors also haven't managed to discuss model overfitting possibility using the collected data. In another investigation, authors on [9] focuses on the application of IoT in healthcare system and advance an original design using IoT concept under fog computing. The main objective of the proposed model is to recognize the fundamental problem and deficiency of clinic-centric health system and change it to a smart patient centric health system. However, the authors with their proposed work, only concentrated on the collection of patients' data and forwarding this data to the health practitioner that have to assess and analyses the situation and then plan the medical intervention. While data is simply sent to the health provider for evaluation, the solution lacks the intelligence required to create such a smart model. In [10] paper, an extensive discussion of technologies that would form a smart healthcare system has been advanced in the aim of creating a blueprint for smart healthcare architecture. The work presented on this paper can surely create a roadmap for governments wishing to adopt smart healthcare as a model. In a separate study, the authors on [11] analyzed the potential applications of IoT in healthcare. In addition to this, the authors tackled the security issue of such new healthcare system model and the importance of forensics and identity access management to protect critical data. This paper shed a light on the criticality of security when it comes to managing and exploiting user's health or medical records. In a separate study, an interesting healthcare architecture has been presented in [12], where the authors presented a new IoT-based element using SDN-enabled WSN and RFID in the vertical domain and big data analytics solutions such as Kafka, Spark, MongoDB, and NodeJS in the horizontal domain. The proposed architecture has shown promising results in relation to health condition detection using logistic regression supervised learning algorithm. Even though the proposed machine learning algorithm demonstrated excellent precision in detecting a medical emergency, the authors didn't manage to discuss other opportunities with other machine learning algorithms or methods.

2 Technologies for Remote Healthcare

The basic purpose of smart home technologies is automation with the intention of making people's life more comfortable [13]. In the same context, healthcare is the service of assisting people with medical conditions in the aim of curing them, ease their symptoms or helping them adapt with a lifelong health situation.

Traditionally, healthcare was always perceived to be given at hospitals or a specialized doctors cabinet. Nevertheless, with the rising numbers of people requiring access to hospitals, classical healthcare services are becoming deprecated, especially with an ongoing crisis like the covid19 pandemic. Although smart healthcare was examined and proposed several times in the past as a solution for remote healthcare, it has never been considered as a necessity until the present times, whereas smart healthcare relies on gathering a patient data remotely and transferring this data to the health physicians for further analysis. In parallel to this, smart home technologies also depend on assembling the user data in order to automate home related tasks. Therefore, smart healthcare and smart home technologies converge on the importance of user data collection and analysis, thus smart home can hold the key to revolutionize healthcare system and make it more sustainable.

2.1 Sensors, Wearables and IoT

The first line of smart home or smart healthcare technologies are sensors which are wireless components that can be responsible for security, access control systems, control appliances, climate, lighting… [14]. Additionally, wearables are special materials or equipment's which can be wore with the capability of sensing, registering, reacting or simply sending a signal to a command center.

Nowadays, Sensors and wearables are used as the main elements of Internet of Things (IoT), which refers to the method of designing and modeling objects connected to the Internet through computer networks [15]. These technologies form the first layer of smart homes and can be extended to answer to smart healthcare model where each element is conceived to monitor and help with the medical situation of a specific user. For example:

- IoT Wearables: Collecting the user vital signals: heart rate, blood pressure, glucose…
- IoT Sensors: Surveillance, movement, fall detection, Smoke, fire detection…

2.2 Cloud Computing

Cloud computing is a computing model that consists on providing remote on-demand dynamic services to clients at all levels anywhere and anytime [16]. Since its creation, cloud computing has solved several problems related to computing infrastructures, platforms and applications for a variety of domains. Hence, smart home and smart healthcare can benefit from this novel computing model by exporting all of their computing processes to the cloud. Therefore, the cloud can host all the required tools to aggregate user's data, store it or prepare it for analysis.

Cloud computing present an outstanding opportunity for healthcare systems that can reduce the costs related to computing infrastructures management (Installation, maintenance, backup, scalability…), consequently concentrating more on studying the medical condition of a patient. Putting this into perspective, designing a smart home for healthcare would be confined to data collection, transmitting and receiving [17]. Afterwards, all of the user data is sent to the cloud for storage and analysis in a client-server architecture.

2.3 Machine Learning

Machine learning is a form of artificial intelligence (AI) that allows a system to learn from data and not through explicit programming [18]. A machine learning model is the output generated when an algorithm is trained with data. After training, the machine learning model gains the ability to return an output result (decision) for each provided input data. Depending on the nature of the problem being addressed, there are different machine learning approaches that vary based on the type and volume of data:

- Supervised learning: It works on labeled data. In this context, the machine already knows the answers that are expected of it and model is trained on several cases of inputs and their corresponding or awaited output (For example: classifications).
- Unsupervised learning: The outputs to predict are not available in the input data. Here, the algorithm uses unlabeled data. The machine model is then asked to create its own responses. Thus, offers answers based on analyzes and data grouping (For example: Clustering).
- Reinforcement learning: It consists of letting the machine learn from their experiences through a reward or penalty system.

3 The Proposed Smart Home Architecture for Healthcare

The proposed smart home architecture for healthcare consists of three layers as illustrated on Fig. 1. The first layer (physical) involves the deployment of specialized IoT devices (Advanced Sensors, Wearables, detectors, cameras etc.), where the main objective of this layer is data collection and transmission to the centralized management system on the cloud. The second layer encompasses a variety of technics for data aggregation, conversion and storage on the cloud data center. The last layer and the most important one, is data management and analysis. This is where advanced analytics is applied to the data collected using machine learning capabilities to take strategic decision regarding the user medical condition.

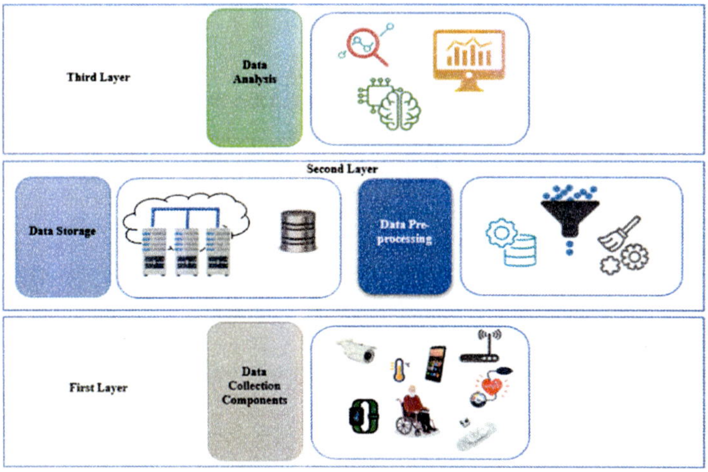

Fig. 1. The proposed three-layer smart home architecture for healthcare.

3.1 Smart Healthcare Application Design

The proposed application is destined for remote healthcare monitoring and comes at the heart of the proposed smart home architecture for healthcare. The application design consists on a backend for frontend model (Fig. 2). In this model, the gateway APIs isolates the clients (Web, mobile) from internal services partitioning and refactoring. This approach offers a lot of flexibility when modifying a backend service, which can be included in the gateway without breaking the client access. The backend for frontend model also is the right fit for a cloud computing architecture, where each client (web or mobile) host only a light weight application (reduce energy consumption) that accesses all the required services on the cloud (backend where the processing is done).

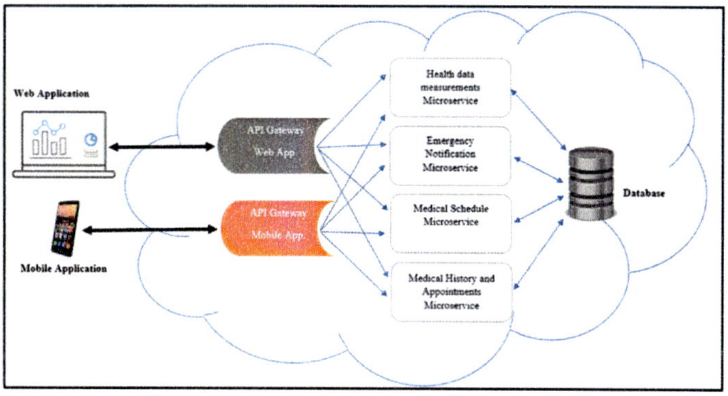

Fig. 2. Backend for Frontend application design

3.2 Data Collection and Analysis Using Machine Learning

After the patient data collection using IoT elements, it gets sent through the mobile application towards the cloud hosting platform where the main application resides. While almost every proposed remote healthcare platform only automates the data collection step and leave the analysis to the practitioner. The advanced smart healthcare model on this paper, introduces a new level of automation, where the collected data is analyzed using a machine learning algorithm and according to the case urgency takes the decision of:

- Sending medical emergency personal to help the patient.
- Notify a family member to check on the patient.
- Notify and remind the patient to take a medication, practice a physical activity or seek medical help.
- Log the information on the patient record.
- Schedule a medical visit.

Essentially, all medical data are already labeled and practically every health condition has a set of outputs for each situation. Taking this into account, supervised learning is the right fit for medical data analysis and specifically classification algorithms. In a previous investigation [18], several machine learning classification algorithms have been compared on a medical dataset case study (heart disease case scenario), and with consideration of this study, "Decision Tree" was the ultimate classification model. All things considered, the proposed algorithm for health condition analysis will take advantage of "Decision Tree" algorithm capabilities to detect a health anomaly and choose the best decision as underlined above.

Before the main algorithm deployment, the health practitioner needs to provide a comprehensive dataset that describes the user health condition based on previous observations of the users' medical history or other patients with the same underlying health condition. This data will represent the training set for the machine learning algorithm and can be organized as follow (Table 1):

Table 1. Example of training data set structure

Input	Output	Intervention
Vital sign measurments	Condition severity	Decision
1.5	1	Send notification
2.3	2	Send Emergency
0.7	0	Log information
...	...	...

- Input: Represents the collected vital signs measurements using the IoT elements and then sent to the cloud for analysis (The input can be an array of values).

- Output: Represents the patient medical condition severity based on the health practitioner knowledge and expertise. The output can be fitted to an integer scaler that displays the case severity: 0 to 5 for example where each value assimilates a way of intervention.

3.3 Case Study: Heart Disease Patient

In pursuance of creating a remote healthcare model (smart healthcare) that manages to assist people with underlying health problems and the healthcare providers, a case study of patients with heart disease is proposed to highlight the importance of the proposed architecture. As underlined on Sect. 3.2, the proposed model uses machine learning (Decision tree classification algorithm) to analyze the patient data, accordingly the dataset used for this case study was established by the University of California, Irvine (UCI) [19]. The dataset consists of 270 entry with the following attributes (Table 2):

Table 2. Dataset attributes

Input	Output
1. Age	Variable to be predicted
2. Sex	-----------------------
3. Chest pain type (4 values)	Absence (1) of heart disease Presence (2) of heart disease
4. Resting blood pressure	
5. Serum cholesterol in mg/dl	
6. Fasting blood sugar >120 mg/dl	
7. Resting electrocardiographic results (values 0, 1, 2)	
8. Maximum heart rate achieved	
9. Exercise induced angina	
10. Oldpeak = ST depression induced by exercise relative to rest	
11. The slope of the peak exercise ST segment	
12. Number of major vessels (0–3) colored by flourosopy	
13. Thal: 3 = normal; 6 = fixed defect; 7 = reversable defect	

The dataset was split into 70% (189 entry) for training and 30% (81 entry) for testing. The decision tree algorithm was tuned using the following parameters (Table 3) in order to dismiss the possibility of overfitting or underfitting the model:

The values of each parameter from the previous Table 3. Was chosen after training the model with several parameter values in order to obtain the finest based on the algorithm accuracy score and precision. The results obtained were as follow (Table 4):

To better explain the results, the algorithm predicted correctly 42 (51.85%) "true positive" cases which represents the actual number of people with a heart disease from the 81 case (30%) of the testing set. Additionally, 25 (30.86%) predicted "true negative"

Table 3. Decision tree tuning parameters

Parameters	Description
criterion = 'gini'	This is a parameter used to minimize the likelihood of misclassification: gini = $1-\sum_j p_j^2$ where p is the probability of class j
max_depth = 6	The depth of the decision tree that allows the model to capture information regarding the data
max_features = 7	This is the features amount to consider each time to make the split decision
max_leaf_nodes = 15	Maximum terminal nodes in a tree
min_impurity_decrease = 0.005	This parameter is used to test the pureness of a split
min_samples_leaf = 3,	A specification for the minimum number of samples required to be at the leaf node
min_samples_split = 15	A specification for the minimum number of samples required to split an internal node

Table 4. Confusion/error matrix

	Actual		
Predicted		Positive	Negative
	Positive	True Positive (TP): 42	False Positive (FP): 6
	Negative	False Negative (FN): 8	True Negative (TN): 25

cases which represents the actual number of people with no concerning heart case. Furthermore, the algorithm predicted 6 (7.40%) "false positive" cases that represents the cases that have no condition but seen by the machine learning algorithm as positive cases. Also, 8 (9.87%) cases were predicted as "false negative" in which the algorithm fails to identify these cases conditions. In consideration to these results, "false positive" cases don't really present an alarming case since the actual case is negative, consequently if we sum these cases percentage with the "true positive" and "true negative" cases the proposed algorithm gives 90.14% correct prediction which is a significant outcome.

In respect to these findings, the following metrics can be extracted:

- Accuracy = $\frac{TP+TN}{TP+FP+TN+FN}$ = 0.82: This metric signifies that the proposed machine learning algorithm can predict a patient having a heart condition in 82% of the time.
- Precision = $\frac{TP}{TP+FP}$ = 0.875: This indicates how many predicted cases are actually positive cases (87.5%).
- Recall = $\frac{TP}{TP+FN}$ = 0.84: This shows how many positive cases were predicted correctly (84%).

- F1 score $= \frac{2 \times Precision \times Recall}{Precision + Recall} = 0.8571$: It is the harmonic mean of precision and recall which is another representation of accuracy when using unbalanced datasets.

The proposed model produced motivating results and in regards to "false negative" cases, the data analysis layer (Fig. 1) will use in parallel to the machine learning classification algorithm, specialized thresholds to monitor the most critical patient metrics. In this case, even if the algorithm fails to identify a patient alarming condition the configured thresholds would fire a notification to signal that something distressing is going on with the patient (For example: if the patient blood pressure is 140/90 mmHg or higher and the "Decision Tree" predicts that the case is negative send a notification to the practitioner to re-evaluate).

4 Discussions and Perspectives

Nowadays, a sustainable healthcare system is not only a research investigation or a simple discussion, it is a necessity. Therefore, a smart healthcare model is essential to reinvent the existing traditional approaches to healthcare. The actual investigation tackles the conception of a smart healthcare system with the service of smart homes as an alternative for hosting patients at hospitals while adopting cloud computing model as a platform for healthcare data management and analysis. The underlined previously mentioned smart home architecture operates using smart IoT devices (sensors, wearables, cameras) for the patient data collection and a mobile application that acts not only as an element to control the smart home aspects but as a portal to the patient health condition follow ups, a notification hub for the patient medical schedule and check-ups. The most valuable aspect of the proposed design is the application of machine learning algorithms (decision tree classification) for patient data analysis. Hence, the proposed model develops into an intelligent system capable of taking strategic decisions in case of a patient health emergency.

The fundamental purpose of healthcare or remote healthcare systems is patient monitoring, diagnosis and eventually referral for hospitalization or treatment [20]. Accordingly, the designed model employs the most trending technologies for healthcare remote monitoring (IoT) combined with machine learning which automatically classifies the patient health condition based on a preconfigured scheme. Consequently, overcrowding a hospital is no more an obstacle for healthcare. A patient can be monitored and diagnosed in the comfort of his home and the proposed smart healthcare model is capable of predicting/detecting an urgency and of course act upon it by sending the emergency team to help the patient or just logs the information on the patient archive. The employment of machine learning is becoming crucial on several emerging models. Hence, the proposed smart home architecture for healthcare apply decision tree classification algorithm on the healthcare collected metrics to predict any patient arising condition and giving the health practitioner real time access to the patient condition by the mean of notifications. By applying the proposed model on a case study of a patients with heart diseases, one can clearly observe the effectiveness of the proposed classification algorithm on predicting with high accuracy (82% of cases) a patient alarming heart condition (90.14% if we consider the "false positive" cases). The proposed model requires further investigation especially on new kind of diseases, thus new health datasets with extensive entries (more than 270 cases used on the case study).

To further advance the proposed smart home architecture, Neural Network Transformer (NNT) can be employed for patient monitoring at home. NNT is one of the most trending machine learning models for natural language processing [21]. Besides, NNT models has been investigated as a solution on several other assignments like for example: image classification and speech recognition. A transformer model doesn't rely on recurrence to perform a task and instead depend on an attention method to get dependencies between input and output [22]. While Recurrent Neural Networks (RNN) and Convolutional Neural Networks (CNN) are time consuming (training) and require extensive resources (Processing and memory), NNT exhibits the best performance [23, 24]. In respect to the problematic in hand, in which a patient should be monitored at home, an NNT model can be used to follow the patient daily activities and behavior. This is where the NNT model intervenes to caption a video sequence and send it for analysis, in this manner the logged in information can be examined to assess if a patient for example is following his in-house sport activity [25] or taking his medications as prescribed.

5 Conclusion

In this paper, we reviewed the most trending technologies that can be used to renovate the healthcare system. We introduced a novel architecture that combines smart home, cloud computing and machine learning capabilities to produce a new design for patient remote diagnosis and monitoring (at home). Moreover, a case study of a patient with heart disease was examined to feature the advantages of using machine learning classification capabilities to analyze the patient medical acquired data and predict the appropriate action. Extensive testing on the case study using the proposed architecture and decision tree classification algorithm demonstrated the strength of the proposed model. Future work would further explore neural network transformers to monitor patient activity and behavior at home by sequencing the surveillance videos and captioning the sequences to inspect if the patient is following his medical prescribed routines.

References

1. Zhou, P., Yang, X.L., Wang, X.G., et al.: A pneumonia outbreak associated with a new coronavirus of probable bat origin. Nature **579**, 270–273 (2020)
2. Lupu, D., Tiganasu, R.: COVID-19 and the efficiency of health systems in Europe. Health Econ. Rev. **12**, 14 (2022)
3. Blomberg, B., Mohn, K. GI., Brokstad, K.A., et al.: Long COVID in a prospective cohort of home isolated patients. Nat. Med. **27**, 1607–1613 (2021)
4. Mammen, A.: Has COVID taught us anything about pandemic preparedness? Nature **596**, 332–335 (2021). doi: https://doi.org/10.1038/d41586-021-02217-y
5. Aziz, S., et al.: Managing ICU surge during the COVID-19 crisis: rapid guidelines. Intensive Care Med. **46**(7), 1303–1325 (2020). https://doi.org/10.1007/s00134-020-06092-5
6. Shuo, T., Wenbo, Y., Jehane, M. L. G. et al. Smart healthcare: making medical care more in-telligent. Global Health J. **3**(3), 62–65 (2019)

7. Jamie, B., Osvaldas, R., Liming, C.: Healthcare in the smart home: a study of past present and future. Sustainability **9**, 840 (2017)
8. Fetjah, L., et al.: Towards a smart healthcare system: an architecture based on IoT, blockchain, and fog computing. Int. J. Healthc. Inf. Syst. Inform. **16**(4), 1–18 (2021)
9. Kharel, J., et al.: An architecture for smart health monitoring system based on fog computing. J. Commun. **12**, 228–233 (2017)
10. Joon-Soo, J., Oakyoung, H., Yen You, Y.: A design characteristics of smart healthcare system as the IoT application. Ind. J. Sci. Technol. **9**(37) (2016)
11. Shaftab, A., Mohammad, I., Yasin, M.A.R.: Internet of things: applications in smart healthcare. In: Proceedings of the 9th International Conference on Society and Information Technologies (ICSIT 2018)
12. Nur, B.O., Mohammed, A., Celal, Ç.: IoT data analytics architecture for smart healthcare using RFID and WSN. ETRI J. **44**(1), 135–146 (2022). wileyonlinelibrary.com/journal/etrij135
13. Mustafa, A.O., Wasaan, K.S., Bashar, J.H., et al.: Survey of various intelligent home applications using IoT and intelligent controllers. Indonesian J. Electr. Eng. Comput. Sci. **23**(1), 490–499 (2021)
14. Nael, A.Al., Sura, A.A., Mohanad, S.A., Ammar, A.S.: Towards secure smart cities: design and implementation of smart home digital communication system. Indonesian J. Electr. Eng. Comput. Sci. **21**(1), 271–277 (2021)
15. Fardin, A.M., Maytham, N.M., Seifedine, K.: Development of an IoT-based and cloud-based disease prediction and diagnosis system for healthcare using machine learning algorithms. IAES Int. J. Artif. Intell. **9**(4), 766–771 (2020)
16. Isaac, O.A., et al.: A systematic mapping study of performance analysis and modelling of cloud systems and applications. Int. J. Electr. Comput. Eng. **11**(2), 1839–1848 (2021)
17. Dalia, R., et al.: A study on cloud computing architectures for smart healthcare services. In: Proceedings of IDDM'2020: 3rd International Conference on Informatics & Data-Driven Medicine, CEUR Workshop Proceedings, 19–21 Nov. 2020
18. Eloutouate, L., et al.: Smart home and machine learning for medical surveillance: classification algorithms survey. J. Theor. Appl. Inf. Technol. **99**(12), 2890–289930 (2021)
19. Janosi, A., Steinbrunn, W., Pfisterer, M., Detrano, R.: Heart disease. UCI Machine Learning Repository (1988)
20. Hawrysz, L., et al.: The research on patient satisfaction with remote healthcare prior to and during the COVID-19 Pandemic. Int. J. Environ. Res. Public Health **18**(10), 5338 (2021). https://doi.org/10.3390/ijerph18105338
21. Vittorio, M., et al.: Action Transformer: A self-attention model for short-time pose-based human action recognition. Pattern Recogn. **124**, 108487 (2022)
22. Selim, R., et al.: A multi-head attention-based transformer model for traffic flow forecasting with a comparative analysis to recurrent neural networks. Expert Syst. Appl. **202**, 117275 (2022)
23. Surafel, M.L., Mauro, C., Marcello, F.A.: Comparison of transformer and recurrent neural networks on multilingual neural machine translation. In: Proceedings of the 27th International Conference on Computational Linguistics, pp. 641–652, Santa Fe, New Mexico, USA, 20–26 Aug 2018
24. Qiang, W., et al.: Learning deep transformer models for machine translation. In: Proceedings of the 57th Annual Meeting of the Association for Computational Linguistics, pp. 1810–1822, Florence, Italy, 28 Jul – 2 Aug 2019
25. Ming, C., et al.: TVT: Two-View Transformer Network for Video Captioning. Proc. Mach. Learn. Res. **95**, 847–862 (2018)

MLOps: Overview of Current State and Future Directions

Anas Bodor[1(✉)], Meriem Hnida[1,2], and Daoudi Najima[1,3]

[1] ITQAN Team, LyRica Lab, Information Sciences School, Rabat, Morocco
anas.bodor@esi.ac.ma
[2] RIME Team, Mohammadia School of Engineers, Mohammed V University in Rabat, Rabat, Morocco
[3] SSLab, ENSIAS, Mohammed V University, Rabat, Morocco

Abstract. This article presents and discuss the industrialization process of ML (Machine Learning) projects with a focus on the principles of MLOps (Machine Learning Operations) and the challenges encountered when putting an ML project into production. The paper also proposes a set of tools used in an MLOps context to facilitate the deployment of ML projects and their production release. This paper is a guide to discover the MLOps domain in its theoretical (MLOps concepts, pipeline and life cycle) and practical (technical and tools) aspects. MLOps must provide answers to the use of ML applications hosted on servers with high performance, also for applications embedded in equipment with minimal sizing, since we are talking about the fourth industrial revolution with the increase of number of sensors in the world generating a mass of data that must be efficiently processed and analyzed.

Keywords: MLOps · DevOps · Pipeline · Machine learning · MLOps tools · Delivery pipeline · Continuous delivery · Industry 4.0

1 Introduction

MLOps is a relatively recent field dating back to 2015 [1] whose founding idea is to accelerate the release process also called the deployment of machine learning (ML) projects. The goal is to overcome some of the difficulties raised during the go-live phase of a ML model. According to [6], an entity may take 6 to 18 months to deploy a single ML model to production. This is highly dependent on the requirements that may occur in both the ML model and the integration environment.

The incorporation, rapid and frequent deployment of a ML model into production [2] becomes a necessity when moving from an exploratory and research environment to the industrialization of ML projects, i.e., into the environment where the ML project will be deployed and operated. The practice of continuous delivery of ML is called MLOps [7].

The model generated within the framework of MLOps will be integrated into a classic software engineering project, having specific characteristics, hence there is a need to see ML projects in their globality for a better setting in service, exploitation, and final use.

M. Ben Ahmed et al. (Eds.): SCA 2022, LNNS 629, pp. 156–165, 2023.
https://doi.org/10.1007/978-3-031-26852-6_14

MLOps is also a set of tools covering the communication flow related to the human resources involved in an ML project [2] in order to facilitate work and knowledge sharing between analysts, data scientists, data engineers, front-end engineers, and production engineers. All these profiles are involved in an ML project and need to develop, test, manage, configure, deploy, monitor and document the ML projects.

The aim of this paper is to provide an overview of the MLOps domain and to highlight its importance in deploying an ML model into production in a reliable and sustainable way over time. The paper also presents a set of tools that can accelerate the ML project production process and ensure a smooth transition between the development environment and the environment in which the ML project will be deployed.

The reminder of this article is as follows: In the Sect. 2 the context and the research problem are presented. In the Sect. 3 the different challenges related to MLOps projects are discussed. The Sect. 4 is devoted to the pipeline and industrialization life cycle of an ML project based on the different technologies and tools. Section 5 presents a discussion of the different concepts addressed in this paper. Followed by a conclusion and research perspectives.

2 Context and Problem Statement

The birth of the ML domain dates to the 50s. Its maturity continues to grow, and the integration of this type of application in the daily life of people and businesses has become a necessity. However, there is a need to run ML code in production while ensuring the non-degradation of the service provided by the model deployed over time.

MLOps is a combination of two words (Fig. 1): Machine Learning and Operations. ML stands for ML Model; Operations stands for production and deployment of solutions in an iterative and incremental process.

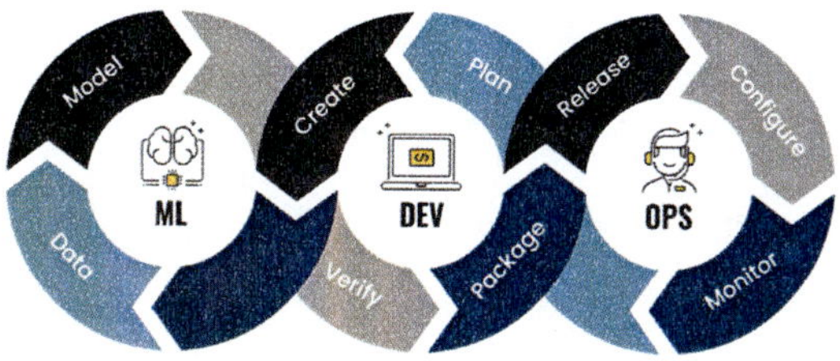

Fig. 1. MLOps decomposition [2]

MLOps is DevOps applied to the ML project. The two main principles of DevOps are continuous integration (CI) with automated testing and validation of code, data, and model and continuous deployment (CD) by trying to deliver frequently through automated deployments. We could add to these last two principles the continuous training (CT) by automating the ML model training pipeline which is a property specific to MLOps [2]. Another interesting and promising concept to add to the list of principles is continuous monitoring (CM) with the need and necessity to continuously monitor the health of models in production in real-time.

This field can also be defined as the set of practices and guidelines for producing ML models, from the analysis phase through model generation to ROI accounting, while reducing the time it takes to develop a service before offering it to the market (Time To Market TTM).

MLOps is the discipline that addresses the problem of deploying ML models in the most reliable and efficient way possible, ensuring a safe transition from the ML model development environment to the ML model operation and deployment environment. It can be seen as a set of tools and techniques performing the functions of the ML pipeline [3], data ingestion, data transformation, continuous (re)training of the ML model, (re)deployment, and presentation of the results, whenever needed.

3 Challenges Related to MLOps

The difficulties encountered when adopting MLOps are directly related to the complexity duc to thc mass of data involved in the ML process [3]. Another aspect is the need to find a way to collaborate and work in a team each with a different profile, level of experience and background.

Among the major problems related to MLOps is versioning, also called version management, initially related to the working environment such as Jupyter notebooks, and the difficulty of keeping versions of the latter and following its evolution over time. In addition, there is the challenge of being able to collaborate in the same notebook and therefore the possibility of tracking the modifications made by the different members of the team. There is also a need to keep data and generated artifacts (hyper parameters) under version control management tool and linked to the generated models. Finally, since the development of ML applications is governed by an iterative process, the pipelines used must also be versioned using version tracking system.

There are major challenges facing ML engineers while rewriting the model with a programming language different from the one in which the model was created by the data scientist in the first phase, to be able to deploy it in production.

In an ML context, the production, delivery, and evaluation of projects are difficult due to their non-deterministic nature, which makes the model complex to test, explain and improve. This brings us back to finding a way to monitor the quality of the information system in real-time.

4 Workflow MLOps and Implementation Tools

The success of a ML project can be measured by the conformity of the expected results of the model created and by the achievement of the objectives defined at the beginning of the project. But also, through the quality of the process and the steps which lead to model being deployed in production and monitored continuously, this process is called pipeline.

4.1 The MLOps Pipeline and Life Cycle

The emergence of ML projects in the world has created an increased need for competence in relation to the benefit that can be derived from this intelligence, as exposed by Google engineers in [1] showing the complexity of managing ML projects compared to managing traditional software. The manifesto (Fig. 2) brings together all the elements that fit into the ML ecosystem, authors of this manifesto argue that The "ML code" block which is the core of data science only takes up a small space and it is very small in size in the whole process. For instance, is a need to deal with ML code in the whole ecosystem. Referring to hidden technical debts related to IT projects.

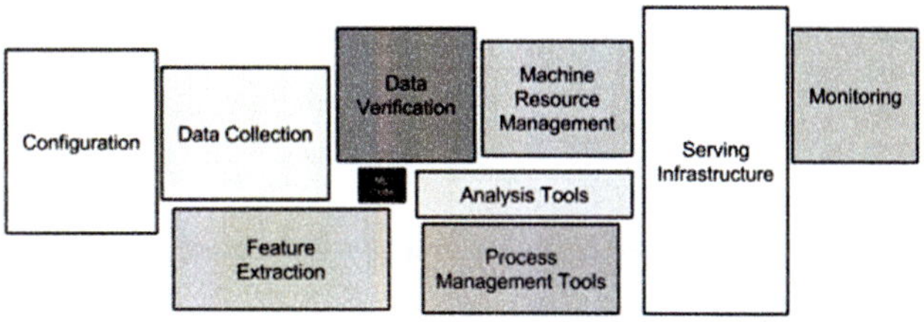

Fig. 2. Google's MLOps manifesto [1]

MLOps can help reduce the technical debt generated by trying to find the right balance between the best and fastest solution to achieve a given ML problem, as it promotes the automation of all steps in building an ML system, from development to deployment [4].

The set of activities, if done manually or if not standardized or even automated, will certainly have an impact on the completion time of an ML project given its multidisciplinary and inter-organizational nature.

There is no standard life cycle for all ML projects. The way these and other practices are adopted in each project may vary from project to another one. Nevertheless, they can generally be placed under the MLOps paradigm which includes the following macro steps (as depicted in Fig. 3):

- **Understanding the needs**: The first step is to define the needs and the expected results of the ML project to be implemented, establish priorities, and consider the risks.
- **Data preparation:** Once the need is well defined, the next step is data collection, analysis, and transformation.
- **Data modeling:** in this step the focus is on training, testing, and validating the model with the setting and choice of hyper-parameters.
- **Deployment of the model:** the model resulting from the previous step is made available for consumption in its final operating environment called also production environment.

There are different techniques for deploying a ML model. The two common used techniques are the "model as a service" and the "integrated model" [3]. In model as a service case, the model is exposed as a service to be consumed by independent applications based on access rights, while in the case of the "integrated model" the model is embedded in an application, which is then published.

- **Model monitoring:** this involves monitoring the performance of the deployed model over time, by measuring the accuracy of the prediction, the detection of outliers and data drift.
- **Action to be taken:** depending on the results obtained in the previous monitoring phase, a decision can be taken. The action to be taken depends mainly on the nature and magnitude of the deployed model results and its impact on the normal course of the beneficiary organization's activity, such as re-training the model with new data.

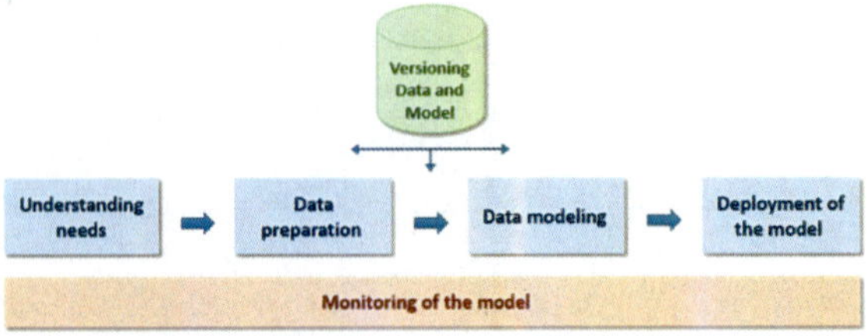

Fig. 3. Macro stages of the ML life cycle

The use of ML projects at an advanced level and the adaptation of the problems encountered to more generic use cases give rise to pipeline or life cycle architectures. Companies such as NVIDIA, Facebook and Google that use Artificial Intelligence in their projects have defined their own architectures [4].

For example, the pipeline proposed by NVIDIA depicted in (Fig. 4) has a cyclical nature. If we unpack this flow proposed by NVIDIA, we notice that the components of the pipeline are strongly correlated with the Google manifesto as shown in (Fig. 2).

The diagram in (Fig. 5) is a concrete example from the company Facebook [4] that shows the importance of pipeline architectures to be able to continuously re-train the models. At the reception of new data, such as the click on Facebook's "Like", "Comment" and "Share" buttons that feed directly into the database to serve as training data to improve the accuracy of the deployed ML models.

Google groups the ML workflow steps into six systems (Fig. 6) (data processing, feature extraction, model creation, model training, model deployment, monitoring and evaluation).

The main feature that characterizes a good pipeline architecture is the interconnection between all the steps. The change of a particular parameter or data in the life cycle automatically triggers the necessary modifications in the whole pipeline based on continuous monitoring.

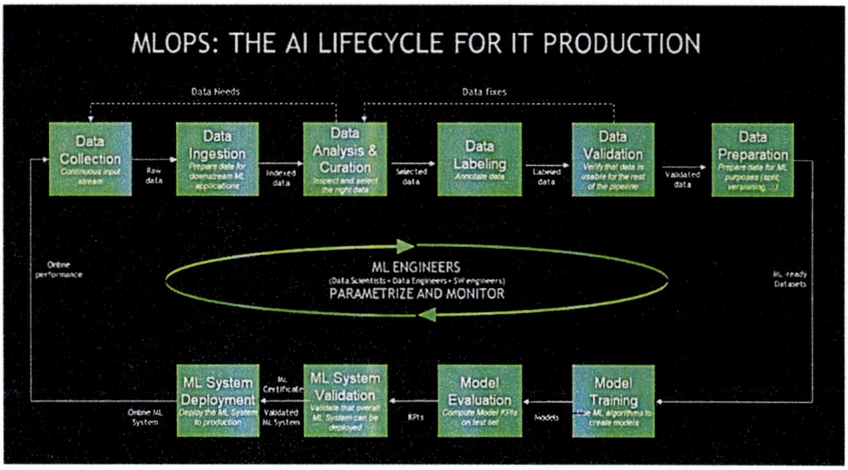

Fig. 4. Pipeline architecture by NVIDIA

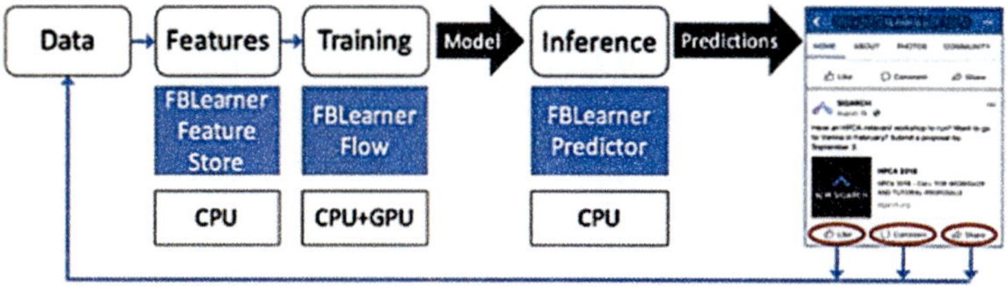

Fig. 5. Example of a pipeline through Facebook [4]

4.2 MLOps Tools and Technologies

Several tools exist on the market that help redeploy, restore or update models according to the specific needs of each project. Still, there is a need to choose the best combination of tools that can deliver the expected results.

The purpose is to address the different challenges and issues mentioned in the Sect. 3.

In the world of MLOps there are open-source platforms that many organizations have contributed to, such as open source mlflow and managed mlflow on Databricks [15], the project has a rapidly growing open source community, with over 50 contributors since its launch in June 2018. Kubeflow is also a “project to make ML stream deployments on Kubernetes simple, portable, and scalable” developed by Google [9] and its list of contributors includes names like IBM, Intel, LinkedIn, Microsoft, Project Jupyter, Shopify, Uber, and many others.

MLFlow: This is an open source agnostic MLOps platform that can be used to manage an end-to-end ML lifecycle. It tracks model experiments, deploys the model, packages the code into reusable components, manages the ML registry, and hosts the results as REST endpoints [8].

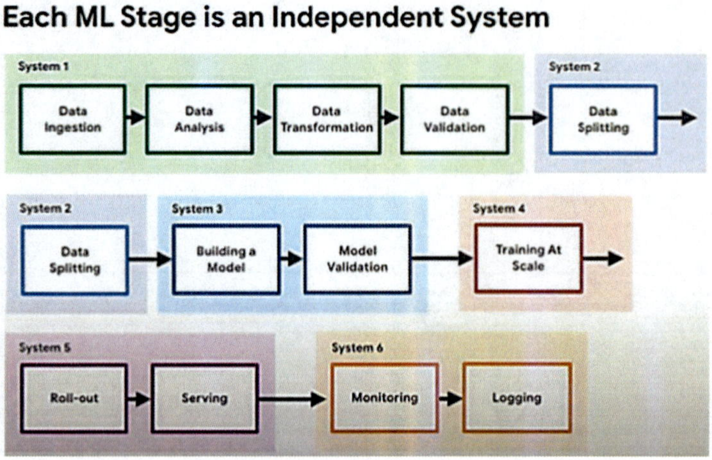

Fig. 6. Google pipeline of ML projects [4]

Kubeflow. [9] started as a simpler way to run TensorFlow jobs on Kubernetes, but has since grown into a multi-architecture, multi-cloud framework for running entire ML pipelines. Given Google's manifesto explained in the first part of the "MLOps Pipeline and Lifecycle" in Sect. 4, they decided to build a tool stack to deploy, scale, and manage these ever-growing systems. This led to the creation of a platform with features such as a centralized dashboard, Kubeflow pipelines for creating ML workflows, integration with JupyterHub servers for simultaneous, multi-user contribution to Jupyter notebooks, distributed training using Tensorflow and Pytorch operators, CLI kfctl.

DVC (Data Version Control): This is an open source version control system for ML projects. It facilitates ML experiment management, project version control, collaboration and makes models shareable and reproducible. Built on top of the popular GitHub version control tool [10], it is designed to overcome some of the very specific challenges of ML model development, such as managing large data files and tracking ML model code and metrics.

CodeReef. [5] is a platform that attempts to facilitate the automation of ML model deployment on various systems. The CodeReef platform is inspired by GitHub and PyPI (Python Package Index): as a collaborative platform to share reusable ML components/systems, portable ML workflows, reproducible ML solutions and associated ML results. Based on open-source technologies with open standards to allow users to exchange all basic blocks (components) needed to have portable MLs without locking them.

ZenML. [12] is an extensible and open-source MLOps framework for creating ML pipelines. Designed for data scientists, it has a simple and flexible syntax. ZenML is tool-independent, it allows writing portable ML code and offers a quick transition from experimentation to production.

AutoML. Is an automatic process that groups the phases of the ML pipeline (data pre-processing, model creation, training, evaluation) in a black box format. It has the advantage of reducing the technical complexity, however it is not the right choice for very specific use cases that require more flexibility to have the necessary openness for model parameterization [2].

The following Table 1 gathers a non-exhaustive list of tools [13] from the MLOps ecosystem with the most dominant category in which each software operates.

Table 1. List of the most popular MLOps tools by application field [13].

	Storage, metadata management	*Data and pipeline versioning*	*Adjustment of hyper parameters*	*Pipeline execution and orchestration*	*Deployment models*	*Monitor the model in production*
Neptune	*					
MLflow	*					
Comet	*					
DVC		*				
Pachyderm		*				
Optuna			*			
Sigopt			*			
Kubeflow pipelines				*		
Polyaxon				*		
Airflow				*		
Kedro				*		
ZenML				*		
BentoML					*	
Cortex					*	
Seldon					*	
CodeReef					*	
Amazon SageMaker Model Monitor						*
Fiddler						*
Hydrosphere						*
Evidently						*

TinyML and IoT: With the emergence of connected objects (IoT), a new opportunity is opening up to involve ML techniques in small-embedded devices, whose resources are limited with the obligation to keep the same quality of service provided. ML requires significant power to operate. The TinyML paradigm [11] is there to facilitate the embedding of ML solutions in devices running on microcontrollers. MLOps is used to cover the industrialization of the production chain of ML projects embedded in connected objects. The use increases with the fourth industrial revolution or Industry 4.0 [16] that implement connected production approach using IoT with a significant requirement for data processing automation and real-time accessibility.

5 Discussion

With the birth of the ML field the focus was on results and on improving the quality of predictions of ML algorithms, but now the focus is shifting to the efficient management of the entire life cycle of an ML project, including the communication flow between the stakeholders of the same project, usually with different profiles and skills. The ultimate goal is to ensure a safe and fluid transition between the exploration environment and the real environment in which the ML project will be deployed. This transition is often complex and requires the intervention and collaboration of several profiles.

There are several proprietary and open source platforms and tools in the ML world, each of these solutions tries to handle and ensure one or more phases or aspects of the MLOps pipeline such as Mlflow, Kubeflow, DVC and many others. These solutions complement each other in some features and overlap in others.

Mlflow facilitates the implementation of the model tracking mechanism using APIs, it integrates with the most used tools in the ML world, while the multi-user mode is not well managed as long as the access is not based on profiles, especially since Mlflow is criticized for not being easy and ideal to deploy models on different platforms [14].

From a version control point of view we find DVC allowing data versioning, model and pipeline tracking, with storage of results for later reference without the need to re-run. DVC is strongly coupled with pipeline management, which means that if you use a different data pipeline tracking tool, you will have redundancy at this level, with a difficulty in managing dependencies.

Kubeflow is a platform capable of deploying ML workflows ensuring portability and scalability based on Docker and Kubernetes containers, this is considered a major advantage but it requires very advanced technical skills to handle.

ZenML on the other hand tries to dissociate the ML pipeline code from the tools and infrastructure needed to run it, with the decoupling of ZenML stacks from the pipelines and their associated code, a principle that allows a quick transition from experimentation to deployment in a production environment.

AutoML, even if it is considered as a black box, can be classified as an advanced tool allowing the automation of ML model training to choose the best one, while reducing technical complexity and processing time.

Future research should focus on proposing an integrative and dynamic architecture of different solutions ensuring the principles of MLOps. As well as to go in depth with the concept of monitoring considered as an essential element in an industrialized ML project, including monitoring in the context of connected and intelligent objects with reference to the approach called TinyML. In an increasingly interconnected world characterized by data-exchange [17] better decisions can be made for value added.

6 Conclusion

In this article, MLOps concepts and tools are presented as well as ML pipeline for creating a good model in ML projects and facilitate its deployment to production. An ML project often requires re-running the code by testing dataset, model type and hyper-parameters. Also to maintain the quality of the models deployed in production. By adopting an MLOps approach, the deployment process is simplified and even automated, thus facilitating the production and operationalization of ML projects in their real context. In the near future, we will deal with the phases of the MLOps pipeline in depth by valuing the different techniques and tools corresponding to each step. Mainly we will focus on the monitoring phase considered as an indispensable element present throughout the life cycle of an ML project in order to ensure the maintenance of quality systems integrating intelligent systems.

References

1. Sculley, D., et al.: Hidden technical debt in machine learning systems. In: NIPS, pp. 2494–2502 (2015)
2. Symeonidis, G., Nerantzis, E., Kazakis, A., Papakostas, G.: MLOps – definitions, tools and challenges, pp. 0453–0460. https://doi.org/10.1109/CCWC54503.2022.9720902 (2022)
3. Ruf, P., Madan, M., Reich, C., Abdeslam, O.D.: Demystifying MLOps and presenting a recipe for the selection of open-source tools. Appl. Sci. **11**, 39 (2021). https://doi.org/10.3390/app11198861
4. Cajamarca, A.: Andres Felipe Cajamarca Mantilla Responsabilidad Social. https://doi.org/10.13140/RG.2.2.35401.06241(2021)
5. Fursin, G., Guillou, H., Essayan, N.: CodeReef: an open platform for portable MLOps, reusable automation actions and reproducible benchmarking (2020)
6. Liu, Y., Ling, Z., Huo, B., Wang, B., Chen, T., Mouine, E.: Building a platform for machine learning operations from open source frameworks. IFAC-PapersOnLine **53**, 704–709 (2020). https://doi.org/10.1016/j.ifacol.2021.04.161
7. Mäkinen, S., Skogström, H., Laaksonen, E., Mikkonen, T.: Who needs MLOps: what data scientists seek to accomplish and how can MLOps help? (2021)
8. Chen, A., et al.: Developments in MLflow: a system to accelerate the machine learning lifecycle, pp. 1–4. https://doi.org/10.1145/3399579.3399867(2020)
9. Bisong, E.: Building machine learning and deep learning models on Google Cloud Platform: a comprehensive guide for beginners. https://doi.org/10.1007/978-1-4842-4470-8(2019)
10. Site officiel dvc [online]. Available on: https://dvc.org/
11. Ray, P.P.: A review on TinyML: state-of-the-art and prospects. J. Sens. **2022**, 11 (2022)
12. ZenML [online]. Available on: https://zenml.io/
13. Neptune [online]. Available on: https://neptune.ai/blog/best-mlops-tools
14. Censius [online]. https://censius.ai/blogs/dvc-vs-mlflow
15. Databricks [online]. https://databricks.com/product/managed-mlflow
16. Chakraborty, A., Mandal, N.: Introduction of industry 4.0 introduction of industry 4.0. Turk. Online J. Qual. Inq. **12**, 8342–8350 (2022)
17. Dharmadhikari, S.: Industry 4.0 Challenges. 40 (2022)

Cloud Services for Smart Farming: A Case Study of the Veracruz Almond Crops in Portugal

Filipe Fidalgo[1,2], Osvaldo Santos[1,2], Ângela Oliveira[1,2], José Metrôlho[1,2], Fernando Reinaldo[1,2], Antonino Candeias[1], Jorge Rebelo[3], Paulo Rodrigues[3], Rodrigo Serpa[3], and Rogério Dionísio[1,2](✉)

[1] Polytechnic Institute of Castelo Branco, 6000-084 Castelo Branco, Portugal
rdionisio@ipcb.pt
[2] R&D Unit DiSAC, 6000-767 Castelo Branco, Portugal
[3] Eracruz, 6060-188 Idanha-a-Nova, Portugal

Abstract. Efficient use of resources is a critical factor in almond crops. Technological solutions can significantly contribute to this purpose. The VeraTech project aims to explore the integration of sensors and cloud-based technologies in almond crops for efficient use of resources and reduction of environmental impact. It also makes available a set of relevant and impactful performance indicators in agricultural activity, which promote productivity gains supported by efficient use of resources. The proposed solution includes a sensor network in the almond crops, the transmission of data and its integration in the cloud, making this data available to be consumed, processed, and presented in the monitoring and alerts dashboard. In the current state of the development, several data are collected by sensors, transmitted over LoRaWAN, integrated using AWS IoT Core, and monitored and analysed through a cloud business analytics service.

Keywords: Smart farming · AWS · Database · LoRaWAN · The Things Network

1 Introduction

Population growth and climate change are the two major challenges for agricultural production in the 21st century. The pressure to produce more with fewer resources implies changes in production methods and in the efficient management of each method. The Food and Agriculture Organization of the United Nations (FAO) recommends that all agricultural sectors equip themselves with innovative techniques and tools, to adopt and take advantage of digital technologies [1].

Thus, there is a growing commitment to precision agriculture, where the objective is to optimize production per agricultural unit, using the most modern tools to achieve better results in a sustainable approach.

The development of wireless sensor network technologies has been opening doors to new ways of developing precision agriculture.

Currently, a multitude of monitoring systems and data generation are installed and operating on many farms in the Beira Interior region, in Portugal. However, such systems

M. Ben Ahmed et al. (Eds.): SCA 2022, LNNS 629, pp. 166–174, 2023.
https://doi.org/10.1007/978-3-031-26852-6_15

operate in a non-integrated way, so the potential of cross-information between image captures and processing, the use of agricultural production factors, the planning of water use, soil, evolution of crop productivity, pests and diseases, meteorology and human resources is not achieved [2].

Current solutions do not supply such level of integration to group multiple data sources and correlate it to extract useful information for supporting and providing decisions. In general, the available solutions provide only closed islands of data (rain sensors, tree counts, stem growth, etc.). Even sensors with a set of APIs available are not, in fact, able to integrate information useful to optimize the agricultural planning solution. Moreover, the amount of data, due to its complexity and multiplicity, is not being accompanied by processes that lead to comprehensive decision support systems for the farmers [3].

The objective of this work is to develop and implement a model and present a solution based on Cloud services platforms for capturing multiple data from different sources, as shown in Fig. 1, so that, through smart farming, precision agriculture, real time monitoring and business intelligence solutions, they can introduce increases in productivity and optimize the usage of resource in the almond cultivation process of the VeraCruz orchards in Portugal.

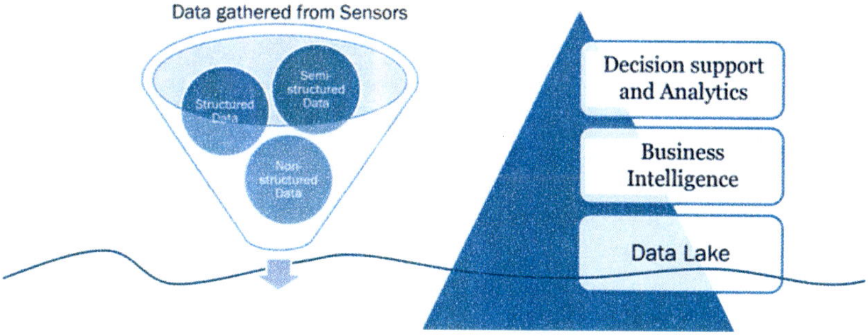

Fig. 1. Integration of several data source into a single decision support platform.

The use of Cloud platforms is a requirement for the development of an integrated decision support service for almond orchards. Given the variety of IoT cloud platforms currently available, difficulties may arise in the process of selecting the one that best suits the necessary requirements. A recent study was made in [4] and concluded that the platforms that best suited a set of defined criteria are commercial solutions, and that open-source platforms generally only provide basic functionalities. We compared the features of commercial IoT cloud platform currently available from the main players in the market: Amazon Web Services (AWS), Google Cloud Platform (GCP) and Microsoft Azure. The results are presented in Table 1.

From the analysis of Table 1, except for GCP that failed to comply with one requirement, the other two platforms have similar services and support and from that point of view, it is difficult to choose. Recent studies have compared and analyzed the performance of the platforms from simulations to identify which ones are more efficient when exposed to different usage situations [5–7]. The results show that AWS is a better performing cloud provider in serverless applications than Azure in constant workload. Thus, AWS is the elected choice to implement our solution.

Table 1. Comparative analysis between three commercial cloud platforms for IoT applications.

Platform/Requirement	AWS	GCP	Microsoft azure
Device management	Yes	Yes	Yes
Device simulation	Yes	No	Yes
Dashboards	Yes	Yes	Yes
Alerts and warning configuration	Yes	Yes	Yes
Development (SDK and programming languages)	Yes AWS IoT Device SDKs and AWS Mobile SDKs (C + +, Java, Javascript, Python, Embedded C)	Yes Node.js, Python, Go, Java,.NET, Ruby and PHP	Yes Azure IoT Hub SDK (C, C#, Java, Node.js, Python) Azure IoT Hub REST API (Go, PHP)
Machine learning	Yes	Yes	Yes
Storage limits	Scalable. According to the subscription	Scalable. According to the subscription	Scalable. According to the subscription
Communication protocols	Yes MQTT, MQTT over WSS – Websockets HTTPS	Yes MQTT, MQTT HTTPS	Yes MQTT and AMQP MQTT and AMQP over WebSockets, HTTPS
Security (Access levels and permissions, device authentication, encryption)	Yes Encryption TLS/SSL of the data transmission; Authentication with X.509 certificate; *endpoints certificate*	Yes TLS 1.2, Support Authentication, access control, security tokens, encrypted connections, and digital certificates	Yes Support Authentication, access control, security tokens, encrypted connections, and digital certificates

Besides the presented introduction, this paper is organized as follows. Section 2 presents the methodology used and details about Architecture and Implementation. Finally, in Sect. 3 the conclusions and further work are presented.

2 Methodology

Considering the goals and requirements previously identified, a set of guidelines was identified for defining the architecture of the solution. Thus, it must include the installation of a sensor network in the almond's plantation, the transmission of data and its integration in the Cloud, and make these data available to be consumed, processed, and presented in a monitoring, recommendation, and alerts dashboard and analyzed using Business Intelligence algorithms.

2.1 Architecture

Four main modules of the architecture were identified, each of which includes several components (see Fig. 2):

1. Sensoring;
2. Integration;
3. KPI identification and generation;
4. Visualization and decision support dashboards.

The architecture must allow the integration of all data from the sensors and devices included in the "sensing network" in the farm. It should also allow the integration of various parameters from external data sources, such as data from farm machinery, data from weather stations, data from and pest and disease detectors, and data about the use of resources and on the characterization of production. This data must be integrated into a data model on the Cloud to allow scalability. Data is later accessed through API by the various application modules of the solution. This data model must be able to store the data and must allow continuous evolution.

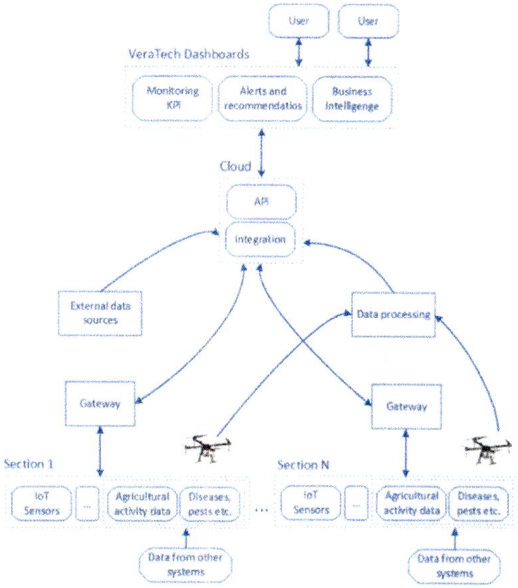

Fig. 2. An overview of the proposed architecture.

2.2 Implementation

Several wireless nodes are currently installed in the Veracruz Almond Crops, as shown in Fig. 3. Sensors are collecting information from soil, air, almond trees, and critical node parameters and sending data to a TTN (The Things Network) server using LoRaWAN.

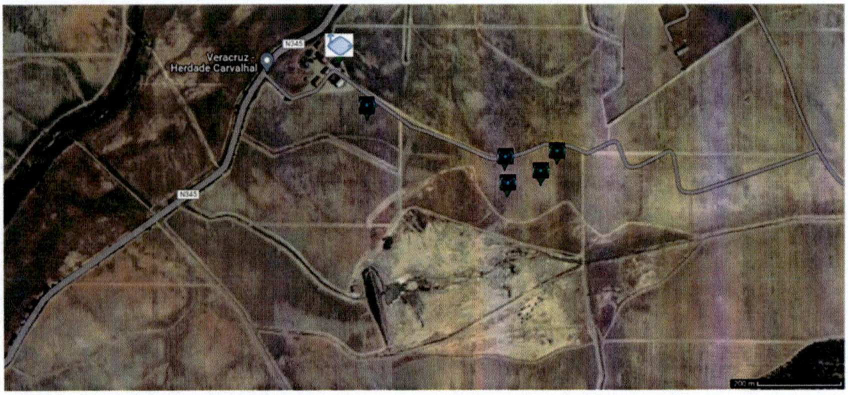

Fig. 3. Location of the sensors and gateway on the crop field.

Payload Encoder and Decoder

Sensing data from a wireless sensor network is an energy and time-consuming process. In the specific case of LoRaWAN sensor connected through the TTN server, for EU863–870 frequency band, the Regional Parameters 1.0.2 Rev B as used by the TTN community network, define duty-cycled limited transmissions to comply with the European Telecommunications Standards Institute (ETSI) regulations. As a rule-of-Thumb, the TTN fair access policy allows for at most 30 s uplink airtime and 10 downlink messages per sensor, per 24 h [8].

For the case of the almond crop wireless device, for sensors with integer output with values greater than 255 (8 bits), the encoding process divides the variable into sections of one byte each, starting by extracting the least significant 8 bits, and proceeding with a bit-shift technique (8 bits) to extract the remaining sections of the variable. For a sensor with analogue (float) outputs, the values are initially multiplied by 100 to store information up to two decimal places, then the value is truncated to an integer, and then follow an approach like the one used for integers. For sensor that may output negative values (e.g., temperature), we set the most significant bit (bit 0) of the most significant byte to represent the value sign: 0 is negative and 1 is positive.

AWS Integration

The TTN is integrated with the AWS IoT Core service, a highly scalable managed IoT broker that can connect billions of IoT devices with cloud applications and other devices [9]. This service is thus the data interface between the IoT devices and the developed applications.

Every time a sensor sends a message to the TTN, a copy of that uplink message is relayed to the AWS IoT Core. Each IoT device publishes its messages in a specific topic related to that device's type. Incoming messages to IoT Core are fed to a set of rules that filter the messages by subscribed topic, and then apply routing policies to them. These rules can be configured with direct connectors to other AWS services such as Dynamo DB, S3, Analytics and others. However, there is not a direct connector to relational databases like Aurora, thus an intermediate layer (Lambda functions) is being used to connect the rule to the Aurora database, as can be seen in Fig. 4.

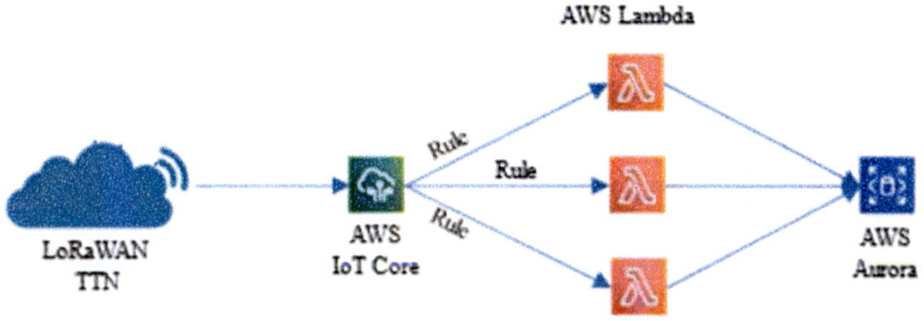

Fig. 4. Message routing path from TTN to Aurora database.

There are three different rules, one for each type of device. Currently the IoT network comprises three types of IoT sensors: Dragino devices (for temperature and humidity), soil devices (for temperature, humidity, and electrical conductivity) and water flow sensors (volume and alarms). Each rule connects to its own lambda function, which intercepts the messages, fetches some fields, and inserts them in the Aurora database. The lambda functions are written in Node.JS and perform some additional pre-processing, such as data quality checks, data format conversion and average calculation. These uplink messages are sent by the TTN in a specific JSON format [10], which includes multiple fields of the TTN itself, and specific fields that come from the IoT device's payload, embedded in the uplink_message section.

Table 2. Data fields fetched by the lambda functions.

JSON element	Description
device_id	TTN end-device identifier
latitude, longitude, altitude	GPS coordinates of the device
received_at	ISO 8601 UTC time
Rssi	Received signal strength indicator (dBm)
Snr	Signal-to-noise ratio (dB)
Airtime	Time-on-air, calculated by the network server

Some of the message fields are common to all the sensors whereas other are device specific. Table 2 enumerates the common fields of the different devices, which are fetched by all the lambda functions and sent to the Aurora database. The dragino IoT devices, which publish message in the aws-dragino topic, send the additional fields shown in Table 3:

Table 3. Data fields specific of the dragino device.

JSON element	Description
BatV	Battery voltage
Bat_status	Battery status
Hum_SHT	Humidity (internal probe)
TempC_SHT	Temperature (internal probe)
TempC_DS	Temperature (external probe)

The soil sensors publish messages in the aws-milesight topic and have the following specific fields (Table 4):

Table 4. Data fields specific of the soil sensor.

JSON element	Description
Ec	Electrical conductivity
Humidity	Humidity
Temperature	Temperature
Battery	Battery level

Finally, the water flow sensors publish messages in the aws-axiom topic and have the additional fields shown in Table 5.

Table 5. Data fields specific of the water flow sensor.

JSON element	Description
volume	Water flow volume
reverseFlow, battery, leak, permanent, burst, dry, lowTemperature	Binary fields used to transmit alarm conditions

A database represents a collection of data core for business, usually stored in digital form. For the work present in this study, Amazon Web Services (AWS) were used. AWS as DBaaS offers more than fifteen engines to support different data models: relational, key-value, document, in-memory, graph, time series, wide column, and ledger databases [11].

After a detailed analysis of the data involved, several relationships between them were identified that should be preserved, and due to the volume of data a relational database – Aurora Database was selected, which made it possible to create meaningful information by joining the tables. Additionally, relational databases allow users to use stored procedures. A stored procedure is an object that can be stored in a database for later use and can be used many times. Several stored procedures were created and were called by other AWS objects that passed them a set of parameters. Stored procedures evaluated these parameters and depending on the result obtained store the received data in specific tables keeping the referential integrity and strong data consistency. This stored data will be later used for creating dashboards (see Fig. 5) and key performance indicators (KPI) to support the organization's decisions. Quicksight [12] was the AWS service selected for creating dashboards with relevant visualizations for users to gain insights from the data.

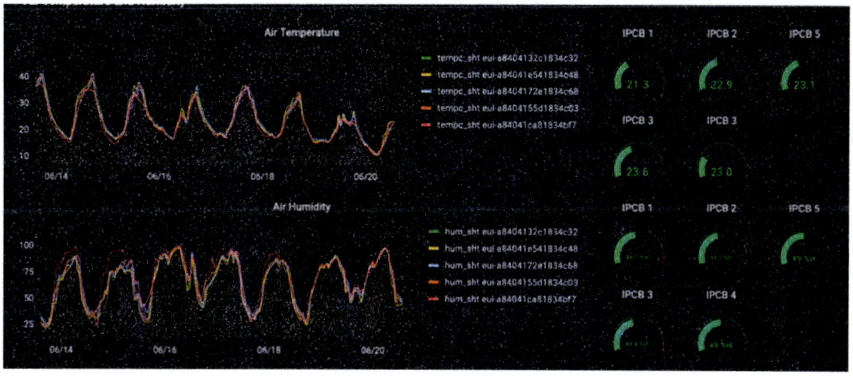

Fig. 5. Section of the dashboard showing temperature and humidity values.

3 Conclusions

This paper presented a case study on the use of Cloud services in the scope of precision agriculture that combines the use of state-of-the-art technological innovations to support the decision-making for the promotion of better working techniques, promotion of better use of natural resources and the protection of the environment in the Beira-Baixa region of Portugal. With the rapid growth in technology, there is a continuous opportunity to manage agriculture data and present it in dashboards to allow an interactive analysis of data, enabling stakeholders to obtain insights to support their decisions. Activating IoT-based alarms can prevent damage in cultures, like a flood in productive areas. Systems accuracy can improve by choosing reliable sensors which fit in the appropriate cloud services.

Acknowledgment. The research leading to these results has received funding from the Portugal2020 (13/SI/2020) under grant agreement No. 113287 [VERATECH]. The authors would like to thank all VERATECH partners for their contributions.

References

1. Hassan, S.I., Alam, M.M., Illahi, U., Al Ghamdi, M.A., Almotiri, S.H., Su'ud, M.M.: A systematic review on monitoring and advanced control strategies in smart agriculture. In: IEEE Access, vol. 9, pp. 32517–32548 (2021). https://doi.org/10.1109/ACCESS.2021.3057865
2. Jahanzad, E., et al.: Orchard recycling improves climate change adaptation and mitigation potential of almond production systems. PLoS ONE **15**(3), e0229588 (2020). https://doi.org/10.1371/journal.pone.0229588
3. Martnez, J.V.Y., Skarmeta, A.F., Zamora-Izquierdo, M.A., Ramallo-Gonzlez, A.P.: IoT-based data management for smart agriculture. Seco. Int. Conf. Embed. Distri. Sys. (EDiS) **2020**, 41–46 (2020). https://doi.org/10.1109/EDiS49545.2020.9296443
4. Oliveira, L., Silva, F.: Análise comparativa de plataformas baseadas em Cloud para o desenvolvimento de aplicações IoT. In: Anais Estendidos do XXXVIII Simpósio Brasileiro de Redes de Computadores e Sistemas Distribuídos, pp. 257–264. Porto Alegre: SBC (2020). https://doi.org/10.5753/sbrc_estendido.2020.12427
5. Marcin, C., Grzegorz, K., Maciej, B., Michal, P., Torsten, H.: SeBS: a serverless benchmark suite for function-as-a-service computing. In: Proceedings of the 22nd International Middleware Conference (Middleware '21), pp. 64–78. Association for Computing Machinery, New York, NY, USA (2021). https://doi.org/10.1145/3464298.3476133
6. Ustiugov, D., Amariucai, T., Grot, B.: Analyzing tail latency in serverless clouds with STeLLAR. IEEE Int. Sympo. Workl. Character. (IISWC) **2021**, 51–62 (2021). https://doi.org/10.1109/IISWC53511.2021.00016
7. Pascal, M., Pascal, F., Peter, K., Valerio, S.: FaaSdom: a benchmark suite for serverless computing. In: Proceedings of the 14th ACM International Conference on Distributed and Event-based Systems (DEBS '20), pp. 73–84. Association for Computing Machinery, New York, NY, USA (2020). https://doi.org/10.1145/3401025.3401738
8. "Airtime calculator for LoRaWAN." https://avbentem.github.io/airtime-calculator/ttn/eu868. Las accessed 22 May 2022
9. AWS IoT Core Homepage: https://aws.amazon.com/iot-core/. Last accessed 22 May 2022
10. The Things Stack Data Formats Reference: https://www.thethingsindustries.com/docs/reference/data-formats/. Last accessed 22 May 2022
11. Amazon Web Services – Databases: AWS Cloud Databases. https://aws.amazon.com/products/databases/
12. Amazon Web Services: "AWS QuickSight," https://aws.amazon.com/quicksight/

Nyon: A Ubiquitous Fall Detection Device for Elders

Cassandra Sofia dos Santos Jesus[1], Ana Rafaela Rosa[1], and Rogério Pais Dionísio[1,2](✉)

[1] Polytechnic Institute of Castelo Branco, 6000-084 Castelo Branco, Portugal
rdionisio@ipcb.pt
[2] DiSAC – Digital Services, Applications and Contents, 6000-767 Castelo Branco, Portugal

Abstract. Falls are one of the main causes of mortality and morbidity in the elderly worldwide. This had let to the research and development of electronic fall-detection systems. We propose a complete fall-detection system, that combines a wearable device (called Nyon) and a message microservice (for email and SMS) to alert caregiver every time a fall occurs. The wearable uses a simple threshold method and has the capability of search and switch between Wi-Fi and Bluetooth, using the available communication technology when a fall occurs. The results have shown that the wearable autonomy is adequate for a daily use and the server microservices are reliable and deliver a message to the caregiver every time a fall alert occurs. Several improvements are planned to increase the autonomy and range of the wearable device.

Keywords: Wi-Fi · Bluetooth · MQTT · Fall detection algorithm · Threshold detection · ROC

1 Introduction

1.1 Scope

The topic of aging in populations has gained prominence around the world, particularly in more developed countries, such as in Europe, where the highest rates of aging are recorded worldwide.

In 2021, Portugal ranked 4th in the list of the oldest countries in Europe, only surpassed by Italy, Greece, and Finland [1], with an aging index of 182.1 [2]. This scenario poses a set of challenges to which Portuguese society must respond effectively.

Aging is a natural, continuous and inevitable process, guided by numerous biopsychosocial changes, which can represent losses and/or gains [3] and lead to a functional decline, making the elderly more vulnerable, which, in turn, increases the probability of developing pathologies, and consequently increases the risk of falling [4].

Falls are one of the main causes of mortality and morbidity in the elderly worldwide and can be defined as an unexpected event, in which the individual moves, unintentionally, from the level where he is or from a higher level to a lower one [5]. It is estimated

M. Ben Ahmed et al. (Eds.): SCA 2022, LNNS 629, pp. 175–184, 2023.
https://doi.org/10.1007/978-3-031-26852-6_16

that in the European Union, the need for health services that respond to falls and consequent hip fracture can cost approximately 37 billion euros, which in Portugal can reach 216 million Euros [6]. In the population over 65 years of age, falls can be the cause of numerous hospitalizations, institutionalization, and premature death, which contributes to the loss of functional capacity, increased consumption of health resources and, above all, loss of quality of life [7]. In Portugal, the prevalence of injuries from falls rises to 76% in the 65–74 age group, reaching 90% in the group of people aged 75 and over [8].

The pandemic caused by the Sars-Cov-2 virus has increased the loneliness and isolation of elderly people, particularly those living in rural areas, two of the most important risk factors for falls in this population [9].

It is in this context, and given the fact that the Castelo Branco district has one of the highest aging rates in the interior of Portugal (213.7) [10] and where 1,826 elderly people living isolated or in a vulnerable situation in the district were counted by the National Security Police [11], that the technological revolution can make a great contribution to improving and facilitating aging in the community, in the sense that they can support the performance of Activities of Daily Living (ADLs), the monitoring of vital parameters while improving safety, mobility, communication and monitoring of the elderly [12]. There are authors who argue that a quick response after a fall considerably reduces the number of consequences inherent to it [13].

1.2 State of the Art

The Internet of Things (IoT) allows a pervasive a seamless interaction between objects, sensors, and computing devices. An important part of the IoT ecosystem are the embedded systems, combining wireless communications and fall-detection sensors.

Recently, many articles have been focusing on the development and testing of fall-detection wearable devices. In [14], the authors review the main hardware and software components involved in a fall-detection system for elderly.

There are two approaches to fall detection using wearable sensors—threshold-based systems and machine learning-based systems. Threshold-based algorithms are typically designed to minimize computational overhead [16]. Most fall detection studies trained their models in an offline mode with a single sensor on personal computers. In studies that used wearable devices, most of them applied wireless methods, such as Bluetooth, which allowed the subject to move unrestricted.

Other research studies have included Electroencephalogram (EEG) in fall detection devices [15]. But there is still much work to do, as EEG systems are normally bulky, mainly used in hospital for seizure detection, which is not portable at all.

Many new and important developments have been made to tackle the best detection algorithm and process for early fall-detection in elderly, but still, wore work must be done to improve the feasibility and reliability of fall-detection devices in real scenarios, especially when elder people are still capable of moving in and out of their house. From a point-of-view of wireless communication ubiquity and battery autonomy, most of the recent works do not include additional input on that question, as mentioned in [17].

1.3 Objectives

With this paper, we propose to combine several wireless technologies into a single fall-detection device, using a 3-axis acceleration sensor and a common algorithm that requires very low energy consumption to detect falls with adequate reliability, and that transmit alerts using the available wireless communication system (indoor or outdoor) when the fall is detected.

The present work is divided into 5 sections: the introduction section presents a brief review of what has been done and developed in recent years, regarding fall detection devices and operation and the objectives of this work; then the methodology used in the development of the Nyon wearable device is described, followed by the description of a ubiquitous prototype in terms of communication technology, implemented with a low consumption device. Finally, the results and respective discussion will be presented, and we finalize with the main conclusions.

2 Fall-Detection Methodology

The fall-detection methodology was divided into two distinct processes. First, we execute several trials with an accelerometer and measuring devices to create a proprietary dataset (not describe in this paper). Then, we implement a fall-detection and alert system, including the design and production of a wearable device incorporating the same accelerometer from the one used in the dataset construction, with information on three independent acceleration axes: x, y and z.

2.1 Fall Detection Features

Fall-detection algorithms can be classified into two types: Simple threshold method or Machine Learning (ML) methods. In the simple threshold method, threshold values of specific parameters calculated from sensor data, such as 3-axial acceleration or gyroscope, are used to detect a fall event [18]. They have low computational requirements and are simple to implement on a wearable device. The ML method is more complex and leads to better detection rates compared to simple threshold methods. However, it is difficult to implement the ML approach in battery powered wearable devices due to the heavy computational and energy resource requirements [19].

To detect a fall event, Nyon wearable uses a simple threshold-based fall detection feature called Differential Sum Vector Magnitude (A_{DSVM}) [20], computed using the following equation,

$$A_{DSVM}(n) = \sqrt{(x_n - x_{n-1})^2 + (y_n - y_{n-1})^2 + (z_n - z_{n-1})^2} \quad (1)$$

where x, y and z are the three independent variables from the acceleration axes and n denotes the sample number.

From a proprietary dataset, several scenarios with fall-event sequences were realized using a common accelerometer sensor and a measurement system (not descried in this paper). Data sequences were collected and stored as a csv file, with x, y, z acceleration measured in g (gravity acceleration) and timestamp measured in second. Figure 1 shows the computed A_{DSVM} of one the sequences with three predefined (black line) fall-events.

Mean duration of a fall-event takes 0.45 s.

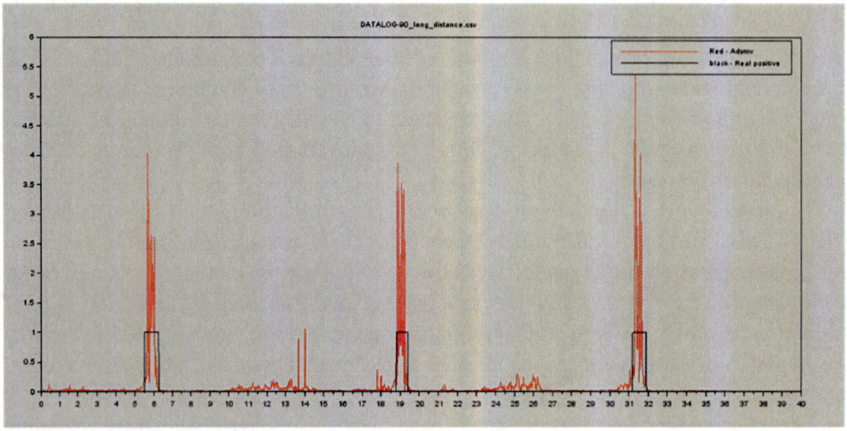

Fig. 1. A_{DSVM} results from one of the time sequences of the dataset, with three fall events: black: True Positive (1) and True Negative (0); red: A_{DSVM} computation.

2.2 Receiver Operating Characteristics Curve

The receiver operating characteristics (ROC) curve shown in Fig. 2.

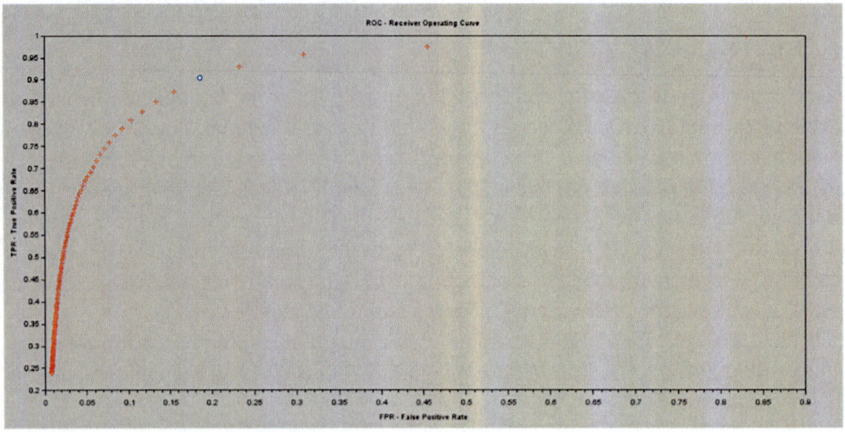

Fig. 2. Receiver operating curve of the fall detection device.

The ROC curve is generated from True positive Rate (TPR) and False Positive Rate (FPR) vectors, calculated for all A_{DSVM} values,

$$TPR = \frac{TP}{TP + FN} \tag{2}$$

$$FPR = \frac{FP}{FP + TN} \tag{3}$$

where TP stands for True Positive, FN is False Negative, FP is False Positive, and TN mean True Negative.

The best threshold is determined from the ROC curve, when the amplitude between TPR and FPR is maximum, as represented by the code snippet from Fig. 3.

```
ampl = TPR-FPR;
best_ampl = find(ampl==max(ampl));
best_th = vect_th(best_ampl);
```

Fig. 3. Pseudo-code of the threshold computation.

The white dot visible in Fig. 2. is the optimal threshold (0.35), when TPR = 90% and FPR = 22%.

2.3 Algorithm

The fall-feature parameters are applied to the simple threshold method to determine whether the computed A_{DSVM} parameter is above a certain threshold (0.35) within a time interval (0,45 s). If any parameter is above a threshold, the sample is determined to be a possible fall indicating a subject fall event or ADL like a fall event. The fall-detection algorithm programmed in the Nyon wearable device is shown in Fig. 4.

```
if A_DSVM value > threshold value
    if elapsed time < Fall interval
        count++
        if count > N_Samples
            count=0
            return fall detection
    else
        reset elapsed time
else
    return no fall detection (ADL)
```

Fig. 4. Pseudo-code of the fall detection algorithm.

3 Implementation

3.1 System Architecture

The proposed System is based on a cloud architecture. The wearable sensor device has two different communications technologies, Wi-Fi and Bluetooth and will switch between them automatically when one of them is unavailable. When a known Wi-Fi access point (AP) is nearby and available, the fall detection device will first try to connect through it to a cloud server, using MQTT protocol. On the other side, when the elder is outside his home, in the garden or farm, and not in reach of a Wi-Fi AP, the fall detection device will try the connection from the elder's mobile phone using a Bluetooth connection. In either case, the Cloud server will send an alert to the caregiver using SMS messages or an e-mail. A diagram of the proposed system is depicted in Fig. 5.

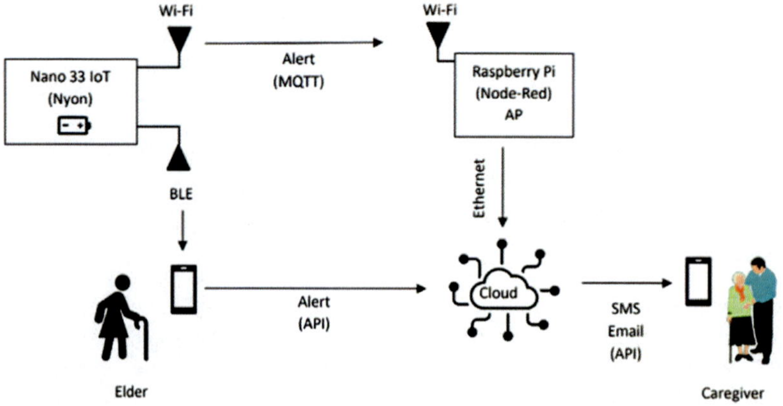

Fig. 5. Diagram of the proposed system for the fall detection device

3.2 Hardware and Software Description

The Nyon fall-detection system consists of an Arduino Nano 33 IoT board, that incorporates a $\pm$ 4 g triaxial accelerometer (LSM6DS3) and a wireless communication module (NINA-W102) to enable Bluetooth and Wi-Fi connectivity; The wearable system is small enough to be weared as a bracelet, a collar, incorporated into hat or clothes. A gateway (Raspberry Pi 3) collects information from one or multiple wireless wearable devices, and a Cloud server (LAMP) send SMS and e-mails to the caregivers. The wearable sensor node is powered by two 12 V batteries (MN21), as shown in Fig. 6.

Fig. 6. Nyon wearable device prototype.

The wearable sensor node was programmed using C language form Arduino IDE, including all the necessary libraries. The server-side application was programmed using Node-Red. The Android App was created with pfodDesignerV3 (V3.0.3774 +), to generate the general purpose BLE connection for the Nyon wearable device, when Wi-Fi is not available.

4 Experimental Results

4.1 Autonomy

Table 1 shows the instantaneous current measurements from different operation modes of the wearable fall-detection sensor and the computed autonomy when the device is powered with two 12 V battery, each with 60 mAh of stored energy.

Considering that the fall detection algorithm running on the wearable device continuously pools the accelerometer sensor, the baseline instantaneous current is 7 mA and the autonomy is approximately 17 h. This value could be increased using a timer interrupt-based approach to measure acceleration, reducing the power consumption of the device, and keeping it periodically in sleep mode between measurements.

Table 1. Current and autonomy for each operating mode of the wearable device

Operation Mode	Current (mA)	Autonomy (hours)
Wi-Fi and Bluetooth disconnected	7	17,14
Connecting to Wi-Fi AP	45	2,67
Connected to Wi-Fi AP	22	5,45
Connecting to Bluetooth (SCAN)	56	2,14
Connected to Bluetooth	10	12,00

The worst case is present when the wearable device searches for a Wi-Fi AP or a mobile phone using BLE to connect. After being connected, either through Wi-Fi or Bluetooth, the device will stay in that operation mode only the time necessary time to send the alert, then returns in a lower power consumption mode. Thus, in general, the wearable must be recharged every day to be kept operational.

4.2 Email and Text Message Microservices

Several microservices were created in the cloud server to receive alerts and send SMS and email messages. The software running the microservices on the cloud server is Node-RED. Figure 7 shows the flow running the microservice that is responsible to receive fall-detection messages from the wearable device, by subscribing to the topic *arduino/fall*. Every time the wearable device publishes a message with the same topic, the MQTT broker (running on the Cloud server) will distribute the information to all the subscribers of that topic.

To use Google Gmail (www.gmail.com) for sending email or Vonage (www.vonage.com) for sending SMS, we have created an account with each of them. When creating an account, we get various credentials needed to configure both nodes (*Send SMS* and *email*) for the first time. Both services are proven being reliable and have delivered the messages in all cases.

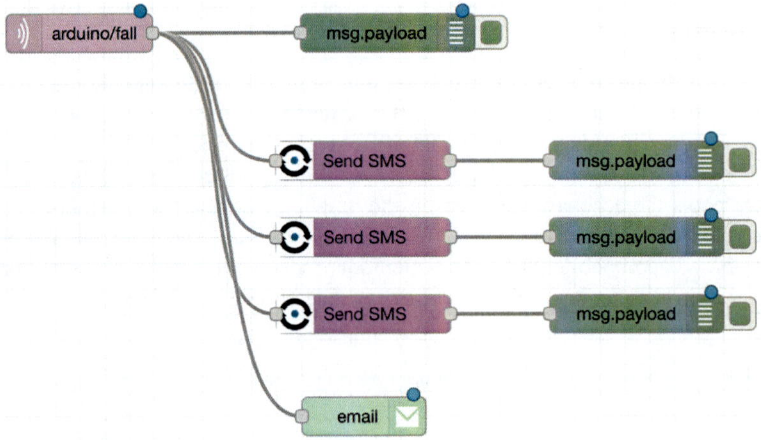

Fig. 7. Node-Red flow for receiving alerts from MQTT and to send SMS and emails.

5 Conclusions

We propose a complete system to detect fall in elder people. The system combines a simple threshold method with a 3-axis accelerometer. Because the proposed algorithm is simple, it can be into an embedded system such as the Arduino Nano 33 IoT.

The proposed wearable device can seamlessly switch between Wi-Fi and Bluetooth, using the communication technology that is available when the fall occurs.

One of the limitations of this work is that the wearable device in continuously pooling data from the accelerometer, draining energy from the batteries, and reducing the autonomy of the device. Further research is required to improve the performance of the proposed wearable device, using an interrupt-based approach to sense fall with the accelerometer. A wireless charging module could also benefit the elder when recharging the device.

Another line of research is to add LoRa communication technology into the wearable device, and increase the possibility of an alert being received by caregiver, when the elder does not have a mobile phone with him and the fall occurs outdoor, or in places out of reach of a Wi-Fi AP.

References

1. EUROSTAT: Archive: Estrutura populacional e envelhecimento. EUROSTAT (2021). https://ec.europa.eu/eurostat/statisticsexplained/index.php?title=Population_structure_and_ageing/pt&oldid=510113. Last accessed 20 December 2021
2. PORDATA: Indicadores de envelhecimento segundo os Censos. PORDATA (2021). https://www.pordata.pt/Portugal/Indicadores+de+envelhecimento+segundo+os+Censos++-525. Last accessed 20 June 2022
3. Berger, L., Mailloux-Poirier, D.: Pessoas Idosas: Uma abordagem global. (Edição revista e corrigida). Lusodidacta (1995)
4. Silva, L., Oliveira, F., Martins, I., Martins, F., Garcia, T., Sousa, A.: Avaliação da funcionalidade e mobilidade de idosos comunitários na atenção primária à saúde. Revista Brasileira de Geriatria e Gerontologia **22**(5), 1 (2019). https://doi.org/10.1590/1981-22562019022.190086
5. World Health Organization: Falls. WHO2018. https://www.who.int/news-room/fact-sheets/detail/falls
6. Venâncio, B., Almeida, A., Filipe, M.: O impacto económico da prevenção de quedas em idosos: uma análise custo-utilidade à intervenção das Equipas de Cuidados Continuados Integrados. Jornal Brasileiro de Economia da Saúde **11**(1), 34–41 (2019). https://docs.bvsalud.org/biblioref/2019/07/1005698/jbes-111-art-05.pdf
7. Santos, J., Arreguy-Sena, C., Pinto, P., Pereira, E., Alves, M., Loures, F.: Social representation of elderly people on falls: structural analysis and in the light of Neuman. Rev Bras Enferm **71**(suppl 2), 851–859 (2017). https://doi.org/10.1590/0034-7167-2017-0258
8. Coimbra, V., Marques, E., Chaves, C.: Prevalência de Quedas em Idosos Residentes numa Comunidade Rural. Millenium **2**(3), 109–116 (2019). https://doi.org/10.29352/mill0203e.09.00218
9. Zeytinoglu, M., Wroblewski, K., Vokes, T., Huisingh-Scheetz, M., Hawkley, L., Huang, E.: Association of loneliness with falls: a study of older US adults using the national social life, health, and aging project. Gerontolology Geriatric Medicine **7**, 1–7 (2021). https://doi.org/10.1177/2333721421989217

10. PORDATA: Índice de Envelhecimento: Onde há mais e menos idosos por 100 jovens?. Disponível em (2020). https://www.pordata.pt/Municipios/%C3%8Dndice+de+envelhecimento-458. Last accessed 20 June 2022
11. Guarda Nacional Republicana: Operação Censos Sénior 2021 – Balanço. GNR (2021). https://www.gnr.pt/comunicado.aspx?linha=4625. Last accessed 20 June 2022
12. Peek, S., et al.: Understanding changes and stability in the long-term use of technologies by seniors who are aging in place: a dynamical framework. BMC Geriatr **19**(236), 1–13 (2019). https://doi.org/10.1186/s12877-019-1241-9
13. Luna-Perejón, F., Muñoz-Saavedra, L., Civit-Masot, J., Civit, A., Domínguez-Morales, M.: AnkFall – falls, falling risks and daily-life activities dataset with an ankle-placed accelerometer and training using recurrent neural networks. Sensors **21**(5) (2021). https://doi.org/10.3390/s21051889
14. Wang, X., Ellul, J., Azzopardi, G.: Elderly fall detection systems: a literature survey. Frontiers in Robotics and AI **7**, 71 (2020)
15. Wang, Z., Ramamoorthy, V., Gal, U., Guez, A.: Possible life saver: a review on human fall detection technology. Robotics **9**(3), 55 (2020)
16. Ramachandran, A., Karuppiah, A.: A survey on recent advances in wearable fall detection systems. BioMed research international (2020)
17. Bet, P., Castro, P.C., Ponti, M.A.: Fall detection and fall risk assessment in older person using wearable sensors: a systematic review. Int. J. Med. Informatics **130**, 103946 (2019)
18. Li, Q., Stankovic, J.A., Hanson, M.A., Barth, A.T., Lach, J., Zhou, G.: Accurate, fast fall detection using gyroscopes and accelerometer-derived posture information. In: 2009 Sixth International Workshop on Wearable and Implantable Body Sensor Networks, pp. 138–143. IEEE (2009 June)
19. Igual, R., Medrano, C., Plaza, I.: Challenges, issues and trends in fall detection systems. Biomed. Eng. Online **12**(1), 1–24 (2013)
20. Kim, N.H., Yu, Y.S.: Fall recognition algorithm using gravity-weighted 3-axis accelerometer data. J. Inst. Elect. Info. Eng. **50**(6), 254–259 (2013)

Design of the Content Part of the Information System in Smart Cities from the Perspective of Regional Governments and Security of Residents

Hana Svecova(✉)

Faculty of Security Management, The Police Academy of the Czech Republic in Prague, Lhotecka 559/4, 143 01 Prague, Czech Republic
svecova@polac.cz

Abstract. Smart cities and smart regions are currently very topical and often discussed topics. Modern technologies are used in the smart cities strategy, primarily intended to improve the quality of life of residents in cities, especially in connection with the security of citizens.

These smart technologies focus on efficiently using resources with the least possible financial costs and the highest benefits for residents. However, the insufficiently resolved implementation and fragmentation of Smart Cities strategies do not provide citizens, cities and municipalities with comprehensive information, which also includes possible problems related to the safety of residents. Cities and municipalities understand the Smart Cities strategy differently, which creates inconsistencies in the implementation of Smart Cities, and there is a non-professional integration of Smart Cities in the Czech Republic.

It is therefore appropriate that information systems (IS) for Smart Cities be designed in a uniform, coordinated manner from the position of higher territorial self-governing units in the form of IS.

These will provide information for citizens, cities and municipalities, which will guarantee a unified, comprehensive integration (implementation) of IS for Smart Cities at the level of regional municipalities, and at the same time, will also contain modern technology/elements for ensuring public order to increase the safety of residents. Therefore, the article's main goal is to analyse and design the content part of IS, focusing on information technology, IoT and smart applications implemented in Smart Cities.

Keywords: Cities · Informational system · IRS · Internet of Things (IoT) · Smart region · IS

1 Introduction

Smart Cities represent the use of different technologies and approaches in many areas, which are often diverse and unique to a given location. For example, each Cities or

M. Ben Ahmed et al. (Eds.): SCA 2022, LNNS 629, pp. 185–195, 2023.
https://doi.org/10.1007/978-3-031-26852-6_17

municipality has a different population, financial budget, area, and overall opportunities for integrating information technology (IT) implemented in Smart Cities.

When selecting and designing implemented IT solutions in Smart Cities, it was necessary to focus on several factors, including defining the area of benefits for the Cities or municipality. The European Commission recommends that the following areas be identified and prioritised in this broad area [1]:

- sustainable urban mobility;
- sustainable construction and housing;
- integrated infrastructures and processes in energy, information and communication technologies and transport;
- citizen focus;
- policy and regulation;
- integrated planning and management;
- knowledge sharing;
- baselines, performance indicators and metrics;
- management of open data;
- standardisation;
- business models, public procurement and financing.

Following the framework of the Innovation Strategy of the Czech Republic, it follows that the use of services to increase efficiency as a reliable source of integrated planning and management according to EU priorities is a significant and action-packed factor in Smart Cities [2]. For the information communicated to citizens and officials to be used effectively for planning and management, the information must be published at the appropriate level through the IS.

The article focuses on the analysis and comparison of existing IS Smart Cities, which are operated by cities and municipalities in the Czech Republic, to design a unified area of IS for Smart Cities operated by regional governments to create a comprehensive IS for Smart Cities for cities, municipalities and citizens. Based on the analysis and comparison according to predetermined criteria (Smart Cities concept), designing the content part of the IS pro-area IT, IoT and Smart solutions was possible.

For the above reasons, the article was structured into chapters:

1. Analysis of IS in Smart Cities in the Czech Republic - this part focuses on analysing existing IS Smart Cities in the Czech Republic operated by cities and municipalities and compares partial parts (selected criteria) according to the Smart Cities strategy.
2. Design of the content part of IS for Smart Cities - this part follows the outputs of the analysis and comparison in the chapter entitled Analysis of IS in Smart Cities and addresses the design of the content part of IS for Smart Cities with a focus on IT, IoT and intelligent technologies.
3. Smart solutions used in Smart Cities - this section summarises the issue of intelligent technologies that can be implemented in cities and towns.

A unified concept of IS Smart Cities was addressed in a study, which aimed to analyse the IS used in Smart Cities, which could be used for everyday use by citizens [3]. Furthermore, with the growing urbanisation and expansion of the Smart Cities strategy, artificial intelligence is being introduced into the transport systems integral to Smart Cities [4]. Based on the above information, it can also be stated that IS in Smart Cities is solved independently, uncoordinated in many cities and municipalities, which results in fragmentation of information provided for citizens and officials, and a single IS for Smart Cities operated by regional governments would solve the problem.

First, an analysis of existing strategies and methodologies for Smart Cities in the Czech Republic was performed, followed by a characterisation of the primary goals, principles and cross-cutting themes for a comprehensive understanding of Smart Cities issues.

Furthermore, the current state of existing Smart Cities information systems (web presentations) was analysed, and criteria for analysis were proposed according to the cross-cutting themes of the Smart Cities strategy. Furthermore, the basic sub-criteria were determined according to which suitable cities involved in Smart Cities were selected.

2 Analysis of IS Smart Cities in the Czech Republic

A criterion was chosen to analyse IS in Smart Cities, which considered selecting suitable cities to analyse IS Smart Cities (size of cities according to population).

In the Czech Republic, therefore, it was a matter of choosing statutory cities according to Act No. 128/2000 Coll., On Municipalities, which contains a list of statutory cities. The list of statutory cities divided into individual higher self-governing units is shown in Fig. 1.

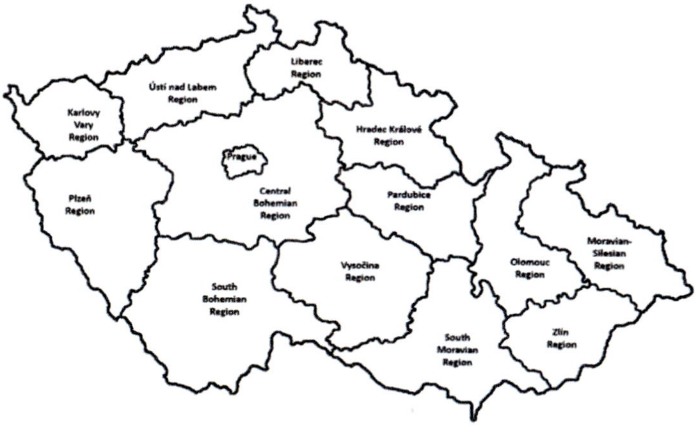

Fig. 1. Regions in the Czech Republic

The list of statutory cities in the Czech Republic is shown in Table 1.

Table 1. List of statutory cities in the Czech Republic

Regions in the Czech republic	Statutory Cities
Central Bohemian Region	Kladno, Mlada Boleslav
Hradec Kralové Region	Hradec Kralove
Karlovy Vary Region	Karlovy Vary
Liberec Region	Jablonec nad Nisou, Liberec
Moravian- Silesian Region	Trinec, Ostrava, Opava, Karvina, Havirov, Frydek-Mistek
Olomouc Region	Olomouc, Prostejov, Prerov
Pardubice Region	Pardubice
Prague	Praha
Plzeň Region	Plzen
South Bohemian Region	Brno
South Moravian Region	Ceské Budejovice
Ústí nad Labem Region	Teplice, Decín, Most, Usti nad Labem, Chomutov
Vysočina Region	Jihlava
Zlín Region	Zlin

The strategy of smart cities is divided into several categories, which are analysed and researched in this work. The categories were chosen with strategies approved by the Government of the Czech Republic entitled "SMART Cities Concept - Resilience through SMART Solutions for Municipalities, Cities and Regions" [6].

This strategy implements more innovation in the Czech Republic to benefit municipalities, cities and regions, especially citizens. It is pretty clear from the presented strategy that it unifies the fragmentation and various existing Smart Cities strategies in the Czech Republic that have emerged in the last five years. SMART Cities strategy - resilience through SMART solutions for municipalities, cities and regions defines the primary goal of implementing new technological tools and innovative solutions in municipalities, cities and regions.

However, the strategy addresses the issue of cities and addresses the issue of Smart Regions. An example of the Smart region can be demonstrated in the Moravian-Silesian region, where this issue is addressed very actively. Deputy for Smart Cities has been established at the Regional Office of the Moravian-Silesian Region [7].

SMART Cities strategy - resilience through SMART solutions for municipalities, cities and regions defines the main goal in the implementation of new technological tools and innovative solutions in municipalities, cities and regions so that it is possible to achieve quality living conditions, good environment and strengthen the competitiveness of the Czech Republic in international competition [8].

We can characterise Smart Cities as a concept of strategic Cities management, resp. Municipalities or regions. The primary goal of Smart Cities is to ensure a quality of life for the inhabitants, where modern technologies are used as tools to influence the quality of life in the Cities and, subsequently, to achieve the economic and social goals of the Cities. At the same time, there are synergies between the various activities and public services that make the Cities work - especially transport, logistics, security, energy, building management, etc. The Smart Cities strategy emphasises both "hard" and "soft" aspects of Cities life management and matches the "grey" and "green" infrastructure of the Cities [9]. The Smart Cities strategy [10] consists of three pillars: People and Communities, Local Economy, and Living Environment. The first pillar is called "People and Communities (Resilient Cities, Municipalities and Regions)" and includes four sections: 1. Education as the basis of the quality of life, 2. Affordable social and health services, 3. Social resilience, 4. Quick reaction.

The second pillar is called "Local Economy (Competitive Cities, Municipalities and Regions)" and includes four parts: 1. Entrepreneurship is a natural part of community, Cities and region life, 2. Citizens and municipalities as partners of energy suppliers, 3. Raw materials and recyclates in the circular economy, developing the bio-economy, 4. ICT infrastructure is an essential prerequisite for the success of digitisation.

The third pillar is called "Environment for life (Green cities, municipalities and regions)" and includes four parts: 1. Environment for quality of life in municipalities, cities and regions, 2. Nature and landscape supporting resilience, 3. Green and blue infrastructure of cities and municipalities of the future, 4. Transport infrastructure and mobility adapted to people. The graphic design of all three pillars is shown in Fig. 2.

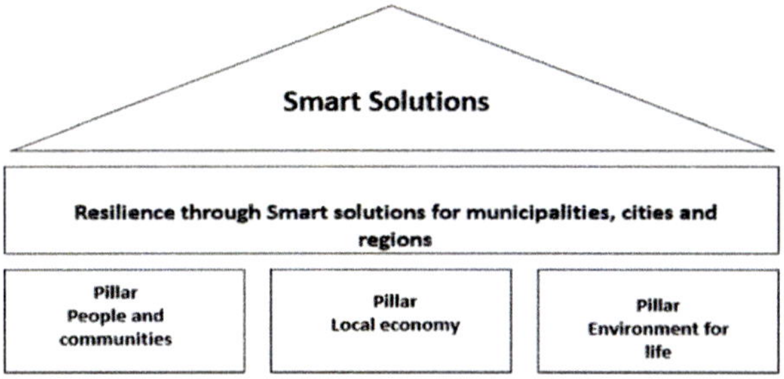

Fig. 2. The basic pillars of the smart cities concept

Intelligent solutions are based on the conceptual approach of regional and municipal governments. These approaches include innovations in existing technological tools, processes, products and services, and just for practical use, the Smart Cities Strategy - resilience through intelligent solutions for municipalities, cities, and regions has defined the basic principles, which are: the principle of change of direction, the principle of one solution with several effects, the principle of short distances, the principle of cooperation and financial sustainability to achieve the solution's effectiveness, the principle of cohesion and complementarity, horizontal and vertical interconnection, the principle of justification-based solutions, the principle of justification-based solutions [1]. The area of inhibitant security is described in the People and Communities pillar, in the section entitled Rapid Response - an integrated approach to resilience. This section develops the objectives:

- Integrated security at the vertical and horizontal level
- Cities, municipalities and regions, organisations established by them and their cooperating associations are technically equipped as best as possible to deal with disasters of anthropogenic and natural origin and other high-impact incidents (terrorist acts, environmental security risks, such as damage to people and the environment) environment toxins).
- There is a systematic and effective integration and sharing of extracted data between municipalities, cities, regions and their organisational units, testing of innovative security technologies and approaches.
- Coordination of municipal, municipal and regional authorities with responsible authorities reduces the risks of disruption of critical infrastructure elements, interruptions in the supply of strategic raw materials or energy.

The comprehensive implementation of the Smart Cities strategy and the integration of individual entities (cities, municipalities, regions) in the Czech Republic is shown in Fig. 3.

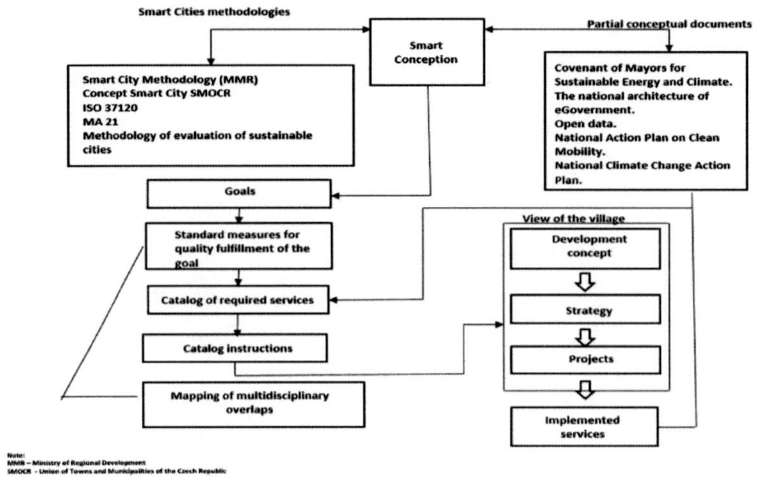

Fig. 3. Comprehensive view of the smart solution, taking into account the specific requirements of the municipality, Cities or region

After analysing and comparing the existing Smart Cities IS operated from the statutory cities, it can be stated that most cities' Cities websites are not ready for the arrival of Smart Cities, and citizens are informed about projects that are implemented or are implemented in cities and municipalities mainly from the media. Or from random articles. Web presentations in the form of information systems contain and display fragmented on-uniform information, often on an already unsatisfactory and unsupported solution from the application point of view.

The Cities of Brno does not operate any information system (web presentation for Smart Cities). Therefore, the only available information dealing with Smart Cities is the Czech Smart Cities Cluster2. The website contains general information such as memoranda, methodologies, and working groups but does not provide clear information for citizens, business entities, or institutions (municipalities, cities).

In České Budějovice, it focuses primarily on sustainable urban mobility, which includes support for non-motorised and low-emission transport, shared transport, increasing the safety of traffic participants, traffic education, and the creation of safe routes for cyclists and pedestrians. In addition to transport, the development of public spaces is also current [11].

Smart Cities for the Cities of Ostrava are healthyOVA7, which informs about the current air quality in individual parts of Ostrava (updated daily). It is also possible to find articles on water, energy, nature, urban transport and waste on the website [13].

The second website that Dopravní podnik Ostrava is Smart Cities Center8, which provides information on exclusions, search options and locations of individual public transport vehicles and the current status of used vehicles (whether it has air conditioning, are ecological, etc.) [14]. The Cities of Jihlava has an IS for Smart Cities focused primarily on mobility, more specifically on car transport, parking policy, public transport and intermodality, bicycle transport, pedestrian transport, primary communication network, urban logistics and freight transport. In addition, mobility is closely linked to the ecology in which Jihlava deals with clean vehicles and public space.

Smart Cities website for the Cities of Prague contains all the categories that Smart Cities should include and is in line with the above-mentioned cross-cutting themes of the Smart Cities strategy, such as future mobility, intelligent buildings and energy, waste-free Cities, attractive tourism, people and urban environment and data area. The opportunity to develop an idea for another project that could take place in the Cities [14].

Based on the performed analysis and subsequent comparison, it can be stated that the problem in the form of unified coordination from the level of regional self-governments has not been solved yet, which shows the current situation shown in Table 2.

Table 2. Readiness of statutory cities for the implementation of IS for smart cities.

Regions in the Czech Republic	Number of cities	Readiness of statutory cities for the implementation of IS for smart cities
Central Bohemian Region	2	No
Hradec Kralové Region	1	No
Karlovy Vary Region	1	Yes
Liberec Region	2	No
Moravian- Silesian Region	6	Partly
Olomouc Region	3	Partly

(continued)

Table 2. (*continued*)

Regions in the Czech Republic	Number of cities	Readiness of statutory cities for the implementation of IS for smart cities
Pardubice Region	1	No
Prague	1	Yes
Plzeň Region	1	Yes
South Bohemian Region	1	No
South Moravian Region	1	Partly
Ústí nad Labem Region	5	Partly
Vysočina Region	1	Yes
Zlín Region	1	Yes

3 Design of the Content Part of IS for Smart Cities

Based on the performed analysis and subsequent comparison of IS for Smart Cities in statutory cities in the Czech Republic, partial content areas were designed to focus on IT, intelligent technologies and IoT. Cities and municipalities in individual regions differ in size, population and, at the same time, budget or other strategic options. It was, therefore, necessary to consider and design information from two perspectives because small municipalities with a population of up to 1,000 have different options than statutory cities or regional cities. At the same time, it was necessary to divide the information in the content part into two parts: information provided for municipalities and cities and the information provided for citizens.

When designing content information, it was necessary to divide the content design for IS in Smart Cities into a part that would include technology and information for cities and municipalities and a part that would include technology and information for citizens.

The distribution of cities according to population is occupied in the Smart Cities Methodology, which the Ministry of Regional Development developed. The population is shown in Table 1 in Table 3 [9].

Table 3. Distribution of cities and municipalities according to population

Category	Population (peoples)	Note
A	Over 150 thousand	Praha, Brno, Ostrava, Plzen
B	40–150 thousand	Statutory cities
C	15–40 thousand	Smaller cities
D	5–15 thousand	Larger villages
E1	1–5 thousand	Small villages
E2	Up to 1 thousand	

One of the most sought-after solutions in Smart Cities is smart technologies, which mainly includes IoT with a focus on increasing digitisation and cyber security. IoT and smart technologies in cities and towns can be found in many areas. The most well-known solutions include:

- regulation of street lighting according to the current number of people, the intensity of sunlight or the time of day;
- regulation of traffic lights according to the presence of people, cars or according to the time of day;
- reporting the status of garbage cans according to the current amount of content for the needs of garbage collectors;
- measurement of air quality according to the values of temperature, humidity, O2, CO2, CO, N2, CH4, radioactivity, etc.;
- activation of alarm systems according to the measurement of critical values or as needed;
- fire risk reporting according to air humidity values, number of precipitation, etc.;
- activation of flood barriers according to water flow or height exceeding the limit;
- updating information signs according to traffic conditions, weather or as needed;
- interconnection of Cities information systems for more comfortable use of citizens;
- optimisation of public transport operation according to the actual number of passengers;
- optimisation of infrastructure for pedestrians and cars according to the movement of people;
- optimisation of business and marketing activities, according to the number and movement of persons, and connection of camera systems with GSM localisation for easy identification of the perpetrator.

3.1 Content Design of IS Information for All Cities and 10,000 Inhabitants and Regional Cities

Some Smart technologies that cities are implementing as part of the Smart Cities strategy are financial, personnel and organizationally expensive and smaller municipalities or cities cannot use them from the above facts. For this reason, more expensive technologies are recommended only to larger cities, mainly statutory and regional cities or cities with more than 10,000 inhabitants.

These technologies include intelligent parking, CCTV, an overview of implemented projects, actualities of individual cities, city's plans for the future, current traffic information, cities weather information, and important links for City administration.

A graphic diagram of the content part of IS Smart Cities for citizens and technologies usable for regional cities and towns with a population of over 10,000 municipalities is shown in Fig. 4.

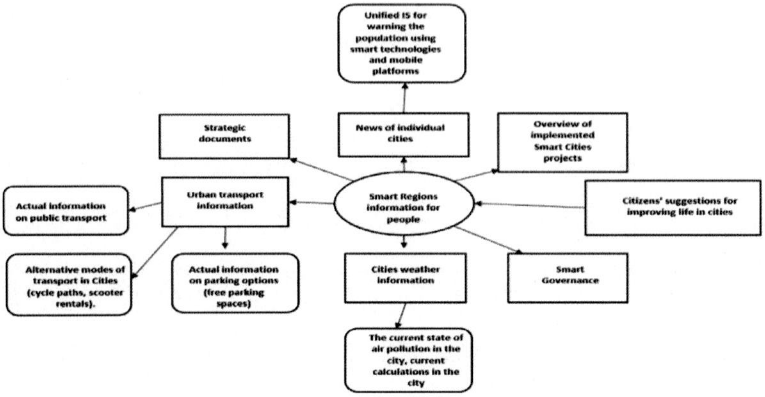

Fig. 4. Information for people

Other sub-sections that could be provided to citizens, cities and municipalities could be the following sections: a complete overview of implemented Smart projects in the cities, news of individual cities, news about intelligent solutions in cities - examples from other cities, the possibility of sharing inspiration for the cities, the ability to communicate important information to citizens quickly, cities plans for the future, information on transport in individual cities, occupancy indicator of individual roads and intersections in the cities, cities weather information, important links to the Cities administration, possibility to use the visit to the office electronically, an opportunity for citizens to submit their idea for the project's solution in Smart Cities.

4 Conclusion

The article focused on the analysis, comparison, and existing IS Smart Cities in the Czech Republic and further on the design of the content part of IS for Smart Cities with a focus on information technology.

Within the framework of the analysis, the existing IS of cities in the Czech Republic were analysed and compared. When comparing the analysed IS, a criterion with a breakdown by local government units was chosen. As a result of the analysis and comparison, it was found that web presentations in the form of information systems contain and display fragmented and inconsistent information, often based on already inadequate and unsupported solutions from the application point of view. Furthermore, it was found that there is no unified concept for implementing Smart Cities that would support all cities or regions.

After analysis and comparison, subcontinents for an IS that would be operated from the level of the regional governments were proposed.

Based on the above proposal of subcontinent areas, a coordinated IS could be created that would provide information for cities, municipalities and citizens, displaying information for cities, municipalities and citizens, especially for the population's security.

The county would provide cities and municipalities with a unified IS concept for Smart Cities. Cities and municipalities would provide citizens with information on important events, especially public safety in emergencies. In the future, it would be optimal for the implementation of the Smart Cities strategy to be supported by the level of regional governments. Smart Cities would be comprehensively integrated into the state-authority cities or smaller cities of higher territorial self-government units, which could, for example, lead to the unification of intelligent parking software.

References

1. Smart cities European Commission-European Commission. https://ec.europa.eu/info/eu-regional-and-urban-development/topics/cities-and-urban-development/Cities-initiatives/smart-cities_en. Accessed 10 Nov 2021
2. Inovační strategie České republiky 2019–2030, https://www.vyzkum.cz/FrontClanek.aspx?idsekce=866015. Accessed 22 Feb 2022
3. Mishbah, M., Sensus, D.I., Noprisson H.: Information system implementation in smart cities based on types, region, sub-area. In: International Conference on Information Technology Systems and Innovation (ICITSI) (2017)
4. Agarwal, P.K., Gurjar, J., Agarwal, A.K., Birla, R.: Application of artificial intelligence for development of intelligent transport system in smart cities. Int. J. Transport. Eng. Traffic Syst. **1**(2) (2015)
5. Jiang, D.: The construction of smart Cities information system based on the Internet of Things and cloud computing. Comput. Commun. 158–166 (2020)
6. SMART Cities concept - resilience through SMART solutions for municipalities, cities and regions. https://mmr.cz/getattachment/3ffecf72-c230-43f6-8c80-b84956fb215d/Koncepce-Smart-Cities-odolnost-prostrednictvim-SMART-reseni-pro-obce,-mesta-a-regiony.pdf.aspx?lang=cs-CZ&ext=.pdf. Accessed 20 Nov 2021
7. Ing. Jakub Unucka, MBA Moravian-Silesian Region. https://www.msk.cz/cs/kraj/zastupitelstvo/ing--jakub-unucka--mba-3277. Accessed 10 Nov 2021
8. 9/21 SMART Cities concept - resilience through SMART solutions for municipalities, cities and regions. https://www.komora.cz/legislation/9-21-koncepce-smart-cities-odolnost-prostrednictvim-smart-reseni-pro-obce-mesta-a-regiony-t8-2-2021/. Accessed 10 Nov 2021
9. Methodology Smart Cities. https://mmr.cz/getmedia/f76636e0-88ad-40f9-8e27-cbb774ea7caf/Metodika_Smart_Cities.pdf.aspx?ext=.pdf. Accessed 10 Nov 2021
10. Cities plan – Ceske Budejovice. http://ceskebudejovice.dobramesta.cz/plan-mesta. Accessed 10 Nov 2021
11. Projects - Innovation Center of the Usti Region. https://icuk.cz/projekty. Accessed 10 Nov 2021
12. ZdravaOVA. https://zdravaova.cz. Accessed 10 Nov 2021
13. Transport Company Ostrava (DPO). https://scc.dpo.cz/dashboard/index.html. Accessed 10 Nov 2021
14. Smart Prague. https://www.smartprague.eu/o-smart-prague. Accessed 10 Nov 2021

Security Approaches for Smart Campus

Srhir Ahmed[1](✉) and Tomader Mazri[2]

[1] Ibn Tofail University - Kénitra-Maroc, Kénitra, Morocco
ahmed.srhir1@uit.ac.ma

[2] National School of Applied Sciences of Kenitra, P.B: 242, 14000 Kénitra, Morocco

Abstract. This paper describes the development of Security Approaches for smart campus using different Methodologies and introduce a global strategy for secure Architecture Especially Secure Internet of Things (IoT) technology. Our main goal is to enhance campus security, minimize costs, and advance towards a smart university campus. We provide not only a methodology that would act as an instant responder, but also provide insight into the evolution of security research for smart campus, therefore with the development of architectures, the more changes are happening, the more we must know the risks we can face, the obligation of security and confidentiality is an essential requirement.

Keywords: Smart campus · IoT sensors · Smart campus · Security · Threats · Attacks · Smart system · Ecosystem · Vulnerabilities · IOT

1 Introduction

The concept of smart campus has recently received a lot of attention from many researchers due to the valuable knowledge gained in smart campus development. In theory, the university campus is a small city that provides different services to different users. There are Many factors motivate researchers to study the smart campus, including the provision of higher quality services, environmental protection, and cost reduction.

The smart campus is an ecosystem, and associated applications, utilities, and use cases, among the key standards that are linked directly to smart campus in particular [1, 2]:

- Education
- Energy
- Urban planning
- Healthcare
- Transport

The Internet of Things is a core part of the smart campus, and it is inevitable that it will be invoked. [3] The IoT (Internet of Things) is a paradigm of communication that has earned its tenacity through its ability to connect a variety of daily reality objects to the internet. These objects comprise (sensors, safety, alarms, drones, robots, home

M. Ben Ahmed et al. (Eds.): SCA 2022, LNNS 629, pp. 196–205, 2023.
https://doi.org/10.1007/978-3-031-26852-6_18

devices, smart grid systems, office equipment, and more...). While the IoT is still in its infancy, there are a lot of applications and standardization that needs to be implemented including: home automation, water and waste control, traffic control, smart vehicles, smart grids, etc. [4].

The IoT is implicated everywhere in the smart campus, including but not limited to:

- Safety
- Classroom Technologies
- Residential Services
- Facilities
- Electrical Systems
- Physical and Mental Health

So, to ensure the security of smart campus it's very important to secure in a first time the IOT technologies used or deployed, this paper provides an overview of security Methodologies threats/risks and preventative measures recommended for a best security, [5–7] among the approaches involved to Secure Smart Campus We worked according to three methods:

1. E2E Trust & Security (E2ET&S)
2. Distributed Big Data & Analytics (DBDA)
3. Internet of Things (IoT)

2 Smart Campus Architecture

The Smart Campus is one of the innovations to be developed at Ibno Tofail University. Among the major aspects considered in the development of the Smart Campus infrastructure; therefore, we are presenting a technical architecture for the Smart Campus and suggesting a local operational model.

To properly secure our campus, we need to implement a convergence infrastructure [8] that covers all aspects and meets the security requirements:

The smart campus architecture described in Fig. 1 defines three fundamental aspects of a smart campus:

1. Smart education: A system which consists of e-learning, Radio Frequency Identification (RFID) tags, virtual classroom.
2. Smart parking: [9–11] a parking system that provides location tracking information for vehicles and also delivers information about the parking space are availability.
3. Smart Building: A system that provides information about building temperature monitoring systems, power systems and illumination systems.

3 Related Work

At present, many research projects have been proposed to develop different approaches to Secure the Connection Between the end device and her components, Security deployment in smart environments is a difficult issue due to their diverse and complex nature

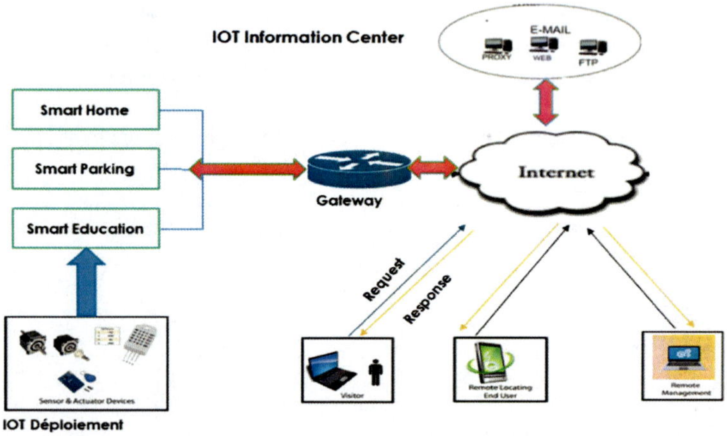

Fig. 1. Smart campus system design

[4–6] environments are composed of various devices with different mechanisms and prerequisites such as evolutivity, scalability, conviviality and level of protection. Also, the implementation of security varies from device to device.

As mentioned earlier that the smart campus has a heterogeneous nature, in many projects, a considerable security mechanism needs to be implemented on the smart environments which is based on four basic security mechanism objectives [1] which are:

1. Authentication,
2. Confidentiality
3. Integrity
4. Availability

in order to counter the security threats, Regarding the protection of system data by IoT devices, Tiburski [8] developed a security architecture for Service-0riented Architecture (SOA) based IoT devices. Middleware, Other proposed research projects include the development of a threat identification model as a first step in risk assessment in the health information domain. System (HIS)Shahri [12], With respect to The fundamental security mechanism monitoring and maintenance systems through IoT devices Florence D. Hudson and Phillip A. Laplante [13] define mechanism to aid in achieving a secure environment for the Internet of Things by developing a TIPPSS framework, a set of recommendations in the context of higher education campus IoT deployments that can be enforced and applied in a standard, uniform way. A TIPPSS framework would include defining vendor requirements, providing a structured approach to identify risks, and informing the lifecycle approach to IoT security risks management. A TIPPSS framework would apply to all levels of devices, applications, data, networks, and users [13, 14]:

- **Trust**: user, data, application, and device
- **Identity**: device, users,

- **Privacy**: application, data, and user

4 Proposed Method

A smart campus collects data from smart devices and sensors that are embedded in it. It exchanges these data via an intelligent communication system that is usually a combination of wired and wireless infrastructure. Then it uses intelligent software engines to make smart decisions to improve service performance [15].

Cybersecurity for a smart campus means the protection of the data, systems, and infrastructure responsible for the operations of the campus and the stability and livability of its inhabitants. That why we should implement a process approach to secure the communication between the end devices and Internet with an isolation for the Distributed Big Data for this purpose We worked according to an approach which respects the different points below:

- **Network access security**: Provide a secure access by using several methods of filtering to the service by the user.
- **Network Domain Security:** Protect the network elements as servers, e-node B or routers and secure the signaling and user data exchange.
- **User Domain Security:** Control the secure access between users and mobile stations.
- **Application Domain Security:** Development of secure applications layer communications.
- **Visibility and configuration of security:** Allows the user to check whether the safety functions are in use.

The U.S. National Intelligence Council (NIC) considers that technological advances, combined with strong market demand, would drive widespread adoption and deployment of IoT [16]. Nevertheless, the biggest fear is that everyday objects will become potential security risks. Worse, widespread IoT penetration would propagate these threats far more widely than today's Internet [17, 18].

Our approach consists in defining three parts of the security Fig. 2, thereafter we will study each part according to specified axes and according to each field.

4.1 Technology Protection

As described in the previous section, which defines our Intelligent Campus Architecture separated into three as seen below:

- Smart Home
- Smart Parking
- Smart Education

For the part of technology protection, we break it down into two parts Network Security and data Security with taking into consideration all the aspects of our smart campus, our proposed architecture consists in securing all the incoming traffic on our

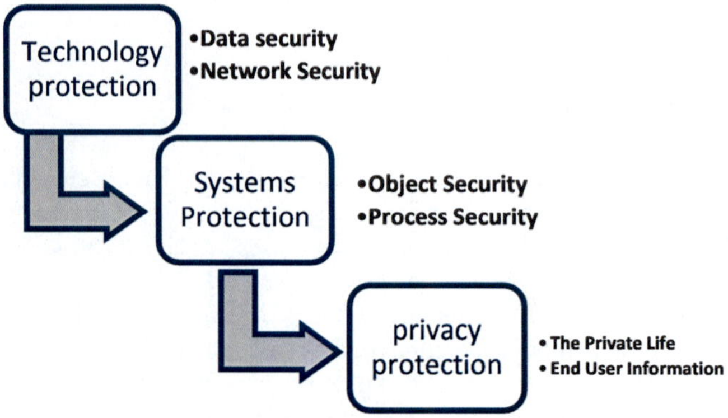

Fig. 2. Security approaches for Smart Campus

end device, either by implementing a Firewall or a Secured virtual private network (Vpn) Site-to-Site communication, and to reinforce the security against malicious attacks we will deploys an intrusion prevention system (IPS).

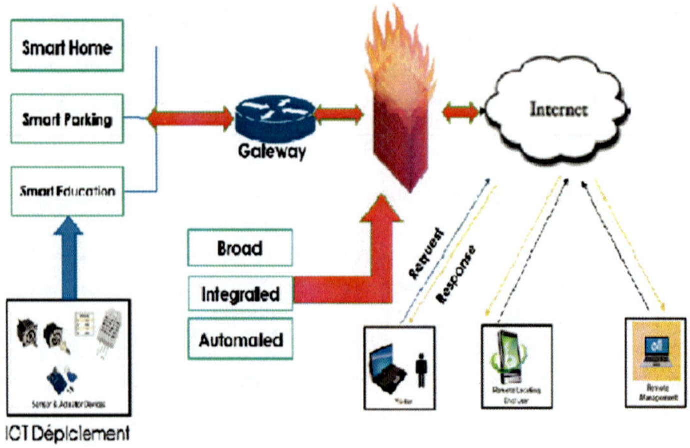

Fig. 3. Secure smart campus architecture

As seen in Fig. 3 we have secured the only entry point that links the internet service with our devices and sensors in our Smart campus. Our approach is to set up a new generation firewall (NGFW)based on three aspects [19, 20], to reinforce the security and reduce the threat from the Iot devices we have also employs many technologies as described in the Fig. 4:

- Broad: Visibility and protection of the digital attack including converged networking and security solutions at the endpoint, network, and cloud. It provides high-performance connection, real-time coordinated threat detection, and policy enforcement across the digital attack surface and lifecycle.

- Integrated: Detection of Advanced threads and the activation of Self-Adaptive Intrusion Detection System (IDS) in each for each rule created in NGFW or an intrusion prevention system (IPS) that permanently scans a network for anomalous network activity and acts to prevent it, including notification, blocking, or aborting it, when it occurs, the implementation of NGFW technologies is incorporated with centralized analysis and automated prevention powered by artificial intelligence (AI). This provides co-efficient and consistent security and simplifies operations across different technologies, locations, and deployments.
- Automated: Response and continuous trust Assessment, to accelerate prevention and operational efficiency, it is a contextual, self-healing security and network posture that leverages cloud computing and advanced artificial intelligence to automatically provide coordinated, nearly real-time.

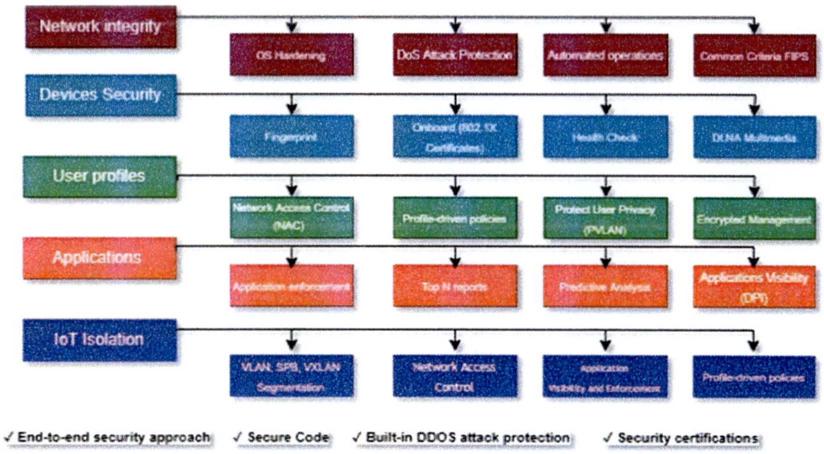

Fig. 4. Security layered approach

After securing the infrastructure and communication between the internet service provider (ISP) and our smart Campus platform as seen in the setwork security section, we will start securing the data as defined in our Approach in the Technology Protection Section, the basic concept is to combine a normal database for storing sensor data with unique reference identifiers on the Application Programming Interface (API) [21]. When the centralized database receives new inputs, a pre-programmed condition triggers that batch of data to be hashed, based in the Fig. 5 that describes how this solution works.

Finally, all data can be audited for whether the hash on the public matches the hash on the file in the centralized database - providing clear evidence of data-tampering and for checking the hashes we should develop a class that can contain all the operations

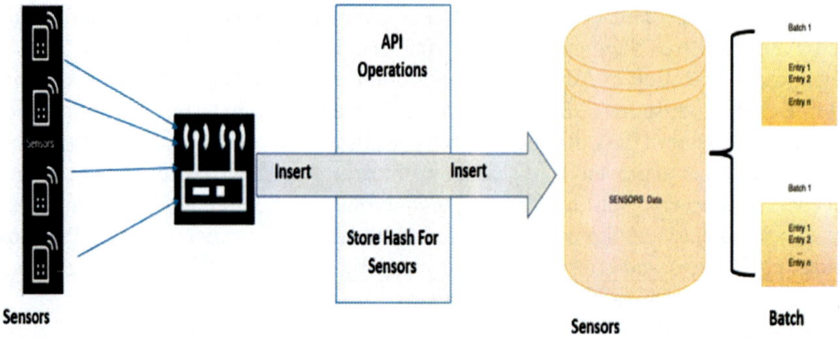

Fig. 5. Database security approach

needed to check that the hashed data of the database and verify that is the same to the hash that was previously calculated and stored on the API.

4.2 System Protection

The protection of interconnected [22] systems and hosted IoT objects concerns the protection of the objects themselves delivered to these systems and the processes they control.

the confidentiality of the data must be accessible not only to approved users but also to approved intelligent devices the data availability and integrity must be guaranteed, the security of the network must be ensuring user confidentiality and preventing the possibility of eavesdropping attacks on the communication channel as well as security at the hardware level must be used to prevent the execution of malicious components or sequences of instructions, Many of the major security considerations for the existing IT systems are similarly applicable to IoT systems, as outlined below:

- Trust management: Trust management is an essential and fundamental aspect to guarantee the trust of devices, entities and data.
- Access control: An access control should be provided, to determine who must be authorized to access smart objects or the data they generate by them This control is the identification of smart object owners, the authentication and the subsequent approval of access.

4.3 Privacy Protection

The protection of persons will concern the protection of the privacy of users [22, 23] which implies, in addition to technological solutions, an appropriate regulation that establishes responsibilities in case of conflicts.

Privacy is a major aspect of the IoT field because of the nature of the data and information involved in it, it's defining the rules by which data about individual users can be accessed [24]. The quantity of data gathered by IoT sensors is incredibly extensive and extremely private in nature; it can be related to private and personal activities,

Global Positioning System (GPS) locations by recorders, and recorded daily activities by connected individuals, Privacy considerations can be divided into two categories: those related to data privacy and those related to access privacy:

- Protection of personal data
- Protection of sensitive data
- Protection of enterprise data
- Data access control

The security of the network must be guaranteed by providing privacy to the user and preventing eavesdropping on the communication channel.

5 Conclusion

The smart campus is a physical or digital environment in which humans and technological systems interact to create more immersive and automated experiences for the different stakeholders of the university. It is an environment that aims to improve the experience of students and staff, ensuring a better quality of services and reliable energy efficiency.

Universities combine a wide infrastructure, including smart devices and modern technologies, that allows them to transform it into a digital space to support the concept of smart campus.

A general overview of the smart campus environment and the communication technologies used has been presented by defining a global smart campus architecture in this paper as well as approaches for securing this environment. In addition, smart campuses are exposed to several security vulnerabilities and attacks through the exploitation of their various component and communication technologies.

These security issues can put at risk the privacy and sensitive data of the university campus.

It is therefore paramount to implement certain security solutions and measures based on the main objectives and good security practices, including authentication, confidentiality, integrity, and availability, in order to successfully circumvent some of the problems and allow the strengthening and implementation of an effective and robust security system for smart campuses.

Among the main security mechanisms to secure networks and data, we have decomposed our campus into several parts for security enhancement as defined in Sect. 4:

1. Technology Protection: We break it down into two parts. Network Security and Data Security by taking into consideration all aspects of our smart campus, which consists of securing all incoming traffic to our terminal, either by implementing a firewall or secure VPN site-to-site communication
2. System Protection: this refers to the security of the objects delivered to these systems as well as the processes they control.
3. Privacy Protection: the protection of users' privacy

Each smart campus needs an adapted and structured computer model and a general architecture that encompasses all the parts involved in the technologies and communication equipment as well as the exposure to several vulnerabilities and security attacks through the exploitation of its different components. It is necessary to study its anomalies and vulnerabilities and circumvent them through reliable security solutions, as well as the implementation of a security strategy and the awareness of the end-users of their risks and criticality.

References

1. Abdullah, A., Thanoon, M.: Anwar AlsulamiToward a Smart Campus Using IoT: Framework for Safety and Security System on a University Campus
2. Hannan, A., et al.: Disaster management system aided by named data network of things: architecture, design, and analysis. Sensors **18**, 2431 (2018)
3. Lee, N., Lee, H., Lee, H., Ryu, W.: Smart home Web of Object Architecture International Conference on Information and Communication Technology Convergence 1200–1216 (2015)
4. Mantas, G., Lymberopoulos, D., Komninos, N.: Security in smart home environment, Wireless Technologies for Ambient Assisted Living and Healthcare: Systems and Applications, Medical Information Science, Hershey, PA, pp. 170–191 (2010)
5. Hager, M., Schellenberg, S., Seitz, J., Mann, S., Schorcht, G.: Secure and QoS-aware communications for smart home services. In: 2012 35th Int. Conf. Telecommunications and Signal Processing (TSP) (2012)
6. Zhai, X., Dong, Y., Yuan, J.: Investigating learners' technology engagement—a perspective from ubiquitousgame-based learning in smart campus. IEEE Access **6**, 10279–10287 (2018)
7. Abomhara, M., Ien, G.M.K.: Cyber security and the internet of things: vulnerabilities, threats, intruders and attacks. J. Cyber Secu. Mobi. **4**, 60–90 (2015)
8. https://edudownloads.azureedge.net/msdownloads/MS-SmartSecureCampus-eBook.pdf. Last accessed 10 August 2021
9. Jurva, R., Matinmikko-Blue, M.: Ville Niemelä & Suvi Nenonen Architecture and Operational Model for Smart Campus Digital Infrastructure
10. Haque, M., Ahamed, S.I.: Security in pervasive computing: current status and open issues. IJ Network Security **3**, 203–214 (2006)
11. Wang, L., Li, K., Chen, X.: Internet of things security analysis of smart campus ICCCS 2018: Cloud Computing and Security, pp. 418–428
12. Qian, L.: Establishment of smart campus based on cloud computing and internet of things. Computer Science **38**(10), 18–21 (2011)
13. Tiburski, R.T., Amaral, L.A., De Matos, E., Hessel, F.: The importance of a standard security architecture for SOA-based IoT middleware. IEEE Communications Magazine **53**(12), 20-26 (2015)
14. https://ieeexplore.ieee.org/document/8338006
15. Defining and Deploying TIPPSS for IoT (Trust, Identity, Privacy, Protection, Safety, and Security for the Internet of Things) Florence D. Hudson - Internet2
16. Song, H., Srinivasan, R.: Tamim Smart Cities: Foundations, Principles, and Applications 396, 14.4
17. Lin, Y.B., Chen, L.K., Shieh, M.Z., Lin, Y.W., Yen, T.H.: CampusTalk: IoT devices and their interesting features on campus applications. IEEE Access **6**, 26036–26046 (2018)
18. Alvarez-Campana, M., López, G., Vázquez, E.V., Villagrá, V., Berrocal, J.: Smart CEI Moncloa: An IoT-based platform for people flow and environmental monitoring on a smart university campus. Sensors **17**, 2856 (2017)

19. Farahat, I.S., Tolba, A.S., Elhoseny, M., Eladrosy, W.: Data Security and Challenges in Smart Cities. In: Hassanien, A.E., Elhoseny, M., Ahmed, S.H., Singh, A.K. (eds.) Security in Smart Cities: Models, Applications, and Challenges. LNITI, pp. 117–142. Springer, Cham (2019). https://doi.org/10.1007/978-3-030-01560-2_6
20. Deogirikar, J., Vidhate, A.: Security attacks in IoT: a survey. In: 2017 International Conference on I-SMAC (IoT in Social, Mobile, Analytics and Cloud) (I-SMAC). Presented at the 2017 International Conference on I-SMAC (IoT in Social, Mobile, Analytics and Cloud) (I-SMAC), pp. 29–40 (2017)
21. Ijaz, S., Shah, M., Khan, A., Ahmed, M.: Smart cities: a survey on security concerns. Int. J. Adva. Comp. Sci. Appl. (2016)
22. Wang, L., Li, K., Chen, X.: Internet of Things Security Analysis of Smart Campus ICCCS 2018: Cloud Computing and Security, pp. 418–428
23. Abdullah, A., Thanoon, M., Alsulami, A.: Toward a smart campus using IoT: framework for safety and security system on a university campus (2019). ASTESJ ISSN: 2415–6698
24. Wang, C., Tam Vo, H., Ni, P.: An IoT application for fault diagnosis and prediction. IEEE Int. Conf. Data Sci. Data Intensive Sys, pp. 726–731 (2015)
25. Miorandi, D., Sicari, S., Pellegrini, F.D., Chlamta, I.: Internet of things: vision, applications and research challenges. Journal of Ad Hoc Networks **10**, 1497–1516 (2012)

Assessing the Implementation of Smart Energy Efficient Street Lighting Projects Within Cities

Tebello N. D. Mathaba(✉) and Moabi K. Manyake

Department of Electrical Engineering, Vaal University of Technology, Vanderbijlpark, South Africa
tmathaba@uj.ac.za

Abstract. The deployment of smart street lighting infrastructure offers cities the benefit of reducing their operating costs, while continuing to offer the citizens the required lighting services and safety. Energy efficient lighting retrofit projects with traffic-adaptive control strategies are some of the attractive projects for cities to consider. This paper proposes a simulation-based simple methodology for assessing the financial viability of smart lighting retrofit projects in smart cities. The methodology is presented and applied on a case study street in South Africa to indicate that the proposed project would result in a payback period of 3.86 years. Sensitivity analysis indicates that project's payback period is mostly sensitive to unit cost price of the newly retrofitted luminaires and electricity price.

Keywords: Smart cities · Energy efficiency · Streetlights · Internet-of-Things

1 Introduction

Streetlights are ubiquitous in cities. They provide for safe movement of vehicular and pedestrian traffic. Due to their high number, they makeup a significant proportion of total energy usage in cities [1]. Retrofitting the currently installed old technology with new smart street lighting infrastructure presents good cost saving and sometimes income generating opportunities for cities [2].

Several technologies have been proposed to make streetlights more energy efficient. These include the use of energy-efficient (EE) streetlights based on the light emitting diode (LED) technology, photo-sensors to automatically turn lights off based on ambient lighting and using photovoltaic solar panels to generate additional energy [2,3]. Beyond these technologies, a number of control techniques have been proposed to lower the amount of energy used by streetlights, such as traffic-adaptive lighting of highways [4,5].

Figure 1 presents typical elements of a smart-street lighting infrastructure. This is an Internet-of-Things (IoT) network of sensor-nodes attached to luminaires, that can be monitored remotely over the Internet to offer multiple services [4,6]. The versatility of this new smart infrastructure enables the cities to also

M. Ben Ahmed et al. (Eds.): SCA 2022, LNNS 629, pp. 206–213, 2023.
https://doi.org/10.1007/978-3-031-26852-6_19

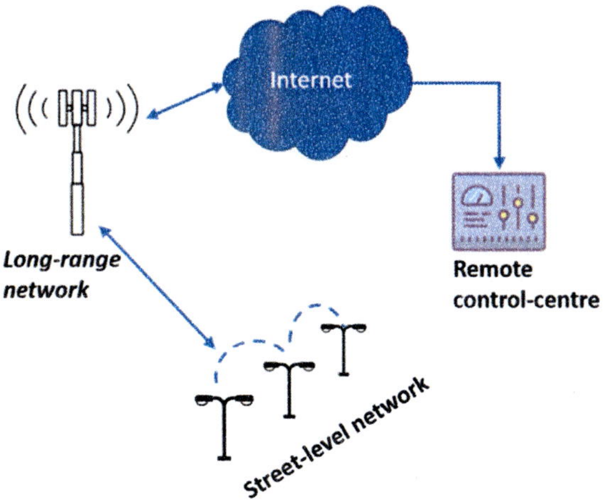

Fig. 1. Elements of smart-street lighting infrastructure

harness the in-built information and communication technologies (ICTs) to offer additional services like internet-connectivity, security surveillance and pollution monitoring [3].

Thus, there is a need for tools to enable swift project appraisal of potential smartlighting projects within cities. This paper proposed a generic methodology for the financial evaluation of retrofitting smart street-lighting in cities. The methodology is then applied to a case-study to show how a city-wide smart-lighting project assessment can be made.

The rest of the paper contains the methodology in Sect. 2, analysis of results in Sect. 3 and a conclusion in Sect. 4.

2 Methodology

Figure 2 summarizes the proposed methodology for assessing the viability of a smart street lighting retrofit project in a city.

The process begins with *Step 1* by modeling the baseline energy usage profile of the current installation of lighting. The purpose of this exercise is to determine the baseline energy usage, the energy cost of the current lights, as well as their other operating and maintenance costs. The current electricity load profile is characteristic by the average power usage, P_i^B measured at regular interval i. The project load profile is a product of the number of luminaries, N_Q, the average energy usage of a single luminaire, P_R and a modification factor f_i as shown in (1).

$$P_i^B := N_Q \cdot f_i \cdot P_R \tag{1}$$

The modification factor f_i is used to account for the average fraction of time in which the luminaries are operating, at a particular time interval i. For example, if in a given location X, the lights are turned off 70% of the time during the year at 6pm because the sun is shining, then $f_i = 0.7$ for that interval.

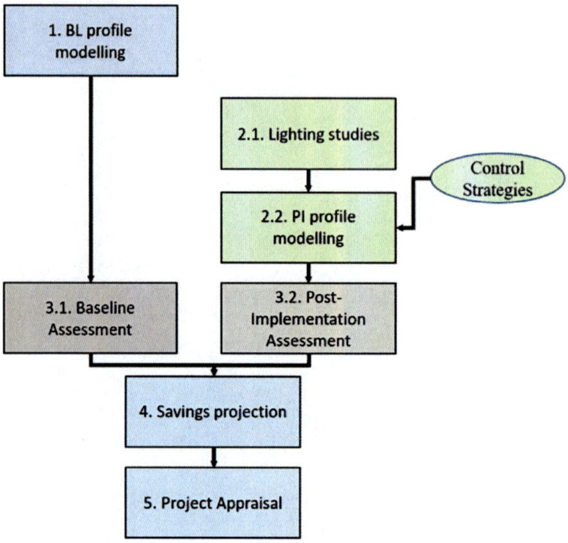

Fig. 2. Proposed methodology of assessment

In *Step 2.1*, lighting studies are performed in order to identify the most suitable and energy-efficient luminaires to retrofit. The purpose of the studies is to ensure that retrofitted lights meet the require lighting and quality requirements as per the applicable standards.

The new load profile, expected after a successful retrofit, is calculated in *Step 2.2*. This post-implementation load profile, P_i^A, will depend on the control strategy to be implemented, as shown in (2).

$$P_i^A[u_i] := N_Q \cdot f_i \cdot u_i \cdot P_R \tag{2}$$

The newly implemented control strategy will determine control inputs, u_is, that regulate the light s on-off condition and hence electricity cost. The control action in lighting application is achieved by simply dimming the lights.

In *Step 3.1* and *Step 3.2*, the baseline and post-implementation profiles are used to evaluate the monthly operating and maintenance costs. Both profiles, are used in *Step 4* to project the potential monthly energy to be saved. The energy savings, E_{sav}, are calculated using (3).

$$E_{sav} := N_Q \cdot \sum_i (f_i - f_i \cdot u_i) \cdot P_R \cdot \Delta t \cdot \tag{3}$$

In *Step 4*, the cost saving C_{sav} are based on both the baseline tariff $B_{tariff}(\cdot)$, baseline price levels ρ_i^B, post-implementation tariff $A_{tariff}(\cdot)$ and post-implementation price levels ρ_i^A. Equation (4) shows the calculation.

$$C_{sav} := B_{tarrif}(P_i^B, \rho_i^B) - A_{tarrif}(P_i^A, \rho_i^A) \tag{4}$$

Step 5: The total investment cost C_{Inv} as well as the regular operations and maintenance costs $C_{O\&M}$ need to be calculated for project appraisal. The total investment cost C_{Inv} is made up of the luminaire's unit price c_{price} and installation price per luminaire, c_{price}, as shown in (6).

$$C_{Inv} := (c_{install} + c_{price}) \cdot N_Q \tag{5}$$

$$C_{O\&M}[\lambda_{avg}] := \frac{c_{replacement}}{\lambda_{avg}} \tag{6}$$

The operation and maintenance cost depends on a cost of replacing a light, $c_{replacement}$, and the average life-time of the luminaire, λ_{avg}, as shown in (6).

Energy-efficiency project viability can be assessed using a variety of methods such as life-cycle cost analysis, internal rate of return and payback period shown in (7) [1]. In this case, the payback period is selected as a quick and easy indicator for the appraisal of the smart-lighting project.

$$N_{payback} := \frac{C_{Inv}}{C_{Sav} + C_{O\&M}[\lambda^B_{avg}] - C_{O\&M}[\lambda^A_{avg}]} \tag{7}$$

Generally, the shorter the payback period the better. According to (7), a smaller payback period is obtained from a smaller investment cost, and larger cost savings from electricity usage, operations and management. The savings on $C_{O\&M}$ depends on average lifetime of the current luminaire technology λ^B_{avg} and new technology λ^A_{avg}.

3 Results and Discussion

For a case-study example, we consider the retrofitting of smart streetlights on Andries Potgieter Boulevard in the city of Vanderbijlpark located within the Gauteng province of South Africa, shown in Fig. 3.

Applying *step 1* of the methodology assessed the area shown in Fig. 3. The stretch of the road considered is currently fitted with, $N_Q = 104$, old high pressure sodium (HPS) lamps rated at 250 W. The measured power consumption of a sample of these lights, including ballasts, gives an average wattage of 289W. The yearly sunrise and sunset data from [7] is used to model annual average modifier values f_i at a 30-min average interval for typical day.

Step 2.1 was applied with the help of lighting design software, DIALux, to establish most suitable replacement LED luminaire for the road, whilst also meeting the lighting standard [8]. The case-study street is an A3 class type of road with minimum lighting specifications given under the SANS 10098-1 standard [9]. The type of road is closely related to the M3 class under the European standard [5]. The results in *step 2.1* indicate that an LED luminaire rated at 178 W is the most suitable replacement. Thus, new LED EE lights can

Fig. 3. Case-study area in Vanderbijlpark

be installed to save the city energy cost.

$$u_i = \begin{cases} 60\% & i \in [1, 11] \cup \in [42, 48], \\ 80\% & i \in [12, 13] \cup \in [37, 41], \\ 100\% & otherwise. \end{cases} \quad (8)$$

However, the advantage of smart lights is that they can be able to further dim the lights based on the traffic flow patterns obtained by their smart sensors. Based on the traffic studies of an average South African city, the smart streetlight will be able to implement the control strategy shown in (8) [10]. The traffic data and lighting studies indicate that the new luminaires can be dimmed to 60% before 5:00am and 8:30pm in the evening. The dimming can then be at 80% just before the morning traffic peak (5:00am–6:30am) and after the evening traffic peak (6:00pm–8:30am).

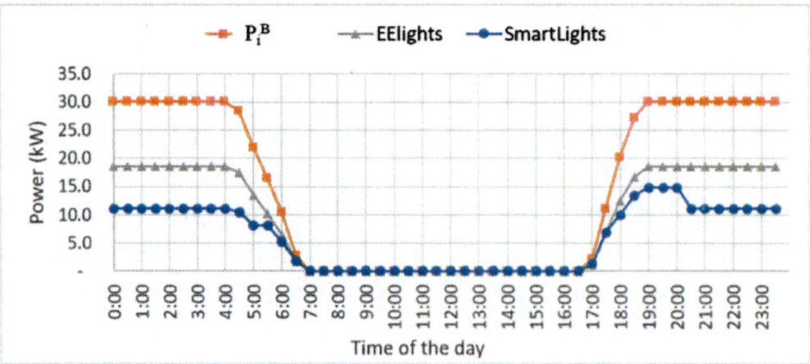

Fig. 4. Simulation energy usage load profiles

The baseline load profile resulting from *step 1*, the post-implementation profiles for EE lighting and smart lighting strategies of *step 2.2* are shown in Fig. 4.

Figure 4 shows the smart-street light load profile that uses the control strategy in (8) as 'SmartLights' and the implementation of EE LED luminaires without dimming control as 'EElights'.

Table 1. Case-study data

Parameter	Value
c_{price}	3,200.00 ZAR
$c_{install}$	480.00 ZAR
ρ^B, ρ^A	1.31 ZAR
$c_{replacement}$-HPS	780.00 ZAR
$c_{replacement}$-LED	3,680.00 ZAR
λ^B_{avg}	24,000 h
ρ^A_{avg}	100,000 h

1 EURO ≈ 16.90 South African Rands (ZAR)

Step 3.1 and *3.2* require the use of lighting market data to obtain the price of LED luminaires and HPS bulbs. The recent market data given in Table 1 is based on the participating suppliers and contractors of the South African municipal Energy Efficiency and Demand-side Engagement (EEDSM) programme [11]. The average life-spans of the two technologies is also noted in Table 1 [12]. The electricity tariff is supplied by Eskom, a South African public utility responsible for supplying the majority of municipal entities in the country [13]. In this case, we assume that the city does not change its electricity supplier post-implementation of the smart streetlight project (Table 2).

Table 2. Results data

Item	Description
Monthly E_{sav} - Smartlights	6,399 kWh
$N_{payback}$ - Smartlights	3.86 years
Monthly E_{sav} - EElights	4,131 kWh
$N_{payback}$ - EElights	6.02 years

Applying Eqs. (1)–(6), results in a $C_{Inv} = 383,720.00$, $C_{O\&M}[\lambda^B_{avg}] = 1,201.00$ and $C_{O\&M}[\lambda^A_{avg}] = 1,360.00$. This gives savings and $N_{payback}$ calculation from (7) shown in Table 1. The payback period of smart lighting is less than 4 years while that of EE lighting is more than 6 years. The results in Table 1 show an added advantage the city gets from implementing smart lighting as opposed to ordinary EE lighting.

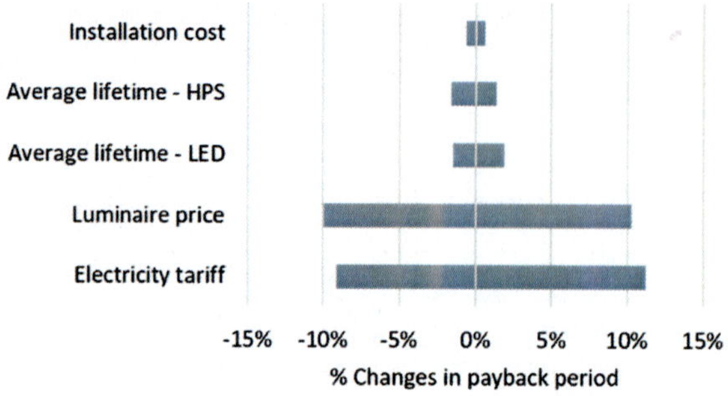

Fig. 5. Sensitivity analysis of the project's payback period

Figure 5 shows the sensitivity of the payback period on a number of parameters with a 10% variation. The $N_{payback}$ is least affected by changes in λ_{avg}^{B}, λ_{avg}^{A} and $c_{install}$. This is because the installation cost forms a very small part, 15%, the total investment. The payback period is mostly affected by c_{price} and electricity price ρ^{A}, resulting in more than 10% variations. Increasing the luminaire price by 10% delays $N_{payback}$ by about 5 months. On the other hand, a 10% increase in the electricity price will result in a positive effect of reducing $N_{payback}$ by slightly over 4 months. However, the project remains attractive in-spite of the 10% variations.

4 Conclusion

This paper presented an easy to use methodology for assessing the viability of implementing smart street-lighting projects in cities. The proposed methodology can be used as a planning tool to conduct preliminary feasibility studies. The methodology is able to estimate the project's payback period using minimal measurements from case-study streets while relying of lighting study simulation software. The methodology is able to illustrate that implementing a smart-streetlight project in a case-study road will result in an attractive payback period of less than 4 years. The analysis in this work also illustrates that cheaper LED luminaire prices and rising electricity cost have a significant positive impact on the project's payback period. Future work on the methodology should focus on applying life-cycle cost assessment which accounts for the time-value of money and on including other benefits offered by smart-streetlights.

References

1. Xia, X., Wu, X., Balamurugan, S., Karuppiah, M.: Effect of environmental and social responsibility in energy-efficient management models for smart cities infrastructure. Sustainable Energy Technol. Assess. **47**, 101525 (2021). https://doi.org/10.1016/j.seta.2021.101525

2. Duman, A.C., Guler, O.: Techno-economic analysis of off-grid photovoltaic LED road lighting systems: a case study for northern, central and southern regions of Turkey. Build. Environ. **156**, 89–98 (2019). https://doi.org/10.1016/j.buildenv.2019.04.005
3. Gorgulu, S., Kocabey, S.: An energy saving potential analysis of lighting retrofit scenarios in outdoor lighting systems: a case study for a university campus. J. Clean. Prod. **260**, 121060 (2020). https://doi.org/10.1016/j.jclepro.2020.121060
4. Petritoli, E., Leccese, F., Pizzuti, S., Pieroni, F.: Smart lighting as basic building block of smart city: an energy performance comparative case study. Measurement **136**, 466–477 (2019). https://doi.org/10.1016/j.measurement.2018.12.095
5. Fryc, I., Czyzewski, D., Fan, J., Galatanu, C.D.: The drive towards optimization of road lighting energy consumption based on Mesopic vision-a suburban street case study. Energies **14**, 1175 (2021). https://doi.org/10.3390/en14041175
6. Pasolini, G., Toppan, P., Zabini, F., De Castro, C., Andrisano, O.: Design, deployment and evolution of heterogeneous smart public lighting systems. Appl. Sci. **9**, 3281 (2019). https://doi.org/10.3390/app9163281
7. Sunrise and Sunset Calculator. http://www.timeanddate.com/sun/
8. DIALux evo 10.1. http://www.dialux.com
9. South African Bureau of Standards. Standard SANS-10098-1: Public lighting Part 1-The lighting of public thoroughfares: Pretoria, South Africa (2003)
10. De Jongh, F., Bruwer, M.: Quantification of the natural variation in traffic flow on selected national roads in South Africa. In: 36th Southern African Transport Conference. University of Pretoria, Pretoria (2017)
11. Energy Efficiency Demand Side Management (EEDSM). http://www.savingenergy.org.za/municipal-eedsm/
12. Paucek, I., Appolloni, E., Pennisi, G., Quaini, S., Gianquinto, G., Orsini, F.: LED lighting systems for horticulture: business growth and global distribution. Sustainability **12**, 7516 (2020). https://doi.org/10.3390/su12187516
13. Eskom. Schedule of standard prices for Eskom Tariffs: Pretoria, South Africa (2021)

Marine and Climatic Impacts on the Hydrogeochemical Functioning of Lake Sidi Boughaba (Kenitra, North West, Morocco)

Mohamed Lachhab[1](✉), Mohamed Najy[1], Fatima Zahra Talbi[2], Hassan Ech-Chafay[1], Mbarek Idoumou[1], Driss Belghyti[1], and Khadija El Kharrim[1]

[1] Natural Resources and Sustainable Development Laboratory, Faculty of Sciences, University Ibn Tofail, BP 133, 14000 Kenitra, Morocco

[2] Faculty of Sciences and Technologies, Laboratory of Biochemistry, Neurosciences, Natural Resources and Environment, Hassan First University of Settat, P. O. Box 577, 26000 Settat, Morocco

Abstract. A seasonal hydrochemical monitoring of the lake Sidi Boughaba (Morocco) was carried out during the years 2020 and 2021 with the aim of characterizing its hydrogeochemical functioning with respect to the climatic and marine impacts. The analyses carried out revealed a basic pH between 7.55 and 9.31, an electrical conductivity in the range of 5 to 39.4 mS/cm and high levels of sodium and chlorides, which can reach the respective values of 9636 mg/L and 15549 mg /L, indicating a marine origin of these waters. The mineralization of the waters of this ecosystem is controlled by four main mechanisms, the mechanism at the base of this high mineralization is the impact of marine spray. The second order processes would be the climatic conditions such as temperature and evaporation, the oxidation-reduction and dissolution-precipitation reactions that modify the hydrochemistry of the waters as well as the biological processes.

Keywords: Lake Sidi Boughaba · Hydrogeochemistry · Sea spray · Kenitra · Morocco

1 Introduction

Morocco is the North African country with the best continental water resources. The great spatial variability of climatic and geological conditions accentuates this richness by creating very varied types of aquatic ecosystems [1]. During the last decades, the number and the surface of these wetlands have experienced a great regression [2, 3]. In order to fight against this degradation, the kingdom has embarked on the path of developing the necessary tools for the implementation of a policy of conservation and sustainable development of its natural resources. In this policy, the National Wetlands Strategy 2015–2024 was published in December 2015, which aimed to establish mechanisms and processes that could ensure sustainable use of wetlands in Morocco [4].

The present work aims to assess the marine and climatic impacts on the hydrogeochemical functioning of Lake Sidi Boughaba, which has been classified in the Ramsar list

M. Ben Ahmed et al. (Eds.): SCA 2022, LNNS 629, pp. 214–225, 2023.
https://doi.org/10.1007/978-3-031-26852-6_20

since 1980 alongside the Moulay Bousselham lagoon located in the same province, Lake Afenourir and Khenifiss lagoon, successively in the provinces of Ifrane and Tan-Tan.

2 Materials and Methods

2.1 Study Area

Lake Sidi Boughaba is located on the northwestern Atlantic coast of Morocco (N 34° 24 ', W 06° 67'), it occupies a coastal interdune basin 6 km long and 300 to 800 m wide parallel to the Atlantic coastline. It is formed by a succession of dunes oriented NNE to SSW [5]. These dunes are covered by a spontaneous vegetation composed mainly of a juniper forest (*Juniperus phoenicea*). At the level of the shallow parts of the lake, one meets a benthic vegetation and macroscopic algae and on its immediate edges, one finds mainly a vegetation composed of tufts of rush, phragmites and typhas [6]. Its fauna is also characterized by its diversity and richness [7–9].

Its geographical location on the Palearctic-Sub-Saharan Africa migratory axis, the importance of its winter population and the nesting of a fairly large number of species has allowed its designation, since 1980, as a "Ramsar" site [10]. It was also selected in the framework of the Master Plan of Protected Areas as a Site of Biological and Ecological Interest (S.I.B.E) of priority I [11].

The surface water level of the lake environment depends on rainfall and groundwater flowing towards the ocean in a northwest direction [12].

Fig. 1. Descriptive map of the wetland of Sidi Boughaba Lake

2.2 Sampling and Physicochemical Analysis

The study site was subject to seasonal physico-chemical monitoring over a period from July 2020 to May 2021 where four water sampling campaigns were conducted. The water sampling sites for the four missions are shown in Fig. 1. Samples were filtered through a 0.45 μm filter and collected in two-liter capacity polyethylene bottles without air bubbles, and those for cation determination were acidified with nitric acid to avoid adsorption and precipitation of chemical elements and stored in portable coolers (+4 °C) [13]. Temperature, pH, salinity, conductivity and dissolved oxygen were measured in situ. While the other parameters (biological oxygen demand (BOD), the total alkalinity (TA), color, cations and anions) were measured at the National Institute of Hygiene. The temperature was measured by a thermometer incorporated in the brand pH meter (WTW pH 330i / SET), the pH by portable field pH meter, type (WTW pH 330i / SET), the salinity by salinometer incorporated in the conductivity meter (WTW cond 330i / SET), Conductivity was measured by a portable conductivity meter, type (WTW cond 330i / SET), dissolved oxygen was measured by an oximeter brand Orion model 607A, the biological oxygen demand was determined using a BOD meter, the color with a Lovibond Nesseleriser comparator, the TA as well as the bicarbonates (HCO_3^-) were analyzed by volumetric dosage with HCl 0,1 N, chlorides (Cl^-) and sulfates (SO_4^{2-}) were measured by ion chromatography version Dionex 2000 and the four cations Na^+, Mg^{2+}, Ca^{2+} and K^+ were determined by atomic absorption spectrophotometer model Varian.

3 Results and Discussions

3.1 Temperature

The values of surface water temperatures recorded in situ during the present study are highly variable, they oscillate between 15.3 °C recorded at S1 and S4 in winter (February 2021) and a maximum of 28.2 °C was recorded at S1 in July 2020, with an average of 21.04 ± 0.78 °C. The recorded values are higher than those found by Ouhmidou *et al.* [14] at Hassan Addakhil Dam in Errachidia (Morocco) and lower than those reported by Allalgua *et al.*[15] at Foum El-Khanga Dam in Algeria. Temperature fluctuations are related to local climatic conditions and more particularly to air temperature and water evaporation phenomena [16] (Fig. 2).

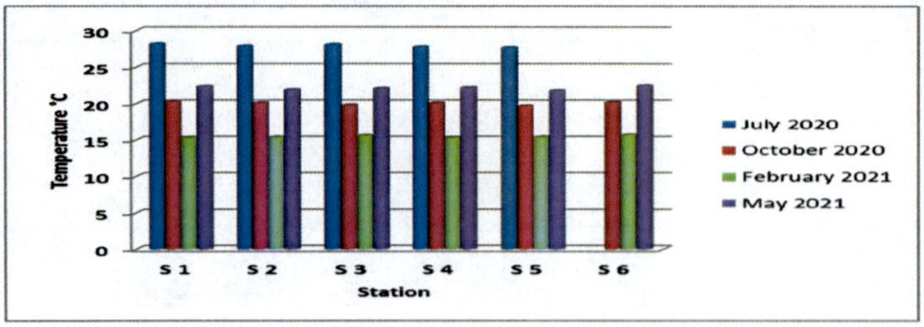

Fig. 2. Spatio-temporal variation in the water temperature of Lake Sidi Boughaba.

3.2 Electrical Conductivity And Salinity

The waters of the lake are highly mineralized, with a maximum of 39400 μS/cm was recorded at station S4 in summer (July 2020) and a minimum of 5000 μS/cm was recorded at the same station in winter (February 2021). These values are lower than those found by Gouasmia *et al.* [17] at Lake Megarine in Algeria. The low values recorded in winter (February 2021 campaign) are related to the beginning of carbonate precipitation. The remarkable increase in electrical conductivity in summer (July 2020 campaign) is due to the evaporation of water and the small decrease in electrical conductivity in autumn (October 2020), resulting from the first inflow of low mineralized rainwater [18] (Fig. 3).

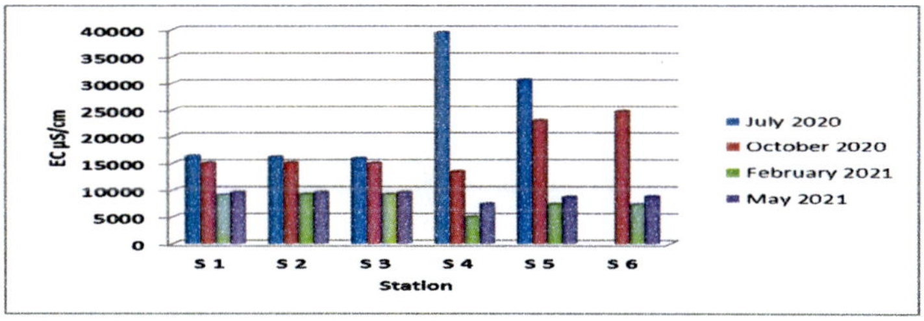

Fig. 3. Spatio-temporal variation of the electrical conductivity of the waters of Lake Sidi Boughaba.

The analysis of the obtained results of salinity, shows that the lake waters are highly saline. We note a maximum of 26.9 g/l in S4 in summer (July 2020) and a minimum of 2.88 g/l in the same station in winter (February 2021), with an average value of about 8.93 ± 5.04 g/l. The strong monthly fluctuations of salinity in Lake Sidi Boughaba are related to the intensity of evaporation, which is more important in hot periods [17] (Fig. 4).

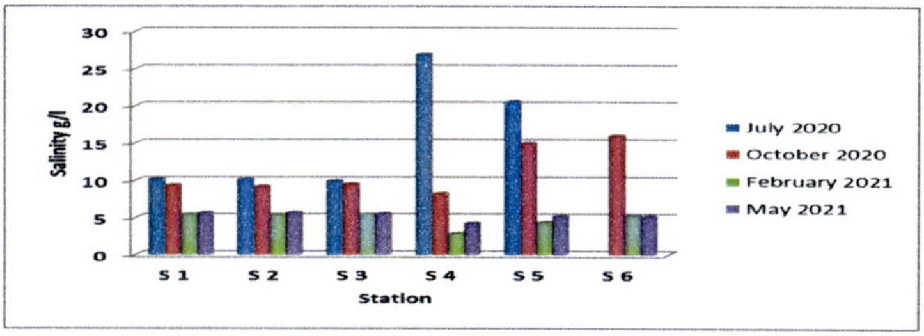

Fig. 4. Spatio-temporal variation in the salinity of the waters of Lake Sidi Boughaba.

3.3 pH and Total Alkalinity

The pH values are generally more or less homogeneous throughout the water column and are alkaline. The maximum pH values were observed in spring (May 2021 campaign) at station S5, without exceeding 9.31. The lowest value of about 7.55 was recorded in winter (February 2021 campaign) at S4. These results are consistent with those found by Mehanned *et al.* [19] at Sidi Chahed dam (Morocco) and those found by Abba et al. [20] at Dayet Aoua (Morocco) and are lower than found by Etebaai *et al.* [18] at Lake Ifrah (Morocco).

The minimum values are recorded during the month of February when the climatic conditions are not favorable for photosynthesis. On the other hand the maximum values are recorded during the month of July, following the intense use of carbon dioxide by aquatic plants through the phenomenon of photosynthesis [20] (Fig. 5).

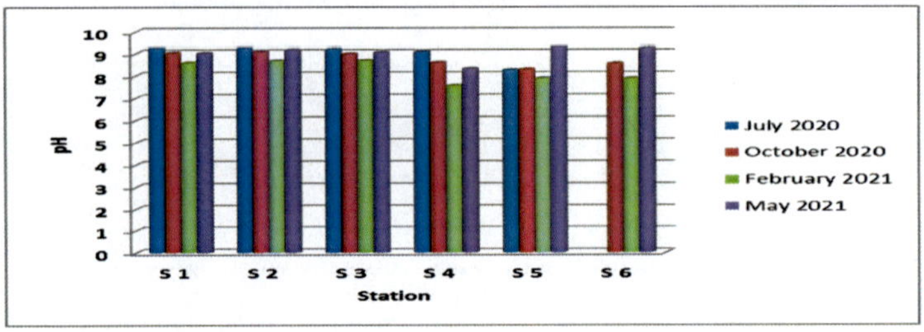

Fig. 5. Spatio-temporal variation of the pH values of the waters of Lake Sidi Boughaba.

The minimum and maximum values of total alkalinity, are respectively around 4.2 recorded in station S5 in autumn (campaign October 2020) and 41.3 meq/l noted in station S4 in summer (campaign July 2020). These results are lower than those found by Lakhdar Idrissi *et al.* [21] at Lake Khnifiss in the Tan-Tan region (Morocco) and are higher than those found by Hakam *et al.* [22] at Lake Fouarate in the Kenitra region (Morocco) The high alkalinity of the lake waters could be due to the fact that it is located on a substrate containing carbonate and bicarbonate [23] (Fig. 6).

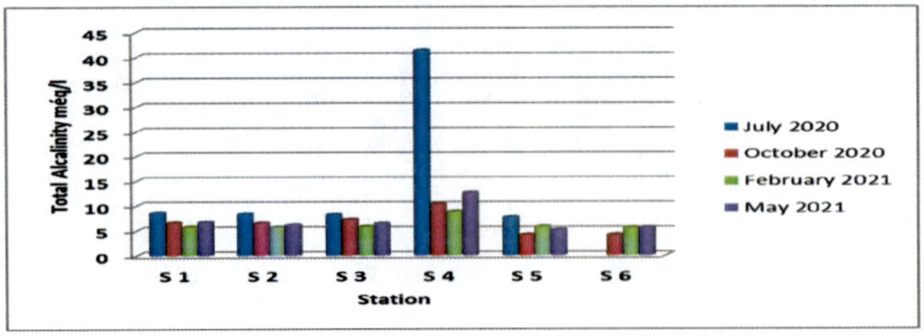

Fig. 6. Spatio-temporal variation of the total alkalinity of the waters of Lake Sidi Boughaba.

3.4 Dissolved Oxygen

The analysis of the profile of dissolved oxygen at the different study sites shows that the temporal evolution of this parameter is very variable from one station to another. We note a maximum of 12.76 mg/l in S5 in spring (May 2021 campaign) and a minimum of 1.73 mg/l in the same station in autumn (October 2020 campaign) with an average of 6.72 ± 2.13 mg/l. Oxygen levels in Lake Sidi Boughaba are similar to those found by Bejaoui *et al.* [24] in Bizerte lagoon in Tunisia and those found by Lakhdar Idrissi *et al.* [21] in Khnifiss lagoon and are lower than those recorded by Lods-Croset *et al.*[25]. The seasonal variation in dissolved oxygen levels is mainly related to plankton dynamics but also to the influence of seasonal climate variability [18] (Fig. 7).

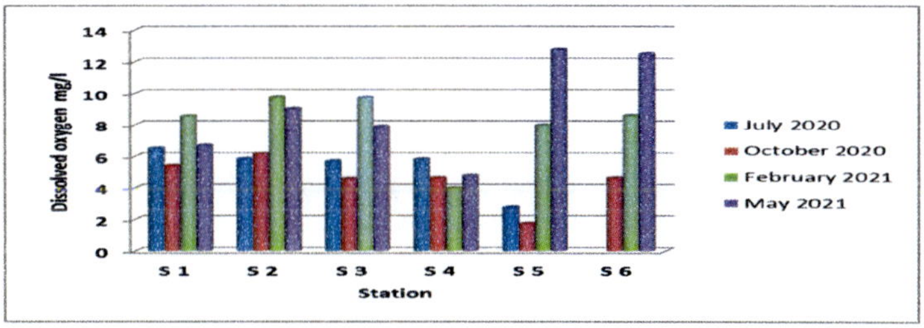

Fig. 7. Spatio-temporal variation in the dissolved oxygen content of the waters of Lake Sidi Boughaba.

3.5 BOD_5

The spatial and temporal evolution of biological oxygen demand is similar at all stations. The minimum and maximum values fluctuate between 0 recorded in stations S2 and S5 in winter (February 2021 campaign) and 11 mg/l recorded in station 3 in autumn (October 2020 campaign). The increase in BOD levels during the dry period (July and October 2020) can be explained by the establishment of conditions for the degradation of organic matter by microorganisms, whose activity intensifies with the decrease in flow velocity and with the warming of the water [26]. Moreover, in wet periods (February 2021) rainwater contributes to the dilution of the organic load [15] (Fig. 8).

3.6 Color

The color of the water is identical in stations S1, S2, S3, S5 and S6. It varies between 50 mg Pt/l recorded in stations S5 and S6 in autumn (October 2020 campaign) and 120 mg Pt/l recorded in the same stations in winter (February 2021). On the other hand, station S4 has values between 350 and 600 mg Pt/l, recorded successively in summer (July 2020) and winter (February 2021). The strong coloration in S4 could be the result of the presence of one or more pigments in the lake water, coming from the vegetation living in this ecosystem (Fig. 9).

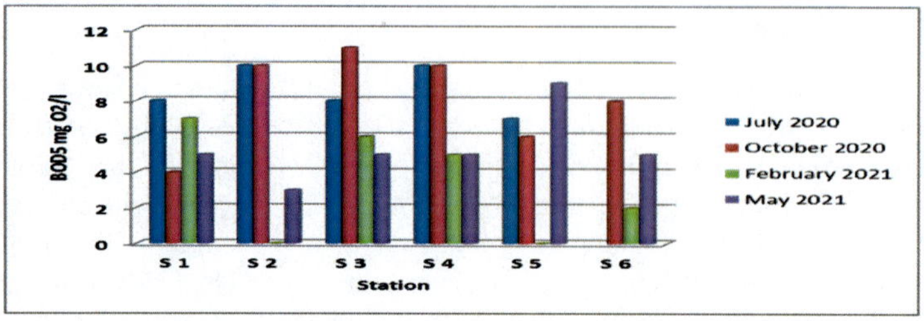

Fig. 8. Spatio-temporal variation of the biological oxygen demand of the waters of Lake Sidi Boughaba.

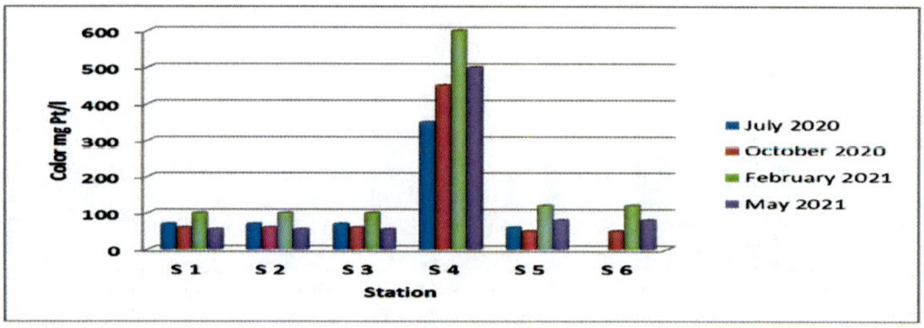

Fig. 9. Spatio-temporal variation in the color of the waters of Lake Sidi Boughaba.

3.7 Cations (Calcium, Magnesium, Potassium and Sodium)

The concentration of calcium ions reaches a maximum value at station S6 during the month of October 2020, which is of the order of 280.3 mg/l, while the minimum content was recorded at station S4 at 8.7 mg/l during July 2020. These values are similar to those found by Loucif *et al.* [27] at Lake Tonga in Algeria. The decrease in calcium content during the dry period (May to September, this is explained by the early precipitation of carbonate minerals, in this case calcite [15]. The temporary stations S5 and S5 do not follow this logic where they experience a great evaporation during the summer (July 2020 campaign), which causes an increase in calcium contents. During the fall (October 2020), precipitation causes the dissolution of minerals and consequently an increase in calcium concentration (Fig. 10).

Magnesium levels range from 781.5 mg/l recorded at station S5 in summer (July 2020) to 125.7 mg/l recorded at station S4 in winter (February 2021). These values are higher than those found by Abba *et al.* [20] at Dayet Aoua (Morocco). The low levels recorded during the wet seasons can be explained by the effect of dilution [16]. On the other hand, the high values noted during dry seasons are due to the evaporation phenomenon [28]. This mineral element is highly valued by bacterial cells, which absorb it from the external environment and concentrate it in the vacuoles of specific resources [29] (Fig. 11).

The concentration of sodium ions reaches a maximum value at station S4 in summer (July 2020) which is of the order of 9636 mg/l, while the minimum content was recorded

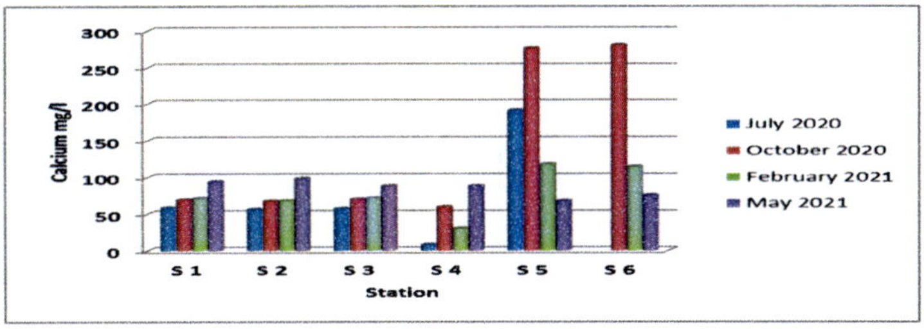

Fig. 10. Spatio-temporal variation in the calcium content of the waters of Lake Sidi Boughaba.

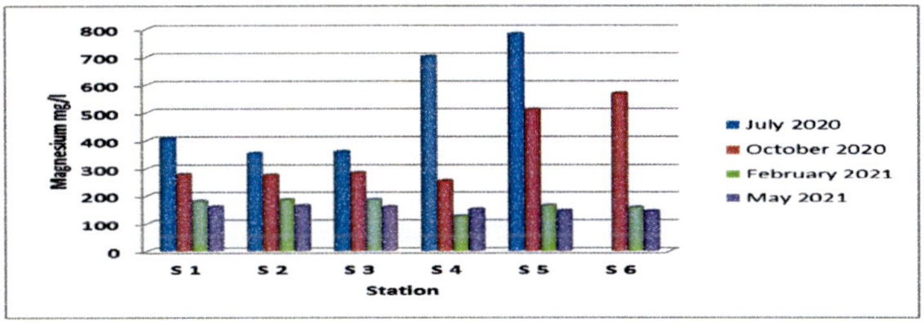

Fig. 11. Spatio-temporal variation in the magnesium content of the waters of Lake Sidi Boughaba.

in the same station at 885 mg/l in winter (February 2021). The high levels recorded during the summer season is due to the decrease in water volume [30] (Fig. 12).

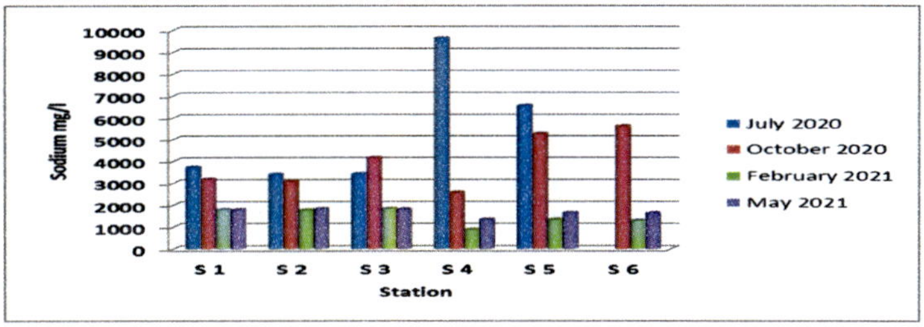

Fig. 12. Spatio-temporal variation in sodium content in the waters of Lake Sidi Boughaba.

The concentration of potassium ions reaches a maximum value at station S6 in winter (February 2021) which is of the order of 97.7 mg/l, while the minimum content of the order of 18.4 mg/l was recorded at the same station in spring (May 2021). The periods of intense growth correspond to massive withdrawals of potassium and phosphorus, mineral salts involved in cell divisions. On the contrary, as the plant ages, the uptake slows down; there may be an outflow of ions from the root [31] (Fig. 13).

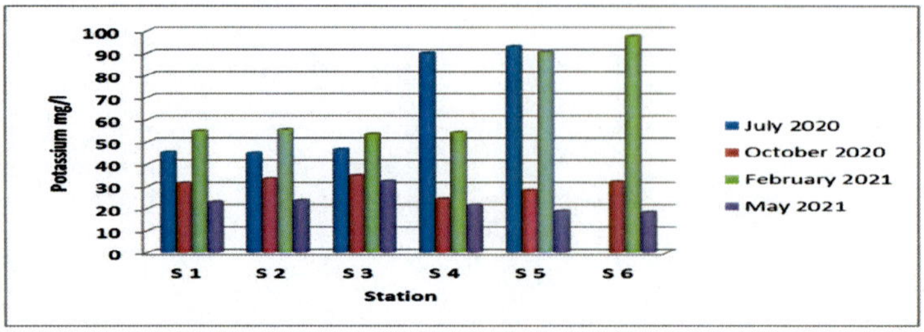

Fig. 13. Spatio-temporal variation in the potassium content of the waters of Lake Sidi Boughaba.

3.8 Anions (Chlorides, Sulfates and Bicarbonates)

The concentration of chloride ions reached a maximum value at station S4 during July 2020, which is of the order of 15549 mg/l, while the minimum content was recorded in the same station at 2225 mg/l during May 2021. The high salinity and high chloride and sodium contents recorded in the waters of Lake Sidi Boughaba reflect typical values of those found in seawater [32, 33], they could be due to the accumulation of sea spray at the level of the lake and its surroundings [34, 35] as well as their concentration by the high number of evaporation cycles that the lake has experienced since its generation [35, 36] which dates to about 4570 ± 110 BC [37] or as a result of marine intrusion [33, 38, 39] (Fig. 14).

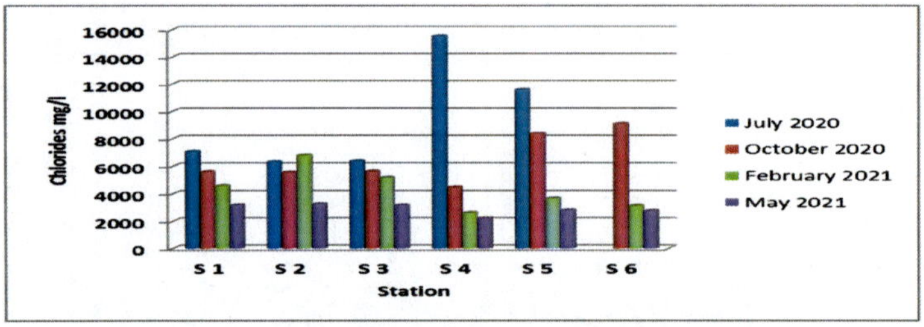

Fig. 14. Spatio-temporal variation of the chloride contents of the waters of Lake Sidi Boughaba.

The concentration of sulfate ions reaches a maximum value in the order of 1070 mg/l at station S4 in spring (May 2021), while the minimum content which is in the order of 63 mg/l was recorded in station S3 in the same season. The high variation of sulfate levels could be explained by a reduction of sulfates [40], and verifying the marine origin of the waters [41]. The high values could be explained by the presence of sea spray in the area.

The concentration of bicarbonate ions reaches a maximum value at station S4 during July 2021 which is of the order of 2065 mg/l, while the minimum content was recorded in station S5 at 256 mg/l during October 2020. The low bicarbonate levels are explained by the precipitation of carbonate minerals. Bicarbonates, calcium and magnesium are controlled by the precipitation of calcite, which is in turn induced by photosynthesis [42] (Figs. 15 and 16).

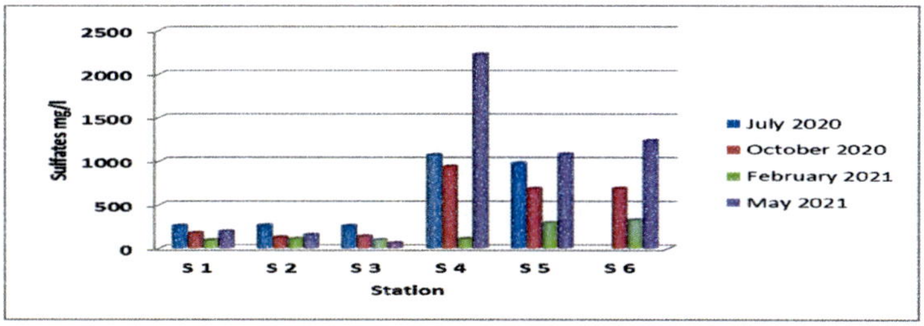

Fig. 15. Spatio-temporal variation in the sulphate content of the waters of Lake Sidi Boughaba.

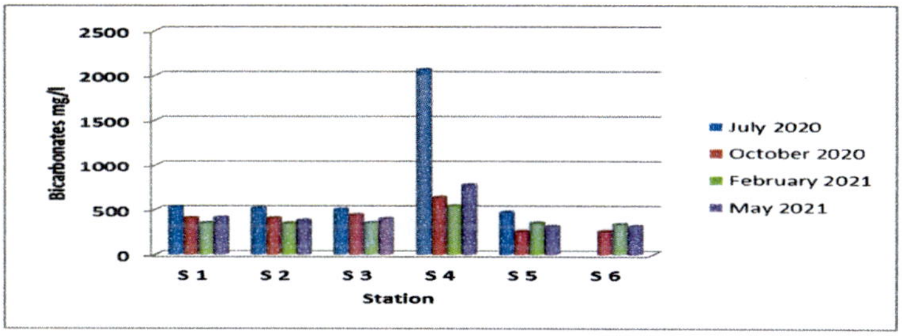

Fig. 16. Spatio-temporal variation in the bicarbonate content of the waters of Lake Sidi Boughaba.

4 Conclusion

The analyses of the quality of the waters of the lake Sidi Boughaba reveal a normal rate in dissolved oxygen, a basic pH, a strong spatio-temporal variation of the mineralization which is accentuated in the temporary parts of the lake environment and especially in summer. The conductivity is generally high and oscillates between 5 and 39.40 mS/cm. The most loaded zones are the northern and southern extremities where the depth of the water table is low. They present high contents of Cl^-, Na^+, SO_4^{2-}, HCO_3^- and Mg^{2+}. The high values of conductivity, chlorides and sodium are probably due to exposure to sea spray, given its location near the sea and the absence of direct or indirect communication (marine intrusion) with the sea, climatic conditions such as temperature and evaporation, oxidation-reduction and dissolution-precipitation reactions that modify the hydrochemistry of the water as well as the biological processes.

References

1. Chillasse, L., Dakki, M., et Abbassi, M.: Valeurs et fonctions écologiques des zones humides du Moyen Atlas (Maroc). Humedales Mediterranios **1**, 139–146 (2001)
2. Thevenot, M.: Les Oiseaux de la réserve de Sidi Boughaba. Bull. Institut Scientifique **1**, 67–99 (1976)
3. Chillasse, L, Dakki, M.: Potentialités et statuts de conservation des zones humides du Moyen-Atlas (Maroc), avec référence aux influences de la sécheresse. Sécheresse **15**(4), 337–345 (2004)

4. Khattabi, A., Benslimane, M.: L'éducation à l'environnement dans la réserve de Sidi Boughaba. Terre et Vie **59**(60), 1–3 (2002)
5. Dakki, M., Menioui, M., et Amhaouch, Z.: Strategie Nationale et plan d'action 2015–2024 pour les Zones Humides du Maroc. Rapport Deutsche Gesellschaft fur Internationale Zusammenarbeit (GIZ) (2015)
6. Cherkaoui, I., Bouchafra, A.: Fiche descriptive sur les zones humides ramsar [online] (2003). https://rsis.ramsar.org/risapp/files/risrep/ma207ris.pdf
7. Ramdani, M., et al.: Open water zooplankton communities in North Africari wetland lakes: the Cassarina Projec1. Aquatic Ecology **35**(3–4), 319–333 (2001a)
8. Ramdani, M., Flower, R.I., El Khiati, N., Birks, H.H., Kraiem, M.M., Fathi, A.A.: Zooplankton (Cladocera-Ostracoda) Chironomidae and other benthic faunal remains in sediment cores from nine wetland lakes: the Cassarina Projec1. Aquatic Ecology **35**(3–4), 389–403 (2001b)
9. El Hamoumi, R., Dakki, M., Michel Thevenot, M.: Etude écologique des larves d'anoures du Maroc. Bulletin de l'Institut Scientifique Rabat, section Sciences de la Vie **29**, 27–34 (2007)
10. Cherkaoui, I., Dakki, M., Lahrouz, S., Hanane, S.: Dix années de suivi des anatidés nicheurs sur le lac de Sidi Boughaba (Nord -ouest Marocain): Situation, tendances d'évolution et perspectives de recherche. Revue d'écologie (Terre & Vie), **68**(2), 167–180 (2013)
11. A.E.F.C.S.: Plan Directeur des Aires Protégées du Maroc vol. 3: Les sites d'intérêt biologique et écologique du domaine littoral. Rapport Inédit, Administration des Eaux et Forêt et de la Conservation des sols (1996)
12. Zouhri, L., Gorini, C., Deffontaines, B., Mania, J.: Relationships between hydraulic conductivity distribution and synsedimentary faults, Rharb-Mamora basin, Morocco; Hydrogeological, geostatistical and modeling approaches. Hydrogeol. J. **12**, 591–600 (2004)
13. Rodier, J.: L'analyse de l'eau : Eaux naturelles, eaux résiduaires et eaux de mer. 7ème Ed. Dunod (2009)
14. Ouhmidou, M., Chahlaoui, A., Kharroubi, A., Chahboune, M.: Étude de la qualité physico-chimique et bactériologique des eaux du barrage Hassan Addakhil d'Errachidia (Maroc). J. Mater. Environ. Sci. **6**(6), 1663–1671 (2015)
15. Allalgua, A., Kaouachi, N., Boualeg, C., Ayari, A., Bensouileh, M.: Caracterisation physico-chimique des eaux du barrage Foum El-Khanga (Région de Souk-Ahras, Algérie). Eur Sci. J. **13**(12), 258–275 (2017)
16. Toumi, A., Reggam, A., Alayat, H., Houhamdi, M.: Caractérisation physico-chimique des eaux de l'écosystème lacustre : cas du Lac des Oiseaux (Extrême NE- Algérien). J. Mater. Environ. Sci. **7**(1), 139–147 (2016)
17. Gouasmia, G., Amarouayache, M., Frihi, H., Kara, M.H.: Caractérisation physico-chimique de trois lacs salés permanents de la vallée d'oued Righ (Sahara septentrional, Algérie nord-est). Revue d'écologie (terre et vie) **71**(4), 330–341 (2016)
18. Etebaai, I., et al.: Impacts climatiques et anthropiques sur le fonctionnement hydrogéochimique du lac Ifrah (Moyen Atlas marocain). Journal des Sciences Hydrologiques **57**(3), 547–561 (2012)
19. Mehanned, S., Chahlaoui, A., Zaid, A., Samih, M., Chahboune, M.: Typologie de la qualité physico-chimique de l'eau du barrage Sidi Chahed-Maroc (Typology of the physic-chemical quality of the waters of the dam Sidi Chahed-Morocco). J. Mater. Environ. Sci. **5**(5), 1633–1642 (2014)
20. Abba, H., Nassali, H., Benabid, M., El Ibaoui, H., Chillasse, L.: Approché physicochimique des eaux du lac dayet Aoua (Maroc). J. Appl. Biosci **58**, 4262–4270 (2012)
21. Idrissi, L., et al.: Organisation et fonctionnement d'un écosystème côtier du Maroc, lagune de Khnifiss. Rev Sci. Eau **17**(4), 447–462 (2004)
22. Hakam, O., et al.: Qualité des eaux naturelles dans la région de la ville Kénitra. Revue marocaine du génie civil **89**, 79–91 (2000)

23. Serghini, A., Fekhaoui, M., El Abidi, A., El Blidi, S., Ben Akkame, R.: Caractérisation hydrochimique d'un site Ramsar: Complexe des zones humides de mohammedia: Maroc. Bulletin de l'institut scientifique Rabat, section Sciences de la vie **32**(2), 133–145 (2010)
24. Béjaoui, B., Ferjani, D., Zaaboub, N., Chapelle, A., Moussa, M.: Caractérisation hydro-biologique saisonnière de la lagune de Bizerte (Tunisie). Rev Sci. Eau **23**(3), 215–232 (2009)
25. Lods-Crozet, B., De La Harpe, M., Reymond, O., Strawczynski, A.: Evaluation de la qualité chimique et Biologique d'un petit lac du plateau suisse (lac du Bret, Canton de vaud). Bull. Soc. vaud. Sc. nat. **91**(4), 363-387 (2009)
26. Bremond, R., Perrodon, C.: Paramètres de la qualité des eaux. Ministère de l'environnement et cadre de vie. 2ème éd, p. 259 (1979)
27. Loucifa, K., Neffard, S., Menasriae, T., Maazib, M.C., Houhamdic, M., Chenchouni, H.: Physico-chemical and bacteriological quality assessment of surface water at Lake Tonga in Algeria. Environmental Nanotechnology Monitoring & Management **13**, 100284 (2020)
28. Nassali, H., Benbouih, H., Sghiri, A.: Influence des eaux usées sur la composition des eaux de surface et des sédiments superficiels du lac Merja Fouarate au Maroc. Afrique Science **01**(1), 145–165 (2002)
29. Schlumberger, O.: Mémento de pisciculture d'étang. 4e édition. CEMAGREF, p. 237 (2002)
30. Verma, P.U., Chandawat, D.K., Solanki, H.A.: Seasonal variation in physico-chemical and phytoplankton analysis of Kankaria lake. Life Sciences Leaflets **19**, 842–854 (2011)
31. Roche, M.A.: Traçage naturel salin et isotopique des eaux du système hydrologique du lac Tchad. Travaux et documents de l'O.R.S.T.O.M. **117**, 383 (1979)
32. Ben Hamouda, M.F., Tarhouni, J., Leduc, C., Zouari, K.: Understanding the origin of salinization of the Plio-quaternary eastern coastal aquifer of Cap Bon (Tunisia) using geochemical and isotope investigations. Environ Earth Sci **63**, 889–901 (2011)
33. Fadili, A., Mehdi, K., Riss, J., Najib, S., Makan, A., Boutayab, K.: Evaluation of groundwater mineralization processes and seawater intrusion extension in the coastal aquifer of Oualidia, Morocco: hydrochemical and geophysical approach. Arab. J. Geosci. **8**(10), 8567–8582 (2015). https://doi.org/10.1007/s12517-015-1808-5
34. Herczeg, A.L., Dogramaci, S.S., Leaney, W.J.: Origin of dissolved salts in a large, semi-arid groundwater system: Murray Basin. Australia Mar. Freshwater Res. **52**, 41–52 (2001)
35. Craw, D., Beckett, S.: Water and sediment chemistry of Sutton Salt Lake, East Otago, New Zealand. NZ J. Mar. Freshwat. Res. **38**, 315–328 (2004)
36. Taheri Tizro, A., Voudouris, K.S.: Groundwater quality in the semi-arid region of the Chahardouly basin. West Iran. Hydrol. Processes **22**(16), 3066–3078 (2007)
37. Reille, M.: Analyse pollinique du lac de Sidi Bou Rhaba, littoral atlantique (Maroc) *Ecologia Mediterranea* **4**, 61–65
38. Batchi, M., Al Karkouri, J., El Maaqili, M., Fenijiro, I.: Cartographie de la vulnérabilité à l'intrusion marine de l'aquifère côtier de Mnasra (Littoral du Gharb, Maroc-Nord-Occidental) Marine Scor Journal **1**, 54–69 (2014)
39. Najib, S., Fadili, A., Mehdi, K., Riss, J., Makan, A., Guessir, H.: Salinization process and coastal groundwater quality in Chaouia. Morocco Journal of African Earth Sciences (2016). https://doi.org/10.1016/j.jafrearsci.2015.12.010
40. Vengosh, A., Rosenthal, E.: Saline groundwater in Israel: its bearing on the water crisis in the country. Journal of Hydrology (Amsterdam) **156**(1–4), 389–430
41. Trabelsi, R., Zaïri, M., Smida, H., Ben Dhia, H.: Salinisation des nappes côtières : cas de la nappe nord du Sahel de Sfax. Tunisie. C. R. Acad. Sci. Paris **337**, 515–524 (2005)
42. Négrel, P.h., Barbier, J., Petelet-Giraud, E.: Caractérisation géochimique et isotopique des relations eaux de surface-eaux souterraines de la zone humide de la loire. Rapport BRGM/RP50668-FR, p. 71 (2001)

Assessment of the Role of Micromobility in ITS by A'WOT Analysis

Rukiye Gizem Öztaş Karlı(✉) and Halil Karlı

Bartın University, Bartın, Turkey
roztas@bartin.edu.tr

Abstract. Transportation is one of the most critical issues affecting the livability of cities. In recent years, micromobility has emerged as an alternative solution to the transportation problem of many cities. The increase in private vehicle ownership, traffic congestion, travel time, emissions and energy consumption in Turkey and worldwide indicates the necessity for the widespread use of micromobility vehicles. Although the A'WOT technique has been used in various fields in the literature, no study has been found on transportation systems. This study aims to assess the potential of micromobility in intelligent transportation systems in Turkey with A'WOT analysis. Data were collected from 5 experts. The findings show that strategy documents for ITS have been created in Turkey, but the strategies for micromobility are insufficient. Furthermore, integrating micromobility vehicles with different transportation modes is a significant opportunity, and inadequate micromobility infrastructure and lack of driving safety are main threats.

Keywords: ITS · Micromobility · Turkey · A'WOT

1 Introduction

Technological developments emerge as a driver of the solution to transportation problems. Technological developments that create change in every field also lead to new trends in transportation, providing various opportunities for transportation systems to be safer, higher quality, convenient, economical, multimodal, environmentally friendly, and sustainable [1]. This situation has emerged the concept of Intelligent Transportation System (ITS). Qi (2008) defines Intelligent Transportation Systems (ITS) as a "broad range of diverse technologies applied to transportation to make systems safer, more efficient, more reliable and more environmentally friendly, without necessarily having to physically alter existing infrastructure" [2]. ITS provides various benefits such as increasing mobility, travel comfort, road and driving safety, reducing traffic congestion, emission, fuel consumption, noise pollution, environmental damage and health problems, shortening travel time and accelerating economic growth [3, 4].

For more environmentally friendly, efficient and safe transportation, it is necessary to use ITS applications and to develop these applications in parallel with the advancing technology. Today, it is aimed to develop and expand the ITS applications by using information communication technologies (ICT) to effectively manage various goals such as

M. Ben Ahmed et al. (Eds.): SCA 2022, LNNS 629, pp. 226–236, 2023.
https://doi.org/10.1007/978-3-031-26852-6_21

reducing emissions, shortening travel time, increasing mobility, and decreasing accident rates [4, 5]. In parallel with this purpose, ITS-related innovations were applied such as "national road network, urban traffic management and control, travel information, ticketing, integrated modes of transportation, logistics and fleet management, C-ITS, connected and autonomous vehicles, data management, road safety, reduced emissions, electric vehicles, communication technologies, mobility and accessibility, smart city and similar ITS components" [2, 6]. With the increase in population in cities, especially developed countries benefit from ITS for primary purposes such as regulation and control of traffic, more efficient transportation, fuel and energy savings and reduced emissions. One of these implementations is micromobility, which has the potential to offer users an alternative mode of transportation and is a new trend in urban mobility.

Micromobility vehicles, a new transportation strategy, have been rapidly increasing worldwide recently, and especially developed countries are using these vehicles to solve the problems experienced in the urban mobility system [7, 8]. The population growth rate and number of private car gradually increase in Turkey [9, 10]. Increases in urbanization and private car ownership increase emissions and energy consumption and cre-ate transportation demand and traffic density. This state makes clear the necessity of the widespread use of micromobility vehicles, one of the ITS applications in Turkey.

In this context, strategies should be created to increase the use of micromobility vehicles in ITS and ensure the sustainability of the urban transportation system. Projects should be carried out within the framework of the strategies created. SWOT analysis is generally used to develop strategies. However, since the SWOT analysis was insufficient to quantitatively determine the importance of the strategies or factors, A'WOT analysis was used.

A'WOT analysis in the literature; tourism [11–13], rural development [14], finance [15] forestry [16], management and strategic planning [17, 18] etc. used in various fields. However, no study using A'WOT analysis in transportation systems has been found in the literature. In this regard, the framework of the study consists of ITS and micromobility.

The study aims to evaluate the potential of micromobility in intelligent transportation systems in Turkey with A'WOT analysis. First, reports and strategy documents in Turkey were examined in detail to provide input to the SWOT analysis. Then, SWOT analysis for ITS in Turkey was created based on micromobility. Finally, the priorities of the SWOT sub-factors were ranked by using the opinions of 5 experts. The ranking process was carried out using the AHP method.

1.1 Micromobility

Large-scale cities, experienced with the various challenges arising from private vehicle use, are working to develop a more sustainable urban transport system globally (Oeschger et al. 2020). Nowadays, cities accept micromobility as a promising urban transportation mode for those struggling with transportation problems [8, 19].

Micromobility is an innovative urban transport solution that enables short-distance journeys, including the first and last mile. Micromobility; is a flexible, sustainable, cost-effective alternative transportation mode [20], reducing traffic congestion [21] and the use of private vehicles, especially for short-distance trips [22, 23] and is integrated with public transportation [24] increases its preferability. On the other hand, the desire of cities to switch to low-carbon and sustainable urban transportation modes also increases the popularity of micromobility vehicles [19].

Micromobility vehicles are generally light vehicles with a maximum speed of 45 km/h and work with human power (bicycle) or electricity (e-bike, e-scooter, etc.). They can also be private or shared (shared bike, shared e-bike, shared e-scooter, etc.) [25]. ICT and the sharing economy are expected to improve the urban transportation system by reducing the use of private motor vehicles [26]. The rapid developments in mobility systems [27] and mobile applications popularize shared micromobility vehicles, supporting this idea.

Efforts to popularize micromobility vehicles around the world are continuing rapidly. In order to be able to compete with global cities and create sustainable urban transportation systems, there is a need to use newly emerging alternative vehicles in the ITS field, such as micromobility vehicles in Turkey.

2 Material and Method

2.1 Material

This study aims to evaluate the potential of micromobility in intelligent transportation systems in Turkey with A'WOT analysis. First of all, national and international literature was searched in different databases. In addition, reports and strategy documents prepared by public institutions and organizations that develop policies for ITS in Turkey have been used. Examined documents can be listed as follows: "National Science Technology and Innovation Strategy and Action Plan, 2003–2023" [28], "9. Development Plan, 2007–2013" [29], "Ministry of Transport, Maritime Affairs and Communications Strategic Plan, 2009–2013" [30], "National Climate Change and Strategy Document, 2010–2020" [31], "Transportation and Communication Strategy Target 2023, 2011–2023" [32], "Energy Efficiency Strategy Document, 2012–2023" [33], "National Intelligent Transportation Systems Strategy Document and Action Plan, 2014–2023" [34], "10. Development Plan, 2014–2018" [35], "Information Society Strategy and Action Plan", 2015–2018 [36], "Current Situation Analysis of Turkey in the Scope of Sustainable Development Goals, 2017" [37], "Ministry of Transport and Infrastructure Strategic Plan, 2019–2023" [38], "11. Development Plan, 2019–2023" [39], "National Smart City Strategy and Action Plan, 2020–2023" [40], "National Smart Transportation Systems Strategy Document and 2020–2023 Action Plan" [6], "Highway Traffic Safety Strategy Document and Action Plan, 2021–2023" [41]. Among these, the National Smart Transportation Systems Strategy Document and Action Plan, 2014–2023, National Smart City Strategy and Action Plan, 2020–2023 and National Smart Transportation Systems Strategy Document and 2020–2023 Action Plan come to the fore.

2.2 Method

Mixed method was used in the study. Mixed method; is defined as the use of quantitative and qualitative research methods together in research [42]. There are different mixed method designs depending on the purpose of the research. In this study, the exploratory sequential mixed research design was used. The exploratory sequential mixed design is use of a qualitative method first and then a quantitative method [43].

In this context, documentation analysis, which is one of the qualitative research methods, was used in the analysis of strategy documents to create the SWOT matrix and to form a base for the AHP. Afterward, SWOT analysis was performed first for A'WOT analysis. Literature reviews, reports and strategy documents on ITS and micromobility were used for SWOT analysis. Then, the SWOT sub-factors were ranked using the AHP method, one of the quantitative research methods. In the ranking phase, firstly, the pairwise comparison matrix of the factors was made. The pairwise comparison scale used in the study is given in Table 1. Pairwise comparison questionnaires were prepared online and sent to 5 experts working in the field of ITS in public institutions and organizations and the private sector in Turkey between 15.03.2022 and 25.03.2022. Then the relative weights were calculated. In this process, the consistency ratio was also calculated. In addition, the consistency ratio was checked in order to measure whether the factors were consistent or not when comparing the factors. A consistency ratio of less than 0.10 makes the matrix reliable.

Table 1. AHP pairwise comparison matrix.

Intensity of importance	Description
1	Equal importance/intensity of two elements
3	Moderate importance of one elements over the other
5	Strong importance of one element over the other
7	Very strong importance of one element over the other
9	Extreme importance of one element over the other
2, 4, 6, 8 are intensities to express intermediates values	

3 Results

In the first stage, the SWOT analysis was applied. Within the scope of the study, SWOT analysis was created by examining ITS and micromobility reports and strategy documents of national institutions and organizations, and the literature review.

Strengths

S1- Availability of micromobility service providers
S2- Supporting the use of micromobility in the National Intelligent Transport Systems Document and the 2020–2023 Action Plan
S3- Widespread use of mobile applications
S4- Being an alternative mode of transport
S5- Beginning of legislative preparations for micromobility vehicles
S6- In ITS vision of Turkey includes "human and environment-oriented transportation system supported by advanced information technologies"

Weaknesses

W1- Insufficient micromobility infrastructure
W2- Insufficient access to shared micromobility vehicles
W3- Insufficient incentives for awareness of the use of micromobility vehicles
W4- Inadequate integration of micromobility vehicles with other transport modes
W5- Lack of strategies for micromobility in transportation policies in Turkey
W6- Insufficient governmental/municipal-supported shared micromobility service providers

Opportunities

O1- Support of micromobility vehicle use by ITS policies around the world
O2- Using micromobility vehicles in various fields such as transportation, recreation, logistics, etc
O3- Use of individual and shared micromobility vehicles and their electric versions in urban transportation
O4- Electric micromobility vehicles require less effort
O5- Increasing awareness of energy efficiency and environmental protection
O6- Access to international funds
O7- Reducing the use of private vehicles and being able to integrate with public transportation

Threats

T1- The impact of weather conditions and topography on micromobility vehicle use
T2- The public's resistance to micromobility vehicle use
T3- High micromobility infrastructure costs
T4- Preference instead of walking
T5- Lack of investment by private service providers for small and medium-sized cities
T6- Driving safety

In the second stage, pairwise comparisons were made by collecting data from 5 experts. This process is aimed to compare the relative importance of the factors. While determining the order of importance of the factors, the scale in Table 1 was considered. The eigenvalues of the matrix were calculated by the AHP method and the SWOT factors were ranked (Table 2).

The weights of strength factors S1, S2, S3, S4, S5 and S6 were 0,07, 0,197, 0,077, 0,194, 0,281 and 0,181 respectively. The weights of weakness factors W1, W2, W3, W4, W5 and W6 were 0,241, 0,093, 0,070, 0,141, 0,353 and 0,102 respectively. The weights of opportunity factors O1, O2, O3, O4, O5, O6 and O7 were 0,102, 0,147, 0,189, 0,075, 0,131, 0,068 and 0,287 respectively. The weights of threat factors T1, T2, T3, T4, T5 and T6 were 0,095, 0,134, 0,218, 0,202, 0,210 and 0,142 respectively.

The consistency ratio is 0.068 for strengths, 0.040 for weaknesses, 0.081 for opportunities, and 0.063 for threats. These calculated consistency ratios indicate that the pairwise comparison matrices are consistent.

Table 2. Weights and consistency ratios obtained as a result of the application.

Factors	Subfactors	Weights	Consistency
Strengths	S1- Availability of micromobility service providers	0,070	0,068
	S2- Supporting the use of micromobility in the National Intelligent Transport Systems Document and the 2020–2023 Action Plan	0,197	
	S3- Widespread use of mobile applications	0,077	
	S4- Being an alternative mode of transport	0,194	
	S5- Beginning of legislative preprations for micromobility vehicles	0,281	
	S6- Turkey's ITS vision is "human and environment-oriented transportation sys-tem supported by advanced information technologies"	0,181	
Weaknesses	W1- Insufficient micromobility infrastructure	0,241	0,040
	W2- Insufficient access to shared micromobility vehicles	0,093	
	W3- Insufficient incentives for awareness of the use of micromobility vehicles	0,070	
	W4- Inadequate integration of micromobility vehicles with other transport modes	0,141	
	W5- Lack of strategies for micromobility in transportation policies in Turkey	0,353	
	W6- Insufficient governmental/municipal-supported shared micromobility ser-vice providers	0,102	
Opportunities	O1- Support of micromobility vehicle use by ITS policies around the world	0,102	0,081
	O2- Using micromobility vehicles in various fields such as transportation, recrea-tion, logistics, etc	0,147	
	O3- Use of individual and shared micromobility vehicles and their electric ver-sions in urban transportation	0,189	
	O4- Electric micromobility vehicles require less effort	0,075	
	O5- Increasing awareness of energy efficiency and environmental protection	0,131	
	O6- Access to international funds	0,068	
	O7- Reducing the use of private vehicles and being able to integrate with public transportation	0,287	

(*continued*)

Table 2. (*continued*)

Factors	Subfactors	Weights	Consistency
Threats	T1- The impact of weather conditions and topography on micromobility vehicle use	0,095	0,063
	T2- The public's resistance to micromobility vehicle use	0,134	
	T3- High micromobility infrastructure costs	0,218	
	T4- Preference instead of walking	0,202	
	T5- Lack of investment by private service providers for small and medium-sized cities	0,210	
	T6- Driving safety	0,142	

4 Discussion and Conclusion

Transportation problems and developing technology have caused a paradigm shift in transportation systems. Today, global and large-scale cities focus on micromobility studies to establish a more sustainable urban transportation system. In this context, implementations of micromobility in cities should be widespread in order to increase Turkey's global competitiveness and ensure sustainability in urban mobility. In this way, transportation problems might be reduced.

The first study in the field of ITS in Turkey was carried out in 1984. However, until the "National Science and Technologies Policies, 2003–2023" strategy document was prepared in 2004, the concept of "intelligent" is not mentioned in any document. This indicates that ITS implementations have started to be used in Turkey, but it takes time to be included in strategic documents. With the emergence of the concept of micromobility in 2017, especially developed countries started to create policies regarding the use of these vehicles. Policy and legislation preparations are of primary importance in the micromobility system, as they include various subjects such as revealing the legal obligations of users and service providers, ensuring users' physical and personal data security, and providing usage information. The vision of the "National Intelligent Transportation Systems Strategy Document and 2020–2023 Action Plan" prepared in Turkey in 2020 includes the human and environment-oriented transportation system. Still, the strategies created for micromobility vehicles are insufficient. For this reason, the necessary importance should be given to micromobility and infrastructure, incentives, etc. should be provided for its widespread use in the strategy documents and strategies should be developed. In addition, the "E-scooter regulation" was prepared in 2021 and legislative arrangements for micromobility vehicles were implemented. The acceleration of regulations on micromobility in recent years reveals that awareness has increased. These regulations may contribute to the widespread use of micromobility.

Another critical element in the use and development of micromobility vehicles is infrastructure. Currently, the infrastructure for micromobility vehicles is insufficient in Turkey. The biggest obstacle to the lack of infrastructure is the high infrastructure costs. However, when the lack of roads is added to the high density of vehicles in cities, the use of micromobility vehicles becomes even more difficult. The drivers' lack of driving safety negatively affects the vehicles' preferability. Furthermore, the integration of micromobility vehicles with different transportation modes and the inadequacy of public/municipal-supported service providers positively affect micromobility vehicle use.

On the contrary, the failure of private service providers to invest in small and medium-sized cities further deteriorates the balance with large-scale cities. Another critical issue that needs attention is the public resistance to micromobility vehicles. Because to provide a sustainable urban transportation system, these vehicles must be used by the city residents. Recently some municipalities' implementing "urban bicycles" and integrating them into the public transportation system, supporting user incentives in strategy documents and carrying out studies to raise awareness are promising studies.

Recently, climate change and global warming problems have led countries, cities and even individuals to be more environmentally friendly. Developed cities support the use of micromobility vehicles in their policies for ITS. Because micromobility vehicles provide energy efficiency and reduce emissions, affecting the preference of use among individuals with increased environmental awareness. Moreover, the potential to replace the use of private vehicles, which cause the most emissions in transportation and its integration with public transportation are essential benefits to be considered in ensuring environmental sustainability.

Micromobility vehicles are used in transportation, recreation, logistics, etc. used, in various fields. It is mainly used as an alternative transportation mode to solve the first mile-last mile problem. These uses reduce emissions and energy consumption caused by transportation, shorten the travel times of individuals in traffic and increase the quality of life. The opportunities offered by micromobility are significant for the widespread use of these vehicles. To acceptance of micromobility vehicles for residents, public/municipal and private investments should increase. With these investments, access to shared micromobility vehicles will be increased. Because in order to create a sustainable urban transportation system, the use of these vehicles should increase not only in transportation but also in all areas that will improve the quality of life of individuals, such as logistics and recreation. Today, especially in developing countries use of micromobility vehicles in their urban mobility plans and the increase in sustainability in transportation support this situation.

In this study, SWOT, a common planning tool, was integrated with AHP, one of the multi-criteria decision-making methods, and used to evaluate the potential of micromobility in ITS in Turkey. Thus, it is aimed to improve the usability of SWOT analysis. The results are presented in an explanatory manner with quantitative information. In addition, performing A'WOT analysis based on expert opinion provides an excellent opportunity to examine the views of different stakeholders regarding the decision-making process. Based on micromobility, this study reveals the factors that should be prioritized in the ITS process in Turkey; It makes an essential contribution to the literature, strengthening the quantitative knowledge bases of micromobility and ITS, and to researchers working on this subject.

Micromobility policies have only recently started to be included in strategy documents in Turkey. Therefore, the results of the policies cannot be observed yet. In future studies,

- It is recommended to evaluate the effects of policies on micromobility.
- Urban development strategies can be developed according to city scales to increase the use of micromobility using methods such as A'WOT analysis.
- Different MCDM can be used for the comparison of results.

References

1. Kenanoğlu, M.E., Aydın, M.: Akıllı ulaşım sistemlerinin dışsallık bağlamında değerlendirilmesi: Seçilmiş ülke uygulamaları. Çanakkale Onsekiz Mart Üniversitesi Uluslararası Sosyal Bilimler Dergisi **3**(2), 363–387 (2018)
2. Qi, L.: Research on intelligent transportation system technologies and applications. In: 2008 Workshop on Power Electronics and Intelligent Transportation System, pp. 529–531. IEEE (2008)
3. Crawford, D.: Benefits of ITS. https://rno-its.piarc.org/en/its-basics/benefits-its. Last Accessed 01 May 2022
4. Yan, X., Zhang, H., Wu, C.: Research and development of intelligent transportation systems. In: 2012 11th International Symposium on Distributed Computing and Applications to Business, Engineering & Science, pp. 321–327. IEEE (2012)
5. Njord, J.,et al.: Safety applications of intelligent transportation systems in Europe and Japan (No. FHWA-PL-06-001). https://international.fhwa.dot.gov/ipsafety/ipsafety.pdf. Last Accessed 11 May 2022
6. Ministry of Transport and Infrastructure: National Intelligent Transportation Systems Strategy Document and 2020–2023 Action Plan. https://www.uab.gov.tr/uploads/announcements/ulusal-akilli-ulasim-sistemleri-strateji-belgesi-v/ulusal-akilli-ulas-im-sistemleri-strateji-belgesi-ve-2020-2023-eylem-plani-eng.pdf. (2020). Last Accessed 13 May 2022
7. Gössling, S.: Integrating e-scooters in urban transportation: problems, policies, and the prospect of system change. Transp. Res. Part D: Transp. Environ. **79**, 102230 (2020)
8. Oeschger, G., Carroll, P., Caulfield, B.: Micromobility and public transport integration: the current state of knowledge. Transp. Res. Part D: Transp. Environ. **89**, 102628 (2020)
9. TÜİK.: Adrese Dayalı Nüfus Kayıt Sistemi Sonuçları. https://data.tuik.gov.tr/Bulten/Index?p=Adrese-Dayali-Nufus-Kayit-Sistemi-Sonuclari-2020-37210. Last Accessed 04 May 2022
10. TÜİK.: Motorlu Kara Taşıtları. https://data.tuik.gov.tr/Bulten/Index?p=Motorlu-Kara-Tasitlari-Aralik-2020-37410. Last Accessed 04 May 2022
11. Akbulak, C., Cengiz, T.: Determining ecotourism strategies using A'WOT hybrid method: Case study of troia historical national park, çanakkale, turkey. Int. J. Sust. Dev. World **21**(4), 380–388 (2014)
12. Baek, I.H., Kim, H.D.: Study on marine tourism development strategy of yeosu city using a A'WOT analysis technique. Journal of Fisheries and Marine Sciences Education **25**(6), 1419–1428 (2013)
13. Heydarzadegan, P., Arslan, M.: Merağa ili (Iran) turizm potansiyelinin A'WOT analizi ile değerlendirilmesi. Düzce Üniversitesi Orman Fakültesi Ormancılık Dergisi **8**(1), 107–117 (2010)

14. Öztaş Karlı, R.G.: Bilgi ve iletişim teknolojilerinin kırsal kalkınmadaki Rolünün A'WOT analizi ile değerlendirilmesi. Yüzüncü Yıl Üniversitesi Tarım Bilimleri Dergisi **31**(2), 494–501 (2021)
15. Çağıl, G., Candemır, G.: Türk bankacılık sektörünün fintek'ler karşısında A'WOT analizi. Maliye ve Finans Yazıları **113**, 207–238 (2020)
16. Güngör, E.: Determination of optimum management strategy for honey production forest lands using A'WOT and conjoint analysis: A case study in turkey. Appl. Ecol. Environ. Res. **16**(3), 3437–3459 (2018)
17. Canto-Perello, J., Curiel-Esparza, J., Calvo, V.: Strategic decision support system for utility tunnel's planning applying A'WOT method. Tunn. Undergr. Space Technol. **55**, 146–152 (2016)
18. Kajanus, M., Kangas, J., Kurttila, M.: The use of value focused thinking and the A'WOT hybrid method in tourism management. Tour. Manage. **25**(4), 499–506 (2004)
19. Abduljabbar, R.L., Liyanage, S., Dia, H.: The role of micro-mobility in shaping sustainable cities: A systematic literature review. Transp. Res. Part D: Transp. Environ. **92**, 102734 (2021)
20. Shaheen, S., Cohen, A., Chan, N., Bansal, A.: Sharing strategies: carsharing, shared micromobility (bikesharing and scooter sharing), transportation network companies, microtransit, and other innovative mobility modes. In: Transportation, Land use, and Environmental Planning (pp. 237–262). Elsevier (2020)
21. Sperling, D.: Three Rrevolutions Steering Automated, Shared, and Electric Vehicles to a Better Future. Island Press (2018)
22. Clewlow, R.: The Micro-Mobility Revolution, https://medium.com/populus-ai/the-micro-mobility-revolution-95e396db3754. Last Accessed 27 Apr 2022
23. Tiwari, A.: Micro-mobility: the next wave of urban transportation in India, https://yourstory.com/journal/micro-mobility-edc6x8f1y1. Last Accessed 27 Apr 2022
24. Møller, T. H., Simlett, J., Mugnier, E.: Micromobility: Moving Cities into A Sustainable Future.London, UK, EY. https://assets.ey.com/content/dam/ey-sites/ey-com/en_gl/topics/automotive-and-transportation/automotive-transportation-pdfs/ey-micromobility-moving-cities-into-a-sustainable-future.pdf. Last Accessed 02 May 2022
25. Dia, H.: Rethinking urban mobility: Unlocking the benefits of vehicle electrification. In: Decarbonising the Built Environment, 83–98. Palgrave Macmillan, Singapore (2019)
26. Machado, C.A.S., de Salles Hue, N.P.M., Berssaneti, F.T., Quintanilha, J.A.: An overview of shared mobility. Sustainability **10**(12), 4342 (2018)
27. Shaheen, S., Chan, N.: Mobility and the sharing economy: potential to facilitate the first-and last-mile public transit connections. Built Environ. **42**(4), 573–588 (2016)
28. The Scientific and Technological Research Council of Turkey: National Science Technology and Innovation Strategy and 2003–2023 Action Plan https://www.tubitak.gov.tr/tubitak_content_files/vizyon2023/Vizyon2023_Strateji_Belgesi.pdf. Last Accessed 15 May 2022
29. State Planning Organization: Ninth National Development Plan, 2007–2013. https://www.sbb.gov.tr/wp-content/uploads/2018/11/Dokuzuncu-Kalk%C4%B1nma-Plan%C4%B1-2007-2013%E2%80%8B.pdf. Last Accessed 15 May 2022
30. Ministry of Transport, Maritime Affairs and Communications of the Republic of Turkey Presidency of Strategy and Budget: Ministry of Transport, Maritime Affairs and Communications Strategic Plan, 2009–2013. https://sgb.uab.gov.tr/uploads/pages/stratejik-planlar/2009-2013-stratejik-plan.pdf. Last Accessed 15 May 2022
31. Presidency of Strategy and Budget: National Climate Change Strategy 2010–2020. http://www.sp.gov.tr/upload/xSPTemelBelge/files/k0fZ2+Ulusal_Iklim_Degisikligi_Strateji_Belgesi_2010-2020.pdf. Last Accessed 15 May 2022
32. Ministry of Transport: Transportation and Communication Strategy Target 2023. http://www.sp.gov.tr/upload/xSPTemelBelge/files/93C5Y+Turkiye_Ulasim_veIletisim_Stratejisi.pdf. Last Accessed 12 May 2022

33. Ministry of Energy and Natural Resources: Energy Efficiency Strategy Document, 2012–2023. https://www.resmigazete.gov.tr/eskiler/2012/02/20120225-7.htm. Last Accessed 13 May 2022
34. Ministry of Transport, Maritime Affairs and Communications of the Republic of Turkey Presidency of Strategy and Budget: National Intelligent Transportation Systems Strategy Document and 2014–2023 Action Plan. http://www.sp.gov.tr/tr/temel-belge/s/175/Ulusal+Akilli+Ulasim+Sistemleri+Strateji+Belgesi+_2014%202023_+ve+Eki+Eylem+Plani+_2014-2016. Last Accessed 13 May 2022
35. Ministry of Development: Tenth Development Plan, 2014–2018. https://www.sbb.gov.tr/wp-content/uploads/2018/11/Onuncu-Kalk%C4%B1nma-Plan%C4%B1-2014-2018.pdf. Last Accessed 16 May 2022
36. Ministry of Development Information Society Department: Information Society Strategy and 2015–2018 Action Plan. http://www.sp.gov.tr/tr/temel-belge/s/109/Bilgi+Toplumu+Stratejisi+ve+Eylem+Plani+_2015-2018. Last Accessed 15 May 2022
37. Ministry of Development: Analysis Report of Turkey's Current Situation within the Scope of Sustainable Development Goals. http://www.ceidizleme.org/ekutuphaneresim/dosya/753_1.pdf. Last Accessed 15 May 2022
38. Ministry of Transport and Infrastructure: Ministry of Transport and Infrastructure Strategic Plan, 2019–2023. https://www.uab.gov.tr/uploads/pages/stratejik-yonetim/uab-2019-2023-stratejik-plani-16-10-2019.pdf. Last Accessed 15 May 2022
39. Ministry of Development: Eleventh Development Plan 2019–2023. https://www.sbb.gov.tr/wp-content/uploads/2019/07/OnbirinciKalkinmaPlani.pdf. Last Accessed 13 May 2022
40. Ministry of Environment, Urbanisation: National Smart City Strategy and 2020–2023 Action Plan. https://www.akillisehirler.gov.tr/wp-content/uploads/EylemPlani.pdf. Last Accessed 15 May 2022
41. General Directorate of Security, Traffic Department: Road Safety Action Plan 2021–2023. http://www.trafik.gov.tr/kurumlar/trafik.gov.tr/01-Haberler/03-2021/2021_2023-Karayolu-Trafik-Guvenligi-Eylem-Plani.pdf. Last Accessed 13 May 2022
42. Creswell, J.W.: A Framework for Design. Research Design: Qualitative, Quantitative, and Mixed Methods Approaches. GE Publications (2003)

Learning Healthcare Ecosystems for Equity in Health Service Provisioning and Delivery: Smart Cities and the Quintuple Aim

Nabil Georges Badr(✉)

Higher Institute of Public Health, Saint Joseph University, Beirut, Lebanon
nabil@itvaluepartner.com

Abstract. Equitable care is about providing care that does not vary in quality because of personal characteristics such as gender, ethnicity, geographic location, and socioeconomic status. We are interested in codifying the capabilities of a learning healthcare ecosystem in smart cities to provide equitable health to reach the goals of the quintuple aim. We look at this problem through the lens of Service-dominant logic in Healthcare. Based on our review we advance a call to action for equitable learning healthcare ecosystems and conceptualize health service provisioning and delivery using digital health technology in a smart health ecosystem, powered by knowledge integration in smart cities.

Keywords: Quintuple aim · Smart city · Smart healthcare · Equitable care · Telehealth · Artificial intelligence

1 Introduction

Long defined as the Triple Aim (better health outcomes, better patient experience, and lower costs), value in healthcare ecosystems has expanded in recent years to include improvements in staff experience [1] and equity [2], bringing the definition of value to encompass all Quintuple Aim.

Equitable care is about providing care that does not vary in quality because of personal characteristics such as gender, ethnicity, geographic location, and socioeconomic status [3]. There is a connection between people's access to and understanding of health services and their own health. An emphasis on health care equity guarantees that a healthcare system is purposefully structured to reduce inequities in healthcare processes and outcomes, particularly for those from low-income families. The ability of clinicians, payers, and other healthcare stakeholders to understand how these nonclinical determinants influence the health trajectories of the patients they care for is critical to effective and equitable population health and cost management.

Equity in healthcare access and service coverage is also about the capacity to make decisions, assess the need, personal economic, social and cultural circumstances influencing access, and physical accessibility [4]. Disease surveillance, healthcare utilization registries, health services statistics, and administrative records are examples of routine

M. Ben Ahmed et al. (Eds.): SCA 2022, LNNS 629, pp. 237–251, 2023.
https://doi.org/10.1007/978-3-031-26852-6_22

data acquisition that provide information for monitoring the population's health status and outcomes, but these records only provide information on individuals who seek healthcare.

The deployment of smart cities provides an opportunity to explore new ways to collect data from all citizens [5]. Medical care is critical for smart city growth, so providing high quality (equitable) medical care is the most difficult goal for city governments to achieve. Decision systems in healthcare must adapt to context aware decision making for healthcare delivery in smart cities [6]. Evidence collected and analyzed must address any challenges to knowledge use including engagement activities to increase awareness [7]. For our paper, we are curious to codify capabilities of a learning healthcare ecosystems in smart cities for the provisioning and delivery of equitable health to reach the goals of the quintuple aim. We do this through a survey of the literature on the Quintuple Aim while mining for evidence of technology in reaching the aim of health equity.

2 Background

2.1 SD Logic and Technologies for Equitable Health Care

The notion of service-dominant logic has been discussed in economics for centuries, with economists such Fredric Bastiat debating it as early as 1848. SD logic is a logic of cooperation in which players use their applicable knowledge and abilities (competences) to benefit others while also benefiting themselves [8]. SD logic is a paradigm for understanding value generation between organizations (here, in the context of Health Systems) and their customers (here, our patient population). Each actor is always learning how to better serve and function in a dynamic environment because of its interactional nature and the constant adjustment of actors and how they appraise value [9]. In our context, we discuss value in the form of "Equitable care" and the exchange of service in forms of service provisioning and service delivery. In SD logic, service is the fundamental basis of exchange. In our healthcare equity context, service provisioning value generation activities encompass the collection of resources, skills and functionality or knowledge, that lead to equitable care, while service delivery is the means of realizing the competitive advantage, here in the form of equitable care. As smart cities encourage further data collection and better focus on the parameters to realize equity in provisioning, they are better prepared to deliver equitable health care services. Therefore, the value proposition is at its peak when the underserved population receives equitable access to care. As in the case of an SD logic, all social and economic actors are resource integrators that transform health service into equitable service when beneficiaries acknowledge the value in form of better health outcome. For example, the care at home model, in which individuals who are functionally or practically homebound receive primary care in the home, are typically otherwise left out of the care value circle if the delivery of care is not at their home, i.e. not cared for equitably. In this case, access to online information through a virtual health portal, usually powered by technology such as AI and Telehealth, is consistent with SD logic as these services focus on community rather than the individual interaction, in a way that information collected from different encounters help shape the response of other future encounters. In this case, patients and providers (actors) are organizing on their own to help each other, therefore establishing the care equity desired. They are cocreating value by integrating their resources and experiences.

2.2 Smart Cities via Smart Healthcare

In smart cities, complex smart service ecosystems are the outcome of systematic methods of continuous learning, timely data collecting, rational innovation, social responsibility, and network governance, rather than basic intuitions [10]. Smart cities need to take advantage of the opportunities that the knowledge-based economy and society can bring to the city. The integration of knowledge and the refinement required must leverage digital health technology components that enable the transformation and development of smart cities, to facilitate the ability to integrate, create and reconfigure both internal and external competences to manage knowledge that originates from within and beyond the ecosystem boundaries. Knowledge, acquired from various actors in the healthcare ecosystem, at the core of its processes, foster a sustainable growth [11].

The idea of learning is particularly important for socalled "intelligent" systems that may be able to selfreconfigure in order to resist / persist over time. While adapting and continuously altering, learning health systems examine how environmental, policy, and system level activities will influence the social determinants of health of individual population groups and modify practices to influence their health behavior [12]. This leads to the nonoptional transformation of the care ecosystem into a selflearning system using technology, data and analytics to maximize the value and equity of care.

2.3 Concepts of Digital Health Connected to the Quintuple Aim

Our survey of the literature isolated digital health technologies such as AI and virtual health (telehealth) as essential means to achieving the quintuple aim. Artificial intelligence (AI) which "*refers to the capacity of computers or other machines to exhibit or simulate intelligent behavior*" (Oxford English Dictionary) has occupied a large presence in the practice of care and enabled its transformation to settings outside the hospital and clinic [13]. AI has disrupted healthcare on many fronts. On the individual level for better personal health, to augment primary care outside of the clinical encounter, or devising population health strategies for improving medication adherence [14].

Inspired by our literature review, in the next section we discuss Artificial Intelligence (AI) and Digital Health (telehealh) as cornerstones for equitable care provisioning and delivery. We use the lens of SD Logic to explore the potential for digital health as an opportunity to bridge the equity gap and reach the quintuple aim. Samples of our findings where the literature was explicit in discussions of health care equity through digital health are in Table 1.

3 Digital Health Aimed at Equity in Smart Cities

The rise of smart cities is mostly due to advances in data science, with the goal of improving people's lives and viability in our densely urbanized population [23]. In our review, we found evidence that AI in primary care [16], is an essential factor in achieving the

Table 1. Concepts of digital health, artificial intelligence and telehealth connected to the quintuple aim

Artificial Intelligence (AI)
- AI enabled its transformation to settings outside the hospital and clinic [13]
- Investments in inclusive primary care research using AI [15]
- Promoting inclusive and moral AI that mitigates rather than health inequities [16]
- Adolescent health, with multidisciplinary developmental science [17]
- Improving personal health and medication adherence through AI [14]
- AI models are required to understand the distinct needs of the populations [18]
Virtual Care or Telehealth (TH)
- Robust, comprehensive telehealth coverage is critical to improving pediatric access and quality of care and services, particularly for under-resourced populations [19]
- Telehealth for at home Care [20]
- Care for kids with severe chronic illnesses is becoming an expanding part of pediatric practice at a distance [21]
- Specialty services offered at a distance - expanded beyond dermatology to include cardiology, nephrology, and haematology [22]

quintuple aim of health care [24]. AI models continue to outpace adoption in existing healthcare for quality care and to fuel concepts of equity in providing care [25–27]. The ability to collect and analyze data for evidence-based approaches to achieve equity would be an equalizer if utilized safely, carefully, and with good intentions. If enough patients from the society provide data to these data analyses, the advances outside of the inequitable clinical research field could help to bridge the existing research equity gap. With technologies such as AI fed by a plethora of sensors through the Internet of Things (IoT) platforms, layers of smartness have been added to hospitals, such as remote monitoring, chronic disease management, medication management, patient selfmanagement, and workflow management [27]. On the other hand, increased social isolation can have negative effects on health outcomes. Interventions designed to mitigate the effects of isolation such as the facilitation of the presence of informal social supports including extended family, neighborhood and community resources have potential to improve health care outcomes. In this, case the role of virtual health technology (Telehealth) as a delivery mechanism for addressing shortcomings in more traditional inperson interactions between clinicians and patients, becomes important. The body of literature connects Telehealth to achieving the quintuple aim for telehealth for at home Care [20], in pediatrics [19, 21], improving access to care, sometimes through asynchronous consultation [22]. Indeed, smart city planning, developments, and growth initiatives emphasize smart healthcare. Smart city implementations of health care tele monitoring systems support vulnerable groups of people (disabled, Alzheimer's disease etc.). Other applications for remote monitoring patient progress in isolated countryside areas can be connected through telemedicine systems for measurements of vital indicators (pressure, sugar, etc.).

These are then integrated to citizens medical records for advice from the hospital/health center doctors [28].

3.1 Equity in Health Service Provisioning

Strategic Lever of Care Transformation. Alami et al. [29] suggest that AI-based systems should be seen as health system transformation levers, rather than discrete sets of technological devices. Indeed, their use could bring significant changes and impacts at several levels: technological, clinical, human and cognitive (patient and clinician), professional and organizational, economic, legal, and ethical, aiming at equity in the healthcare ecosystem. Lower costs, improved clinician wellbeing, and a focus on health equity and inclusion are all priorities for the sustainable use of AI in healthcare. Further, the growing affordability of remote monitoring devices, reduced reliance on brick and mortar healthcare infrastructure, and realtime feedback mechanisms of these tools position AI as a critical component in achieving the Quintuple Aim. The affordability of remote monitoring devices, for example, boosted web based care and real-time feedback tools, putting AI on the map as a critical component of attaining integrated and equitable care [13]. Rather of replacing human caregivers or manipulating the human interactions that are at the heart of healing, AI can be designed to augment human capabilities and support humandriven care delivery models [30].

Fairness in Data Collection and Coding. The application of advanced analytics and AI to medical practice offers a lot of promise. However, the ability to use these technologies comes with significant difficulties. High quality data are critical to healthcare research and are required to meet the quintuple goal [31]. This necessitates increasing clinical engagement to improve data collection and coding [32]. The increased use of algorithms in numerous fields of medicine raises the possibility of health disparities. Those in charge of AI in healthcare must make sure that the algorithms are not just accurate, but also objective and fair. Because many clinical trial guidelines and diagnostic tests consider a patient's race and ethnicity, a dispute has erupted about whether or not the selection of these characteristics is evidence-based. The application of AI can evaluate enormous amounts of data maintained by healthcare organizations in the form of digital scans, clinical research trials, and medical claims, and can spot patterns and insights that are typically missed by humans. In order to reduce health inequities, AI systems can be trained to lessen or eliminate potential prejudice through encouraging data transparency and diversity. Without such efforts, tools may represent the many different ways that data, computational, and analytic biases can be created, thereby widening inequities based on race, ethnicity, gender, and other sociodemographic characteristics linked to unequal health outcomes [33].

Ethics - Justifiable Use of Information The importance of anticipating and addressing social inequalities and health disparities in AI development and implementation [34], was evidenced by the recent COVID19 pandemic's disproportionate impact on racial/ethnic minority and socially disadvantaged people, for example. The pandemic has presented a number of opportunities to use AIdriven tools and applied treatments to better analyze, mitigate, and manage the pandemic and its impacts. Some ethical and social considerations were forgotten in the scope of development, and the use of modern technologies in health care settings has become more prevalent as the public health crisis has progressed. Unless they incorporate ethical AI principles into the development process in ways that intentionally promote beneficence (do only good), nonmaleficence (do not harm), autonomy (ability to decide), justice (promoting prosperity and preserving solidarity), and explicability, hastily deployed AI tools may exacerbate structural inequities that can lead to disparate health outcomes [35]. Explicability attempts to enable the other principles through transparency and accountability, while the first four are essential bioethics principles. Solutions that emphasize actor engagement and implement methods for providing methods of data privacy, feedback, flow tuning, and a justifiable use of information are necessary [36].

Digital Compassion and Empathy. There are potentials to construct and focus on AI to lower costs and improve population health, care team wellbeing, patient experience, equity, and inclusion; but, as AI in healthcare evolves, we should challenge ourselves to use the technology to provide compassionate, high quality care. We must strike a balance between innovative digital technologies for enhancing health and equitable inclusivity, because AI has the potential to worsen current imbalances in health and health care. For example, incorporating ethnography that may help sidestep the individualization of structural inequities as well as increase partnerships to improve both individual health and broader social conditions, in line with both health equity and social justice [17]. It is critical that healthcare organizations and their stakeholders have the cognitive, psychomotor, and affective abilities needed to use AI to improve and optimize care delivery [37]. Evidence of health inequities abounds, whether caring for cancer patients [38], the elderly or newborn [39]. Practitioners and technology businesses must collaborate to cocreate, promote, or support innovation and insights that can potentially enhance decisionmaking and healthcare equity, in order to make sense of the multitude of data accessible to fuel AI technologies. As AI work may ameliorate, increase, or perpetuate existing disparities, especially in times of pandemic, the concept of digital compassion and digital empathy should evolve [40].

3.2 Equity in Health Service Delivery

Smart cities, empowered by advances in technology, aim at realizing the competitive advantage, here in the form of equitable care. Through the lens of SD logic, competitive advantage is a resource integration capability that centers on serving the underserved, through the virtues of virtual health for achieving equity in care. Virtual care refers to "ways in which healthcare providers communicate with their patients at a distance". In addition to treating patients via telemedicine, doctors can communicate with them remotely using live video, audio, and instant messaging. Virtual Health programs are being developed by care centers, considering a postpandemic situations of increased awareness of equity in care, encouraging joy in practice, and bending the expense curve in particular [41]. Service delivery for rural and remote areas, integrated care, and hospital at home care and outreach services can greatly benefit from the ubiquity of virtual care. Patients and member networks want constant access to their healthcare team as well as materials that are tailored to their specific requirements. From new patient onboarding to followup care, resilient organizations are extending virtual care services and improving patient engagement. Equitable care is also about providing care with equal access; "Care that does not vary in quality because of personal characteristics such as gender, ethnicity, geographic location, and socioeconomic status". Telemedicine has connected hardtoreach rural areas and populations who are underserved [42]. The combination of highspeed internet and widespread smartphone use will enable telehealth consultations and rapid deployment of video teleconsultations from a patient's home [43, 44]. The number of patients having to attend facetoface consultations has decreased thanks to teleaudiology [45] and teledermatology [46]. The use of telehealth as a treatment option for various contagious diseases improves the continuity of care while avoiding physical contact, allowing all sorts of lionesses to participate, and improving patient outcomes [47, 48]. The underlying enabler for patientclinician interaction in the majority of these solutions is communications technology which, given its relative ease of use, low cost and ubiquitous nature, has the potential to deliver health care solutions that transcend socioeconomic class. The potential positive impact that certain interventions can have on health care outcomes by addressing patient resource, communication and navigation barriers; these interventions are often facilitated by technology and dependent upon (1) Disease or condition type; (2) phase of the care process and (3) accessibility of technology [49]. The capacity to use these technologies, however, is fraught with obstacles.

Funding for Virtual Health Infrastructure. That said, we still have an arduous task to democratize virtual care delivery. About half of the world's population is still without access to the internet. Households still lack internet connectivity at home, with the majority of the offline population concentrated in Africa (71.8%) and Asia Pacific (51.6%) [50]. Funding telemedicine infrastructure is critical, as is ensuring that telemedicine technologies are compatible and simple to use for those with impairments, the elderly [50], the children [21] and by the majority of the care providers [25]. In some parts of the world, poverty and a lack of education are the most significant barriers to telemedicine adoption [51].

Bridging the Technological Gap. While Virtual Health may improve access to health care for many people, it may exacerbate equity issues for those with limited access to advanced technologies or a lack of technological competence. Technology advances faster than firms can accept and adapt to it, necessitating "extensive employee training or retraining [25] as well as professional growth in related science, engineering, and math literacy. " Health care providers must have the skills and ability to initiate change and define the future of their discipline and practices for developing highquality care inside the digital ecosystem in order to help shape these new practices [24]. The patient population's technological gap must also be addressed. Some patients, for example, may not have smartphones but can still communicate with care providers via SMS. Conversely, due to poor socioeconomic level and computer capabilities, older persons have limited access to internetbased services; this disparity around telemedicine may exacerbate mental health problems and extend global health disparities [50]. Trust in technology, design, cognitive impairment, and physical constraints such as poor vision, hearing, or sensory impairment are all challenges to telemedicine for the elderly [52].

Consideration for Contextual Factors. Clinical consultations must address patient technology needs for virtual visit preparation, as well as piling on new components to understand the perspectives of all participants. Further, integrated healthcare interventions including patient counseling, education, alert and monitoring, and communication, should be tailored by contextual factors [49]. Equitable access must equally consider various age groups. For instance, children present unique obstacles during examinations, such as a child's level of participation during video conferencing and the necessity to rely on another person to elicit physical indications so that correct diagnoses may be made [53]. The literature identifies disparities in telehealth's impact on the workforce treating children, as well as its societal consequences, indicating that there are chances to develop a more equitable care delivery paradigm. Complex considerations such as prudence, acceptability, design, interoperability, effectiveness, diffusion, costeffectiveness, adaptability, monetizability, and security can assist ensure equity as telehealth for pediatric populations becomes more widely implemented [21].

3.3 Learning Health System for Equitable Care

A learning health system is an essential foundation to achieve the quintuple aim [32, 54]. A learning care practice's main approach is to use predictive modeling in conjunction with evidencebased best practices to achieve and maintain populationlevel health gains [41]. A learning pharmacy practice for example, continually generates datapowered discoveries as a byproduct of pharmacists' patient care process (PPCP) interactions, information from patients on drug interactions, to supplement clinical and research findings [55], transforming information for the service of equitable care. For example every encounter generates data, which is then merged with existing data and analyzed by teams of pharmacists and data scientists. The new knowledge gained is subsequently applied to all future contacts in the form of forecasts and actionable suggestions. The cycle of knowledge transformation creates a type of Information Common Goods that aid in the democratization of information and its availability for longterm societal growth [11]. As a result, such models must account for improving care to demonstrate value, investing in infrastructure to enable enhanced care, and building and leveraging partnerships to achieve health disparity reduction [56, 57]. Patients and families must be involved in equity assessments, as well as members of affected groups and communities, in the planning, design, and delivery of interventions to ensure that they are inclusive, culturally sensitive, and structurally appropriate. Decision support systems for equitable care must consider rules that identify the vulnerable populations so that disparities can be monitored. Clinical decision support would then be applied regardless of patient demographic, ethic and socioeconomic status. Care models would evolve into a sophisticated system with adaptive feedback loops that alter payment model features, payment levels, and risk adjustments in real time to maximize value and equity. This necessitates taking into account both health and social complexity, as well as the patient's/and family's community's socioeconomic situation. For example, using models to adjust risk evaluation and cost alignment to incentivize equity in meaningful outcomes (e.g., obesity reduction, smoking cessation, and avoidable hospitalizations, etc.) that may necessitate more resources and must serve a population with varying levels of affordability, access, and education.

The ability to create and integrate the knowledge base for exchange, management, re-use, sharing of information and responsible management of common data are key drivers of the evolution of societies, the progress of humanity and competitiveness. The purpose of smart cities is to use technology to combine various infrastructures and services, as well as to understand social factors in order to build human infrastructures and governance for institutional improvement and citizen involvement. Smart cities are known to have three core objectives to improve productivity, sustainability, and livability [58]. Solutions such as Artificial intelligence (AI) and telehealth support a learning healthcare system in achieving healthcare equity by closing care gaps that have been created by current care models and hence ensuring continuity of care.

4 Conclusion

Our review identified value propositions and key principles that bolster the contribution of digital health technologies (DHT) to achieving the quintuple aim (Table 2).

Table 2. Key Principles for DHT for equitable care - connected to the quintuple aim

Provide Equitable Care
• Apply fairness in data collection and coding. Accurate and timely collection measures include underlying causes of inequities in AI Models
• Focus on reducing poverty and a promoting awareness and education through AI
Enhance Patient Experience of Care
• Virtual clinical consultations must address patient technology needs literacy gaps
• Respect ethical boundaries - Justifiable use of information
Improve Population Health Outcome
• Train AI systems to lessen potential prejudice (data source diversity and transparency)
• Factor in complex considerations for equitable access (age groups and disabilities)
Ensure Practitioner Well-Being
• Emphasize digital compassion and empathy outside of the clinical encounter to improve and optimize care delivery
• Bolster training of practitioners to raise technology competence and trust
Reduce Per Capita Cost of Care
• Maintain essential priorities for sustainable use of AI - Lower costs
• Funding of telemedicine infrastructure is critical

Such principles emphasize that for care to be equitable, learning healthcare systems must apply fairness in data collection and coding with the timeliness and accuracy focused on promoting awareness and education through AI. Digital health technology can be a strategic lever to improve access, quality, and efficiency of care, however, virtual clinical consultations must address patient technology needs and technology literacy gaps and respect ethical boundaries, through justifiable use of information. AI systems must be trained to lessen potential prejudice through the inclusion of diverse data sources and transparency in analysis, while factoring in considerations for equitable access for different age groups and disabilities. Practitioners must be trained to raise technology competence and trust especially among the vulnerable patient population. Finally, government agencies and private partnerships ought to maintain essential priorities for sustainable use of AI to lower overall costs and consider funding of telemedicine infrastructure as a critical.

4.1 Contribution to Equitable Care Practice – Call to Action

The Health Information Technology for Economic and Clinical Health (HITECH) Act of 2009 promoted the adoption of meaningful use emphasizing the integration of social determinants of health in the value delivered through clinical health practice [59]. As defined by the center for disease control (CDC), the five Pillars of Meaningful Use support equitable health outcomes. In addition to outcomes of improving public health and ensuring privacy for personal health information, meaningful use principles support improving quality, safety, and efficiency of care while reducing health disparities; which requires the design and implementation of evidencebased interventions. Our review crystallizes these principles with equity and equitable care in mind.

Based on our review we advance a call to action for equitable learning healthcare ecosystems to focus on funding for the required infrastructure, reducing poverty and a promoting awareness and education to bridge the technology gap among practitioners, patients and their care circle. Virtual clinical consultations must address patient technology needs, and complex considerations for equitable access for different age groups and disabilities. Smart cities are true learning organizations, capable of not only creating, growing, and gaining information, but also of using new knowledge and skills to influence people's and organizations' behavior [60]. In the context of smart cities and under the lens of SD logic, stakeholders are required to prepare for the provisioning of a smart care system that can address equity parameters in the modeling of service provision and delivery. Practitioners in digital health ecosystems must levy essential priorities for sustainable use of AI and emphasize on lowering costs, improved clinician wellbeing, and a focus on health equity and inclusion, while augmenting primary care outside of the classical, in person clinical encounter. To achieve true health equity, however, we must reconsider the basics of integrated healthcare and technology, rather than simply digitizing old healthcare delivery models. Smart cities are true learning organizations, capable of not only creating, expanding, and receiving information, but also of influencing people and organizations' behavior with new knowledge and abilities. AI and telehealth facilitate the use of research knowledge by several stakeholders of smart cities, such as practitioners, policymakers, patients and the public. Evidence collected and analyzed through AI engines and Telehealth tools provide knowledge to service integration capabilities to synthesize and translate that evidence into interventions designed to adapt knowledge to the local context [7]. Therefore, we must assess and comprehend the demands of underrepresented communities in order to develop custom-made digital solutions that give precise, targeted health interventions that address their urgent issues. One approach is to identify the common needs of certain groups and curate a solution utilizing digital tools to address the issues and improve the efficacy and efficiency of care. Nevertheless, for equitable care delivery, an intricate understanding of the interaction of social, cultural and economic realities is needed to overcome barriers to technology access that aggravate health inequalities. We need to address connectivity, affordability and accessibility with appropriate investments, innovative and inclusive solutions, and policy support in order to bridge the gaps and create sustained holistic healthcare solutions though a complex and equitable health ecosystem.

Smart cities need Smart Healthcare. Learning health ecosystems in smart cities use knowledge co-creation practices for the integration of provisioning and delivery of equitable care (Fig. 1).

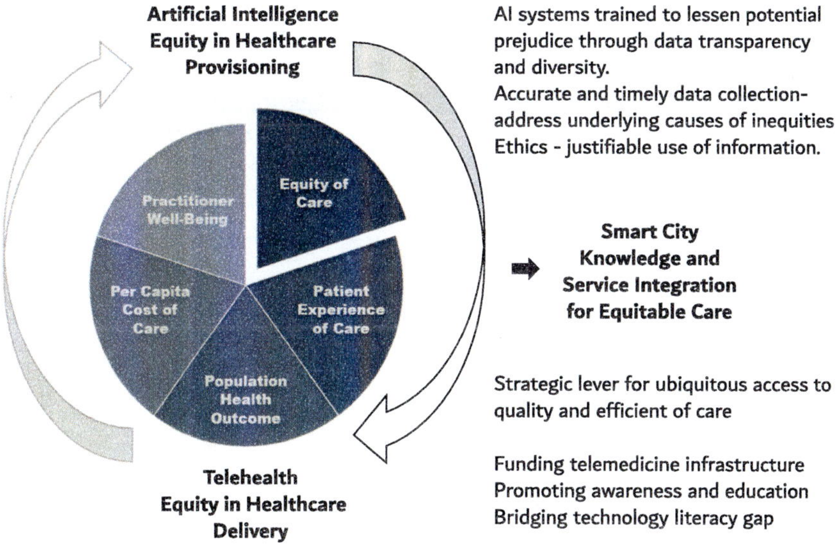

Fig. 1. Learning ecosystems for equitable health service provisioning and delivery – conceptualization

This work has demonstrated that, in order to be equitable, learning health systems must be powered by artificial intelligence (AI) in the provisioning of equitable care, that enable the collection of data that illustrate all factors of equity in a population and develop models to address the provisioning of health services that serve the quintuple aim. Correspondingly, collaboration, facilitated by digitalization have significantly increased equity in healthcare delivery using digital technology such as telemedicine for ubiquitous access to affordable care anywhere and strategic levers to improve access, quality, and efficiency of care.

Digital technology is making healthcare proactive rather than reactive, expanding and enabling access to quality healthcare for previously underserved or marginalized areas. In the future, digital healthcare will enable the delivery of value based care while also enhancing health and yielding real results. It may even pave the road for universal healthcare to become a reality. Our work provides a basis for further research in digital health in a conceptual model that aligns service provisioning ad delivery to "Equity" as value delivered from a smart healthcare ecosystem. Follow on research could reference this model to expand on the constraints, viability and sustainability of the cyclical concept of knowledge creation to advance the learning of equitable healthcare ecosystems.

4.2 Further Exploration

We learned that AI and Telehealth are strategic levers to improve access, quality, and efficiency of care. Equitable evidence based interventions must rely on artificial intelligence based on fair, accurate and timely data collection and coding, with measures of the underlying causes of inequities. AI systems must be trained to lessen potential prejudice through data transparency and diversity. In order to give the greatest potential health outcomes, the AI sector must build healthcare systems and technologies that promote fairness and equality, both in data science and clinical investigations. It is the obligation of all stakeholders to ensure that equitable principles are incorporated into its implementation and use. The problematic that AI technologies should be linked to an important outcome and the construct that will be used to prioritize and measure performance. Alignment aids in securing leadership and resource support [47].

References

1. Sikka, R., Morath, J.M., Leape, L.: The quadruple aim: care, health, cost and meaning in work. BMJ Qual. Saf. **24**(10), 608–610 (2015)
2. Nundy, S., Cooper, L.A., Mate, K.S.: The quintuple aim for health care improvement a new imperative to advance health equity. JAMA. **327**(6), 521–522 (2022)
3. Orbuch, T.L., Fine, M.A.: The context of race/ethnicity in interpersonal relationships: crossing the chasm. JSPR **20**(2), 147–152 (2003)
4. Carroll, C., Sworn, K., Booth, A., Tsuchiya, A., Maden, M., Rosenberg, M.: Equity in healthcare access and service coverage for older people: a scoping review of the conceptual literature. Integrated Healthcare Journal **4**(1), e000092 (2022)
5. Pacheco Rocha, N., Dias, A., Santinha, G., Rodrigues, M., Queirós, A., Rodrigues, C.: Smart cities and healthcare: a systematic review. Technologies **7**(3), 58 (2019)
6. Aborokbah, M.M., AlMutairi, S., Sangaiah, A.K., Samuel, O.W.: Adaptive context aware decision computing paradigm for intensive health care delivery in smart cities—A case analysis. Sustain. Cities Soc. **41**, 919–924 (2018)
7. Graham, I.D., Tetroe, J.M.: The knowledge to action framework. Models and Frameworks for Implementing Evidence-Based Practice: Linking Evidence to Action, 207, 222 (2010)
8. Vargo, S.L., Lusch, R.F.: Evolving to a New Dominant Logic for Marketing, pp.2146. Routledge (2014)
9. Joiner, K., Lusch, R.: Evolving to a new servicedominant logic for health care (2016)
10. Demirkan, H., Spohrer, J.C., Krishna, V. (eds.): The Science of Service Systems. Springer US, Boston, MA (2011)
11. Badr, N.G., Dragoicea, M., Crihana, I.: What do we know about renewable knowledge and sustainable societal growth? a scoping review. In: 17th conference of the Italian Chapter of AIS (Association for Information Systems), October 16th–17th, 2020, Pescara, Italy (October 2020)

12. Carrubbo, L., Clarizia, F., Hysa, X., Bilotta, A.: New'smarter'solutions for the healthcare complex service system. Gummesson, E., Mele, C., Polese, F.(a cura di). System Theory and Service Science: Integrating Three Perspectives In a New Service Agenda, Giannini, Napoli, 14–17 (2013)
13. Aggarwal, N., et al.: Advancing artificial intelligence in health settings outside the hospital and clinic. NAM Perspectives (2020)
14. Esmaeilzadeh, P.: Use of AIbased tools for healthcare purposes: a survey study from consumers' perspectives. BMC Med. Inform. Decis. Mak. **20**(1), 119 (2020)
15. Bierman, A.S., Tong, S.T., McNellis, R.J.: Realizing the dream: the future of primary care research. The Annals of Family Medicine (2022)
16. Lin, S.: A clinician's guide to artificial intelligence (AI): why and how primary care should lead the health care AI revolution. JABFM **35**(1), 175–184 (2022)
17. Israni, S.T., Matheny, M.E., Matlow, R., Whicher, D.: Equity, inclusivity, and innovative digital technologies to improve adolescent and young adult health. J. Adolesc. Health **67**(2), S4–S6 (2020). https://doi.org/10.1016/j.jadohealth.2020.05.014
18. Itchhaporia, D.: The evolution of the quintuple aim: health equity, health outcomes, and the economy. JACC **78**(22), 2262–2264 (2021)
19. Curfman, A.L., et al.: Telehealth: improving access to and quality of pediatric health care. Pediatrics **148**(3) (2021)
20. Allard, B.L.: INSPIREd Healthcare: A Valuebased Care Coordination Model. Sigma Theta Tau (2020)
21. Cahan, E.M., Mittal, V., Shah, N.R., ThadaneyIsrani, S.: Achieving a quintuple aim for telehealth in pediatrics. Pediatr. Clin. **67**(4), 683–705 (2020)
22. Liddy, C., Moroz, I., Mihan, A., Nawar, N., Keely, E.: A systematic review of asynchronous, providertoprovider, electronic consultation services to improve access to specialty care available worldwide. Telemed. E Health **25**(3), 184–198 (2019)
23. Jokanović, V.: Smart healthcare in smart cities. Towards Smart World: Homes to Cities Using Internet of Things, pp. 45 (2020)
24. Wiljer, D., et al.: Accelerating the appropriate adoption of artificial intelligence in health care: protocol for a multistepped approach. JMIR Res. Protoc. **10**(10), e30940 (2021)
25. Matheny, M.E., Whicher, D., Israni, S.T.: Artificial intelligence in health care: a report from the national academy of medicine. JAMA **323**(6), 509–510 (2020)
26. Smith, M., Sattler, A., Hong, G., Lin, S.: From code to bedside: implementing artificial intelligence using quality improvement methods. J. Gen. Intern. Med. **36**(4), 1061–1066 (2021)
27. Singh, V.K., Singh, N.: Redefining healthcare of tomorrow in smart city. In: Singh, V.K., Paul, L., Singh, V.K., Lillrank, P. (eds.) Planning and Designing Healthcare Facilities: A Lean, Innovative, and Evidence-Based Approach, pp. 141–155. Productivity Press (2017). https://doi.org/10.1201/b21814-10
28. Badr, N.G., Sorrentino, M., De Marco, M., Fugini, M.: Improving interaction in integrated chronic care management. In: 2019 IEEE 28th International Conference on Enabling Technologies: Infrastructure for Collaborative Enterprises (WETICE), pp. 265–270. IEEE (June 2019)
29. Alami, H., et al.: Artificial intelligence and health technology assessment: anticipating a new level of complexity. JMIR **22**(7), e17707 (2020)
30. Williams, M., et al.: Summary of fourth annual MCBK public meeting: mobilizing computable biomedical knowledge—metadata and trust (2022)
31. Filliettaz, S.S., Berchtold, P., Koch, U., PeytremannBridevaux, I.: Integrated care in switzerland: strengths and weaknesses of a federal system. IJIC **21**(4) (2021)

32. McGuckin, T., Crick, K., Myroniuk, T.W., Setchell, B., Yeung, R.O., CampbellScherer, D.: Understanding challenges of using routinely collected health data to address clinical care gaps: a case study in Alberta, Canada. BMJ Open Qual. **11**(1), e001491 (2022)
33. Clark, C.R., et al.: Health care equity in the use of advanced analytics and artificial intelligence technologies in primary care. J. Gen. Intern. Med. **36**(10), 3188–3193 (2021). https://doi.org/10.1007/s11606-021-06846-x
34. DankwaMullan, I., et al.: A proposed framework on integrating health equity and racial justice into the artificial intelligence development lifecycle. J. Health Care Poor Underserved **32**(2), 300–317 (2021)
35. Floridi, L., et al.: AI4People—An ethical framework for a good AI society: opportunities, risks, principles, and recommendations. Minds Mach. **28**(4), 689–707 (2018). https://doi.org/10.1007/s11023-018-9482-5
36. Badr, N., Walletzký, L., Carrubbo, L., Dragoicea, M., Toli, A.M.: Modelling for ethical concerns for traceability in time of pandemic "do no harm" or "better safe than sorry!". In: 54th Annual Hawaii International Conference on System Sciences, HICSS 2021, pp. 1779–1788 (2021)
37. Charow, R., et al.: Artificial intelligence education programs for health care professionals: scoping review. JMIR Med. Educ. **7**(4), e31043 (2021)
38. Wiljer, D.: To the lighthouse: embracing a grand challenge for cancer education in the digital age. J. Cancer Educ. **35**(3), 425–427 (2020)
39. Reichman, Vicky, Brachio, Sandhya S., Madu, Chinonyerem R., Montoya-Williams, Diana, Peña, Michelle-Marie.: Using rising tides to lift all boats: Equity-focused quality improvement as a tool to reduce neonatal health disparities. Semin. Fetal Neonatal Med. **26**(1), 101–198 (2021). https://doi.org/10.1016/j.siny.2021.101198
40. Röösli, E., Rice, B., HernandezBoussard, T.: Bias at warp speed: how AI may contribute to the disparities gap in the time of COVID19. JAMIA **28**(1), 190–192 (2021)
41. Srinivasan, M., et al.: Qualitative assessment of rapid system transformation to primary care video visits at an academic medical center. Ann. Intern. Med. **173**(7), 527–535 (2020). https://doi.org/10.7326/M20-1814
42. Qureshi, I., Raza, H., Whitty, M., Naqvi, Z.: Telemedicine implementation and benefits for quality and patient safety in Pakistan. Knowledge Management & ELearning: An International Journal 367–377 (2015)
43. Ohannessian, R., Duong, T.A., Odone, A.: Global Telemedicine implementation and integration within health systems to fight the COVID19 pandemic: a call to action. JMIR Public Health and Surveillance **6**(2), e18810 (2020)
44. Knight, P., et al.: Positive clinical outcomes are synergistic with positive educational outcomes when using telehealth consulting in general practice: a mixed-methods study. JMIR **18**(2), e31 (2016)
45. Singh, G., PichoraFuller, M.K., Malkowski, M., Boretzki, M., Launer, S.: A survey of the attitudes of practitioners toward teleaudiology. Int. J. Audiol. **53**(12), 850–860 (2014). https://doi.org/10.3109/14992027.2014.921736
46. Snoswell, C., Smith, A.C., Scuffham, P.A., Whitty, J.A.: Economic evaluation strategies in telehealth: obtaining a more holistic valuation of telehealth interventions. J. Telemed. Telecare **23**(9), 792–796 (2016)
47. Smith, W.R., Atala, A.J., Terlecki, R.P., Kelly, E.E., Matthews, C.A.: Implementation guide for rapid integration of an outpatient telemedicine program during the COVID19 pandemic. JACS **231**(2), 216–222.e2 (2020)
48. Monaghesh, E., Hajizadeh, A.: The role of telehealth during COVID19 outbreak: a systematic review based on current evidence. BMC Public Health **20**(1) (2020)

49. Knighton, A.J., Stephenson, B., Savitz, L.A.: Measuring the effect of social determinants on patient outcomes: a systematic literature review. J. Health Care Poor Underserved **29**(1), 81–106 (2018)
50. Zhai, Y.: A call for addressing barriers to telemedicine: health disparities during the COVID19 pandemic. Psychother. Psychosom. **90**(1), 64–66 (2020)
51. Ashfaq, A., et al.: Knowledge and attitude regarding telemedicine among doctors in Karachi. Cureus. Published (2020)
52. Tan, L.F., Ho Wen Teng, V., Seetharaman, S.K., Yip, A.W.: Facilitating telehealth for older adults during the COVID-19 pandemic and beyond: Strategies from a Singapore geriatric center. Geriatr. Gerontol. Int. **20**(10), 993–995 (2020)
53. Gattu, R., Teshome, G., Lichenstein, R.: Telemedicine applications for the pediatric emergency medicine. Pediatr. Emerg. Care **32**(2), 123–130 (2016)
54. Bazemore, A.W., Ireland, J., Cattoi, R., Newton, W.P.: Fostering Generalist Leaders in a Subspecialized World: Congratulations to an Expanded Cohort of New Pisacano Scholars (2022)
55. Lester, C.A., Coe, A.B., Dorsch, M.P., Farris, K.B., Flynn, A.J.: A learning pharmacy practice enabled by the pharmacists' patient care process. JAPhA **60**(6), e66–e72 (2020)
56. Fiscella, K., Carroll, J.K.: Re: how evolving United States payment models influence primary care and its impact on the quadruple aim: the need for health equity. JABFM **32**(1), 118–118 (2019)
57. Maxwell, J., et al.: The Road Ahead: A Model for Advancing High Performance in Primary Care and Behavioral Health under Value Based Payment. The Delta Center, Berkeley, CA (2019)
58. Kim, H.M., Sabri, S., Kent, A.: Smart cities as a platform for technological and social innovation in productivity, sustainability, and livability: a conceptual framework. In: Smart Cities for Technological and Social Innovation, pp. 9–28. Elsevier (2021). https://doi.org/10.1016/B978-0-12-818886-6.00002-2
59. HHS.ORG Homepage. https://www.hhs.gov/hipaa/for-professionals/special-topics/hitech-act-enforcement-interim-final-rule/index.html
60. Kofman, F., Senge, P.M.: Communities of commitment: the heart of learning organizations. Organ. Dyn. **22**(2), 523 (1993)

Comparative Study of Optimal Tuning PID Controller for Manipulator Robot

Faiza Dib[1](✉), Nabil Benaya[1], Khaddouj Ben Meziane[2], and Ismail Boumhidi[3]

[1] LRDSI Laboratory, FSTH, Abdelmalek Essaadi University, Tetouan, Morocco
dib.fayza@yahoo.fr
[2] Higher Institute of Engineering and Business (ISGA), Fez, Morocco
[3] FSDM, Sidi Mohammed Ben Abdellah University, Fez, Morocco

Abstract. In this article, two advanced optimization techniques in command by metaheuristic algorithms are applied to optimally tune the controller for a robot manipulator 2 DOF.The performance of the Differential Evolution (DE) Algorithm and the Teaching-Learning Based Optimization (TLBO) Algorithm are compared for the optimal operation of the robot manipulator. These algorithms are applied to adjust the tuning parameters of PID controller Ki, Kd and Kp in order to minimize the error and assure a good tracking of the desired value of the joint which guarantees the precision of the movements. The robot manipulator has been simulated on Matlab/Simulink with the application of the two algorithms on the PID controller of each articulation. To prove the practicability of the proposed control, a simulation diagram is made by applying the control for each of the joints of the robot. The results of the study indicate that both DE and TLBO algorithms performed nearly equally and provide a good tracking performance in joint movement; however, the DE Algorithm reached its result more quickly than the TLBO algorithm.

Keywords: Differential evolution · Teaching-learning based optimization · Metaheuristic · PID · Robot manipulator · Two-link robot

1 Introduction

Robotics remains a promising market with the development of personal and service robotics making it possible to respond to major societal challenges, this most certainly explains the development of research activities around robotics in recent years [1].

One of the important points in robotics is the command whose function is to obtain the required movement of a robot manipulator to perform a task. The object of the command relates to the dynamic behavior of the robot in its workspace by acting via its actuators. The dynamic modeling of a manipulator robot is an important step for the synthesis of its control law [2]. The use of a rigid dynamic model (bodies assumed to be non-deformable) makes it possible to establish efficient control laws, in particular to follow a trajectory precisely. Such a model makes it possible to take into account the nonlinear behavior of the robot and the couplings between its joints [3]. Nevertheless,

M. Ben Ahmed et al. (Eds.): SCA 2022, LNNS 629, pp. 252–261, 2023.
https://doi.org/10.1007/978-3-031-26852-6_23

the use of such a model for control suffers from several limitations. A main limitation comes from the complexity of the generated nonlinear models, which makes the control synthesis and its implementation difficult (computational cost). A second limit is linked to the risk of obtaining a control whose performance is not robust to structured and unstructured modeling uncertainties as well as to disturbances. Moreover, given the significant modeling uncertainties and the external disturbances, the robustness of the performances is difficult to guarantee [1]. The challenge is therefore to propose control synthesis methods which make it possible to achieve the highest possible performances while guaranteeing their robustness (to modeling uncertainties and disturbances) and mastering the complexity of the law to be implemented.

In this article, two advanced optimization techniques in command by metaheuristic algorithms are presented. The nature of real physical systems (nonlinear systems by nature) requires special attention to nonlinearity when designing controllers and more specifically in the choice of parameters Ki, Kd and Kp. These parameters for various modified structures, can be estimated and optimized using TLBO and DE, which are adopted as offline optimization techniques. A new swarm intelligence metaheuristic called Teaching and Learning Based Optimization (TLBO) is reported to be an efficient way for solving numerous optimization issues, although, Differential Evolution (DE) is a population-based metaheuristic search technique that enhances a candidate solution through an evolutionary process in order to optimize a problem.

Many researches apply metaheuristic algorithms to optimize the gains of the PID controller or of the parameters in the control device. Sudarshan and Madhusudan have applied the adaptive particle swarm optimization algorithm and the multi-objective bio inspired algorithm to adjust the parameters of two type of PID [4]. Javier et al. have compared the performance of metaheuristic algorithms for solving the problems of inverse kinematics of robot manipulators [5]. Hao et al. have presented an optimized PID based on improved PSO for trajectory tracking [6]. In [7] the authors have applied a controller based of the TLBO algorithm and the adaptive network-based fuzzy inference system in order to optimize the Path navigation of multiple robots. The type-II fuzzy logic with the Teaching Learning Based Optimization algorithm is designed to control a STATCOM in power systems [8]. The authors in [9] have proposed a PID control based on DE algorithm for a planar parallel robot. In [10, 11] the authors have used the TLBO algorithm and the DE algorithm to tune optimally the three parameters of PID to have an optimal surface sliding mode in the wind turbine system. A Hybrid algorithm DE–TLBO was proposed by [12] for an optimal control in the multi-machine power system.

In this work, we focused on determining the function allowing to correct the command according to the initial instruction and the measured error, while focusing on the commands; in order to achieve better performance and more robustness to generate smooth motion in low cost robotic arms. We have contribute to the development of optimal control laws allowing a manipulator robot to follow a desired trajectory using a control methodology based on the use of PID controllers and two metaheuristic optimization algorithms the DE and the TLBO algorithm. We have therefore carried out a comparative study on the effectiveness of each algorithm in order to improve the precision of movements. The simulation results show that the DE and TLBO algorithms exhibit almost equal performance, and offer good performance in terms of tracking and convergence and robustness.

2 Proposed Algorithms

2.1 Differential Evolution Algorithm

The differential evolution algorithm (DE) is a simple yet powerful, outperforming many existing global stochastic search techniques. It is generally considered an accurate, fast and robust optimization tool. The main advantage of DE is its simplicity, thus allowing an easy implementation for global optimization in continuous spaces. These characteristics encouraged us to adopt this algorithm in our research work. An initial population xi, 0, i ∈ {1,..., N} is created by uniform random drawing [13]. The algorithm then performs the evolution of the population until convergence. Evolution is reflected iteratively by creating a new generation from the individuals of the current generation, by applying mutation and crossover operations. The selection is then made to keep only the best individuals for the next generation. The main steps of the differential evolution algorithm are given below [14]:

Mutation
For each target individual $X_{i,G}$, a mutant vector $V_{i,G}$ is generated according to:

$$V_{i,G} = X_{r1^i,G} + F.\left(X_{r2^i,G} - X_{r3^i,G}\right) \tag{1}$$

where r1, r2, r3 ∈ {1,…,N} are three different integers between them and different from i and F ∈ [0, 2], a constant called amplification factor controls the magnitude of the difference $\left(X_{r2^i,G} - X_{r3^i,G}\right)$.

Crossover
After the mutation, a crossing operation between the individuals $V_{i,G}$ and $X_{i,G}$ is carried out in order to introduce diversity.

$$U_{j,i,G} = \begin{cases} V_{j,i,G} \text{ } if \text{ } rand_j \leq CR \text{ } or \text{ } j = j_{rand} \\ X_{j,i,G} \text{ } if \text{ } rand_j \succ CR \text{ } or \text{ } j \neq j_{rand} \end{cases} \tag{2}$$

Here, randj is a random number ∈ [0, 1] drawn according to a law of uniform distribution for dimension j. CR ∈ [0, 1] is a constant called crossing rate and j_{rand} is randomly chosen dimension index ensuring that at least one parameter of the mutant individual $V_{j,i,G}$ is transmitted to $U_{j,i,G}$.

Selection
In order to determine if the generated individual will belong to the next generation (G + 1) of the population, a selection criterion is applied. For a minimization problem, we have:

$$X_{i,G+1} = \begin{cases} U_{i,G} \text{ } if \text{ } J(U_{i,G}) \prec J(X_{i,G}) \\ X_{i,G} \text{ otherwise} \end{cases} \tag{3}$$

This algorithm therefore has a restricted number of parameters: the size of the population N as well as the coefficients F and CR. The general process is described by the following algorithm:

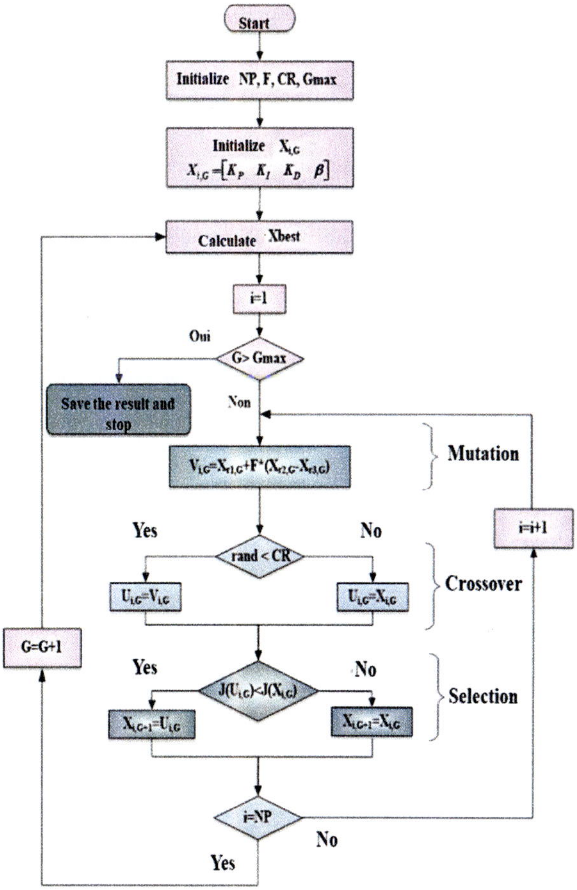

2.2 Teaching Learning Based Optimization Algorithm

The Teaching Learning Based Optimization algorithm (TLBO) is a population-based metaheuristic optimization technique that simulates a classroom environment to optimize a given objective function and it has been proposed by RV Rao et al. in 2011 [15]. The TLBO algorithm basically consists of two phases: the Teaching phase and the Learning phase. Learners gain knowledge not only from the instruction given by a teacher, but also from their mutual interaction through group discussions, presentations, and formal communications. Each individual in the population is considered a learner and the variables are considered subjects to be learned. After evaluating the learners, the best learner is selected to become a teacher for the next phase of teaching [16].

The Teaching Phase

In the teaching phase, learners learn through the teacher. The latter tries to improve the results of his learners. Each Yi learner learns from the most competent individual in the current population (teacher), denoted by YTeacher, taking into consideration the average level of all *mean* learners. During this phase, a new learner *new*Yi is obtained using the following equation

$$newY_i = Y_i + rand(0, 1) * (YTeacher - Tr * Mean) \tag{4}$$

Tr is the teaching factor ϵ {1,2} to emphasize the importance of the average quality of learners.

The Learning Phase

After the completion of the teaching phase, the refreshed learners will participate in the learning phase. A learner interacts with another learner, to increase his knowledge if the latter has more knowledge than him. For each learner Yi, a colleague Xk is randomly chosen to interact with him following equations:

$$newY_i = \begin{cases} Y_i + rand(0, 1) * (Y_i - Y_k) \ if \ f(Y_i) < f(Y_k) \\ Y_i + rand(0.1) * (Y_k - Y_i) \ if \ f(Y_k) < f(Y_i) \end{cases} \tag{5}$$

The flowchart of TLBO algorithm is given as follow.

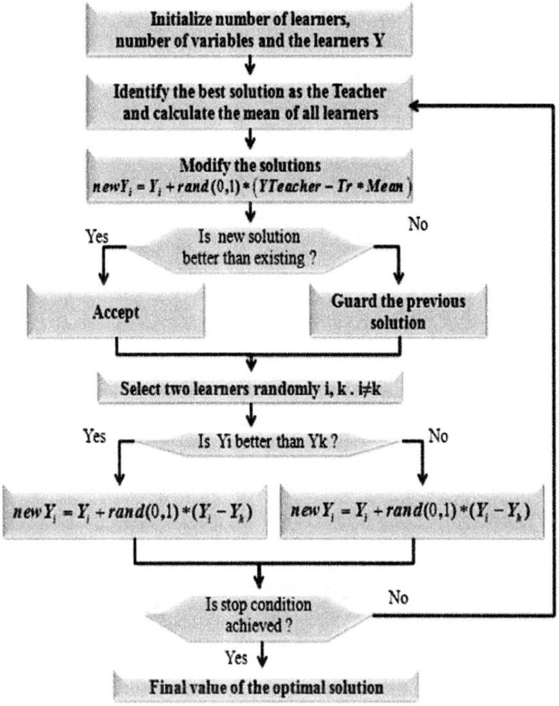

F is the objective function that must be minimized using the integral time absolute error function (ITAE). The mathematical equation that describes the ITAE objective function is given by the following equation:

$$F = ITAE = \int_0^{\infty} t|\ e(t)\ |\ dt \tag{6}$$

3 System Modeling of Two-Link Robot Manipulator

The manipulator robot we are interested in in our study is a robot with 2 degrees of freedom. It is characterized by two articulation with respect to 2 horizontal and parallel axes. The position of the terminal organ is completely determined using its two coordinates in space (x, y, z) which are called operational coordinates. The configuration of the robot is determined using the angles (q1, q2), which are called joint coordinates. The mathematical model which represents the dynamics of the robot above is presented as follows:

$$D(q)\ddot{q} + C(q, \dot{q})\dot{q} + G(q) = u \tag{7}$$

$$\begin{bmatrix} u_1 \\ u_2 \end{bmatrix} = \begin{bmatrix} D_{11}(q)\ D_{12}(q) \\ D_{21}(q)\ D_{22}(q) \end{bmatrix} \begin{bmatrix} \ddot{q}_1 \\ \ddot{q}_1 \end{bmatrix} + \begin{bmatrix} C_{11}(q)\ C_{12}(q) \\ C_{21}(q)\ C_{22}(q) \end{bmatrix} \begin{bmatrix} \dot{q}_1 \\ \dot{q}_2 \end{bmatrix} + \begin{bmatrix} G_1(q) \\ G_2(q) \end{bmatrix} \tag{8}$$

where $u = [u1\ u2\ u3]^T$ is the input torque vector of the joint, $q = [q1\ q2]^T$ is the vector of joint position, and $\dot{q} = [\dot{q}_1,\ \dot{q}_2]^T$ is the joint velocity vector

D(q) is a positive definite symmetric matrix of inertial accelerations given by the following equation:

$$\begin{aligned} D_{11}(q) &= I_1 + I_2 + 4m_2 l_1^2 + 4m_2 l_1 l_2 \cos(q_2); \quad D_{12}(q) = I_2 + 2m_2 l_1 l_2 \cos(q_2) \\ D_{21}(q) &= I_2 + 2m_2 l_1 l_2 \cos(q_2); \quad D_{22}(q) = I_2 \end{aligned} \tag{9}$$

C(q, $\dot{q}$) is the matrix of Coriolis and centrifugal forces described by:

$$\begin{aligned} C_{11}(q) &= -2m_2 l_1 l_2 \dot{q}_2 \sin(q_2); \quad C_{12}(q) = -2m_2 l_1 l_2 (\dot{q}_1 + \dot{q}_2) \sin(q_2) \\ C_{21}(q) &= 2m_2 l_1 l_2 \dot{q}_1 \sin(q_2); \quad C_{22}(q) = 0 \end{aligned} \tag{10}$$

$G(q)$ is a state variable vector of gravity terms given by:

$$\begin{aligned} G_1(q) &= m_2 g l_2 \sin(q_1 + q_2) + m_1 g l_1 \sin(q_2) \\ G_2(q) &= m_2 g l_2 \sin(q_1 + q_2) \end{aligned} \tag{11}$$

The control law applied in this study is based on the PID controller optimized by the metaheuristic optimization algorithms DE and TLBO. This ensures good tracking of the output towards the setpoint, which cancels the tracking error and guarantees good performance of the robot manipulator. The PID controller is a closed loop control and it is most popular in industries because it is easy to implement while achieving good performance. The mathematical representation of the PID controller is of the form:

$$u_{pid} = K_P e(t) + K_I \int e(t)dt + K_D \frac{de(t)}{dt} \tag{12}$$

The position error in radians of each joint; is given by:

$$e(t) = q(t) - q_d(t) \tag{13}$$

$q_d(t)$ is the requested position in radians.

4 Simulations Results

The control method developed with the DE and TLBOA optimization algorithms were carried out by simulation using the MATLAB/Simulink environment in order to verify the efficiency and the capacity achieved for the tracking performance of the two joints of the robot manipulator (Fig. 1).

In the first scenario, the following joint trajectories are chosen as qd = [qd1 qd2]^{T}:

$$\begin{aligned} qd1 &= 0.3 + 0.1 * sin(t) + 0.3 * sin(1.7 * t) + 0.2 * sin(0.9 * t) \\ qd2 &= 0.4 + 0.1 * cos(t) + 0.3 * sin(0.8 * t) + 0.2 * sin(1.7 * t) \end{aligned} \tag{14}$$

As shown in Figs. 2, 3, 4, 5, the responses in red relating to the PID controller optimized by the DE algorithm and the responses in blue relating the PID controller optimized by the TLBO algorithm are significantly better than those obtained by the classical PID (in green).

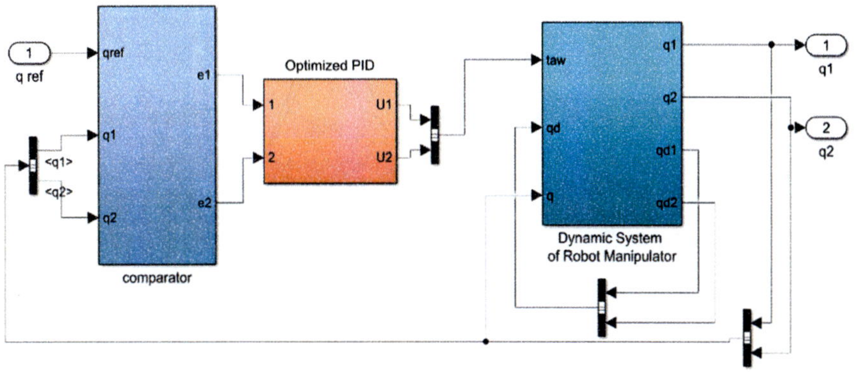

Fig. 1. The model of robot manipulator with optimized PID in Matlab/Simulink

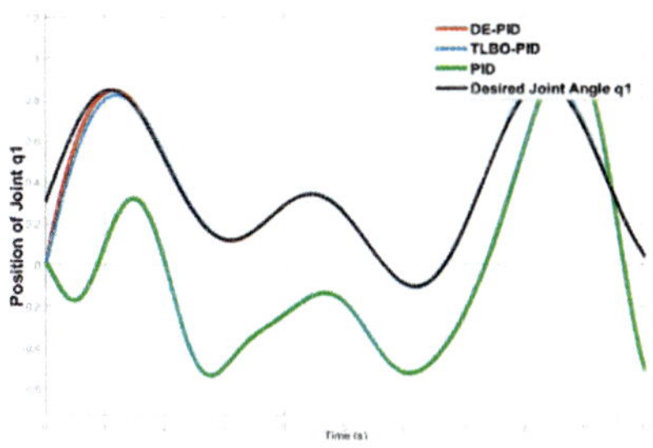

Fig. 2. The position of joint q1.

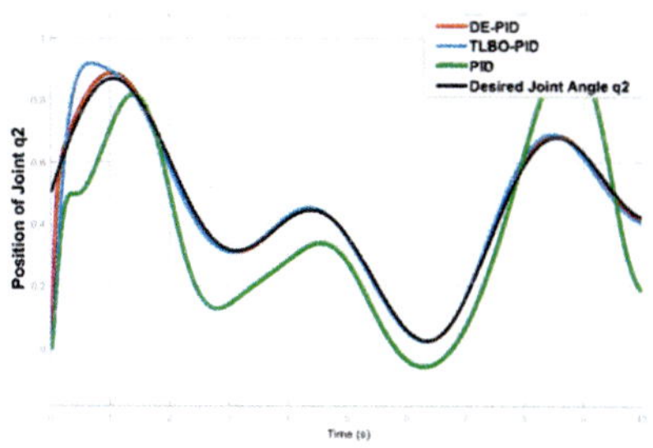

Fig. 3. The position of joint q2.

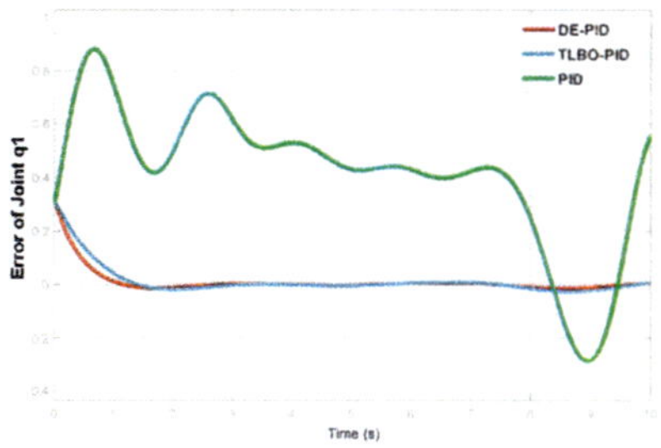

Fig. 4. The error of joint q1.

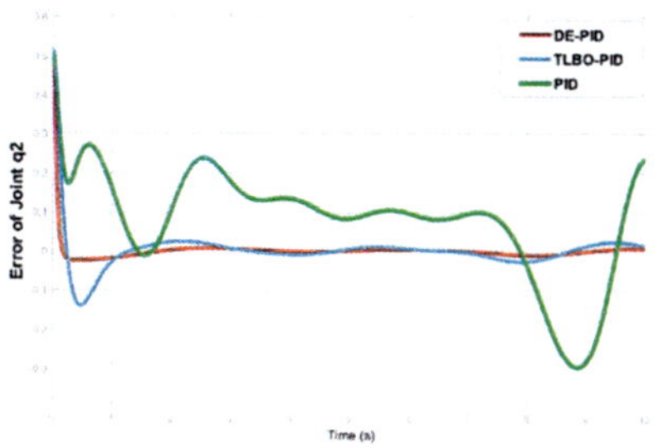

Fig. 5. The error of joint q2.

In the second scenario, the following joint trajectories are chosen as $qd = [qd1\ qd2]^T$:

$$\begin{aligned} qd1 &= 0.5 * sin(t) \\ qd2 &= sin(t) \end{aligned} \tag{15}$$

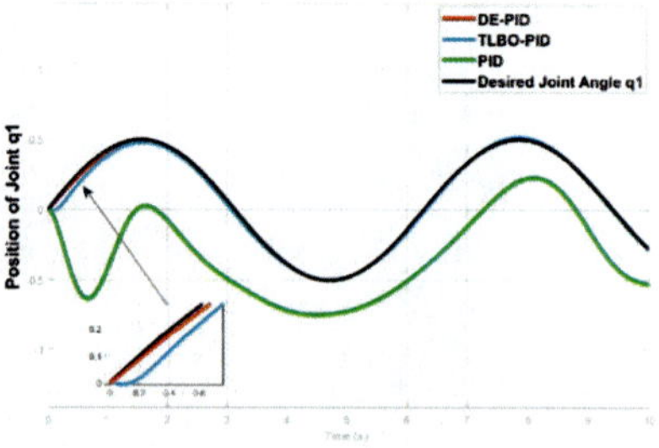

Fig. 6. The position of joint q1.

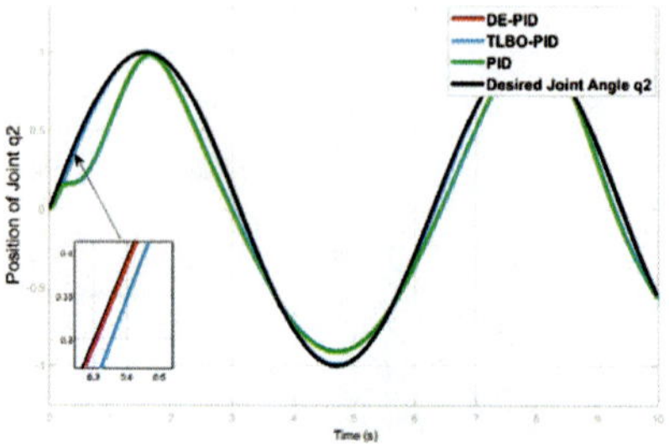

Fig. 7. The position of joint q2.

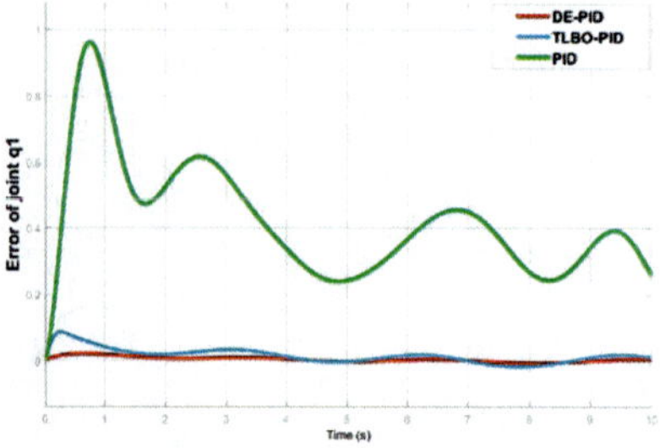

Fig. 8. The error of joint q1.

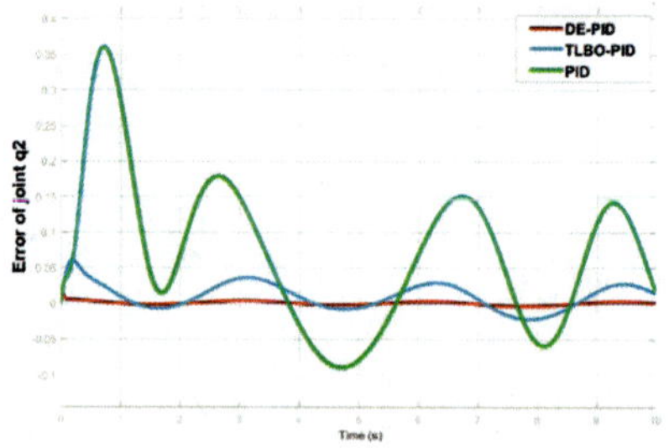

Fig. 9. The error of joint q2.

The results in Figs. 6, 7, 8, 9 show that the PID optimized using the two algorithms DE and TLBOA does not present, in any scenario, an overshoot, in opposition to the PID without adaptation, where the overshoots were observed. According to the results obtained on the response time, we notice that the controller acts quickly to eliminate the error and ensures a good tracking of the reference trajectory. The test's results indicate that both the DE and TLBO algorithms offered good tracking performance in joint motion and performed almost equally efficiently; however, the DE Algorithm arrived at its result more faster than the TLBO algorithm.

5 Conclusion

This work proposed a design of the intelligent PID controller for the manipulator robot. we carried out a comparative study on the effectiveness of two optimization algorithm DE and TLBO for adjusting the gains of the PID controller in order to improve the precision of the movements. The purpose of the proposed command is to allow the robot to follow the imposed trajectory. For this reason, the joints are controlled by the optimized PID controller to satisfy the requirement exact trajectory followed.

By using the advantages of DE and TLBO algorithms, the convergence has been improved and its steady-state error has been reduced compared to the application of classical PID. The simulations results indicate that both DE and TLBO algorithms performed nearly equally and provide a good tracking performance in joint movement; however, the DE Algorithm reached its result more quickly than the TLBO algorithm.

References

1. Mark, W.S., Seth, H., Vidyasagar, M.: Robot Modeling and Control. John Wiley & Sons (2020)
2. Shafei, A.M., Shafei, H.R.: Dynamic modeling of planar closed-chain robotic manipulators in flight and impact phases. Mech. Mach. Theory **126**, 141–154 (2018)
3. Benaya, N., Dib, F., Ben Meziane, K., Boumhidi, I.: Type 2 fuzzy PID for robot manipulator. In: Ben Ahmed, M., Abdelhakim, B.A., Ane, B.K., Rosiyadi, D. (eds.) Emerging Trends in Intelligent Systems & Network Security. NISS 2022. Lecture Notes on Data Engineering and Communications Technologies, vol. 147. Springer, Cham (2023). https://doi.org/10.1007/978-3-031-15191-0_46
4. Sudarshan, K.V., Madhusudan, S.: Optimization strategy of bio-inspired metaheuristic algorithms tuned PID controller for PMBDC actuated robotic manipulator. Procedia Comput. Sci. **171**, 2040–2049 (2020)
5. Abdor-Sierra, J.A., Merchán-Cruz, E.A., Rodríguez-Cañizo, R.G.: A comparative analysis of metaheuristic algorithms for solving the inverse kinematics of robot manipulators. Results Eng. **16**, 100597 (2022). https://doi.org/10.1016/j.rineng.2022.100597
6. Feng, H., Ma, W., Yin, C., Cao, D.: Trajectory control of electro-hydraulic position servo system using improved PSO-PID controller. Autom. Constr. **127**, 103722 (2021)
7. Kashyap, A.K., Pandey, A., Parhi, D.R., Sharma, A.: Path optimization for multiple humanoid robot using TLBO based ANFIS controller in obscure environment. Mater. Today: Proc. **47**, 2677–2684 (2021)
8. Tavana, M.R., Khooban, M.H., Niknam, T.: Adaptive PI controller to voltage regulation in power systems: STATCOM as a case study. ISA Trans. **66**, 325–334 (2017)
9. Villarreal, C.M.G., Alvarez, G.J.: Off-line PID control tuning for a planar parallel robot using DE variants. Expert Syst. Appl. **64**, 444–454 (2016)
10. Dib, F., Boumhidi, I., Akchioui, N.E., Benaya, N.: Sliding mode control optimized by teaching learning-based optimization algorithm for variable speed wind turbine system. In: 2019 International Conference on Wireless Technologies, Embedded and Intelligent Systems (WITS), pp. 1–6. Fez, Morocco (2019)
11. Ben Meziane, K., Naoual, R., Dib, F., Boumhidi, I.: Intelligent sliding mode control based on optimized PID by DE algorithm for wind turbine system. In: 2021 Fifth International Conference On Intelligent Computing in Data Sciences (ICDS), pp. 1–8. Fez, Morocco (2021)
12. Dib, F., Boumhidi, I.: Hybrid algorithm DE–TLBO for optimal H∞ and PID control for multi-machine power system. Int. J. Syst. Assur. Eng. Manag. **8**, 925–936 (2017)
13. Mohamad, F.A., Ashidi, N.M.I., Wei, H.L., Koon, M.A.: Differential evolution: A recent review based on state-of-the-art works. Alex. Eng. J. **61**, 3831–3872 (2022)
14. Dib, F., Boumhidi, I.: Optimal H∞ control without reaching phase with the differential evolution PID based on PSS for multi-machine power system. In: 2015 Intelligent Systems and Computer Vision (ISCV), pp. 1–8 (2015)
15. Rao, R.V., Savsani, V.J., Vakharia, D.P.: Teaching–learning-based optimization: a novel method for constrained mechanical design optimization problems. Comput. Aided Des. **43**, 303–315 (2011)
16. Venkata, R.R.: Teaching-learning-based optimization (TLBO) algorithm and its engineering applications. Springer International Publishing, Switzerland (2016)

Real-Time Hand Gesture Recognition for Humanoid Robot Control Using Python CVZone

Muhammad Yeza Baihaqi[1](✉), Vincent[2], and Joni Welman Simatupang[3]

[1] Graduate Institute of Automation and Control, College of Engineering, National Taiwan University of Science and Technology, Taipei, Taiwan
M11012803@mail.ntust.edu.tw

[2] School of Electrical Engineering and Informatics, Institut Teknologi Bandung, Bandung, Indonesia

[3] Study Program of Electrical Engineering, President University, Bekasi, Indonesia

Abstract. Humanoid robot is able to mimic repetitive activities done by human. By having them, humans can focus on tasks that require extra focus, while leaving the simple one to be done by humanoid robot. To extend the functionality of the 3D-printed upper-body humanoid robot, a camera along with a machine learning algorithm is implemented. In this case, their responses can be categorized based on the gesture it sees. In this paper, the authors evaluate the performance of an upper-body humanoid robot that implements CVZone. The method of evaluation is conducted under three different distances, and each distance has three different brightness levels. It is found that the success rate is highest at the shortest distance which is 50 cm, and under medium-lighted environment. As the distance increases, the hand gesture gets harder to differentiate. At the same time, low or high brightness environment makes the image that contain hand gesture hard to interpret, thus lowering the success rate. Finally, hand gesture that involves third and little finger has the lowest success rate of 75%, while the movement that involves three fingers simultaneously has the second lowest success rate of 80.9%. Meanwhile, simple hand gesture such as clenched-fist or lifting a thumb has 100% success rate under all condition.

Keywords: CVZone · Hand gesture · Humanoid robot · Machine learning

1 Introduction

As a terminology, "Robot" was first introduced to describe a helper role that mimics human being [1]. In other word, it can do various tasks that humans can do [2]. Several possible movements are greeting, accompanying, as well as providing information performing a certain gesture. It is very helpful, especially by mimicking certain monotonous tasks, thus allowing humans to focus on the important one [3]. A lot of improvements have been made for this, and it has been implemented in hotel industry, such as in China, Malaysia, and also Japan since 2016 [4–6]. It is also granted a name of "humanoid robot" as it mimics humans' tasks and having human-like posture.

M. Ben Ahmed et al. (Eds.): SCA 2022, LNNS 629, pp. 262–271, 2023.
https://doi.org/10.1007/978-3-031-26852-6_24

Humanoid robot come in different size, starting from the one smaller than human, similar size, up to a huge one. It is normally built by servo motor as a joint. It is possible to construct multiple joints by having multiple servos. Besides, it is also possible to add sensors or camera, especially to accommodate the robot in completing its mission. The mission itself may be a monotonous task, of which the robot may be programmed to do certain movements only. The task can be based on certain inputs, such as from a cable-connected controller, Bluetooth, Wi-Fi, or even the camera [7–11].

For instance, author in [7] focused in controlling a mobile robot via Bluetooth connection. Even so, it was meant for motor controlling, such that the robot can move forward, backward, and to turn left or right. The accuracy was found to be dependent to the distance. It degraded significantly as the distance got longer than 5 m. On the other hand, author in [8] implemented the Bluetooth-based control for humanoid robot. The controller was able to provide several commands, such as to move forward. Beside the distance problem, a mobile application was required to send the command. Author in [9] implemented Internet-based communication, where the distance problem was solved. However, the system required a web-based application to send the command. Finally, author in [12] focused in the implementation of Bluetooth-based communication, where the user controlled the humanoid robot by doing gestures. The gesture itself were recognized by sensors installed in gloves, which data was sent to the humanoid robot via microcontrollers. Even though the system did not require any mobile or web-based application, the cost was higher compared to the other system, as multiple microcontrollers and sensors were necessary.

Seeing from a wider perspective, each communication module has its own advantages and disadvantages. For instance, cable-connected one will have very low bias, but space-constrained. It can only be utilized as long as the cable is connected. Besides, a controller provides limited combination of inputs, making the output to be as limited as it is. For wireless module such as Bluetooth or Wi-Fi, the basic principle is still using a controller, although it is not connected physically with the robot. It extends the range of communication, but still has limited input. Meanwhile, when using a camera, the robot is basically doing task based on recognized gestures. For instance, the robot will be pre-trained to comprehend several gestures. Then as its camera reads image, it will evaluate the image and see if the gesture is present or not. If yes, then it will perform an output. In addition, having gesture as the input for the robot creates infinite possibilities, where the user can simply make a gesture or a combination of it, and the robot can be programmed to response each gesture.

In this work, the Authors focus on the implementation of Open-Source Computer Vision Library (OpenCV). It is an open-source software providing a machine learning library. It is helpful in training a robot to recognize sets of gestures [13]. This way, a controller is not required anymore, and the user just have to perform a certain gesture in front of the robot, then it will give a reaction. This type of control is also suitable for disabled and/or mute people [14]. Since the controlling does not require any device, hand gestures allow easy and interactive method, as long as the user stays within the camera view. Therefore, in this paper, we perform hand-gesture-controlled robot. Also, we evaluate its performance with variable distance and light intensity.

The rest of this paper is organized as follow. The next section elaborates important components in this work. Then it is followed by the result and its discussion. Finally, the last section concludes this paper.

2 Methodology

The explanation of computer vision in Python, including its well-known library and package, begins in this section. The upper-body humanoid robot is discussed next to it. The topic then moves on to the system's block diagram. Finally, many hand motions employed in this research are discussed, as well as the robot's expected response.

2.1 Computer Vision in Python

Computer vision is a field of study that combines image processing and pattern recognition to produce specific results [15, 16]. This output may take the form of models, data extracts, or information. It is now used in a variety of fields [17]. Initially, its software was widely developed in C/C++ and MATLAB [18]. MATLAB is a powerful platform for computer vision, but its performance was lower than that of a well-designed and optimized C/C++ implementation [19]. Python was discovered in 2009 as a new valuable solution for computer vision implementations [20]. Also, it was discovered that computer vision environment is easier to set up in Python. This study will primarily concentrate on the computer vision usage in Python to recognize hand gesture.

There are several libraries used in developing computer vision implementations in Python, but OpenCV, NumPy, Time, and OS are the heavily used one. OpenCV is an abbreviation for open-source computer vision. OpenCV was made by Intel Company in 1999 specialized in real time processing. Where, NumPy is specialized in multidimensional array manipulation. Time library enables tasks related with time, such as to extract the duration of time since the code has started. To interact with the operating system, such as accessing file, OS has the functions stored in its library. But in this research, the authors will use a package called CVZone. It compiles essential libraries in Python to do hand gesture recognition, including the acquisition of the hand gesture image, as well as its processing.

CVZone was created by a private company at first. Its primary libraries are OpenCV and Mediapipe. It was designed to recognize gestures or patterns such as body estimation and hand gestures, especially by utilizing existing libraries. Common code for hand gesture recognition requires a hundred lines of code. Compared to this work, CVZone utilization is able to reduce it to just forty lines. In short, hand gesture recognition will be simpler and require fewer lines of code if this package is used.

2.2 Upper-Body Humanoid Robot

A humanoid robot, in essence, imitates the human body, which is separated into two major parts: the upper and lower bodies. The upper-body humanoid robot, in particular, has a shape that is comparable to a human's upper-body, covering the head, body, and both arms.

To enable a movement, humanoid robot has several servos throughout its body. For instance, one servo can be installed on each shoulder to move the whole arm. Another servo can be installed to mimic elbow's movement. By using multiple servos, the movement can be more detailed. Thus, making the movement more accurate. The 3D-printed upper-body humanoid robot used in this paper is shown in Fig. 1. It was constructed with 3 Degree of Freedom (DOF) in each hand, and 1 DOF for its head, making it to have 7 DOF in total.

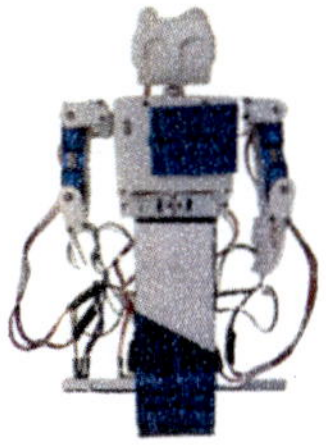

Fig. 1. Upper-body humanoid robot

2.3 Block Diagram

Figure 2 depicts the system's block diagram. It has two sections on a grand scale. The first is Real-Time Hand Gesture Recognition. It is made up of Python libraries (i.e., CVZone) for gesture detection and PySerial for serial communication to send data to the Arduino. The system immediately processes the image taken by laptop's webcam. The serial communication links the first and second section, which is the Upper-body Humanoid Robot. The Arduino UNO controls five servos that serves as the joint of the robot. These servos move the robot's arms to perform movements.

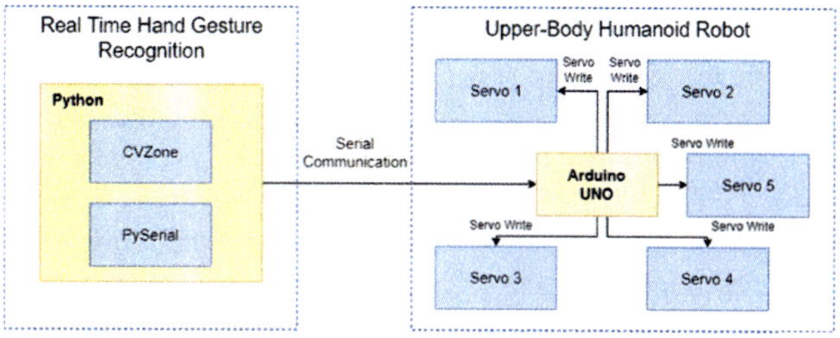

Fig. 2. Block diagram of the system

2.4 Flowchart

In addition to the block diagram, a flowchart is shown in Fig. 3. The system has two main sections. The first one is the laptop, where the OpenCV runs. As the camera reads an image, it tries to identify the hand gesture. It keeps reading with a rate of 20–42 fps until a recognized gesture is present. When it is, the laptop sends a command through a serial communication to the second section, which is the Upper-body Humanoid Robot. As the command is received, it responses by performing a specific gesture. Otherwise, it will provide no response.

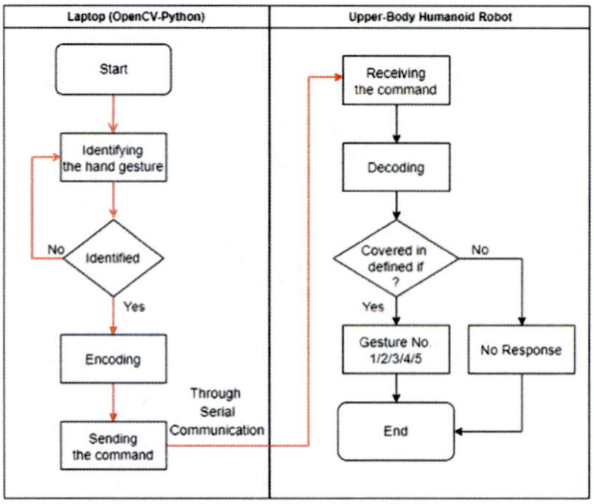

Fig. 3. Flowchart of the system

2.5 Hand Gestures

This work focuses on implementing six inputs using hand gesture. As a response, each gesture corresponds to a specific robot's movement. Table 1 shows all the gestures and its response. The first gesture is fist-clenching. In this case, the robot will lower both hands. It will be referred as movement 1 for the rest of this paper. The second one is done by lifting the thumb. In this case, the robot will lift its right hand. It will be referred as movement 2 for the rest of this paper. The next one is done by lifting the index finger. In this case, the robot will lift its left hand. It will be referred as movement 3 for the rest of this paper. After that, by lifting both thumb and index finger, the robot will lift both hands as the response. It will be referred as movement 4 for the rest of the paper. Next, by lifting the third and little finger, the robot will lift both hands to point to its left-hand side. Last, by lifting the thumb, index, and middle finger at the same time, the robot will lift both hands to point to its right-hand side. These movements are then referred as movement 5 and movement 6, respectively.

3 Result and Discussion

After implementing the block diagram as discussed in previous section, the system is evaluated based on its ability in recognizing the gestures correctly. Besides the number of hand gestures, the evaluation extends to two variables. The first one is distance. In this work, there are three distances used, 50 cm, 75 cm, and 100 cm. Each distance is also evaluated on three different room brightness, i.e., 40–43 lx, 50–53 lx, and 60–63 lx, which will be written as low, medium, and bright environment for the rest of this paper. Thus, there are nine cases for each gesture. Each case is evaluated 50 times, making each gesture to be evaluated 450 times. Table 2 shows the success rate of each hand gesture under every condition.

Table 1. Hand gesture and its corresponding response

No	Hand Gesture	Response and Description	No	Hand Gesture	Response and Description
1		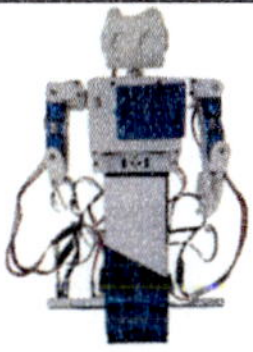 (Lower both hands)	4		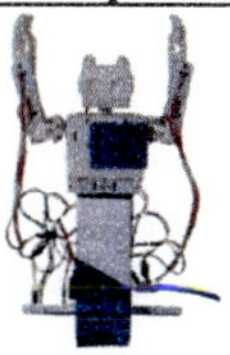 (Lift both hands)
2		 (Lift right hand)	5		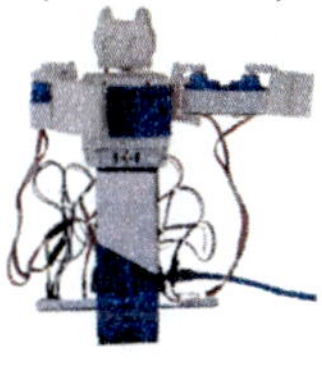 (Point left)
3		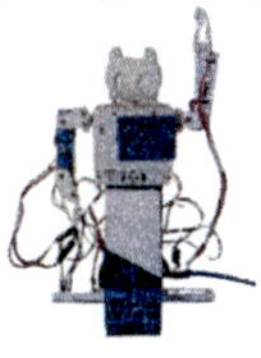 (Lift left hand)	6	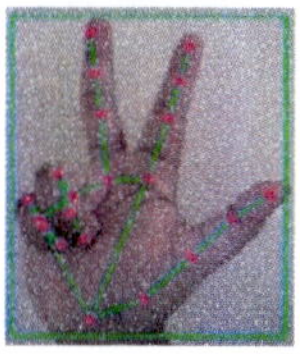	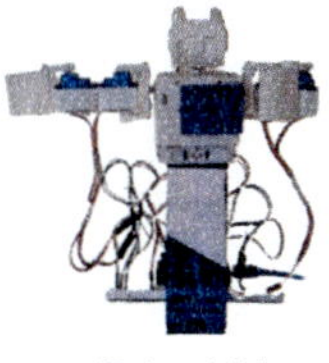 (Point right)

Figure 4(a) shows the evaluation on movement 1. The robot is able to recognize the movement perfectly within all conditions. In fact, it is the simplest one, making the recognition becomes easier. In this case, the success rate of gesture 1 is 100%.

Figure 4(b) shows the evaluation of movement 2. It performs similarly to the previous one, where the perfect recognition was achieved under all conditions. In this case, the gesture is also simple, with only the thumb lifted. Its success rate is also 100%.

Figure 4(c) shows the evaluation on movement 3. The robot is unable to have a perfect recognition. At distance of 50 cm, the success rate is 100% at all brightness level. As the distance increases to 75 cm, the success rate is 100% only at medium-lighted and bright environment. Since the recognition is based on a camera-taken image, it is common to have lower success rate at longer distance for low light environment. As for the longer distance of 100 cm, the success rate is even lower. Even though the bright environment has higher success rate compared to the low one, the highest one is still achieved at medium light level. For this movement, the success rate is 94.9%.

Table 2. Performance evaluation based on different distance and brightness

No	Distance (cm)	Light intensity	Success rate (%)	No	Distance (cm)	Light intensity	Success rate (%)
1	50	Low	100	4	50	Low	84
	50	Medium	100		50	Medium	92
	50	Bright	100		50	Bright	90
	75	Low	100		75	Low	84
	75	Medium	100		75	Medium	92
	75	Bright	100		75	Bright	88
	100	Low	100		100	Low	76
	100	Medium	100		100	Medium	84
	100	Bright	100		100	Bright	76
2	50	Low	100	5	50	Low	80
	50	Medium	100		50	Medium	88
	50	Bright	100		50	Bright	82
	75	Low	100		75	Low	74
	75	Medium	100		75	Medium	80
	75	Bright	100		75	Bright	78
	100	Low	100		100	Low	60
	100	Medium	100		100	Medium	72
	100	Bright	100		100	Bright	70
3	50	Low	100	6	50	Low	84
	50	Medium	100		50	Medium	90
	50	Bright	100		50	Bright	84
	75	Low	98		75	Low	78
	75	Medium	100		75	Medium	84
	75	Bright	100		75	Bright	80
	100	Low	82		100	Low	72
	100	Medium	90		100	Medium	80
	100	Bright	84		100	Bright	76

Figure 4(d) shows the evaluation on movement 4. In any distance, it is obvious that the robot has higher recognition success rate at brighter environment, but it performs best in medium light environment. Besides, as the distance increases, the success rate decreases. For this movement, the success rate is 85.1%.

Figure 4(e) shows the evaluation on movement 5. Its performance has similar pattern with the previous one, where the success rate decreases as the distance increases. At the same time, the robot has recognized the hand gesture better in brighter environment, but it performs the best at medium light environment. For this movement, the success rate is 75%.

Figure 4(f) shows the evaluation on movement 6. The performance over this hand gesture is also similar the previous movements, where the success rate decreases as the distance increases. Concurrently, brighter environment improves the recognition success rate. However, the robot best performs in medium light environment compared to the brightest one. For this movement, the success rate is 80.9%.

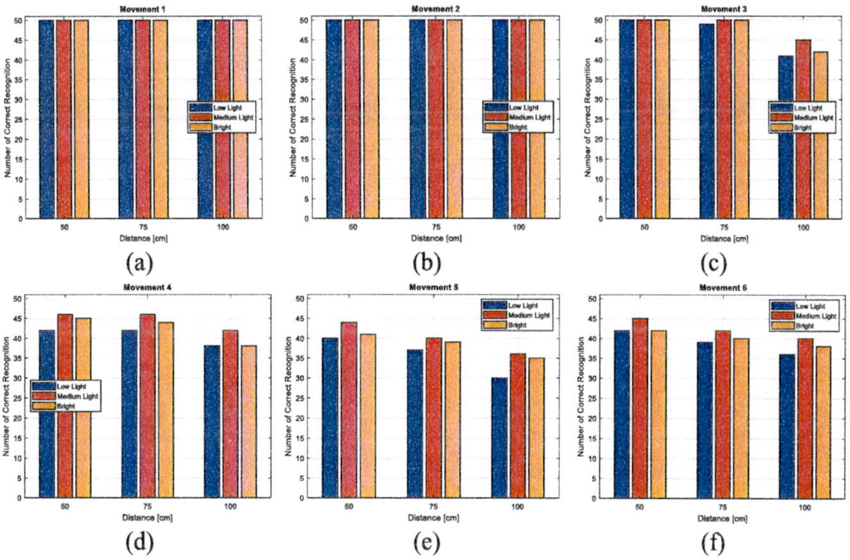

Fig. 4. Evaluation of (a) Movement 1, (b) Movement 2, (c) Movement 3, (d) Movement 4, (e) Movement 5, (f) Movement 6.

Within the six gestures, the best result is achieved when the distance is taken to be 50 cm as the lowest one in the scenario. Since the recognition is done by taking an image using camera, the robot will find it hard to differentiate between one gesture and another at longer distance, i.e., 75 cm or 100 cm. The success rate also decreases as the brightness is low. However, bright environment does not guarantee the best performance. The captured image under bright environment makes the hand gesture to be unclear as in the low light one. Even though brighter one makes the robot to have higher success rate, the best performance is achieved under the medium-lighted environment. Apart from the distance and brightness, the hand gesture also plays a significant role on the recognition rate. Among the six gestures, the robot shows the worst performance on movement 5 with success rate of 75%. There are at least three reasons behind this. First, movement 5 is hard to perform, especially because it involves the third finger, which is hard to lift. Besides, using multiple nearby fingers may confuses the recognition process. In addition, the system may have taken the image sample even before we have lifted all fingers. Thus, making the system to perform wrong output, and lowering the success rate simultaneously. In fact, movement 6 – which involves three fingers at the same time – has the second lowest success rate at 80.9%. Additionally, Table 3 shows the confusion matrix of all six movements, with one additional column of NR, which denotes the No Response condition. In that case, the robot did not make any response, while at the same time, the user had already given a command. For instance, when the user performs movement 3 but the robot recognized it as the middle finger – which did not exist in the movement list–, so the robot gave no response.

Table 3. Confusion matrix

Actual Movement	1	2	3	4	5	6	NR
1	450	0	0	0	0	0	0
2	0	450	0	0	0	0	0
3	0	9	427	0	0	0	14
4	0	32	35	383	0	0	0
5	0	0	0	0	342	0	108
6	0	34	30	22	0	364	0

Predicted Movement

4 Conclusion

To sum up, this paper evaluates the performance of hand gestures recognition using CVZone. The output is shown as a movement performed by an upper-body humanoid robot, where each hand gesture corresponds to a specific robot's movement. The evaluation is done by considering three parameters. The first one is variable distance between the user and the robot, i.e., 50 cm, 75 cm, and 100 cm. In addition, the brightness of the environment is adjusted to three levels. Last, six distinct hand gestures are used as the input. Each hand gesture is evaluated under the three different distances, and each distance covers three different brightness levels. To have better result, each condition is evaluated 50 times, and the robot is found to have the highest success rate under medium-lighted environment at close distance of 50 cm. Since the recognition is done by evaluating the camera-taken image, longer distance will make the hand gesture hard to differentiate. Concurrently, the gesture is hard to be evaluated by the system under low brightness environment, as well as the one that is too bright. Finally, the movement involving third and little finger has the lowest success rate of 75%, while the movement that involves three fingers at the same time has the second lowest success rate of 80.9%. Meanwhile, simple hand gesture such as clenched-fist or lifting a thumb has 100% success rate under all condition.

In this research, the camera that was used was a webcam. In the next development, the camera may be attached to the robot. So, there is no need to bring the laptop while controlling the robot. Also, it was found that the current hand gesture algorithm still needs an improvement in terms of accuracy. In the future research, the new algorithm may be developed. Among the possibilities, the next research may apply the Kalman filter to reduce the noise. In addition, it is also possible to modify the features to improve accuracy of the hand gesture recognition.

References

1. Behnke, S.: Humanoid robots – from fiction to reality? KI-Zeitschrift **4**, 5–9 (2008)
2. Denny, J., et al.: Humanoid robots – past, present and the future. Eur. J. Adv. Eng. Technol. **3**, 8–15 (2016)
3. What is Humanoid Robot. https://gfi.world/tn-en/news/parole_expert/139-how-humanoid-robotics-re-invents-customer-experience. Accessed 1 Jul 2021
4. Wang, E.: Humanoid Robot Service China (2019). https://www.chinamoneynetwork.com/2018/09/20/alibabasai-unit-to-launch-hospitality-robots-in-october. Accessed 1 Jul 2021
5. Puteri, I.: Humanoid Robot Service Malaysia (2019). https://www.thesundaily.my/local/ecoworld-to-unveil-malaysias-first-robothotel-YF546100. Accessed 1 Jul 2021
6. Verge, T.: Humanoid Robot Service Japan (2019). https://www.theverge.com/2019/1/15/18184198/japans-robot-hotel-lay-off-work-forhumans. Accessed 1 Jul 2021
7. Vincent, V., Chiu, E., Tae, K.S., Bukhori, I., Setiyadi, S.: Automatic wireless mobile weather station. J. Electr. Electron. Eng. **2**, 54–59 (2018)
8. Baihaqi, M.Y., Simatupang, J.W.: Aplikasi Metode Kontrol Bluetooth Pada Robot Humanoid Berbasis Arduino Mega 2560. **88**, 49–56 (2016)
9. Simatupang, J.W., Yosua, M.: A remote controlled car using wireless technology. J. Electr. Electron. Eng. **1**, 56–61 (2016)
10. Baihaqi, M.Y., Vincent, V., Simatupang, J.W.: Humanoid robot application as COVID-19 symptoms checker using computer vision and multiple sensors. ELKHA J. Tek. Elektro. **13**, 105–112 (2021)
11. Sitompul, E., Baihaqi, M.Y.: Quasi-dynamic walking optimization of humanoid robot using genetic algorithm. Int. J. Artif. Intell. **20**, 24–45 (2022)
12. Iliukhin, V.N., Mitkovskii, K.B., Bizyanova, D.A., Akopyan, A.A.: The development of motion capture system based on kinect sensor and bluetooth-gloves. Procedia Eng. **176**, 506–513 (2017)
13. What is OpenCV. OpenCV (2021). https://opencv.org/about/. Accessed 9 Sep 2021
14. Sabuj, B., Islam, M.J., Rahaman, M.A.: Human robot interaction using sensor based hand gestures for assisting disable people. In: 2019 Int. Conf. Sustain. Technol. Ind. 40, STI (2019). https://doi.org/10.1109/STI47673.2019.9068087
15. Cosido, O., et al.: Hybridization of convergent photogrammetry, computer vision, and artificial intelligence for digital documentation of cultural heritage – a case study: The Magdalena Palace. In: Int. Conf. Cyberworlds. IEEE, pp. 369–376 (2014)
16. Long, J., Shelhamer, E., Darrell, T.: Fully convolutional networks for semantic segmentation. In: IEEE Conf. Comput. Vis. Pattern Recognit. IEEE, pp. 3431–3440 (2015)
17. Patel, K.K., Kar, A., Jha, S.N., Khan, M.A.: Machine vision system: a tool for quality inspection of food and agricultural products. J. Food. Sci. Technol. **49**, 123–141 (2012)
18. Kovesi, P.D.: MATLAB and Octave Functions for Computer Vision and Image Processing (2000). http://www.peterkovesi.com/matlabfns/
19. Doak, J., Prasaf, L.: Python and computer vision. In: Tenth Int. Python Conf. (2002)
20. Thorne, B.: Introduction to Computer Vision in Python. Python Pap. Monogr. (2009)

RobVRL@bs: Web-Based Application for Teaching Robotics

Meriem L. Aarizou(✉) and Nasr Eddine Berrached

Intelligent Systems Research Laboratory, Université des Sciences et de la Technologie d'Oran, Oran, Algeria
meriem.aarizou@univ-usto.dz

Abstract. Remote teaching platforms offer to students the possibility of using real systems and devices in their practical sessions through the Internet. This allows students who are away from campus or with special needs to benefit from practical training in engineering, especially in robotics. Virtual and remote robotics laboratories are web-based telerobotic platforms (Networked Telerobotic System (NTS)) that allow real-time manipulation from web pages. The implementation of such a system may raise issues related to the often limited networks.

In this paper we propose different practical sessions to teach robotics to engineer students. Our objective is to propose a technical solution to Algerian universities in order to address economic issues such providing expensive educational material. Our solution addresses as well the issues of the Algerian Internet network, i.e. adapting remote manipulations on real robots to a limited bandwidth Internet network. Moreover, our tool can be used in many other applications as a prototype for optimized telerobotic applications.

Keywords: Virtual laboratories · Remote laboratories · Networked telerobotic systems · E-learning · Telerobotics · Telemanipulations · Practical sessions

1 Introduction

Teaching robotics to engineer student requires combining practical activities to theory classes. Some of practical activities need educational material and equipment which could be expensive and difficult to provide. Virtual and Remote Laboratories (VRLs) are particularly useful as they offer tools to instructors and learners so they can remotely do their practical activities on real robotic devices or on realistic simulated environments. Furthermore, if these platforms are available from the Web, they will provide accessibility, mobility and diverse software tools to learners. Networked Telerobotic Systems (NTS) are web platforms from which users can perform teleoperations on real robots from a web browser. In this work, we developed a NTS-based VRL for teaching robotics to

M. Ben Ahmed et al. (Eds.): SCA 2022, LNNS 629, pp. 272–279, 2023.
https://doi.org/10.1007/978-3-031-26852-6_26

engineer students. Therefore, we are dealing with three main issues: 1) issues in relation with web-based e-learning systems, 2) issues related to web-based telerobotic systems and 3) visual feedback and web-based Human Machine Interfaces (HMI) issues. We present in this paper RobVRL@bs a NTS-based VRL for teaching robotics in E-learning program. Our main goal is to develop a system that has the following features:

1. Real-time teleoperation of robotic devices
2. Distributed and shared environments
3. Suitable tools and dedicated environments for E-Learning
4. Manipulating robotic systems over a limited bandwidth network
5. Real-time visual feedback without compromising the quality of service.

The remaining sections are outlined as follows: In Sect. 2 we present an overview of the different NTS applications then a technical evaluation for NTS-based VRL for teaching robotics in E-learning program. Sections 3 describes the architecture we used in our work. And our different practical sessions are showed and described in Sect. 4.

2 State of the Art

NTS use the two revolutionary technologies of the World Wide Web, which are the HTTP protocol and the HTML language. These technologies provide users the ability to manipulate remote robotic systems from a web browser, without installing any additional software. Only keyboards and mouses can be used to perform the remote manipulations from a HTML page. In a previous work [2] we presented a state of the art describing the evolution of NTS systems from their first implementation until 2018 (the year of the study). The first generation began in 1995 when the mercury project [7], the first NTS, went online. This generation is a great achievement for Telerobotic Engineering as it defined the main features of NTS-based applications. Furthermore, different technologies, that were new at the time and used for developing the NTS were not necessarily suitable for real-time robotic telemanipulation (such as the first version of HTTP, Common Gateway Interfaces (CGI) and static HTML pages). Even though the manipulations were brief and short, researchers could carry out remote manipulations with static images or videos as visual feedback.

The second generation started in 1997 [4] when researchers added more interactivity to web pages in order to improve the experience and reduce time-delay. Dedicated protocols, such as TCP and RTP, were used for manipulation and visual feedback independently from HTTP. Java Applet technology were used on web pages to implement those protocols, add interactivity to the GUI and to stream live videos of the remote manipulations.

The third generation has implemented distributed architectures on the remote sites. Therefor multiple and heterogeneous robots and devices were used in NTS-based applications. Works, such as [9] [3] and [15], have implemented the CORBA technology. Communication, manipulation commands and even feedback visual data are transmitted via IIOP protocol, and the CORBA client is implemented by Java Applet on the client side. Nevertheless, only a few works have been published in this generation. As a matter of fact, the third generation didn't last long and was quickly dropped by researchers as Java plugin, and other NPAPI-based plugins, were slowly banned from web browsers. The fourth generation has therefore come to overcome the NPAPI restrictions and to address time-delay issues for real-time remote control. The first application in the fourth generation is the PR2 Remote Lab [11] from ROS developers who used three main technologies: JavaScript-based interactive interfaces on the client side, ROS-based distributed architectures on the remote site and WebSocket protocol for real-time communication and transmission.

Despite the fact that NTS have evolved a great deal during this period, NTS-based VRL for teaching robotics has only appeared in the second generation [16], and only a few articles related to real education programs can be found in the literature. In the following we select research papers that presents a NTS-based VRL that have been used for real practical activities, from 2003 and 2018. Table 1 presents the technical evaluation we performed for each of the selected papers. From our evaluation we can deduce the following:

1. Reducing time-delay has always been related to the choice of the transmission protocol.
2. The only strategy used to reduce time-delay is based on closed-loop control.
3. No high level protocol is implemented for data transmission
4. No work has addressed the problem of reducing the latency related to visual feedback
5. No experience has been tested on congested web networks

3 RobVRL@bs Architecture

Based on the above conclusions, we propose to develop a NTS-based VRL in which we implement a high-level protocol to reduce time-delay and latency. Our application must also meet the requirements of the fourth generation, by having:

1. a distributed architecture in the remote site,
2. a WebSocket transmission,
3. JavaScript-based interfaces on the client side, and
4. a 3D/2D based modelization for visual feedback.

In our previous work [2], we developed custom_tf2_republisher, a software tool that can reduce bandwidth consumption during a telemanipulation. Our module, which can be found in [1] under GPL license, is a *rospy* node that can be used in any 4th generation NTS. We tested our tool in different scenarios and on two types of robots. We resume our results in Table 2.

Therefor, we propose to develop RobVRL@bs, a 4th generation NTS-based VRL in which we implement custom_tf2_republisher to reduce time-delay during the manipulation and for visual feedback. The software architecture of RobVRL@bs is described in Fig 1.

Table 1. Technical evaluation of state of the art

Project	VR lab [13]	IVL [10]	Robo-Lab [12]	Automat-L@bs [14]	RoboUA-Lab [8]	RFCP [5,6]
Generation	2	N/A	2	2	2	2
Manipulation protocol	TCP	N/A	TCP	TCP and	N/A	N/A
Visual feedback protocol	RTP	N/A	N/A	N/A	N/A	N/A
Communication	Full-Duplex	N/A	N/A	Half-Duplex	Half-Duplex	N/A
Using NPAPI	Yes	N/A	Yes	Yes	Yes	Yes
Used API	Java Applet	N/A	Java Applet	Java Applet Flash, ActiveX	Java Applet	Java Applet
Distributed remote system	No	N/A	No	N/A	N/A	N/A
Remote middleware	None	N/A	None	LabView	N/A	N/A
Implementing control strategy	N/A	N/A	N/A	Yes	N/A	N/A
Limiting length of transmitted data	N/A	N/A	No	N/A	N/A	N/A

Table 2. Results of reducing bandwidth consumption with custom_tf2_republisher

	Bandwidth reduction rate	Packet reduction rate	Time reduction rate
Remote visualization	82%	29.31%	–
LAN manipulation	82%	56.42%	12.52%
Internet manipulation	–	–	64%

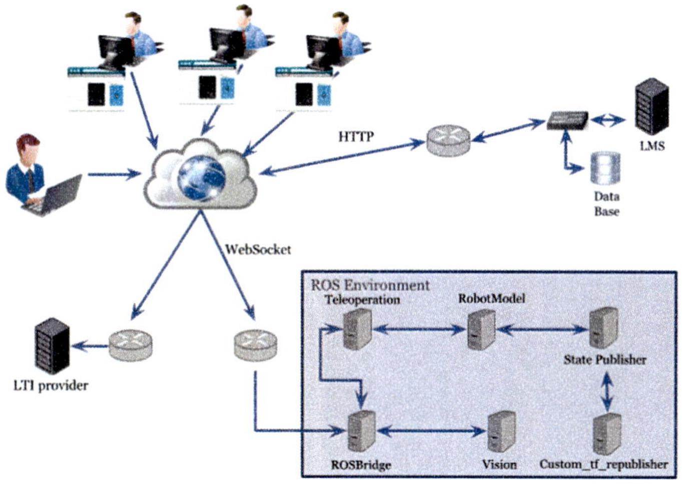

Fig. 1. RobVRL@abs architecture.

4 RobVRL@bs Sessions

In this section we will describe the remote practical sessions we developed for RobVRL@bs. The devices that were used during the sessions are: Pioneer3-DX mobile robot, Pioneer3-AT mobile robot and cyton-gamma-300 robot arm.

4.1 Practical Session 1: Initiation to the Platform

The first session is a Virtual Lab (VL) platform designed to introduce RobVRL@bs to students and to get them used to remote manipulations. Each registered student has a virtual mobile robot Pioneer3-AT. All the robots are located in a indoor environment, in which they are operated from a starting point to a finishing point, as presented in Fig 2a. The remote virtual environment is completely unknown from students. They only have a navigation map that describes the virtual site (see Fig. 2b).

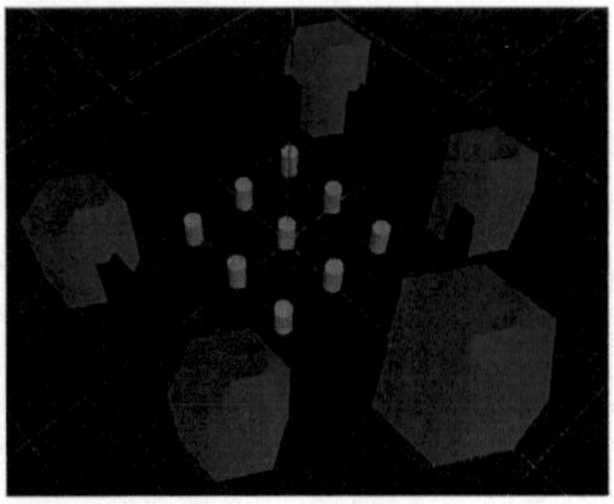

(a)

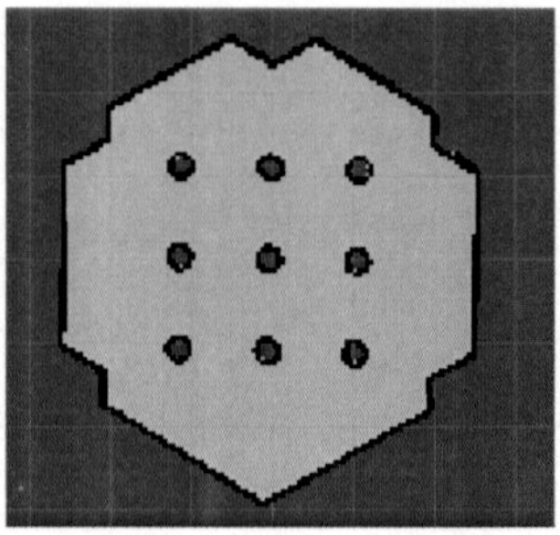

(b)

Fig. 2. First session of RobVRL@abs. **a** The simulated environment. **b** The navigation map.

The starting positions of each robot are already preset on the map, but no student knows the finishing points of the other robots on site. As a result, their respective paths may cross. In this session, students will be able to identify the challenges of robot navigation in shared environments such as avoiding mobile and fixed obstacles. Students will also be able to notice how it is important to have navigation maps and to know the exact positions and localization of any robot in real-time on the map.

4.2 Practical Session 2: Navigation Map

Like in the first session, students are dealing with a virtual environment in which they have to manipulate a mobile robot Pioneer3-DX, as showed in Fig. 3. Each one of them must navigate a robot with a lidar in a large and unknown facility (Willow Garage labs) in order to build the navigation map. Real-time data from lidars are used for the supervised remote navigation. Such a difficult task will help students to understand the difficulty of mapping a large and an unknown environment. They will notice how it is hard to locate other robots in their own map and will understand how important is to collaborate multiple robots to carry out different tasks.

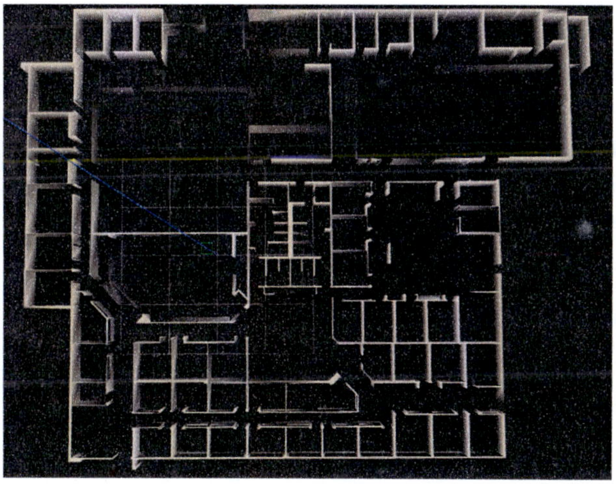

Fig. 3. Simulated environment for the second session.

4.3 Practical Session 3: Path Generation

In the third session we propose a platform from which students can perform manipulations on a 07 DoF robot arm. The interface is highly interactive. Users are able to chose any feasible position for the robot arm. The path will be automatically generated and tested on a virtual arm. If positions are valid, the commands will be sent to the real remote robot, and immediately visualized on the web page (see Fig. 4). In this session, students will be able to understand robotic basics such as forward and inverse kinematics and path generation.

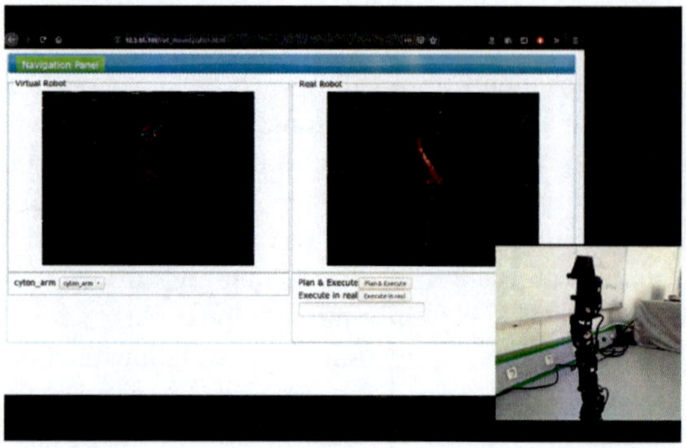

Fig. 4. Third session's interface with a real-time visual feedback of the remote robot arm.

5 Conclusion

In this paper, we presented our remote robotics training tool. We proposed an architecture that meets the requirements of the 4th generation. The implementation of RobVRL@bs has been done using the ROS package we developed to reduce the bandwidth consumption during the telemanipulation custom_tf2_republisher. It is important to point out that only virtual mobile robots were used during the practical sessions. Nevertheless, due to the software architecture we can easily replace the virtual devices by their corresponding real robots, without changing the software programs nor the architecture. Therefore, if enough mobile robots are available in the laboratory, it will be easy to use them instead of the virtual models.

References

1. ROS package for publishing optimal TF over the web. https://github.com/MerAARIZOU/py_tf2_web. Accessed 9 Feb 2022
2. Aarizou, M.L., Berrached, N.E.: ROS-based telerobotic application for transmitting high-bandwidth kinematic data over a limited network. Int. J. Control Autom. Syst. **17**(2), 445–453 (2019)
3. Amoretti, M., Bottazzi, S., Reggiani, M., Caselli, S.: Evaluation of data distribution techniques in a corba-based telerobotic system. In: Proceedings 2003 IEEE/RSJ International Conference on Intelligent Robots and Systems (IROS 2003)(Cat. No. 03CH37453). vol. 2, pp. 1100–1105. IEEE (2003)
4. DePasquale, P., Lewis, J., Stein, M.R.: Java interface for asserting interactive telerobotic control. In: Telemanipulator and Telepresence Technologies IV. vol. 3206, pp. 159–169. SPIE (1997)
5. Fabregas, E., Farias, G., Dormido-Canto, S., Guinaldo, M., Sánchez, J., Dormido Bencomo, S.: Platform for teaching mobile robotics. J. Intell. Robot. Syst. **81**(1), 131–143 (2016)

6. Galán, D., et al.: Online virtual control laboratory of mobile robots. IFAC-PapersOnLine **51**(4), 316–321 (2018)
7. Goldberg, K., et al.: Beyond the web: manipulating the real world. Comput. Netw. ISDN Syst. **28**(1–2), 209–219 (1995)
8. Jara, C.A., Candelas, F.A., Puente, S.T., Torres, F.: Hands-on experiences of undergraduate students in automatics and robotics using a virtual and remote laboratory. Comput. Educ. **57**(4), 2451–2461 (2011)
9. Jia, S., Takase, K.: A corba-based internet robotic system. Adv. Robot. **15**(6), 663–673 (2001)
10. Noguez, J., Sucar, L.: Intelligent virtual laboratory and project-oriented learning for teaching mobile robotics. Int. J. Eng. Educ. **22**(4), 744 (2006)
11. Pitzer, B., Osentoski, S., Jay, G., Crick, C., Jenkins, O.C.: PR2 remote lab: an environment for remote development and experimentation. In: 2012 IEEE International Conference on Robotics and Automation, pp. 3200–3205. IEEE (2012)
12. Torres, F., Candelas-Herías, F.A., Puente Méndez, S.T., Pomares, J., Gil, P., Ortiz Zamora, F.G.: Experiences with virtual environment and remote laboratory for teaching and learning robotics at the university of alicante. Int. J. Eng. Educ. **22**(4), 766–776 (2006)
13. Tzafestas, C.S., Palaiologou, N., Alifragis, M.: Virtual and remote robotic laboratory: comparative experimental evaluation. IEEE Trans. Educ. **49**(3), 360–369 (2006)
14. Vargas, H., Sánchez, J., Jara, C.A., Candelas, F.A., Torres, F., Dormido, S.: A network of automatic control web-based laboratories. IEEE Trans. Learn. Technol. **4**(3), 197–208 (2010)
15. Wang, D., Ma, X., Dai, X.: Web-based robotic control system with flexible framework. In: IEEE International Conference on Robotics and Automation, 2004. Proceedings. ICRA2004. 2004. vol. 4, pp. 3351–3356. IEEE (2004)
16. Yu, L., Tsui, P.W., Zhou, Q., Hu, H.: A web-based telerobotic system for research and education at essex. In: 2001 IEEE/ASME International Conference on Advanced Intelligent Mechatronics. Proceedings (Cat. No. 01TH8556). vol. 1, pp. 37–42. IEEE (2001)

Implementing Machine Learning-Based Simulation in Physics Virtual Laboratory

Fatima Cheddi(✉)

ACSA Laboratory Mohammed I University in OUJDA, Oujda, Morocco
Fatima.chedi@gmail.com

Abstract. Integrating virtual labs and machine learning is one way to use the ICT in education. With the use of virtual laboratories, students may perform theoretical tasks without having to enter a physical science lab. However, creating such software code is a time-consuming procedure that calls for a skilled developer and is still challenging to do. The first stage in designing an application is to create drawings that serve as prototypes. In order to convert a hand sketch into a code simulation, this work provides a deep learning-based solution together with a Laboratory as a Service (LaaS) for user-drawing sketch detection and identification. Our method relies on feature extractions from object detectors SSD (Single Shut Detector) and Alex-Net with the deployment of a SOA to produce simulation code. Our approach has been tested and trained on the "ImageNet Sketch" dataset. The results have demonstrated that the suggested methodology may be utilized realistically to generate code in a variety of fields.

Keywords: Sketch · Virtual laboratories · User interface · Simulation · CNN

1 Introduction

Communication technologies (ICT) are applied in several domains such as agriculture, industry, health, and education field, one of the methods to integrate ICT in education is the use of virtual laboratory and machine learning methods. Virtual laboratories are also needed in various levels of higher education since they can be helping more convenience to the teaching process and offer students new learning opportunities.

A virtual laboratory is a set of simulations where a user interacts and controls the system via a graphical user interface (GUI). Moreover, the hand-drawn image also called a sketch is used as a tool for representing ideas and transmitting thoughts and opinions. With the evolution of technology and smart device, sketches become easy and universal. Sketch has gotten much attention from researchers and various novel tasks are analyzed in this field including sketch recognition [1, 2], sketch-based image retrieval [3], etc.

The initial stage for developing a design is the classification of sketch elements these elements represent the real instruments, in order to implement each element. With the development of CNNs, significant progress has been achieved in the field of object detection and classification. Many deep neural network-based object detectors have been

M. Ben Ahmed et al. (Eds.): SCA 2022, LNNS 629, pp. 280–290, 2023.
https://doi.org/10.1007/978-3-031-26852-6_27

proposed, such as R-CNN [4], Fast R-CNN [5], Faster R-CNN [6], R-FCN [7], SSD [8], and YOL [9]. These detectors have been successful in different applications.

In this article, we present a novel approach based on a convolutional neural network combined with Laboratory as a service (LaaS) [10, 11], in order to transform a sketch (design of experience) into simulation code, That allows any teacher or student with no programming skills to create a personalized simulation, generate an experience from a simple image (sketch). To achieve this our proposed solution is divided into two essential phases:

- The first phase: this phase is divided into two parts the objects recognition based on a conventional neural network technology to predict the objects that exist in image drawing by users, we choose Single Shot Detector (SSD) [8] with AlexeNe [12] and the second part is an optical character recognition (OCR) technique which is able to recognize and classify the words, texts, numbers…
- the second phase: is regeneration, this part is based on the paradigm Laboratory As a Services (LAAS) [10, 11] and modular components [13] that allow calling all web services desired and generate a code of graphical user interface.

This paper is structured as follows: in Sect. 2 related works we present an overview of existing in the creation of online labs side clients. Section 3 presents the proposed method. In Sect. 4 we implement our system for evaluating our approach. Finally, a conclusion in Sect. 5, gives some future work on the item.

2 Related Work

In recent years, the implementation of virtual laboratories has gotten much consideration from researchers. Several solutions were proposed in this context.

Ejss [14] is a system for the creation of remote and virtual laboratories lab user interfaces. Moreover, they give an answer that is reusable and keeps application engineers from building UIs without any preparation for every lab, this system just backings the age of UIs for labs that are viable with their execution of the introduced application developer. Another solution for the development of the simulation user interface named LabVIEW [15] is an Application alluded to as virtual instruments. Is an incredible and flexible graphical programming climate that was implemented to work with instrumentation control and information acquisition. Besides AJAX [16] is a Web technology for the creation of the Graphical user interface that combines many Web languages (HTML, Javascript, CSS…). This technology allows Javascript to call XML Web Services asynchronously.

On the other side, many studies have attempted to employ sketch-based user interfaces using the classical computer vision techniques in several fields such as Electric Circuits [17], Chemistry [18], Mechanical Engineering, Maths [19], and Computer Science [20]. Physics domain [21] using Sketch-Based Tutoring Systems.

In the field of generation of code automatically from sketches using deep learning methods is a relatively new area of study in following some related works: pix to code [22] is a new method created in 2018 for transforming a screenshots into code basing on deep learning. Also in 2019 sketch2code [23] created by Microsoft, is to transform a sketch into code HTML by merging the object and the text. In 2019 another approach named Sketch2Code [24], this approach based on Faster-RCNN permits the Transformation of Sketches to UI in Real-time. Another method [25] created in 2021 using AlexNet, SSD, RNN that permit to detect of graphical user interface elements and generate an HTML code.

3 Methods and Materials

3.1 System Overview

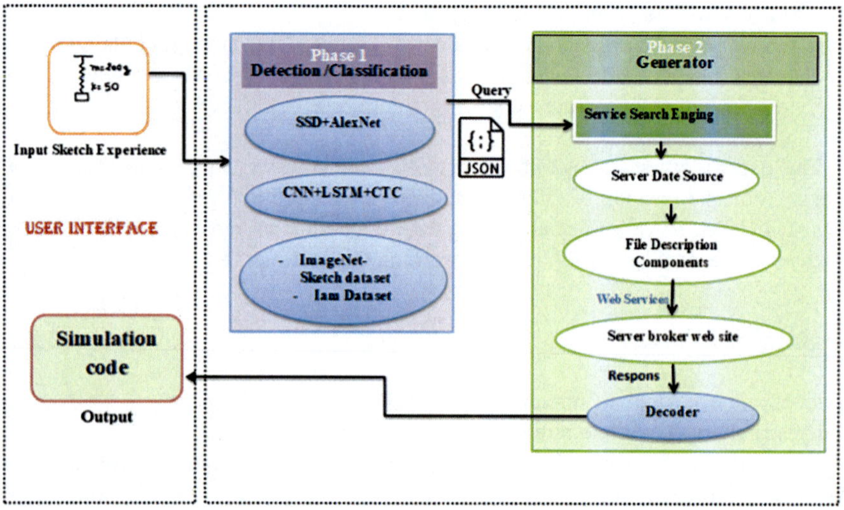

Fig. 1. The flowchart of our approach

The proposed method is a novel approach that allows transforming a sketch as input into a simulation Code of simulation. Our Solution is composed of two phases detection/classification and a Generator that is responsible for calling all services corresponding and building an output code of simulation. The Flowchart of our approach is shown in Fig. 1. We explain each phase in the following sections.

A sketch-based simulation needs to recognize drawn experience schema, and process them in table elements. Our input image is a hand-drawn sketch containing the representation of Graphical User Interface, which is resized to the size of 224 * 224. Then is transmitted it through the network. The starter is the block Detection/classification based on a convolutional neural network to identify and predict all objects that existed in the sketch. The output of this branch is a file JSON that contains pieces of information of all objects predicted. The file JSON is considered a starter of the second branch. Where a communication process (request-response) is enabled between the "list" of elements detected and the web laboratory. We adopting the concept of Laboratory as a service (LaaS) and modular laboratory paradigms [10, 11]. For developing a laboratory as separated Elements and each element in the laboratory as exposed as a service. Finally based on the web services called a parser generates the output code of GUI.

3.2 Input Image

Our input image is a hand-drawn consisting of one or more elements of UI on a white background. The input images can be of any dimension and are re-sized to 224 * 224, before being fed into the neural network.

3.3 Classification

In this work, a sketch represents the design of the experience. Any sketch can include hand-drawn elements and text. For this reason, we propose a neural network separated into two stages that combines Object recognition and character optic recognition. It's important to mention that two neural networks have been trained separately. For the first stage, we used a convolutional object detector SSD [8] with feature extractor Model "AlexNet" [12] for object recognition. And for the second stage, we utilize LSTM Recurrent Neural Network layers combined with CTC for text Recognition. Furthermore, we combine their outputs to construct a merging representation.

3.3.1 Object Recognition

CNN models are becoming the most used method for detecting objects. In the first branch of the proposed system, we used object detectors SSD [8], with AlexNet [12] model as a feature extractor. The output of this part is a list of elements predicted.

1) **Single Shut Detector:** SSD

Is an object detector based on a convolutional neural network that split the output space of bounding boxes to default box sets by locating all feature maps with different sizes according to various aspect ratios.

SSD network generates a set of fixed-size bounding boxes with scores for the existing items class instance in the bounding boxes, then outputs the final detections using a non-maximum suppression function. SSD has the following characteristics to improve the precision: a) predictors of object classes for detection with convolutional filters; b) for managing aspect ratio, the network partitions filters, and default boxes. c) Locating items using multi-scale feature maps.

2) **AlexNet**

It was considered a more efficient CNN model than LeNet and achieves the most difficult on (ILSVRC). Against all classic machine learning and computer vision techniques, AlexNet obtained state-of-the-art recognition accuracy. It was a watershed moment in the history of machine learning and computer vision for visual identification and classification techniques, and it marked the beginning of a surge in interest in deep learning. AlexNet architecture comprises four convolutional layers, three max-pooling layers, two normalization layers, two fully connected layers, and one SoftMax layer. The major key features of AlexNet make the success of this model as follows.

3) **SSD/AlexNet**

Our contribution is the combination of the single-shot detector (SSD) and Alexnet model. We changed the basic architecture of SSD; the size of the input is changed to 224 * 224. In AlexNet network architecture, for the second layer, the output of the Conv layers is 3 * 3 Conv kernels to make the first category of the default box. Moreover, AlexNet's last two fully connected layers are replaced with convolutional layers, and the network architecture is the same as SSD's, with four convolutional layers. In this proposed method the parameters are considerably reduced. The architecture of our proposed network is illustrated in Fig. 2

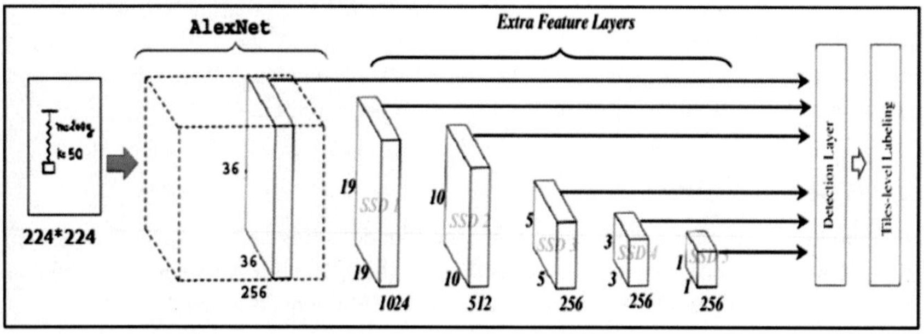

Fig. 2. The architecture of our proposed method SSD/AlexNet

3.3.2 OCR Detection

Text detection stage: is conceived to identify text in sketch images, such as words and numbers. The architecture is composed of three steps: the first is the convolutional layers, the second concerns the bidirectional LSTM recurrent layers and the use of the connectionist temporal classification (CTC) function. The CNN layers to extracts feature representations from the input sketch. The second part incorporates these representations. The sequences will be labeled by this LSTM layer. The CTC decoder receives these outputs at various time steps in order to extract text data from sketches. The proposed method is depicted in Table 1.

Table 1. Architecture of our OCR

Layers composition	Dimension
Input sketch	224 × 224 × 1
Conv 2D	3 × 3
Max pooling	2 × 2
Conv 2D	3 × 3
Max pooling	2 × 2
Resizing	
Dense Layer	
Bidirectionnel LSTM	
Sum	
Bidirectionnal LSTM	
Concatenation	
Dense Layer	
CTC Decoder	

The network of our proposed architecture for the text recognition is described below: The size of the input hand-drawn sketch is 224 * 224 * 1. The network starts with two convolution layers, sequentially, a maximum pooling follows every convolutional layer. After that, a resize operation is applied to the output of the previous convolutional layer in order to become a two-dimensional matrix; then, a dense layer follows this operation for reducing the matrix dimension. A bidirectional LSTM recurrent neural layer with concatenation function of the two output matrices. Followed by a second bidirectional LSTM layer. Before the final CTC output layer, a dense layer is applied.

3.3.3 Merging Branch

This sub-section explains the technique utilized to merge the results of two previous steps computer vision and OCR to build an output representation of the input sketch's features. The output is a sequence of information that describe the sketch's annotations. We utilized the JavaScript object notation (JSON) format.

3.4 Second Step: Generator

This part explains the second phase "Generator" of our proposed model which represents a communication procedure between three essential elements: result obtained by the previous branch, web server, and Parser. Consider a virtual laboratory for implementing experiments. The file JSON created by the previous phase is considered a user Query that contains all parameters such as (Elements, Pos x, Pos y, height…). In this phase, we integrate the LaaS paradigm into the system. In this case, we propose services retrieval in a virtual laboratory system. Indeed, the main goal of this technique is to find the corresponding service for a given request from a hand-drawn sketch. The virtual laboratory is exposed as a set of services. And the laboratory provider creates the "service description file" of this laboratory. Where the "service description file" lists all services. The provider transmits the file to the service broker Web site and indexes it with the appropriate metadata in order for the user may simply access it. This communication builds via API to send a query and receive a response. Therefore the service broker Web site produces a range of personalized services based on the user's requirements. Each service is a set of functions that is ready to be used. And finally, a parser based on the NodeJS server will build the output of the Graphical User Interface code. In this case, we choose Html/javascript language.

4 Experimental Results and Discussion

4.1 Evaluation Metrics

The performance of object detectors used in our system is evaluated by using four metrics: Accuracy, Precision, Recall, and F-measure. We will collect the True Positive (TP), False Positive (FP), False Negative (FN), and True Negative (TN) values to calculate the evaluation criteria. The number of experience sketches successfully recognized is referred to as True Positive (TP)

$$Recall = TP/TP + FN \tag{1}$$

$$Precision = TP/TP + FP \tag{2}$$

$$Accuracy = TP + TN/TP + TN + FP + FN \tag{3}$$

$$F1-measure = 2 * P * R/P + R \tag{4}$$

4.2 Dataset

Our proposed method is trained on sketch datasets: *ImageNet* dataset and Iam for OCR.

- *ImageNet-Sketch* [26]*:* There are 50000 photos in this collection, including 50 images for each of the 1000 ImageNet classes. The data set is built using Google Image queries such as "sketch of," where is the typical class name. We only look for things that are black and white.
- *Iam* [27]*:* This dataset is composed of over 1500 scanned text pages handwritten by 657 different writers. This dataset is divided into three parts: for training set has 747 documents (6,482 lines), the validation set contains 116 documents (976 lines) and the test set has 336 documents (2,915 lines).

Table 2. Some examples of GUI simulation generated by our system

Simulation title	Experience sketch	Generated simulation
Conical pendulum	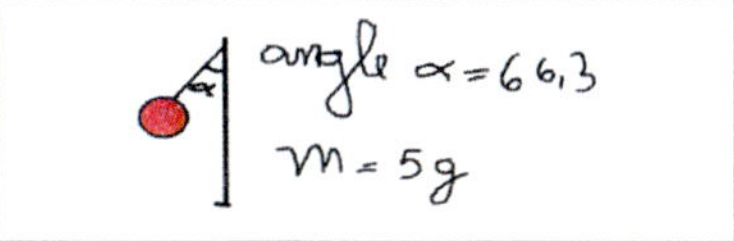	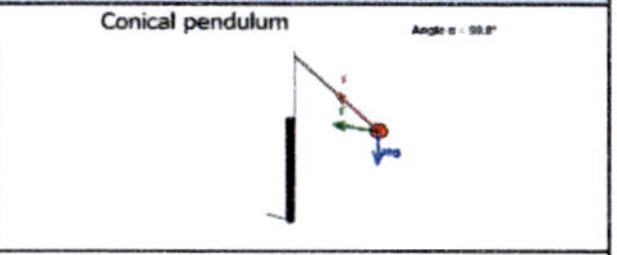
the power applied to the ballistic Pendulum	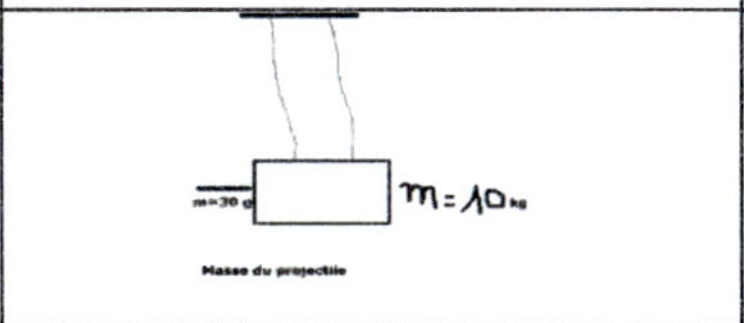	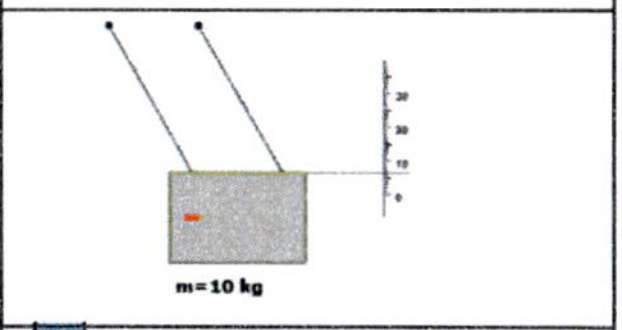
the fall of a ball with a limit speed	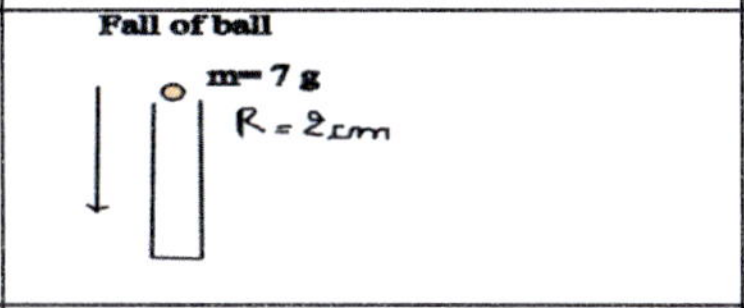	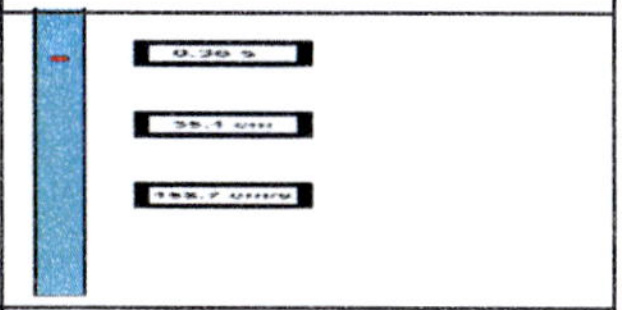
Newton's Pendu-lum	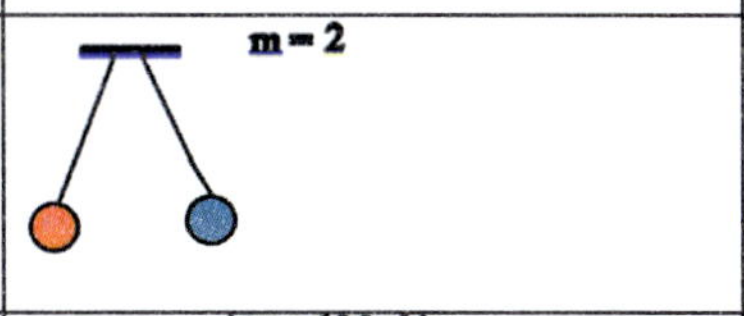	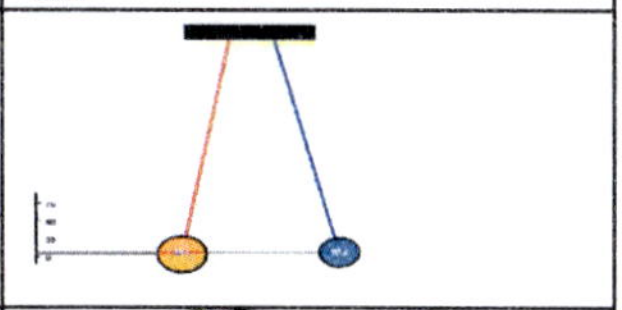
Atwood Machine	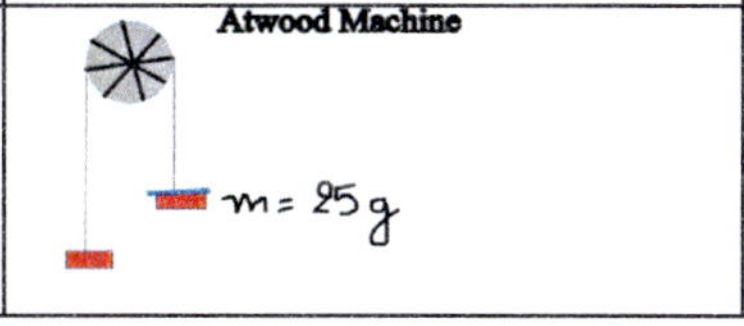	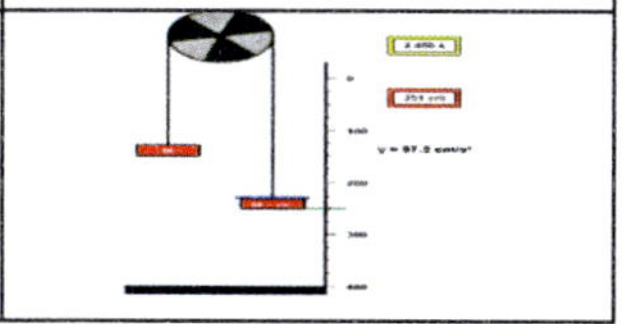

4.3 Qualitative Results

To test and validate our previous architecture we have developed our first system for some basic experiments in MECHANICS OF MATERIALS and power domain (pendulum, Fall of the ball, rotation event ...). Table 2 presents some simulations that the system can realize.

4.4 Quantitative Results

The proposed system implements deep learning-based SSD and AlexNet combined with CNN + RNN + CTC for regeneration of a simulation from a sketch image. We evaluate The performance of our system in terms of the most known data evaluation measures. For our study and as mentioned previously, we proceed to comparison using only four measures that are accuracy, precision, recall, and the F-Measure (F1 Score). The performance result is shown in Table 3.

In this section, we describe the results of our proposed method. We choose six experiments examples to effectuate the tests (ballistic Pendulum, Conical pendulum, the

Table 3. Performance evaluation

Object detector	Simulations	Precision	Recall	Accuracy	F1 score
SSD + AlexNet	Ballistic Pendulum	95,24%	94,12%	94,44%	94,67%
	Conical pendulum	96,51%	95,84%	96,10%	96,14%
	the fall of a ball	96,80%	94,18%	95,88%	95,91%
	Newton's Pendulum	95,39%	93,62%	94,33%	95,47%
	Machine atwood	94,05%	93,25%	93,33%	94,05%
	Spring	91,95%	90,96%	91,11%	91,67%
	Calorimeter	94,90%	93,29%	94,87%	94,89%
	Displacement of object	96,35%	95,62%	96,23%	96,27%

fall of a ball, Newton's Pendulum, Machine Atwood, Spring, calorimeter, Displacement of object). To obtain the evaluation measures; we need to calculate the average of the values obtained from 150 sketches to determine each value (TP, TN, FP, FN).

The models used have been implemented by php5, javascript, JSON using the TensorFlow and ML5 API and run on a Core i7 computer with a Windows 10 (64 bits) 12.00 GB operating system.

The performance results illustrate that the used models have obtained over 96.23% accuracy and over 95% on F1-score results. The proposed object detectors clearly provide efficient and effective results, as shown by similar results.

4.5 Discussion

In Table 4, we present the evaluation metrics comparison. This table Includes Comparisons in terms of average recall, F1- score, accuracy, and specificity.

Table 4. Performance comparaison

Object detector	Average performances %			
	Precision	Recall	Accuracy	F1 score
SSD + Inception v2 [28]	91,58%	90,80%	91,27	91,33%
Faster R-CNN + ResNet [29]	92,22%	90,85%	91,78	91,14%
This work SSD + Alexnet	94,85%	96,34%	93,97%	95,56%

It is observed that SSD with AlexNet offers higher accuracy (93,97%) and recall (95.56%) performances than Faster-RCNN with ResNet followed by SSD with ResNet. High recall indicates that fewer false positives were detected. SSD with alexNet achieve the highest F-measure (96.85%), followed by SSD with Inception v2 (91.33%) and Faster R-CNN with ResNet (91.14%). Finally, we can see that the proposed method was successful to generate simulations from sketches efficiently in many subjects taking into account the difficult scenarios.

5 Conclusion

In this work, we present a method based on deep learning which generates a Graphical User Interface for simulation in virtual laboratories from sketch images. Our system tackles the recognition of Elements in input hand-drawn image and generates a UI code. We evaluate our system on the ImageNet-Sketch dataset, the performance evaluations illustrate that method has achieved important performance. In future works, we try to test the proposed methods on a large dataset and generate GUI for another platform like an android.

References

1. Li, Y., Song, Y., Gong, S.: Sketch recognition by ensemble matching of structured features. In: British Machine Vision Conference (BMVC), pp. 1–11 (2013)
2. Sarvadevabhatla, R.K., Babu, R.V.: Freehand sketch recognition using deep features. Computer Science (2015)
3. Zhang, Y., Qian, X., Tan, X.: Sketch-based image retrieval using contour segments. In: IEEE International Workshop on Multimedia Signal Processing (2015)
4. Girshick, R., Donahue, J., Darrell, T., Malik, J.: Rich feature hierarchies for accurate object detection and semantic segmentation. arXiv:1311.2524v5 [cs.CV] (2014)
5. Girshick, R.: Fast R-CNN. arXiv:1504.08083v2 [cs.CV] (2015)
6. Ren, S., He, K., Girshick, R., Sun, J.: Faster R-CNN. Towards real-time object detection with region proposal networks. Advances in neural information processing systems (2015)
7. Dai, J., Li, Y., He, K., Sun, J.: R-FCN: object detection via region-based fully convolutional networks. arXiv preprint arXiv:1605.06409 (2016)
8. Liu, W., et al.: SSD: single shot multibox detector. In: European Conference on Computer Vision, pp. 21–37. Springer (2016)
9. Redmon, J., Divvala, S., Girshick, R., Farhadi, A.: You only look once: unified, realtime object detection. In: Proceedings of the IEEE Conference on Computer Vision and Pattern Recognition, pp. 779–788 (2016)
10. Salzmann, C., Govaerts, S., Halimi, W., Gillet, D.: The SmartDevice specification for remote labs, pp. 199–208. IEEE (2015)
11. Tawfik, M., et al.: Laboratory as a Service (LaaS): a novel paradigm for developing and implementing modular remote laboratories. Int. J. Online Eng. **10**(4), 13–21 (2014)
12. Krizhevsky, A., Sutskever, I., Hinton, G.E.: Imagenet classification with deep convolutional neural networks. Neural Inf. Process. Syst. **25** (2017). https://doi.org/10.1145/3065386
13. Tobarra, L., et al.: Creation of customized remote laboratories using deconstruction. IEEE Revista Iberoamericana de Tecnologias del Aprendizaje **10**(2), 69–76 (2015). https://doi.org/10.1109/RITA.2015.2418011
14. de la Torre, L., Sanchez, J., Andrade, T.F., Restivo, M.T.: Easy creation and deployment of javascript remote labs with EJSS and moodle. In: 2016 13th International Conference on Remote Engineering and Virtual Instrumentation (REV), pp. 260–261. IEEE (2016)
15. Wang, Z., Shang. Y., Liu. J., Wu, X.: A LabVIEW based automatic test system for sieving chips. Measurement **46**, 402–410 (2013)

16. Paulson, L.D.: Building rich web applications with Ajax. Computer **38**(10), 14–17 (2005)
17. Gennari, L., Kara, L.B., Stahovich, T.F., Shimada, K.: Combining geometry and domain knowledge to interpret hand-drawn diagrams. Comput. Graph **29**(4), 547–562(2005)
18. Ouyang, T.Y., Davis, R.: Chemink, a natural real-time recognition system for chemical drawings. In: Proceedings of the 16th International Conference on Intelligent User Interfaces, IUI 2011, pp. 267–276. ACM, New York, NY, USA (2011)
19. Anthony, L., Yang, J., Koedinger, K.R.: A paradigm for handwriting-based intelligent tutors. Int. J. Hum. Comput. Stud. **70**(11), 866–887 (2012)
20. Buchanan, S., Laviola, J.J.: Cstutor a sketch-based tool for visualizing data structures. Trans. Comput. Educ. **14**(1), 3:1–3:28 (2014)
21. Cheema, S., LaViola, J.J.: Using animation to enrich learning experience in sketch-based physics tutoring systems. In: Kapros E., Koutsombogera M. (eds.) Designing for the User Experience in Learning Systems. Human–Computer Interaction Series (2018)
22. Beltramelli, T.: pix2code: generating code from a graphical user interface screenshot. In: Proceedings of the ACM SIGCHI Symposium on Engineering Interactive Computing Systems, p. 3. ACM (2018)
23. https://www.microsoft.com/en-us/ai/ai-lab-sketch2code
24. Vanita, J., et al.: Sketch2Code: Transformation of Sketches to UI in Real-time Using Deep Neural Network. ArXiv abs/1910.08930 (2019)
25. Cheddi, F., Tahiri, A., Serrhini, M.: Regenerating a graphical user interface using deep learning. In: Saeed F., Al-Hadhrami T., Mohammed E., Al-Sarem M. (eds.) Advances on Smart and Soft Computing. AISC, vol. 1399. Springer, Singapore (2022). https://doi.org/10.1007/978-981-16-5559-3_28
26. Wang, H., Ge, S., Lipton, Z., Xing, E.P.: Learning robust global representations by penalizing local predictive power. Adv. Neural Inf. Process. Syst. **32** (2019)
27. Marti, U.-V.: Horst Bunke: the I am-database: an English sentence database for offline handwriting recognition. Int. J. Doc. Anal. Recogn. **5**(1), 39–46 (2002)
28. Szegedy, C., Vanhoucke, V., Ioffe, S., Shlens, J., Wojna, Z.: Rethinking the inception architecture for computer vision. In: Proceedings of the IEEE Computer Society Conference on Computer Vision and Pattern Recognition, vol. 2016-Decem, pp. 2818–2826 (2016)
29. He, K., Zhang, X., Ren, S., Sun, J.: Deep residual learning for image recognition. In: Proceedings of the IEEE Conference on Computer Vision and Pattern Recognition, pp 770–778 (2016)

ALF - Ambient Assisted Living for Healthcare Framework Based on IoMT and Big Data

Kanwal Zahoor(✉) and Narmeen Zakaria Bawany

Center for Computing Research, Department of Computer Science and Software Engineering, Jinnah University for Women, Karachi 74600, Pakistan
{kanwal.zahoor,narmeen.bawany}@juw.edu.pk

Abstract. A rapid technology change has significant opportunities for almost every industry, especially for healthcare. By using advanced network technologies and Internet of Medical Things (IoMT) devices, health professionals can monitor remotely and respond to emergency cases efficiently. Ambient Assistive living (AAL) aims to improve living conditions for the elderly and disabled people. Data from heterogeneous sensors allow AAL systems to continuously monitor the health status of the elderly. The main contribution of this paper is to propose an Ambient Assisted Living Framework (ALF) for health care. The framework - ALF for health care comprises four layers namely, data sources, data integration, data analytics and ambient assisted living applications for healthcare.

Keywords: Ambient assisted living · Healthcare · Big data · IoMT devices · HDFS · Spark · Storm

1 Introduction

According to the Global Health and Aging report published by the World Health Organization (WHO), the number of individuals aged 65 years or more were estimated around 524 million with the prospect of increasing up to 1.5 billion by 2050 [1]. Challenges brought forth by aging are not restricted to individuals but spans over caregivers, government, and healthcare. In individuals, the effects of aging are numerous from physical to psychological. It includes obesity, loss of hearing or vision, diabetes, reduced physical activity, back pain, arthritis, dementia, depression etc. [2]. Due to physical and physiological weaknesses, the rate of falls resulting in fatal injuries in the elderly is increased and often leads to their admission in hospitals. Research studies indicate that 25% of the hospitalized elderly live for at most one year after the fall whereas 40% are rendered incapable of living on their own [3]. Healthcare sectors in developing countries are often insufficient for the increasing demands of the growing population due to economic barriers and thus, unable to offer sufficient or good quality care to the elderly who require keen attention. As healing mechanisms of the body weaken in old age, the elderly become prone to severe damages even with a minor disease or a fall [4]. On analysis of a survey, it was revealed that 90% of the elderly population prefers to stay at their own homes instead of nursing homes or hospitals. Therefore, there is a need to develop

M. Ben Ahmed et al. (Eds.): SCA 2022, LNNS 629, pp. 291–298, 2023.
https://doi.org/10.1007/978-3-031-26852-6_28

technology supported environments that enable their preference and address their need for the comfort of their homes [5]. Ambient Assisted Living utilizes modern technology to improve the value of life for both, young and old people, thereby improving the quality of life on a global scale. It uses technologies like smart devices and standalone assisted devices to monitor an individual's daily routine and work environment [6, 7]. Through assistive devices, elderly people's health can be easily monitored continuously at home. Wearable sensors can be incorporated to track the daily activities of older adults and offer them the assistance of caretakers and health professionals. IoMT based AAL systems help senior citizens to increase their quality of life by managing and preventing chronic conditions, assisting them in remaining socially active, independent in their daily activities and autonomy, and encouraging them to remain active in their communities [8]. Significant progress made in wireless technology, processing and computing power are the powerful force for innovation in various sectors, including leaps of expansions in healthcare. The development of connected medical devices has enabled the capture, generation, analysis, and transmission of data as per requirement. Connected devices combined with data form the IoMT a connected infrastructure of healthcare medical devices, software applications, and computing systems for healthcare services. IoMT connected devices provide real-time monitoring to protect lives in case of medical emergencies like cardiac arrest, diabetes issues, stress, and any other chronic disease like Alzheimer's, depression, and fall etc. [9]. These linked devices sensors improve patient safety by enabling healthcare systems to streamline their clinical operations and workflow management, even from remote locations [10]. Research studies [11, 12] estimate a drastic change in the number of connected medical devices from 10 billion to 50 billion. Because of Mobile Health (MHealth), sensor technology and wearable's technology, the amount of biological data is increasing day by day such that a recent Cisco report [13] presents that the data created by IoMT devices is around 847 Zettabytes (ZB). Analysis, storage, and management of such voluminous data by traditional methods is a huge challenge. Cutting-edge technologies of big data analytics and machine learning techniques accelerate the process of data analysis, storage and management when incorporated with healthcare IoT. Big Data offers the analysis of large amounts of data. Big data analytics techniques in healthcare are beneficial for disease diagnosis, drug discovery, predicting epidemic, and improving the quality of medication.

Machine Learning techniques are used to evaluate data, detect patterns, learn from it, and then prepare results. The analysis of large amounts of medical data combined with pattern detection by machine learning algorithms streamlines the process healthcare. This enables healthcare personnel to avoid problems by detecting them early and ensure patient's safety. Combining the Big Data analytics and IoMT devices we propose an Ambient Assisted Living framework (ALF) in healthcare for remote monitoring of the patients and providing various health related services including emergency care.

The structure of the paper is as follows: Sect. 1 presents the introduction followed by related work in Sect. 2. In Sect. 3, the four-layered Ambient Assisted living Framework (ALF) for healthcare is discussed and Sect. 4 concludes the paper.

2 Related Work

Ambient assisted living is based on data collected using various sensors which includes connected sensors, wearable sensors, smart watches, combined with non-invasive wearable electrochemical sensors for patient monitoring. To monitor and collect data of everyday activities of elderly people is a huge challenge. Some of the most popular use cases of AAL are Diabetes prevention, cardiac activities monitoring, chronic disease management etc.

In this section a brief survey of existing work related to AAL is discussed. Ali et al. 2017 [14] proposes smart healthcare solution based on Ambient Assistive Living for elderly people. Hossain and Muhammad [15] developed a cloud-based patient monitoring system for collecting and sending patients' ECG as well as other health related data to the cloud. The cloud infrastructure is utilized to allow doctors, nurses, and other health-care professionals and careers to have seamless access to their patients' data. Another research study by Hossain [16] focuses on utilizing a different facial expressions and speech to recognize patients' states, such as normal, pain, and tensed. The accuracy of around 98% was reported, however it's not clear how much this knowledge of patients' state will help in understanding the scenario in an AAL situation. Buthelezi et al. [17] proposed a New Tree Routing Protocol (NTRP) for collecting Heart Rate data signals from a human body in healthcare monitoring systems. NTRP enhances Zigbee Tree Routing (ZTR) by using Kruskal's minimum spanning tree to identify the shortest routes on a ZigBee network. In NTRP, neighbor tables are used instead of the parent–child approach used in ZTR. NTRP divides nodes into clusters to reduce end-to-end delay, and cluster heads employ neighbor tables to send heart rate data to the target node. The NS-2 simulation tool is used to assess the performance of the NTRP. Moreover, Maskeliunas et al. [18] discuss cloud-based data centers can help with patient state recognition by offering data storage, analytics, and audio-visual data analysis, as well as storing data and granting safe access to healthcare professionals.

3 ALF – Ambient Assisted Living Framework for Healthcare Description

To monitor and perform health analysis for the elderly, we need to develop a framework that is modular enough to handle a wide range of data such as numerical, signals, environmental and contextual data. Each of these require different big data analytics techniques. This study proposes Ambient Assisted Living Framework (ALF) for healthcare as shown in Fig. 1. The first layer takes input and is known as the data source layer. Second layer is known as a data integration layer that performs filtering and data processing methods to reduce the quantity of redundant data while keeping the most significant one. Third layer is known as data analytics that executes the machine learning algorithms. The fourth layer is known as Ambient Assisted application layer which is shown on the user interface of applications.

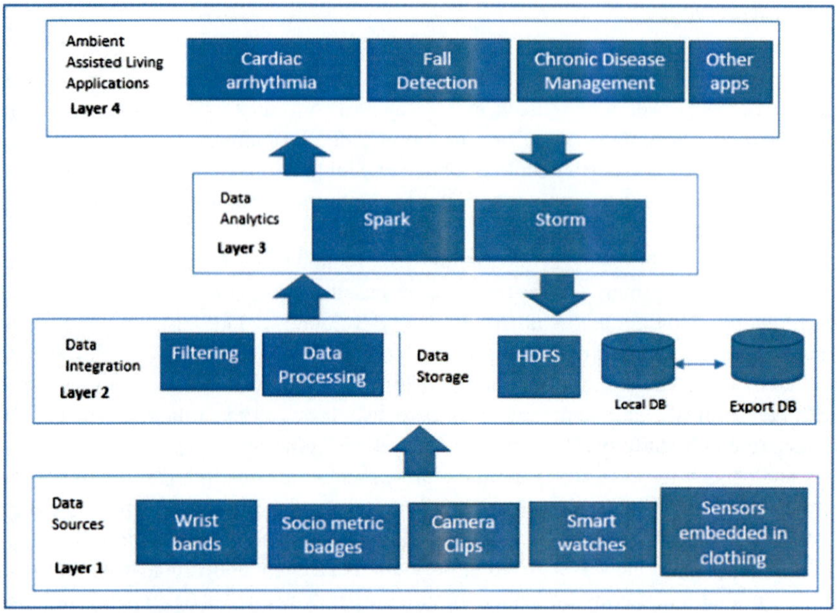

Fig. 1. Ambient Assisted Living Framework (ALF) for healthcare

3.1 Ambient Assisted Living (AAL) for Healthcare Applications

AAL applications provide users with various health monitoring services. This research is focused on several applications, such as cardiac arrhythmia, fall detection, and Chronic Disease Management etc. Some of data sources for ambient assisted application for healthcare [9] are mentioned in Table 1.

Cardiac Arrhythmia. Cardiac Arrhythmia happens when the electrical impulses that arrange pulses does not work as expected, making the heartbeat excessively fast, excessively sluggish, or sporadically. Cardiac arrhythmia may feel like a vacillating or dashing heart and might be innocuous. Notwithstanding, some heart arrhythmias might cause trouble here and there are even dangerous signs and side effects [19]. A constant heart monitoring sensor is required to ensure any emergency is handled timely.

Fall Detection. A fall detection can be defined as an assisted gadget whose standard target is to notify to health professionals when a fall event has occurred. In a real circumstance, it has the potential to relieve some of the ominous aftereffects of a fall. Specifically, fall indicators can clearly influence the reduction in the fear of falling and the quick provision after a fall [20]. Remote monitoring of the patient is required such that immediate help can be provided in case of accident.

Chronic Disease Management. AAL provides a considerate way to deal with diseases. The application incorporates screenings, registration, checking and organizing treatment, and patients training. It can work to the users' personal satisfaction while decreasing medical care costs on the off chance that individuals fall sick frequently by preventing or limiting the impacts of an infection [21].

Table 1. Ambient Assisted Living applications for healthcare

Application	Data source
Cardiac arrhythmia	Smart watch, camera clips
Fall detection	Wearable sensors (e.g., wrist band, smart watches), camera
Chronic disease management	Wearable sensors, sensors embedded in clothing, camera

3.2 Data Analytics

This layer receives physical activity data from various IoMT devices and wearable sensors and delivers it to cloud platforms such as Google Cloud Platform (GCP) and Amazon Web Services (AWS) where it is stored and subsequently processed by a big data processing technique.

Spark. Spark is an open-source platform used for large scale data processing. It has many other libraries like Spark SQL, Spark Streaming, GraphX, and MLib Spark SQL. It is a database engine that can restore data from a variety of sources and update it using SQL languages. Spark Streaming allows users to use this library to create applications that analyze and present data in real time. GraphX comprises of built-in libraries. It provides graph-parallel computation algorithms, whereas MLib provides machine learning algorithms [22].

Storm. Storm analyzes data in real time, and it is made up of spout and bolts. Bolt examines input streams and creates output streams, whereas Spout may produce data or load data from an input stream [23].

3.3 Data Integration

Hadoop MapReduce uses HDFS, a large data processor, to analyze and perform daily activities. Hadoop MapReduce processes many cores in parallel, allowing it to process massive volumes of input data. Two main processes of data integration are filtering and data processing. Filtering process can remove redundancy from the incoming data and maximize the useful data. Filtering is a type of lossy compression that decreases the amount of data while also deleting some of it. Data processing is also a vital process because without this step, data cannot be processed and classified. Table 2 presents some methodologies for data processing [24].

Table 2. Applications processing methodologies

Task	Machine learning algorithms	Sensors
Cardiac arrhythmia	SVM, DT, KNN, NB	ECG (Electrocardiography)
Fall detection	LSTM, SVM, and KNN	Wearable Sensors: Accelerometer, Gyroscope and Magnetometer At Wrist And Chest
Chronic disease management	CNN, RF, KNN	Wearable's Accelerometer

* Where: SVM: Support vector machine, DT: Decision Tree, NB: Naïve Bayes, RF: Random Forest, KNN: K-nearest neighbors, CNN: Convolutional neural networks, LSTM: Long Short-Term Memory.

3.4 Data Sources

To gather the information of environment and different health-related activities, various sensing devices are used through many deployment methods. Wearables and sensing devices, for example, smart devices such as wristband, sociometric badges, camera clips, smart watches, and sensors are embedded in clothing and its measures and sensors [9] are mentioned in Table 3.

Table 3. Data sources sensors

Data sources	Measure	Sensors
Smart watches	Physical activity, sleep pattern, heart rate, oxygen level	Oximeter, accelerometer
Wrist band	Stress	Electro dermal sensor
Socio metric badges	Level of social interactions	Proximity detection
Sensors embedded in clothing	Muscles activity	Electromyography sensor

4 Conclusion

In this research, ALF for healthcare is proposed by using big data analytics and machine learning techniques. ALF comprises four layers which are data source, data integration, data analytics, and ambient assisted living applications for health care. This paper concludes with some challenges in the field of Ambient Assisted Living applications. ALF for healthcare is limited because we are not addressing security and privacy layer but, in the future, it will be extended by incorporating a security layer.

References

1. Colombo, P.J., Crawley, M.E., East, B.S., Hill, A.R.: Aging and the Brain. Encyclopedia of Human Behavior: Second Edition, pp. 53–59 (2017). https://doi.org/10.1016/B978-0-12-375000-6.00006-9
2. Jaul, E., Barron, J.: Age-related diseases and clinical and public health implications for the 85 years old and over population. Front. Public Health **5**, 1–7 (2017). https://doi.org/10.3389/fpubh.2017.00335
3. AgeSafeAmerica: Home Safety for Seniors (2019). https://agesafeamerica.com/home-safety-seniors-statistics-solutions/
4. Azimi, I., Rahmani, A.M., Liljeberg, P., Tenhunen, H.: Internet of things for remote elderly monitoring: a study from user-centered perspective. J. Ambient. Intell. Humaniz. Comput. 1–17 (2016). https://doi.org/10.1007/s12652-016-0387-y
5. Ganesan, B., Gowda, T., Al-Jumaily, A., et al.: Ambient assisted living technologies for older adults with cognitive and physical impairments: a review. Eur. Rev. Med. Pharmacol. Sci. **23**, 10470–10481 (2019). https://doi.org/10.26355/eurrev_201912_19686
6. Delgado, J.M.P.Q., Azevedo, A.C., Guimarães, A.S.: State-of-the-Art. In: Interface Influence on Moisture Transport in Building Components. SAST, pp. 5–15. Springer, Cham (2020). https://doi.org/10.1007/978-3-030-30803-2_2
7. Singh, D., Kropf, J., Hanke, S., Holzinger, A.: Ambient assisted living technologies from the perspectives of older people and professionals. In: Holzinger, A., Kieseberg, P., Tjoa, A.M., Weippl, E. (eds.) CD-MAKE 2017. LNCS, vol. 10410, pp. 255–266. Springer, Cham (2017). https://doi.org/10.1007/978-3-319-66808-6_17
8. Syed, L., Jabeen, S., Manimala, S., Alsaeedi, A.: Smart healthcare framework for ambient assisted living using IoMT and big data analytics techniques. Future Gener. Comput. Syst. **101**, 136–151(2019). https://doi.org/10.1016/j.future.2019.06.004
9. IOT Veteran: Internet of Medical Things (IoMT) (2019). https://iotveteran.com/what-is-iomt/
10. Haughey, J., Taylor, K.: Medtech and the Internet of Medical Things (2019). https://www2.deloitte.com/global/en/pages/life-sciences-and-healthcare/articles/medtech-internet-of-medical-things.html
11. Dimitrievski, A., Zdravevski, E., Lameski, P., Trajkovik, V.: A survey of Ambient Assisted Living systems: challenges and opportunities. In: Proceedings - 2016 IEEE 12th International Conference on Intelligent Computer Communication and Processing, ICCP 2016, pp. 49–53 (2016). https://doi.org/10.1109/ICCP.2016.7737121
12. Zeadally, S., Siddiqui, F., Baig, Z., Ibrahim, A.: Smart healthcare. PSU Res. Rev. **4**, 149–168 (2019). https://doi.org/10.1108/prr-08-2019-0027
13. Cisco: Global Cloud Index: Forecast and Methodology (2018). https://www.CiscoCom/C/En/Us/Solutions/Collateral/Service-Provider/Global-Cloud-Index-Gci/White-Paper-C11-738085Pdf 2016–2021
14. Ali, Z., Muhammad, G., Alhamid, M.F.: Situation awareness in ambient assisted living for smart healthcare. IEEE Access **5**, 19835–19841 (2017)
15. Hossain, M.S., Muhammad, G.: Cloud-assisted Industrial Internet of Things (IIoT) - enabled framework for health monitoring. Comput. Netw. **101**, 192–202 (2016). https://doi.org/10.1016/j.comnet.2016.01.009
16. Hossain, M.S.: Patient state recognition system for healthcare using speech and facial expressions. J. Med. Syst. **40**(12), 1–8 (2016). https://doi.org/10.1007/s10916-016-0627-x
17. Buthelezi, B.E., Mphahlele, M., DuPlessis, D., et al.: ZigBee healthcare monitoring system for ambient assisted living environments. Int. J. Commun. Netw. Inform. Secur. **11**, 85–92 (2019)

18. Maskeliunas, R., Damaševicius, R., Segal, S.: A review of internet of things technologies for ambient assisted living environments. Future Internet **11** (2019). https://doi.org/10.3390/FI11120259
19. Röcker, C.: Designing ambient assisted living applications : an overview over state-of-the-art implementation concepts. In: International Conference on Information and Digital Engineering (ICIDE 2011) pp. 167–172 (2011)
20. Sarabia-Jácome, D., Usach, R., Palau, C.E., Esteve, M.: Highly-efficient fog-based deep learning AAL fall detection system. Internet Things **11**, 100185 (2020). https://doi.org/10.1016/j.iot.2020.100185
21. Patel, A., Shah, J.: Real-time human behaviour monitoring using hybrid ambient assisted living framework. J. Reliable Intell. Environ. **6**(2), 95–106 (2020). https://doi.org/10.1007/s40860-020-00100-7
22. Hassan, M.K., El Desouky, A.I., Elghamrawy, S.M., Sarhan, A.M.: Big data challenges and opportunities in healthcare informatics and smart hospitals. In: Hassanien, A.E., Elhoseny, M., Ahmed, S.H., Singh, A.K. (eds.) Security in Smart Cities: Models, Applications, and Challenges. LNITI, pp. 3–26. Springer, Cham (2019). https://doi.org/10.1007/978-3-030-01560-2_1
23. El Aboudi, N., Benhlima, L.: Big data management for healthcare systems: architecture, requirements, and implementation. Adv. Bioinform. **2018** (2018). https://doi.org/10.1155/2018/4059018
24. Qureshi, M.A., Qureshi, K.N., Jeon, G., Piccialli, F.: Deep learning-based ambient assisted living for self-management of cardiovascular conditions. Neural Comput. Appl. 1–19 (2021). https://doi.org/10.1007/s00521-020-05678-w

Leveraging Moroccan Arabic Sentiment Analysis Using AraBERT and QARIB

Ghizlane Bourahouat(✉), Manar Abourezq, and Najima Daoudi

ITQAN Team, LyRICA Laboratory, ESI, Rabat, Morocco
{ghizlane.bourahouat,mabourezq,ndaoudi}@esi.ac.ma

Abstract. Given the significance of Sentiment Analysis, NLP emerged with studies focusing on this task. The studies being conducted to investigate Arabic Natural Language Processing are not as significant as those in other non-Latin alphabet languages are. This lack of research is due to the rich and complex nature of the Arabic language, as well as the lack of free lexical resources. When it comes to dialectal Arabic, the studies become even more limited, revealing additional challenges. As a result, our paper focuses on Moroccan Arabic Sentiment Analysis by incorporating pre-trained Arabic BERT models -AraBERT and QARIB- throughout the process with various combinations such as SVM, CNN, and fine-tuning the pre-trained model itself. Our proposed system achieved the highest accuracy of 93% while using QARIB model. The research results obtained using our approach are compared with those of other leading methods, showing that our approach is highly effective and can provide valuable guidance for future improvements.

Keywords: ANLP · Arabic Sentiment Analysis · Moroccan dialect · Transformers · QARIB · AraBERT · SVM · CNN

1 Introduction

People all over the world are interacting more through social media, which they regard as a valuable tool for sharing knowledge on a variety of topics and expressing their ideas freely and openly. This engagement and exchange generate a large amount of data. Once thoroughly studied, decision-makers can use this huge amount of data to gain a better knowledge of a situation and, as a result, make the best decision possible. It's practically hard to analyse and process this many data without the correct technologies. This is where Natural Language Processing (NLP) and Sentiment Analysis (SA) are needed.

SA is an area of study within NLP that is concerned with identifying the mood or opinion of subjective elements within a text [1]. In recent years, SA has received great attention not only among researchers but also among businesses, governments, and organizations [2]. Many users express their views and opinions through various online resources. Therefore, to continuously monitor public opinion and aid decision-making, we must analyse user-generated data.

Multiple challenges are faced while dealing with SA such as informal writing style, sarcasm, irony, and language-specific challenges. Many words in different languages

M. Ben Ahmed et al. (Eds.): SCA 2022, LNNS 629, pp. 299–310, 2023.
https://doi.org/10.1007/978-3-031-26852-6_29

have different meanings and orientations depending on the context and domain in which they are used. As a result, there are few tools and resources available for all languages [3]. The Arabic language belongs to this category of languages. Therefore, researchers struggle while working in Arabic Sentiment Analysis (ASA).

The Arabic language is ranked as the fifth most widely spoken language in the world with a community that reaches billions of people. The Arabic language is characterized by a set of specifics, which however present challenges in the context of NLP. This later is conducted by following multiple steps starting with the data collection and arriving at the task performed which, in our research, is the task of Sentiment Analysis (SA).

From the various steps followed in the NLP process, we find that word embedding impacts the final performance of the implemented model. Several papers and works were interested in this step and tested different Machine Learning and Deep learning models, but few are the ones that implemented the Arabic pre-trained BERT models and especially to deal with the Moroccan dialect. For this reason, we aim to test different combinations of Arabic pre-trained BERT models such as AraBERT and QARIB, as an embedding layer and a baseline model, in order to study the impact of their structures.

The organization of the remainder of this paper is as follows: Sect. 2 presents the general context of this work and the statement of the problem that was addressed, namely the application of SA techniques to the Arabic text, the challenges faced while performing ASA, the use of Machine Learning and Deep Learning algorithms, and the Transformers architecture with the BERT model and its Arabic pre-trained based on it. Section 3 presents the related works. Section 4 presents the proposed approach. Finally, Sect. 5 presents and discusses the results and some insights into future research directions.

2 Context and Problem Statement

2.1 Arabic Sentiment Analysis

Sentiment Analysis (SA), also called opinion mining, is the field of study that analyses people's opinions, sentiments, evaluations, appraisals, attitudes, and emotions towards entities such as products, services, organizations, individuals, issues, events, topics, and their attributes [4]. A more thorough definition of SA is given by [5], where it is stated that there are many different names for the same thing-- SA, opinion mining, opinion extraction. However, both terms, SA and opinion mining, are more commonly to represent the same field of study.

SA automates the extraction or classification of sentiment from sentiment reviews using NLP, text analysis, and computational techniques. Its applications have spread to nearly every possible domain, including consumer products, services, healthcare, and financial services, as well as social events and political elections. Besides Named Entity Recognition (NER), SA is one of the most challenging tasks in NLP [4].

Since the Arabic language is a non-Latin language spoken by more than 422 million speakers, researchers have increasingly been interested in applying SA to Arabic texts, yielding to Arabic Sentiment Analysis (ASA).

Performing ASA faces various challenges related to the nature and structure of the Arabic language, which are different from other languages, such as English. Besides its specific structure, the Arabic language has three main varieties: The Classical Arabic (CA) used in religious and literary contexts. The Modern Standard Arabic (MSA), which is the official language used in education and formal communications. Finally, the Arabic dialect (AD) where there are no spelling standards in spoken for it. Also, AD is mostly divided into five main groups: Egyptian, Levantine, Gulf, Iraqi, Maghrebi [6]. In the category of Maghrebi dialect, we mention the Moroccan Dialect (MD). The Moroccan dialect is a spoken form of Arabic. In Morocco, it is often considered an informal language, used mainly in daily conversations and media. In the MD text, different colloquial words of the same text can be distinguished: MSA, French, Spanish and Tamazight [7].

Research in the field of ASA isn't as substantial as in other languages. They are even fewer when dealing with AD and especially with Moroccan dialect. The complexity and the challenges related to ASA increases while dealing with Arabic dialects such as Moroccan one [7]. In ASA, many challenges are faced. For instance, when collecting data, the available sources are not only limited in number, but they often contain misspelled words and transliterated especially in the case of Moroccan dialect. To handle these challenges, researchers turned to Machine Learning and Deep Learning.

2.2 Machine Learning and Deep Learning

In the field of NLP, various research use Machine Learning models to deal with SA. As investigated with [8, 9] and [10] Naïve Bayes, decision tree and Support Vector Machine (SVM) were implemented. Moreover, [11] showed that one of the mostly preferred method deployed for classification is SVM. It is a supervised learning approach that uses Machine Learning algorithms to perform effective regression analysis and data classification [11].

Furthermore, in ASA, several Machine Learning methods have been used. However, three methods have been consistently showing superior performances. They are SVM, K-nearest neighbour (K-NN), and Naïve Bayes. In [12–14] and [15] we see that SVM achieved the highest accuracy for ASA. SVM is a very prevalent supervised classifier [11]. It is non-probabilistic and uses hyperplanes to segregate labels. SVM supports linear and nonlinear models [11].

SA has also taken advantage of the performance of Deep Learning techniques. In [16–19] various models were used such as Convolution Neural Network (CNN) and Long Short-Term Memory (LSTM). In most cases, the CNN model achieved the best performance in SA. CNN are unique classes of neural network models that are designed to identify hidden patterns and relationships in large data sets. CNN is a feedforward neural network composed of a 1-dimensional convolutional layer and a 1-dimensional pooling layer [20]. CNN's role is to learn various features through many filters [21]. CNN has been proved to be effective for NLP tasks and has achieved remarkable performance in sentence classification [22].

Moreover, in [23–26] and [27], the use of CNN model were worthy since it achieved highest performance.

2.3 Transformers

The Transformer is a new architecture in NLP that aims to solve sequence-to-sequence tasks while handling long-range dependencies with ease. It completely relies on self-attention to compute representations of its input and output, with no use of sequence-aligned RNNs or convolution [28]. The Transformer's architecture is based on two elements: encoder and decoder. It uses stacked self-attention and pointwise, fully connected layers for both the encoder and decoder. The architecture of Transformers is not based on attention [28], which decides what sequences are important in each computational step. The encoder maps the input to a higher dimensional space vector. Moreover, it uses the important keywords as additional input to the decoder. This improves the decoder because it now has more information, such as which sequences are important and which keywords provide context for the sentence [29]. Transformers has shined in several NLP tasks such as SA, and BERT is one of its typical representatives.

BERT stands for Bidirectional Encoder Representations from Transformers. It is a language representation model, released in late 2018, advances the state of the art in a variety of NLP tasks. ERT is the first deeply bidirectional and unsupervised language representation model developed. BERT's model architecture is a multi-layer bidirectional Transformer encoder based on the original implementation described in [28]. BERT has different models for several languages. As our research is interested in Arabic language and especially the Moroccan dialect, we'll investigate Arabic pre-trained models based on BERT.

Indeed, the last few years have seen a massive revolution in the development of various bidirectional transformer-based models, particularly for Arabic. They serve as effective transfer learning tools, assisting in the improvement of a wide range of NLP tasks such as text classification, question answering, named entity recognition (NER), and so on. Several Arabic pre-trained based on BERT model are available, one of the main differences is the training corpus. We can categorize them to three classes. Firstly, models for MSA such as AraBERT, alBERT, AraELECTRA. Secondly, models trained on Arabic dialects such as QARIB, MARBERT, ArabicBERT and CAMelBERT. Finally, models trained on multilingual corpus that are XLM and mBERT.

3 Related Work

The objective of this study is to investigate SA in Moroccan dialect, especially when used with Arabic Transformers models. Based on this objective, we've focused on BERT models used for Moroccan or Maghrebin SA and compared between the models based on the effectiveness in classifying Moroccan/Maghrebin dialect. We've also examined already tested combination with Transformers and Deep Learning or Machine Learning models. Based on our findings, we notice that there are several techniques used in ANLP to solve the problem, however there were fewer one concerning the Maghrebi dialect. In this section, we will present some work in this field concisely. [26] has processed the Algerian dialect and the CNN model has achieved the best accuracy combined with several embedding techniques, and the highest accuracy was 82% when used with word2vec. With Algerian dialect, we present [30] that implemented the CNN model for SA and get an accuracy of 89.5%.

Works are even fewer when dealing especially with Moroccan dialect SA. In [31], various Machine Learning models were implemented and compared such as Support Vector Machine, Naïve Bayes and Logistic Regression. The best obtained result for was achieved by SVM, combined with TF-IDF, for an accuracy of 84.31%. In [32], three classification algorithms; Naive Bayes, Maximum Entropy and Support Vector Machines are used to improve the SA of Moroccan tweets. The best classifier was SVM with an accuracy of 82.5%.

As seen in this section most of the approaches applied to the Maghrebi dialect SA observed are based on traditional ML and DL. In the implementation of BERT models, we observe that very few works have been done in relation to other languages. In the following, we quote some of viewed research. In [33], AraBERT model was implemented and tested on a DA dataset and obtained an accuracy value of 92.5%. AraBERT was also used in [34] on multi-dialect dataset and got an accuracy of 89.6%. In [35], AraBERT was used with SVM on a dataset of MSA and Algerian dialect, the model had an accuracy of 87%. [36] has also employed the AraBERT model to perform SA and got an accuracy of 82.5%. QARIB was implemented as a classifier and embedding model in [37] and achieved an accuracy of 70% applied on a multi-dialect dataset. In [38], QARiB was also implemented for sentiment detection and achieved 75% as F1-score. QARIB has achieved an accuracy of 95,8% in [36] while used as an embedding model and classifier. Moreover, in [39], QARIB achieved an accuracy of 92% applied on a mixed corpus (dialectal Jordanian and MSA tweets). We notice that QARIB was also used in [35] with SVM and achieved an accuracy of 90% applied to a combination of MSA and Algerian dialect.

Concerning the AraBERT and QARIB model, we note that there is little work done in relation to other languages. Moreover, we've seen that AraBERT and QARIB were combined, in one article, only with SVM. For this reason, in our approach, we'll going to use both AraBERT and QARIB to perform SA. After we'll going to use the embedding provided by these models and combine them with the SVM and then with CNN. The use of SVM and CNN is justified by the high performance, in various cases, in ASA as seen earlier.

4 Our Approach

Our approach, as shown in Fig. 4, is based on Arabic BERT pre-trained models, namely AraBERT and QARIB. First we'll use the models, AraBERT and QARIB, to perform the SA taking advantage of the whole architecture of the model.

As stated earlier, we wanted to see the effectiveness of using Arabic Transformers model on Moroccan dialect. We also want to compare the transformer models to a baseline that uses an SVM classifier and CNN with embeddings generated by the Arabic Transformers as illustrated in Fig. 1.

AraBERT and QARIB

Since Arabic BERT-based models can be adapted to different down-stream tasks according to different needs, we will be using it to achieve ASA. The core of the architecture is trained on particularly large text corpora and, consequently, the parameters of the

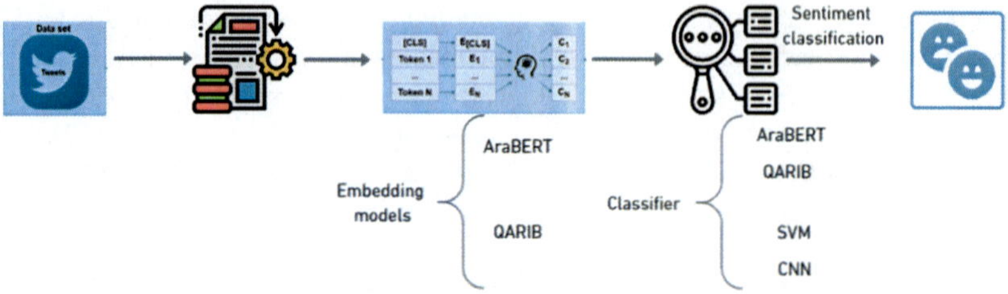

Fig. 1. Overall proposed approach

internal layers of the architecture are frozen. The outermost layers are instead those that adapt to the task and on which the so-called fine-tuning is performed. Therefore, we'll be using it as an embedding and a classifier model.

The next step is designed to combine the Arabic BERT pre-trained models as an embedding layer with SVM and CNN respectively. For the encoding phase it uses, respectively, AraBERT's and QARIB's embedding model and then the classifier is trained using SVM (see Fig. 2). Once the system classifier is trained, it is ready for an input as test data. The output is in the form of a confidence level, represented as a number between 0 and 1, where 1 is the highest confidence possible. The intents are encoded through the encode() call which takes a list of string and converts it into a list of vectors. After we can extract the pretrained embeddings. BERT provides contextual embeddings, meaning, the embeddings of each word depends on its neighbouring words. However, since it's contextual embedding, we can make an assumption that the fist token which is '[CLS]' captures the context can be treated as sentence embedding as be used as input to 'SVM' for Machine Learning approach and CNN for the Deep Learning one.

For our model, we will be applying the embedding using the pre-trained Arabic based model. After applying the embedding, the embedded sentence is fed to the CNN model layers as used in [40]. The first layer of the CNN is the convolution layer which contains multiple filters for 16 different regions to extract features of the sentences. Those filters convolve the input and generate feature maps. The second layer is the max-pooling layer which is responsible for capturing the most important features from the previous feature maps. The third layer is the fatten layer that merges all features into unified matrix. The last layer is a fully connected layer of type sigmoid function used as an activation function. This layer generates the output of the network, indicating that whether the input sentence is positive or negative.

We then proceeded with the implementation of our approach, as detailed in the next section.

5 Experiments and Results

The dataset is the baseline for any NLP research. The choice of the dataset was challenging since we needed a Moroccan dialect dataset designed for SA with sufficient data. We noticed that there are few datasets, for Moroccan dialect, that are open source. Therefore, in our work, we choose the dataset [41]. The data was collected from the Twitter social network and consists on +50K tweets in five (5) national dialects [41] including the

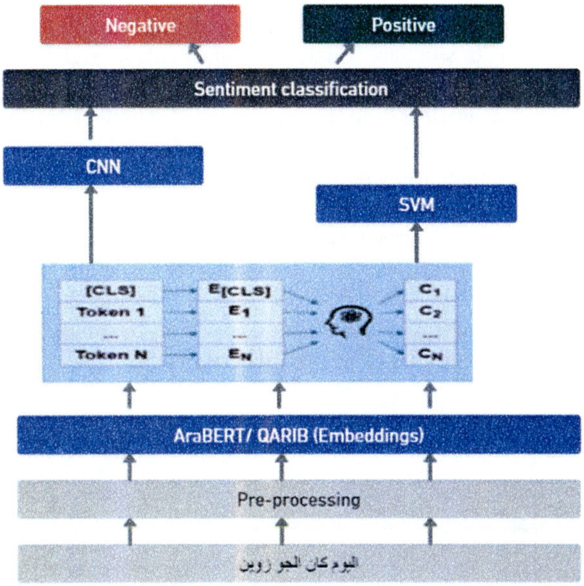

Fig. 2. Machine learning and deep learning approach based Arabic BERT pre-trained model

Moroccan dialect. From the first analysis, we found that the data is highly imbalanced, as shown in Fig. 4, which will be a blocker for an optimal performance (Fig. 3).

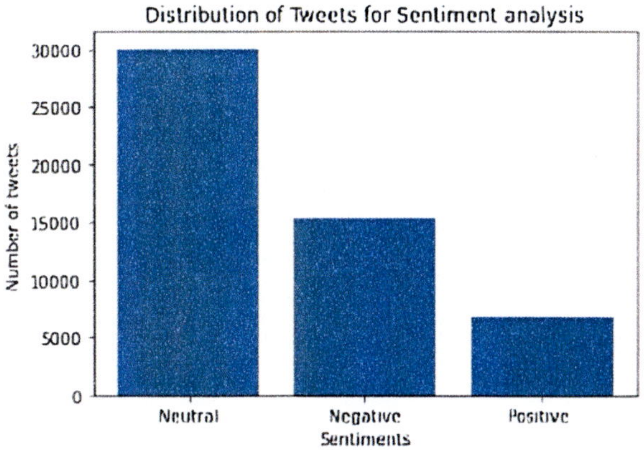

Fig. 3. Distribution of tweets for sentiment analysis

For that, we decided to perform SA based on two classes: negative and positive. Regarding these two classes, we've performed an under sampling to eliminate data point from the negative class randomly until the classes are balanced.

We then proceeded to test different combination of Arabic pre-trained BERT model to study the impact of its structure. First, we implemented, AraBERT and QARIB, respectively to perform the SA taking advantage of the whole architecture of the model.

AraBERT is based on the BERT model, a stacked Bidirectional Transformer Encoder. It is a multi-layer bidirectional transformer encoder based on the implementation of [60]. AraBERT is pre-trained using Modern Standard Arabic (MSA) data [36]. It was trained on Arabic corpora consisting of internet text and news articles of (8.6 B tokens) [42].[1]

QARiB (QCRI Arabic and Dialectal BERT) is a dialectal BERT model. It was trained on a collection of 420 Million tweets, 14B of tokens, 64k amount of vocabulary [39].[2]

To use the pre-trained Arabic based models, one of the parameters to specify is the maximum length of the sequence. For our case, the specific maximum length is less than 150, as shown in Fig. 4, but we used 160 for more precision.

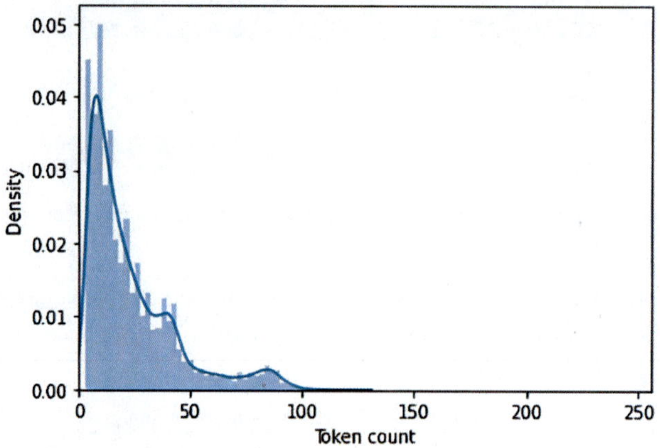

Fig. 4. Maximum length of tokens selected for the models

Table 1. Accuracy of the AraBERT and QARIB with themselves

Embedding	Algorithm	Maximum sequence length	Number of epoch	Metric	Learning rate	Performance
AraBERT	AraBERT	160	3	Accuracy	1e−6	92%
Qarib	Qarib	160	3	Accuracy	1e−6	93%

While using only QARIB and AraBERT, we found that the QARIB outperformed the AraBERT with only 1% to get an accuracy of 93%.

For the next step, we'll try the combination of AraBERT and QARIB respectively with SVM and CNN as classifiers. For these Arabic transformers, we'll only use the

[1] URL: https://github.com/aub-mind/araBERT.

[2] URL: https://github.com/qcri/QARiB.

generated embeddings and use them as an input for SVM and CNN respectively (see Tables 2 and 3). The results showed that the QARIB embeddings combined with SVM and CNN has also achieved the highest accuracy with 91% and 82% respectively.

Table 2. Accuracy of the AraBERT and QARIB and SVM

Embedding	Algorithm	Maximum sequence length	Metric	Performance
AraBERT	SVM	160	Accuracy	85%
Qarib	SVM	160	Accuracy	91%

Table 3. Accuracy of the AraBERT and QARIB and CNN

Embedding	Algorithm	Maximum sequence length	Number of epoch	Metric	Batch size	Performance
AraBERT	CNN	160	10	Accuracy	6	75%
Qarib	CNN	160	10	Accuracy	6	82%

The results shown in Tables 1, 2 and 3 demonstrate that the transformer-based models generally outperformed the standard Machine Learning Deep Learning models, which are based on SVM and CNN. This can be interpreted by the huge language knowledge gained by the Transformers while training them on language modeling objectives. This result emphasizes the importance of training a language model on various topics and indicates the superiority of Transformers over embeddings for Arabic dialect. For transformer-based models, it was difficult to determine which model worked best because different training methods could lead the model to behave differently. However, we found that QARiB obtained high score exceeding an accuracy of 93% but the difference between it and AraBERT was not huge as expected.

6 Conclusion and Future Works

Arabic Sentiment Analysis is still a challenge for researchers in the field of ANLP. In the present article, we have presented our approach grounded on the AraBERT and QARIB models. In addition, we have used diagrams and illustrations to clarify and detail the key steps of the proposed architecture. Our approach begins with the input of our model into a pre-processed text from the Moroccan dialect dataset. Then, we implemented the AraBERT/QARIB respectively to perform SA. After, we used the generated embeddings from ARABERT/QARIB and classify them using SVM/CNN respectively. Moreover, our evaluation is based on a Moroccan dataset, which contains more than 50,000 comments and feedback. We obtained an accuracy value of 93% using well-tuned parameters of the QARIB model, which is a very competitive result. We also achieved an accuracy of 82% using CNN and QARIB embedding. In the future, we hope to look into other Arabic Transformers for tasks other than SA, such as named entity recognition.

References

1. Bhadane, C., Dalal, H., Doshi, H.: Sentiment analysis: measuring opinions. Procedia Comput. Sci. **45**, 808–814 (2015). https://doi.org/10.1016/j.procs.2015.03.159
2. Fernando Sánchez-Rada, J., Iglesias, C.A.: Social context in sentiment analysis: formal definition, overview of current trends and framework for comparison. Inform. Fus. **52**, 344–356 (2019). https://doi.org/10.1016/j.inffus.2019.05.003
3. Wankhade, M., Rao, A.C.S., Kulkarni, C.: A survey on sentiment analysis methods, applications, and challenges. Artif. Intell. Rev. **55**(7), 5731–5780 (2022). https://doi.org/10.1007/s10462-022-10144-1
4. Hussein, D.M.E.D.M.: A survey on sentiment analysis challenges. J. King Saud Univ. Eng. Sci. **30**(4), 330–338 (2018). https://doi.org/10.1016/j.jksues.2016.04.002
5. Liu, B.: Sentiment Analysis and Opinion Mining. Springer, Cham (2012)
6. Guellil, I., Saâdane, H., Azouaou, F., Gueni, B., Nouvel, Damien: Arabic natural language processing: an overview. J. King Saud Univ. Comput. Inform. Sci. **33**(5), 497–507 (2021). https://doi.org/10.1016/j.jksuci.2019.02.006
7. Alsayat, A., Elmitwally, N.: A comprehensive study for Arabic sentiment analysis (challenges and applications). Egypt. Inform. J. **21**(1), 7–12 (2020). https://doi.org/10.1016/j.eij.2019.06.001
8. Hasan, A., Moin, S., Karim, A., Shamshirband, S.: Machine learning-based sentiment analysis for twitter accounts. Math. Comput. Appl. **23**(1), 11 (2018). https://doi.org/10.3390/mca23010011
9. Agarwal, B., Mittal, N.: Machine learning approach for sentiment analysis. In: Prominent Feature Extraction for Sentiment Analysis. SC, pp. 21–45. Springer, Cham (2016). https://doi.org/10.1007/978-3-319-25343-5_3
10. Krishna, A., Akhilesh, V., Aich, A., Hegde, C.: Sentiment analysis of restaurant reviews using machine learning techniques. In: Sridhar, V., Padma, M.C., Rao, K.A.R. (eds.) Emerging Research in Electronics, Computer Science and Technology. LNEE, vol. 545, pp. 687–696. Springer, Singapore (2019). https://doi.org/10.1007/978-981-13-5802-9_60
11. Yi, S., Liu, X.: Machine learning based customer sentiment analysis for recommending shoppers, shops based on customers' review. Compl. Intell. Syst. **6**(3), 621–634 (2020). https://doi.org/10.1007/s40747-020-00155-2
12. Duwairi, R.M., Qarqaz, I.: Arabic sentiment analysis using supervised classification. In: Proceedings - 2014 International Conference on Future Internet of Things and Cloud, FiCloud 2014, pp. 579–583 (2014). https://doi.org/10.1109/FiCloud.2014.100
13. Bhatt, S., Patwa, F., Sandhu, R.: An access control framework for cloud-enabled wearable Internet of Things. In: Proceedings - 2017 IEEE 3rd International Conference on Collaboration and Internet Computing, CIC 2017, vol. 2017, pp. 328–338 (2017). https://doi.org/10.1109/CIC.2017.00050
14. Mohamed, A.: SVM and Naive Bayes for Sentiment Analysis in Arabic (2022)
15. Dehghani, M., Noughabi, E.A.: Sentiment Analysis of Persian Political tweets using Machine Learning Techniques. In: Researchgate.Net (2022). https://www.researchgate.net/profile/Mohammad-Dehghani-23/publication/358461419_Sentiment_Analysis_of_Persian_Political_tweets_using_Machine_Learning_Techniques/links/62037402c8b46c1ad9757c34/Sentiment-Analysis-of-Persian-Political-tweets-using-Machine-Le

16. Dang, N.C., Moreno-García, M.N., De la Prieta, F.: Sentiment analysis based on deep learning: a comparative study. Electron. **9**(3) (2020). https://doi.org/10.3390/electronics9030483
17. Mulyo, B.M., Widyantoro, D.H.: Aspect-based sentiment analysis approach with CNN. In: International Conference on Electrical Engineering, Computer Science and Informatics, vol. 2018, pp. 142–147 (2018). https://doi.org/10.1109/EECSI.2018.8752857
18. Cui, H., Lin, Y., Utsuro, T.: Sentiment Analysis of Tweets by CNN utilizing Tweets with Emoji as Training Data. Wisdom, pp. 1–8 (2018). https://sentic.net/wisdom2018cui.pdf
19. Sitaula, C., Shahi, T.B.: Multi-channel CNN to classify nepali covid-19 related tweets using hybrid features (2022). http://arxiv.org/abs/2203.10286
20. Li, H., Ma, Y., Ma, Z., Zhu, H.: Weibo text sentiment analysis based on bert and deep learning. Appl. Sci. **11**(22) (2021). https://doi.org/10.3390/app112210774
21. El Rifai, H., Al Qadi, L., Elnagar, A.: Arabic text classification: the need for multi-labeling systems. Neural Comput. Appl. **34**(2), 1135–1159 (2021). https://doi.org/10.1007/s00521-021-06390-z
22. Kalchbrenner, N., Grefenstette, E., Blunsom, P.: A convolutional neural network for modelling sentences. In: 52nd Annual Meeting of the Association for Computational Linguistics. ACL 2014 - Proceedings Conference, vol. 1, pp. 655–665 (2014). https://doi.org/10.3115/v1/p14-1062
23. Darwish, K., et al.: A panoramic survey of natural language processing in the Arab world. Commun. ACM **64**(4), 72–81 (2021). https://doi.org/10.1145/3447735
24. Alayba, A.M., Palade, V.: Leveraging Arabic sentiment classification using an enhanced CNN-LSTM approach and effective Arabic text preparation. J. King Saud Univ. Comput. Inform. Sci. **34**(10), 9710–9722 (2022). https://doi.org/10.1016/j.jksuci.2021.12.004
25. Alharbi, A.I., Lee, M.: Multi-task learning using a combination of contextualised and static word embeddings for {A}rabic sarcasm detection and sentiment analysis. In: Proceedings of the Sixth Arabic Natural Language Processing Workshop, pp. 318–322 (2021). https://aclanthology.org/2021.wanlp-1.39
26. Guellil, I., et al.: A semi-supervised approach for sentiment analysis of Arab(ic+izi) messages: application to the algerian dialect. SN Comput. Sci. **2**(2), 1–18 (2021). https://doi.org/10.1007/s42979-021-00510-1
27. Mhamed, M., Sutcliffe, R., Sun, X., Feng, J., Almekhlafi, E., Retta, E.A.: Improving Arabic sentiment analysis using CNN-based architectures and text preprocessing. Comput. Intell. Neurosci. **2021**, 1–12 (2021). https://doi.org/10.1155/2021/5538791
28. Vaswani, A., et al.: Attention is all you need. Adv. Neural Inf. Process. Syst. **2017** 5999–6009 (2017)
29. Hoang, M., Rouces, J.: Aspect-Based Sentiment Analysis Using BERT (2019)
30. Soumeur, A., Mokdadi, M., Guessoum, A., Daoud, A.: Sentiment analysis of users on social networks: overcoming the challenge of the loose usages of the algerian dialect. Procedia Comput. Sci. **142**, 26–37 (2018). https://doi.org/10.1016/j.procs.2018.10.458
31. Adam, E.L.B.: Sentiment Analysis for Moroccan Dialect (2020)
32. Oussous, A., Lahcen, A.A., Belfkih, S.: Improving sentiment analysis of moroccan tweets using ensemble learning. In: Tabii, Y., Lazaar, M., Al Achhab, M., Enneya, N. (eds.) BDCA 2018. CCIS, vol. 872, pp. 91–104. Springer, Cham (2018). https://doi.org/10.1007/978-3-319-96292-4_8
33. El Moubtahij, H., Abdelali, H., Tazi, E.B.: AraBERT transformer model for Arabic comments and reviews analysis. IAES Int. J. Artif. Intell. **11**(1), 379–387 (2022). https://doi.org/10.11591/ijai.v11.i1.pp379-387
34. El Mahdaouy, A., El Mekki, A., Essefar, K., El Mamoun, N., Berrada, I., Khoumsi, A.: Deep Multi-Task Model for Sarcasm Detection and Sentiment Analysis in Arabic Language (2021). http://arxiv.org/abs/2106.12488

35. Chowdhury, S.A., Abdelali, A., Darwish, K., Soon-Gyo, J., Salminen, J., Jansen, B.J.: Improving Arabic text categorization using transformer training diversification. In: Proceedings of the Fifth Arabic Natural Language Processing Workshop, pp. 226–236 (2020). https://www.aclweb.org/anthology/2020.wanlp-1.21
36. Al-Yahya, M., Al-Khalifa, H., Al-Baity, H., Alsaeed, D., Essam, A.: Arabic fake news detection: comparative study of neural networks and transformer-based approaches. Complexity **2021** (2021). https://doi.org/10.1155/2021/5516945
37. Farha, I.A., Magdy, W.: Benchmarking transformer-based language models for Arabic sentiment and sarcasm detection. Arabic Natural Language Processing, pp. 21–31 (2021)
38. Abuzayed, A., Al-Khalifa, H.: Sarcasm and sentiment detection in Arabic tweets using BERT-based models and data augmentation. In: Proceedings of the Sixth Arabic Natural Language Processing Workshop, pp. 312–317 (2021). https://www.aclweb.org/anthology/2021.wanlp-1.38)
39. Abdelali, A., Hassan, S., Mubarak, H., Darwish, K., Samih, Y.: Pre-Training BERT on Arabic Tweets: Practical Considerations (2021). http://arxiv.org/abs/2102.10684
40. Mohammed, A., Kora, R.: Deep learning approaches for Arabic sentiment analysis. Soc. Netw. Anal. Min. **9**(1), 1–12 (2019). https://doi.org/10.1007/s13278-019-0596-4
41. Boujou, E., Chataoui, H., El Mekki, A., Benjelloun, S., Chairi, I., Berrada, I.: An open access NLP dataset for Arabic dialects : data collection, labeling, and model construction (2021). http://arxiv.org/abs/2102.11000
42. Antoun, W., Baly, F., Hajj, H.: AraBERT: Transformer-based Model for Arabic Language Understanding (2020). http://arxiv.org/abs/2003.00104

Transfer Learning for Automated Melanoma Classification System: Data Augmentation

Dahdouh Yousra[1,2](✉), Anouar Boudhir Abdelhakim[1,2], and Ben Ahmed Mohamed[1,2]

[1] Faculty of Sciences and Techniques, Abdelmalek Essaâdi University, Tangier, Morocco
dahdouhyousra@gmail.com

[2] LIST Laboratory FSTT UAE, Tangier, Morocco

Abstract. Skin cancer, and especially melanoma, is the most dangerous diseases today because it is one of the fastest and active growing cancers worldwide [1]. It can be diagnosed and analyzed by specialist dermatologists, using Dermoscopic imaging techniques. Early detection is an important factor in reducing death rates related with this type of skin cancer. This is the cause why many researchers in this domain wanted to obtain accurate computer-aided diagnosis systems based with Artificial Intelligence (AI) algorithms to simplify the early detection of such diseases. Thanks to wide applications of Artificial Intelligence and especially Deep Learning in healthcare, including Dermatology.

In this paper, we concentrate on the problem of skin lesion classification particularly melanoma, where we proposed an effective, accurate and fast model to solve the problem of Dermoscopic image classification containing a cutaneous melanoma as malignant or benign. This model based on different pre-trained deep learning algorithms with fine-tuned to maximizing accuracy. The proposed model includes a preprocessing step for data organization, normalization and data augmentation; after all, we employed transfer learning algorithms with fine-tuning, where we first change the last layer to match the classes in our dataset, then, used the different strategies of freezing layer's techniques of fine-tuning, add other layers and finally retrain the layers of the network. We used six algorithms for state-of-the-art models previously developed for the ImageNet Large Scale Visual Recognition Challenge (ILSVRC): (EfficientNetB0) / (Mobile-Net) / (InceptionV3) / (XceptionV3) / (VGG19) / (ResNet152V2), Finally, we have evaluated our performances models using the following three metrics: Accuracy, Precision, F1 Score.

Our experimental study and the results prove that transfer learning is efficient in skin melanoma cancer classification. Mobile-Net, XceptionV3, Inceptionv3 and VGG19 obtained the highest accuracy of 87%, 80%, 80%, 82%, respectively. Resnet152v2 obtained the lowest accuracy of 70%. However, Efficient-Net outperformed all other models in melanoma detection with an accuracy 92%. The outstanding performance of these pre-trained model can significantly improve the speed, efficacy and accuracy of skin cancer diagnosis.

Keywords: Skin cancer · Melanoma · Dermoscopy · Deep learning · Transfer learning · Data augmentation

M. Ben Ahmed et al. (Eds.): SCA 2022, LNNS 629, pp. 311–326, 2023.
https://doi.org/10.1007/978-3-031-26852-6_30

1 Introduction

Cancer is the world's main cause of death if not diagnosed and treated in a timely fashion. [2] Skin cancer is one of the most prevalent and deadly types of cancer. As the skin is the body's largest organ, the point of considering skin cancer as the most common type of cancer among humans is understandable [3].

Skin cancer is the most common form of cancer because of its ability to spread faster to all organs, caused by growth of abnormal cells in the upper layer of the skin due to DNA damage that causes mutations which cause tumor formation due to skin cells grow rapidly. These tumors may be malignant or benign [4]. Among the main types of skin cancer are Melanoma. It is a hazardous, rare, and deadly type of skin cancer. It grows rapidly and has the ability to affect and spread to any area of the human body. A melanoma develops from cells called melanocytes on the areas exposed to sun rays, such as on the hands, face, neck, lips, etc. [5]. Early diagnosis and treatment are necessary where can be treated more easily if detected in its early stage before it develops into a full-blown skin cancer or reaches the inner layers of the skin.

Dermoscopy is technique primarily used for the examination of skin lesions, is able to give illuminated and magnified images to increase the clarity of the spots of skin lesions. However, automatic recognition of melanoma Dermoscopy images is a challenging task due to the high similarity between melanoma and non-melanoma lesions [6]. Recently, for overcoming this challenge the new automated systems have been developed based on Artificial Intelligence can be aided in the early detection of skin cancer, lowering the incidence of disease and also the associated deaths. In addition to reducing the workload for the doctors, AI-based systems can also help directly in better skin lesion diagnostics.

Deep Learning, is a subdivision of Machine Learning supported by mass of algorithms. This is a novel computer vision technique that develops and extends traditional machine learning models. Deep Learning is applied effectively to various domains such as speech recognition [7], text classification [9], object detection [8]…., and has proved promising for the field of medical imaging. Where Deep Learning algorithms have recorded huge success in the early detection and diagnostic of skin cancer. Among various Deep Learning neural networks, the Convolutional Neural Network (CNN) [10] is most frequently used in the image classification task in medical domain. Because of insufficiency the large dataset, Transfer Learning is suggested use as solution.

Transfer Learning is a machine learning method where reuse a pre-trained model such as Inception [12], Xception [13] and Mobile-Net [11]… as the starting point for a model on a new and different task, it very effective in the case of deep learning to speed up training, improves the adaptation ability, and to enhance the model performance.

In this paper, we evaluate six different pre-trained CNN algorithms ((Efficient-Net B0) / (Mobile-Net) / (InceptionV3) / (XceptionV3) / (VGG 19) / (ResNet152V2)) with fine-tuned to maximize detection accuracy. The dataset used in this study is collected by the ISIC archive. After preprocessing step for data organization, normalization and data augmentation, we have employ transfer learning algorithms with fine-tuning techniques, where we freezing layers, adding other and retrain the layers of the network.

This paper is organized as follows: Sect. 2 reviews related work; Sect. 3 described the proposed solution and associated methods and materials; we have introduced the theory of transfer learning and the CNN algorithms pre-trained used, and then describes the proposed method in detail including the introduction to the dataset used. Section 4 displays the results of experiments and discusses their significance. In the same section we have presented the experimental evaluation of the proposed model. Finally, Sect. 5 offers concluding observations, and instructions for future work.

2 Related Works

Skin cancer is by far the most common dangerous types of cancers that affect large number of people every year in the world. The detection and classification of skin cancer using deep learning and computer vision techniques and especially CNN algorithm is a new challenge for all researchers and dermatologists. There has been a masses of work published in medical domain especially for skin cancer classification using deep learning algorithms. These works use a different approach.

This section presents some most related works on skin cancer using different techniques for detecting and classifying.

Hossin and al. [14] proposed a multilayered CNN architecture using dropout and Batch Normalization as regularization techniques for melanoma classification. This system has provided as accuracy of 93.5%.

In another research, Noortaz Rezaoana and al. [15] proposed an automated technic based with Convolution Neural Network for skin cancer classification (9 types of skin cancer). This methodology uses notion of image processing, image augmentation techniques and deep learning.in addition, the authors used transfer learning approach to further improve the accuracy. Approximately 79.45 percent accuracy.

In [16], S. Alagua and K. Bhoopathy Bagan proposed an architecture based of convolutional neural network incorporated with optimization technique to identify the melanoma cells. After preprocessing step and data augmentation method the images are given as input to convolutional neural network. They adopted the DenseNet architecture with transfer learning for skin cancer classification. They are found that the classification accuracy of 95% is obtained using CNN with stochastic gradient descent optimization technique.

In [32], the authors proposed a novel deep transfer learning model for melanoma classification using MobileNetV2. Various data augmentation techniques were applied to tackle the class imbalance issue and add diversity to the dataset. This architecture works effectively and has a diagnostic accuracy of 98.2 percent.

Further, in [17] Laila Moataz and al presents the transfer learning classification technique to classify and diagnosis skin cancer. They used a Xception pre-trained model with fine-tuning method by adding a some of layers after the basic ones of the Xception model. These layers contain a dense layer followed by a dropout layer then a Batch Normalization layer and ends with a ReLU activation function. And finally added a softmax layer as a classifier. The model achieved a better accuracy of 96% using a balanced dataset.

Following this context, this article proposes an advanced deep learning architectures based with transfer learning to contribute for early diagnosis of skin cancer. These models show superior performance for detection and classification of skin cancer using

Dermoscopy images. We have used six a fine-tuning on pre-trained deep CNN models –Efficient-Net B0) / (Mobile-Net) / (Inception V3) / (Xception V3) / (VGG 19) / (ResNet152V2) --. We have freezing, adding a group of layers after the basic ones of the model and also retrain the layers of the networks.

3 Materials and Methods

In this section, we present a brief introduction to Transfer Learning and pre-trained algorithms used, and also present the description of dataset used will be made first, and then the proposed model will be detailed.

3.1 Dataset Description

In this work, to train and test the models, a dataset of images is required. We collected a public dataset from ISIC website [18] for skin cancer classification task. We used the ISIC Archive Downloader to download the images of Dermoscopic which allow better diagnostic accuracy as they obtain the micro-structures in the lesion and provide a clear view. A dataset consisting of malignant and benign images of skin cancer.

We used 3297 images separated into 1800 of benign images and 1497 of malignant images as shown in Fig. 1. After collecting the dataset, we preprocessed images by resized into 224x224 pixel our dataset because all the images were not the same size.

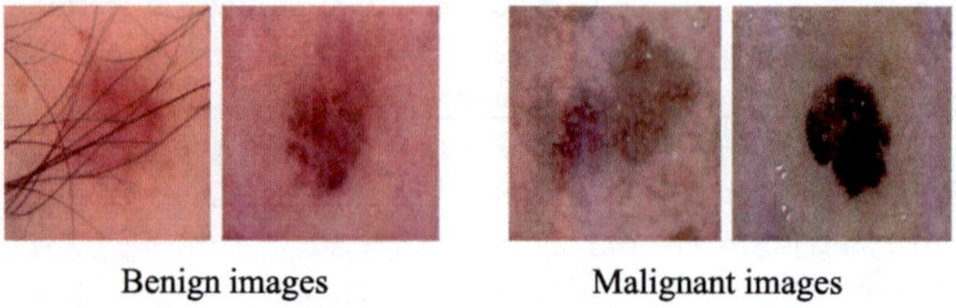

Fig. 1. Sample images.

3.2 Transfer Learning Nets

In this section, we focus is on transfer learning and we discuss briefly the models used in the research and the steps of development of individual model.

Transfer Learning is a famous method in computer vision that allows us to build fastly accurate models using pre-trained neural networks. Where a model developed for a task is reused to another task. There are two popular common ways to apply transfer learning:

Fixed Feature Extractor: the use the pre-trained model as a feature extraction system. It works is by frozen the weights in the feature extraction layer after removing the last fully connected layer.

Fine-tuning: the weights in the feature extraction layer are not frozen. But some fine adjustments to augment performance further.

In our case the advantages of using transfer learning are:

- Is the more efficient in classifying Melanoma and benign lesions because these latter contain a high similarity.
- Transfer Learning Networks are trained on large datasets and their model weights are frozen. The advantage is the last few layers are changed according to different dataset.

3.2.1 VGG19

VGG19 [19] proposed by Simonyan and Zisserman (2014) it is an improvement of the model VGG-16. it is a simple convolutional neural network that comprises 19 layers with 16 convolution layers are used for feature extraction and 3 fully connected work for classification. The input image into this model of size 224 × 224 pixel (Fig. 2).

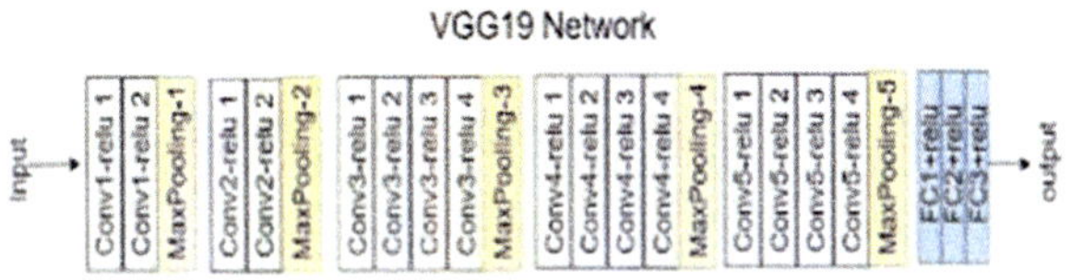

Fig. 2. VGG19 architecture.

Development of Individual Model:
In our case, we used fine-tuning model by pre-trained model (VGG19). we have tried to make fine-tuning model in better manner for classification task, especially classification the Dermoscopy images. With fine-tuning, with fine-tuning, we first change the last layer to match the classes in our dataset and besides, we also retrain the layers of the network.

Firstly, we have added the new layers to the pre-trained model. we have get layers and add average pooling layer and fully-connected layer and output layer with activation function Softmax for classification, after adding new layers, we have train those layers to update the weights pre-trained model layers have. Finally, we chose the layers to train and not to train by testing many techniques to fine-tune:

- Just last layers are updated by training and the others are remained as they are.
- Train just the part that we added.
- Consider 90% layers are not trained.

3.2.2 Resnet152

ResNet is one of the deepest deep learning architectures proposed by Microsoft researchers. Resnet introduces a structure called residual learning unit to alleviate the degradation of deep neural networks. This unit's structure is a feedforward network with a shortcut connection which adds new inputs into the network and generates new outputs. The main merit of this unit is that it produces better classification accuracy without increasing the complexity of the model. [21] (Fig. 3).

Development of Individual Model:
Fine tuning on a pre-trained model involves training some of the layers instead of freezing

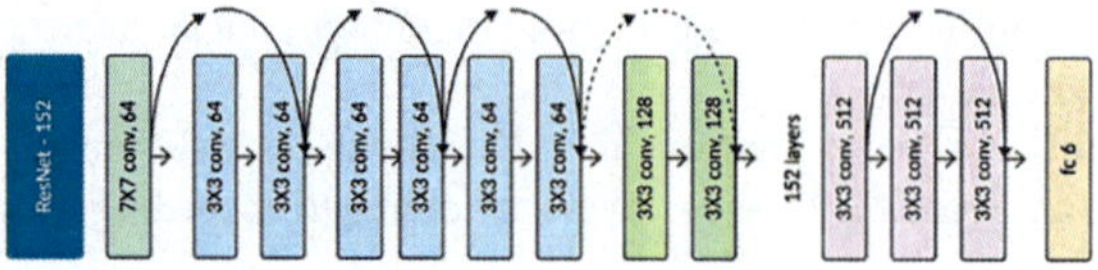

Fig. 3. Resnet architecture.

all the weights.in our case, we have used the pre-trained ResNet152 for fine tuning, by doing the following:

- We've set all layers in the model to be untrainable, except for the last ten.
- Adding our average pooling layer and two own fully connected layer with 512 and 256 neurons.
- Adding a single neuron dense layer with Softmax activation as our classifier.

3.2.3 InceptionV3

Inception models are a type of Convolutional Neural Networks. The inception models differ from the ordinary CNN in the structure where the inception models are inception blocks, that means lapping the same input tensor with multiple filters and concatenating their results. Inception-v3 [12] is an extended network of the popular GoogLeNet [22] which has achieved a good performance in different task in several domains (Fig. 4).

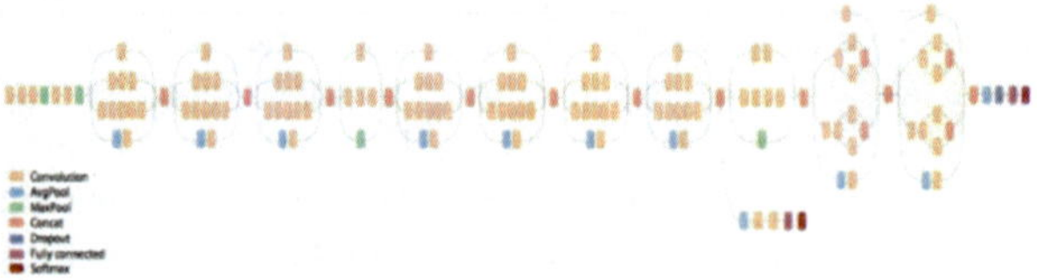

Fig. 4. InceptionV3 architecture.

Development of Individual Model:

On the fine tuning phase, we have added fully connected layers, selected training target layers, and trained.

Firstly, we have get layers of based model and add average pooling layer and fully-connected layer and output layer with function Softmax as our classifier. After, we have freeze pre-trained model area's layer and update the weight that are added, after that, we have chosen the layers which are updated by training before update the weights. we have freeze the first 279 layers and unfreezes the remaining layers of the InceptionV3 part. Finally, we have start the training with best hyper-parameter.

3.2.4 Xception

Xception, is one of the advanced and efficient deep learning model, it is an ameliorated version of Inception architecture and involves depthwise separable convolutions. It replaces the standard Inception modules with depthwise separable convolutions [13].

The Xception architecture involves Depthwise Separable Convolutions with residual connections. is divided into three major phases: Entry flow, Middle Flow, and Exit flow. The default input image size for this model is 299x299 (Fig. 5).

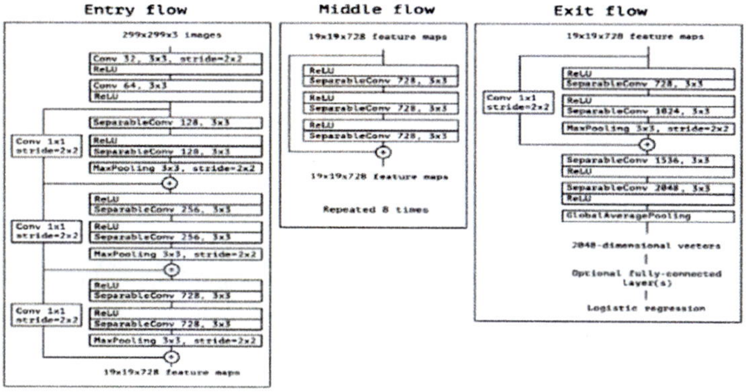

Fig. 5. Xception architecture.

Development of Individual Model:
In our case, we have make fine-tuning Xception model, we have adding the new layers to the pre-trained model. The layers added is one average pooling layer and fully-connected layer and output.

After adding new layers, we have train those layers, to say concretely the weights. we have chosen the layers to train and not to train. we just set that first 80% layers are not trained and next 20% are trained. After that we have train and evaluate the model.

3.2.5 MobileNetV2

MobileNetV2 is a convolutional neural network architecture developed by Google provides to perform well on mobile devices and embedded vision applications.

The MobileNets [11] is a concise form of the Xception and Inception nets with less parameters and complexity. It is based on an inverted residual structure where the residual connections are between the bottleneck layers and applies depthwise convolution a single filter to each input channel. The pointwise convolution then applies a 1×1 convolution to combine the outputs of the depthwise convolution (Fig. 6).

Development of Individual Model:
MobileNetV2 is a powerful classification model. In our case to fine tune the model, we have unfreeze the base model and freeze all the layers before 100th layer. Finally; we have recompile and start training the model.

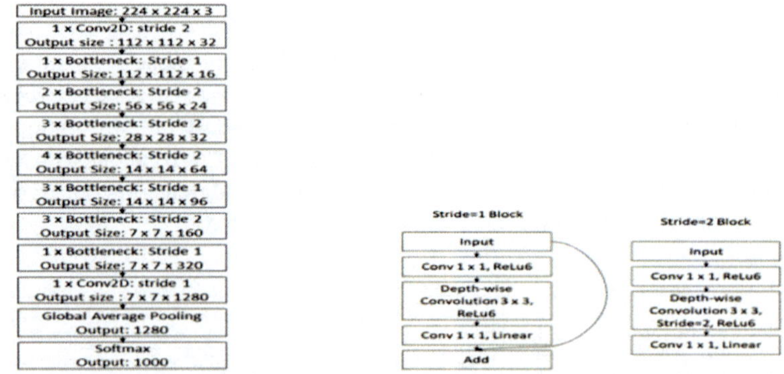

(a) MobileNet-V2 Architecture (b) MobileNet-V2 Bottleneck Stride Block

Fig. 6. MobileNet architecture.

3.2.6 Efficient-Net

Efficient-Net [23] is an image classification model family based with scaling method uniformly scales network width, depth, and resolution with a set of fixed scaling coefficients. The network has different versions, ranging from B0 to B7.

Efficient-Net are faster, contain a few parameter and generalize well to obtain higher accuracy on other datasets.

In our case we have used the base EfficientNet-B0 network is based on the inverted bottleneck residual blocks of MobileNetV2, in addition to squeeze and excitation blocks (Fig. 7).

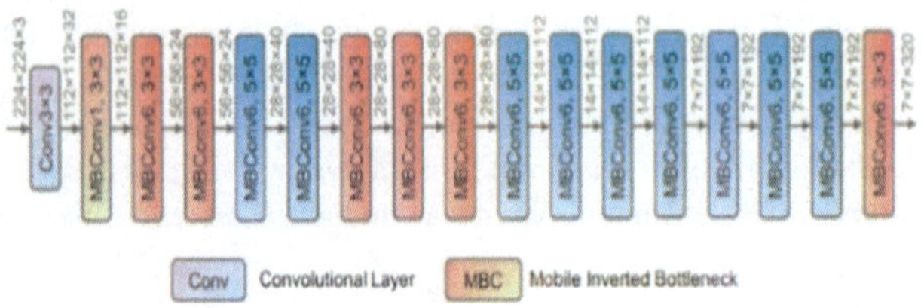

Fig. 7. Efficient-net architecture.

Development of Individual Model:
To resolve our problem, we have used fine-tuning. The steps for fine-tuning a network we have used:

- Added our custom network on top of an already trained base network.
- Applied freeze to the base network.
- We have training the part added (average pooling layer and fully-connected layer and output layer with function Softmax).
- Unfreezed some layers in the base network.

– Jointlying train both these layers and the part added.

3.3 Proposed Methodology

In this section, we will shine a light the key components of our proposed classification model for detecting skin cancer using Dermoscopie image.

In this research, we have a try to build different architectures use transfer learning with fine-tuning method for compare the performance of models in classification Dermoscopie images task. We implemented: Efficient-Net B0, Mobile-Net, InceptionV3, XceptionV3, VGG19 and ResNet152 models and used the different strategies and methods of freezing/Unfreezing layers of fine-tuning to get various pre-trained weights to increase accuracy. As illustrated in Fig. 8, our proposed model contains a preprocessing step including data augmentation, resizing images and contrast enhancement to clean images and obtain the best results. Additionally, we employ transfer learning algorithms with fine-tuning, where we first freezing layers and change the last layer to match the classes in our dataset, and also retrain the layers of the network with optimal hyper-parameters.

The main steps of our architecture are as follows:

1) Collecting the Dermoscopie images for the dataset from skin cancer case benign and malignant images.
2) Applying image preprocessing techniques, such as reading, resizing images, image quality enhancement: remove of noise, glare, rules, hair..., and applying One Hot Encoding on the labels of dataset.
3) Using Train_Test_Split to splitting the dataset into two sets: a training set and a test set.
4) Using Keras ImageDataGenerator to apply data augmentation methods for training a CNNs models properly and solving the overfitting problem.
5) Extracting prominent features from images and Classifying the Dermoscopie images using the CNNs networks pre-trained with fine tuning techniques.
6) Reducing overfitting and local response normalization [24] to reduce error rates using Dropout.
7) Choosing the optimal Hyper-parameter.
8) Evaluating and testing the performance of our model.

Preprocessing: is the most essential task for analysis before applying any feature extractor and classification methods [25] [26].is necessary be performed on images and especially images acquired by Demoscopy.

To produce correct outcomes is necessary to applied some procedures to remove noise and imperfection from each images.

In our work, after reading the data all images were resized to 224x 224 pixels to make the computation and training faster. The main steps of the preprocessing are image enhancement, and noise removal. we have used median filter the is a simple and powerful non-linear filter is used for reducing the amount of intensity variation between one pixel and the other pixel. The pixel value replaced with median value, and the median is calculated by first sorting all the pixel values into ascending order and then replace the pixel being calculated with the middle pixel value. [27].

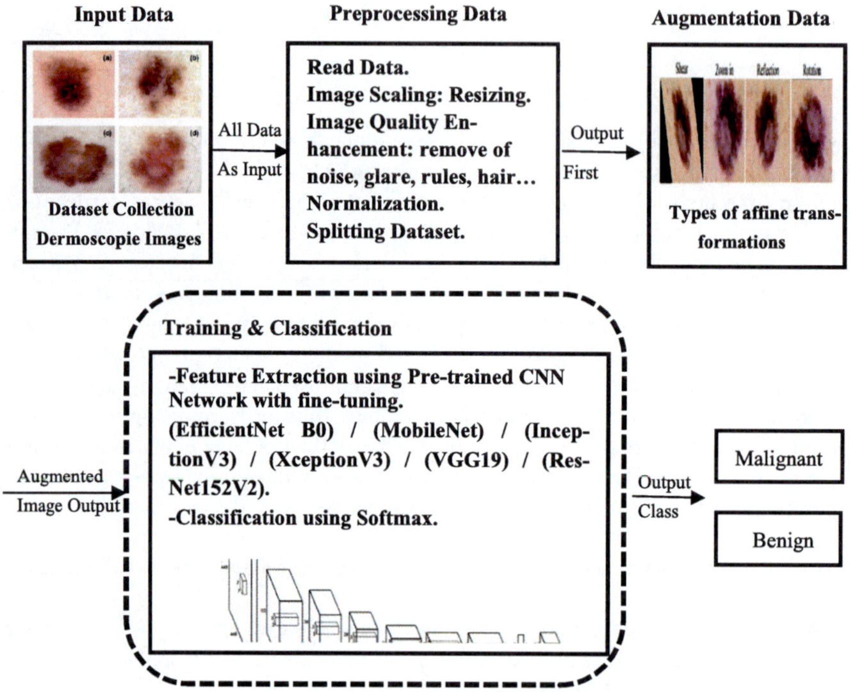

Fig. 8. The architecture of the proposed method.

Then, we have normalizing the pixel values to a [0,1] range to reduce the space of variation of the values of a feature. Finally, we apply the Train_Test_Split technique to randomly splitting data. Where 80% of the images are used for training, and 20% are used for testing.

Data augmentation: is an AI method for increasing the size and the diversity of training sets. Data augmentation methods are commonly used in deep learning for training a CNN model properly and increase the accuracy of the model, in addition to address class imbalance problems, reduce overfitting problem, and improve convergence [28].

In the medical domain, owing to privacy concerns, obtaining large amounts of data is a significant challenge [29]. That's why we have augmented our data collected via a number of random transformations. The selected data augmentation techniques were: rotation_range: input data generates with the rotation from -10 to 10, width_shift_range: image is randomly shifted horizontal direction by 0.1, height_shift_range: image is randomly shifted vertical direction by 0.1,zoom_range: zomme in or out 0.1 from the center, horizontal_flip: randomly flips the input horizontally,vertical_flip: randomly flips the input vertically.These methods are applied through the keras library with the ImageDataGenerator function.

The settings for image augmentation used in our experiment can be illustrated in Table 1.

Table 1. Image augmentation parameters.

Augmentation Method	Value
Rotation_range	10
Width_shift_range	0.1
Height_shift_range	0.1
Zoom_range	0.1
Horizontal_flip	True
Vertical_flip	True

Training & Classification: After Preprocessing step, we used transfer learning, which reduce the neural network's training time and minimize the generalization error to improve the accuracy of the diagnosis. we have using pre-trained CNNs with proper fine-tuning because it works better with medical image of type Dermoscopy.

We trained, fine-tuned the state-of-the-art deep learning models such as Inceptionv3, effecientnetB0, MobileNetV2, Xceptionv3, vgg19 and ResNet152V2 to extract features and classify the skin cancer if the is Melanoma or Non-melanoma.

Each algorithm will modify by different techniques, we first freezing layers, add the pooling layers and change the last layer to match the classes in our dataset, we used activation function Softmax as classifier and finally, we retrain the layers of the network with optimal hyper-parameters (see details on Sect. 3.2).

4 Experiments and Results

4.1 Evaluation Metrics

For measuring our model performance, we evaluated it to observe which algorithm gives the highest accuracy by predicting the sample data. We calculated accuracy, precision, recall, F1 score by using the True Positive (TP), True Negative (TN), False Positive (FP), False Negative (FN) values. The classification performance has been evaluated using the following quality measures:

$$Accuracy = \frac{TP + TN}{TP + TN + FP + FN} \tag{1}$$

$$Precision = \frac{TP}{TP + FP} \tag{2}$$

$$Recall = \frac{TP}{TP + FN} \tag{3}$$

$$F1Score = 2 \times \frac{(Recall \times Precision)}{(Recall + Precision)} \tag{4}$$

4.2 Results

In this section we present our results. To train our proposed models, we used Google Colab GPU platform using python environment.

For training, we used technique "Train_Test_Split" to randomly splitting data. Where 80% of the images are used for training, and 20% are used for testing. We compiled our models with different optimizers such RMSprop [30], Adam [30], and SGD [30]; batch size, learning rates and other hyper-parameters. After that, we train our model for different values of epochs until we obtained the best accuracy with "categorical cross-entropy" as the loss function. The following table shows the results obtained (Table 2):

Table 2. Results of the performance models.

Methodology	Optimizer	F1 Score	Precision	Accuracy
Resnet152	Adam	0.71	0.70	0.70
InceptionV3	RMSProp	0.71	0.80	0.80
VGG19	Adam	0.82	0.80	0.82
Xception	SGD	0.82	0.81	0.80
Mobile-Net	SGD	0.87	0.86	0.87
Efficient-Net Bo	**SGD**	**0.89**	**0.90**	**0.92**

We compared the pre-trained algorithms with fine tuning method, we have found the best result for Efficient-Net B0 with the **accuracy = 92%** with the following hyper-parameters: **Dropout: 0.5, Optimizer: Adam, Learning rate: 1e-5, Batch size: 32, Epochs: 120** (Fig. 9).

Epochs: 120.

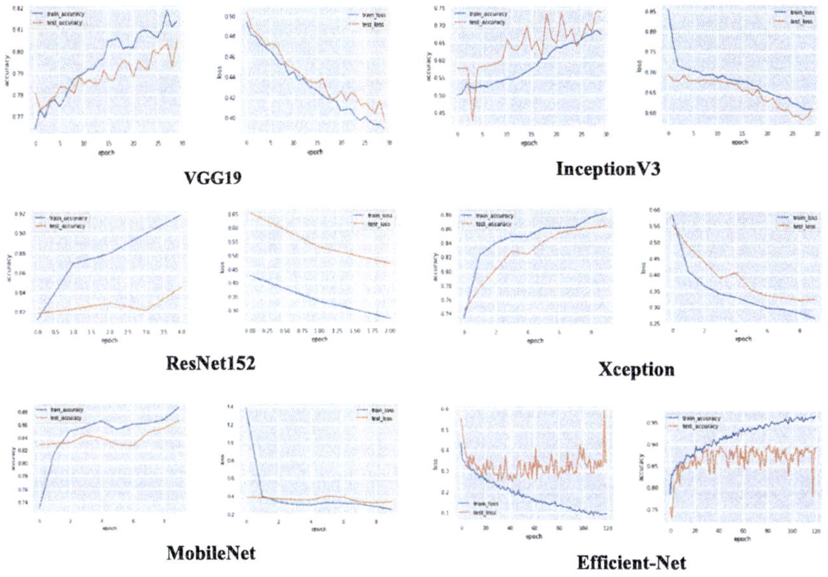

Fig. 9. Training and validation accuracy vs. loss.

5 Conclusion

Skin cancer is one of the most prevalent and severe cancers. The detection and diagnosis of skin cancer including skin cancer have improved over the years. Due to the development of artificial intelligence algorithms has always played an important part in this domain [31].

The goal of this paper was to applied a fine-tuned deep learning which uses the different pre-trained models to classify skin cancer into benign and malignant. We present a comparison on the comparison of six different state-of the-art CNN architectures on the classification task (VGG19, InceptionV3, Res-Net153V2, Efficient-Net, Xception, Mobile-Net). This comparison is establish from the F1-score, Precision, accuracy.

Our models are trained with ISIC dataset, we applied augmented dataset increased accuracy rates and avoided overfitting. Through various experiments, we have found the best result for Efficient-Net B0 with the accuracy = 92%. And there is still room be improved if we provide more data using larger datasets and advanced other techniques for preprocessing step.

To demonstrate the performance of the presented approach, we compared the performance of the presented model with state-of-the-art techniques. The presented our deep learning model was observed to outperform state-of-the-art approaches. The performance of the developed method was evaluated in comparison with other previously published strategies for classifying melanoma and benign tumors. The study results showed that the presented method had the highest accuracy compared with other current studies, as shown in Table 3.

Table 3. Comparison with state-of-the-art models.

Ref	Methodology	Diseases	Dataset	Accuracy
[33]	InceptionV3, ResNet and VGG19	Melanoma, Benign	ISIC 2019–2020	86.90%
[34]	Alex-Net	Melanoma, Benign	HAM10000	84%
[35]	CNN	Melanoma, Benign	ISIC 2018	0.8318%
	Proposed Method	Melanoma, Benign	ISIC Archive	**0.92%**

The proposed Method outperformed the existing studies, as Maad M. Mijwil [33] obtained 86.90% accuracy in classifying the melanoma and benign skin cancer diseases on the ISIC dataset. Ameri A. [34] reported 84% accuracy using Alex-Net model on HAM10000 dataset. Walaa Gouda et al. [35] used different models and achieved performed 0.8318% accuracy of CNN model using the ISIC 2018 dataset. We can say our proposed method dominated the existing techniques, it obtained the highest accuracy compared to other models and thus achieved 0.92% accuracy.

References

1. Ashraf, R., et al.: Region-of-interest based transfer learning assisted framework for skin cancer detection. IEEE Access **8**, 147858–147871 (2020)
2. Munir, Khushboo, Hassan Elahi, Afsheen Ayub, Fabrizio Frezza,and Antonello Rizzi. : Cancer diagnosis using deep learning:a bibliographic review.Cancer 11.no.9 (2019):1235
3. Byrd, A.L., Belkaid, Y., Segre, J.A.: The human skin microbiome. Nat. Rev. Microbiol. **16**, 143–155 (2018)
4. Skincancer.org. Melanoma - SkinCancer.org (2022). https://www.skincancer.org/skin-cancer-information/melanoma/
5. Khan, M.Q., et al.: Classification of melanoma and nevus in digital images for diagnosis of skin cancer. IEEE Access **7**, 90132–90144 (2019)
6. Lee, Y.C., Jung, S.-H., Won, H.-H.: WonDerM: Skin Lesion Classification with Fine-tuned Neural Networks. arXiv:1808.03426v3 [cs.CV] (10 May 2019)
7. Rashid, H., Tanveer, M.A., Aqeel Khan, H.: Skin lesion classification using GAN based data augmentation. In: Proceedings of the 2019 41st Annual International Conference of the IEEE Engineering in Medicine and Biology Society (EMBC), Berlin, Germany, pp. 916–919 (2019)
8. Dai, J., Li, Y., He, K., Sun, J. (2016). R-fcn: Object detection via region-based fully convolutional networks. In: Advances in Neural Information Processing Systems, pp. 379–387
9. Minaee, S., Kalchbrenner, N., Cambria, E., Nikzad, N., Chenaghlu, M., Gao, J.: Deep Learning Based Text Classification: A Comprehensive Review. arXiv:2004.03705v3 [cs.CL] (2021)
10. Bardou, D., Zhang, K., Ahmad, S.M.: Classification of breast cancer based on histology images using convolutional neural networks. IEEE Access **6**, pp. 24680–24693 (2018)
11. Howard, A.G., et al.: MobileNets: Efficient Convolutional Neural Networks for Mobile Vision Applications. ArXiv, abs/1704.04861. (2017)
12. Szegedy, C., Vanhoucke, V., Ioffe, S., Shlens, J., Wojna, Z.: Rethinking the inception architecture for computer vision. In: Proceedings of the IEEE Conference on Computer Vision and Pattern Recognition (CVPR), Las Vegas, NV, USA, pp. 2818–2826 (2016)

13. Chollet, F. :Xception: deep learning with depthwise separable convolutions. In: Proceedings of the 2017 IEEE Conference on Computer Vision and Pattern Recognition (CVPR), Honolulu, HI, USA, pp. 1800–1807 (2017)
14. Hossin, M.A., Rupom, F.F., Mahi, H.R., Sarker, A., Ahsan, F.,Warech, S.: Melanoma skin cancer detection using deep learning and advanced regularizer. In: Proceedings of the 2020 International Conference on Advanced Computer Science and Information Systems (ICACSIS), pp. 89–94 (2020)
15. Rezaoana, N., Hossain, M.S., Andersson, K.: Detection and classification of skin cancer by using a parallel CNN model. In: 2020 IEEE International Women in Engineering (WIE) Conference on Electrical and Computer Engineering (WIECON-ECE) (2020)
16. Alagua, S., Bagan, K.B.: Skin cancer classification in dermoscopy images using convolutional neural network. In: AIP Conference Proceedings 2336, 040013 (2021); https://doi.org/10.1063/5.0045757 Published Online (26 March 2021)
17. Moataz, L., Salama, G.I., Abd Elazeem, M.H.: Skin cancer diseases classification using deep convolutional neural network with transfer learning model. In: this article: Laila Moataz et al 2021 J. Phys.: Conf. Ser. 2128 012013 (2021)
18. ISIC Archive https://challenge.isic-archive.com/data/ Accessed 2020
19. Simonyan, K., Zisserman, A.: Very Deep Convolutional Networks for Large-Scale Image Recognition. arXiv 2014 arXiv:1409.1556 (2014)
20. He, K., Zhang, X., Ren, S., Sun, J.: Deep residual learning for image recognition. In CVPR (2016)
21. He, K., Zhang, X., Ren, S., Sun, J.: Identity mappings in deep residual networks. In: Leibe, B., Matas, J., Sebe, N., Welling, M. (eds.) ECCV 2016. LNCS, vol. 9908, pp. 630–645. Springer, Cham (2016). https://doi.org/10.1007/978-3-319-46493-0_38
22. Szegedy, C., Liu, W., Jia, Y., Sermanet, P., et al.: Going deeper with convolutions. In: Proceedings IEEE Computer Soc. Conference Computational Vision Pattern Recognition, pp. 1–9 (2015)
23. Tan, M., Le, Q.: EfficientNet: Rethinking Model Scaling for Convolutional Neural Networks (2019)
24. Deng, J., Dong, W., Socher, R., Li, L.J., Li, K., FeiFei, L.: ImageNet: a large-scale hierarchical image database. In: 2009 IEEE Conference on Computer Vision and Pattern Recognition, Miami, FL, USA (2009)
25. Fan, C., et al.: A review on data preprocessing techniques toward efficient and reliable knowledge discovery from building operational data. Front. Energy Res. **9**, 652801 (2021)
26. Khan, M.S.I., et al.: Water quality prediction and classification based on principal component regression and gradient boosting classifier approach. Journal of King Saud University-Computer and Information Sciences (2021)
27. Pimpale, A.N., Khambra, P.A.: Review on median filter in image filtration. International Journal on Emerging Technologies **7**(1), 52–56 (2016)
28. Islam, M.Z., Hossain, M.S., ul Islam, R., Andersson, K.: Static hand gesture recognition using convolutional neural network with data augmentation. In: 2019 Joint 8th International Conference on Informatics, Electronics & Vision (ICIEV) and 2019 3rd International Conference on Imaging, Vision & Pattern Recognition (icIVPR). IEEE, pp. 324–329 (2019)
29. Shah, A., Ahirrao, S., Pandya, S., Kotecha, K., Rathod, S.: Smart cardiac framework for an early detection of cardiac arrest condition and risk. Front. Public Health 2021, **9**, 762303 (2021)
30. Ruder, S.: An Overview of Gradient Descent Optimization Algorithms. arXiv:1609.04747v2 [cs. LG] (15 Jun 2017)
31. Park, C.-W., et al.: Artificial intelligence in health care: current applications and issues. Article in Journal of Korean medical science (2020)

32. Rashid, J., et al.: Skin cancer disease detection using transfer learning technique. Appl. Sci. (2022)
33. Mijwil, M.M.: Skin cancer disease images classification using deep learning solutions multimedia tools and applications. Multimed. Tools Appl. (2021)
34. Ameri A: A deep learning approach to skin cancer detection in dermoscopy images. J. Biomed. Phys. Eng. (2020)
35. Gouda, W., Sama, N.U., Al-Waakid, G., Humayun, M., Jhanjhi, N.Z.: Detection of skin cancer based on skin lesion images using deep learning. Healthcare (2022)

Towards a Formal Specification for a Knowledge Model of a Multi-agent System Using Ontologies

Noureddin Amaigarou[1](✉), Mouna Boulaajoul[1], and Mohamed Khaldi[2]

[1] Faculty of Science, AbdelmalekEssaadi University, Tetouan, Morocco
Noureddin.amaigarou@gmail.com
[2] Higher Normal School, AbdelmalekEssaadi University, Tetouan, Morocco

Abstract. This study is about ontology-based knowledge and its use in multi-agent systems. It is commonly understood that knowledge is extremely valuable in all sorts of businesses, in people's daily lives, and in the field of e-learning [1]. Ontology and MAS: describes how ontology can be used in a Multi-Agent System, the specification of an agent's architecture, the definition of the agent's knowledge model, the definition of the agent's knowledge methodology, and the description of the Java libraries developed to manipulate an ontology in our intelligent system.

Keywords: Multi-agent systems · Knowledge · Knowledge · Ontology · CommonKADS · RDF

1 Introduction

The concept of knowledge management/capitalization (KM: Knowledge Management) has two dimensions. On the one hand, it raises concerns about human resource management, organizational and managerial practices, and corporate culture. On the other hand, information and communication technologies are critical to many areas of KM; hence, a KM approach is fundamentally interdisciplinary. KM technologies are typically built on what are known as corporate memories.

The construction of a corporate memory allows one to "preserve in order to utilize them later or as soon as feasible, reasoning, conduct, knowledge, even in their contradictions and in all their variety" [2]. Knowledge management (KM) and multi-agent systems (MAS) are two areas of research that are still in their early stages. Agent technology seems to be very promising since it provided a new paradigm for software technologies, particularly those devoted to learning and decision support [3].

Recent technologies such as SOAP, WSDL, UDDI, and Peer-to-Peer are also outcomes of MAS (Multi-agent Systems), however they are only partial realizations of MAS's potential. Knowledge Management and Ontologies, on the other hand, have already achieved success in the Semantic Web arena. The work discussed in this study attempts to strengthen the link between KM and MAS, primarily by incorporating work done at MAS in the semantic web. All domains of artificial intelligence, including as learning, negotiation, and knowledge management, are used and exploited by agent technology. This work focuses on the usage of KM by agent systems, allowing for the development of an architecture that can be used to a wide range of application fields [4].

M. Ben Ahmed et al. (Eds.): SCA 2022, LNNS 629, pp. 327–334, 2023.
https://doi.org/10.1007/978-3-031-26852-6_31

2 Motivation, Research Context and Goals

In the modern world, knowledge has definitely a huge asset. The so-called knowledge economy, in which knowledge is more valuable and in demand than any other resource, is where the world economy is headed. The majority of knowledge management solutions available today are centralized, which is insufficient to support solutions operating in diverse and distributed contexts. Agents are a good option in this situation, however MAS-based solutions lack knowledge and intelligence. Similar to how individuals grasp formalized knowledge, agents can as well [5].

In this chapter, we will go through the following goals:

- Design an agent architecture that employs an ontology-based knowledge model.
- Create a software development method for creating agents using an ontology-based knowledge model.
- Create a software library for creating intelligent agents with an ontology knowledge model and the ability to connect agents to existing commercial technologies.

3 Ontologies: Knowledge Design and Representation

A knowledge representation language can be used to explicitly express an ontology. Ontologies, formalisms, and tools are prominent in Semantic Web research disciplines, where various ontology representation languages have been established and are actively described and standardized. The most well-known languages are those based on the RDF (Resource Description Framework) standard language and its RDFS (RDF Schema) extension, as developed by the W3C consortium. In addition to RDF (S), several other standards, such as DAML + OIL, which extends RDF (S) by enforcing certain implications and established language constraints [6] or OWL [7], have been introduced. Berners-Lee proposed a layered architecture of a set of Semantic Web technologies within the framework of the languages used for the representation of Web data (Fig. 1) [8].

Any computer system that is created must be able to work with various types of data. Most information systems now employ relational database management systems (RDBMS) [9]. RDBMS are most appropriate when the information system's problem area is properly characterized and does not change over time. Object Database Management Systems (ODBMS) are better suited to information systems that require any combination of data extraction. However, updating indexing and other database operations in OODBMS is much slower than in RDBMS, and these solutions are not very stable. The information system is adaptable and adjustable thanks to the use of ontology and ODBMS. Such an information system accurately reflects knowledge facts and its information structure and can be built on an ontology or RDF-based solutions, such as RDF stores or the Jena library.

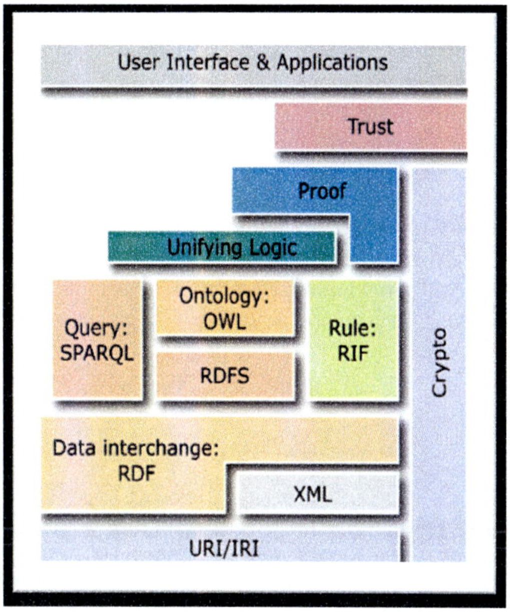

Fig. 1. Architecture of Semantic Web technologies

4 A Formal Specification for Our Agent Knowledge Model

Our model is comprised of five major components: resources, actions, actors, contexts, and events.

The set R contains all of the agent environment's resources. It can be partitioned or made of subsets Ri that represent the many types of resources.

$$R_i \subset R\, r \in R$$

$A^R$is a special subset of R. It is a set of actors in the environment.

$$A^R \subset R$$
$$a^r \in A^R$$

The A^Ractor can take an action year from the set of A^N actions.

$$a^n \in A^N$$
$$C^A_{New} = f_c\left(e^a, C^A_{old}\right)$$
$$R^A_{new} = f_R\left(C^A_{new}, R^A_{old}\right)$$

We begin by updating the actor context based on an old actor context and a new/received actor event. Then, we update the actor resources based on the actors' current/new context and old resources.

Without explanation, the actor context is understandable. It is obvious that the actor or agent requires a state of the relevant context or environment. The f_c function or algorithm maintains the context. Function or algorithm f_R updates actor resources. These resources are frequently extended by a more complicated structure, but we find that such resources are required for the actor/agent to take action or for actions to result from actions made.

5 Formal Specification of the Model Using Description Logic

In this section, we use description logic to specify the same model more precisely (DL). Figure 2 displays a formal graphical representation of the concept described in this paper.

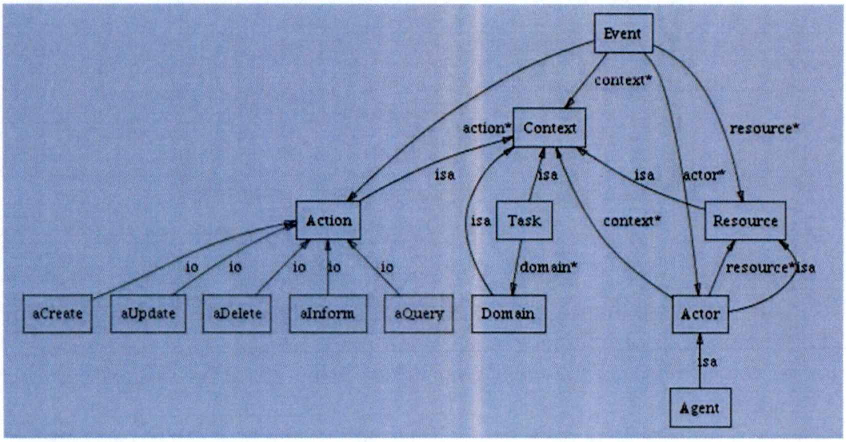

Fig. 2. Basic ontology for knowledge modeling

All of the resources in the agent environment are represented by the resource class. As part of the model modification, many subclasses representing different types of resources can be defined. Actor is an important subclass of Resource.

$$Actor \subseteq Resource$$
$$\{actor\} \in Actor$$

Actors in the environment are represented by the actor class. Actors can perform actions, which are members of the Action class.

$$\{action\} \in Action$$

The task class represents tasks that have been completed or must be completed in the environment, but depending on the application, it can also represent difficulties. The domain class corresponds to the ontology's application domain extension, and all extensions must be subclasses of this type. The context class represents the actors', the environment's, or other contexts.

$$\begin{gathered}
Resource \subseteq Context \\
Action \subseteq Context \\
Domain \subseteq Context \\
\{domain\} \in Domain \\
Task \subseteq Context \\
Context \supset Resource \cup Action \cup Domain \cup Task \\
\{context\} \in Context
\end{gathered}$$

Important attributes of the actor class include context. Actor and resource.Actor.

The context.Actor class represents the actor's current context. Based on saved events, the system or application environment can be found. The current state of the event model's environment. New relevant occurrences alter the current state of the environment or the environment or context linked with the actor. The actor's current resources are represented by the resource.Actor property. These resources are the product of the actors' intentions or goals. These resources are thus dependent on the current actors' status/context of the environment (context.Actor). The functions/algorithms for updating context and resources have already been defined.

$$\begin{gathered}
\text{Actor} \subseteq \\
\text{resource.Actor}(\{\text{resource}\}) \cap \\
\text{context.Actor}(\{\text{actor}\}) \cap \\
\text{Resource} \\
\text{contextActor}(\{\text{actor}\}) = \\
\text{fC}(\forall \{\text{event}\};\ \text{actor.Event}(\{\text{actor}\}) \in \{\text{event}\}) \\
\text{resource.Actor}(\{\text{resource}\}) = \\
\text{f}_{\text{R}}(\text{contextActor}(\{\text{actor}\}))
\end{gathered}$$

This is where the system's brain is located. This model has the advantage of allowing improved results to be obtained when such algorithms are updated in the future while using the same model and data. Because all events are saved, we can model the environment at any moment in the past and process it from any starting point using enhanced algorithms for updating context and resources.

6 Specifying the Method of Ontology Creation and the Agent Model

When building the knowledge model, We must enhance the generic agent model (Fig. 3) to include new elements and linkages. We will utilize a graphical representation of the models in this study to help us understand them better.

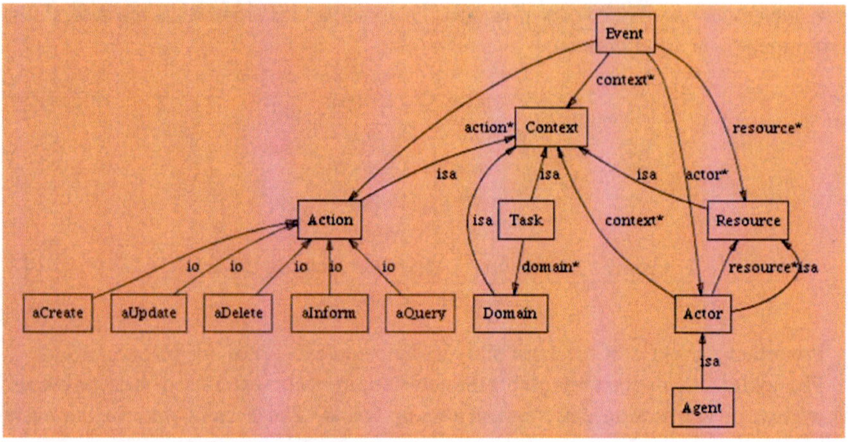

Fig. 3. Basic ontology for knowledge modeling

7 Design and Specification of the Agent Software Library

The developed library is based on the JADE agent system and the Jena library. It addresses the elements that we identified as lacking in existing agent architectures such as JADE [10]. These missing components are as follows:

- Agent knowledge model based on RDF/OWL;
- Action resource as a basis of communication and included in the OWL ontology;
- Sending ACL-based RDF/OWL messages;
- Receipt of RDF/OWL messages based on ACL;
- XMLRPC receives messages;
- XMLRPC returning RDF and XML;
- Inference engine;
- Management of RDQL messages;
- The library includes the following classes: Ontology, Memory and Message.
- The Ontology class (Fig.4) has basic constants related to the OWL ontology and the ontology used.

Figure 5 displays a functionality diagram of the agent library that makes use of this library. This library can be used to construct a JADE-based agent that supports the Jena OWL model as agent memory. It is also possible to include XML-RPC functionality in order to present knowledge or receive events from external systems as RDF messages based on an ontology. It also enables communication amongst RDF/OWL agents [11].

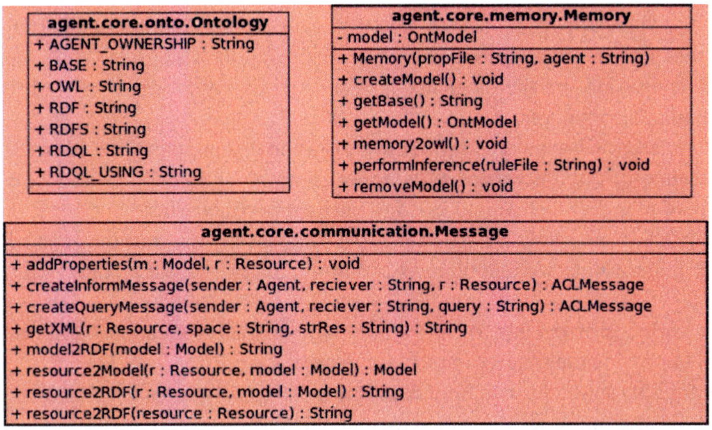

Fig. 4. Library classes with main properties and methods, UML class diagram

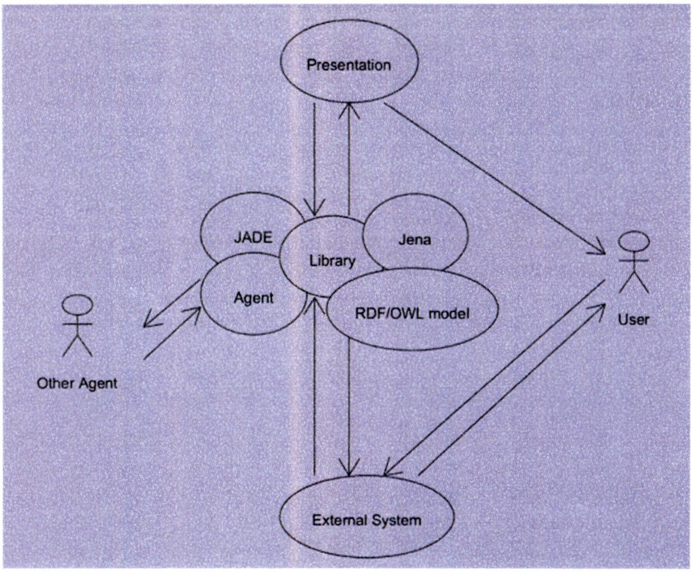

Fig. 5. Agent library functionality diagram

8 Conclusion

In this article, we designed an agent architecture based on an ontology-based knowledge model. A software development methodology for the creation of agents with an ontology-based knowledge model was also designed. This developed methodology combines elements from other methodologies. It adheres to the CommonKADS methodology. CommonKADS, on the other hand, is not tied to any modeling tool, knowledge representation, or ontologies.

The third section is concerned with the design and development of a software library for the creation of intelligent agents based on an ontology knowledge model, with the possibility of plugging agents into existing business technologies.

The library produced is based on the JADE agent system and the Jena library. It covers the features that we identified as lacking in existing agent architectures such as JADE.

The knowledge model and agent memory models, each with a brief description, are presented to provide tangible examples of the methods utilized.

Several UML diagrams with brief descriptions are presented in this article to provide a concrete illustration of the described methodology.

References

1. Sanfilippo, E.M., Belkadi, F., Bernard, A.: Ontology-based knowledge representation for additive manufacturing. Comput. Ind. **109**, 182–194 (2019). https://doi.org/10.1016/j.compind.2019.03.006
2. Pomian, J.: Mémoire d'entreprise: techniques et outils de la gestion du savoir. Éd. Sapientia. proposal, Znalosti 2004, Brno, Czech Republic (2004)
3. Balakrishnan, S., Janet, J.: An overview of agent based intelligent systems and its tools. CSI Communications Magazine **42**, 15–17 (2019)
4. Amaigarou, N., Khaldi, M.: S@J-Wiki: a collaborative semantic wiki to manage juridical knowledge. Int. J. Computer Appl. **104**, 44–49 (2014)
5. Amaigarou, N., Khaldi, M.: Modeling a collaborative multi-agent system using a process for complex engineering systems. International Journal of Advance Research in Science and Engineering (IJARSE) (ISSN: 2319–8354) **04** (2015)
6. Connolly, D., et al.:. Daml+ oil URI: http://www.daml.org/2001/03/daml+oilindex. Accessed 8 June 2022 (2001)
7. Bechhofer, S., et al.: OWL web ontology language reference. W3C recommendation, 10 (2004)
8. Berners-Lee, T., Brickley, D., Heflin, J.:. Daml+ oil (march 2001). URI: http://www.daml.org/2001/03/daml+oilindex. Accessed 8 June 2022 (2001)
9. Faizul, H., Mahesh, K., Raja, M.K.: The use of relational database management systems (DBMS) for information retrieval in a group technology (GT) environment. Computer Industrial Eng. **26**(2), 253–266 (1994)
10. Komma, V.R., Jain, P.K., Mehta, N.K.: An approach for agent modeling in manufacturing on JADE™ reactive architecture. Int. J. Adv. Manuf. Technol. **52**, 1079–1090 (2011). https://doi.org/10.1007/s00170-010-2784-2
11. Amaigarou, N., Khaldi, M.: Design of a smart tool for knowledge management in R&D projects. Communications on Applied Electronics **7**(22), 1–6 (2018)

Usage of Close-Range Photogrammetry to Obtain Digital Twin of Human Tooth

Iniyan Thiruselvam[1,2(✉)], Ashish M. Bhalkikar[1], Dhananjay M. Kulkarni[1,2], Kiran D. Mali[1,2], Savio D. S. A. Lourenço[1,2,3], and Sakthivel Jayabal[1,2]

[1] BITS PILANI K. K. Birla Goa Campus, Zuarinagar, Goa 403726, India
iniyann@goa.bits-pilani.ac.in
[2] BIONIC Imprints Private Limited, BGIIES, Zuarinagar, Goa 403726, India
[3] Leslie Memorial Dental Clinic, Madgaon, Goa 403601, India

Abstract. A digital twin of the human tooth provides an option for patient-specific dental services. Close-range Photogrammetry (CP) is a low-cost alternative to Cone Beam Computerized Tomography (CBCT) in preparing a digital twin. Though CP is generally known to be less accurate than CBCT, a comparative analysis of their scan output has not been reported in literature so far. A CP-based generation of a digital twin of the human tooth and a comparative evaluation of the digital twins (DTs) obtained by CP and CBCT are presented in this paper. CP-based DTs look noisier than CBCT-based DTs. They displayed a maximum error of 400–600 microns. Highlights of this paper are: 1. a standardized workflow for generating CP-based DTs has been proposed and 2. the factors that increase the accuracy of such digital twins have been identified. Future work will be carried out to ensure the repeatability, accuracy and efficiency in creating digital twins of the human tooth using close-range photogrammetry.

Keywords: Digital twin · Close-range photogrammetry · Human tooth

1 Introduction

Smart systems have gained popularity in the recent times owing to the arrival of Industry 4.0 technologies. In general, smart systems employ latest advancements in technologies, which include but are not limited to big data, smart factory, cyber-physical systems, interoperability, Internet of Things (IoT), and digital twins [1]. Usage of a digital twin (DT) is increasing in engineering applications related to healthcare [2, 3]. DT is a digital representation of real-world object/system/process, which is further used to perform tasks that add some business value. For personalized healthcare, It is generated by teamwork between engineers and medical professionals [3]. DT of a real-world entity is used in structural health monitoring, failure prediction and other value-added activities.

In the recent years, smart health care has become an active area of research due to the latest advancements in bioinstrumentation, a branch of science that focuses on the tools and mechanics used to analyze, diagnose, and treat biological systems. Its prime focus is on the application of sensors to track physiological features of a human

M. Ben Ahmed et al. (Eds.): SCA 2022, LNNS 629, pp. 335–343, 2023.
https://doi.org/10.1007/978-3-031-26852-6_32

being. For dental procedures, dentists are increasingly turning to digital dentistry, which involves the use of digital or computer-controlled components rather than traditional mechanical or electrical tools. Recent advances in Digital Twin technology, one of the important aspects of Industry 4.0, has the potential to advance the current state of the art in digital dentistry. Some commonly used sensors in digital dentistry are X-ray imaging systems that employ cone beam computed tomography (CBCT) technology (for example, Papaya 3D, Genoray, Republic of Korea) and structured light scanning systems (for example, Tupel 3D, Oxford, England). Digital twins are generated using these sensors for several purposes. Some of them are: 1. To record the patient information, 2. To scan models, impressions and dentures, and 3. To fabricate dental holders designed for models, impressions, dies and dentures.

The human tooth has unique mechanical properties and microstructure [4]. Hardness and elastic modulus would decrease gradually from the surface of the enamel to the dentin-enamel junction, thereby making its shape important. Each tooth in a human skull has a unique shape and size, and no two teeth are of identical characteristics. Therefore, it is essential to recreate the shape of the human tooth as accurately as possible for both aesthetic and structural needs. In [5], machine learning was used to identify dental illness of a patient using photos. Close-range photogrammetry (CP), an alternative to other expensive scanning technologies, can also be used with machine learning to recreate the surface of the tooth.

A digital twin would be essential when human tooth loss necessitates replacement by fixed dental prostheses. It provides useful information required to fabricate fixed dental prostheses [6]. Digital twins (DTs) of the human tooth are invaluable in a patient-specific approach to dental implantology. For example, DTs of the human tooth and the jawbone could be used to predict cervical stress profile through a finite element analysis [7]. They are useful in anthropological applications as well. They are obtained by scanning the original specimen, which helps preserve shape information [8, 9]. Some scanning techniques based on light rays are illustrated in Fig. 1.

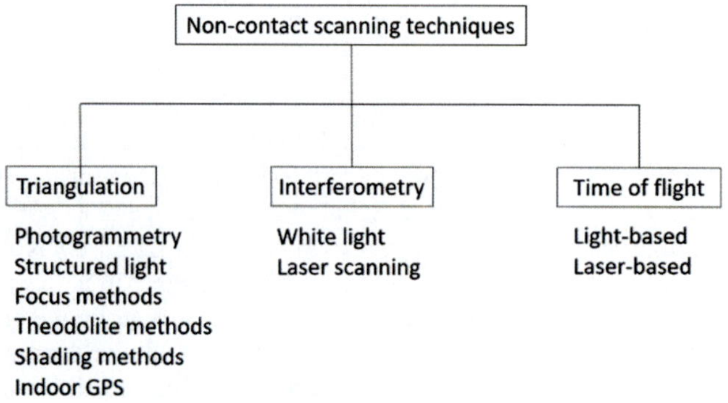

Fig. 1. Non-contact scanning/measurement techniques.

Apart from cone beam computed tomography (CBCT), 3D scanning techniques like structured light scanning and photogrammetry are used in dentistry. In structured light scanning, a pattern of light is projected onto the object and a camera collects the images. The pattern of light is then altered while collecting images of the tooth. Finally, digital twin of the tooth is obtained by processing the distorted patten of light [10]. In dentistry, CBCT is considered as the standard for generating digital twin of the human tooth [11]. In this technique, cone-shaped X-ray beam is used instead of structured light beam to obtain an accurate anatomical scan of tooth and surrounding bone. However, CBCT is an expensive technique, which costs more than 100 USD. Close-range Photogrammetry using commodity cameras or smartphones is less expensive than structured light scanning and CBCT. It is popular in many engineering applications due to its low cost, versatility, ease of use and relative accuracy. However, use of this technique in dentistry to scan the entire retrieved/extracted human tooth is limited. The present work focuses on using Close-range Photogrammetry (CP) to obtain digital twin (DT) of the human tooth.

2 Overview of Close-Range Photogrammetry

Close-range Photogrammetry (CP) is a non-contact image-based 3D Scanning technique used to compute 3D models of real-world objects. In this technique, first, commodity cameras (even smartphones recently) collect images of an object from multiple view-points. These images are processed through self-calibrating bundle adjustment [12] to compute point cloud, from which 3D model of the object is obtained [13]. Photogrammetry, generally finds usage in aerial applications, ground surveying and large-scale metrology [14]. Important industrial applications of CP include among others:

- reverse engineering of design models
- material testing
- robot calibration
- heritage conservation, and
- production planning

Output of CP. CP offers both coordinate and graphical information of real-world objects (Fig. 2). Distances, surface data, process control data, and deviation map (comparison with other data) are the coordinate information obtained using CP. Drawings (maps), 3D model (CAD data), image rectifications and orthophotos are the graphical information obtained using CP. Among these output, deviation map (comparison with other data) is useful to evaluate the measurement accuracy of CP. Cloud to Cloud distance (C2C) measurement is the most commonly used method for computing deviation maps. C2C is defined as the nearest neighbour Euclidean distance between two dense point clouds.

CP of Small Objects. CP has rarely been used to scan small objects until recently [15]. Since small objects have features in sub-millimeter length scale, prime/zoom lenses used for portrait/landscape photography are inadequate. This problem can be averted by using macro lenses or 2x/3x zoom lenses with extension tubes [16] or a combination thereof [17]. However, obtaining accurate measurements with macro/zoom lenses in CP can pose difficulties [18]. One difficulty is that the entire object will not be in focus in some

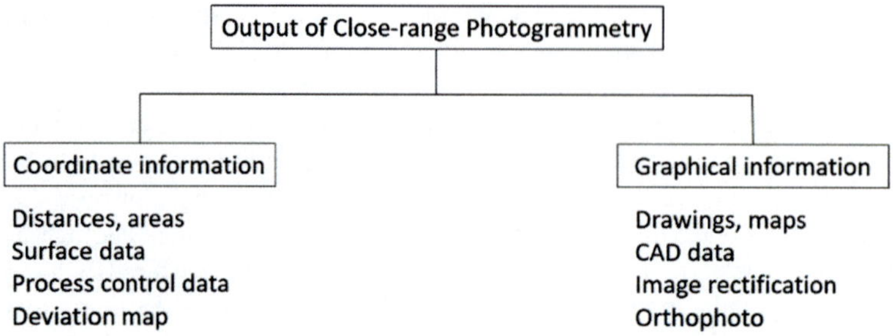

Fig. 2. Output of close-range photogrammetry.

viewpoints (shallow depth of field). This can be overcome by focus-stacking method [19], which is computationally expensive. Alternative solutions include fusing the images by specialized software [15] or tilt-shift method [20]. Another difficulty with accurate 3D digitization of the sub-millimeter-sized features is calibration of high-magnification imaging systems [18]. Recently published guidelines alleviated these difficulties [12].

In the present work, a digital twin (DT) of the human tooth is obtained by computing 3D point cloud using Close-range Photogrammetry (CP). This DT is evaluated by computing C2C between the output of CP and CBCT, which is considered as the standard in dentistry.

3 Methodology

Periodontally-compromised maxillary Premolar and Molar teeth were extracted during implant surgery at LMDC, Madgaon, Goa, India. Each tooth was placed on a flat plate (Fig. 3), which was later rotated to generate the different viewpoints required for CP. The results and summary are later presented in the next section.

Fig. 3. Flat textured plate on which tooth was placed.

The flat plate was textured to increase the accuracy of camera calibration [12]. A coaxial lighting arrangement was used to illuminate the tooth. A smartphone suitable for macro imaging was mounted on a gorilla tripod (Fig. 4) to collect the images of the tooth. Specifications of the smartphone camera are listed in Table 1.

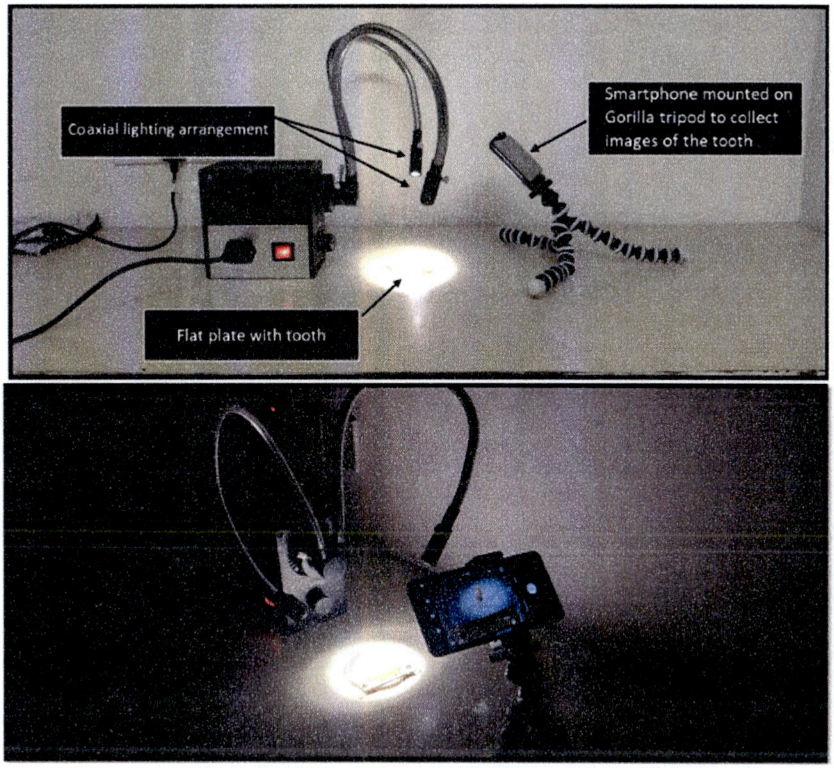

Fig. 4. Setup for image collection and Close-range Photogrammetry.

Table 1. Camera specifications.

Details of camera lens	Wide lens	Macro lens
Resolution (MP)	64	2
Aperture (f-stop)	f/1.9	f/2.4
Shutter speed (s)	1/60	1/60
Pixel size (μm)	0.8	0.8
Focal length (mm)	26	13

The images of the tooth collected from multiple viewpoints were processed through photogrammetry pipeline, which comprises of feature tracking and self-calibrating bundle adjustment algorithms [12]. For photogrammetric 3D model reconstruction (CP data), Agisoft Metashape software (St. Petersburg, Russia) was used. The 3D model computed by using CBCT was obtained from the imaging center. To compare CP data with CBCT data, C2C (deviation map) between the data obtained using CP and CBCT was calculated using the open-source Cloudcompare software.

There are three phases in the CP pipeline adapted in the present work to obtain the 3D model of the entire tooth. The first phase is 3D modelling of the crown of the tooth; the second phase is 3D modelling of the root of the tooth; and, the third phase is merging the crown and root parts by using the existing features in the tooth. To compensate for the inadequate natural features, artificial features were later added by using red and blue coloured markers to aid CP (Fig. 8). The angle of incidence for the camera were set to be 10°, 45° and 80°. For each angle, 20 photos were collected from all sides. The shutter speed was set to 1/60, the focus was set to AUTO, and the ISO level to 100. The coaxial light was set at an angle of 45° to the flat plate (Fig. 4).

The full tooth was reconstructed using CP protocol (Fig. 5). The 3D model of the tooth (in STL format) obtained by using CBCT was finally imported in Cloudcompare software. By using "sub-sampling" option, the deviation maps i.e. the cloud-to-cloud distance (C2C) between the point clouds obtained by using CP and CBCT were calculated and analyzed.

1. Place extracted tooth on flat plate with scale visible in image.
2. Align coaxial light at about 45° angle to the plate.
3. Align the smartphone camera at about 10°, 45° and 80° to the tooth.
4. Collect 2 photo sets (one is crown and another is root). Repeat for all 3 angles.
5. Align camera in Agisoft Metashape for both photo sets (crown and root).
6. Generate dense point cloud for both photo sets in Agisoft Metashape.
7. Export both dense point clouds to Cloudcompare.
8. Merge and export dense cloud from Cloudcompare to Agisoft Metashape.
9. Generate STL of full tooth using the merged dense cloud in Agisoft Metashape.

Fig. 5. Close-range photogrammetry protocol.

4 Results and Discussion

The digital twin of the human tooth computed using Close-range Photogrammetry (CP) is visibly noisier (Fig. 6) when compared to the one computed using cone-beam computed tomography (CBCT).

The C2C (deviation) maps between CP and CBCT for the three teeth under study is illustrated in Fig. 7. The maximum C2C distance in Premolar, First Molar and Second Molar teeth are 428 μm, 600 μm, and 547 μm respectively. The digital twin of the First Molar tooth shows the highest error. The regions of digital twins where maximum error is

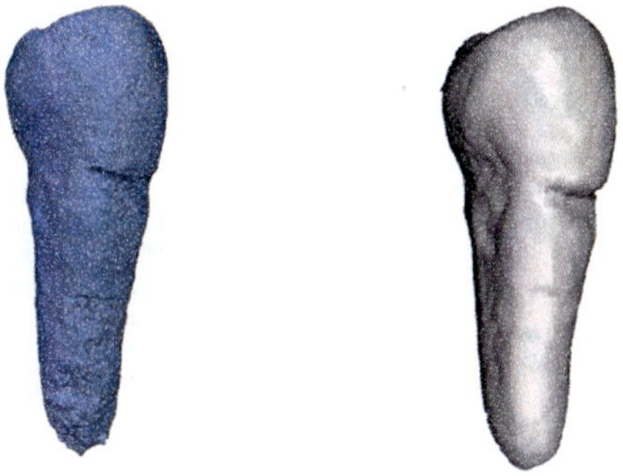

Fig. 6. Digital twins of Premolar tooth obtained by using CP (left) and CBCT (right).

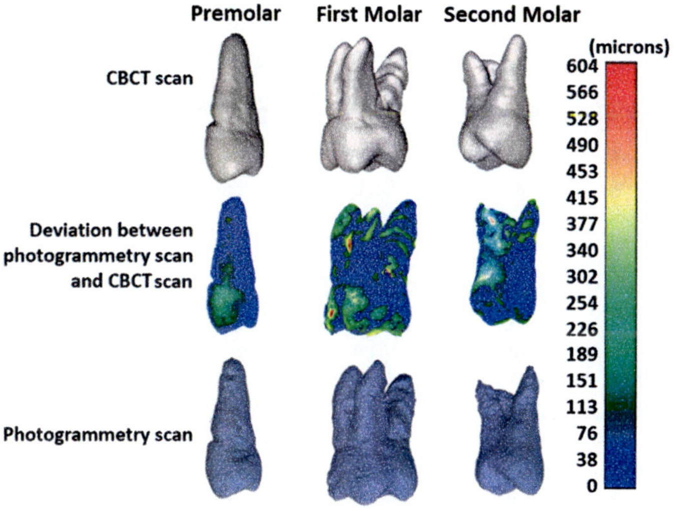

Fig. 7. C2C distance/deviation maps (middle) between the digital twins computed by using CBCT (top) and CP (bottom).

seen are either shadow regions, reflective regions or low texture/feature regions. In these regions, the image features are not accurately recognised or tracked across viewpoints (Fig. 8).

Figure 7 shows that that the greatest possible error in the digital twins generated using Close-range Photogrammetry is 600 microns. In [4], the average elasticity modulus on the chewing surface of a tooth, known as enamel, is 70 GPa, which suggests that the observed error is insignificant in terms of mechanical performance. Therefore, it is evident that close-range photogrammetry (CP) is a less expensive alternative to CBCT that still provides a reasonable accuracy. The cost of creating a digital twin of the human tooth is reduced by using close-range photogrammetry. Such digital twins can be used to document dental health of a patient for future reference in the context of preventative smart healthcare, as reported in [5].

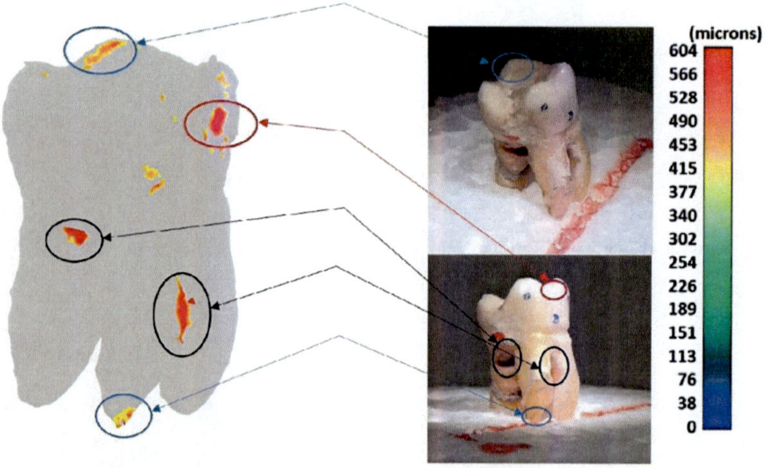

Fig. 8. C2C distance/deviation maps (middle) between the digital twins computed by using CBCT (top) and CP (bottom).

Photogrammetry scans show loss of data/details in low texture/feature regions. In this work, this problem is overcome by painting red/blue colored dots on these regions of the tooth. The reflective regions can be avoided by diffusing the coaxial light. The shadow regions can be avoided by altering the angle of incidence. Image collection from smartphone requires touching screen, which causes vibration. To avoid this, a 5-s time delay between images is used. These changes would allow an accurate digital twin of the human tooth at low cost. Implementation of the aforementioned improvements to the existing photogrammetry solution is deferred for future work.

5 Conclusion

Close-range Photogrammetry (CP) has been used for the first time to generate a digital twin of the human tooth. An accuracy of 400–600 μm has been achieved when compared to cone beam computed tomography (CBCT). We have demonstrated that CP has the potential to replace the expensive CBCT process in generating the digital twins of human tooth. A standardized workflow for generating CP-based digital twins has been proposed and factors which increase the accuracy of such digital twins have been identified. Future work will be conducted to improve the generation of CP-based digital twins of the human tooth to ensure repeatability, accuracy and efficiency.

References

1. Büchi, G., Cugno, M., Castagnoli, R.: Smart factory performance and Industry 4.0. Technological Forecasting and Social Change **150**, 119790 (2020)

2. Alloghani, M., Al-Jumeily, D., Hussain, A., Aljaaf, A.J., Mustafina, J., Petrov, E.: Healthcare services innovations based on the state of the art technology trend industry 4.0. In: 11th International Conference on Developments in eSystems Engineering (DeSE), Cambridge, England, pp. 64–70 IEEE (2018)
3. Angulo, C., Ortega, J.A., Gonzalez-Abril, L.: Towards a healthcare digital twin. In: Artificial Intelligence Research and Development, pp. 312–315. IOS Press (2019)
4. Zhang, Y.R., Du, W., Zhou, X.D., Yu, H.Y.: Review of research on the mechanical properties of the human tooth. Int. J. Oral Sci. **6**(2), 61–69 (2014)
5. Liu, L., Xu, J., Huan, Y., Zou, Z., Yeh, S.C., Zheng, L.R.: A smart dental health-IoT platform based on intelligent hardware, deep learning, and mobile terminal. IEEE J. Biomed. Health Inform. **24**(3), 898–906 (2019)
6. Shengli, W.: Is human digital twin possible? Computer Methods and Programs in Biomedicine Update **1**, 100014 (2021)
7. Thiruselvam, N.I., Kulkami, D.M., Lourenço, S.D.: Pre-operative simulation of implant angulation effect on mandibular masticatory stresses-a finite element study. Procedia Engineering **64**, 815–824 (2013)
8. Fiorenza, L., et al.: Technical note: The use of 3D printing in dental anthropology. Am. J. Phys. Anthropol. **167**, 400–406 (2018)
9. Silvester, C.M., Hillson, S.: A critical assessment of the potential for structure-from-motion photogrammetry to produce high fidelity 3D dental models. Am. J. Phys. Anthropol. **173**(2), 381–392 (2020)
10. Geng, J.: Structured-light 3D surface imaging: a tutorial. Advances in Optics and Photonics **3**(2), 128–160 (2011)
11. Ozcan, G., Sekerci, A.E., Cantekin, K., Aydinbelge, M., Dogan, S.: Evaluation of root canal morphology of human primary molars by using CBCT and comprehensive review of the literature. Acta Odontol. Scand. **74**(4), 250–258 (2016)
12. Thiruselvam, N.I., Subramanian, S.J.: On improving the accuracy of self-calibrated stereo digital image correlation system. Meas Sci Technol **32**(2), 025201 (2020)
13. Nguyen, C.V., Lovell, D.R., La Salle, J.: Capturing natural-colour 3D models of insects for species discovery and diagnostics. PLoS ONE **9**(4), e94346 (2014)
14. Luhmann, T.: Close range photogrammetry for industrial applications. IS- PRS J. Photogrammetry and Remote Sensing **65**(6), 558–569 (2010)
15. Gallo, A., Muzzupappa, M., Bruno, F.: 3D reconstruction of small sized objects from a sequence of multi-focused images. J. Cult. Herit. **15**(2), 173–182 (2014)
16. González, M.Á.M., Yravedra, J., González-Aguilera, D., Palomeque-González, J.F., Domınguez-Rodrigo, M.: Micro-photogrammetric characterization of cut marks on bones. Journal of Archaeological Science **62**, 128-142 (2015)
17. Percoco, G., Guerra, M.G., Sanchez Salmeron, A.J., Galantucci, L.M.: Experimental investigation on camera calibration for 3D photogrammetric scanning of micro-features for micrometric resolution. Int. J. Advanced Manufacturing Technol. **91**(9–12), 2935–2947 (2017)
18. Shortis, M.R., Bellman, C.J., Johnson, G.W.: Stability of zoom and fixed lenses used with digital SLR cameras. In: Proceedings of the ISPRS Commission V Symposium of Image Engineering & Vision Metrology, pp. 285–290 (2006)
19. Galantucci, L.M., Lavecchia, F., Percoco, G.: Multistack close range photogrammetry for low cost submillimeter metrology. J. Comput. Inf. Sci. Eng. **13**(4), 044501 (2013)
20. Nocerino, E., Menna, F., Remondino, F., Beraldin, J.A., Cournoyer, L., Reain, G.: Experiments on calibrating tilt-shift lenses for close-range photogrammetry. Int. Arch. Photogramm. Remote. Sens. Spat. Inf. Sci. **5**, 99–105 (2016)

Eye-Tracking Technology in Smart System for Monitoring of Human's Psychophysiological States

Vitaliy Pavlenko(✉) and Tetiana Shamanina

Odessa Polytechnic National University, Av. Shevchenko, 1, Odessa 65044, Ukraine
pavlenko_vitalij@ukr.net

Abstract. Project methods of human neuropsychological research based on experimental data using innovative Eye-tracking technology and computational means of their processing allow monitoring and diagnosing the state of cognitive processes. Instrumental algorithmic and software tools for building a nonparametric dynamic model of the human oculo-motor system (OMS) taking into account its inertial and nonlinear properties based on the data of experimental studies "input-output" using technology tracking are developed. The Volterra model is used in the form of multidimensional transient functions. Classifiers of human cognitive states are built on the basis of the studied heuristic features that are resistant to computational errors.

Keywords: Eye-tracking technology · Oculo-motor system · Volterra polynomial · Nonlinear dynamic identification · Multidimensional transient functions · Psychophysiological states · Machine learning

1 Introduction

The study of the trajectory of human eye movements reveals the structure of the relationship of the individual with the environment. Knowledge of eye movement is of great theoretical and applied importance, expanding the possibilities of studying the specifics of many professions to increase the efficiency of the subject of labor [1–4].

The aim is to develop software tools for building a non-parametric dynamic model of human OMS considering its inertial and nonlinear properties based on experimental input-output data using test visual stimulus and innovative Eye-tracking technology; introduction of the received information models in diagnostic practice of states of cognitive processes.

The identification process is based on the use of test visual stimuli that are displayed on the computer monitor screen at different distances from the starting position (Fig. 1).

The developed software enables support of the following tasks:

1. The relationship study of mental states and cognitive processes in educational activities [5–9].

M. Ben Ahmed et al. (Eds.): SCA 2022, LNNS 629, pp. 344–353, 2023.
https://doi.org/10.1007/978-3-031-26852-6_33

2. The interaction of mental states and cognitive processes during the educational activities of students, an objective assessment of their cognitive development level, assessment of the effectiveness of training to improve mental processes and for psychological correction of personality [2].
3. Extension of the individual's creative life due to the early diagnosis of degenerative processes of cognitive functions of the brain. Identification of a gifted personality (building a psychological model of the personality) and evaluation of its abilities. Professional selection (the identification and education of leaders) [10].
4. The assimilation of scientific knowledge and their respective skills serves as the main goal and the main result of educational activities. The process of mastering knowledge is the central part of the learning process. Managing this process implies the existence of effective objective indicators for assessing an individual's intellectual abilities [11].

The methods of psychological identification of an individual proposed in the project, based on obtaining experimental data using Eye-tracking technology and computing means of processing them, allow monitoring and diagnostics of the state of cognitive processes during the educational activities of students.

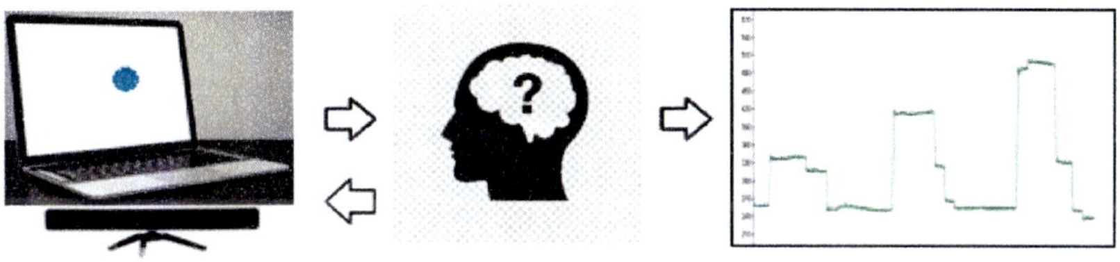

Fig. 1. The eye-tracking process

2 Volterra Model and the Method of the Identification OMS

The basis for creating a mathematical (informational) model of the researched object is the results of measurements of its input and output variables and solving the problem of identification is connected with obtaining experimental data and their processing taking into account measurement noise.

To describe objects of unknown structure, it is advisable to use the most universal non-linear non-parametric dynamic models – Volterra models [12–14]. At the same time, the nonlinear and dynamic properties of the object under study are unambiguously described by a sequence of multidimensional weight functions invariant to the type of input signal – Volterra kernels.

The "input-output" ratio for nonlinear dynamic systems (NDS) with an unknown structure (like a "black box") with one input and one output can be represented by a discrete cubic Volterra polynomial ($N = 3$) in the form [15]:

$$y[m] = \sum_{n=1}^{N=3} \hat{y}_n[m] = \sum_{k_1=0}^{m} w_1[k_1]x[m-k_1] + \sum_{k_1=0}^{m}\sum_{k_2=0}^{m} w_2[k_1,k_2]x[m-k_1]x[m-k_2]$$
$$+ \sum_{k_1=0}^{m}\sum_{k_2=0}^{m}\sum_{k_3=0}^{m} w_3[k_1,k_2,k_3]x[m-k_1]x[m-k_2]x[m-k_3], \qquad (1)$$

where $\hat{y}_n[m]$ is the n-th partial component of the NDS model response; $w_1[k_1]$, $w_2[k_1,k_2]$, $w_3[k_1,k_2,k_3]$ are discrete weight functions of the first, second and third orders; $x[k]$, $y[k]$ are input (stimulus) and output (response) functions of the system being modeled, respectively; k is the time variable.

The task of identification consists in the selection of test effects $x[m]$ and the development of an algorithm that makes it possible to isolate partial components $y_n[m]$, ($n = 1, 2, 3$) based on the measured reactions $y[m]$ and to determine on their basis Volterra kernels $w_1[k_1]$, $w_2[k_1,k_2]$, $w_3[k_1,k_2,k_3]$ [15].

3 Eye Movement Tracking to Identify OMS

The information technology of construction of non-parametric dynamic model of human OMS taking into account its inertial and nonlinear properties on the basis of data of experimental research "input-output" is developed. The OMS Volterra model is used in the form of multidimensional transient functions (MTF) [16, 17].

Methods and tools for the identification of OMS have been developed using technology tracking. The obtained MTF are used to construct the space of diagnostic signs and to carry out the optimal classification of neurophysiological states of personality for research in neuroinformatics and computational neurology. Experimental studies of individuals' OMS were performed using the Tobii Pro TX300 eye-tracker (frame rate 300 Hz) and the corresponding software in the Laboratory of Motion Analysis and Ergonomics of Interfaces of the Lublin University of Technology (Lublin, Poland) [15].

Considering the specifics of the studied OMS, test step signals are used for identification. If the test signal $x[m] = \theta[m]$, where $\theta[m]$ is a unit function (Gaviside function), then the partial components of the response $\hat{y}_1[m]$, $\hat{y}_2[m]$, $\hat{y}_3[m]$ will be equal to the transient function of the first order $\hat{h}_1[m]$ and the diagonal intersections of the transient functions of the 2nd and 3rd orders $\hat{h}_2[m,m]$, $\hat{h}_3[m,m,m]$ respectively [15]:

$$\hat{y}_1[m] = \hat{h}_1[m] = \sum_{k_1=0}^{m} w_1[m-k_1],$$
$$\hat{y}_2[m] = \hat{h}_2[m,m] = \sum_{k_1=0}^{m}\sum_{k_2=0}^{m} w_2[m-k_1, m-k_2], \qquad (2)$$
$$\hat{y}_3[m] = \hat{h}_3[m,m,m] = \sum_{k_1=0}^{m}\sum_{k_2=0}^{m}\sum_{k_3=0}^{m} w_3[m-k_1, m-k_2, m-k_3].$$

Multistep test signals [14, 15] with different a_j ($j = 1,2,\ldots,L$; L – number of experiments, $L \geq N$) $x_j(t) = a_j\theta(t)$ are used for identification. The feedbacks of the OMS, which are measured at the same time, denoted as $y_1[m]$, $y_2[m]$,..., $y_L[m]$. Determining the partial response components of the model $\hat{y}_1[m]$, $\hat{y}_2[m]$, $\hat{y}_3[m]$ leads to the estimation of the transient functions of the first order $\hat{h}_1[m]$ and the diagonal intersections of the transient functions n-th order ($n = 2, 3$) $\hat{h}_2[m,m]$, $\hat{h}_3[m,m,m]$ (2).

The responses of the Volterra polynomial model are equal

$$\tilde{y}_i[m] = a_i\hat{y}_1[m] + a_i^2\hat{y}_2[m] + a_i^3\hat{y}_3[m],\ i = 1, 2, \ldots, L. \tag{3}$$

To determine the transient functions $h_1[m],\ h_2[m, m],\ h_3[m, m, m]$ is used the method of least squares (LSM) is used, which provides a minimum of the root mean square error of the deviation of the model responses from the OMS responses to the same stimulus:

$$J_N = \sum_{j=1}^{L}\left(y_j[m] - \sum_{n=1}^{N} a_j^n\hat{y}_n[m]\right)^2 \to \min \tag{4}$$

The minimization of the criterion (4) is reduced to the solution of the system of normal Gaussian equations, which can be written in the vector-matrix form as

$$\mathrm{A}'\mathrm{A}\hat{\mathrm{y}} = \mathrm{A}'\,\mathrm{y}, \tag{5}$$

where

$$\mathrm{A} = \begin{bmatrix} a_1 & a_1^2 & a_1^3 \\ a_2 & a_2^2 & a_2^3 \\ \ldots & \ldots & \ldots \\ a_L & a_L^2 & a_L^3 \end{bmatrix},\quad \mathrm{y} = \begin{bmatrix} y_1[m] \\ y_2[m] \\ \ldots \\ y_L[m] \end{bmatrix},\quad \hat{\mathrm{y}} = \begin{bmatrix} \hat{y}_1[m] \\ \hat{y}_2[m] \\ \hat{y}_3[m] \end{bmatrix}.$$

Having solved the system of Eqs. (5) with respect to $\hat{y}_1[m],\ \hat{y}_2[m],\ \hat{y}_3[m]$, we obtain estimates of the multidimensional transient functions of the OMS $h_1[m],\ h_2[m, m],\ h_3[m, m, m]$ at each time moment m of the observation interval. From Eq. (5) we get

$$\hat{\mathrm{y}} = (\mathrm{A}'\mathrm{A})^{-1}\mathrm{A}'\mathrm{y}. \tag{6}$$

After performing matrix operations in (6), we get

$$\begin{bmatrix} \hat{y}_1[m] \\ \hat{y}_2[m] \\ \hat{y}_3[m] \end{bmatrix} = \begin{bmatrix} \sum_{j=1}^{L} a_j^2 & \sum_{j=1}^{L} a_j^3 & \sum_{j=1}^{L} a_j^4 \\ \sum_{j=1}^{L} a_j^3 & \sum_{j=1}^{L} a_j^4 & \sum_{j=1}^{L} a_j^5 \\ \sum_{j=1}^{L} a_j^4 & \sum_{j=1}^{L} a_j^5 & \sum_{j=1}^{L} a_j^6 \end{bmatrix}^{-1} \cdot \begin{bmatrix} \sum_{j=1}^{L} a_j y_j[m] \\ \sum_{j=1}^{L} a_j^2 y_j[m] \\ \sum_{j=1}^{L} a_j^3 y_j[m] \end{bmatrix}. \tag{7}$$

In this way, we obtain estimates of the transient functions of the first, second, and third orders [15].

The system of normal Gaussian Eqs. (6) gives good results for the approximation of functions if the number of measurements L is large enough (much greater than the degree of the approximating polynomial N) or the measurement errors are small. Otherwise, the determinant of the system turns out to be close to zero, and the system becomes indeterminate. In this case, large errors in the estimation of the parameters of the approximating polynomial are possible.

The regularization method of A.N. Tikhonov [18], which is based on the variational method of constructing the regularizing operator, is used to obtain the System of linear algebraic equations (SLAE) solution (6) that is resistant to measurement errors. This method boils down to finding an approximate solution vector that minimizes a certain smoothing functional. A single vector that satisfies the condition of the minimum of the smoothing functional can be determined from the SLAE solution:

$$(\mathrm{A}'\mathrm{A} + \alpha \mathrm{I})\hat{y}_{\alpha} = \mathrm{A}'\ \mathrm{y}, \tag{8}$$

where A' is the transposed matrix; I is the identity matrix; α is the A.N. Tikhonov regularization parameter.

The approximate solution obtained on the basis of (8) corresponds to the zeroth order of regularization (weak regularization).

When implementing this algorithm, the regularization parameter α is chosen sufficiently small from the analysis of the available information about the error of the initial data and the calculation error. The appropriate value of the regularization parameter α is determined by selection, i.e. repeated calculations, for different values of α. The quasi-optimal value of the parameter $\alpha = \alpha_0$ is selected from the condition:

$$||\hat{y}_{\alpha_{i+1}} - \hat{y}_{\alpha_i}|| \to \min, \tag{9}$$

where $\alpha_{i+1} = \mu\alpha_i,\ 0 < \mu < 1,\ i = 0, 1, 2, \ldots$ It should be noted that different methods of determining the regularization parameter can give different results and, as a result, different regularized solutions.

4 Experimental Research of the OMS

When conducting experimental studies, such actions are carried out:

1. The test subject is placed in front of the computer so that his eyes are at the center of the monitor at a distance of 40–50 cm from him.
2. The subject's head is fixed in order to prevent its movements during the study and to ensure the same experimental conditions.
3. On the subject's readiness, the Signal Manager of the test visual stimulus program is launched. A red circle appears in the center (or from its edge) – of the screen in the starting position.

4. After a short pause (2–3 s), the circle in the starting position disappears and a circle of a different color appears at the point with the specified coordinates – a visual stimulus (test signal), which is displayed in this position for a specified duration (1–2 s) – the action makes the eye move in the direction of the visual stimulus. Then this stimulus circle disappears and a red circle appears in the starting position – this makes the eye move in the opposite direction to the starting position, after these actions the experiment ends.
5. Using the eye-tracker, the coordinates of the pupil of the eye are determined during its movement (reaction to the visual stimulus) in the period between the starting positions and the coordinate values are stored in the xls-file.

In the studies of each respondent, three experiments were successively implemented for three amplitudes of test signals in the horizontal direction. The distance between the starting position and the test incentives is equal to: 0.33 l_x, 0.66 l_x, 1.0 l_x, where l_x is the length of the monitor screen. Coordinates of the starting position ($x = 0, y = 0.5\ l_y$), l_y – mean the width of the monitor screen.

The experiments were organized in order to classify subjects by the state of fatigue. The data for constructing the model – the OMS responses to the same test signals, were obtained using the Tobii Pro TX300 eye-tracker at different times of the day: "In the morning" (before work) and "In the evening" (after work).

In Fig. 2 and Fig. 3 presents graphs of experimental data at different amplitudes of the test signals "In the morning" and "In the evening", received from the eye tracker.

The average values of the OMS obtained from the eye-tracker at various amplitudes of the test signals "In the morning" and "In the evening" are shown in Fig. 4.

Graphs of transient functions for the states of the respondent "In the Morning" and "In the Evening" at $N = 1$ are presented in Fig. 5, at $N = 2$ – in Fig. 6.

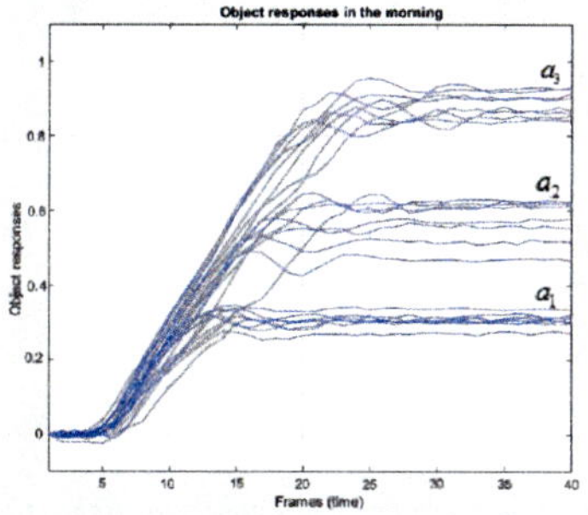

Fig. 2. OMS responses at different amplitudes of test signals "In the morning"

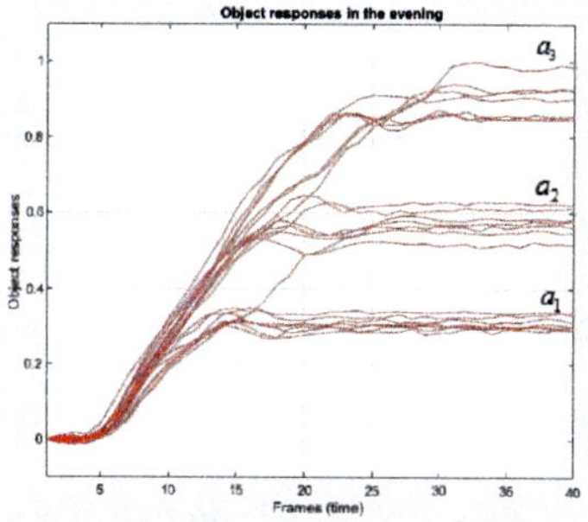

Fig. 3. OMS responses at different amplitudes of test signals "In the evening"

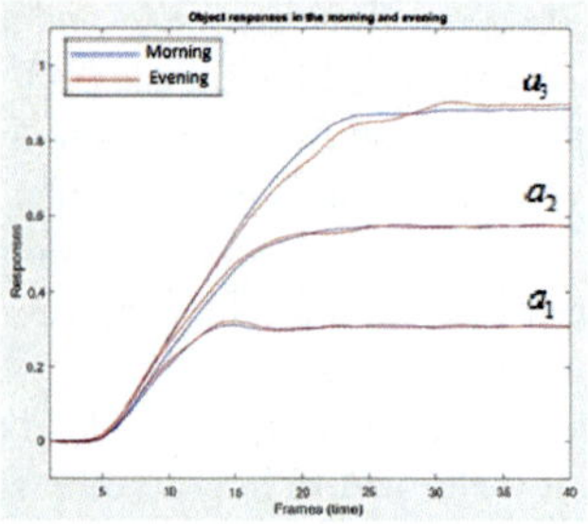

Fig. 4. Averaged OMS responses at various amplitudes of test signals "In the morning" and "In the evening"

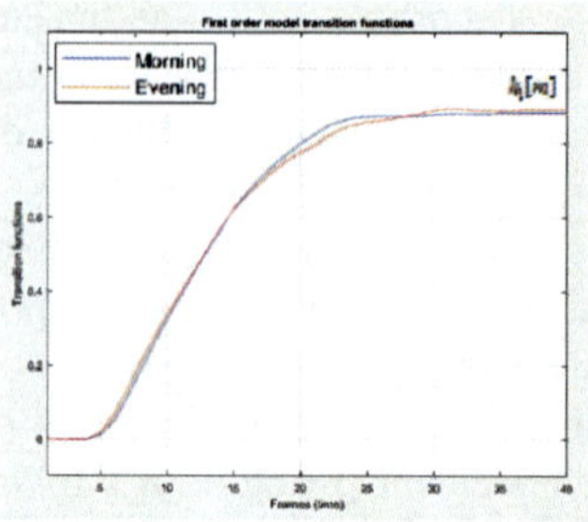

Fig. 5. The transient functions estimates of the 1st order ($N = 1$) for states of the test subject "In the morning" and "In the evening"

According to averaged data of OMS responses on visual stimuli with a different distance from the start position on the basis of formula (5) the functions of the OMS were defined when approximation models of degrees $N = 3$ were used. Graphs of the transient functions estimates for the "In the Morning" and "In the Evening" states of the subject based on model (1) are shown in Fig. 7.

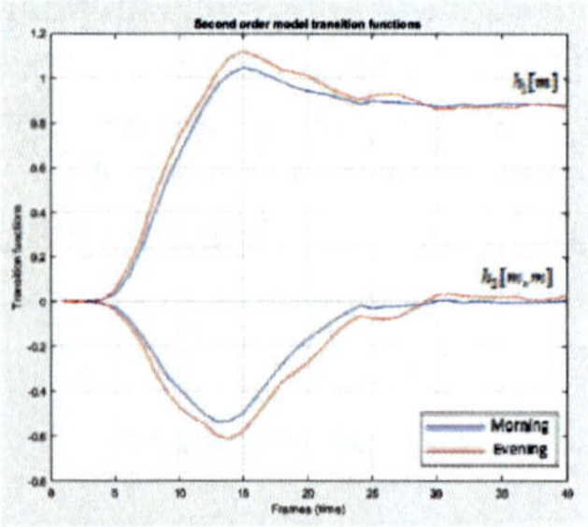

Fig. 6. The transient functions estimates of the 1st and 2nd orders ($N = 2$) for states of the test subject "In the morning" and "In the evening"

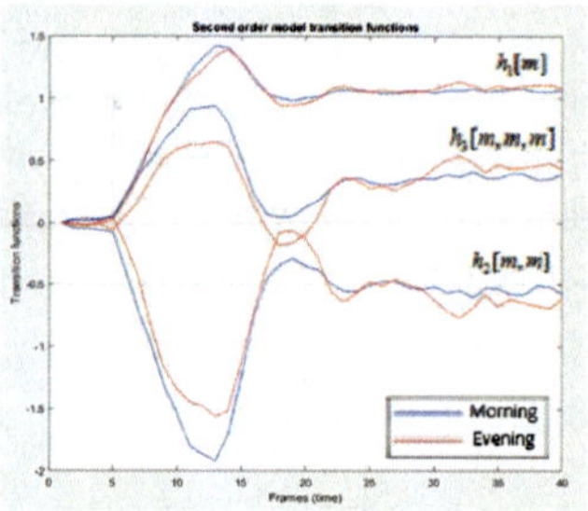

Fig. 7. The transient functions estimates of the 1st, 2nd, and 3rd orders ($N = 3$) for states of the test subject "In the morning" and "In the evening"

5 Building a Classifier of the Fatigue

To assess the psycho-physiological state of a person based on the OMS model, studies were carried out:

1. Getting a feature space for building a human state classifier using machine learning.
2. Building classifiers using deterministic and statistical learning methods for pattern recognition based on the data obtained using Eye-tracking technology.

The discriminant function $d(x)$ is sequentially calculated based on training datasets for object classes **A** ("In the Morning"), **B** ("In the Evening").

Gaussian classifier is built for separate the two classes (dichotomy case) a discriminant function of the form is used:

$$d(\mathrm{x}) = \frac{1}{2}\mathrm{x}'(\mathrm{S}_2^{-1} - \mathrm{S}_1^{-1})\mathrm{x} + (\mathrm{S}_1^{-1}\mathrm{m}_1 - \mathrm{S}_2^{-1}\mathrm{m}_2)'\mathrm{x} + \frac{1}{2}(\mathrm{m}_1'\mathrm{S}_1^{-1}\mathrm{m}_1 - \mathrm{m}_2'\mathrm{S}_2^{-1}\mathrm{m}_2 + \ln\frac{|\mathrm{S}_2|}{|\mathrm{S}_1|}) + \lambda_{\max} \tag{10}$$

where $\mathrm{x} = (x_1, x_2, \ldots, x_n)'$ – features vector, n – features space dimensionality, m_i – mathematical expectation vector for a features of class i, $i = 1, 2$; $\mathrm{S}_i = \mathrm{M}[(\mathrm{x}\text{-}\mathrm{m}_i)(\mathrm{x}\text{-}\mathrm{m}_i)']$ – covariance matrix for class i (M[] – mathematical expectation operation). S_i^{-1} – matrix inverse to S_i, $|\mathrm{S}_i|$ – matrix determinant S_i, $\lambda_{\max}$ – classification threshold providing the highest criterion probability of correct recognition training sample objects.

Gaussian classifier of a person's fatigue state in the two-dimensional space of heuristic features provided with the maximum recognition reliability ($P = 0.9375$) with combinations of the following features:

$$\left(x_3 = \sum_{m=0}^{M} |h_3[m, m, m]| \ \& \ x_{12} = \min_{m \in [0,M]} h_2'[m, m]\right), \text{ or} \tag{11}$$

$$\left(x_3 = \sum_{m=0}^{M} |h_3[m, m, m]| \ \& x_{14} = \min_{m \in [0,M]} h_3'[m, m, m]\right), \text{ or} \tag{12}$$

$$\left(x_8 = \max_{m \in [0,M]} h_3'[m, m, m] \ \& \ x_6 = \max_{m \in [0,M]} h_2'[m, m]\right), \text{ or} \tag{13}$$

$$\left(x_9 = \arg\max_{m \in [0,M]} h_3'[m, m, m] \ \& x_6 = \max_{m \in [0,M]} h_2'[m, m]\right), \text{ or} \tag{14}$$

$$\left(x_{13} = \arg\min_{m \in [0,M]} h_2'[m, m] \ \& \ x_6 = \max_{m \in [0,M]} h_2'[m, m]\right), \text{ or} \tag{15}$$

$$\left(x_{14} = \min_{m \in [0,M]} h_3'[m, m, m] \ \& x_{12} = \min_{m \in [0,M]} h_2'[m, m]\right). \tag{16}$$

Separately, the PCR features: x_9 or x_{13} – $P = 0{,}625$; x_3 or x_8 – $P = 0{,}688$; x_{12} – $P = 0{,}75$; x_6 or x_{14} – $P = 0{,}813$.

Support Vector Machine (SVM) [19] classification gives close results ($P = 0{,}875$) in two-dimensional feature space x_6, x_9 or x_6, x_{13}. The classifier is built using a 2nd order polynomial kernel:

$$K(\mathrm{x},\mathrm{x}') = (\langle \mathrm{x},\mathrm{x}' \rangle + 1)^d \tag{17}$$

where d is specified by parameter degree, $d = 2$.

Scikit-learn library was used (class sklearn.svm.SVC) to apply SVM.

6 Conclusion

Instrumental algorithmic and software tools for building a nonparametric dynamic model of the human oculo-motor system considering its inertial and nonlinear properties based on the data of experimental studies "input-output" using technology tracking are developed. The Volterra model is used in the form of multidimensional transient functions.

The following tasks are solved:

1. The application of the OMS identification method based on the Volterra model in the form of multidimensional transient functions using test visual stimuli with different distances from the starting position - step functions of different amplitude is substantiated.
2. The information technology of obtaining experimental data for the identification of OMS with the help of test visual stimuli and the use of eye-tracking to track the corresponding eye movements has been developed.
3. Developed in the Matlab system software for the identification of OMS based on Volterra polynomials in the form of multidimensional transient functions according to eye-tracking.
4. Experimental studies of OMS were performed with the help of Eye-tracking technology and the transient functions of the first, second and third orders were determined based on oculographic studies. Studies of local self-government with the help of the obtained transient functions by means of computer modeling confirm the adequacy of the constructed approximation model to the real system.
5. Classifiers of human cognitive states are built based on the studied heuristic features that are resistant to computational errors. The features were calculated on multidimensional transient functions obtained from the integral Volterra models of human OMS within a new approach to diagnosing human conditions.

The analysis of variability of transient functions corresponding to different psychophysiological states of the individual (states of fatigue) is carried out. It is established that the diagonal intersections of the transient functions of the second and third order with respect to the transient functions of the first order for fatigue states change significantly in magnitude. Thus, they can be used as a data source in the formation of spaces of diagnostic features for the construction of classifiers of human psychophysiological states.

References

1. Wamain, Y., Corveleyn, X., Ott, L., Coello, Y.: Does the motor system contribute to the perception of changes in objects visual attributes? The neural dynamics of sensory binding by action. Neuropsychologia 132, 107121 (2019)
2. Daoudi, M., Coello, Y., Descrosiers, P.-A., Ott, L.: A new computational approach to identify human social intention in action. In: 2018 13th IEEE International Conference on Automatic Face & Gesture Recognition, FG 2018, Xi'an, China on May 15–19, pp. 512–516 (2018)

3. Quesque, F., Mignon, A., Coello, Y.: Cooperative and competitive contexts do not modify the effect of social intention on motor action. Conscious. Cogn. **56**, 91–99 (2017)
4. Lanata, A., Sebastiani, L., Di Gruttola, F., Di Modica, S., Scilingo, E.P., Greco, A.: Nonlinear analysis of eye-tracking information for motor imagery assessments. Front. Neurosci. **13** (2020). https://doi.org/10.3389/fnins.2019.01431
5. El Haj, M., Coello, Y., Kapogiannis, D., Gallouj, K., Antoine, P.: Negative prospective memory in Alzheimer's disease: "Do not perform that action". J. Alzheimer's Dis. **61**(2), 663–672 (2017)
6. Corveleyn, X., et al.: Body-centred and object-centred motor imagery in Alzheimer's disease. Curr. Alzheimer Res. **15**(3), 247–258 (2018)
7. Jansson, D., Medvedev, A., Axelson, H., Nyholm, D.: Stochastic anomaly detection in eye-tracking data for quantification of motor symptoms in Parkinson's disease. In: Sun, C., Bednarz, T., Pham, T.D., Vallotton, P., Wang, D. (eds.) Signal and Image Analysis for Biomedical and Life Sciences. AEMB, vol. 823, pp. 63–82. Springer, Cham (2015). https://doi.org/10.1007/978-3-319-10984-8_4
8. Jansson, D., Rosen, O., Medvedev, A.: Parametric and nonparametric analysis of eye-tracking data by anomaly detection. IEEE Trans. Control Syst. Technol. **23**(4), 1578–1586 (2015)
9. Bro, V., Medvedev, A.: Nonlinear dynamics of the human smooth pursuit system in health and disease: model structure and parameter estimation. In: 2017 IEEE 56th Annual Conference on Decision and Control (CDC), 12–15 December 2017, Melbourne, Australia, pp. 4692–4697 (2017)
10. Guillot, A., Louis, M., Collet, C.: Neurophysiological substrates of motor imagery ability. In: Guillot, A., Collet, C. (eds.) The Neurophysiological Foundations of Mental and Motor Imagery, pp. 109–124. Oxford University Press (2010)
11. Guillot, A., Collet, C.: Construction of the motor imagery integrative model in sport: a review and theoretical investigation of motor imagery use. Int. Rev. Sport Exerc. Psychol. **1**(1), 31–44 (2008)
12. Doyle, F.J., Pearson, R.K., Ogunnaike, B.A.: Identification and control using Volterra models. Springer, London (2002)
13. Pavlenko, V., Pavlenko, S., Speranskyy, V.: Identification of systems using Volterra model in time and frequency domain. In: Haasz, V., Madani, K. (eds.) Advanced Data Acquisition and Intelligent Data Processing, pp. 233–270. River Publishers (2014)
14. Pavlenko,V., Pavlenko, S.: Deterministic identification methods for nonlinear dynamical systems based on the Volterra model. Appl. Aspects Inf. Technol. **01**(01), 9–29 (2018)
15. Pavlenko, V., Milosz, M., Dzienkowski, M.: Identification of the oculo-motor system based on the Volterra model using eye tracking technology. J. Phys. Conf. Ser. **1603**(012011), 1–8 (2020). https://doi.org/10.1088/1742-6596/1603/1/012011
16. Pavlenko, V., Salata, D., Dombrovskyi, M., Maksymenko, Yu.: Estimation of the multidimensional transient functions oculo-motor system of human. In: Mathematical Methods and Computational Techniques in Science and Engineering, vol. 1872, pp. 110–117. AIP Publishing, Melville, New York (2017). https://doi.org/10.1063/1.4996671
17. Pavlenko, V., Ivanov, I., Kravchenko, E.: Estimation of the multidimensional dynamical characteristic eye-motor system. In: 2017 9th IEEE International Conference on Intelligent Data Acquisition and Advanced Computing Systems: Technology and Applications (IDAACS), 21–23 September 2017, Bucharest, Romania, pp. 645–650 (2017)
18. Tikhonov, A., Goncharsky, A., Stepanov, V., Yagola, A.: Numerical Methods for the Solution of Ill-Posed Problems. Springer, Netherlands, Netherlands (2018)
19. Vapnik, V.: The Nature of Statistical Learning Theory. Springer-Verlag, New York (2010)

Toward a Smart City: Reinforcement Learning for Traffic Light Control

Asma Ait Ouallane[1](✉), Ayoub Bahnasse[1,2], Assia Bakali[1,3], and Mohamed Talea[1]

[1] TI Lab, Faculty of Science Ben M'sik, Hassan II University of Casablanca, BP 7955 Casablanca, Morocco
asmaaaitouallane@gmail.com

[2] ENSAM Casablanca, University Hassan II of Casablanca, Casablanca, Morocco

[3] Royal Naval School, Boulevard Sour Jdid, B.P 16303, Casablanca, Morocco

Abstract. For decades, the optimization of traffic congestion problem has presented a significant issue for the transportation field. The continuous increase in the number of vehicles and the impossibility of building new high-capacity road infrastructures in big cities, which are densely populated and limited in space, make the alleviation of this risk a challenge for the scientific research community. The transportation researchers are called upon to deal with the complexities of ensuring effective management of intersections, which are in fact the nodes of road traffic congestion. But the real-world decision-making problems always involve uncertainty and indeterminacy, which lead to a paralyzed traffic light control system in some scenarios. Policemen address the traffic directly in an efficient way relying on their knowledge. This motivated the researchers to use Reinforcement Learning to create intelligent intersection traffic light management systems that can learn over time how to handle the intersection traffic flow based on real-time traffic conditions. This paper presents a brief overview of the three most used traffic data sources of a Traffic Management System and the most recently used Reinforcement Learning technics to handle the traffic light control problem. Finally, various experimental parameters that can affect the evaluation of traffic signal management methods are discussed.

Keywords: Smart city · Traffic light control · Intelligent urban traffic management system · Reinforcement learning · Artificial intelligence

1 Introduction

Cities are the principal hubs of people and economic activity. They are capable of establishing connections that will create excellent growth opportunities for their population. However, as the rate of urban population growth grows, as does the size and complexity of cities [1], current urban transportation systems become less effective in accommodating the expanding number of commuters [2]. The concept of a smart city is inextricably

M. Ben Ahmed et al. (Eds.): SCA 2022, LNNS 629, pp. 354–364, 2023.
https://doi.org/10.1007/978-3-031-26852-6_34

linked to the use of information and communication technology to improve different city operations, boost economic growth, and improve inhabitants' quality of life [3], in order to construct more effective and sustainable communities. The six main urban development fields that constitute the smart city are Smart Governance, Smart Economy, Smart Living, Smart Environment, Smart People, and Smart Mobility, which includes urban road traffic management as illustrated in Fig. 1. Transport researchers have long been interested in urban road traffic management [4]. Congestion is a major problem affecting road users and officials alike. Despite significant efforts and research to minimize traffic congestion [5], this serious problem continues to worsen. The primary reason for this is the growing number of cars on the road and the slow development of road infrastructure and capacity. Therefore, traffic congestion is a serious matter that should be earnestly addressed in order to offer a comfortable and safe environment for people [6]. The clear solution is to build more high-capacity road infrastructure. However, this strategy is expensive and, in most situations, impractical in major cities with dense populations and limited space [7]. As a result, Intelligent Transportation Systems (ITSs) were created in order to provide innovative intelligent solutions that make the greatest use of existing infrastructure while improving road users' safety and comfort [8]. ITSs serve various purposes, including traffic management, emergency management, electronic payment, smart parking [9], traffic safety, and traffic law enforcement. Among these services, traffic management is the most challenging subject for academics and transportation engineers to tackle, since inefficient traffic control results in a city full of bottlenecks, traffic delays, energy waste, increased pollution and accident rates. As a result, it has a negative influence on almost all ITS applications. Intelligent Traffic Light Control Systems are crucial in traffic management. Taking real-time traffic information as an input and dynamically modifying traffic signal length is required to increase efficiency. Real-world decision-making difficulties, on the other hand, frequently incorporate ambiguity and imprecision, resulting in an intelligent traffic light system that is stuck in some scenarios. The intervention of the policemen in these instances is usually good. They handle traffic fast and effectively since they are familiar with the intersection. Based on various Reinforcement Learning approaches, this witness inspired scientific researchers to consider developing a smart intersection traffic light management system that learns directly from its environment how to manage the intersection efficiently without making any unrealistic assumptions. The rest of the paper is organized as follows. In Sect. 2, we present the two categories of traffic congestion, In Sect. 3, we provide a brief overview of three commonly used traffic data sources in a Traffic Management System, and the recently employed Reinforcement Learning techniques for dealing with the traffic light control problem. In Sect. 4, several experimental characteristics that might influence traffic signal management technique evaluation are examined. Finally, in Sect. 5, we draw our conclusion.

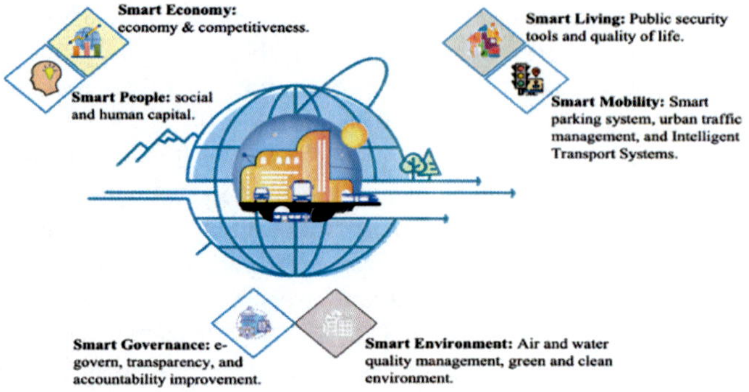

Fig. 1. The six urban development fields that make up the smart city.

2 Categories of Road Traffic Congestion

The fast development in traffic demand, along with the limited resources and capacity of urban road networks, has resulted in a significant increase in traffic congestion, affecting the comfort of road users. Congestion not only generates general social unease, but it also has a significant negative influence on national economic development [9]. There are two kinds of traffic congestion. The first is recurrent traffic congestion. It is typically produced by regular traffic. It occurs every day at the same time and location. The second is irregular traffic congestion. It is related to uncommon events such as traffic accidents [10]. To reduce traffic congestion, the origins of the problem must be recognized. The Federal Highway Administration in the United States identified the six sources of traffic congestion as depicted in Fig. 2.

Fig. 2. The six sources of road traffic congestion

To alleviate traffic congestion and ensure smooth traffic flow on urban roads, smart cities rely on traffic management systems [11]. To improve overall traffic efficiency and thus reduce traffic congestion, the following factors must be considered when designing a traffic management system: the road users and the traffic management systems should have access to accurate real-time traffic data, the traffic management system must be able to make quick decisions at junctions, the emergency vehicles must be prioritized, the security of smart cities must be ensured through traffic management systems, public transportation should be prioritized especially during peak hours, the traffic management systems must be able to handle many tasks at the same time, multiple junctions must be controlled and synchronized by traffic management systems, and last but not least, traffic incidents must be detected.

3 Reinforcement Learning for Traffic Light Control

As we already detailed in our previous survey [4], the Urban Traffic Management System is composed of three subsystems, including the data collection subsystem, the traffic light control subsystem, and the safety subsystem (as illustrated in Fig. 3). Traffic Light Control (TLC) is a traditional and centralized method of controlling traffic at intersections. It enables the calculation and synchronization of the different phases' duration for each direction in a static manner independent of road conditions [12]. To address the shortcomings of this inflexible approach, researchers have developed intelligent TLC systems that adapt to current road conditions based on the data collection subsystem, which consists of three main data sources: Wireless Sensor Networks (WSNs), Cameras, and Vehicular Ad hoc Networks (VANETs).

Regarding the WSNs, they are the most often utilized sources in data gathering systems to regulate and manage traffic congestion due to their counting precision and insensitivity to environmental traffic circumstances [13, 14]. Regarding the camera-based computer vision technology, it can provide a more detailed description of the traffic situation [15]. For example, it can identify irregularities and traffic accidents, which is very useful for traffic officers and road traffic management. The real-time data gathering module is included in traffic management in this situation. Then comes the data processing module, followed by the transfer of important information to the decision-making module [16, 17]. Finally, there is the VANET, which is a promising technique in the ITS field. It provides the Traffic Management System with real-time insight into the traffic condition. In a VANET, each vehicle serves as a wireless router. It has an On-Board Unit (OBU) that comprises a collection of sensors, a processing unit, and a wireless interface that allows cars to connect with other vehicles and with central entities via Vehicle-to-Vehicle (V2V) and Vehicle-to-Infrastructure (V2I) communication, respectively [18, 19].

Reinforcement Learning for Traffic Light Control: The TLC System makes critical decisions to manage traffic on the road. However, in many other situations, such as a high-traffic hour scenario, the TLC system fails. As a result, policemen regulate traffic promptly by waving signals. They may assess real-time traffic conditions in junctions and intelligently select the duration of the permitted passing time for each direction based on their long-term knowledge and awareness of the intersection [20]. This process is quite efficient. This has inspired transportation researchers to create a smart traffic signal control system that can learn how to handle intersections depending on real-time traffic circumstances. To that purpose, Reinforcement Learning, a machine learning

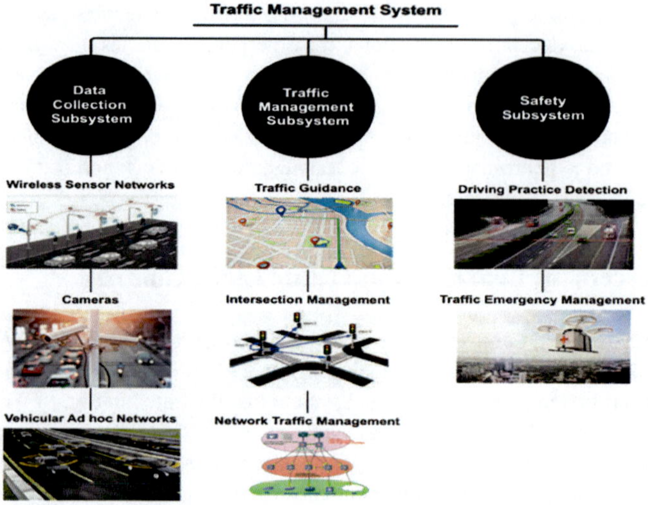

Fig. 3. The three subsystems of a Traffic Management System

method, appears to be a potential answer [2, 21, 22]. Through numerous steps, it learns how to attain a difficult task or optimize along a certain dimension. A reinforcement learning system's goal is to teach an agent the optimum strategy for interacting with the environment in order to maximize the reward [23] as illustrated in Fig. 4.

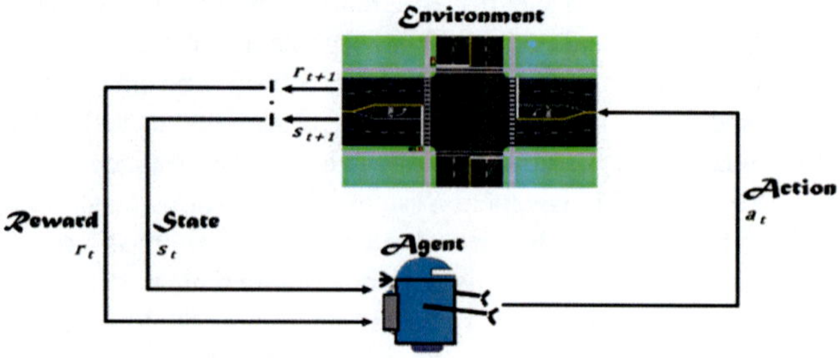

Fig. 4. Reinforcement Learning cyclicality.

In the case of a traffic light control system, detectors (e.g., induction loop, traffic camera image processing…) monitor the environment to determine its status. In this situation, the action is the traffic light control action, such as extending or shortening the green period, switching to another management plan, etc.

Q-learning: Q-learning is one of the most notable model-free reinforcement learning algorithms studied for traffic light control due to its simplicity and performance. It investigates the complicated and stochastic world through trial and error and selects the optimal behavior based on experience [24]. Using the Q-function, the agent estimates the efficiency of each action in simulation states.

$$Q^{new}(s_t, a_t) \leftarrow \underbrace{Q(s_t, a_t)}_{\text{old value}} + \underbrace{\alpha}_{\text{learning rate}} \cdot \overbrace{\Big(\underbrace{\underbrace{r_t}_{\text{reward}} + \underbrace{\gamma}_{\text{discount factor}} \cdot \underbrace{\max Q(s_{t+1}, a_t)}_{\text{estimate of optimal future}}}_{\text{new value (temporal difference target)}} - \underbrace{Q(s_t, a_t)}_{\text{old value}}\Big)}^{\text{temporal difference}} \quad (1)$$

Each state-action pair and the associated Q-value are stored in a Q-table. The convergence and speed of the Q-learning algorithm are affected by the learning rate and the discount factor, as indicated in Eq. (1). The Q-learning algorithm has been fully explored and used for traffic light control at an isolated intersection [25, 26]. However, the state space should be simple due to technical limitations [23]. The Deep Q Network, a fusion of Q-learning and Deep Learning, can offer a much more flexible traffic light management system by determining the state space of traffic if the simulation parameters become more complicated and hence require more memory and computational resources.

Deep Q-Network (DQN): The cyclicality of the agent and environment is the same in DQN as in Q-learning, but the Q-values are estimated using a deep neural network rather than storing them in the Q-table as illustrated in Fig. 5.

Q-values are adjusted based on a Q-function similar to the one illustrated in Eq. (1):

$$Q(s_t, a_t) = R_t + \gamma \cdot \max Q(s_{t+1}, a') \quad (2)$$

Q (st, at) is predicted by the neural network based on the state space, Q (st + 1, a′) is predicted by the neural network for the future state space, and γ is the discount factor.

The replay memory, the main network, and the target network are the three basic components of an agent. First, the replay memory is an agent's experiences dataset that is accumulated over time when the agent interacts with the environment. During the training process, the experiences are applied. The main network is utilized to estimate the Q-values at iteration k. It picks a specific action a_t for a specific state s_t seen in the environment in order to obtain the best potential reward $r_{t+1}(s_{t+1})$ and the following state s_{t+1} at the next time instant t + 1. The target network is a copy of the main network that is used to estimate Q-values after the kth iteration. The main network is utilized for action selection and training, whereas the target network is solely used for training in order to enhance training stability [27].

Double DQN: Hasselt et al. [28] created Double DQN to address the issue that arises when the network continuously assigns a higher Q-value to non-optimal actions, leading the algorithm to become stuck in a local optimum and hence fail to find the optimal action during training. Double DQN's target value is periodically constant. The same structure as the base neural network is used to create a second neural network (target model, Q^t) with periodically constant weights. The learning is continually frozen, and the original model' weights are copied after a certain number of steps. The target value is calculated using the following formula:

$$Q(s_t, a_t) = R_t + \lambda \cdot Q^t(s_{t+1}, \arg\max Q(s_{t+1}, a')) \quad (3)$$

Dueling DQN: When all actions in a given state produce the same Q-value, it is assumed that none of the actions produces a more advantageous position than the others. Hence Wang et al. [29] created Dueling DQN that can filter out states in which it is not essential to learn the value of the actions corresponding to them. In environments where the actions do not always have a significant impact, Dueling DQN is particularly useful. It is actually the case of the traffic control environment where it is not essential to adjust the traffic light phases during periods of low traffic. The neural network structure differs from previous ones in that we generate two independent threads at the end of the network. The first thread estimates Q-values in the same manner as previously, while the second thread defines the advantage of taking an action in a particular state. It is called the advantage function. The two threads are then grouped together based on:

$$Q(s, a) = V(s) + A(s, a) - \frac{1}{|A|}\sum_{a'} A(s, a') \tag{4}$$

The state value and the benefit of each action are represented by V(s) and A(s, a) respectively.

Double Dueling DQN (3DQN): The principles of Dueling DQN and Double DQN may be merged without issue, resulting in Double Dueling DQN.

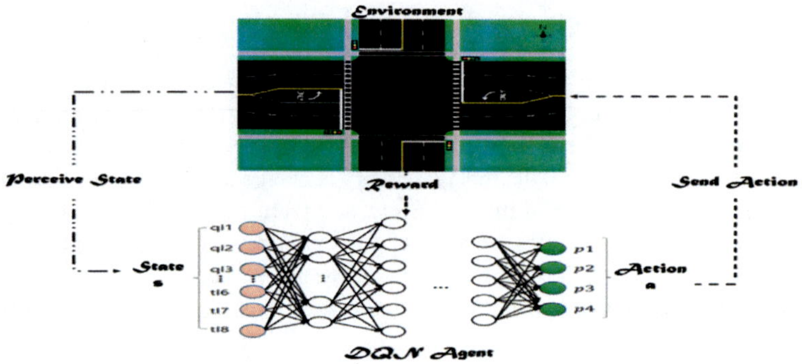

Fig. 5. DQN

4 Evaluation

The evaluation of the various traffic lights management reinforcement learning approaches can be affected by different experimental parameters. In this section, we discuss these parameters.

Simulation Platforms: Testing and deploying novel reinforcement learning algorithms for traffic light management in a realistic traffic network scenario is both expensive and challenging because real-world traffic is highly dynamic over time [2]. As a result, before real implementation, transportation researchers employ simulators as auxiliary substitutes. The most commonly used simulators in the literature are microscopic simulators because traffic light control simulations often involve heterogeneous road traffic and are concerned with the behavior of each vehicle in detail including speed, location, destination, and lane change. The most used open-source microscopic simulator is SUMO (Simulation of Urbain MObility). It was created at German Aerospace Centre. This simulator allows vehicles to travel freely and considers the lane width and vehicle width fixed. Thus, the various sorts of vehicles are not considered. SUMO can simulate collisions and accidents and multimodal traffic systems including pedestrians and public transportation. For a thorough comparison of road traffic simulators see [30].

Road Network: To evaluate suggested reinforcement learning methods, researchers generally use synthetic or real-world road networks. Regarding the synthetic grid network, it consists of edges that represent roads and nodes that represent intersections. Regarding the real-world network, it is more intricated concerning its features for instance the speed limit of each vehicle, the traffic movement, the number of intersections and their construction and combination, and the number of traffic lights. The number of traffic lights in the network has the greatest effect on experiment outcomes of all road network features since the larger the road network, the larger the scale of explorations for reinforcement learning to take. Actually, most research is conducted on small road networks compared to the size of a city, which can comprise hundreds or even thousands of traffic lights.

Performance Evaluation: To evaluate their algorithms' performance, the majority of researchers compare them with a pre-scheduled traffic light system. The latter is adequate for examining the viability of a suggested technique and its concept. However, it is ineffective in assessing reinforcement learning approaches performance since it is incompetent to adapt to real-time traffic flow. Thus, any comparison with a fixed-time method will be useless in terms of performance evaluation and it is obvious that the results of the proposed approach are going to be better than those of a pre-scheduled traffic light system.

Traffic Density: The simulation's traffic flow density can impact the evaluation of the proposed reinforcement learning methods. As input, the simulator consumes traffic data. Each vehicle is defined as (o, t, d), where o represents the origin position, t represents time, and d represents the destination position. Generally, the reinforcement learning approaches have many difficulties to learn the optimal policy in environments with dynamic and heavy traffic.

Evaluation Parameter: Using different state, action, and reward elements in RL methods is essential to achieve effective results [31, 32]. Regarding the state, the Queue length, the waiting time, the volume, the delay, the speed, the phase duration, the congestion, and the phase were used in the literature. Regarding the action, keep or change, choose the next phase, set phase split, and set current phase duration were used in the literature. Regarding the reward, Queue length, waiting time, change of delay, speed, number of stops, throughput, accident avoidance, and pressure were all used in the literature to improve the learning of reinforcement learning algorithms and evaluate their efficiency.

To evaluate their approaches, researchers relied on various measures among them we found the average waiting time of vehicles, the average delay of vehicles, the travel time of vehicles, the queue length, and the throughput of the road network.

5 Conclusion and Future Work

In this paper, we provided a brief overview of the smart city applications and we underlined the traffic management problem, which threatens smart cities. Afterward, we investigated the latest used Reinforcement Learning approaches for traffic light control and discussed various experimental parameters that can affect the evaluation of traffic light control methods. Based on this brief overview, we conclude that the DQN approach and its variants are promising in solving the problem of traffic light control. In our next work, we will compare these algorithms and propose our approach to solving the traffic signal control problem.

Acknowledgment. This work was supported by the National Center for Scientific and Technical Research (CNRST) in Morocco as part of the Research Excellence Scholarship Program.

References

1. Monzon, A.: Smart cities concept and challenges: bases for the assessment of smart city projects. In: 2015 International Conference on Smart Cities and Green ICT Systems (SMARTGREENS), pp. 1–11, May 2015
2. Noaeen, M., et al.: Reinforcement learning in urban network traffic signal control: a systematic literature review. Expert Syst. Appl. **199**, 116830 (2022). https://doi.org/10.1016/j.eswa.2022.116830
3. Belli, L., et al.: IoT-enabled smart sustainable cities: challenges and approaches. Smart Cities **3**(3), Article no. 3 (2020). https://doi.org/10.3390/smartcities3030052
4. Ait Ouallane, A., Bakali, A., Bahnasse, A., Broumi, S., Talea, M.: Fusion of engineering insights and emerging trends: Intelligent urban traffic management system. Inf. Fus. **88**, 218–248 (2022). https://doi.org/10.1016/j.inffus.2022.07.020
5. Ait Ouallane, A., Broumi, S., Ayele, E.T., Bakali, A., Bahnasse, A., Talea, M.: Towards intelligent road traffic management based on neutrosophic logic: a brief review October 2022. https://doi.org/10.5281/zenodo.7135256
6. Ait Ouallane, A., Bahnasse, A., Bakali, A., Talea, M.: Overview of road traffic management solutions based on IoT and AI. Procedia Comput. Sci. **198**, 518–523 (2022). https://doi.org/10.1016/j.procs.2021.12.279
7. Jabbarpour, M.R., Zarrabi, H., Khokhar, R.H., Shamshirband, S., Choo, K.-K.: Applications of computational intelligence in vehicle traffic congestion problem: a survey. Soft. Comput. **22**(7), 2299–2320 (2017). https://doi.org/10.1007/s00500-017-2492-z
8. Bui, K.-H., Camacho, D., Jung, J.E.: Real-Time traffic flow management based on inter-object communication: a case study at intersection. Mob. Netw. Appl. **22**(4), 613–624 (2017). https://doi.org/10.1007/s11036-016-0800-y
9. Ma, C., Zhou, J., (Daniel) Xu, X., Xu, J.: Evolution regularity mining and gating control method of urban recurrent traffic congestion: a literature review. J. Adv. Transp. **2020**, e5261580 (2020). https://doi.org/10.1155/2020/5261580

10. Olayode, I.O., Tartibu, L.K., Okwu, M.O., Uchechi, U.F.: Intelligent transportation systems, un-signalized road intersections and traffic congestion in Johannesburg: a systematic review. Procedia CIRP **91**, 844–850 (2020). https://doi.org/10.1016/j.procir.2020.04.137
11. de Souza, A.M., Brennand, C.A., Yokoyama, R.S., Donato, E.A., Madeira, E.R., Villas, L.A.: Traffic management systems: a classification, review, challenges, and future perspectives. Int. J. Distrib. Sens. Netw. **13**(4), 1550147716683612 (2017). https://doi.org/10.1177/1550147716683612
12. El Hamdani, S., Benamar, N., Younis, M.: Pedestrian support in intelligent transportation systems: challenges, solutions and open issues. Transp. Res. Part C Emerg. Technol. **121**, 102856 (2020). https://doi.org/10.1016/j.trc.2020.102856
13. A Hilmani, A., Maizate, A., Hassouni, L., Ahmed, G.: Automated real-time intelligent traffic control system for smart cities using wireless sensor networks. Wirel. Commun. Mob. Comput. **2020** (2020). https://doi.org/10.1155/2020/8841893
14. Sarrab, M., Pulparambil, S., Awadalla, M.: Development of an IoT based real-time traffic monitoring system for city governance. Glob. Transit. **2**, 230–245 (2020). https://doi.org/10.1016/j.glt.2020.09.004
15. Makhmutova, A., Anikin, I.V., Dagaeva, M.: Object tracking method for videomonitoring in intelligent transport systems. In: 2020 International Russian Automation Conference (RusAutoCon), September 2020, pp. 535–540 (2020). https://doi.org/10.1109/RusAutoCon49822.2020.9208032
16. Gong, S., Kumar, R., Kumutha, D.: Design of lighting intelligent control system based on OpenCV image processing technology. Int. J. Unc. Fuzz. Knowl. Based Syst. **29**(Supp. 01), 119–139 (2021). https://doi.org/10.1142/S0218488521400079
17. Suseendran, G., Akila, D., Balaganesh, D., Elangovan, V.R., Vijayalakshmi, V.: Incremental multi-feature tensor subspace learning based smart traffic control system and traffic density calculation using image processing. In: 2021 2nd International Conference on Computation, Automation and Knowledge Management (ICCAKM), January 2021, pp. 87–91 (2021). https://doi.org/10.1109/ICCAKM50778.2021.9357743
18. Kshirsagar, M., et al.: GREE-COCO: green artificial intelligence powered cost pricing models for congestion control. Presented at the ICAART 2021 - Proceedings of the 13th International Conference on Agents and Artificial Intelligence, vol. 2, pp. 916–923 (2021)
19. Contreras, M., Gamess, E.: An Algorithm based on VANET technology to count vehicles stopped at a traffic light. Int. J. Intell. Transp. Syst. Res. **18**(1), 122–139 (2019). https://doi.org/10.1007/s13177-019-00184-3
20. Liang, X., Du, X., Wang, G., Han, Z.: A deep reinforcement learning network for traffic light cycle control. IEEE Trans. Veh. Technol. **68**(2), 1243–1253 (2019). https://doi.org/10.1109/TVT.2018.2890726
21. Genders, W., Razavi, S.: Evaluating reinforcement learning state representations for adaptive traffic signal control. Procedia Comput. Sci. **130**, 26–33 (2018). https://doi.org/10.1016/j.procs.2018.04.008
22. Yau, K.-L.A., Qadir, J., Khoo, H.L., Ling, M.H., Komisarczuk, P.: A survey on reinforcement learning models and algorithms for traffic signal control. ACM Comput. Surv. **50**(3), 34:1–34:38 (2017). https://doi.org/10.1145/3068287
23. Pálos, P., Huszák, Á.: Comparison of Q-learning based traffic light control methods and objective functions. In: 2020 International Conference on Software, Telecommunications and Computer Networks (SoftCOM), pp. 1–6, September 2020. https://doi.org/10.23919/SoftCOM50211.2020.9238290
24. Pandey, D., Pandey, P.: Approximate Q-learning: an introduction. In: 2010 Second International Conference on Machine Learning and Computing, pp. 317–320, February 2010. https://doi.org/10.1109/ICMLC.2010.38

25. Liao, Y., Cheng, X.: Study on traffic signal control based on Q-learning. In: 2009 Sixth International Conference on Fuzzy Systems and Knowledge Discovery, vol. 3, pp. 581–585 August 2009. https://doi.org/10.1109/FSKD.2009.539
26. Joo, H., Ahmed, S.H., Lim, Y.: Traffic signal control for smart cities using reinforcement learning. Comput. Commun. **154**, 324–330 (2020). https://doi.org/10.1016/j.comcom.2020.03.005
27. Rasheed, F., Yau, K.-L.A., Noor, R., Wu, C., Low, Y.-C.: Deep reinforcement learning for traffic signal control: a review. IEEE Access **8**, 208016–208044 (2020). https://doi.org/10.1109/ACCESS.2020.3034141
28. van Hasselt, H., Guez, A., Silver, D.: Deep reinforcement learning with double Q-learning. In: Proceedings of the Thirtieth AAAI Conference on Artificial Intelligence, Phoenix, Arizona, pp. 2094–2100, February 2016
29. Wang, Z., Schaul, T., Hessel, M., van Hasselt, H., Lanctot, M., de Freitas, N.: Dueling network architectures for deep reinforcement learning. arXiv, arXiv:1511.06581, April 2016. https://doi.org/10.48550/arXiv.1511.06581
30. Saidallah, M., Fergougui, A.E., Elalaoui, A.E.: A comparative study of urban road traffic simulators. MATEC Web Conf. **81**, 05002 (2016). https://doi.org/10.1051/matecconf/20168105002
31. Cabrejas-Egea, A., Zhang, R., Walton, N.: Reinforcement learning for traffic signal control: comparison with commercial systems. Transp. Res. Procedia **58**, 638–645 (2021). https://doi.org/10.1016/j.trpro.2021.11.084
32. Wei, H., Zheng, G., Gayah, V., Li, Z.: A survey on traffic signal control methods. arXiv:1904.08117 [cs, stat], January 2020. http://arxiv.org/abs/1904.08117. Accessed 23 Feb 2022

Speech Emotion Recognition Using Pre-trained and Fine-Tuned Transfer Learning Approaches

Adil Chakhtouna[1](✉), Sara Sekkate[2], and Abdellah Adib[1]

[1] Team Computer Science, Artificial Intelligence and Big Data, MCSA Laboratory, Faculty of Sciences and Technologies of Mohammedia, Hassan II University of Casablanca, Mohammedia, Morocco
adilchakhtouna10@gmail.com, abdellah.adib@fstm.ac.ma
[2] Higher National School of Arts and Crafts of Casablanca, Casablanca, Morocco
sarasekkate@gmail.com

Abstract. Speech Emotion Recognition (SER) seems to be an exciting task due to its promising uses in many areas. A large number of studies have extracted acoustic descriptors from input speech and used them as features to train and test emotion recognition models. On the other hand, dimensional representations of speech signal such as spectrograms have been explored to appraise deep neural networks and more specifically the Convolutional Neural Network (CNN). To address this last issue, the current study aims to take advantage of the implementation of transfer learning architectures by using Mel spectrograms feature extraction mechanism to transform speech cues into images and then forward them to the pre-trained models. As well as make a comparative analysis of the performance using four different pre-trained architectures (VGG-16, VGG-19, EfficientNet B0 and EfficientNetV2 B0) both including and excluding the fine-tuning process. The experimental results were performed with two different configurations, pre-trained and fine-tuned models, the VGG-19 has obtained the highest accuracy of 79.83% on the held-out test. An evaluation carried out on the Italian Database of Elicited Mood in Speech (DEMoS) proves the effectiveness of the suggested model, which outperforms the state-of-the-art in SER.

Keywords: Transfer learning · Speech emotion recognition · Mel spectrogram · VGG-16 · VGG-19 · EfficientNet B0 · EfficientNetV2 B0 · DEMoS

1 Introduction

One of the primary forms of communication between humans around the world is speech. Emotions can be characterized throughout speech and they play an influential part in the human communication process. Speech Emotion Recognition

M. Ben Ahmed et al. (Eds.): SCA 2022, LNNS 629, pp. 365–374, 2023.
https://doi.org/10.1007/978-3-031-26852-6_35

(SER) appears on the horizon as a promising task by virtue of giving computers the ability to understand and recognize various emotional aspects in speech. The future advances in human-computer communication technologies are anticipated to enhance emotional competence through the diversity of its multiple pathways, such as body language [1], facial expressions, electroencephalography (EEG) [2] and speech [3].

· So far, SER has created a growing worldwide interest from researchers and scientists as a result of its significance in exploring the emotional aspects of humans and their relationship towards the development of a variety of multidisciplinary fields including psychological sciences, knowledge systems and health informatics. The critical issue facing investigators in the field of SER relates to the way of extracting reliable and discriminative features from the vocal content of a spoken utterance. This task has been approached by a large number of studies, each in its own way. Spectral and prosodic features including fundamental frequency, energy, zero-crossing rate, Mel Frequency Cepstral Coefficients (MFCC) and Linear Predictive Coefficients (LPC) are part of the most widely used ones in SER [4]. Furthermore, there are several kinds of machine learning techniques that are applied to understand the connection between the extracted features of speech and the predefined emotion labels. Regarding modeling approaches in SER, Support Vector Machine (SVM) [5], ensemble learning [6], Hidden Markov Model (HMM) [7] and, more recently, the involvement of deep neural networks such as Recurrent Neural Network (RNN) [8] and Convolutional Neural Network (CNN) [9] have been addressed.

New investigative approaches have been recently proposed to analyze speech signals based on the transfer learning technique. The primary goal of transfer learning is to take the learning gained from the baseline task and use it to help the target task learn. the current study aims to take advantage of the implementation of well-known transfer learning architectures and addresses the SER task as an image classification problem by using Mel spectrograms feature extraction mechanism to transform speech cues into images and then forward them to the pre-trained and fine-tuned models.

The body of this document is structured in the following manner: Sect. 2 briefly reviews the existing works in SER. Section 3 outlines the proposal's methodology. Section 4 explains in detail the different experiences and presents the main results achieved in this study. Finally, in Sect. 5, we summarize the most relevant findings and give future guidelines for investigation in SER.

2 Related Works

Over the last few decades, several surveys have been conducted on the SER, each according to its own point of view. The traditional pioneer approaches start by extracting key features from the speech signals and then ranking various classes of emotions depending on the feature values. Alternatively, emerging deep learning methods accomplish the SER task by merging the two steps into a single process. Authors in [10], introduce a multi-resolution feature extraction based on MFCC features, which are obtained from the Discrete Wavelet

Transform (DWT) sub-band coefficients and mixed with standard MFCC and pitch features. The feature set was reduced using Linear Discrimination Analysis (LDA) before its input fed into a naive Bayes classifier. The proposed methodology is evaluated on the German Emotional Speech Database (EMO-DB) with a variety of noise types such as restaurant, train, exhibition, street, car, babble and airport. The average accuracies of 80.10% and 89.09% were achieved in the speaker-independent and speaker-dependent configurations, respectively, at a Signal-to-Noise-Ratio (SNR) of 30 dB.

In [11], the author conducted a combination of deep features extracted from deep learning architectures including VGG-16, ResNet18, ResNet50, ResNet101, SqueezeNet, and DenseNet201 and acoustic features such as Root Mean Square energy (RMS), MFCC and zero-crossing rate. The ReliefF feature selection algorithm was used to choose the most effective features from the combined feature set. The chosen attributes were given to SVM-based modeling to perform the classification task. The highest accuracies achieved with ResNet101 deep features + acoustic features were 79.41% and 90.21% for the RAVDESS and EMO-DB datasets, respectively. While for the IEMOCAP database, an accuracy of 85.37% with VGG-16 deep features + acoustic features was achieved. In [12], an end-to-end trained model to recognize continuous emotions was employed, it consists of a stacked CNN with 2 layers of Long Short-Term Memory (LSTM). The idea is mainly based on retrieving features from the raw signal rather than on extracting hand-created features. The presented model outperforms the state-of-the-art on the RECOLA database having the concordance correlation coefficients of 0.815 and 0.502 for the excitation and valence dimensions, respectively, on the validation set.

The SER system proposed by Sun and Zou [13] was built with a DNN-decision tree SVM framework, the authors' contribution is based on the calculation of the degree of emotional confusion. At each leaf of the tree, different DNNs were trained to extract deep characteristics to train every SVM classifier in the decision tree. The experimental performance conducted on the Chinese CASIA database indicates that the overall recognition rate has increased by 2.91% and 6.25% compared to the DNN-SVM and conventional SVM classification schemes, respectively. To improve the efficiency of SER systems in cross-language and cross-corpus situations, a transfer learning method based on Deep Belief Network (DBN) has been proposed in [14]. The eGeMAPS [15] feature set extracted with openSMILE[1] toolbox were used. Different experimental setups including within corpus and multi-language schemes, were undertaken to test the performance of the DBN in comparison with the sparse autoencoder (AE) with SVM. Five corpora of German, English and Italian languages were investigated. The proposed method will prove helpful for developing strong SER system involving data from multiple languages.

The main contribution of this study is manifested by taking benefit of well-known CNN architectures such as VGG and EfficientNet, which are pre-trained on the large ImageNet database, and also addresses the issue of SER as an image

[1] https://www.audeering.com/research/opensmile/.

classification task. The impact of transfer learning in SER is demonstrated by performing the fine-tuning process.

3 Methodology

In this section, we propose a SER system that relies on the use of Mel spectrogram to model the speech signal as an image. The proposed SER framework has two major stages, as depicted in Fig. 1: it starts by transforming all speech utterances into Mel spectrogram representations and apply then transfer learning configurations to recognize several emotions. Further details are provided in the following subsections.

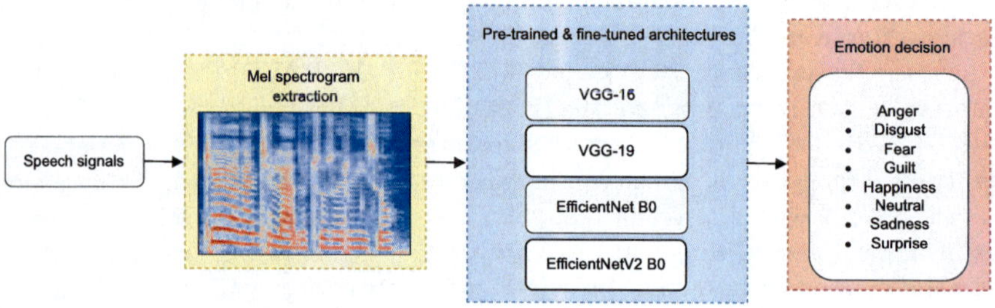

Fig. 1. Diagram of the proposed SER system.

3.1 Emotional Database

The use of database of Elicited Mood in Speech (DEMoS) [16] was proposed in this study. It is an extensive new database with an Italian emotional speech. The access to the data requires the signature of an academic license agreement. DEMoS was generated by 68 native speakers, including 23 females and 45 males, comprising 9697 induced emotional samples from eight emotions, the 'big six' according to Ekman [17]: happiness, anger, sadness, surprise, fear, disgust, the secondary emotion guilt and the neutral state.

3.2 Mel Spectrogram Extraction

From all previous work related to SER, we can infer that finding the appropriate input features can make a difference in the performance of emotion recognition. Speech properties can be modeled using a one-dimensional vector or two- or three-dimensional representations, depending on the type of considered features. Among the most common ways to simultaneously observe frequency, time, and intensity variations is via speech spectrogram, which maps the energy of speech over time and frequency. Mel spectrogram provide the time-frequency representation of speech signal.

To obtain the Mel spectrogram, the magnitude spectrogram is first computed, and then mapped onto the Mel scale. Furthermore, each speech signal is divided into equal number of frames, and each frame is produced by overlapping the preceding frame by half. The windowing process is applied to each frame in order to prevent the discontinuities arising at the extremities of each frame. The widely recommended Hamming window function is employed in this study. Hence, the Fourier spectrum is revealed on each shorter segment to transform it from time domain to frequency domain. Finally, the power spectrograms of each signal processed by the Fourier transform are extracted. In order to get the same dimension for all Mel spectrograms, short speech signals must be padded by a segment of their part to reach the maximum length of the longer statement in the dataset. Because the speech utterances in the DEMoS database have not the same length.

3.3 Transfer Learning Architectures

In our study, the aim of representing speech signals as Mel spectrograms (images) is to take advantage of the use of advanced deep learning algorithms dedicated to the field of computer vision, in particular CNN architectures. In this section, we discuss four existing CNN architectures: VGG-16, VGG-19, EfficientNet B0 and EfficientNetV2 B0.

A- VGG-16 and VGG-19

VGG-16 [18] and VGG-19 [18] are two commonly known CNN architectures used in a variety of image classification tasks. The principal concept behind the VGG model is to increase the overall performance of the network by expanding the depth of its layers, therefore its strategy relies on the application of convolution kernels to transform the input into several small volumes of kernels, thereby minimizing the number of parameters in the model, and allowing the network to be more discriminative.

The structure of the pre-trained VGG-16 architecture used on the ImageNet dataset, comprising a total of 21 main layers, includes 13 convolutional layers, 5 max pooling layers and 3 Fully Connected (FC) layers. In addition, a deeper network than the VGG-16, called the VGG-19, is introduced. The VGG-19 possesses 24 fundamental layers, with 16 convolutional layers, five max pooling layers and three FC layers.

B- EfficientNet B0 and EfficientNetV2 B0

EfficientNet B0 [19] provides a rule-based approach for scaling patterns and increasing the performance of the CNN. It is generally accepted that improving CNN performance necessitates dimensional scaling in terms of depth, width, or resolution. The basic EfficientNet B0 is then scaled using a compound scaling approach to produce a family of patterns from B1 to B7. EfficientNetV2 B0 [20], a novel class of CNN that exhibit faster learning speed and better parameter effectiveness than earlier models.

The two architectures contain 7 blocks. These blocks also vary in the number of sub-blocks, which increases as we move from EfficientNet B0 to EfficientNetV2 B0. As a result, the total number of layers in EfficientNet B0 is 237, whereas in EfficientNetV2 B0 is 255.

In this paper, we explore how well the four architectures including VGG-16, VGG-19, EfficientNet B0, and EfficientNetV2 B0 transfer to the SER classification task by comparing their performance with other related studies.

4 Experiments and Results

4.1 Implementation Details

In this subsection, we provide further details on the implementation parameters used in the different experimental setups. We employ the Google Colab environment to train several fine-tuned models. To secure the validity of our results, the Mel spectrogram images extracted from the DEMoS dataset were divided into 70% for training and 30% for testing. The architecture of the used FC layers added in the top of all the four pre-trained models. The FC layers block consists of three dense layers and and dropout layer, the detailed parameters of the FC layers are shown in Table 1.

Table 1. Parameters of the FC layers.

Layer	Parameter value	Activation function
Dense	512 (nodes)	Relu
Dropout	0.5 (probability)	–
Dense	256 (nodes)	Relu
Dense	8 (nodes)	Softmax

4.2 Fine-Tuned Models

The fine-tuning technique is the main part in the transfer learning approach. It allows to carefully choose the layers to be fine-tuned in the model to obtain a better result. To improve the accuracy of our pre-trained models, we conducted two configurations for fine tuning using our Mel spectrogram images for the SER classification task. In the first configuration (VGG-16-C1, VGG-19-C1, EfficientNet B0-C1, and EfficientNetV2 B0-C1), we froze all pre-trained layers of the used architectures and replaced the last FC layers used on the ImageNet dataset with our FC layers. Alternatively, a part of the pre-trained model can be considered in the fine-tuning stage together with the FC layers. In the second configuration (VGG-16-C2, VGG-19-C2, EfficientNet B0-C2, and EfficientNetV2 B0-C2), in addition to the FC layers, we have fine-tuned the fourth and fifth convolutional blocks of the VGG-16-C2 and VGG-19-C2. Furthermore, we fine-tuned 32 layers for both EfficientNet B0-C2 and EfficientNetV2 B0-C2.

4.3 Results and Discussion

In this section, we provide all the results obtained using our pre-trained models. Table 2 displays the results obtained on the test set using several measures, including precision, recall, F1-score, and accuracy. According to Table 2, the VGG-19-C2 architecture outperforms the other architectures in all evaluation metrics. It achieved the accuracy of 80% followed by the VGG-16-C2, EfficientNet B0-C2, EfficientNetV2 B0-C1, EfficientNet B0-C1, EfficientNetV2 B0-C2, VGG-19-C1 and VGG-16-C1 with respective accuracies of 71%, 68%, 66%, 65%, 64%, 57% and 55%. We can confirm that for all the pre-trained models, the second fine-tuning configuration gives better results compared to the first one, except for the EfficientNetV2 B0 model which has the opposite case.

Table 2. The obtained results of the pre-trained models using both fine-tuning configurations.

Model	Precision	Recall	F1-score	Accuracy
VGG-16-C1	0.57	0.53	0.54	0.55
VGG-16-C2	0.70	0.69	0.69	0.71
VGG-19-C1	0.59	0.55	0.56	0.57
VGG-19-C2	**0.79**	**0.80**	**0.79**	**0.80**
EfficientNet B0-C1	0.66	0.62	0.63	0.65
EfficientNet B0-C2	0.69	0.66	0.67	0.68
EfficientNetV2 B0-C1	0.70	0.63	0.65	0.66
EfficientNetV2 B0-C2	0.66	0.63	0.64	0.64

The confusion matrix of the best architecture among the used models (VGG-19-C2) is presented in Table 3. The most recognized samples for all emotions are placed on the diagonal of the confusion matrix. The recognition rates of the majority of the emotions were anticipated with remarkable percentages. Recall values of 79%, 85%, 78%, 73%, 79%, 80%, 83% and 81% were reached for anger, disgust, fear, guilt, happiness, neutral, sadness and surprise, respectively. The highest misclassified part comes from guilt emotion, in which 9% of its samples are incorrectly classified as sadness.

4.4 Comparison with Other Studies

In this section, the only related work we found for the DEMoS dataset was proposed by [21], whose model was trained on the DEMoS dataset without the

Table 3. The confusion matrix of the VGG-19-C2 architecture.

Emotion	Anger	Disgust	Fear	Guilt	Happiness	Neutral	Sadness	Surprise
Anger	**0.79**	0.06	0.05	0.01	0.02	0.01	0.03	0.02
Disgust	0.03	**0.85**	0.02	0.02	0.03	0.01	0.03	0.01
Fear	0.03	0.05	**0.78**	0.03	0.04	0.03	0.03	0.02
Guilt	0.01	0.07	0.03	**0.73**	0.06	0.01	0.09	0.01
Happiness	0.02	0.06	0.02	0.03	**0.79**	0.02	0.04	0.02
Neutral	0.01	0.02	0.03	0.05	0.04	**0.80**	0.01	0.04
Sadness	0.04	0.05	0.02	0.03	0.02	0.01	**0.83**	0.01
Surprise	0.03	0.03	0.03	0.01	0.03	0.01	0.04	**0.81**

neutral emotion. To make a fair comparison, we repeated all the experiments without considering the neutral state, then we chose the best pre-trained architecture among the four used models for comparison with the existing work.

Gerczuk et al. [21] proposed to train the ResNet architecture from scratch on Mel spectrograms extracted from the DEMoS database. They attained a 73.8% recognition rate on the test set. In contrast, we obtained a superior performance of 79.7% in our work. It was better if the authors had employed the fine-tuning technique on the ResNet architecture, this might improve the recognition rate of their study (Table 4).

Table 4. Results obtained by our proposed method and other state-of-the-art work for SER.

Work	Model	Training technique	Dataset	Feature	Accuracy
[21]	ResNet	Trained from scratch	DEMoS without neutral	Mel spectrogram	73.8%
Our	VGG-19-C2	Transfer learning	DEMoS without neutral	Mel spectrogram	**79.7%**

5 Conclusion

In this study, we approached the SER task based on the use of Mel spectrogram feature represented as an image. This approach allowed us to overcome the SER challenge by applying the transfer learning approach that enables us to leverage the use of pre-trained models such as VGG-16, VGG-19, EfficientNet B0 and EfficientNetV2 B0. Depending on the experimental results, by using the fine-tuning technique, the second fine-tuned configuration of the VGG-19 architecture achieved the highest recognition rate compared to the other architectures.

Acknowledgements. This work was supported by the Ministry of Higher Education, Scientific Research and Innovation, the Digital Development Agency (DDA) and the CNRST of Morocco (Alkhawarizmi/2020/01).

References

1. Noroozi, F., Marjanovic, M., Njegus, A., Escalera, S., Anbarjafari, G.: Audio-visual emotion recognition in video clips. IEEE Trans. Affect. Comput. **10**(1), 60–75 (2017)
2. Tan, Y., Sun, Z., Duan, F., Solé-Casals, J., Caiafa, C.F.: A multimodal emotion recognition method based on facial expressions and electroencephalography. Biomed. Signal Process. Control **70**, 103029 (2021)
3. Sekkate, S., Khalil, M., Adib, A., Ben Jebara, S.: An investigation of a feature-level fusion for noisy speech emotion recognition. Computers **8**(4), 91 (2019)
4. Chakhtouna, A., Sekkate, S., Adib, A.: Improving speech emotion recognition system using spectral and prosodic features. In: Abraham, A., Gandhi, N., Hanne, T., Hong, T.-P., Nogueira Rios, T., Ding, W. (eds.) ISDA 2021. LNNS, vol. 418, pp. 399–409. Springer, Cham (2022). https://doi.org/10.1007/978-3-030-96308-8_37
5. Al Dujaili, M.J., Ebrahimi-Moghadam, A., Fatlawi, A.: Speech emotion recognition based on SVM and KNN classifications fusion. Int. J. Electr. Comput. Eng. **11**(2), 1259 (2021)
6. Zehra, W., Javed, A.R., Jalil, Z., Khan, H.U., Gadekallu, T.R.: Cross corpus multilingual speech emotion recognition using ensemble learning. Complex Intell. Syst. **7**(4), 1845–1854 (2021)
7. Schuller, B., Rigoll, G., Lang, M.: Hidden markov model-based speech emotion recognition. In: 2003 IEEE International Conference on Acoustics, Speech, and Signal Processing, 2003. Proceedings.(ICASSP2003). vol. 2, pp. II–1. IEEE (2003)
8. Li, D., Liu, J., Yang, Z., Sun, L., Wang, Z.: Speech emotion recognition using recurrent neural networks with directional self-attention. Expert Syst. Appl. **173**, 114683 (2021)
9. Qayyum, A.B.A., Arefeen, A., Shahnaz, C.: Convolutional neural network (CNN) based speech-emotion recognition. In: 2019 IEEE International Conference on Signal Processing, Information, Communication & Systems (SPICSCON), pp. 122–125. IEEE (2019)
10. Sekkate, S., Khalil, M., Adib, A., Ben Jebara, S.: A multiresolution-based fusion strategy for improving speech emotion recognition efficiency. In: Renault, É., Boumerdassi, S., Leghris, C., Bouzefrane, S. (eds.) MSPN 2019. LNCS, vol. 11557, pp. 96–109. Springer, Cham (2019). https://doi.org/10.1007/978-3-030-22885-9_10
11. Er, M.B.: A novel approach for classification of speech emotions based on deep and acoustic features. IEEE Access **8**, 221640–221653 (2020)
12. Tzirakis, P., Zhang, J., Schuller, W.: End-to-end speech emotion recognition using deep neural networks. In: 2018 IEEE International Conference on Acoustics, Speech and Signal Processing (ICASSP), pp. 5089–5093. IEEE (2018)
13. Sun, L., C, B., Fu, S., Chen, J., Wang, F.: Speech emotion recognition based on DNN-decision tree SVM model. Speech Commun. **115**, 29–37 (2019)
14. Latif, S., Rana, R., Younis, S., Qadir, J., Epps, J.: Transfer learning for improving speech emotion classification accuracy. arXiv preprint arXiv:1801.06353 (2018)
15. Eyben, F., et al.: The geneva minimalistic acoustic parameter set (GeMAPS) for voice research and affective computing. IEEE Trans. Affect. Comput. **7**(2), 190–202 (2015)
16. Parada-Cabaleiro, E., Costantini, G., Batliner, A., Schmitt, M., Schuller, B.W.: DEMoS: an italian emotional speech corpus. Lang. Resour. Eval. **54**(2), 341–383 (2020)

17. Ekman, P.: Expression and the nature of emotion. Approaches Emot. **3**(19), 344 (1984)
18. Simonyan, K., Zisserman, A.: Very deep convolutional networks for large-scale image recognition. arXiv preprint arXiv:1409.1556 (2014)
19. Tan, M., Le, Q.: EfficientNet: rethinking model scaling for convolutional neural networks. In: International Conference on Machine Learning, pp. 6105–6114. PMLR (2019)
20. Tan, M., Le, Q.: EfficientNetv2: smaller models and faster training. In: International Conference on Machine Learning, pp. 10096–10106. PMLR (2021)
21. Gerczuk, M., Amiriparian, S., Ottl, S., Schuller, B.W.: EmoNet: a transfer learning framework for multi-corpus speech emotion recognition. IEEE Trans. Affect. Comput. arXiv:2103.08310 (2021)

A New Modelization of Concentrated/Simple Industrial InGaP/InGaAs/Ge and Silicon Solar Cells: Insights on High Leverage Variables

Zineb Aqachmar[1](✉), Tarik Bouhal[2], Hamid El Qarnia[1], Mohammed Boukendil[1], ElAlami Ibnouelghazi[1], and Abdelkader Outzourhit[1]

[1] Faculty of Sciences Semlalia, Department of Physics, Cadi Ayyad University, 40000 Marrakesh, Morocco
zineb.aqachmar@uca.ac.ma, zaynab.aqachmar@gmail.com
[2] ALTES Energy, Alternative Energy Solutions, Zagora, Morocco

Abstract. This paper aims plotting P-V curves of industrial solar cells (Concentrated InGaP/InGaAs/Ge; Simple InGaP/InGaAs/Ge; Concentrated Silicon; Silicon) under different parameters using the most updated mathematical modelization of a solar cell which is tackled in the first time in the literature. The second objective is finding high leverage parameters that influences the most the power at the output of the considered solar cells. It was found that the concentration ratio impacts the most concentrated solar cells while the irradiance influences the most solar cells without concentrators. An increase of 10% of the concentration ratio makes the power at the output of the concentrated InGaP/InGaAs/Ge/ Concentrated Silicon solar cells increasing by 3.7 W and 1.06 W respectively. Besides, an increase of 10% of the irradiance makes the power of the triple junction without concentrator rising by an average of 0.05 W. It is also shown that all the studied solar cells are the less sensitive to the series resistance.

Keywords: Photovoltaics · Silicon · CPV · P-V curves · Optimization

1 Introduction

Solar energy has the potential to play a pivotal role in the global energetic transition worldwide due to the huge amount of available solar resources. Besides, electrical demand is constantly increasing due to the growing usage of electrical devices such as electrical cars and processes using electricity as desalination. Moreover, solar energy has the advantage to fulfill the second objective of decreasing greenhouse gases caused by conventional fossil sources of energy. Furthermore, solar energy allows avoiding the risks associated to nuclear energy as the major accident of Fukushima (Japan) that caused the death of 18000 people [1]. The challenge nowadays is to develop low-cost operational devices for solar energy conversion and complementary technologies [2].

Many numerical studies assessing and optimizing the maximal power produced by solar cells are available in the literature. Rezk et al. [3] proposed a model to evaluate the I-V and P-V curves of triple junction solar cells using one diode modelization of the solar

M. Ben Ahmed et al. (Eds.): SCA 2022, LNNS 629, pp. 375–387, 2023.
https://doi.org/10.1007/978-3-031-26852-6_36

cells. Their model was able to improve the converted energy per day from 3.37 kWh to 3.75 kWh. Another study was conducted by Anaty et al. [4] to modelize and simulate a concentrated triple junction solar cell using one diode modelization of the solar cell. They plotted J-V curves and P-V curves under different temperatures and concentration ratios and found that the maximal power point of their device drops from 19.26 W to 18.58 W when the temperature increases from 25 °C to 65 °C. Besides, Hussain et al. [5] developed a model of a simple and triple solar junction (InGaP/InGaAs/Ge) using one diode modelization. They found that changing the irradiance from 1 to 100 suns makes the open circuit voltage change by 11.6%. Furthermore, Das et al. [6] simulated a single solar cell, double solar cell and triple solar cell without concentrator using two diode modelization and assesses the effect of the temperature of the cell and the irradiance on the I-V curves and P-V curves. Moreover, Segev et al. [7] proposed mathematical models to evaluate the current produced by a concentrated triple junction solar cell using one diode and two diode modelizations. They made an experimental validation and found a RMSE lower than 2.5%. Another study was conducted by Aqachmar et al. [8] to study the effect of different parameters on a concentrated silicon solar cell. They used a single diode model and found that an increase of the temperature of the cell from 20 °C to 90 °C decreases the maximal power produced from 8.89 W to 6.79 W. A shaded solar cell was studied by Hofer et al. [9] using one diode modelization of a solar cell. They found that the orientation of the tracking axis strongly influences timing and spatial distribution of shading patterns, and thus total performances of the solar cell.

In this work, the latest mathematical model of a solar cell is used on the following industrial solar cells:

- Concentrated InGaP/InGaAs/Ge: inspired from Arzon solar that installed the most important capacity of concentrated photovoltaic power plants [10];
- InGaP/InGaAs/Ge: inspired from Arima group, a major manufacturer of InGaP/InGaAs/Ge [11];
- Concentrated silicon solar cell: where the expensive triple junction InGaP/InGaAs/Ge solar cell is replaced by the available and easy to manufacture silicon. It is used in low cost concentrated projects as LOUCOUM project in Morocco and studied in many research centers as the IRESEN in Benguerir in Morocco;
- A simple silicon solar cell to validate the model. This cell is inspired from Sunpower, a pioneer and leader company in the manufacturing of solar cells.

The P-V curves of the described solar cells are plotted under different irradiances, temperature of the cells, series resistances, shunt resistances, and concentration ratio to compare their performances. Furthermore, another objective of the paper is to make a sensitivity study to find high leverage variables that affect the most the power produced by the cells which may help scientists focusing their research axis on the criteria that achieve optimum performances.

The studied solar cells can be used in the international strategy about smart grids as the process of manufacturing of these cells is already mastered by the cited companies in the previous paragraph. Moreover, through this work, smart energy can be converted in the most efficient way by the identification of the most impacting parameters.

2 Materials and Methods

2.1 Material Used

The solar cells used for the study are InGaP/InGaAs/Ge, Si, Concentrated InGaP/InGaAs/Ge, and concentrated Si.

The studied InGaP/InGaAs/Ge is inspired from Arzon Solar, as this company installed the most important capacity of concentrated InGaP/InGaAs/Ge powerplants [2]. Moreover, it has one of the best efficiencies among industrial concentrated solar cells. The top junction (InGaP) has a bandgap of 1.86 eV, the middle junction (InGaAs) has a bandgap of 1.4 eV and the bottom junction (Ge) has a bandgap of 0.65 eV. The differences of bandgaps allow the maximization of absorbed photons. Furthermore, these three junctions are lattice-matched (the atomic spacing of the used materials is almost constant which achieves a maximal conductivity of current). The National Renewable Energy Laboratory (NREL) mentioned that $\pm 0.01\%$ mismatch on the lattice matching engenders an important degradation of the quality of the photovoltaic solar cell, and therefore, to energy conversion [6].

The physical characteristics of the InGaP/InGaAs/Ge are detailed in Table 1.

The used Silicon solar cell is manufactured by Sunpower as it is a pioneering company in the field of photovoltaic solar cells and still has a big share of the market [12]. The details of this cell are presented in Table 2.

In this study, we use a second diode in the modelization of the solar cell as it represents the recombination in the junction at low voltages [13], therefore results are found in better accuracies. The equivalent circuit of the InGaP/InGaAs/Ge and the Silicon junctions are displayed on Fig. 1.

Table 1. Characteristics of the InGaP/InGaAs/Ge solar cells [3]

Physical characteristic	InGaP	InGaAs	Ge
E_g(eV) at 298 K	1.976	1.519	0.744
Isc(mA)	6.7522	7.7126	10.094
K (A/cm^{2}K^4)	$1.86.10^{-9}$	$1.288.10^{-8}$	$10.5.10^{-6}$
n	1.97	1.75	1.96
γ	2	2	2
α	$7.5.10^4$	$5.405.10^4$	$4.774.10^4$
β	500	204	235

2.2 Mathematical Modelization

Concentrated/Simple Triple-Junction Solar Cell

In this study, we use the most updated and most accurate modelization of a concentrated solar cell that is described in what follows. The simulations are made under Matlab/Simulink environment.

Table 2. Characteristics of the silicon solar cell used [14]

Physical characteristic	Value
Open circuit voltage	0.665 V
Short circuit current	5.75 A
Energy gap of the crystalline Silicon	1.12 eV
Quality factor of the diode	1.2
Temperature coefficient of the short circuit current	0.07%/K
Temperature coefficient of the open circuit voltage	−0.29%/K
Temperature coefficient of the maximal power point	−0.39%/K

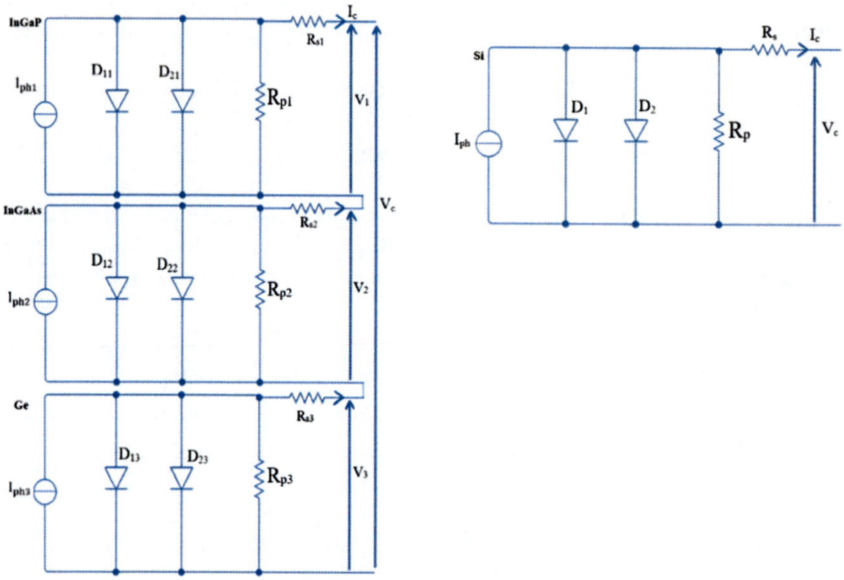

Fig. 1. Double diode modelization of the studied triple-junction solar cell and single solar cell

Referring to Fig. 1, the current produced by the solar cell I_c can be represented by the following equation [15]:

$$I_c = I_{phi} - I_{si}(T) \times \left(e^{\frac{(V_i+I_c\times R_{si})}{N_i\times V_{ti}}} - 1\right) - I_{s2i}(T) \times \left(e^{\frac{(V_i+I_c\times R_{si})}{N_{2i}\times V_{ti}}} - 1\right) - \left(\frac{V_i + I_c \times R_{si}}{R_{shi}}\right) \quad (1)$$

where i = 1 for InGaP, i = 2 for InGaAs, and n = 3 for Ge, I_{phi} is the solar induced current by the cell number i and is detailed in Eq. 2, $I_{si}(T)$ is the saturation current of the first diode depending on the temperature detailed in Eq. 3, Vi is the voltage at the output of the cell, R_{si} is the series resistance of the studied cell, N_i is the diode emission coefficient (quality factor) of the first diode, V_{ti} is the thermal voltage detailed in Eq. 4,

$I_{s2i}(T)$ is the saturation current for the second diode that is detailed in Eq. 5, N2i is the diode emission coefficient (quality factor) of the second diode, and R_{shi} is the shunt resistance (in parallel).

The solar induced current by the cell is given by the following Eq. 2:

$$I_{phi} = C \times I_{ph0} \times \frac{I_r}{I_{r0}} \times (1 + T_{ci} \times (T_i - T_{ref})) \tag{2}$$

where C is the concentration ratio (is taken equal to one for a the InGaP/InGaAs/Ge junction without concentrator), I_{ph0} is the measured solar-generated current for the irradiance for the reference irradiance I_{r0} (1000 W/m^2), I_r is the irradiance (W/m^2) received by the cell, T_{ci} is the temperature coefficient of the solar induced current I_{phi}, T_i is the temperature of the cell, and T_{ref} is the reference temperature (298K).

The saturation current of the first diode is expressed by Eq. 3:

$$I_{si}(T) = I_{si} \times \left(\frac{T_i}{T_{ref}}\right)^{\left(\frac{T_{cis}}{N_i}\right)} \times e^{\left(\frac{E_{gi} \times \left(\frac{T_i}{T_{ref}} - 1\right)}{N_i \times V_{ti}}\right)} \tag{3}$$

where I_{si} is the saturation current of the first diode, T_{cis} is the temperature coefficient for I_s, and E_{gi} is the energy gap of the cell and is detailed in Eq. 6.

The thermal voltage is given by the following Eq. 4:

$$V_{ti} = \frac{k \times T_i}{q} \tag{4}$$

where *k* is the Boltzmann constant, T_i is the cell's temperature, and *q* is the elementary charge on an electron.

The saturation current of the second diode is expressed by Eq. 3:

$$I_{s2i}(T) = I_{s2i} \times \left(\frac{T_i}{T_{ref}}\right)^{\left(\frac{T_{cis2}}{N_{2i}}\right)} \times e^{\left(\frac{E_{gi} \times \left(\frac{T_i}{T_{ref}} - 1\right)}{N_{2i} \times V_{ti}}\right)} \tag{5}$$

where I_{s2i} is the saturation current of the second diode, and T_{cis2} is the temperature coefficient for I_{s2}.

The energy gap of a cell is given by the following Eq. 6 [3]:

$$E_{gi}(T) = E_{gi}(298) + \frac{\alpha_i \times T_i^2}{T_i + \beta_i} \tag{6}$$

where $E_{gi}(298)$ is the energy gap of the cell at standard temperature of 298 K (corresponding to 25 °C), α_i and β_i are material dependent constants described in Table 1.

The voltage at the output of whole cell is given in the next Eq. 7:

$$V_c = \sum\nolimits_i V_i \tag{7}$$

The power at the output of the cell is given by the following Eq. 8:

$$p_c = I_C \times V_C \tag{8}$$

Concentrated/Simple Silicon Solar Cell
By modifying the factor i used in the previous equations to Silicon, we obtain the modelization of the concentrated silicon cell. In this case, there is no need to sum the voltages as there is only one voltage.

3 Results and Discussion

3.1 Parametric Study

Based on the described mathematical model, performances of the described cells can be drawn in terms of generated current, voltage and power. To produce a comparative study, disparate P-V curves are analyzed. The subsequent results demonstrate how different temperatures of the cell, irradiances, series resistances, shunt resistances, and concentrated ratios influences the concentrated InGaP/InGaAs/Ge, simple InGaP/InGaAs/Ge, concentrated silicon, and silicon solar cells.

Figure 2 is the comparative study of the concentrated InGaP/InGaAs/Ge solar cell. It shows that increasing the temperature of the cell from 20 °C to 60 °C increases the decreases the voltage. One can observe the inverted correlation between the power produced and the temperature of the cell. The power generated by the considered solar cell decreases from 5.9 W to 5 W while increasing the cell's temperature. Moreover, increasing the irradiance from 50 W/m^2 to 250 W/m^2 enhances the voltage and power produced. Concretely, the power produced is augmented from 1.27 W to 7.39 W. This late effect is observed for the concentrated ratio while enhancing the concentration ratio from 50 to 1000 increases the generated power from 55.53 W to 145.3 W. Furthermore, increasing the value of the parallel resistance (named also shunt resistance) from 10 Ω to 1010 Ω augments the power generated from 5.8 W to 5.9 W. Besides, there is an inverted relation between the series resistance and the power produced as increasing the series resistance from 0 to 0.04 Ω reduce the power from 5.9 W to 4.6 W.

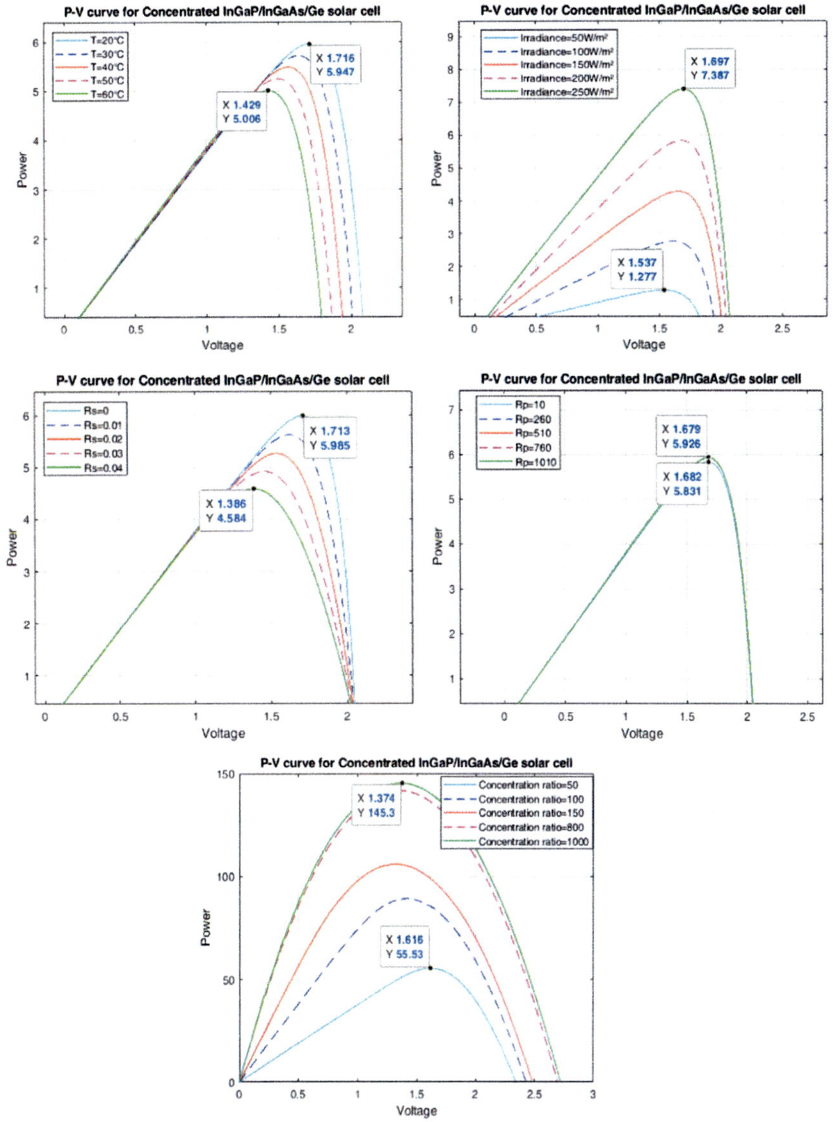

Fig. 2. P-V curves of concentrated InGaP/InGaAs/Ge under parameters

Figure 3 exhibits the effects of the temperatures of the cell, irradiances, series resistances and shunt resistances on the P-V curves of a InGaP/InGaAs/Ge without concentrator. Rising the cell's temperature from 20 °C to 60 °C decreases the voltage from 1.86 V to 1.52 V and the generated power from 1.013 W to 0.83 W. Additionally, P-V curves are plotted for irradiance from 50 W/m^2 to 250m^2. A shy P-V curve is illustrated for 100 W/m^2 with a maximal power of 0.4 W. At 250 W/m^2, more confirmed P-V curves are remarked. Furthermore, increasing R_s from 0 Ω to 0.04 Ω decreased the produced power from 0.99 W to 0.94 W. While rising R_p from 10 Ω to 1010 Ω augments power from 0.99 W to 1.12 W.

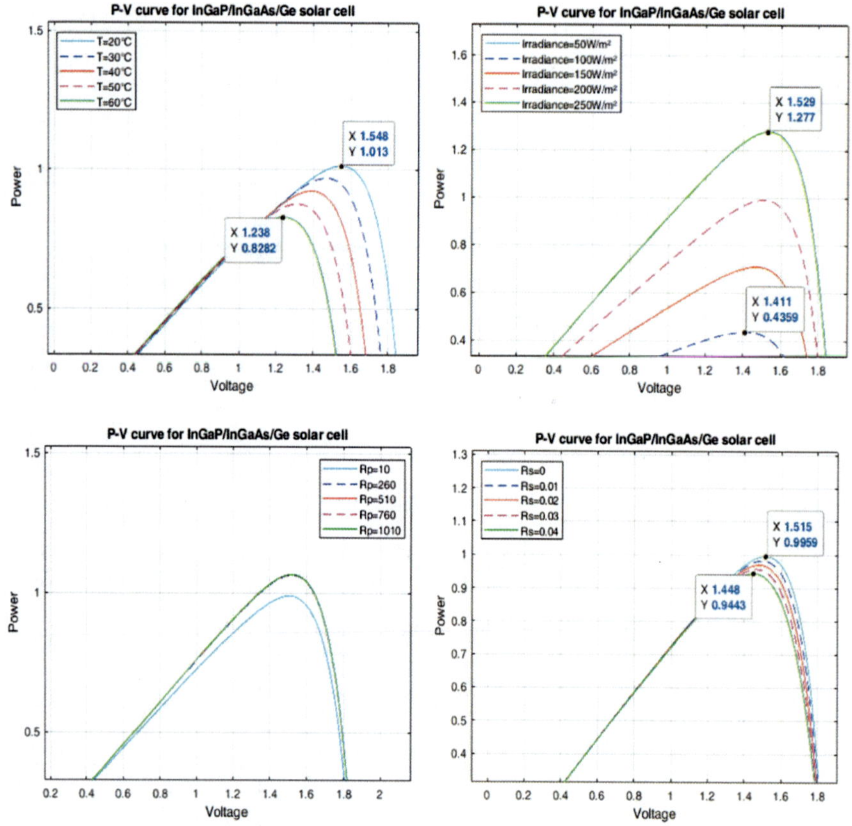

Fig. 3. P-V curves of InGaP/InGaAs/Ge under different parameters

Figure 4 illustrate the effect of the variation of the considered parameters on a low cost concentrated photovoltaic solar cell (with Silicon solar cell) while Fig. 5 shows the behavior of a simple silicon solar cell to highlight the differences between using or not a concentrator. Increasing the irradiance from 50 W/m^2 to 250 W/m^2 increases the voltage and the power. It is shown that the power of the concentrated silicon solar cell rises from 0.39 W to 2.25 W while the power of the silicon cell goes from 0.056 W to 0.39 W. From the one hand, varying the series resistance from 0 to 0.04 Ω decreases the power of the concentrated silicon solar cell from 1.8 W to 1.37 W and the power of simple silicon cell from 0.3 W to 0.29 W. From the other hand, increasing the value of the shunt resistance (or parallel resistance) from 10 Ω to 1010 Ω rises the value of the produced power of the concentrated silicon solar cell from 1.779 W to 1.806 W and the produced power from the silicon solar cell from 0.3 W to 0.32 W. Finally, varying the concentration ratio from 50 to 1000 sunbeams, improves the power of the concentrated silicon solar cell from 16.55 W to 40.7 W.

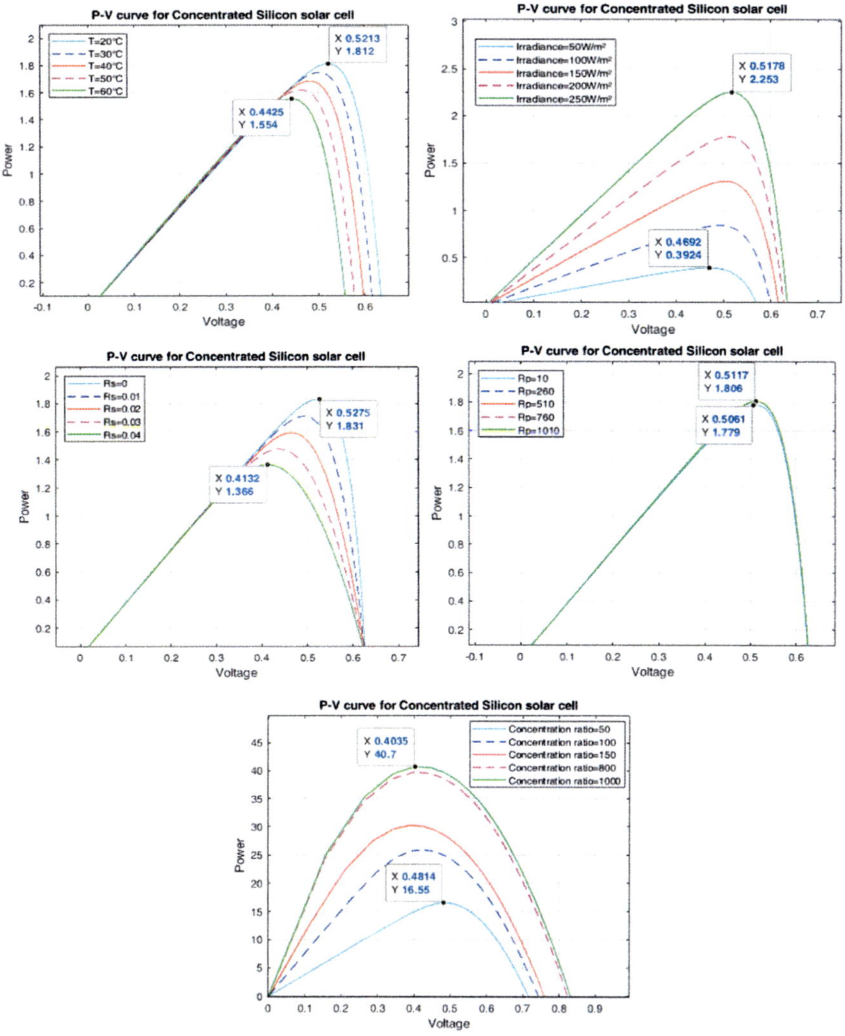

Fig. 4. P-V curves of concentrated Silicon cells under different parameters

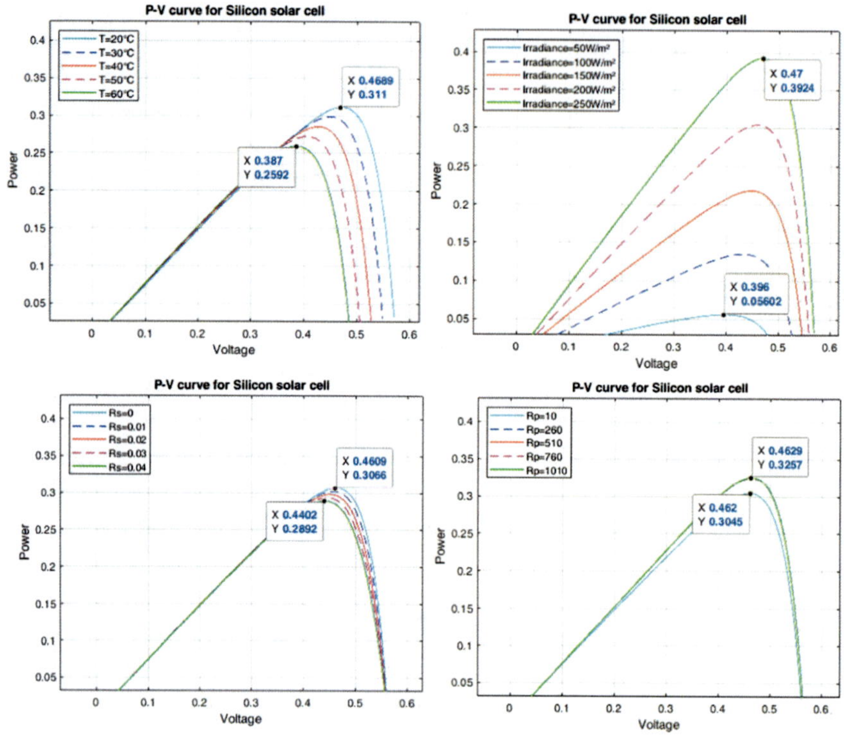

Fig. 5. P-V curves of Silicon cells under different parameters

3.2 Sensitivity Study

A sensitivity study is a mighty methodology to assess the influence of the considered parameters on the uncertainty of the indicator of performance. It exhibits the impact of the modification of uncorrelated inputs on correlated parameters under a set of suppositions. By using this powerful calculation, distinguishing the most sensitive parameters is realizable. A sensitivity study permit detecting the high leverage variables from low leverage variables. Choosing carefully high leverage variables allow optimizing the parameters that impacts the most the solar cell.

In this part, the sensitivity of the produced power to the studied criteria is emphasized. The previous section show that the maximal power produced by a solar cell depends on the temperature of the cell, the irradiance, the series resistance, the shunt resistance and the concentration ratio. Therefore, the sensitivity of the produced power to these parameters is investigated and displayed on Fig. 6 by varying the input parameters by 10%. The results obtained demonstrate that the concentrated triple junction is the most sensitive to the concentration ratio, as an average variation of 10% of the concentration ratio changes the produced power by 3.7 W while it is the less sensitive to the variation of the series resistance as a variation of the series resistance by 10% make the power of the solar cell varying by 0.03 W. The simple triple junction is the most sensitive to

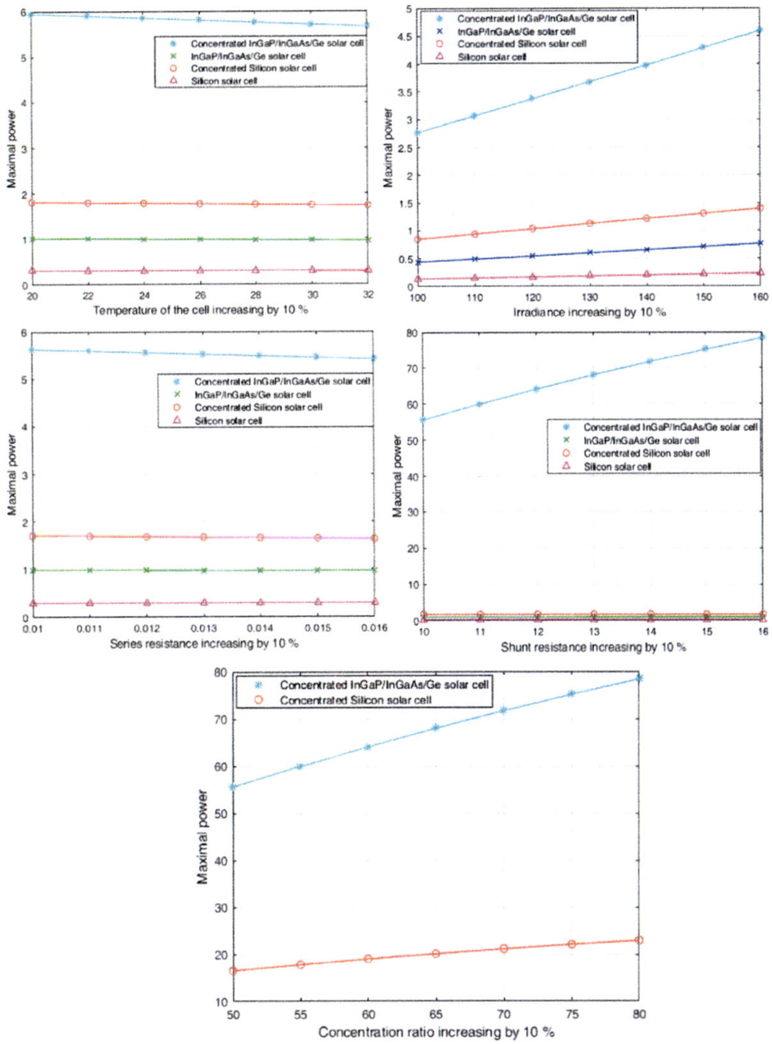

Fig. 6. Sensitivity of the maximal power produced by the studied solar cells with a variation of temperatures of the cell, irradiances, series resistances, shunt resistances, and concentration ratios of 10%

the irradiance, a rise of 10% of this criterion makes the power of the cell increasing by an average of 0.05 W and is the less sensitive to the series resistance. The concentrated silicon solar cell is also the most sensitive to the concentration ratio as an increase of 10% of the concentration ratio rise the power produced by an average of 1.06 W, the series resistance is the less impacting parameter the concentrated silicon solar cell. Finally, the simple silicon solar cell is the most sensitive to the irradiance and the less sensitive to the series resistance. Hence, the manufacturers of concentrated triple junction solar cells and low cost concentrated solar cells should undertake carefully the optimization and the maximization of the concentration ratio. While people involved in the industrialization of simple triple junction solar cells and silicon solar cells should work on the maximization of the irradiance received by the cells by choosing the sites with the highest irradiances and the materials that absorbs the most sun beams.

4 Conclusion

The latest mathematical modelization of a solar cell is used in this work to plot P-V curves of industrial Concentrated InGaP/InGaAs/Ge, Simple InGaP/InGaAs/Ge, Concentrated Silicon and Silicon. The effect of irradiance, series resistance, parallel resistance, concentration ratio and temperature of the cell were all considered. Moreover, a sensitivity study was conducted in order to find the most influencing parameters. This last analysis makes it possible to deduce that concentration ratio affects the most the concentrated solar cells while the irradiance influences the most solar cells without concentrator. Furthermore, series resistance is the less impacting parameter on the power produced by all studied industrial solar cells. The study makes possible to deduce that the irradiance and the concentration ratio should be studied rigorously studied during the industrialization and installation of PV and CPV power plants. This sensitivity study is useful for smart energy conversion. Moreover, the studied solar cells are applicable in the smart grids.

References

1. Greenpeace: Catastrophe nucléaire de Fukushima: quelles conséquences? https://www.greenpeace.fr/catastrophe-nucleaire-de-fukushima-quelles-consequences/. Accessed 14 Aug 2022
2. Aqachmar, Z., et al.: Electrification of Africa through CPV installations in small-scale industrial applications: energetic, economic, and environmental analysis. Renew. Energy. **197**, 723–746 (2022). https://doi.org/10.1016/j.renene.2022.07.106
3. Rezk, H., Hasaneen, E.-S.: A new MATLAB/Simulink model of triple-junction solar cell and MPPT based on artificial neural networks for photovoltaic energy systems. Ain Shams Eng. J. **6**, 873–881 (2015). https://doi.org/10.1016/j.asej.2015.03.001
4. Anaty, M.K., Belkasmi, M., Bouziane, K., Aggour, M., El Ouahabi, M.: Modeling and simulation of a C3MJ+ triple junction solar cell using Matlab/Simulink. In: 2016 International Renewable and Sustainable Energy Conference (IRSEC), pp. 752–757. IEEE (2016)
5. Hussain, A.B., Abdalla, A.S., Mukhtar, A.S., Elamin, M., Alammari, R., Iqbal, A.: Modelling and simulation of single- and triple-junction solar cells using MATLAB/SIMULINK. Int. J. Ambient Energy **38**, 613–621 (2017). https://doi.org/10.1080/01430750.2016.1181567
6. Das, N., al Ghadeer, A., Islam, S.: Modelling and analysis of multi-junction solar cells to improve the conversion efficiency of photovoltaic systems. In: 2014 Australasian Universities Power Engineering Conference (AUPEC), pp. 1–5. IEEE (2014). https://doi.org/10.1109/AUPEC.2014.6966482
7. Segev, G., Mittelman, G., Kribus, A.: Equivalent circuit models for triple-junction concentrator solar cells. Sol. Energy Mater. Sol. Cells **98**, 57–65 (2012). https://doi.org/10.1016/j.solmat.2011.10.013
8. Aqachmar, Z., et al.: Modelization and simulation of a low cost concentrated photovoltaic solar cell: parametric and sensitivity study under MATLAB. Instrum. Mesure Métrologie **21**, 1–6 (2022). https://doi.org/10.18280/i2m.210101

9. Hofer, J., Groenewolt, A., Jayathissa, P., Nagy, Z., Schlueter, A.: Parametric analysis and systems design of dynamic photovoltaic shading modules. Energy Sci. Eng. **4**, 134–152 (2016)
10. Wiesenfarth, M., Philipps, S.P., Bett, A.W.: Current status of concentrator photovoltaic (CPV) technology (2017)
11. Arima-Group: Triple Junction Solar Cell (InGaP/InGaAs/Ge). http://www.arima.com.tw/en/group1-detail.php?index_m1_id=1&index_m2_id=6&index_id=3. Accessed 14 Aug 2022
12. Share market Photovoltaic. https://www.mordorintelligence.com/industry-reports/solar-energy-market. Accessed 14 Aug 2022
13. PV Education: Double Diode Model. https://www.pveducation.org/pvcdrom/characterisation/double-diode-model#:~:text=The%20single%20diode%20equation%20assumes,factor%20is%20close%20to%20one. Accessed 14 Aug 2022
14. Sunpower: Sunpower solar cell. https://us.sunpower.com/sites/default/files/sp-gen2-solar-cell-ds-en-a4-160-506760e_0.pdf. Accessed 14 Aug 2022
15. MATLAB: Solar cell modelization. https://fr.mathworks.com/help/physmod/sps/ref/solarcell.html?searchHighlight=solar%20cell&s_tid=srchtitle_solar%20cell_1. Accessed 15 Aug 2022

Learning Theories and Technology Adoption Models: A Review

Lubabalo Mbangata[1] and Abdultaofeek Abayomi[2(✉)]

[1] Department of Information Systems, Durban University of Technology, Durban 4000, South Africa
lubabalom1@dut.ac.za

[2] Department of Information and Communication Technology, Mangosuthu University of Technology, Jacobs, Durban 4026, South Africa
abayomi.abdultaofeek@mut.ac.za

Abstract. E-learning concepts and strategies in smart learning and smart education are very important elements in smart city paradigm. The adoption of technologies including those related to e-learning has been studied for decades to understand their adoption in different contexts. There exist different theorists that construct the classifications that were founded on three basic approaches while dealing with learning theories. The aim of this current study is to critically review and compare technology adoption models and learning theories as well as their classifications with regards to technology adoption for a learning process to occur. The pros and cons of the technology adoption theories are also discussed. The technology adoption models are classified according to three categories including the models focusing on the antecedents or determinants of the adoption of e-learning, the models focusing on e-learning usage continuance behaviour, and the models focusing on e-Learning adoption outcome.

Keywords: Technology adoption · Adoption models · e-Learning · Learning theories · Smart education

1 Introduction

Over the past decades the learning process has been investigated by numerous studies with various theories and models being used to define and describe it. In addition, several theories and models such as the technology acceptance (TAM) model [1] have also been developed in the field of information systems (IS) to explain technology adoption and acceptance. However, the adoption of technology in the education sector can be mainly for teaching and learning particularly in the context of e-Learning.

As the specific contexts of learning become even more crucial in academia, how learning phenomena could be abstracted into a learning theory is a matter for further exploration [2]. Thus, the pertinent question is how learning can be viewed from the perspective of classification of learning theories and models especially in the context of technology adoption in e-Learning. Learning theorists construct the classification founded on three basic approaches while dealing with learning theories. These three

M. Ben Ahmed et al. (Eds.): SCA 2022, LNNS 629, pp. 388–403, 2023.
https://doi.org/10.1007/978-3-031-26852-6_37

classifications are behaviourist, cognitive, and constructivism constructs. On the other hand, it is difficult to classify these theories in this context, with some other theories that can be involved in more than one class in different ways [3]. Therefore, the aim of this current study is to critically review how technology adoption models and learning theories can be classified with regards to technology adoption for learning to enable learning process to occur. To achieve this, the existing learning theories and models used for technology adoption will be reviewed as well as the antecedents for technology adoption while making recommendations on how these theories and models can be classified based on the determinants of technology adoption.

2 Methodology

This study reviewed different existing learning theories and technology adoption models with the objective of comparatively analysing them. The study used digital articles and journals to select the relevant literature that was analysed by looking at the pros and cons of the selected learning theories and technology adoption models. For the selection of the literature, we applied the following composite keywords, namely "learning theories"; "technology acceptance model"; and "technology adoption". This search led to several research studies showing models linking e-learning adoption antecedents or determinants, e-learning adoption behaviours, and e-learning adoption outcomes [4]. In this regard, we sub-divided the review into three sections including describing models on the antecedents or determinants of technology adoption; the models explaining people's behaviours when using technology; and the models regarding the outcomes or consequences of the use and adoption of technology. In addition, we also included learning theories with respect to technology adoption.

3 Technology Adoption Antecedents' (Determinants') Models

The main models on the identification of the antecedents of technology adoption especially in the context of e-Learning, include Theory of Reasoned Action (TRA), the Theory of Planned Behaviour (TPB), the Decomposed Theory of Planned Behaviour (DTPB), the Technology Acceptance Model (TAM), and the Diffusion of Innovations Theory (DIT). All these models focus on the following constructs [4] as the main determinants of technology adoption in the context of e-Learning: ease of use, usefulness, enjoyment, system quality, information quality, service quality, self-efficacy, usability, and playfulness.

Before describing these theories, it is important to note that, the adoption and use of ICT for the purpose of smart education is in fact not so much influenced by education administration authority, but by the teachers and students themselves. It is therefore, the teachers who ultimately decide how to implement new technologies in the classrooms, and they are also likely to use ICT as an effective teaching tool for smart education when they are comfortable with the technology.

3.1 Theory of Reasoned Action (TRA)

The Theory of Reasoned Action (TRA) [5] depicted as Fig. 1 shows that, the intention to adopt a particular technology which is represented by behavioural intention in the Figure depends on the user's attitude towards that technology and on the receptivity to external subjective norms in relation to the use of that technology [6]. Subjective norms are referred to as one's assumption or perceptions and normative beliefs of importance about others' expectations to learn about technology or performing a specific behaviour. Other models perceive subjective norms as a consequence of awareness, and other view it as antecedent of awareness, while the behaviour is the action of the user to either adopt or reject the technology for use.

One of the short comings of the TRA is that, as much as it has been an extensively used theory which explains the determinants of consciously intended behaviours, it has however not been effectively used to show the impact it has on technology adoption rather than that of behavioural intention [7]. The initial TRA refers to an intention to partake in a certain behaviour is perceived the best determinant of whether a person actually partake in that behaviour or not [8]. Even in this case, the adoption or rejection behaviour is not explored.

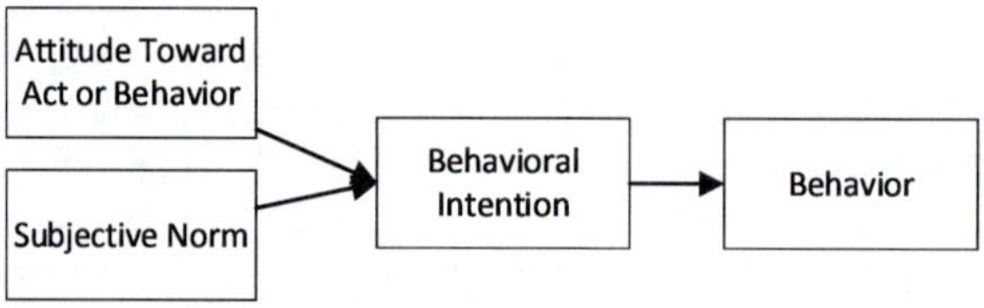

Fig. 1. The steps involved in the theory of reasoned action (TRA).

3.2 Theory of Planned Behaviour (TPB)

The Theory of Planned Behavior (TPB) [5] is an expansion of the Theory of Reasoned Action (TRA) as shown in Figs. 1 and 2. According to the TPB, the intention to adopt technology certainly depends on the attitude towards the technology and on the receptivity to external subjective norms, but it also depends on the extent to which one perceives to be able to master the control or use of that technology. Moreover, TPB posits that attitude, subject norms, perceived behavioral control, behavioral intention, and behavior are all inter-connected.

Amongst other key pros of the TPB are that: it is widely used in different context like social and behavioral science; well-established methodological tools have accompanied it to enhance its reliability and validate its theoretical constructs measures; its structural model that provides conceptual framework; and it can be applied for any type of behaviour investigation.

The TPB in its explicit definition with regards to behaviour, is based on three factors that include the target, the action convoluted, the context it occurs on and the time frame [9]. It is argued that the target is mostly ballooned to include a comprehensive

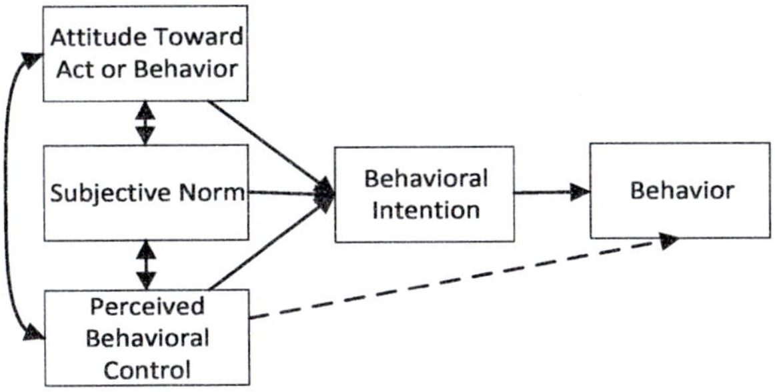

Fig. 2. The steps involved in the theory of planned behaviour (TPB)

choice of technologies to be adopted, however, the context is left unspecified. Also, measuring actual behavioural control is difficult as compared to perceived behavioural control, which in fact has led to several studies to rely on perceived behaviour over the actual behaviour control itself. The behavioural definition adopted determines how all constructs in the TPB are to be formulated and measured.

3.3 Decomposed Theory of Planned Behaviour (DTPB)

The Decomposed Theory of Planned Behaviour (DTPB) [10] is an extension of TPB as depicted in Figs. 2 and 3. According to the DTPB, the intention to adopt a technology certainly depends on the attitude, subject norms, and on the receptivity to external subjective norms, but each of these constructs has its own determinants. DTPB therefore indicate that attitude towards a given technology depends on the perception as to whether that technology provides an added advantage, and it is perceived as complex as well as compatible with one's activities. However, DTPB relates perceived behavioural control to the perceived efficacy or efficiency of the technology, and to the existence of suitable conditions for the deployment and the use of the technology.

According to [11], one of the strengths of DTPB rely on the fact that it has relatively managed to be used to predict the performance of behaviour and adoption decision to a products' trials. DTPB also adds on the construct perceived behaviour control as an extract component to those of TRA by pointing out the possible specific behaviour factor to be applied to different scenarios [12].

On the other hand, DTPB is widely used to determine the determinacies like those of TPB and TRA, however, it has not been widely used in the context of technology adoption for e-Learning.

The author of [13] notes that as much as DTPB has been used to study technology adoption especially in the context of Wb 2.0 adoption, it hasn't been applied much in other technology application. Also, the theory needs to be applied more in the e-Learning field to identify other factors that influence learners' information technology adoption. Furthermore, [13] suggests the integration of DTPB with other theoretical models to study technology adoption, as they have not been investigated in the perspective of technology adoption in e-Learning.

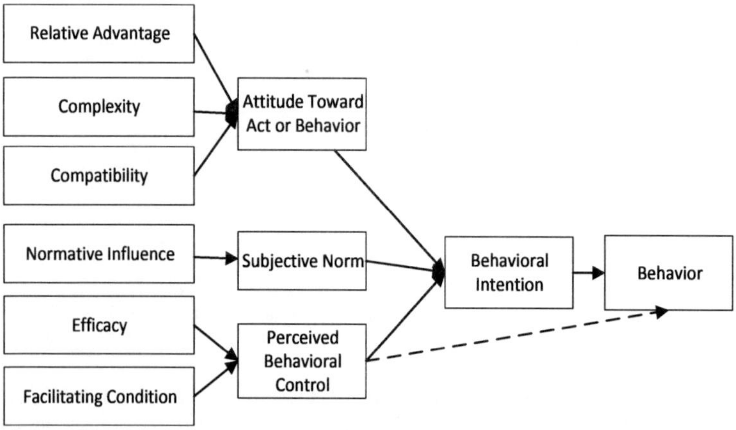

Fig. 3. The steps involved in the decomposed theory of planned behaviour (DTPB).

3.4 Technology Acceptance Model (TAM)

The technology adoption model (TAM) [14] shown in Fig. 4 indicate that, the intention to adopt a technology depends on the extent to which that technology is perceived as useful and easy to use as well as on external variables such as the characteristics of the technology, the user involvement in its design, roll out, and training.

According to [15], TAM is a widely utilised model to study technology embedded education research and renown to be strong in perceiving user's intentions to adopt technology for education. TAM precisely validate the determining factor of computer acceptance and accomplished explaining user behaviour through a distinct range of end-user computing technologies and the user groups.

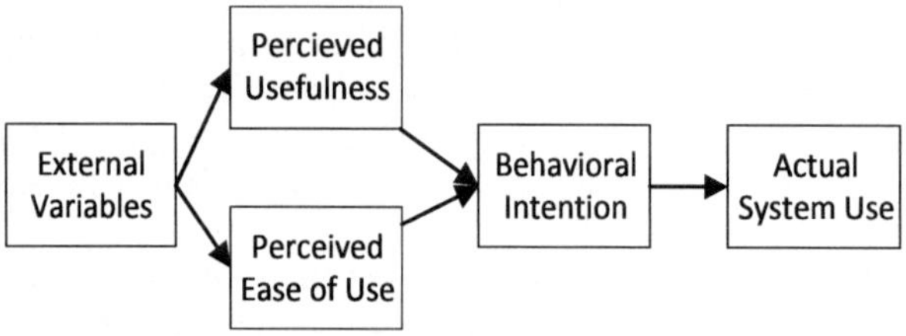

Fig. 4. The steps involved in the technology acceptance model (TAM).

In addition, TAM has been modified in many studies to suit a distinct perspective of technology being studied. Some research studies have recommended the extension and refinement of the model, by incorporating extra determinants to enable the justification of over 40% of technology adoption and usage, and to improve its generalization.

3.5 Diffusion of Innovation Theory (DIT)

The Diffusion of Innovation Theory (DIT) [16] depicted in Fig. 5 is a process by which an innovation is conveyed through a variety of ways over time amongst the participants of the social system. In the DIT theory, there are five main categories of technology users according to the timing of their adoption of technology. These include innovators, early adopters, early majority adopters, late majority adopters, and laggards. Innovators are usually recruited from technology developers and testers, and they constitute a low number (2.5%) of technology adopters. The innovators require shortest time for adoption, appreciate technology, can understand and apply complex technical knowledge. The early adopters are people that start using the technology as soon as it is available on the market, and they constitute about 13.5% of technology adopters. The early adopters tend to be opinion leaders, trend setters and adventurous with technology on the technology adoption hierarchy. While the early majority adopters are the first group of people to adopt a technology based on the success stories they heard from their early adopters' peers and this group of adopters and late majority constitute about 34% each of the adopters in the technology adoption hierarchy. The late majority adopters only adopt a technology when its reputation is completely established, and the laggards are those that usually avoid technology and only adopt it after they are certain that it will not fail. They constitute only about 16% of technology adoption because of their suspiciousness of innovations and maintaining the status quo.

Amongst the many advantages of DIT is its ability to look at the broader adoption process which include how the technology innovation is shared and disseminated by individuals between and/or within the context as these factors will impact on the successful adoption of the technology.

The DIT has been widely utilized to study technology adoption, however very limited studies have carefully analyzed if it is arguable to expand it to explain the diffusion of technology adoption especially in the context of education.

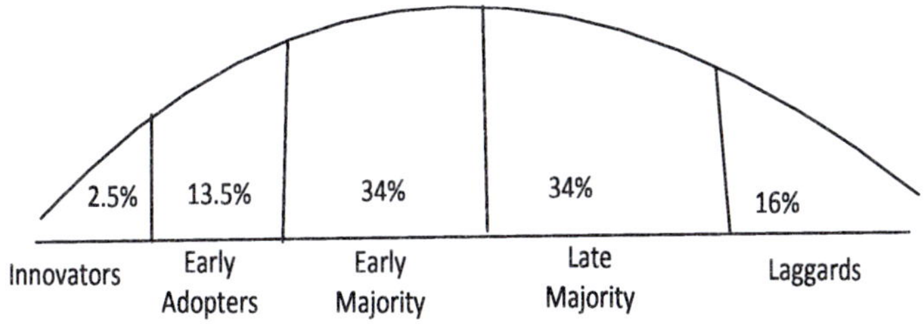

Fig. 5. The categories of technology users in the diffusion of innovation theory (DIT) model

4 Adoption Behaviour or Technology Use Theories and Models

The TAM, TPB, DIT, and DTPB have been identified as being appropriate for the modelling of the antecedents of technology adoption. However, they can also be considered as forming part of the technology adoption behaviour models because they are relevant to people's behaviour when they are using technology. Other adoption behaviour models as described by [4] in terms of technology usage, continuous use, individual feature use, extended use, and habitual use include the Information Systems Continuance Model (ISCM), Expectation-Confirmation Theory (ECT), Expectation Disconfirmation Theory (EDT) and Unified Theory of Acceptance and Use of Technology (UTAUT). These are hereby discussed as follows.

In higher education institutions worldwide, blended learning environments and e-learning integration are getting more attention with traditional course delivery systems [32]. In order for these blended learning environments to foster smart education, adoption behaviors need to be studied, the use of adoption behavior or technology use theories and models would assist in this regard.

4.1 Information Systems Continuance Model (ISCM)

The Information System Continuance Model (ISCM) [17] is depicted by Fig. 6. It revealed that the intention to continue or to discontinue the use of a given technology directly depends on the extent to which one perceives that technology as useful and the extent to which one is satisfied with the technology. According to this model, technology users perceive a given technology as useful and satisfactory when the use of that technology confirms their expectations.

ISCM theorises the Expectation-Confirmation Theory (ECT) [18] that has extensively been utilised in the Marketing to examine the impact of consumers' satisfaction on their intention to continue using an adopted technology. A perceived usefulness for ex-post anticipation is especially significant for information systems products or services, where early expectations often change with time [17]. In the primary adoption context, it is also possible to be present in the continuance context because individual continuance intention can be perceived as a series of usage assessments that are objective of timing or behavioural stages.

On the other hand, one of the ISCM cons is that it focuses on personal continuance acceptance of technology, whereas the technology adoption sometimes is at the organizational level.

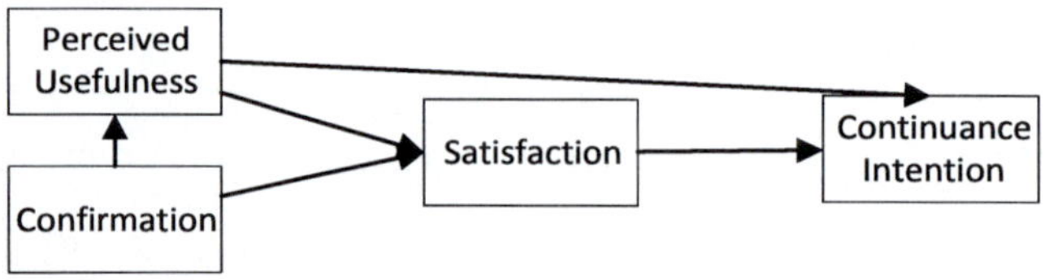

Fig. 6. The process flow of information systems continuance model (ISCM).

4.2 Expectation-Confirmation Theory (ECT)

The Expectation-Confirmation Theory (ECT) [18] is an expansion of the Information System Continuance Model (ISCM) as visible in Figs. 6 and 7. According to the ECT, the intention to repurchase a technology fully depends on the extent to which one is satisfied with that technology and users are satisfied with a given technology when the performance of that technology confirms their expectations.

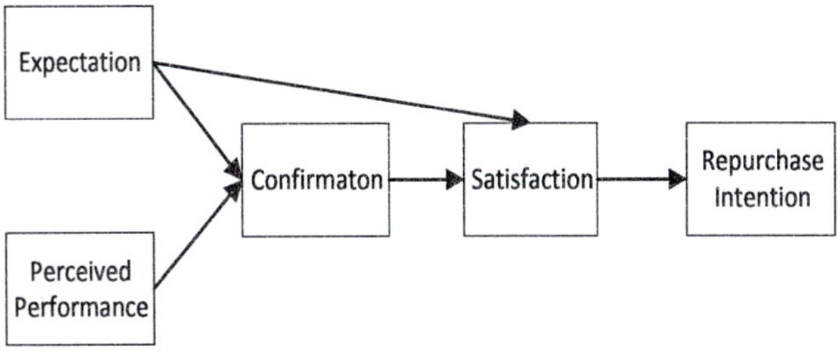

Fig. 7. The process flow of expectation confirmation theory (ECT).

4.3 Expectation Disconfirmation Theory (EDT)

The Expectation Disconfirmation Theory (EDT) [18] as shown in Fig. 8 indicate that users' satisfaction is particularly disconfirmed by expectations and perceived performance of the technology or Information System (IS). Where expectations are attributes or characteristics of a person predicted from technology or IS, product or service and this expectation has direct influence on both disconfirmation of beliefs and perceptions of performance. Whereas disconfirmation is a decision or assessment that a person takes with respect to a technology, product or service and one's evaluation is based on the expectations and perceptions of the individual.

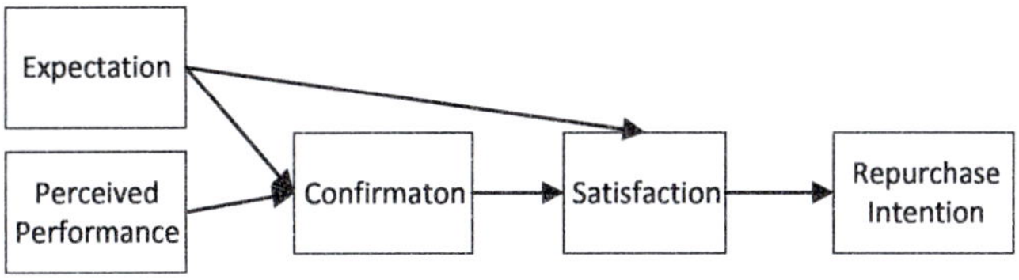

Fig. 8. The process flow of expectation disconfirmation theory (EDT).

Both ECT and EDT are characterised with expectation or desire and experience or perceived performance variables. The difference between initial expectation or desire and perceived experience or performance is known as disinformation of expectation or desire [19] thus indicating that disconfirmation of expectation or desire can be a positive or negative value.

4.4 Unified Theory of Acceptance and Use of Technology (UTAUT)

Venkatesh et al. 2003 [20] proposed the Unified Theory of Acceptance and Use of Technology (UTAUT) shown in Fig. 9. UTAUT indicate that technology use directly depends on the user's intentions and on the conditions under which the technology is being used. UTAUT assembles nearly all the technology adoption constructs from the EDT, ECT, ISCM, DIT, and TAM models, in addition to gender, age, experience, and voluntariness of use, which are considered as the mediating factors affecting the influence of these constructs.

UTAUT developed from a collaboration of eight well developed and popular models including: The Theory of Reasoned Action (TRA), Technology Acceptance Model (TAM), Motivational Model (MM), Theory of Planned Behaviour (TPB), Model Combining the Technology Acceptance Model and Theory of Planned Behaviour (C-TAM-TPB), Model of PC Utilization (MPCU), Innovation Diffusion Theory (IDT), and Social Cognitive Theory (SCT). Moreover, this model has been examined in a variety of research of user behaviour for technology adoption, for instance, e-government, mobile banking, multipurpose smartcard and mobile shopping.

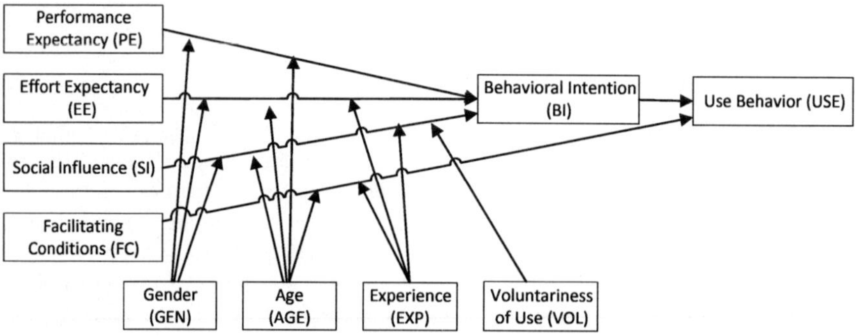

Fig. 9. The unified theory of acceptance and use of technology (UTAUT) model.

5 Adoption Outcomes Theories and Models

The main theories and models that explain the outcomes of technology adoption are the Technology-to-Performance Chain (TPC), the Information System Success Model (ISSM), and the Task-Technology Fit Model (TTFM). The focus of all these theories and models can be tailored to the following set of characteristics of e-Learning adoption outcomes [4]: learning assistance, community building assistance, and academic performance. This is towards aligning with the aim of this current study of critically reviewing the learning theories and models with respect to e-Learning adoption.

The adoption and use or continuation of use of technology for smart education is based on the outcomes that are brought about by the adopted technology. These outcomes can either be positive or negative. According to the results of a study conducted by [31], teachers' beliefs were consistent with the five factors of immediateness, interest, interactivity, instability, and inconvenience. Based on this study, teachers' concerns over smart mobile devices' instability or discomfort may be the primary barrier to using technology for smart education.

5.1 Technology-to-Performance Chain (TPC)

The Technology-to-Performance Chain (TPC) [21] is shown in Fig. 10 that the performance of a technology adoptions directly depends on how it is being utilised and on how it fits with the task at hand. In addition, the task-technology fit depends on the characteristics of the task, the technology being adopted, and of the user of the technology being adopted while the precursors of utilisation include technology adopted factors such as user's attitude, social norms, habits, and facilitating conditions under which the technology is being utilised.

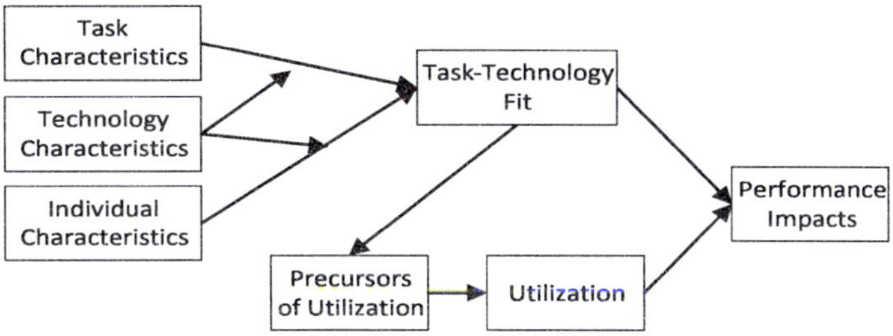

Fig. 10. Technology to performance chain (TPC) model.

Furthermore, TPC is the way in which adopted technologies lead to performance impacts at the individual level. The model thus gives a more accurate picture of the way in which adopted technologies, user tasks, and utilization relate to changes in performance. However, as it may be the case also with TTF, utilization of the adopted technology is not voluntary as the performance impacts will differ progressively upon how task-technology fit rather than it is being utilised [21]. TPC has been used to determine how task-technology fit impact performance on Learning Management Systems (LMS) and found a strong sustenance for the importance of task–technology fit that impact perceived influence on learning either directly and/or indirectly depending on the level of utilization. Whilst task–technology fit had a robust impact on perceived influence of the LMS on learning, however, it had a low influence on student grades' outcomes.

5.2 Information System Success Model (ISSM)

The Information System Success Model (ISSM) [22] is shown in Fig. 11. The model states that the net benefits of using a given technology directly depends on the extent to which one is using that technology and the extent to which one is satisfied with it. In addition, the intention to use a given technology depends on the technology being deemed beneficiary and satisfactory to the user. The other relationships which include use, intention to use and user satisfaction, between the ISSM constructs really focus on the antecedents of technology adoption.

Many studies have been conducted using this model and expanded it by adding "service quality" and "use" to "intention to use", while "individual impact and organizational impact" were exchanged with "net benefits". However, the model was decried by researchers for involving the construct "use". [23] also argued that this model is confusing because of the combination of process and variance in the same model.

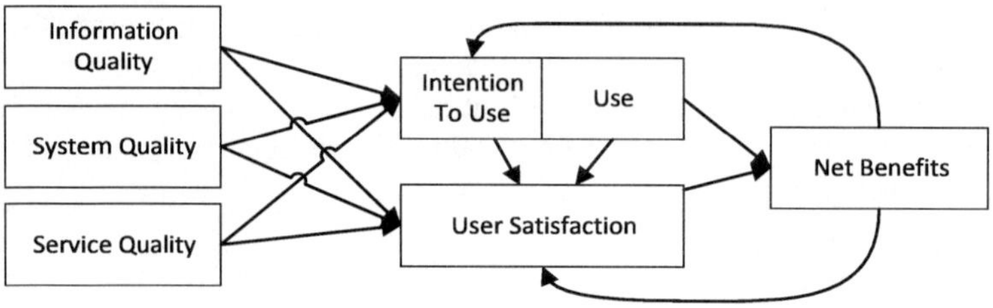

Fig. 11. The information system success model (ISSM).

5.3 Task-Technology Fit Model (TTF)

The Task-Technology Fit model (TTF) proposed by [21] is depicted by Fig. 12 and indicate the performance of a technology and its utilisation directly depends on how it fits with the task at hand. The TTF also posits that task-technology fit depends on the characteristics of the task and of the technology.

Fig. 12. The task-technology fit (TTF) model.

Amongst the shortcomings of TTF is that utilisation is not always voluntary such that several user's utilisation of the technology is highly reliant on how the job is intended to be achieved, rather than the quality or how the technology is useful, or the users' attitude about utilising the technology to be adopted. Furthermore, there is limited unambiguity to recognise that increased utilization of a technology will not automatically take the lead to better performance [21].

6 Learning Theories

Learning theories can be classified into three categories [24] which include the associationist, the cognitive, and the situative perspectives. According to [33] these three categories of learning theories have different links to smart education in the context of e-Learning. Associationist learning theories use a metaphor of learning as activity which require a smart education that allow interactive content delivery system based on assessment and feedback. With cognitive learning theories, they use a metaphor of learning as achieving understanding which require a smart education that allow adaptive and intelligent learning system, and asynchronous and synchronous dialogue. Conversely, the situative learning theories use as metaphor of learning as social practice which require a smart education that allow different tools that create opportunities for social interaction like social network to facilitate learning process. These are hereby briefly discussed.

6.1 Associationist Learning Theories

In the associationist perspective, knowledge is constructed and developed by the associations of mechanisms through activities. Learning is therefore defined as the process of connecting the elementary mental or behavioural units through sequences of activity. According to [25], there are three main types of associationist learning theories which include the associationism, behaviorism, and connectionism or neural networks, and are briefly explained thus:

Associationism. In the associationism theory, learners acquire knowledge by associating what is being acquired with what is already known. The primary idea is that mental events and their underlying brain processes can be analysed by psychological and neural structures. Associationism further outlines that human memory is made of thoughts and these thoughts are linked in an associative network. For example, if action A and action B regularly occur either in parallel or in sequence, then the representations of these events in one's brain become 'associated', such that the thought of A calls to brain the thought of B. Therefore, future occurrences of A trigger the brain to expect the occurrence of B.

Behaviourism. The behaviourism learning theory [26] focuses on observable objectives and its concession psychological activities. It proposes that learners acquire a given behaviour by observing, repeating, adjusting to how the instructor achieves that behaviour, and by reinforcing it. Reinforcement can either be positive or negative. Positive reinforcement is used to encourage the continuation of a behaviour, while negative reinforcement is used to discourage a behaviour. Behaviourists believe that all pupils can learn if they are placed in the right environment under the instructions of an educator. For behaviorism, learning is nothing more than the acquisition of new behaviour transmitted from a teacher, and learners' motivation is kept alive through incentives such as rewards and punishment.

Connectionism. Connectionism learning theory [27] posits that learning is the result of associations formed from the repetition of sequences of action and responses. Learning occurs here according to the following three laws: the law of effect, the law of readiness, and the law of exercise. The law of effect stipulates that actions that attract rewards tend to be perpetuated as opposed to the ones that do not attract rewards. This means that the human mind is good at connecting rewarding actions to their responses. On the other hand, the law of readiness stipulates that human tend to be ready to answer to unexpected actions as a sequence of responses in an attempt to reach their own goals. However, the law of exercise stipulates that the connection between an action and response sequence is strengthened by the frequency of the repetition of that action and response sequence.

6.2 Cognitive Learning Theories

In the cognitive learning theory perspective, knowledge is acquired and developed by understanding concepts and theories through reasoning, planning, problem solving, and language comprehension. Consequently, learning is defined here as the process of acquiring reasoning abilities derived from metacognitive processes and problem solving. There are three main types [25] of cognitive learning theories which include Gestalt psychology, constructivism, and symbolic information processing.

Gestalt Psychology. The idea behind Gestalt theory is that humans structure or analyse problems according to the following four "groupings": *proximity, similarity, closure, and simplicity*. Gestalt have five Gestalt principles [28] that are commonly known and used in designers. These include similarity, continuity, proximity, closure, and figure and ground thus indicating that when analysing problems, humans tend to group concepts or objects that they perceive as close or as similar. They also seek to understand which aspects are simple as opposed to the complex ones, as well as the ones that are complementing other aspects.

Constructivism. Constructivism learning theory focuses on the environment in which learning occurs and on mental reflection. It proposes that learning happens because of the replication of previous experiences, as well as the cultural, physical and social settings, by the learners. Constructivism believes that teaching and learning are conditioned by self-reflection and by self-understanding. Here, learners are more involved in the teaching process as knowledge acquisition relies more on individual experiences. The constructivist view of learning suggests that knowledge is personally formed and socially constructed by a learner during his or her interaction with the world. Constructivism agrees that mental activities are an important aspect of learning because they are the ones that filter the learner's mind in order to create his or her own specific knowledge.

Information Processing Theory. In the information processing theory, learning happens within a test-operate-test-exit (TOTE) repetitive process where the operate phase involves the "chunking" of the short-term acquisition of knowledge into small units. Once this short-term knowledge is acquired, it is then tested, and the learning process is seen as conclusive in case of a positive test. Otherwise, the operate phase is triggered again for another test, over-and-over again, until when the learning process is either perceived as conclusive or it is abandoned.

6.3 Situative Learning Theories

In the situative learning theory perspective, knowledge is transmitted to individuals from their settings and the societies they live in. Learning is therefore the consequence of the intellectual interactions with other people either physically or through technological systems. There are three main types of situative learning theories [25]: *ethnography, ecological psychology, and situation theory.*

Ethnography. Ethnographic teaching and learning make use of "ethnomethods" for the construction of learning through social artefacts such as proverbs, languages, crafts, games, and other social activities [29]. The prefix "ethno" indicates that ethnographic teaching and learning activities "are peculiar to particular people in specific places who are dealing with the artifacts mentioned". Ethnography indicate that individuals learn better when they are taught from the premises of their historical understanding and experiences within their own.

Ecological Psychology. Ecological psychology focusses on the mutual communication between an object such as person or an animal and its surrounding environment. Learning occurs here when one becomes a different person in his or her interactions with others and with the environment. This theory posits that "individuals cannot be understood or understand without reference to their specific environments". This means that people's actions and their learning capabilities can be improved by the analysis of their relationships with their environment.

Situation Theory. The situation theory posits that humans acquire knowledge within their environment and "all sorts of information about the world are organised in terms of situations" [30]. The theory indicates that learning is made possible because of situations and because of infons. Situations are the representation of structured parts of the world that constitute the context for the communication between people and for their behaviours [30]. On the other hand, infons are the representation of different pieces of information about the world. This means that one's mind always contains an elementary representation of one's surrounding environment, and it is that elementary representation that is associated to the production of information.

7 Conclusion

This current study critically reviewed technology adoption models and the learning theories that can be used for the examination of adoption process of technology especially in the context of e-Learning.

The technology adoption models were classified according to three categories including the models focusing on the antecedents or determinants of the adoption of e-learning; e-learning usage continuance behaviour; and e-Learning adoption outcomes [4]. The first category of the models explains how a person decides to adopt or not to adopt a given technology while the second describes how one decides to continue or discontinue an already adopted technology. The e-Learning adoption outcomes category of the models elucidate how to explore the possible positive or negative consequences of the use of an adopted technology. The models identified as determinants of the adoption of e-Learning are the TRA, TPB, DTPB, TAM, and DIT. The e-Learning usage continuance behaviour models identified include the ISCM, ECT, EDT, and UTAUT; while the e-Learning usage continuance behaviour models are TPC, ISSM, and TTFM.

References

1. Peng, G., Wang, Y., Kasuganti, R.: Technological embeddedness and household computer adoption. Inf. Technol. People (2011)
2. Kuk, H.S., Holst, J.D.: A dissection of experiential learning theory: alternative approaches to reflection. Adult Learn. **29**(4), 150–157 (2018)
3. Çeliköz, N., Erisen, Y., Sahin, M.: Cognitive learning theories with emphasis on latent learning, gestalt and information processing theories. Online Submiss. **9**(3), 18–33 (2019)
4. Islam, A.N.: Investigating e-learning system usage outcomes in the university context. Comput. Educ. **69**, 387–399 (2013)
5. Ajzen, I.: The theory of planned behavior. Organ. Behav. Hum. Decis. Process. **50**(2), 179–211 (1991)
6. Taherdoost, H.: A review of technology acceptance and adoption models and theories. Procedia Manuf. **22**, 960–967 (2018)
7. Buabeng-Andoh, C.: Predicting students' intention to adopt mobile learning: a combination of theory of reasoned action and technology acceptance model. J. Res. Innov. Teach. Learn. (2018)
8. LaCaille, L.: Theory of reasoned action. In: Encyclopedia of Behavioral Medicine, pp. 2231–2234 (2020)
9. Ajzen, I.: The theory of planned behavior: frequently asked questions. Hum. Behav. Emerg. Technol. **2**(4), 314–324 (2020)
10. Taylor, S., Todd, P.: Decomposition and crossover effects in the theory of planned behavior: a study of consumer adoption intentions. Int. J. Res. Mark. **12**(2), 137–155 (1995)
11. Khasawneh, M.H.A., Irshaidat, R.: Empirical validation of the decomposed theory of planned behaviour model within the mobile banking adoption context. Int. J. Electron. Mark. Retail. **8**(1), 58–76 (2017)
12. Tao, C.C., Fan, C.C.: A modified decomposed theory of planned behaviour model to analyze user intention towards distance-based electronic toll collection services. Promet-Traffic Transp. **29**(1), 85–97 (2017)
13. Lai, H.J.: Examining civil servants' decisions to use Web 2.0 tools for learning, based on the decomposed theory of planned behavior. Interact. Learn. Environ. **25**(3), 295–305 (2017)
14. Davis, F.D.: Perceived usefulness, perceived ease of use, and user acceptance of information technology. MIS Q. **13**(3), 319–340 (1989)
15. Rabaa'I, A.A.: Extending the technology acceptance model (TAM) to assess students' behavioural intentions to adopt an e-learning system: the case of Moodle as a learning tool. J. Emerg. Trends Eng. Appl. Sci. **7**(1), 13–30 (2016)
16. Rogers, E.M.: Diffusion of Innovations. Simon and Schuster (2010)
17. Bhattacherjee, A.: Understanding information systems continuance: an expectation-confirmation model. MIS Q., 351–370 (2001)
18. Oliver, R.L.: A cognitive model of the antecedents and consequences of satisfaction decisions. J. Mark. Res. **17**(4), 460–469 (1980)
19. Elkhani, N., Bakri, A.: Review on "expectancy disconfirmation theory" (EDT) model in B2C E-commerce. J. Inf. Syst. Res. Innov. **2**(12), 95–102 (2012)
20. Venkatesh, V., Zhang, X.: Unified theory of acceptance and use of technology: U.S. vs. China. J. Glob. Inf. Technol. Manag. **13**(1), 5–24 (2010)
21. Goodhue, D.L., Thompson, R.L.: Task-technology fit and individual performance. MIS Q., 213–236 (1995)
22. DeLone, W.H., McLean, E.R.: The DeLone and McLean model of information systems success: a ten-year update. J. Manag. Inf. Syst. **19**(4), 9–30 (2003)

23. Seddon, P.: A re-specification and extension of the DeLone and McLean model of IS success. Inf. Syst. Res. **8**(3), 240–253 (1997)
24. Mayes, T., De Freitas, S.: Review of e-learning theories, frameworks and models (2004)
25. Greeno, J.G., Moore, J.L., Smith, D.R.: Transfer of situated learning (1993)
26. Watson-Gegeo, K.A.: Ethnography in ESL: defining the essentials. TESOL Q. **22**(4), 575–592 (1988)
27. Thorndike, E.L.: A fundamental theorem in modifiability. Proc. Natl. Acad. Sci. U.S.A. **13**(1 Pt 1), 15 (1928)
28. Mingyang, W.: Augmented reality as an appropriate technology to learn user experience design (2021)
29. Henning, P.H.: Everyday cognition and situated learning. In: Handbook of Research on Educational Communications and Technology: A Project of the Association for Educational Communications and Technology, pp. 829–861 (2004)
30. Akhras, F.N., Self, J.A.: System intelligence in constructivist learning. Int. J. Artif. Intell. Educ. **11**(4), 344–376 (2000)
31. Leem, J., Sung, E.: Teachers' beliefs and technology acceptance concerning smart mobile devices for SMART education in South Korea. Br. J. Edu. Technol. **50**(2), 601–613 (2019)
32. Gunasinghe, A., Abd Hamid, J., Khatibi, A., Azam, S.F.: The adequacy of UTAUT-3 in interpreting academician's adoption to e-Learning in higher education environments. Interact. Technol. Smart Educ. (2019)
33. Hammad, R., Khan, Z., Safieddine, F., Ahmed, A.: A review of learning theories and models underpinning technology-enhanced learning artefacts. World J. Sci. Technol. Sustain. Dev. (2020)

Check for updates

On the Management of Universities Exchange Programs: A Decision Aid Case Study

Basma Jaafar[1,2](✉), Hamza Kamal[1,2], Fatima Zahra Belhadj[1,2], Yousra Chtouki[1,2], and Nabil Benamar[1,2,3,4]

[1] Al Akhawayn University in Ifrane, Ifrane, Morocco
{B.jaafar,Ha.kamal,F.Belhadj,y.chtouki,N.Benamar}@aui.ma
[2] School of Science and Engineering, Rabat, Morocco
[3] Moulay Ismail University of Meknes, Meknes, Morocco
[4] IMAGE Laboratory, Rabat, Morocco

Abstract. Study abroad programs are frequently omnipresent in the specific context of universities and common colleges. Exchange programs offer a significant number of benefits for the considerable category of students who opt for it. It helps improving foreign language abilities and contributes to an imperforate university graduates' profile. Going on exchange is reputable as one of the most relevant averments behind the choice of university. However, Moroccan university students substantially struggle while applying to these study abroad programs. Existing literature does not highlight the problematic of exchange course offerings that African university students endure. To the best of our knowledge, this paper is the first to address the issue of exchange courses in Morocco. It exposes the inequality of exchange opportunities in terms of courses among the diverse majors and schools present in a university. This paper analyzes the prevailing issues of exchange programs present et al. Akhawayn University in Ifrane (AUI) to assist educational institutions in managing a proper decision in this regard. We adopt quantitative research methods including surveys and pre-existing statistical data in the context of AUI. The findings reveal the dominance of business administration courses in comparison to courses of other disciplines which is a pertinent basis to draw a decision aid conclusion on the exchange issue. In particular, courses pertaining to finance, management, and marketing were favored above others. Moreover, the findings note the preference of students opting to take courses in the fall semesters, along with an excess of students who take courses with three credits in comparison to the other credits available.

Keywords: Exchange program management · Education · University · Study abroad program · Decision aid · Student life · Curriculum

1 Introduction

For presumably collegiate purposes, it is often an ascertained referral to adduce "exchange studies as an inclusive and integrated part of students' education aimed to make it easier for the students to move and study abroad. It is intended that studying abroad

M. Ben Ahmed et al. (Eds.): SCA 2022, LNNS 629, pp. 404–414, 2023.
https://doi.org/10.1007/978-3-031-26852-6_38

would be an open opportunity for students" [1]. In close correlation with this assertion, this paper explores data analysis of statistics collected by AUI related to studies abroad and exchange programs offered by numerous other universities. The objective is to scrutinize the extant relationship between the student's major and the choice of courses studied abroad. The significance of this study lies in its interpretation of research in the academic field to predict the pattern of potential courses, universities, and countries chosen, all in accordance to the semester's season in the year. The findings of this investigation will contribute greatly in the management of exchange programs. The conclusions drawn in this paper constitute an important asset in decision aid procedures related to study abroad programs.

Remarkably, "the number of university students participating in exchange programs has risen sharply over the last decade" [2]on a global scale. In fact, we aim to break down this increase and its specific occurrences on a smaller level and in a measurable method. The data collected in this study is further analyzed to be put in hand for students interested in exchange opportunities and for managers of university international programs to make a thoroughly based and information-oriented decision. The latter will be achieved through data manipulation of courses taken on exchange in line with its respective school (SSE, SBA, SHSS, or electives). Henceforth, the methodological approach of this paper exploits data analysis and visualization to collect accurate interpretations and inferences. In this context, we classified host universities' data by country to showcase the popular countries chosen for exchange programs. In addition, we compared recent data to past data so as to assess the increase or the decrease between data collected before and data collected recently.

As per the research substratum of this paper, an exhaustive scrutiny was conducted on the exchange opportunities offered at AUI. Our findings indicate the existence of two types of exchange programs: bilateral exchange, which can be divided into 3 sub-types: Full Exchange which is the most competitive as the fees are equal to those paid to the home university. Partial exchange is the sort of program where students do not fully pay their home university. Whereas tuition only exchange consists of the students paying the same tuition fees to AUI but with an additional amount for books, housing and the meal plan to the host institution. The second type of exchange is Multilateral exchange, also called "ISEP exchange"; it also includes Full Exchange and Direct Exchange. In the multilateral exchange, students have the option to choose between a range of 400 institutions in over 50 countries around the world. Moreover, we discovered the global affiliations which is a database that contains all institutions that have affiliations with AUI. "The following database represents all institutions that have a connection to AUI either through a shared membership such as the Global Liberal Arts Alliance, signed MoUs or specific agreements detailing cooperation and responsibilities such as the exchange of students." [3].

The prospective interpretations of this research have set an ambitious goal of reaching a subequal potential that enables AUI and other universities to further develop and enhance its exchange programs through proper decision making and adequate exchange programs management. As a matter of fact, this study derived its data from realistic prospects and purposeful statistics. In fact, the "internationalization of higher education develops an open perspective, establishes open channel communication, facilitates understanding, and encourages people to broaden their experience and their knowledge" [4]. Thus, our vision is to exploit data analysis tools to monitor the patterns involved in students' decisions related to study abroad. We set the fundamental basis for our investigation which is data obtained from the Office of International Programs (OIP) and First Year Experience program (FYE) by virtue of data manipulation on visualized graphs and figures.

The remainder of this paper is organized as follows. Section 2 presents the most related works to the study abroad programs topic. Section 3 describes the methodology and underlying technologies adopted throughout this study, and Sect. 4 presents the results and findings identified. Finally, Sect. 5 concludes the paper and presents our future works.

2 Related Works

Substantial research efforts were carried out in the topic of study abroad programs. This section gives an insight on some of the existing research articles tackling this topic.

A. Chwialkowska identified the components of exchange study abroad programs (ESP) that facilitate student cross-cultural learning (CCL) [6]. Building on transformative learning theory (TLT), we propose and test a conceptual model of relationships between different components of exchange programs and student CCL. The data collected from more than 700 students participating in a semester and two-semester-long programs are analyzed through logistic regression. This research contributes to the literature on the effectiveness of ESP by identifying the key components that maximize positive outcomes for students. By building on TLT, it reveals the importance of getting out of one's comfort zone and providing students with support during the ESP. This study bears practical implications as it provides academic institutions and students with important insights that help maximize student CCL.

Given the increasing emphasis on evaluating the quality of higher education internationalization, the authors in [7] explored the motivations for, and outcomes of, Maltese nursing students' Erasmus+ exchanges against the program's intended outcomes. In this vein, Sixty-five Erasmus+ mobility participants completed an online questionnaire (response rate = 44.8%), and 16 participated in a 2-h focus group. Results found that participants overwhelmingly agreed that intended outcomes of the Erasmus+ program were achieved. Thematic analysis revealed five themes: Enhanced Employability as a Nurse; Exposure to Nursing Beyond the National Shores; Personal Growth; Context-Sensitivity of Nursing Care Delivery; Language and Citizenship. These positive experiences confirm that nurse educators should continue encouraging and facilitating study abroad experiences, and that such opportunities should be offered to even more students. However, more research is required to explore any negative outcomes of student mobility and whether the participants' perceptions change over time.

The acquisition of intercultural skills by studies abroad is often considered as desirable. But although we can observe a steady increase of studies abroad in the last two decades, the vast majority of students can, obviously, compete on the labor market also without study abroad experience. This leads to the consideration that it could be increasingly a socially expected and thus normative behavior to study abroad, which develops only in specific social and professional contexts. In [8], both the conditions and effects of a social norm to study abroad are discussed theoretically and empirically. Data of a cross-sectional survey among students of economics and engineering at a German university are used. The direct mobility experience is the strongest predictor of a social norm to study abroad and this norm, in turn, determines the intention to study abroad most, compared to expected personality development and career success. The results are finally discussed in terms of possible effects on individual mobility biographies and social inequality.

This study in [9] examines students' perceptions of the impact of their experiences studying abroad. The authors used narrative inquiry methodology to obtain in-depth personal profiles detailing the perceptions of five undergraduate students at Penn State regarding their experience and how to make meaning of it. The study specifically focuses on what motivates students to participate in study abroad, how students perceive the personal impacts of their international experience, what factors shape these perceptions and the short and long-term impacts of this perception. The results of the study showed that all the interviewed students expressed intrinsic and extrinsic factors which they believe motivated them to study abroad. Each student also shared at least one goal they strived towards during their study abroad. Students' perceived expectations of their study abroad experiences were extremely varied and influenced the outcomes of their trip. Every student shared at least one short-term effect and believed that their experience abroad is impacting their life and will continue to do so in the future. This research highlights important themes for future analysis of the impacts of studying abroad.

[10] presents a potential research study that explored how students conceive of and make meaning out of their international education experience in a study abroad context. The data are based on detailed semi-structured interviews conducted with a sample of 28 undergraduate students at an American university who engaged in a variety of different study abroad opportunities. Guided by Variation Theory of Learning, the analysis of the data resulted in a typology of student conceptions of international experience (SCIE) that identified four distinct categories of conceptions of international experience (observing, interacting, participating and embracing) described across three constitutive features (being in the other culture, relating to the other culture and learning in the other culture). Hierarchically related through nine critical aspects of variation, the typology provides a unique and useful framework against which to map the "messiness" of students' complex, often complicated understandings of their experience. It offers study abroad programmers' opportunities to better understand and design student learning experiences and assessment instruments that go beyond competencies, skills and other learning outcomes.

The authors in [11] explored how postsecondary institutions with high and low study abroad participation rates communicate with students about study abroad opportunities. We were especially interested in identifying institutional policies and practices that focused on inequalities in access to study abroad. We addressed this purpose using a mixed-methods research design. We first conducted a quantitative regression analysis to identify two groups of institutions with especially high or low study abroad participation rates. We then applied a comparative qualitative content analysis to these institutions' study abroad websites. Results suggested ways in which institutional policies and practices either perpetuate exclusion or promote access among diverse student populations.

Findings reported in [12] revealed that exchange programs can enrich student teachers immensely with an awakening awareness of other cultures, of events occurring and resources used by other countries for quality teaching. There is a clear need for institutions to introduce and foster international collaboration with institutions abroad to establish enduring relationships that will present students with opportunities to increase their subject knowledge and to comprehend its relevance in the greater context of the world. In the given case, South African students were granted the opportunity to learn and to teach at schools in the Netherlands. They could observe and participate in the machinations of a well-organized and effective educational system. They returned with boundless enthusiasm to improve the local systemic dysfunctional structures encountered in the schools where they contributed their efforts during prescribed experiential training.

3 Methodology

The manipulation of data analysis tools and techniques enabled us to display a count of the total of all our columns (institutions, course code at host, courses at host intuitions, credits hours at AUI, course code at AUI and first semester approved.). Hereupon, we used numeric computations to filter out the number of courses taken in exchange for each of the three schools (SBA: school of business and administrations, SSE: school of science and engineering, SHSS: school of humanities and social sciences). This numeric computation further allowed us to generate a graph showcasing the differences and contrasts between the 3 schools and elective courses.

This graph explicitly displayed that SBA has more classes taken on exchange compared to the other schools (Fig. 1). Furthermore, we used Matplotlib to create a graph that showcases all the courses that were taken abroad in order to compare them and deduce which are the most popular amongst students (Fig. 2).

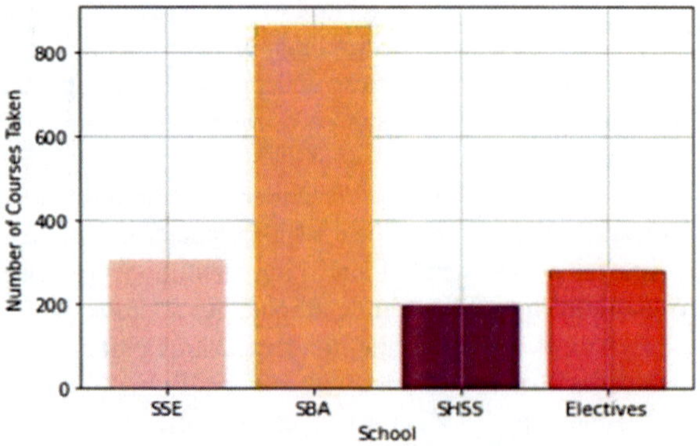

Fig. 1. A comparison of courses taken abroad per school

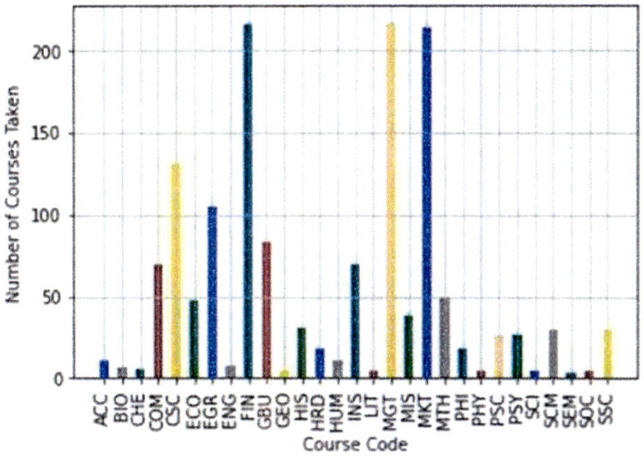

Fig. 2. A comparison of all courses taken abroad per course code

For each respective school in AUI, we implemented graphs that display the courses taken on exchange based on the school that they belong to. These three graphs showed that for SBA, the courses with the following course codes MGT, MKT, FIN are the most popular (Fig 3).

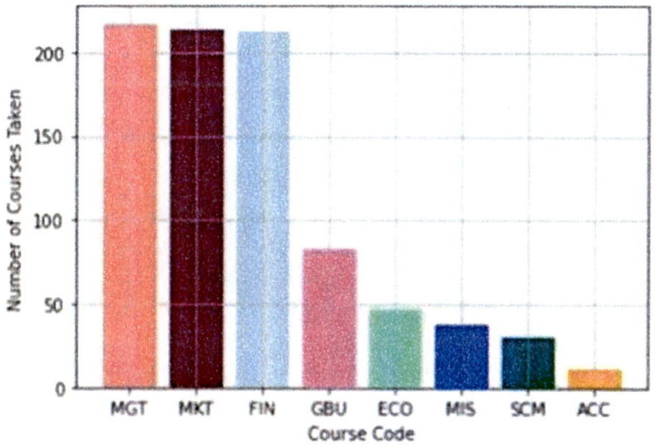

Fig. 3. A comparison of SBA courses taken abroad per course code

As per the School of Science and Engineering, it is CSC and EGR, and INS for the School of Humanities and Social Sciences that are considered popular course codes for students' course choice. These 3 graphs were designed in order to give the reader of our code a more visual representation of the numeric values that our data offers. And to opt for a comparative perceptive approach when it comes to finding the most popular courses for exchange programs in AUI. All of these graphs were designed thanks to the use of so many functions for example: (plt.ylabel, plt.bar, plt.xlabel, plt.grid, plt.label…) (Figs. 4 and 5).

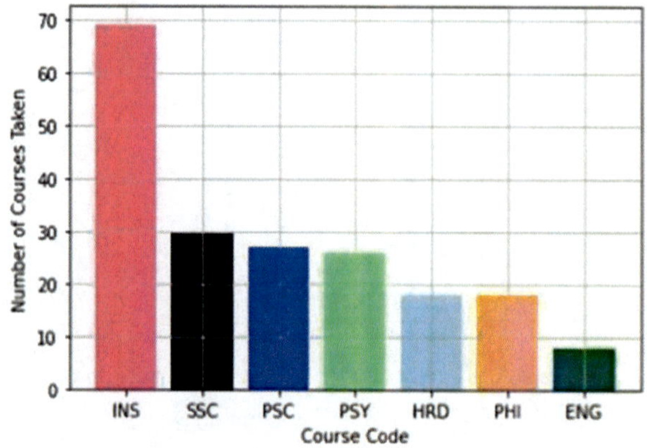

Fig. 4. A comparison of SHSS courses taken abroad per course code

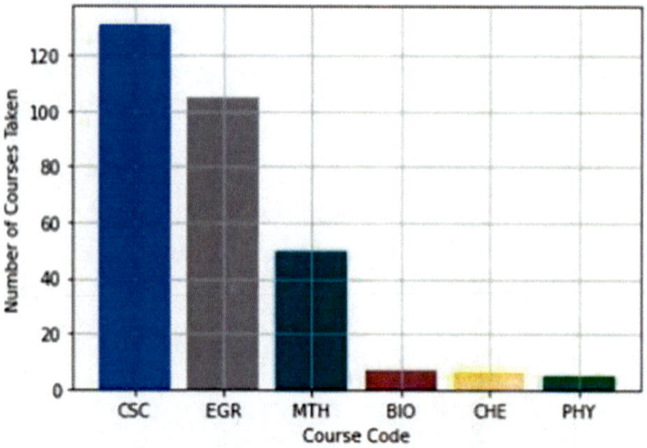

Fig. 5. A comparison of SSE courses taken aborad per course code

Using scattering, we created a graph where we separated our data with the use of the function: plt.scatter so as to display the courses in function of how many times they were used on exchange. Right after this graph, we plotted another scatter plot that scattered the X and Y coordinates of the location of the universities. This plot was displayed with X in function of Y and represents the location of our sample universities. In order to get an easier understanding of the location that seems to be more popularized by AUI students when it comes to choosing their exchange location (Fig. 6).

The second part of the analysis focused on exchange in function of shorter time frames also known as academic semesters. We the used a function to filter out and only display rows that have specific semesters for example fall 2018 so that we can use the length function in order to calculate how many classes were taken in this specific semester. This process was repeated for each semester or summer session. The data

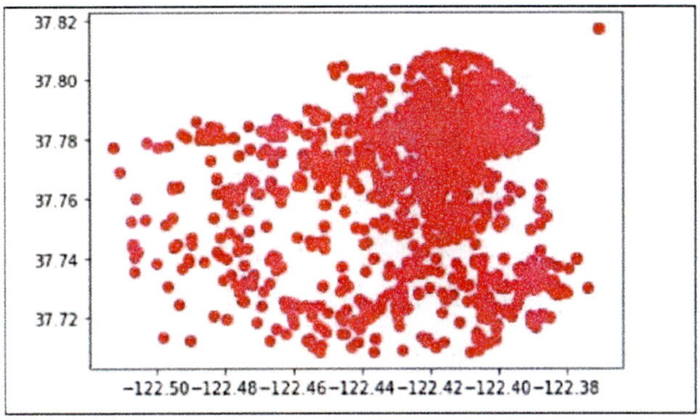

Fig. 6. Scatter plot of universities location taken during exchange programs by AUI students

collected was later reused in order to create new graphs that compare and contrast between semesters in order to conclude which semester has the most courses (Fig. 7).

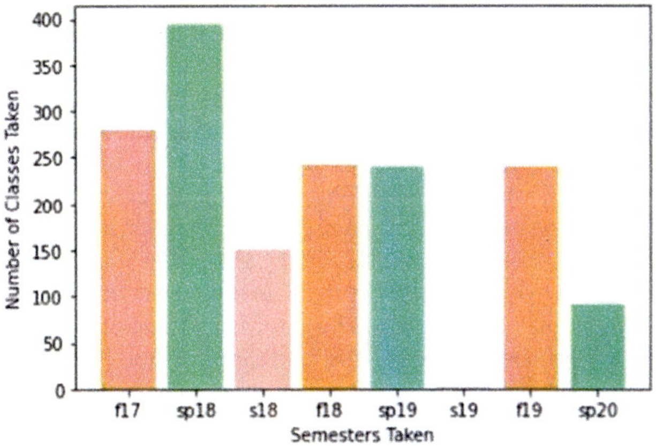

Fig. 7. A comparison of approved semesters over the years

The same process of study was used to filter out classes based on credits. We had 4 types of credits to filter out four, three, two, one, and zero. The same code was utilized in all four categories and the data collected was later repurposed in the making of two other graphs one that represented the number of courses taken based on credits and a second one with the same concept but with different x-axis values (Fig. 8).

We concluded our analysis with simple indexing in order to display the 50 most popular courses and another one in order to display the 10 top universities when it comes to the number of classes that they offer.

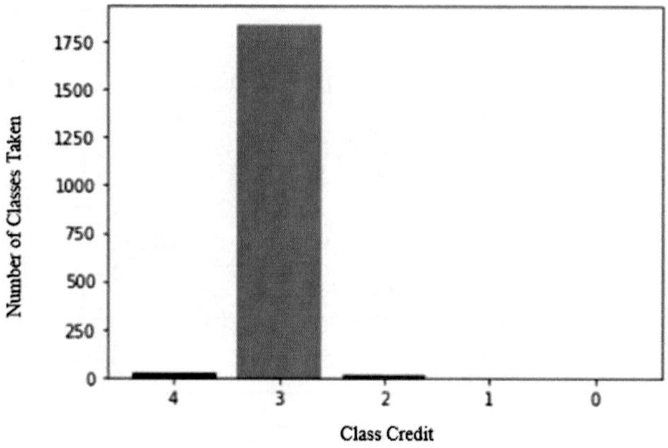

Fig. 8. A comparison of the number of classes taken abroad based on credits

4 Results

Following the graphical analysis, the data led us to a few conclusions:

Firstly, we have attained a confirmation of our initial hypothesis. The latter is that SBA classes are the most available courses for students to take on an exchange, which is strongly reinforced by the graphs assembled. The reason for this is due to the diversity and quantity of the classes offered in the school of business and administration being of higher quality than AUI's other schools, SHSS and SSE.

Moreover, the analysis has established that the fall semester is consistently the popular season of choice for students to go on exchange, hence the numbers being almost consistently consonant for the years 2017, 2018, and 2019. The pyplot figures have revealed that SBA students possess the luxury of an assortment of opportunities in special distinctive countries, such as Japan. This pushes students majoring in the schools of SSE and SHSS to pursue a minor in SBA to benefit from the same various opportunities that SBA students have. Furthermore, students in AUI tend to reserve their three credits elective courses for exchange as there are many non-traditional unique courses offered abroad, like yoga for example, that are not typically offered by AUI. Our data has shown that most of these exchange courses are taken in Europe and America, due to the localization and appeal that the market provides for students who have an interest in relocating to another country.

5 Conclusions

In this paper, we explored data about current extant study abroad programs et al. Akhawayn University. Following precise methodological manners, we cogently designed specific code to ensure suitable data analysis processes in the frame of higher academia opportunities. The graphs and their implication enabled us to deduce reinforcement of our first premise, which is that SBA classes are the most readily available courses for students to take on an exchange, which is well supported by the graphs generated. This is because the range and number of classes given in the school of business and administration are of greater quality than those offered in AUI's two schools, SHSS and SSE. The research determined that fall is the most popular semester season. According to the numbers, SBA students enjoy the luxury of a diverse range of possibilities in special-interest countries. This thoroughly encourages students majoring in SSE and SHSS to obtain a minor in SBA to benefit from the same diverse options as SBA students.

Furthermore, AUI students choose to reserve their three credits, optional classes, for exchange because there are many non-traditional unique courses provided abroad. According to our research, the preponderance of these exchange courses is taken in Europe and America, owing to the localization and attractiveness that the market gives for students interested in transferring to another nation. For future prospective work, and upon delving into seemingly endless research on machine learning, we discovered what our research could truly amount to. With the correct implementation of machine learning, we can create software that could predict and recommend the courses a student could take for studies abroad. This software could ask for the input of school, major, current semester, GPA, and even if you hold a scholarship. It would then be able to match the student with the universities that best fit their profile, along with a list of courses that the student should take. This software would facilitate the procedure of choosing where to study abroad and which course to take as it is a tedious and rather stressful process. Furthermore, as more data is imputed into the database, as in the more students that go study abroad, the more accurate the software would become and the more precise the suggestions would be.

Acknowledgment. The completion of this research would not have been possible without the help and assistance of many people, and their respective functions. Their contributions are tremendously appreciated, and their efforts are significantly recognized. We would like to express our gratitude, and sincere appreciation to Dr. Amine Bensaid, president of Al Akhawayn University, Dr. Tajjeedine Rachidi, dean of the School of Science and Engineering, Abderrafie El Mektoub, administrative assistant from the Office of International Programs who provided the data used in this paper, Omar Bougamza from the First Year Experience Program who also provided a significant amount of statistical referrals that we analyzed, and implemented in the form of graphs.

References

1. To participate or not to in an exchange program: a study on the Swedish students' attitudes on exchange programs in Higher Education. SwePub. http://swepub.kb.se/bib/swepub:oai:DiVA.org:liu-60493?vw=short&tab2=abs
2. Messer, D., Wolter, S.C.: Are student exchange programs worth it? SSRN Electron. J. (2005)
3. Taoufiq, A.: Global affiliations and maps. Al Akhawayn University in Ifrane. http://www.aui.ma/en/international/international-cooperation/partner-institutions.html
4. Globalization and education: an economic perspective. https://www.researchgate.net/publication/232495220_Globalization_and_Education_An_Economic_Perspective. Accessed 07 Jan 2022
5. Chwialkowska, A.: Maximizing cross-cultural learning from exchange study abroad programs: transformative learning theory. J. Stud. Int. Educ. **24**(5), 535–554 (2020)
6. Trapani, J., Cassar, M.: Intended and actual outcomes of study abroad programs: nursing students' experiences. J. Nurs. Educ. **59**(9), 501–505 (2020)

7. Petzold, K., Peter, T.: The social norm to study abroad: determinants and effects. High. Educ. **69**(6), 885–900 (2014)
8. Harrell, A., Sterner, G., Alter, T.R., Lonie, J.: Student perceptions of the impact of their diverse study abroad experiences. NACTA J. **61**(1), 56–65 (2017). https://www.nactateachers.org/attachments/article/2510/14%20%20Harrell.pdf
9. Streitwieser, B.T., Light, G.J.: Student conceptions of international experience in the study abroad context. High. Educ. **75**(3), 471–487 (2017). https://doi.org/10.1007/s10734-017-0150-0
10. Whatley, M., Stich, A.E.: From exclusive to inclusive: a mixed- methods investigation of study abroad participation and practices. J. High. Educ. **92**(1), 140–167 (2020)
11. Mphahlele, L.K., Coetzer, L.J., Coetzee, I.E.: Professional teacher development: a case study on a student exchange program. J. Soc. Sci. **53**(1), 27–35 (2017)

Smartphones Usefulness in Mobile Learning: The Moroccan University Case - Students of High School of Education and Training, Agadir

Tilila Mountasser(✉)

High School of Education and Training, Ibn Zohr University, Agadir, Morocco
t.mountasser@uiz.ac.ma

Abstract. The development of digital technology has transformed our lifestyle and habits in all areas especially since the Covid-19 pandemic. Higher education is a field where digital pedagogy has impacted the practices of teachers and students on mobile devices (smartphones, tablets, etc.). Mobile learning has become one of the information and communication technologies in education (ICTE) used to search, communicate, study and disseminate information through several learning modalities. According to several studies, the latter is widely used among students. The aim of the present paper is to identify the use of mobile learning in the learning process of teacher training. More precisely, we question the contribution of smartphones in the learning process of students from the High School of Education and Training of Agadir. The data we have presented comes from a quantitative survey along with an interview-based research to analyze the effects of mobile technology. The research shows that students use their cell phones to search for educational resources, and also to interact with the instructors and exchange files with their peers, and they are highly motivated to have opted for this type of learning. Ideally, this article is intended to stimulate reflections on the subject of mobile learning and its integration in education in Morocco.

Keywords: Mobile learning · Higher education · Smartphone · Moroccan University · Teacher training

1 Introduction

Morocco is oriented towards technological development. This is evidenced by the digital transformation of the city of Casablanca into a Smart city, an ongoing project for the city of Agadir. As far as the education sector is concerned, the need has been pressing since the Covid-19 pandemic, which forced education players to work or study remotely for several months.

Since the global pandemic, the uses of the information and communication technologies in education (ICTE) have expanded even in the developing countries. The strategies of education are mostly oriented to evolving the e-learning methods and devices.

M. Ben Ahmed et al. (Eds.): SCA 2022, LNNS 629, pp. 415–425, 2023.
https://doi.org/10.1007/978-3-031-26852-6_39

Among these digital transformations, a spontaneous behavior started to become more and more common. As a professor in a new School specialized in the training of future teachers in Morocco, we have noticed the increase of uses of smartphones by students in classrooms. Therefore, we intended to question this behavior and the attitudes of students regarding mobile learning.

Smartphones are an all in one tool that combines communication and the functions of a laptop. Even though the device is limited in terms of storage capacity and size of the screen in comparison with a computer, its portability makes it the most used device by the university students in Morocco.

Many recent studies have shown the efficiency of mobile learning and its expanse (Qureshi et al., 2020) [1], especially through smartphones (Klimova, 2019) [2], a tool commonly spread among students in higher education (Goksu, 2021) [3]. The studies conducted in Moroccan universities are mostly optimistic concerning the integration of cell phones as a mobile learning tool in the educational system (Rouah et al., 2021) [4] and more studies explore the effects of these uses on the learning process and students' motivation and involvement (Daoudi et al., 2022) [5].

The aim of the present paper is to identify the uses of mobile learning in the learning process of teacher training relaying on smartphones. In order to answer our main question: how is the mobile learning using smartphones contributes in the process of learning of students from the High School of Education and Training of Agadir?

The results will be detailed and discussed after a brief review of the concept of mobile learning and the description of the methodology adopted for this research.

2 Literature Review and Hypothesis Development

2.1 Defining Mobile Learning

The concept of "mobile learning" finds its first appearances in the 70s and 80s in the United States, when Alan Kay and his colleagues from the Learning Research Group of the PARC laboratories (Palo Alto Research Center) proposed Dynabook: a laptop with the size of a book that children can carry everywhere displaying activities related to reading, writing, mathematics, music, art and entertainment (Löwgren and Stolterman, 2004) [6].

In the 1990s, the "Apple Classrooms of Tomorrow" project (1985–1995) created, in 1991, Wireless Coyote, allowing students to use laptop computers to record and share environmental information (Grant, 1993) [7].

In 1996, through the "Anytime, Anywhere Learning initiative", launched by Microsoft Corporation in the United States in collaboration with Toshiba America, the first practical advantages of lighter devices were brought to light. Students from 29 public and private schools received laptops to use in the classroom while having the ability to take them home. Both teachers and students were positive about the features of these computers because they found it easier to move from home to school and from room to room. A few students even mentioned that they took the opportunity to do homework while waiting for transport or when their parents were shopping (Rockman et al., 1997) [8].

In Europe, it was not until the beginning of the 21st century that the European Union and Ericson Education Ireland launched the "From e-learning to M-learning" project in 2001 [9]. The aim of this project was to create a new approach to the virtual learning environment, moving from the technologies of the 90s to the new wireless technologies of the beginning of this century.

In African countries and other developing countries, the projects and frameworks of mobile learning are very recent, mostly from the last decade, and the empirical studies are still low (Okai-Ugbaje, Ardzejewska, and Imran, 2022) [10].

Many researchers have defined the Mobile learning or M-learning according to their fields or angle of interest. M-learning refers in general to the learning using wireless mobile devices such as laptops, smartphones, tablets, PDAs, etc. The learner can move those devices and take them to any places and at any time (O'Malley, Vavoula, Glew, Taylor, Sharples, and Lefrere, 2003; Georgiev, Georgieva, and Smrikarov, 2004; Kukulska-Hulme, 2005). One of the first definitions of Mobile learning, and which in our view sums up the concept perfectly, is from O'Malley et al. (2003: 6) [11]:

> "Any sort of learning that happens when the learner is not at a fixed, predetermined location, or learning that happens when the learner takes advantage of the learning opportunities offered by mobile technologies."

Göth & Schwabe (2008: 152) [12] added a specification regarding the learning process, focusing on the learning issues rather than the technical aspect:

> "Mobile learning is learning of mobile actors. In contrast to other mobile activities (e.g., for pleasure or work), mobile learning activities are embedded in a didactic framework. Computer science research is interested in those mobile learning approaches that are supported with mobile technologies."

Bidin & Ziden (2013) [13] synthesized, on one hand, the factors that motivate both instructors and learners to use mobile learning:

- For starters, the mobile devices are flexible and can be carried out anywhere and anytime;
- They are also quick and functional and induce a continuous and spontaneous use;
- The learner is more active and motivated because they own the device and have their privacy;
- The learning can be self-regulated, flexible, life-long and fun;
- Mobile devices can support collaborative, blended, interactive, experiential and problem-based learning.

One the other hand, adopting mobile learning comes with some issues and challenges like the small screen size, the compatibility of the design schemes, the lack of connectivity, the draining of the battery's energy, theft or loss, the cost of the device and other expenses, the blockage from school policy's and the quick obsolescence of older versions.

2.2 Hypothesis

The learner caaccess the training and information from where he wants and when he wants through mobile learning. He learns at his own pace and according to his own needs.

The research question of our contribution is to know how the Mobile learning contributes in the learning process of Moroccan students in the School of education and training of Agadir.

To answer our research question, we formulated the following hypotheses:

- The discussions via social networks are what make the mobile learning efficient;
- Mobile Learning promotes student motivation;
- Exchanges and communication via smartphones are considered beneficial and rich learning for students;
- The pandemic enhanced mobile learning.

2.3 Methodology

This work aims to identify the use of mobile learning in the learning process of students of teacher training at the High School of Education and Training of Agadir. The school was created 3 years ago in September 2019 to meet an increased need for qualified teachers in Morocco at the primary (multidisciplinary) and secondary (specialized) education. The school counted 864 students in June 2022. This preliminary study will allow us to clear the ground in order to discover the different uses of mobile phones in student learning and to be able to orient and target future research.

To do so, we opted for a combination of quantitative and qualitative approach to collect the data in the form of, respectively, a questionnaire and an interview addressed in June 2022 to 110 students, aged between 18 and 25 years, 81% females and 19% males (as the school counts more female students), from the eight specialties provided by the training (Primary teaching, French, English, Mathematics, Physics, Life Sciences, Computer Science, Economics and management) in order to analyze the uses and the effects of mobile learning. These methods were chosen because the survey gives us general data on the uses of mobile learning, and interviews make it possible to discuss the particular experience of each student.

Questions asked in the questionnaire:

- Do you use the Smartphone in your studies?
- How do you use the Smartphone in your studies?
- If you use social networks in your studies, which ones?
- How do you use social networks?
- Do you use the ESEFA-cours Moodle app?
- Does the Smartphone help you in your studies?
- What are the advantages of using the Smartphone in your studies?
- What are the disadvantages of using the Smartphone in your studies?

Interview questions:

- What is your favorite or most frequent use of the smartphone for your studies?
- Can you tell us more about your particular experience of mobile learning?

3 Results

The results are presented in the following in the form of a synthesis combining quantitative and qualitative research.

3.1 The Use of Smartphones

All of the students that were questioned have smartphones and use them in their studies in a daily basis. Within students in Morocco, the equipment in mobile phones is much easier and affordable than any other equipment; therefore, M-learning based on smartphones is very common in Moroccan universities.

Smartphones can be carried easily and safely while a laptop for example is heavier and occupy more space. Furthermore, smartphones are an all in one device that can be used as a cell phone and as a little computer; which is why it can be the only device owned by students who come from poor areas.

The majority of the students' constituting our sample use smartphones both in the classroom and outside (68.5%) while 30.6% use it only outside of the classrooms. They tend to use it in classrooms without asking the instructor for permission, and consider this behavior completely normal because the device is a tool for their education.

The device is used for several activities, the main ones being:

- Google search (94.6%);
- Translation (73.9%) as the Moroccan education system uses different languages;
- Consultation of the instructor's course (69.4%);
- Consultation of dictionaries (67.6%);
- Moodle app of the school (52.3%);
- Reading literature and books (48.6%);
- And the consultation of other courses besides their instructors' (46.8%).

The discussions between students collected only 36% of the responses. The first hypothesis on the social aspect provided by smartphones does not constitute the main purpose of mobile learning from the students. What interests them is getting the information easily and quickly, like this quote from one student when questioned about his preference in mobile learning:

- "Ease of getting information quickly, dictionary in hand to explain difficult words, help in research and preparation of presentations, use of study applications such as Le Conjugueur…"

However, the uses are different whenever they happen in the classroom or outside class hours. During class, students need to define words, concepts, look for the correct information and read the books when they don't have the paper version. While outside class hours, students prepare their lessons, revise them, consult courses or watch videos on their subjects.

- "I think the phone helps me a lot, for example when I want to read a novel and I can't find it on paper I use the smartphone to read it in PDF form, or during exams I would like to revise some notions before sleeping so I skim the information easily on Google, in class also when I'm faced with a difficult word I use the dictionary, and especially when I find history lessons in the form of a podcast so before sleeping I listen."
- "In fact I find that the use of the smartphone is very useful in order to study and this for a variety of reasons: first in class, personally, I often use the digital dictionary it helps me a lot as a student in the French language specialty, then outside class there are Google and Youtube very interesting sites and capsules to complete the understanding of the course, then there is Whatsapp which allows a discussion with our comrades or with our teachers it ensures this permanent contact with them, without forgetting Facebook there are in fact groups that allow contact with other ESEF and ENS of Morocco which allows to exchange information and books it's really interesting. The smartphone is easy to use especially to read the documents shared by our teachers I can read in any time or place."

To do so, students use some apps like social networks and a Moodle app created by the school where they can find their instructors' courses. Social networks used in educational purposes are mostly Whatsapp (88.3%) and Youtube (81.1%). Whatsapp, usually used daily, allows the students to be part of group chats, used to share information and documents. Therefore, students can keep up with the class and the homework and can contact their professors in an informal way, more spontaneous than e-mail and considered by students more efficient because of the instant messages. On the other hand, Youtube is a platform containing videos of courses in various formats (podcasts, short videos, documentaries…) that is less frequently used because it takes time to learn about a subject.

As for the Moodle app, only 20.7% of the students use it frequently and find it useful. Others either use it occasionally (37.8%), or prefer the Moodle website "http://esefa-cours.uiz.ac.ma/" (22.5%), or do not use it at all because they find it not interesting (8%) or don't even know its existence (11%). Since the courses are face-to-face, except for the period between March 2019 and June 2021 when the global pandemic was still at its peak, the Moodle app and website are less frequently used.

Smartphones practices as a mobile learning device are multiple, as explained by the testimony of a student:

- "Practical use of the smartphone for translation, dictionary, search in Google, video course in Youtube discussion in Whatsapp group with the teacher or with the class, Gmail to get the lessons and question the teacher, Moodle to get the lessons, tutorials and tests online and consult the school site to receive new announcements, download books and read in pdf, use of study applications such as Duolingo and Zamericanenglish to improve my French and English languages, as well as the Conjugueur for the conjugation of irregular verbs, then use notepad to list the activities and homework that must be done without forgetting it (To do list)."

3.2 Advantages and Limitations of Mobile Learning

The majority of the students interviewed (92.8%) view the use of smartphones in mobile learning very useful. And only 7.2% of them find this device useful but not significantly. Not one of them answered the question by a negative response.

- "The phone helped me a lot during my studies, especially when I took class photos with it, and I also do research on it during my presentations. I don't forget that thanks to it, we studied a group of classes on line."
- "Mobile Learning represents a new trend in didactics called "Connectivism", which involves all the learning methods instilled through social networks as well as technological devices. It is a new approach that reinforces rather than replaces the traditional method of teaching which is based on the use of papers and lectures. Mobile Learning also allows the student to be autonomous and enrich his knowledge by referring to several sources instead of depending on a single one, which is in class, which could identify him on limited knowledge."

As for the advantages, the most cited are the practicality and functionality of the smartphones when used for learning purposes. Smartphones are easy (84.7%) and fast (82%) to use, and the information is available and reachable (77.5%). Additionally, these mobile devices allow them to contact professors (68.5%).

These popular opinions have been noted by some students:

- "For me the smartphone helps me a lot since it facilitates access to information."
- "We all know that the smartphone in general is very important for our studies (it facilitates access to websites - we find a lot of information easily without wasting time) and even in our daily lives. Generally the use of smartphone facilitates the task."

Moreover, smartphones are also convenient for storage (43.2%), appreciated for their portability (40.5%) and suitable to discuss with their peers (38.7%). Approximately the third of the interviewed students (31.5%) are more motivated to study through smartphones.

Some students shared only positive experiences regarding the use of smartphones in their studies:

- "As soon as I discovered the usefulness of the smartphone in studies, I did not refrain from using it constantly and everywhere because it is the most developed source allowing learning with perfection and maximization."
- "It helps me a lot because it's very useful for understanding a difficulty during a lesson even outside when you don't have access to the computer you can use your phone."

- "The phone helped me a lot during my studies, especially when I took class photos with it, and I also do research on it during my presentations. I don't forget that thanks to it, we studied a group of classes on line."

The issues or limitations are not so significant in comparison. The main issue is the lack of internet connection (56%); there is no free Wi-Fi at the school and public places and its cost is not affordable for students. Likewise, smartphones are tools used for entertainment and sociability which leads to the loss of time (52.3%) and loss of concentration (41.4%). Less than the third of the students prefer the paper format (28.8%) or the use of the laptop (27%), and less than the quarter mentioned the addiction problem related to smartphones (24.3%).

Few testimonials detailed disadvantages of mobile learning using smartphones:

- "Regarding my experience of mobile learning, I found myself faced with several obstacles, including moving from one site to another, from one window to another during research, the abundance of information... so well that I am in a state of loss of concentration, sometimes the course was disturbed, and I do not manage to understand well. Generally, it can be said that learning using the computer is more effective than the smartphone."
- "I find that mobile learning is very important in all that concerns my studies especially in the search for information... But in the revision of the courses which are in PDF format I think that the paper formats or the computer are better."
- "Easy, fast and comfortable in the hand, but difficult in the case of study: whoops a message is sent and hence loss of concentration... I uninstalled Whatsapp and therefore I don't get documents sent by the teacher... They must use email so everyone benefits."
- "The use of the Smartphone has many contributions such as portability, access to information. But it is necessary to use applications to hide the things which will lead you to a loss of concentration during research."
- "I think it is a good experience that promotes self-learning but you can easily get distracted."

Finally, these testimonials summarize most of the stories about the personal experiences regarding mobile learning of the interviewed students:

- "It is a very important tool nowadays since success today relies more on the diversity of information present in each student in addition to the spread of several researches in several fields with different biases provides a multiplicity of open doors for innovative ideas which then leads to thinking and asking questions and for me that's the usefulness of mobile in my learning."
- "My mobile learning experience is summed up in exchanges with other students through forums, emails and messages. His discussions added value to our knowledge and experiences."
- "In my experience, I find that mobile learning; as an indispensable tool for training generally; really paved the way for new ways of working outside the walls, etc. This will contribute to a more fun and educational learning that allows anyone to progress

at its own pace, without constraints relating to face-to-face. And it is this point that creates the difference and makes mobile learning beneficial for me despite its flaws."

4 Discussion

Throughout this research, we have noted several potentials of smartphones as a tool for mobile learning:

- Quick and easy search for information during and outside class.
- A significant potential for exploration since the student leaves the classroom to discover, search, deepen and revise knowledge.
- Ease to access educational applications.
- Use of other applications like social networks for educational purposes: reading, searching for difficult words, translation, conjugation of verbs, interaction with professors or peers, revising and learning through videos and podcasts…
- Autonomy for learners: Students use their device according to their needs with their own learning pace.
- Different features of the device to save the information: taking pictures, recording…
- Switching from a teaching-focused mode of learning to collaborative learning between students.

Even though all these great abilities for learning, smartphones are not a device created initially for this purpose, so the entertaining and calls or messages make it difficult for students to stay focus. Another main issue for the students of the High School of Education and Training of Agadir, and for most students in Morocco, is the affordability of mobile devices and internet connection to benefit from mobile learning.

Nevertheless, the use of smartphones is favored because this tool is already owned by young people and it contains several practical features.

The hypotheses formulated for our work have been confirmed or invalidated:

- The discussions via social networks are what make the mobile learning efficient: according to the attitudes noted with the students, this hypothesis can be invalidated in the sense that the interactions are important for them in their mobile learning but without being the most important benefit.
- Mobile Learning promotes student motivation: this hypothesis can also be invalidated because Mobile learning contributes to motivation of most students, not all of them, while it also decreases it with distractions.
- Exchanges and communication via smartphones are considered beneficial and rich learning for students: this hypothesis is confirmed because students benefit from the exchange of documents and interactions with instructors and peers.
- The pandemic enhanced mobile learning: While the use of mobile phones as a learning tool is definitely a new habit of today's students, on line courses and Moodle apps are mostly used with the lack of face-to-face classes. So this hypothesis is confirmed.

5 Conclusion

Throughout this research we have found that Mobile learning is an inevitable and promising new method in the student curriculum.

Mobile technologies are opening new perspectives for mobile learning, and are currently becoming essential for educational purposes, since the young generations are already equipped and use these devices daily and in every field.

The research question presented took an interest in the contribution of Mobile learning through smartphones in the learning process of Moroccan students in teacher training. The study revealed that Mobile Learning is a very important mean in the learning of students because of the availability of information and knowledge anytime and anywhere, as well as allowing the interaction, exchange and sharing of information.

This experience allowed us to see the motivation of students through their autonomy in learning and the collaboration with each other and how this mode of learning builds a flexible atmosphere and becomes rooted in the students' routine.

The results obtained after this preliminary survey will allow us to deepen with further investigations certain questions that arise, like the integration of mobile learning by teachers, the generational and social gaps that could impact the uses, … We also intend to enlarge the number of responders and do comparisons based on different parameters like gender, year of studying and specialty.

Finally, the rapid development of new information and communication technologies requires the university to follow the trend in order to catch up with the teaching methods of the ultra-connected era. Some functions were added recently or are currently being added, namely the possibility of easily connecting via WIFI and mobile Internet, as well as the possibility of following free online training courses such as MOOCs.

In the case of teacher training, priority should be given to e-learning and mobile learning so that these current students and future teachers can themselves develop competences to train the next generations.

References

1. Qureshi, M.I., Khan, N., Gillani, S.M.A.H., Raza, H.: A systematic review of past decade of mobile learning: what we learned and where to go. IJIM **14**(6), 67–81 (2020). https://doi.org/10.3991/ijim.v14i06.13479,lastaccessed2022/09/11
2. Klimova, B.: Impact of mobile learning on students' achievement results. Educ. Sci. **9**(2), 90 (2019). https://doi.org/10.3390/educsci9020090,lastaccessed2022/09/07
3. Goksu, I.: Bibliometric mapping of mobile learning. Telematics and Inf. **56** (2021). https://doi.org/10.1016/j.tele.2020.101491. Accessed 08 Sept 2022
4. Rouah, I., Bourekkadi, S., Khoulji, S., Slimani, K., Kerkeb, M. L.: Mobile learning driving the development of higher education through a new vision of teaching methods thanks to educational technology. Ilkogretim Online - Elementary Education Online 20(5), 976–987 (2021). http://ilkogretim-online.org. Accessed 11 Sept 2022
5. Daoudi, M., Lebkiri, N., Ouali, Y., Oumaira, I.: Student Involvement in Mobile-Learning: Case of Ibn Tofail University. Stat. Optim. Inf. Comput. **10**, 59–74 (2022). www.IAPress.org. Accessed 11 Sep 2022
6. Löwgren, J., Stolterman, E.: Thoughtful Interaction Design: A Design Perspective on Information Technology. MIT Press, Cambridge MA (2004)

7. Grant, W.C.: Wireless coyote: a computer-supported field trip. Commun. ACM **36**(5), 57–59 (1993)
8. Rockman, et al.: Report of a Laptop Program Pilot. San Francisco, CA: Microsoft Corporation & Toshiba America Information Systems (1997). https://rockman.com/docs/downloads/microsoft-anytime-anywhere-evaluation.pdf. Accessed 14 June 2022
9. Keegan, D.: The future of learning: From eLearning to mLearning. Hagen: Fernstudienforchung, Germany (2003). E-published version available: http://learning.ericsson.net/mlearning2/project_one/book.html
10. Okai-Ugbaje, S., Ardzejewska, K., Imran, A.: A mobile learning framework for higher education in resource constrained environments. In: Education and Information Technologies, Springer, Heidelberg (2022). https://doi.org/10.1007/s10639-022-11094-5. Accessed 15 June 2022
11. O'Malley, C., Vavoula, G., Glew, J. P., Taylor, J., Sharples, M., Lefrere, P.: MOBIlearn WP4 – Guidelines for learning/teaching/tutoring in a mobile environment. (2003). http://www.mobilearn.org/download/results/guidelines.pdf. Accessed 15 June 2022
12. Göth, C., Schwabe, G.: Designing tasks for engaging mobile learning. In: MLearn2008, pp. 152–159. MLearn2008, Telford, UK (2008)
13. Bidin, S., Ziden, A.A.: Adoption and application of mobile learning in the education industry. In: 6th International Conference on University Learning and Teaching (InCULT 2012), Procedia - Social and Behavioral Sciences 90, pp. 720–729. Elsevier, Amsterdam (2013).

A Survey of the Security and DDoS Attacks in the Software Defined Network

Abdoulkarim Tahirou(✉) and Karim Konate

Université Cheikh Anta Diop UCAD, Dakar, Senegal
{abdoul.tahiroudjambeidou,karim.konate}@ucad.edu.sn

Abstract. The Software Defined Network (SDN) is a new network paradigm aimed at making it easier to manage and operate computer networks. SDN is becoming more and more emerging because of the ease and flexibility it offers to companies working in CLOUD and TELECOM. Unlike the traditional network where the control plane and the data plane are integrated together, the SDN network separates the two planes and then makes the network more agile and scalable through programming. In SDN, all network management logic relies on the controller. The OPENFLOW protocol is one of the most widely used protocols in SDN. It offers an opportunity for researchers and industries to develop their own ideas on innovative network protocols or solutions. However, SDN is still immature and has many security holes. Most of its vulnerabilities are related to the fact that the architecture is based on software but also the centralized nature of the controller. Indeed, all the security problems encountered in traditional networks remain valid in SDN, but Distributed Denial of Service (DoS/DDoS) attacks remain the heaviest on SDN, linked to the separation of the control plane from the data plane. On one hand, this paper focuses on the various security threats from to DDoS attacks that weigh on the controller and transfer equipment. On the other hand, it reviews the existing solutions of DDoS attacks on the controller and data plane. Finally, it introduces our research plan in the field.

Keywords: Software Defined Network · Security · DoS · DDos · Control plane · Data plane · Application plane · OpenFlow

1 Introduction

IT has evolved exponentially through the Cloud, Internet of Things (IoT), Big Data, virtualization, block chain, etc. The number of devices connected to the internet has increased considerably, to more than 74,44 billion devices in 2025 [1]. In addition, modern applications make further demands on the availability of bandwidth. Despite this, traditional computer networks have remained static for decades. This is because its network architectures operate hierarchically and vertically [2]. The design, evaluation and deployment of a protocol can take 5 to 10 years in today's networks [3] as the IPV4 to IPV6 transition takes place. The difficulty of updating existing network security policies lies in the CLI command line [4]. These limitations have resulted in a lack of flexibility and agility in traditional networks.

M. Ben Ahmed et al. (Eds.): SCA 2022, LNNS 629, pp. 426–441, 2023.
https://doi.org/10.1007/978-3-031-26852-6_40

The SDN was born in this context to overcome the monotony of traditional networks. SDN is a network architecture where the control plane is dissociated from the transfer plane and is directly programmable. All the intelligence of the SDN network is logically centralized on the software controller. The router and switches become simple packet transfer devices. The SDN has lots of strengths but also still faces security issues. Indeed, the centralized nature of the controller, its programmability, the decoupling of the control from the data planet and the lack of intelligence of the SDN transfer equipment [5] amplify its vulnerability and introduce new security flaws [6]. Research in the field of performance, virtualization, loadbalacing and the network supervision has been carried out in the SDN. One of the broadest fields of research concerns SDN security in general and more specifically DoS/DDoS attacks, to assure the transition between traditional network and the SDN.

Most of the DoS/DDoS attacks perpetrated in the SDN network are issues out on the controller, or on the transfer equipment [7]. This paper presents security in the SDN network in general and more specifically in DoS/DDoS attacks. We will focus on the different types of DoS/DDoS attack on the SDN communication layers and interfaces. This document is organized as follows: Sect. 2 presents the SDN network; Sect. 3 describes security in SDN. Section 4 classifies the modus operandi and different types of DoS/DDoS attacks. Section 5 discusses the different security threats and the corresponding mitigation mechanisms. Section 6 proposes the contribution of the authors and opens the perspectives of research in the field.

2 Presentation of the SDN Architecture

The SDN architecture has three main layers which are the Infrastructure, Control, and Application.

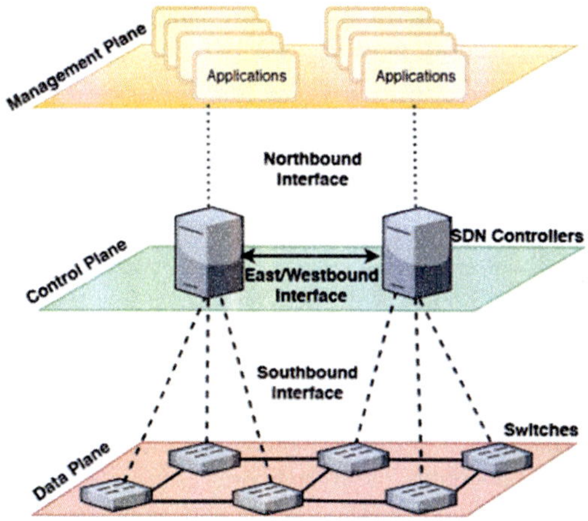

Fig. 1. Layered view of SDN architecture

2.1 Infrastructure Layer or Data Transfer Layer

The data layer is made up of various network devices that form the underlying network to carry traffic. Switches or routers and are linked by either physically wired or wireless media. Each switch is a simple packet transfer device, which has one or more flow tables within it. Each switch flow table contains a set of flow entries with packet routing rules [2].

2.2 Control Layer

The control layer is the intelligent part of the SDN network [3], because it maintains the entire structure of the network. All network management decisions are made by one or more controllers on the SDN network. The interaction between the controller and the deployed switches is called the southbound interface. The OpenFlow protocol is the most widely used between the control and transfer layer. This protocol was proposed by researchers at CleanState at Stanford University because of its flexibility and programmability. Its success is due to the support of certain technology giants such as Microsoft and Google [9], who quickly adopted it in their datacenters. A consortium called ONF has been set up for the standardization of the OpenFlow protocol [10]. On the opposite side, we have the Northbound API [11], which links the communication of business applications to the controller. There are two other interfaces in the case of a multi-controller network called Eastbound and Westbound respectively. Since the controller is the brain of the SDN network, its centralized nature creates a single point of failure in the network. To avoid network downtime most recent controller implementations offer the possibility of having multiple and distributed controllers such as NOX, POX [12], FLOODLIGHT [13], OPENDAYLIGHT [14].

2.3 Application Layer

It is the layer where all innovative network applications are developed. These applications can be used in the field of security, network load balancing, quality of service, etc. [15]. The application layer communicates with the control layer through the Northbound interface.

3 Generality on SDN Security

DDoS attacks are increasing continuously with sophisticated characteristics. These attacks concern all digital services with heterogeneous devices (computers, phones, cameras, etc.). This motivates researchers and equipment manufacturers namely CISCO, IBM, DELL, HUAWEI to focus more on this question. According to Yao et al. [16], there are three types of security issues: SDN intrusion, denial of service, and application trust management. A network intrusion is any action aimed at compromising the availability, integrity and confidentiality of any network resource or service [17]. There are other types of security issues in SDN. Those are applications, developed by third parties which make the application layer vulnerable. According to Li et al. [18], the security challenges in the SDN architecture based on OpenFlow are security issues on switches, controllers, applications and communications interfaces.

Table 1. The various security issues by layer and interface.

SDN Vulnerabilities	Attack target
Application Layer	**Application**
	Unauthenticated access to applications; Fraudulent rule insertion; Policy Application
Application-Controller Infterface	**Application-Controller Interface**
	Unauthenticated access to applications; Fraudulent rule insertion; Policy Application
Control layer	**Controller**
	Unauthorized and Unauthenticated access to controller; Modification of flow rules; Insertion of fraudulent rules; Hijacking of controllers; Saturation of the Controller-Switch communication; Saturation of the Controller-Switch communication
Controller-Data Interface	**Controller-Data Interface**
	Unauthorized access to the controller; Modification of flow rules; Hijacking of controllers; Controller-Switch link saturation
Data Layer	**Data**
	Unauthorized access to the controller; flow rule discovery; Discovery of transfer rules; Modification of flow rules to modify packets; controller hijacking

3.1 Fields of Attack in the Architecture of SDN

In the SDN all layers and interfaces can be attacked. There are six DDoS attack points according to Shu et al. [19].

- **Attacks on SDN switches:** A SDN switch is typically a separate device made up of related hardware and software, which are vulnerable to DDoS attacks. An example of a vulnerability is the small size of the SDN switch flow table and the resources of CPU, memory CAM/TCAM (Ternary Content Address Memory) [20].
- **Attacks on links between SDN switches:** the Packets transmitted between SDN switches are not encrypted and may contain sensitive user information. These packets can be intercepted by attackers easily, especially when the links between switches are wireless [21].
- **Attacks on the SDN controller:** all network management is done at the level of the controller, so it constitutes a potential target for attackers. Any attack on the control layer will have a direct impact on the operation of the SDN. Since each new flow is sent to the controller for decision, this leads to the saturation of the controller resources [20].
- **Attacks between SDN controller's links:** In a multi-controller environment, communication between different controllers is necessary to maintain the consistent state of

the entire SDN. In the event of an attack, packets can be intercepted via the Eastbound and Westbound APIs.

- **Attacks between Controller and links:** SDN centralizes all network intelligence on the controller. All new transfer rules are inserted into the switches by the controller via the Southbound API [19]. Data packets that contain these rules can be modified or tampered with by a malicious attacker who listens on the link between the controller and the switch.
- **Attacks on SDN applications:** It constitutes the added value in SDN, because it allows the development of innovative applications of the network. When applications solicit the controller through the Northbound malicious code can be embedded in the controller. The lack of a security mechanisms to ensure a relationship of trust between controllers and applications is at the root of these kinds of attacks [19].

3.2 Discussions on SDN Security

The SDN network offers more flexibility and programmability compared to the one in traditional network but the security issue in the SDN is almost identical to traditional network. We can have several types of attacks in SDN (see Table 1). The data layer is presented as the weakest link of the SDN chain in terms of security. There are many Southbound API solutions like OpenFlow, Open vSwitch Database Management Protocol (OVSDB), Path Computation Element Protocol (PCEP), Interface to the Routing System (I2RS) that exist [18]. However, not all these APIs are reliable and address security concerns. Several types of attacks can occur at this level, compromising switches or hosts. According to Dayal et al. [22] these are mainly denial of service, Man in the Middle (MIM), data modification, repudiation and side channel attacks.

4 Overview of DoS/DDoS Attacks in the SDN Security

A DDoS attack aims to make a server, service unavailable to legitimate users [4]. We will speak of a distributed DoS attack, any attack that is carried out remotely from several sources consisting of several hundred or even thousands of devices [23]. According to Saman et al. [24], the DDoS attack aims to disrupt the connectivity of legitimate users by depleting network resources. In this section, we describe the classification of DDoS attacks on the different layers and interfaces.

4.1 DDoS Security Threats on the Infrastructure Layer

This consists of networking devices that control the forwarding and data processing capabilities for the network. The attack against SDN data plane takes place at two levels, either on the CAM/TCAM memory of the switch by its overload on the one hand, and on the other hand, the saturation of its flow table. In the process of the Openflow switches, each new flow without correspondence in the flow table is sent to the controller for decision. The new rules can be inserted in the flow table via the interface southbound. In the event of a DDoS attack, the attacker can easily fill the flow table with new flow and saturate the switch. In addition, if the switch memory is saturated, instead of sending just the packet header, the entire packet will be sent to the controller [25].

4.1.1 Detection Mechanisms and Threats Against DDoS Attacks on the Infrastructure Layer

Several detection and defense mechanisms against DDoS attacks have been developed by researchers and industry. According to R. Swami et al. [4], these mechanisms can be classified into three main categories which are mechanism based on statistics, Machine Learning, or specific applications. Statistics-based detection mechanisms are statistical analyzes, which collect, and exploit data samples based on network traffic in order to make a decision on DDoS attacks. The work of several authors [26, 27] are algorithms based on statistics such as entropy, chi-square, which are used to detect DDoS attacks in the SDN.

These statistical techniques commonly used in the work are based on adaptive correlation analysis, standard deviation, probability, and entropy measures. Network features such as source IP address, destination IP address, and port numbers are used to calculate entropy with predefined thresholds to identify the presence or absence of DDoS attacks [28]. Another technique based on machine learning are used recent years due to their effectiveness in detecting DDoS attacks in SDN. These algorithms can be used to detect malicious traffic from legitimate traffic in the SDN. The algorithms used are artificial neural networks [41], Bayesian networks [42], self-organizing map (SOM) [29],and fuzzy logic [30, 31]. Other techniques use the intrinsic characteristics of SDN with specific applications for detection and defense against DDoS attacks.

According to Chen et al. [32], they proposed Flexprotect to protect data centers. They used the intrinsic features of SDN and NFV to protect the data center network from DDoS attacks. Two modules are offered by Flexprotect, the first one for detection and the second for mitigation, that are deployed separately in the system. A solution to defend against DDoS attacks with the monitoring tool named sFlow was proposed by Aizuddin et al. [33]. The proposed system used SDN features to support DDoS attacks against DNS amplification. It collects and processes header flows to check whether they originate from a DNS server. Zheng et al. [34] have implemented a solution to mitigate DDoS attacks by applying a real time adaptive correlation analysis called RADAR (Reinforcing Anti-DDoS Actions in Realtime). It consists of three main modules which are the collector, detector, and locator. RADAR can identify several types of attacks in real time, such as SYN flood, UDP flood and DNS amplification. The FlowTrApp tool [35] is an SDN-based DDoS defense mechanism for protecting data centers. FlowTrApp uses two parameters which are the rate and the duration of a flow. The characteristics of OpenFlow [36] are coupled with those of sFlow [37] for the collection of flow statistics. A tool called Woodpecker [38] is proposed by L. Wang et al. to detect and mitigate the type of DDoS flood attack using the characteristics of the SDN. Several selected ordinary switches are upgraded to SDN compatible switches. With the help of the global view provided by the SDN controller, Woodpecker locates the location of the bottleneck and identifies whether the congestion is really caused by link flooding. Woodpecker uses heuristic traffic engineering as an application on the controller to mitigate the impacts of the attack. The work of J. Liu et al. [39] focused on a modular tool called Floodlight Guard, which was implemented for detection and defense against DDoS attacks in SDN. FLGUARD applies dynamic IP address binding to solve the problem of IP address spoofing and uses the C-SVM algorithm to detect attacks. According to the authors Q. Niyaz et al. [40] their work is based on machine learning, the SAE (Stacked Autoencoder) to detect multi vector attacks in SDN. The packet headers are extracted classified in an unsupervised way by machine learning. SAE can detect DDoS attacks on control and data plane.

4.1.2 Summary Table of the Detection and Defense Mechanisms Against DDoS on the Infrastructure Layer

Security at the infrastructure layer is summarized as mentioned in the table below, DDoS attacks are the pure enemies of SDN. Because in the event of successful attacks on this layer, the entire network is crippled. This amplification of DDoS attacks on this layer could be explained by the non-intelligence of the switches. Several efforts are being deployed by researchers and industry to ensure the security of the SDN against DDoS attacks, but we mention some shortcomings in the work carried out.

Table 2. Detection and defense mechanisms against DDoS attacks on data layer

Ref.	Features	Intrusion	Target	Benefits	Limitations
[32]	FP-SYN, NFV, SDN, Snort, Pfsense	SYN-Flodding	Switch memory	Reduce traffic cost, bandwidth consumption	TCP-SYN DDOS only
[33]	SFlow, SDN	DNS amplification	Switch memory	Reduce controller response time	Only for DNS amplification
[34]	SDN, Adaptive correlation technique	Link flood, SYN flood, UDP/DNS amplification	Switch memory and control/data rate	Detects several types of DDoS attacks	Detection/defense mechanism for large network topologies
[35]	Openflow et Sflow-RT	DDoS ICMP-UDP Flood	Switch memory and control/data rate	Reduce controller load and false negatives	Too much information exchange with the controller
[38]	SDN, Openflow	Attack UDP flood	Bandwidth between controller and switches	Exact location of bottleneck Reduces and bandwidth usage	Use too much memory and CPU for controller resources
[39]	SDN, et SVM (Support Vector Machine)	Attack ICMP flood	Bandwidth between controller and switches	Flexibility, accuracy in DDoS Flooding attacks	Tool not tested in a real environment
[40]	SDN, SAE, Deep Learning	Attack TCP/UDP/ICMP flood	Controller memory Control/data link	Reduce controller, switch bandwidth	High memory and CPU controller resource usage

4.2 Security Threats at the Control Plane in SDN

Control plane security has a direct impact on the data and application layer. If one controller is compromised, the entire network, including switches, is affected. This is because when an Openflow switch can no longer receive forwarding rules from the controller, it will not know how to handle packets. Therefore, due to its important role, the controller is a key target for attackers. Hence the need to protect the control layer to preserve the security of SDN. There are several types of DDoS attack threats that still weigh heavily on the control layer. DDoS attacks generate an enormous amount of packets to overwhelm the resources of the controller in order to make network services or interface bandwidth unavailable to legitimate users [53]. In the SDN architecture, all packets management commands are concentrated on the controller. Most DDoS attacks on the controller try to saturate the controller with the arrival of new packets to increase the CPU workload and creat a botlneck between the control plane and the data plane. This separation of control over packet forwarding is one of the strengths of SDN [57], yet it is also one of the weaknesses of the SDN architecture.

4.2.1 Security Threat Detection Mechanisms Against DDoS Attacks on the Control Layer

The control plane is the most critical in the SDN architecture, as all the intelligence of the network is focused on it. Several tools have been developed to protect this plane against DoS/DDoS attacks in order to avoid its downtime. A tool called DBA (DDoS Application Blocking) has been proposed by Lim et al. [53], which consists of blocking DDoS attacks from the abnormal traffic. The architecture requires communication between the DDoS blocking application running on the SDN controller and the server to be protected. The other exchanges are carried out through the standard interfaces of Openflow. L. Dridi et al. [58] focuses on the development of an effective tool called SDNGUARD. This tool helps to protect SDN networks from DDoS attacks simultaneously and mitigate DoS impact on SDN controller and bandwidth between data layer and switch control. SDNGUARD also dynamically manages flow routes, rule entry delays, and aggregate flow rule entries. This tool performs well in protecting switch controller and bandwidth during DDoS attacks. To solve the bottleneck problem between the data plane and the control plane, the authors of [59] introduce Avant-Guard, an extension of the Openflow data plane called "connection migration". The purpose of this connection migration is to add intelligence to the switches in the data plane, in order to prevent the TCP based DDoS attacks. The objective of the Avant-Guard is to reduce the spoofing of IP addresses, by effectively delete a quantity of data to be transferred to the control plane during a DDoS attack.

FloodDefender [60] is an effective, protocol independent defense tool for SDN / Openflow networks that helps mitigate DDoS attacks. It sits between the control plane and the other controller applications, and can protect both the data layer, memory and CPU resources of the control plane using three new techniques that are: "table engineering-miss", "packet filtering", and finally "management of the flow rule". According to Celyn et al. [24], they implemented a tool for intrusion detection and prevention against certain types of TCP-based DoS attacks in the SDN network. Two connection techniques were used for IPDS namely CB-TRW (Credit-Based Threshold Random Walk) and RL (Rate Limiting), a port scanning detection technique that the authors call Port Bingo (PB), and a QoS technique that relies on throughput statistics to mitigate DoS attacks.

4.2.2 Summary of the Table of the Mechanisms for Detecting and Defending Threats Against DDoS Attacks on the Control Layer (See Table 3)

In the SDN network, the controller is the intelligent part of the SDN. All network management décisions are made by the controller. This makes the controller a potential point of attack. In addition, in the case of an SDN network with a single controller, the latter is a single point of failure of the network. These attacks are mainly focused on saturating the resources of the controller. Another DDoS attack angle on the control layer is the saturation of the bandwidth of the Openflow channel that links the data layer to the controller. Several techniques are used to ensure the security of the controller against DDoS attacks, mainly detection and defense mechanisms. These mechanisms used are IDS, IPS and NIDS tools, for the detection and mitigation of DDoS attacks. Other techniques are used to strengthen the detection and mitigation of DDoS attacks on the controller and interfaces (Table 4).

Table 3. Different sources of SDN attacks on the SDN control and data layer

Réf.	Features	Intrusion	Target	Benefits	Limitations
[41]	SDN, Openflow	Http-Floo	Controller memory; control/data bandwidth	Reduces traffic costs and bandwidth consumption	Only handles TCP-SYN attacks
[42]	SDN, OpenFlow	DoS Flood	Saturation link Control and Data plan	Controller memory and CPU protection Control/data bandwidth reduction	Additional hardware to store data
[43]	SDN, Openflow, connection migration	All DoS Flood	Saturation link Data to Control	two modules used connection migration and trigger for statistics	Adding modules to SDN architecture, Controller overload
[44]	SDN, Openflow	ICMP spoof, IP, UDP/TCP flood	bandwidth and memory saturation of the controller	Detects several types of attacks	Controller overload, not tested to scale

(*continued*)

Table 3. (*continued*)

Réf.	Features	Intrusion	Target	Benefits	Limitations
[45]	SDN, Openflow, Snort, Wireshark, TCP-replay CB-TRW, Rate Limit and port scan	DoS TCP, UDP ICMP	Controller to switch bandwidth	Creation of rules to counter the attack	Controller overload, To many false positives
[46]	SDN, Openflow	TCP/UDP/ICMP flood DoS attack	Bandwidth controller and switch	Allows to protect the controller saturation	Adding additional hardware as a cache to the data plane
[47]	SDN, Openflow	TCP, UDP	Bandwidth between controller and switch	Controller overload, control-data link saturation reduction	Packet loss between controller and data plane

Table 4. The various security issues by layers and interfaces

SDN	SDN Sources of DoS/DDoS attacks	Defense mechanisms
Data layer	Non-intelligent switch	Adding intelligence to switches
	Saturation of CAM/TCAM memory	CAM/TCAM memory protection
		Adding additional equipment
Controller-Data Interface	Bandwidth saturation between the contact layer and the data layer	Dynamic flow management
		Delegation of flow processing locally on the switches
		Connection migration
		Proactive analysis of flows
Controller Layer	Centralized control of the SDN network	Minimum processing on the packets by the switches before transfer to the controller
	Single point of failure (SPOF)	Adding multiple controllers in SDN networks
	Limited controller resources (CPU, RAM)	Adding NIDS to protect resources

4.3 Security Threats on the Application Layer

In the SDN architecture, the application layer is a critical point of DDoS attack. This is due to the variety of applications supported by this layer. A DDoS attack can occur either on different applications or on the Northbound API between the control plane and the application. The diversification of applications and their design by several stakeholders lead to a notorious security breach on the application layer, application can contain malicious code, which can spread throughout other applications.

4.3.1 Security Threat Detection Solutions Against DDoS Attacks on the Application Layer

Nowadays, computer applications are diverse and multiple in all areas. This exposes the SDN network to face DDoS attacks from them. Thus, several studies have been carried out in the context of securing SDN applications. The FLOVER [48] is a model verification system that verifies the global disconnection policies instantiated in OpenFlow network do not violate the network security policy. Another tool VERICON [49] is a system which verifies that the SDN network is indeed properly configured. Adel et al. have proposed an OrchSec tool [50], an orchestrator module developed in the application layer which uses the functionalities of the Openflow controller (redirects or blocks packets) and SFlow-RT for monitoring. It aims to improve network security by reducing overloads on SDN controllers by decoupling the control and monitoring functions. Another technique using WILDCARD, to access the information level by sampling a packet and keep its visibility on the Openflow network has been proposed by Sajad et al. FleXam [51]. Flexam collects network statistics efficiently.

5 Panoply of Open Issues on the Control Layer, Data Layer, and Southbound Interface in SDN

The SDN network presents a lot of openness in terms of vulnerabilities for attackers in general and more specifically for DDoS attacks. This is how several researches are carried out or underway to mitigate DDoS attacks in SDN. Most of these DDoS attacks happen on the control plane, the bandwidth of the interface between the controller and the switches.. Most of the research for the detection and mitigation of DDoS attacks has been carried out for the control layer, data and their link interface. It is clear that these DDoS attacks are constantly being improved by attackers in order to bypass detection and defense measures. The most widespread attacks are above all TCP, ICMP, UDP floods (see Tables 2 and 3).

6 Contribution of This Document and Future Research Perspectives

In this paper we have seen the security challenges that abounds in the SDN network in general and more specifically against DoS/DDoS attacks. We have reviewed the security issues by layer and by interfaces between layers. Most detection mechanisms use the intrinsic characteristics of the openflow based SDN in order to collect statistics and information based on the openflow protocol. Once a certain number of conditions are met, mitigation tools such as IDS, NIDS or IPS are mobilized to mitigate attacks. Our future research will focus on the detection and mitigation of DDoS attacks in the two SDN layers as well as the link interface between them.

7 Conclusion

SDN is the latest trend in computer networking, its adoption as a new paradigm is shaking up the habits of traditional networking. The separation of the control plane from the data plane is the most interesting advantage of the SDN. The centralization of control makes it possible to better secure the SDN from different types of attacks, in general and to bring innovations in its management through network programming. In this article we have reviewed the architecture of the SDN network and DDoS attack. This showed that the separation of the control plane from the data plane, which is one of the biggest strengh of the SDN, can also be at the same time, the biggest weakness of the SDN. In recent years, several researchers have initiated research with a particular focus on fighting DDoS attacks, but much research remains to be done in this area. In this paper we have focused on the two planes of the SDN architecture namely the control plane and the data plane. We have noticed that most DDoS attacks take place on these two planes on the one hand, and on the other hand between the communication interface between these two planes.

In addition, most of the proposed detection and defense solutions to mitigate DDoS attacks in the SDN manage the rates of the inbound flows to determine the attack. Other solutions combine input streams with either NIDS, IDS, IPS, packet filtering, machine learning, or statistical tools to define the attack. Some proposed mitigation tools also make it possible to add a little intelligence at the level of data plane or use additional equipment for the detection of DDoS attacks.

In this context we will propose a tool, which can use the intrinsic characteristics of SDN, Openflow protocol and sFlow for the detection and mitigation of DDoS attacks. The combination of SDN and sFlow makes this possible whitout the need for additional equipment, just the intrinsic characteristics of SDN.

References

1. «IoT : nombre d'appareils connectés dans le monde 2015–2025», *Statista*. https://fr.statista.com/statistiques/584481/internet-des-objets-nombre-d-appareils-connectes-dans-le-monde--2020/ (consulté le 4 mars 2022)
2. Nunes, B.A.A., Mendonca, M., Nguyen, X.-N., Obraczka, K., Turletti, T.: A survey of software-defined networking: past, present, and future of programmable networks. IEEE Commun. Surv. Tutorials **16**(3), 1617–1634 (2014). https://doi.org/10.1109/SURV.2014.012214.00180
3. Kreutz, D., Ramos, F.M.V., Esteves Verissimo, P., Esteve Rothenberg, C., Azodolmolky, S., Uhlig, S.: Software-defined networking: a comprehensive survey. Proc. IEEE **103**(1), 14–76 (2015). https://doi.org/10.1109/JPROC.2014.2371999

4. Swami, R., Dave, M., Ranga, V.: Software-defined networking-based DDoS defense mechanisms. ACM Comput. Surv. **52**(2), 1–36, avr. 2019. https://doi.org/10.1145/3301614
5. Devi, B.S.K., Subbulakshmi, T.: A comparative analysis of security methods for DDoS attacks in the cloud computing environment. Indian J. Sci. Technol. **9**(34) (2016). https://doi.org/10.17485/ijst/2016/v9i34/93175
6. Liu, Y., Zhao, B., Zhao, P., Fan, P., Liu, H.: A survey: typical security issues of software-defined networking. China Commun. **16**(7), 13–31, juill. 2019. https://doi.org/10.23919/JCC.2019.07.002
7. Cui, J., He, J., Xu, Y., Zhong, H.: TDDAD: time-based detection and defense scheme against DDoS attack on SDN controller. In: Susilo, W., Yang, G. (eds.) ACISP 2018. LNCS, vol. 10946, pp. 649–665. Springer, Cham (2018). https://doi.org/10.1007/978-3-319-93638-3_37
8. Latif, Z., Sharif, K., Li, F., Karim, M.M., Wang, Y.: A comprehensive survey of interface protocols for software defined networks. arXiv:1902.07913 [cs], févr. 2019, Consulté le: 28 mars 2022. [En ligne]. Disponible sur: http://arxiv.org/abs/1902.07913
9. Jain, S., et al.: B4: Experience with a Globally-Deployed Software Defined WAN, p. 12
10. SDN Technical Specifications|Open Networking Foundation. https://www.opennetworking.org/software-defined-standards/specifications/ (consulté le 27 mars 2019)
11. «What are SDN Northbound APIs (and SDN REST APIs)?», SDxCentral. https://www.sdxcentral.com/networking/sdn/definitions/north-bound-interfaces-api/ (consulté le 5 mars 2022)
12. Lunagariya, D., Goswami, B.: A comparative performance analysis of stellar SDN controllers using emulators. In: 2021 International Conference on Advances in Electrical, Computing, Communication and Sustainable Technologies (ICAECT), Bhilai, India, févr. 2021, pp. 1–9. https://doi.org/10.1109/ICAECT49130.2021.9392391
13. Daha, M.Y., Zahid, M.S.M., Husain, K., Ousta, F.: Performance evaluation of software defined networks with single and multiple link failure scenario under floodlight controller. In: 2021 International Conference on Computing, Communication, and Intelligent Systems (ICCCIS), Greater Noida, India, févr. 2021, pp. 959–965 (2021). https://doi.org/10.1109/ICCCIS51004.2021.9397125
14. Xiaohua, Y., Canhui, H.: Design and implementation of OpenDayLight manager application. In: 2020 13th International Congress on Image and Signal Processing, BioMedical Engineering and Informatics (CISP-BMEI), Chengdu, China, pp. 977–982, October 2020. https://doi.org/10.1109/CISP-BMEI51763.2020.9263553
15. Mubarakali, A., Alqahtani, A.S.: A survey: security threats and countermeasures in software defined networking. In: 2019 IEEE 2nd International Conference on Information and Computer Technologies (ICICT), Kahului, HI, USA, mars 2019, pp. 180–185. https://doi.org/10.1109/INFOCT.2019.8711319
16. Yao, Z., Yan, Z.: Security in Software-Defined-Networking: A Survey. In: Security, Privacy, and Anonymity in Computation, Communication, and Storage, vol. 10066, G. Wang, I. Ray, J. M. Alcaraz Calero, et S. M. Thampi, Éd. Cham: Springer International Publishing, 2016, pp. 319–332 (2016). https://doi.org/10.1007/978-3-319-49148-6_27
17. Abhilash, G., Divyansh, G.: Intrusion detection and prevention in software defined networking. In: 2018 IEEE International Conference on Advanced Networks and Telecommunications Systems (ANTS), Indore, India, déc. 2018, pp. 1–4 (2018). https://doi.org/10.1109/ANTS.2018.8710141
18. Li, W., Meng, W., Kwok, L.F.: A survey on OpenFlow-based Software Defined Networks: Security challenges and countermeasures. J. Network Comput. Appl. **68**, 126–139, juin 2016. https://doi.org/10.1016/j.jnca.2016.04.011
19. Shu, Z., Wan, J., Li, D., Lin, J., Vasilakos, A.V., Imran, M.: Security in software-defined networking: threats and countermeasures. Mobile Networks Appl. **21**(5), 764–776 (2016). https://doi.org/10.1007/s11036-016-0676-x

20. Mladenov, B.: Studying the DDoS attack effect over SDN controller southbound channel. In: 2019 X National Conference with International Participation (ELECTRONICA), Sofia, Bulgaria, mai 2019, pp. 1–4 (2019). https://doi.org/10.1109/ELECTRONICA.2019.8825601
21. Kuka, M., Vojanec, K., Kucera, J., Benacek, P.: Accelerated DDoS attacks mitigation using programmable data plane. In: 2019 ACM/IEEE Symposium on Architectures for Networking and Communications Systems (ANCS), Cambridge, United Kingdom, pp. 1–3, September 2019. https://doi.org/10.1109/ANCS.2019.8901882
22. Dayal, N., Maity, P., Srivastava, S., Khondoker, R.: Research trends in security and DDoS in SDN. Secur. Commun. Networks **9**(18), 6386–6411 (2016). https://doi.org/10.1002/sec.1759
23. Bhaya, W., Manaa, M.E.: Review clustering mechanisms of distributed denial of service attacks. J. Comput. Sci. **10**(10), 2037 (2014)
24. Zargar, S.T., Joshi, J., Tipper, D.: A survey of defense mechanisms against distributed denial of service (DDoS) flooding attacks. IEEE Commun. Surv. Tutorials **15**(4), 2046–2069 (2013). https://doi.org/10.1109/SURV.2013.031413.00127
25. Dong, S., Abbas, K., Jain, R.: A survey on distributed denial of service (DDoS) attacks in SDN and cloud computing environments. IEEE Access **7**, 80813–80828 (2019). https://doi.org/10.1109/ACCESS.2019.2922196
26. Mishra, A., Gupta, B.B., Perakovic, D., Yamaguchi, S., Hsu, C.-H.: Entropy based Defensive Mechanism against DDoS Attack in SDN-Cloud enabled Online Social Networks. In: 2021 IEEE International Conference on Consumer Electronics (ICCE), Las Vegas, NV, USA, janv. 2021, pp. 1–6 (2021). https://doi.org/10.1109/ICCE50685.2021.9427772
27. Li, R., Wu, B.: Early detection of DDoS based on φ-entropy in SDN networks. In: 2020 IEEE 4th Information Technology, Networking, Electronic and Automation Control Conference (ITNEC), Chongqing, China, juin 2020, pp. 731–735 (2020). https://doi.org/10.1109/ITNEC48623.2020.9084885
28. Bawany, N.Z., Shamsi, J.A., Salah, K.: DDoS attack detection and mitigation using SDN: methods, practices, and solutions. Arab. J. Sci. Eng. **42**(2), 425–441 (2017). https://doi.org/10.1007/s13369-017-2414-5
29. Hung, S.-C., Iliev, N., Vamanan, B., Trivedi, A.R.: Self-organizing maps-based flexible and high-speed packet classification in software defined networking. In: 2019 32nd International Conference on VLSI Design and 2019 18th International Conference on Embedded Systems (VLSID), Delhi, NCR, India, janv. 2019, pp. 545–546 (2019). https://doi.org/10.1109/VLSID.2019.00128
30. Xu, Y., Muqing, W., Guohao, Y.: An effective routing mechanism based on fuzzy logic for software-defined data center networks. In: 2020 IEEE 6th International Conference on Computer and Communications (ICCC), Chengdu, China, déc. 2020, pp. 1793–1798 (2020). https://doi.org/10.1109/ICCC51575.2020.9344964
31. Novaes, M.P., Carvalho, L.F., Lloret, J., Proenca, M.L.: Long short-term memory and fuzzy logic for anomaly detection and mitigation in software-defined network environment. IEEE Access **8**, 83765–83781 (2020). https://doi.org/10.1109/ACCESS.2020.2992044
32. Chen, M.-H., Ciou, J.-Y., Chung, I.-H., Chou, C.-F.: FlexProtect: a SDN-based DDoS attack protection architecture for multi-tenant data centers. In: Proceedings of the International Conference on High Performance Computing in Asia-Pacific Region - HPC Asia 2018, Chiyoda, Tokyo, Japan, 2018, pp. 202–209 (2018). https://doi.org/10.1145/3149457.3149476
33. Aizuddin, A.A., Atan, M., Norulazmi, M., Noor, M.M., Akimi, S., Abidin, Z.: DNS amplification attack detection and mitigation via sFlow with security-centric SDN. In: Proceedings of the 11th International Conference on Ubiquitous Information Management and Communication - IMCOM '17, Beppu, Japan, 2017, pp. 1–7 (2017). https://doi.org/10.1145/3022227.3022230

34. Zheng, J., Li, Q., Gu, G., Cao, J., Yau, D.K.Y., Wu, J.: Realtime DDoS defense using COTS SDN switches via adaptive correlation analysis. IEEE Trans. Inform. Forensic Secur. **13**(7), 1838–1853 (2018). https://doi.org/10.1109/TIFS.2018.2805600
35. Buragohain, C., Medhi, N.: FlowTrApp: an SDN based architecture for DDoS attack detection and mitigation in data centers. In: 2016 3rd International Conference on Signal Processing and Integrated Networks (SPIN), Noida, Delhi NCR, India, févr. 2016, pp. 519–524 (2016). https://doi.org/10.1109/SPIN.2016.7566750
36. Alsaeedi, M., Mohamad, M.M., Al-Roubaiey, A.A.: Toward adaptive and scalable OpenFlow-SDN flow control: a survey. IEEE Access **7**, 107346–107379 (2019). https://doi.org/10.1109/ACCESS.2019.2932422
37. Panchen, S., Phaal, P., McKee, N.: InMon Corporation's sFlow: A Method for Monitoring Traffic in Switched and Routed Networks. https://tools.ietf.org/html/rfc3176 (consulté le 25 juin 2019)
38. Wang, L., Li, Q., Jiang, Y., Jia, X., Wu, J.: Woodpecker: detecting and mitigating link-flooding attacks via SDN. Comput. Networks **147**, 1–13 (2018). https://doi.org/10.1016/j.comnet.2018.09.021
39. Liu, J., Lai, Y., Zhang, S.: FL-GUARD: a detection and defense system for DDoS attack in SDN. In: Proceedings of the 2017 International Conference on Cryptography, Security and Privacy - ICCSP '17, Wuhan, China, 2017, pp. 107–111 (2017). https://doi.org/10.1145/3058060.3058074
40. Niyaz, Q., Sun, W., Javaid, A.Y.: A deep learning based DDoS detection system in software-defined networking (SDN). EAI Endorsed Trans. Secur. Saf. **4**(12) (2017), Consulté le: 29 octobre 2020. [En ligne]. Disponible sur: https://eudl.eu/doi/10.4108/eai.28-12-2017.153515
41. Lim, S., Ha, J., Kim, H., Kim, Y., Yang, S.: A SDN-oriented DDoS blocking scheme for botnet-based attacks. In: 2014 Sixth International Conference on Ubiquitous and Future Networks (ICUFN), Shanghai, China, pp. 63–68, juill. 2014. https://doi.org/10.1109/ICUFN.2014.6876752
42. Dridi, L., Zhani, M.F.: SDN-guard: DoS attacks mitigation in SDN networks. In: 2016 5th IEEE International Conference on Cloud Networking (Cloudnet), pp. 212–217, October 2016. https://doi.org/10.1109/CloudNet.2016.9
43. Shin, S., Yegneswaran, V., Porras, P., Gu, G.: AVANT-GUARD: scalable and vigilant switch flow management in software-defined networks. In: Proceedings of the 2013 ACM SIGSAC Conference on Computer & Communications Security - CCS '13, Berlin, Germany, 2013, pp. 413–424 (2013). https://doi.org/10.1145/2508859.2516684
44. Shang, G., Zhe, P., Bin, X., Aiqun, H., Kui, R.: FloodDefender: protecting data and control plane resources under SDN-aimed DoS attacks. In: IEEE INFOCOM 2017 - IEEE Conference on Computer Communications, Atlanta, GA, USA, mai 2017, pp. 1–9. https://doi.org/10.1109/INFOCOM.2017.8057009
45. Birkinshaw, C., Rouka, E., Vassilakis, V.G.: Implementing an intrusion detection and prevention system using software-defined networking: Defending against port-scanning and denial-of-service attacks. J. Network Comput. Appl. **136**, 71–85 (2019). https://doi.org/10.1016/j.jnca.2019.03.005
46. Wang, H., Xu, L., Gu, G.: FloodGuard: a dos attack prevention extension in software-defined networks. In: 2015 45th Annual IEEE/IFIP International Conference on Dependable Systems and Networks, juin 2015, pp. 239–250. https://doi.org/10.1109/DSN.2015.27
47. Wu, P., Yao, L., Lin, C., Wu, G., Obaidat, M.S.: FMD: a DoS mitigation scheme based on flow migration in software-defined networking: FMD: A DoS mitigation scheme based on flow migration in software-defined networking. Int. J. Commun. Syst. **31**(9), e3543 (2018). https://doi.org/10.1002/dac.3543

48. Son, S., Shin, S., Yegneswaran, V., Porras, P., Gu, G.: Model checking invariant security properties in OpenFlow. In: 2013 IEEE International Conference on Communications (ICC), Budapest, Hungary, juin 2013, pp. 1974–1979. https://doi.org/10.1109/ICC.2013.6654813
49. Ball, T., et al.: VeriCon: towards verifying controller programs in software-defined networks. In: Proceedings of the 35th ACM SIGPLAN Conference on Programming Language Design and Implementation, New York, NY, USA, pp. 282–293 (2014). https://doi.org/10.1145/2594291.2594317
50. Zaalouk, A., Khondoker, R., Marx, R., Bayarou, K.: OrchSec Demo: Demonstrating the Capability of an Orchtestrator-based Architecture for Network Security, p. 3
51. Shirali-Shahreza, S., Ganjali, Y.: Traffic statistics collection with FleXam. .In: Proceedings of the 2014 ACM conference on SIGCOMM - SIGCOMM '14, Chicago, Illinois, USA, pp. 117–118 (2014). https://doi.org/10.1145/2619239.2631441

Recommender System Using Machine Learning and BLE Beacons

Dalal Zaim(✉), Aziza Benomar, and Mostafa Bellafkih

Networks, Informatics and Mathematics Department, INPT, Rabat, Morocco
dallalzaim@gmail.com

Abstract. Nowadays, Recommender system is present in our daily activities, people need the help of computer systems and machine learning to simplify their regular activities. Grocery shopping is an everyday essential task of people's life so the retailers must exploit the advancement of technology to know their customers better and to improve their marketing strategies. This paper aims to apply machine learning algorithms to the development of a recommender system capable of preparing personalized grocery shopping lists. The proposed framework exploits K-means algorithm and Apriori association rules to understand the customer's behavior. The K-means algorithm was used to identify groups of customers based on transactional data and movements of customers; it also divides customers into ten groups of customers. Apriori association rule was used to characterize the groups of customers by creating customer profiles. The analysis of customer behavior is helpful for the decision marketing managers to send the right promotion to the right consumer at the right time.

Keywords: Datamining · Recommendation systems · Location-based services · Trajectory-ticket · K-means algorithm · Apriori association rule

1 Introduction

Over the past decades, prediction of customer's purchases has been recognized as one of the most significant topics in customer behavior studies, because the retailers need to detect the purchasing habits and to understand the customer behavior in order to offer adequate promotions and maximize the satisfaction of the customer. Recommender system is one of important innovation in this field of research. It helps the retailer to boost the sales and adapt the services by providing customized advertisements and offers to targeted customers. Most recommendation systems are based on purchase history not on the movements of customer inside the mall and staying time at every selling area. Few recommender systems are based on similarity of purchase processes, like customers' paths without integrating the purchases history. This gave birth to the idea of developing a recommender system that exploits the history purchases and customer's paths in order to offer the right promotion to the right customer at the right time.

By using big data tools and machine learning algorithms, the marketing decision-makers can extract valuable data which will be the key to apply efficient marketing

M. Ben Ahmed et al. (Eds.): SCA 2022, LNNS 629, pp. 442–451, 2023.
https://doi.org/10.1007/978-3-031-26852-6_41

strategies that meet the needs of the customers. Analyzing customer behavior is a very challenging task to choose which strategy should be adopted. Understanding the preferences of customers why, when, and how purchasing is done and discovering the factors that influence the buying decisions of customers is the key to build efficient marketing strategies.

Clustering or segmentation of customers [1] is usually the first step in the process of analyzing the purchasing behavior in retail. But most approaches construct models based on parameters such as the recency, the frequency and the monetary briefly the RFM Model [2] this model captures only partial information of a real user state. As a result, we must exploit other parameters in order to extract meaningful knowledge and understand the customer purchasing habits.

In this paper, we propose a framework that exploits the data collected about the movements and the purchases of the customers to create a trajectory-ticket. This data will be used to cluster customers into groups by k-means algorithm, then it uses Apriori algorithm to find the representative basket for each cluster and discover the purchasing habits of the customers. So, it will be an efficient tool for the marketing decision-makers to achieve better knowledge of purchasing behavior and to apply the adequate marketing strategies. In the remainder of this paper, we first provide a brief overview of the framework, we next use k-means algorithm to cluster customers into homogeneous groups. Then we exploit the Apriori algorithm to discover applied rules in the data. Finally, we conclude this paper and outline the future work.

2 The Description of the Geomarketing System

2.1 The Proposed Framework

The purpose of our Recommender System is to derive meaningful product recommendations addressed to smartphone application customer, so as to draw the attention of customers to products they would potentially be interested in. The core Recommender System algorithm employs a hybrid approach, combining association rules and K-means algorithm. To build a recommendation model, three phases are required:

1. Information collection phase: Collect the data about the customers, their movements and their purchases.
2. Learning phase: Apply the machine learning algorithm to extract meaningful knowledge from the data obtained in the previous phase.
3. Prediction/recommendation phase: Apply the algorithm to send recommendation or prediction based on the customer's preferences.

The recommender system proposed [3, 4] consists of Beacons, Server Application, and Smartphone Application. We use BLE beacons (small devices using Bluetooth low energy to communicate) placed at both ends of each aisle to build an indoor positioning system able to detect the location of customer inside the mall in order to construct the shopper paths. By matching the time, it passed through the cash register with the time of purchase history data, we can obtain the history of purchases of the customer and their movements inside the mall and then it is possible to analyze customer behavior as well.

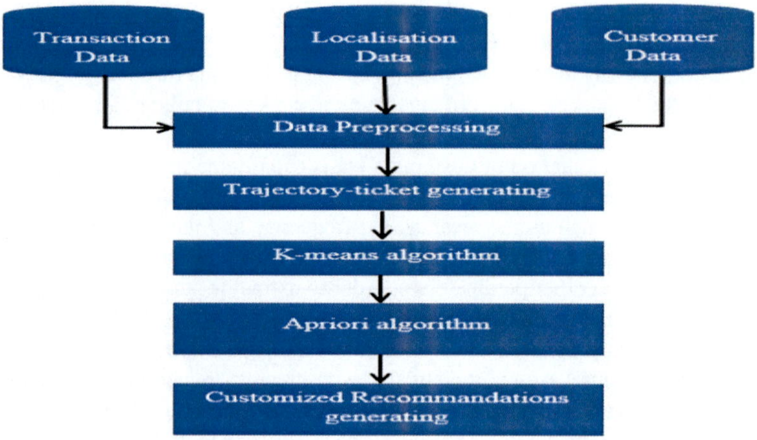

Fig. 1. The proposed approach for the Recommender System.

Next, our recommender system [5] launches the processing of the data collected in order to send customized recommendations via the smartphone application of the customer. The generation of recommendations should head based on purchases history and historical shopping path data of the customer. As shown in Fig. 1, The recommender system proposed presents a novel approach for customer behavior analysis using profile data of the customers, their movements, and their transactions. The key feature of our proposed approach is a cascade involving K-means and Apriori algorithm to generate marketing strategies that respond to the customer's needs.

The K-means algorithm is an efficient algorithm used to divide customers into homogeneous groups that have similar preferences. K-means was employed to segment the customers of the mall into ten clusters based on the customers' transactions and their movements. Once we identified the groups of customers by using K-means, we employ Apriori algorithm to extract association rules and find out the relationship between products purchased by each cluster. Based on these rules, we can create representative baskets for each cluster that characterize the transactions of each cluster. As a result, the store managers can understand the behavior of customer. Hence, they can establish marketing strategies that respond to customer's needs.

To understand the consumer behavior most of studies [6] focus on some parameters such the recency (this parameter refers to when the customer did the most recent transaction, the frequency (refers to how often customers do transactions), the monetary value refers to how much does a customer spend in our store and group the customers accordingly. But these parameters capture only partial data about the customer's habits, we can detect two customers with the same values of recency, frequency and monetary but they don't have similar purchasing behavior. To overcome this problem, we exploit more significant parameters to discover the customer's purchasing habits. In this context, we calculate the average of amount paid at each aisle and the average of duration by aisle to extract the purchasing habits during the observation period. So, we use the calculated values to classify the customers into homogeneous groups that have similar behavior and to predict which customer belongs to which group in order to offer personalized promotions.

2.2 The Databases of Our System

For our analysis, the key of success is the quality and the integrity of data, so we need to process the data and extract useful knowledge in order to ensure the consistency in customer behavior model. To collect the data, we use our Geomarketing system, it consists of three components: beacons, the smartphone application, and the server application. The data of this system are stored in databases, the positions of customers are stored in database of localizations, the history of purchases and the account of customers are stored in the database of transactions, and the organization of the aisles is stored in the database of areas.

The first step is to organize and integrate the data of these databases: the database of areas and the database of localizations were joined using beacon ID to create the trajectory of the customer. Then, the database of transactions and the table of trajectories were joined using customer ID to create the trajectory-ticket of the customer. The extracted data will be used to classify and segment the customers of the mall.

When we process the data and extract the useful information for our analysis. The second step is to select the season because the consumer changes the behavior according to the season, so we choose for our case the summer. We extract the purchases of the customers of our mall during the last summer to understand their purchasing habits and extract the representative basket for each cluster during summer.

3 Clustering the Customers by K-Means

3.1 The Processing of Collected Data

Over the decades, many marketing tools were developed and used in various stages of marketing to create an effective strategy of marketing. But these tools may be limited by the poor quality of collected data. The key of success of a new marketing strategy is the customer clustering, it helps the marketing decision makers to divide customers into homogeneous groups and adjust the promotions according to their preferences.

To understand the customer habits and offer personalized promotions, we assume that the customer behavior follows a pattern like his past patterns, so we need to use the past transactions to understand the customer behavior and divide the customers into homogeneous groups. Due to the innovation in wireless communication technologies and huge evolution of big data, it is the right time to develop a new marketing strategy based on the customer behavior analysis. The decisions to be made include how to promote products, estimate the probability of acceptance new products according to the preferences of each group in order to rich the customer satisfaction and encourage direct marketing. To rich the goals of our recommender system, we need to collect valuable data and preprocess it in order to extract meaningful knowledge.

For our system, we consider transaction and localization data as input data. Firstly, we use k-means algorithm to cluster the customers into homogeneous groups. Then, we exploit Apriori algorithm to find out the representative set of products purchased by each group in order to understand the customer behavior. For our case, we divided our store into 10 aisles: Dairy products M, Frozen products FP, cleaning products C, Cereal and Snacks CS, Deli D, Bakery B, Meats ME, Fruits and Vegetables FV, Fish F and Personal Care P. We used the details of the database of localizations, the database of tickets and the database of areas to generate the trajectory-ticket. For example, we have the purchases of the customer C02, and the movements of this customer during the summer. Then, we construct the trajectory-ticket of the customer C02 as shown in Table 1.

Table 1. The trajectory-ticket of the customer C02

Aisle	M	FP	C	CS	D	B	ME	FV	F	P
Amount	40	80	30	50	0	20	60	42	0	0
Staying time (mn)	5	6	3	4	1	6	10	7	0	0

We calculate the percentage of purchases by using (1):

$$\sigma : \begin{cases} P_i = \frac{Amount_i}{Staying\ time_i} & \textbf{if the staying time} > 0 \\ P_i = 0 & \textbf{otherwise} \end{cases} \quad (1)$$

We use the scales (Table 2) to prepare the trajectories-tickets for the datamining process. For example, the customer C02(2, 3, 3, 3, 1, 2, 2, 2, 1, 1).

Table 2. The scales

Scale	≥30	20 ≤ Pi < 30	10 ≤ Pi < 20	0 < Pi < 10	0
P	5	4	3	2	1

3.2 Dataset

We implement the proposed scheme on a transactions and localizations dataset of 5900 customers, this dataset contains 45200 transactions occurring between 01/06/2014 and 01/09/2014 for a retail store. Table 3 shows the sample of the trajectory's data obtained using our BLE localization system. When the customer comes to the checkout register and makes a purchase, data are recorded and entered into our dataset. The dataset contains the shopping details as shown in Table 4.

3.3 K-Means Clustering

K-Means is a popular algorithm used to cluster the customers into homogenous groups, it takes the parameters of clustering and the number of clusters as inputs and partitions the

Table 3. Sample of the paths data.

Customer ID	Date Time	Area	Elapsed Time
C02	2014/11/12 11:01:12	Entrance	1
C02	2014/11/12 11:02:51	Meat	2
C02	2014/11/12 11:04:51	Meat	2
C02	2014/11/12 11:04:51	Fish	2

Table 4. Sample of the transactions data.

Customer ID	Date Time	Product	Category	Amount
C02	2014/11/12 11:31:12	Banana	Fruit	10
C02	2014/11/12 11:31:12	Potatoes	Vegetable	9
C02	2014/11/12 11:31:12	Chicken	Meat	45
C02	2014/11/12 11:31:12	Orange	Fruit	11

data into the defined number of clusters such that the inter-cluster similarity is low and the intra-cluster similarity is high. This algorithm is iterative, it uses an iterative approach which computes the value of centroids before each iteration. The data points are moved among different clusters depending on the centroids calculated at each iteration. The process is repeated until the objective function cannot be decreased any more.

Algorithm [7]:

- Input: Customer Dataset, k: the number of clusters
- Output: k clusters

1. Initially, depending on the value of k, k random points are chosen as initial centroids.
2. The distances of each data point from the centroids chosen earlier are evaluated using the Euclidian distance.
3. The obtained values are compared and the data point is assigned to the centroid which has shortest Euclidian distance value.
4. Recalculate the new center points for all clusters, then the previous steps are repeated. The process is stopped if the clusters obtained are same as that of the previous step.

After the processing of data, and the generation of the vectors for all customers, we move to another step the clustering of customers by using clustering algorithm. The most useful clustering algorithm is K-means [4, 8], it is an iterative clustering algorithm used to form homogeneous clusters. Firstly, we choose to cluster our customers into ten groups, and we assign randomly ten customers to the ten groups. To classify our customers k-means algorithm aims at minimizing an objective function. The objective function of our case is:

$$D(X, Y) = \sum_{I=1}^{10} (Xi - Yi)^2 \tag{2}$$

For each vector, we calculate the distance between the vector and the center point of each cluster, and we choose the closest cluster. When we classify all the customers into groups, we need to recalculate the center point of each cluster. We obtain new center points for all clusters; we recalculate the distance between each vector and the new center point of all clusters to choose the closest one. We repeat these steps until no vector moves from one cluster to another. By the use of K-Means algorithm, we obtain the clustering shown in Table 5.

Table 5. Clustering result using k-means.

Cluster	Size %	Number	M	FP	C	CS	D	B	ME	FV	F	P
1	35,5	21030	4,8	3,6	4,5	4,7	2,3	3,5	4,6	4,7	4,5	3,8
2	4,33	2565	3,1	2,2	1,2	2,3	1,4	1,2	2,5	1,3	2,12	0,3
3	7,5	4443	1,5	0,6	1,7	0,8	0,2	1,7	2	1,3	1,4	0,5
4	0,06	36	2,3	1,2	0,8	1,1	0,9	1,2	0,7	1	0,5	0,2
5	9,93	5883	1,4	4,7	0,4	0,5	0,2	1,1	5	4,4	4,8	0,2
6	17,67	10468	3,2	2,3	1,54	1,37	2,6	1,2	0,4	2,1	1,3	0,4
7	4,7	2784	0,4	0,7	0,8	0,5	0,2	0,3	0,7	0,4	0,3	0,1
8	1,3	770	5	4,5	4,8	4,1	2,2	3,3	4,9	4,8	5	3,5
9	15,2	9004	1,2	2,5	2,1	1,4	0,3	1,15	4,5	4,7	5	1,5
10	3,81	2257	2,2	3,2	1,4	4,5	4,9	5	0,9	1,2	0,3	0,7

3.4 Discussion

Our approach uses the K-means algorithm to divide the shoppers of our store into 10 segments based on their purchases for different aisles and their paths inside the store. The next paragraph will summarize the characteristics of some clusters.

Cluster 1 relates to 21030 consumers, composed of 35,5% of the whole dataset. This cluster shows high purchase rate in all areas of the store, it contains loyal customers because they visited quite often the store and as a result, spent a quite high amount of money. It seems that the customers of this cluster do all their purchases in our store.as a result, they deserve a special treatment since they are very loyal an represent high profit.

Cluster 5 is a small group relates to 5883 consumers. It shows high purchase rate in areas of Frozen products, Meats, Fruits and Vegetables and Fish. It looks like these customers often visit our store and they appreciate the products of these areas. As a result, the store managers can offer personalized promotions based on their preferences in these areas.

Cluster 6 is large group contains to 10468 consumers, which represents 17,67% of the whole dataset. This cluster shows low purchase rate in all areas of the store. It seems that the customers of this cluster do most of their purchases in other stores but they often visit all the areas of our store. As a result, the store managers must offer the adequate promotions that motivate them to do their shopping in our store.

Cluster 9 is a group with moderate size relates to 9004 consumers. It shows high purchase rate in areas of Meats, Fruits and Vegetables and Fish. It looks like these customers often visit our store, but they don't appreciate the products of the other areas, so they purchase them in another store. As a result, the store managers must exploit the frequency of visits of this group and offer more personalized offers to strength the relationship with them and boost the profit of this cluster.

Cluster 10 is a small group relates to 2257 consumers. It shows high purchase rate in areas of Cereal and Snacks, Deli and Bakery B. It looks like these customers prefer the products of these three areas and they often visit our store to purchase the products of these areas. The store managers must offer promotions that link the products of these areas with the products of other areas in order to satisfy their needs and increase the profit of this cluster.

4 Generating Recommendations

4.1 Apriori Algorithm

For our case, the customers with similar purchasing habits are grouped into ten clusters by using k-means algorithm. Then, for each cluster, we will use the Apriori algorithm to identify the products that are frequently bought together by the customers in order to define representative basket for each cluster. There are different tools to analyze the behaviors of customers and predict their purchasing habits, but the most useful one is association rules. It is an efficient tool to discover the products which more frequently purchased together. The Apriori algorithm [9, 10, 11] is the most well-known association rules mining algorithm. It involves two stages, the first one is to generate the large itemsets and then to generate association rules.

For our case, we have ten clusters of customers, and we will generate representative baskets for each cluster in order to tailor the promotions and satisfy the customer's needs. For each cluster, we find the frequent itemsets whose occurrences exceed the predefined support threshold; we call those itemsets representative baskets. Firstly, we will create a representative basket for each customer; it will include all the products purchased by the customer during the last summer.

Map Reduce Apriori Algorithm

- Step 1: scan the dataset and we start with itemsets containing just a single item.
- Step 2: calculate min _supp = number of transactions*60%.
- Step 3: determine the support of each itemset.
- Step 4: keep the itemsets that meet min_supp and remove itemsets that do not.
- Step 5: Using the itemsets kept from step 3, generate all possible itemset combinations.
- Step 6: repeat steps from 3 to 5 until there are no more itemsets

4.2 Results

For our case, we apply the Apriori algorithm to generate representative basket for each cluster. As a result, we obtain ten representative baskets as shown in Table 6.

Table 6. Representative baskets of the clusters.

Cluster	Representative basket
Cluster 1	{bread, pasta, breakfast snacks, yogurt, meat, fish, orange, potatoes, onions, shampoo, tomatoes, tomato sauce, toothpaste}
Cluster 2	{bread, milk, coffee, meat, onions, snacks, fish, tomatoes, potatoes, toothpaste}
Cluster 3	{butter, cereal, apples, tomatoes, chocolate, potatoes, milk, carrots}
Cluster 4	{bread, milk, cheese, eggs, tomato sauce, milk, tea, potatoes, coffee}
Cluster 5	{bread, butter, milk, frozen meat, frozen fish, potatoes, onions, apples, oranges, fish}
Cluster 6	{bread, milk, pasta, breakfast snacks, yogurt, meat, onions, bananas, shampoo, fish, tomatoes, tomato sauce}
Cluster 7	{bread, milk, coffee, meat, onions, shampoo, potatoes}
Cluster 8	{apples, tomatoes, chocolate, bananas, oranges, potatoes, butter, meat, chicken, eggs, milk, carrots}
Cluster 9	{bread, milk, cheese, eggs, tomato sauce, milk, tea, coffee}
Cluster 10	{bread, butter, milk, potatoes, breakfast snacks onions, apples, oranges, shampoo}

The store's managers can use the results of our solution to send accurate recommendations, to optimize the organization of the shelf space and to place the associated items close together such that customers would find the products easily and not forget to purchase both items. On the other hand, if the store's managers stocked the associated items far apart, then the customers would spend more time browsing aisle by aisle in order to satisfy their needs. Moreover, they can use the basket analysis to design effective marketing strategies and cross-coupon program where customers purchasing an item X get a promotional discount for an item Y. Finding the association rules of a large dataset is an efficient tool to establish customer buying patterns with regard to which products have been purchased together frequently by which customer or cluster. Due to k-means algorithm and Apriori algorithm, we can understand the behavior of the customers, extract the preferences, group similar customers, generate representative baskets and find the products tend to buy together by customers in order to decide which products should be cross-sold or promoted together.

5 Conclusion and Future Work

The analysis of the customer behavior enables the marketing decision makers to apply efficient marketing strategies which aim to maintain good relationship with customers and improve their loyalty.

In this paper, Data analytics and machine learning have been employed to study customer's past data in order to extract the purchasing patterns of customers.

With the help of k-means algorithm, we can group the customers into homogeneous clusters with similar purchasing behavior. As a result, the retailers tailor promotions according to the customer's needs. The distinct customer groups characterized can help the retailers better understand its customers in terms of their profitability, and accordingly, adopt appropriate marketing strategies for different consumers. Moreover, by using the association rules we detect the representative basket for each group and establish customer buying patterns in order to offer the adequate promotions, optimize the organization of the store and maximize the customer satisfaction.

In the future research, we plan to enhance our solution by implementing its proto-type in a real supermarket in order to evaluate its efficiency and detect its drawbacks.

References

1. Phrabu, S., Venatesan, N.: Data Mining and Warehousing. New Age International Publisher, pp. 34–40 (2007)
2. Cho, Y.S., Jeong, S., Ryu, K.H.: Implementation of personalized u-commerce recommendation system using preference of item category based on RFM. In: the 6th International Conference on Ubiquitous Information Technologies & Applications, pp. 109–114, December 2011
3. Zaim, D., Bellafkih, M.: Bluetooth Low Energy (BLE) based geomarketing system. In: 2016 11th International Conference on Intelligent Systems: Theories and Applications (SITA), Mohammedia, pp. 1–6 (2016)
4. Zaim, D., Benomar, A., Bellafkih, M.: Developing a geomarketing solution. In: The second International Conference on Intelligent Computing in Data Science (ICDS 2018), Fez (2018)
5. Zaim, D., Benomar, A., Bellafkih, M.: Analyzing the trajectories of customers by using LCSS approach. In: 2020 Fourth International Conference on Intelligent Computing in Data Sciences (ICDS), 2020, pp. 1–4. https://doi.org/10.1109/ICDS50568.2020.9268721
6. Haiying, M., Yu, G.: Customer segmentation study of college students based on the RFM. In: 2010 International Conference on E-Business and E-Government, Guangzhou, pp. 3860–3863 (2010). https://doi.org/10.1109/ICEE.2010.968
7. MacQueen, J.B.: Some methods for classification and analysis of multivariate observations (1967)
8. McCarty, J.A., Hastak, M.: Segmentation approaches in data-mining: A comparison of RFM, CHAID, and logistic regression. J. Bus. Res. **60**(6), 656–662 (2007)
9. Agrawal, R., Srikant, R., et al.: Fast algorithms for mining association rules. In: Proceedings of 20th International Conference Very Large Data Bases, VLDB, pp. 487–499 (1994)
10. Agrawal, R., Imielienski, T., Swami, A.: Mining association rules between sets of items in large databases. In: Proceedings of Conference on Management of Data, 207–216. ACM Press, New York (1993)
11. Agrawal, A., Mannila, H., Srikant, R., Toivonen, H., Verkamo, A.: Fast discovery of association rules. In: Fayyad et al. (1996), pp. 307–328 (1996)

Check for updates

Driver Fatigue Detection via Eye State Analyses Based on Deep Learning Approach

Burcu Kır Savaş(✉) and Yaşar Becerikli

Kocaeli University, Umuttepe Campus, 41001 Kocaeli, Turkey
{burcu.kir,ybecerikli}@kocaeli.edu.tr

Abstract. Eye state analysis systems are generally developed in different fields such as driver fatigue detection system, human-computer interaction systems and security systems, etc. This paper proposes a detection system based on Deep Learning (DL) model to detect eye states on dynamic images. This system models in DL method for the classification on eye state information by using the SGD method with different adaptive parameters in daylight conditions. To determine the optimum parameter values in this proposed Convolutional Neural Network (CNN) model, we generated 2 different models as M-1 and M-2 changing the hyper-parameters of convolution steps. The generated M-1 model has been designed into 3 different submodels by M-1(a), M-1(b) and M-1(c) as updating the filter number in the convolution layers. In the experiments on these parameters, the average accuracy obtained using the ZJU database is 97.3%, the average accuracy obtained using the YawDD database is 97.9%, the average accuracy obtained using the NTHU-DDD database is 98.4%, and the average accuracy obtained using the KOU-DFD (our own dataset) database is 98.7%. These results prove that our approach is comparable to other state-of-the-art approaches. Moreover, the proposed DL method is more stable and more robust for detecting driver fatigue.

Keywords: Fatigue detection · Eye state classification · CNN · Human-computer interaction

1 Introduction

In our modern world, various means of transportation are widely driven by people in daily life. Therefore, human behavior is accepted to be important for detecting driver fatigue. Drivers can be affected by fatigue before they even notice a change in their driving style or fatigue. When they have fatigue, their reaction time to any event is prolonged. When these reaction times are delayed, drivers will be delayed for any maneuvers. Due to driver fatigue, such accidents are more severe. Traffic accidents due to fatigue based on driver behavior have become a subject pursued by researchers today. Also, automobile suppliers such as Bosch [1] and NVIDIA [2] have begun to develop driver fatigue detection systems based on behavioral-based (BM) measurement (BM) [3]. In the next few years, it is expected that autonomous systems, which leave the task of driving the vehicles to automatic driving systems, will be gradually put into use. Therefore,

M. Ben Ahmed et al. (Eds.): SCA 2022, LNNS 629, pp. 452–462, 2023.
https://doi.org/10.1007/978-3-031-26852-6_42

robust driver fatigue detection is a challenging problem for automated driving systems requiring further research. BM measurement is based on the human facial features [4–6]. It is accepted that eye state information is one of the significant facial features in humans [7–10] to be able to detect the physiological state of humans precisely. The eye state information is generally classified into two categories: open and closed eyes. Automatic eye state classification systems are widely used as security technologies, human–computer interaction systems and driver drowsiness recognition structures etc. Given the significance of these fields, automatic eye state classification systems attract the interest of researchers. There are many machine learning (ML) methods that have been researched such as Artificial Neural Network (ANN), Adaboost classification, Support Vector Machine (SVM) [11] etc. These ML methods are used to determine best feature extraction distribution and classification algorithm pair [12]. However, these traditional methods depend on manual feature extraction. Nowadays, DL models are mostly preferred to image and pattern recognition tasks because they contribute more in terms of time and performance compared to existing methods and detect the features automatically.

In this study, we detect driver fatigue in real time by using a BM measurement. A single camera monitors the driver's physical behavior and analyze the eye state information. The experimental results demonstrate that the CNN model obtains state-of-the-art results on popular benchmarks. This paper is organized as follows: Sect. 2 demonstrates related works. Section 3 gives an overview of the proposed fatigue detection approach and illustrates experimental results. Finally, Sect. 4 presents the conclusion and future works.

2 Related Works

In this part, some fatigue detection methods are detailed based on eye state analysis. Fatigue is associated with a natural physiological need, so it cannot be eliminated [13]. Therefore, researchers are still working on methods of monitoring and warning whether the driver is drowsy, instead of preventing driver fatigue [14, 15]. Lalonde et al., [16] in their study, detected the blinking movement of people in real time. In their system they used a blink detector on contrast images obtained in near infrared illumination. With this detector, they performed data collection and analysis. Coetzer et al. [17] tested the different classification algorithms for eye detection and compared the success rate of these algorithms to determine which technique will be the most sufficient for the driver fatigue detection system. Kir Savas et al. [18] detected driver face and found eye location for driver fatigue monitoring. Also, they used TBA for eye detection and analysis. Finally, they adopted Adaboost algorithm for eye classification. In recent years, deep learning algorithms has come into prominence to obtain and classify features related to driver fatigue. Zhang et al. [19] proposed a driver fatigue detection system with eye tracking based on CNN. They tested their system on IRF and ZJU dataset (total 4864 data). Kir Savas et al. [20], examined eye, mouth and head information to detect driver fatigue and classified them on their own dataset with the SVM classification algorithm. Dwivedi et al. [21] proposed the CNN model to take several hidden facial features and non-linear feature interactivity. They used Softmax to classify the driver as fatigue or non-fatigue.

Miah [22] et al. carried out their work to determine driver fatigue, which considers the state of the eye. And their system depended on eye locations and the optimum distance of the eye locates. Kir Savas et al. [23] proposed a Multi-task ConNN model to determine the driver fatigue/non-fatigue. The characteristics of the eye and mouth are used in their system. Driver fatigue is determined by computing PERCLOS and FOM metrics and also the model is tested on YawdDD and NTHU-DDD datasets. Zhao [24] et al. proposed a Multitasking Cascade Convolutional Network (MTCNN) architecture to detect driver drowsiness. First, they extracted a Region of Interest (ROI) for using feature point and they recommended EM-CNN network to detect eyes and mouth using ROI images.

3 System Overview

The system overview in proposed model is shown in Fig. 1. ZJU, YawDD, NTHU-DDD and KOU-DFD (Kocaeli University Driver Fatigue Dataset) data sets are generated for the training phase. Pseudocode for the system overview is shown in Fig. 2.

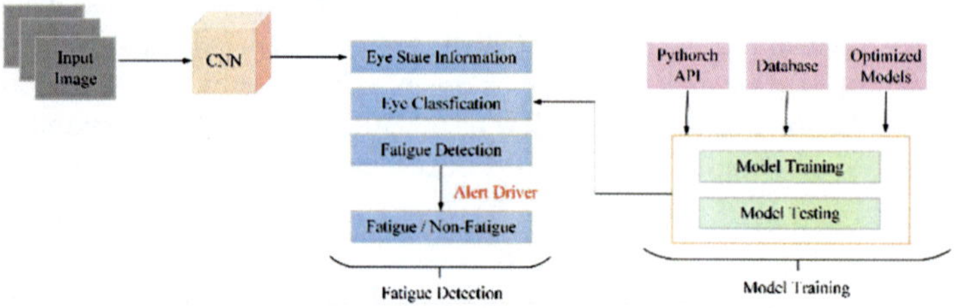

Fig. 1. System overview.

"Without any pre-processing step, the change of behavioral attribute in the driver's face area by monitoring the driver with fixed camera(s) in daylight conditions; By classifying them with the original models developed, the driver's fatigue/sleepiness can be detected with high performance on real-time systems with the help of fatigue parameters" is the hypothesis on which this project is based, and in this regard, the driver is monitored with two fixed cameras and the driver's face area is determined through the images obtained by cameras. Then, this system uses DL model for classification of eye state information. With the developed system, the proposed models CNN were tested with data sets in the literature and with a new and original image data set called KOU-DFD. The SGD optimization algorithm is employed while training the data in training phase. Many experiments have been determined the best optimal hyperparameters of the study. For this study, batch size 32, learning rate 0.01, momentum value 0.6, and weight decay 0.0005 have been determined. We proposed a cross-entropy loss by monitoring the loss function. This study was developed with python programming language and Nvidia CUDA. In the experimental stage, we used NVIDIA GeForce GT540M, Intel Core i7, 4GB DDR3, Ubuntu 16.04.

```
Begin
    Input fixed camera video
    frames_count :=0
    For frames_count==120 Do
    EyeStateInformation= {0, state < 0.24
                         {1,        else
    Check state of eyes
    Load Eye Classification
    For c=1 to 3 Do
            Training CNN(K×L×C)
                For k=1 to C Do
                    conv(:,:,k)
                        For j=1 to L Do
                            For i=1 to K Do
                                conv(i,j,k)
                            end for i
                        end for j
                end for k
                pooling(max)
                fullyconnected(pooling)
        Load Fatigue detection
        If EyeStateInformation == fatigue then
        Driver alert
End
```

Fig. 2. Pseudocode for the system overview.

3.1 Fatigue Determination

This system will forecast if the eye is open/closed and calculate every progress from open to closed/closed to open state. According to the literature [25], a normal blink can change in duration from less than 10 ms to 80 ms, thus a blink rate of 20 blinks per minute for an ordinary person. This paper determines the driver's fatigue state using the count of blinking and closed time of driver.

3.2 Experimental Results

In this part, we discuss the driver fatigue detection system in real time. What might this system need? In addition, we present the architecture of our proposed method and CNN network structure of the designed models. The architecture of our proposed approach can be partitioned into two parts; model training and fatigue detection. During the model training part, CNN models are trained and tested with different adaptive parameters on different datasets. On the other hand, during the fatigue detection part, the eye state information is classified and the driver is warned according to the fatigue condition.

In real-time drivers' fatigue systems, running in a fixed frequency range makes it easier to detect fatigue parameters without loss. The fixed frequency range is determined as 4 s (total number of frames: 120) and this interval is kept constant in the study. In other words, every time a new frame is processed, the last frame is deleted and continuity is ensured. In the system, a 120-length vector whose elements are 0 or 1 is processed. It is detected that the eye is closed in "0" frames; "1" indicates that the eye is open in the frame. Thus, the input consisting of 120 frames is converted into a vector with 120 elements. In Fig. 3, the process of getting the image to the frame queue as a sample image in the KOU-DFD dataset has been given in detail.

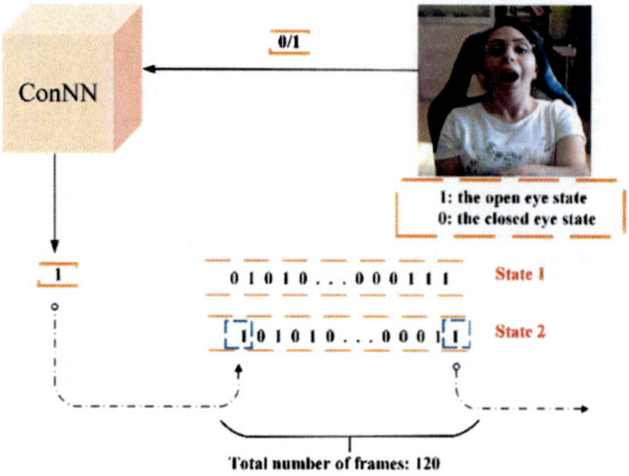

Fig. 3. Image run into frame queue.

In the problem-solving phase, the selection of modifiable model hyperparameters (kernel/kernel dimensions/parameters weight, layer, etc.) is one of the critical conditions in mathematical model-based approaches. The selection of model parameters is very challenging to determine the best optimal parameter. Therefore, various model parameters have been tried many times to detect the most suitable parameter values. Therefore, it would be beneficial to enhance adaptive parameter substitution approaches. Therefore, different hyperparameters including filters count in the convolution layers, striding count and padding count have been updated and multiple tests have been conducted to select the optimum parameter values. To determine the optimum parameter values in this proposed CNN model, we generated 2 different models as M-1 and M-2 updating the hyperparameters of convolution steps. The generated M-1 model have been designed into 3 different submodels as M-1(a), M-1(b) and M-1(c) by updating the filter number in the convolution layers. This proposed CNN architecture is indicated in Table 1 and the modified hyperparameters are indicated in bold.

Table 1. The proposed models' network structure

Model	Layers name	Hyper parametres
M-1 (a)	1(Convolution, A, Pooling) 2(Convolution, A, Pooling) 3(Convolution, A, Pooling)	5 × 5, 3, str = 1; ReLU; 2 × 2, Pooling(max), str = 2
		5 × 5, 4, str = 1; ReLU; 2 × 2, Pooling(max), str = 2
		4 × 4, 5, str = 1; ReLU; 2 × 2, Pooling(max), str = 2
M-1(b)	1(Convolution, A, Pooling) 2(Convolution, A, Pooling) 3(Convolution, A, Pooling)	5 × 5, 6, str = 1; ReLU; 2 × 2, Pooling(max), str = 2
		5 × 5, 8, str = 1; ReLU; 2 × 2, Pooling(max), str = 2
		4 × 4, 10, str = 1; ReLU; 2 × 2, Pooling(max), str = 2

(continued)

Table 1. *(continued)*

Model	Layers name	Hyper parametres
M-1 (c)	1(Convolution, A, Pooling)	5 × 5, 12, str = 1; ReLU; 2 × 2, Pooling(max), str = 2
	2 Convolution, A, Pooling)	5 × 5, 16, str = 1; ReLU; 2 × 2, Pooling(max), str = 2
	3(Convolution, A, Pooling)	4 × 4, 20, str = 1; ReLU; 2 × 2, Pooling(max), str = 2
M-2	1(Convolution, A, Pooling)	5 × 5, 3, str = 2; ReLU; 2 × 2, Pooling(max), str = 2
	2 Convolution, A, Pooling)	5 × 5, 4, str = 2; ReLU; 2 × 2, Pooling(max), str = 2
	3(Convolution, A, Pooling)	4 × 4, 5, str = 2; ReLU; 2 × 2, Pooling(max), str = 2

* str = stride

The performance metric calculation is given in Eq. 1. N is data size, y is target values, $\hat{y}$ is predicted values.

$$Accuracy = \frac{1}{N}\sum_{i}^{N} 1(y_i == \hat{y}) \quad (1)$$

3.3 Experimental Results

Experimental results indicate that the best accuracy rate is in M-l-(a). For M-1(a) models, accuracy and loss graphs on ZJU data are demonstrated in Fig. 3, accuracy and loss graphs on YawDD data are demonstrated in Fig. 4, accuracy and loss graphs on NTHU-DDD are demonstrated in Fig. 5 and accuracy and loss graphs on KOU-DFD are demonstrated in Fig. 6.

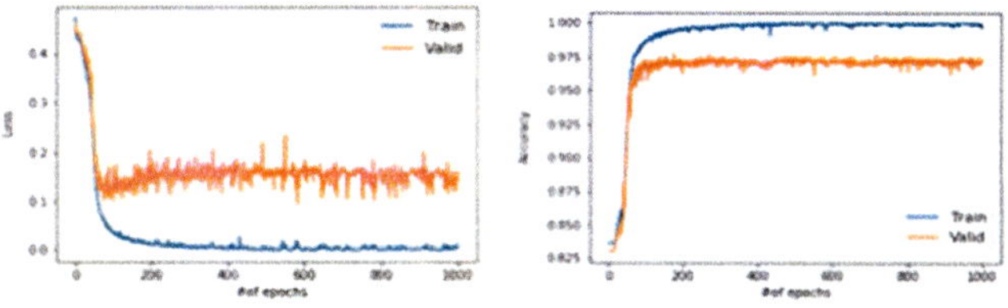

Fig. 4. Training/validation loss and accuracy graphics on ZJU - 1000 epoch

The training and validation errors tend to diminish in the same direction, based on the error function. The proposed model has been observed in up to 1000 epochs and it has been concluded that though the fluctuations continue in the graph at the end of 1000 epochs, the fluctuations remain the same and nothing changed, so the model learning is finished. Similarly, based on the accuracy graph, the training and validation accuracy has been observed and they tend to go up rapidly and simultaneously. At the end of 1000 epochs, it is concluded that even if the fluctuations continue, the fluctuations remain

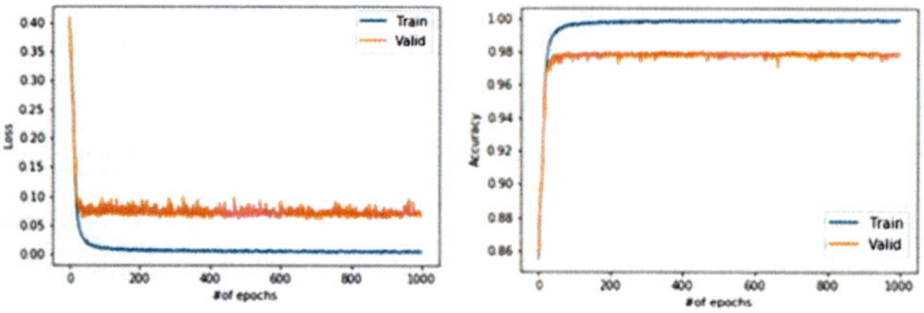

Fig. 5. Training/validation loss and accuracy graphs on YawDD - 1000 epoch

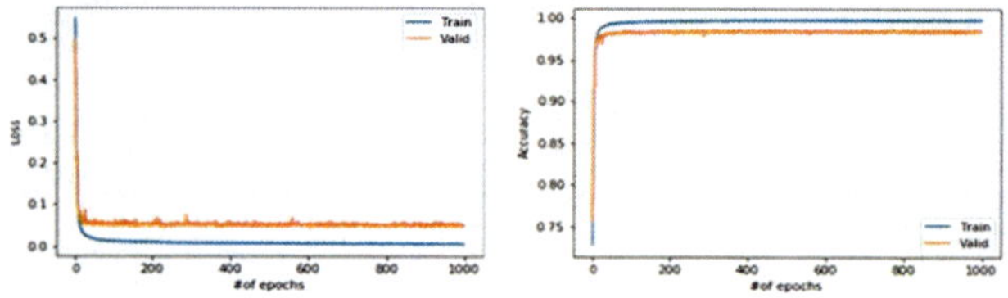

Fig. 6. Training/validation loss and accuracy graphs on NTHU-DDD -1000 epoch

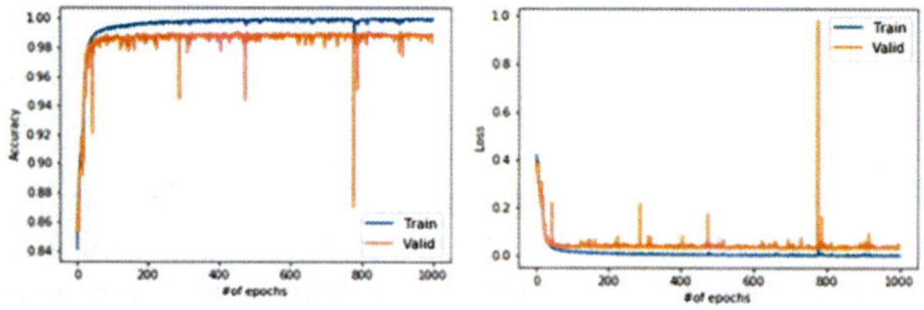

Fig. 7. Training/validation loss and accuracy on KOU-DFD -1000 epoch

the same and nothing changed, so the model learning is finished. Table 2 indicates the comparison of success rates based on the eye state information.

During the end of 1000 cycles, test accuracy on the ZJU dataset is 98.75%, test accuracy on the YawDD dataset is 97.73%, test accuracy on NTHU-DDD is 98.55%, and test accuracy on the KOU-DFD dataset is 98.81%. The experimental results prove that the proposed model is successful and works stably. Comparison of success rate of the proposed model with other approaches is given in Table 3. Also, accuracy rates show that the proposed approach has higher accuracy rate and is more successful than other state-of-the-art approaches.

A comparison of different studies is given in Table 3. This paper outperforms a better result than many studies in the literature using the ZJU, YawDD, NTHU-DDD, KOU-DFD datasets. These results proves this study robust.

Table 2. The success rates comparison based on the eye state information

CNN M-1 (a)	Epoch	ZJU		YawDD		NTHU-DDD		KOU-DFD	
		F-score	Accuracy	F-score	Accuracy	F-score	Accuracy	F-score	Accuracy
Training accuracy	100	98,68%	97,96%	99,60%	99,40%	99,46%	98,46%	99,25%	99,35%
	250	99,66%	98,62%	99,76%	99,68%	99,63%	98,52%	99,70%	99,64%
	500	99,89%	98,65%	99,80%	99,76%	99,70%	98,72%	99,89%	99,76%
	750	99,81%	99,93%	99,86%	99,84%	99,76%	98,66%	99,96%	99,82%
	1000	99,64%	99,78%	99,82%	99,82%	99,72%	99,05%	99,88%	99,84%
Test accuracy	100	96,72%	97,72%	97,80%	97,38%	98,35%	97,25%	98,64%	98,04%
	250	97,12%	97,28%	97,84%	97,27%	98,40%	97,35%	98,69%	98,22%
	500	97,22%	98,17%	97,70%	97,18%	98,36%	98,64%	98,92%	98,48%
	750	96,87%	98,97%	97,81%	97,67%	98,39%	98,45%	98,82%	98,76%
	1000	97,31%	98,75%	97,90%	97,73%	98,48%	98,55%	98,71%	98,81%

Table 3. The success rate compared with other approaches

Literature	Year	Method	Dataset	Accuracy (%)
Luo et al. [26]	2013	AdaBoost	Their own dataset	92%
Punitha et al. [27]	2014	SVM	Their own dataset	93.5%
Yan et al. [28]	2015	CNN	Their own dataset	95.56%
Zhang et al. [19]	2017	CNN	IRF, ZJU	98.83%
Li et al. [29]	2018	Kalman Filter &Mean Shift	Their own dataset	93.19%
Huang et al. [30]	2018	FDCN	ZJU, CEW	94.9%
Ma et al. [31]	2018	SVM	Their own dataset	91%
Gu et al. [32]	2018	MTCNN	EMD, ZJU, CEW	98.85%
Sharan et al. [33]	2019	CNN	Their own dataset	96.56%
Liu et al. [34]	2019	CNN	DFII, CEW	96.2%
Zhang et al. [35]	2019	Facial Key Points	Their own dataset	92%

(*continued*)

Table 3. *(continued)*

Literature	Year	Method	Dataset	Accuracy (%)	
Savaş et al. [23]	2019	SVM	YawDD	97.93%	
Sheikh et al. [36]	2021	SVM	DROZY, NTHU-DDD	90%	
Liu et al. [37]	2021	WBCNNTL	ZJU, CEW	97.15%	97.41%
Saurav et al. [38]	2022	CNN	ZJU, CEW, MRL	97.99%	
Proposed Model	2022	CNN	ZJU, YawDD	97.3%	97.9%
			NTHU-DDD, KOU-DFD	98.4%	98.7%

4 Conclusions

This study presents eye state analysis used in fatigue detection fields in daylight conditions. This detection system uses the DL model to determine the eye states on dynamic images and classifies the datasets. CNN models have been developed using an SGD optimizer by updating different adaptive parameters on the ZJU, YawdDD, NTHU-DDD and KOU-DFD video datasets. The experimental results show that CNN accuracy outperforms most state-of-the-art methods. Also, experimental results exhibit that our DL method is greatly stable and robust for detecting driver fatigue. In the future, we are planning to use different mixed DL methods. And also, we would like to develop a segmentation method to increase the real-time testing accuracy.

References

1. Bosch, CES® 2017: Bosch is showing these smart solutions in Las Vegas - Bosch Media Service. http://www.bosch-presse.de/pressportal/de/en/ces-2017-bosch-is-showing-these-smart-solutions-in-las-vegas-81792.html. Accessed 16 Nov 2020
2. Nvidia, NVIDIA DRIVE IX Self-Driving Vehicles|NVIDIA Automotive. https://www.nvidia.com/en-us/self-driving-cars/drive-ix/. Accessed 16 Nov 2020
3. VentureBeat, Autonomous cars could end drowsy driving, but don't hold your breath. https://venturebeat.com/2017/11/19/autonomous-cars-couldend-drowsy-driving-but-dont-hold-your-breath/. Accessed 14 Jan 2020
4. Pratama, B.G., Ardiyanto, I., Adji, T.B.: A review on driver drowsiness based on image, bio-signal, and driver behavior. In: IEEE International Conference on Science and Technology-Computer, Yogyakarta/Indonesia (2017)
5. Kang, H.B.: Various approaches for driver and driving behavior monitoring: a review. In: Proc. IEEE Int. Conf. Comput. Vis., Aralık (2013)
6. Q. Ji, Z. Zhu, P. Lan., "Real-time Nonintrusive monitoring and prediction of driver fatigue", IEEE Transactions on Vehicular Technology, 53(4), 1052–1068, 2004
7. D. C. Ciresan, U. Meier, L. M. Gambardella, J. Schmidhuber, "Convolutional Neural Network Committees for Handwritten Character Classification," International Conference on Document Analysis and Recognition, Beijing/China 2011

8. Junaedi, S., Akbar, H.: Driver drowsiness detection based on face feature and PERCLOS. J. Phys. **1090**, 12037 (2018)
9. Darshana, S., Fernando, D., Jayawardena, S., Wickramanayake, C., Silva, D: Efficient PERCLOS and gaze measurement methodologies to estimate driver attention in real time. In: Proc. - Int. Conf. Intell. Syst. Model. Simulation, Langkawi/Malaysia (2015)
10. Xie, J.F., Xie, M., Zhu, W.: Driver fatigue detection based on head gesture and PERCLOS. In: Int. Conf. Wavelet Act. Media Technol. Inf. Process. (ICWAMTIP), Chengdu/China (2012)
11. Akdeniz, F., Becerikli, Y.: Performance comparison of support vector machine, K-nearest-neighbor, artificial neural networks, and recurrent neural networks in gender recognition from voice signals. In: 2019 3rd International Symposium on Multidisciplinary Studies and Innovative Technologies (ISMSIT), pp. 1–4. IEEE, October 2019
12. Kayikcioglu, I., Akdeniz, F., Köse, C., Kayikcioglu, T.: Time-frequency approach to ECG classification of myocardial infarction. Comput. Electr. Eng. **84**, 106621 (2020)
13. Čolić, A., Marques, O., Furht, B.: Driver drowsiness detection: systems and solutions. Springer International Publishing (2014)
14. Kim, J., Shin, H.: Algorithm & SoC design for automotive vision systems. Springer (2014)
15. Hong, T., Qin, H.: Drivers drowsiness detection in embedded system. In: 2007 IEEE International Conference on Vehicular Electronics and Safety, December 13 (2007)
16. Lalonde, M., Byrns, D., Gagnon, L., Teasdale, N., Laurendeau, D.: Real-time eye blink detection with GPU-based SIFT tracking. In: Fourth Canadian Conference on Computer and Robot Vision (CRV'07), May 28 (2007)
17. Coetzer, R.C., Hancke, G.P.: Eye detection for a real-time vehicle driver fatigue monitoring system. In: 2011 IEEE Intelligent Vehicles Symposium (IV), June 5, 2011
18. Savas, B.K., Becerikli, Y.: Development of driver fatigue detection system by using video images. Akıllı sistemlerde Yenilikler Ve Uygulamalar Konferansı, Alanya/Turkey (2017)
19. Zhang, F., Su, J., Geng, L., Xiao, Z.:Driver fatigue detection based on eye state recognition. In: Proc. - 2017 Int. Conf. Mach. Vis. Inf. Technol. C., Singapore/Singapore (2017)
20. Savas, B.K., Becerikli, Y.: Real time driver fatigue detection based on SVM algorithm. In: Proceedings of 6th International Conference Control Eng. Inf. Technol. (CEIT), 25–27 October, İstanbul/Turkey (2018)
21. Dwivedi, K., Biswaranjan, K., Sethi, A.: Drowsy driver detection using representation learning. In: 2014 IEEE International Advance Computing Conference (IACC), 21–22 Şubat, Gurgaon, Hindistan (2014)
22. Miah, A.A., Ahmad, M., Mim, K.Z.: Drowsiness detection using eyeblink pattern and mean eye landmarks distance. In: International Joint Conference on Computational Intelligence (IJCCI 2018), 4–15 Aralık, Bangladeş (2018)
23. Savas, B.K., Becerikli, Y.: Real Time Driver Fatigue Detection System Based on Multi-Task ConNN. IEEE Access **8**, 12491–12498 (2020)
24. Zhao, Z., Zhou, N., Zhang, L., Yan, H., Xu, Y., Zhang, Z.: Driver fatigue detection based on convolutional neural networks using EM-CNN. In: Computational Intelligence and Neuroscience, 18 November 2020
25. Boucetta, Z., El Fazziki, A., El Adnani, M.: Deep learning based driver's fatigue detection framework. In: Innovations in Smart Cities Applications Volume 4: The Proceedings of the 5th International Conference on Smart City Applications, pp. 370–380. Springer International Publishing (2021)
26. Luo, X., Hu, R., Fan, T.: The driver fatigue monitoring system based on face recognition technology. In: Fourth International Conference on Intelligent Control and Information Processing (ICICIP), 9–11 Haziran, Beijing, Çin (2013)
27. Punitha, A., Geetha, M.K., Sivaprakash, A.: Driver fatigue monitoring system based on eye state analysis. In: 2014 International Conference on Circuits, Power and Computing Technologies [ICCPCT-2014], 20–21 Mart, Nagercoil, Hindistan (2014)

28. Yan, C., Jiang, H., Zhang, B., Coenen, F.: Recognizing driver inattention by convolutional neural networks. In: International Congress on Image and Signal Processing (CISP), 14–16 Oct., Shenyang/China (2015)
29. Li, H., Xiaoyu, Z., Qisong, Z.: Design of human eye information detection system. In: 2018 Chinese Control And Decision Conference (CCDC), 9–11 Haziran, Shenyang, Çin (2018)
30. Huang, R., Wang, Y., Guo, L.: P-FDCN based eye state analysis for fatigue detection. In: IEEE 18th International Conference on Communication Technology (ICCT), 8–11 Ekim, Chongqing, Çiin (2018)
31. Ma, J., Zhang, J., Gong, Z., Du, Y.: Study on fatigue driving detection model based on steering operation features and eye movement features. In: 2018 IEEE 4th International Conference on Control Science and Systems Engineering (ICCSSE), 24–26 Ağustos, Wuhan, Çin (2018)
32. Gu, W.H., Zhu, Y., Chen, X.D., He, L.F., Zheng, B.B.: Hierarchical CNN-based real-time fatigue detection system by visual-based technologies using MSP model. IET Image Proc. **12**(12), 2319–2329 (2018)
33. Sharan, S.S., Viji, R., Pradeep, R., Sajith, V.: Driver fatigue detection based on eye state recognition using convolutional neural network. In: 2019 International Conference on Communication and Electronics Systems (ICCES), 17–19 Temmuz, Coimbatore, Hindistan (2019)
34. Liu, S., Yu, L., Hou, M.: An efficient method for driver fatigue state detection based on deep learning. In: 2019 2nd International Conference on Safety Produce Informatization (IICSPI), pp. 28–30 Kasım, Chongqing, Çin (2019)
35. Zhang, N., Zhang, H., Huang, J.: Driver fatigue state detection based on facial key points. In: 2019 6th International Conference on Systems and Informatics (ICSAI) (2019)
36. Sheikh, A.A., Mir, J.: Machine learning inspired vision-based drowsiness detection using eye and body motion features. In: 2021 13th International Conference on Information & Communication Technology and System (ICTS), pp. 146–150. IEEE, October 2021
37. Liu, Z.T., Jiang, C.S., Li, S.H., Wu, M., Cao, W.H., Hao, M.: Eye state detection based on weight binarization convolution neural network and transfer learning. Appl. Soft Comput. **109**, 107565 (2021)
38. Saurav, S., Gidde, P., Saini, R., & Singh, S.: Real-time eye state recognition using dual convolutional neural network ensemble. J. Real-Time Image Process., 1–16 (2022)

Real Time Multi-digit Number Recognition System Using YOLOv3 and YOLOv5

Muhammed Ali Irmak, Hakan Akgün, Emirhan Ekşi, Sefa Öztürk, Fulya Akdeniz, Burcu Kır Savaş(✉), and Yaşar Becerikli

Kocaeli University, Umuttepe Campus, 41001 Kocaeli, Turkey
{215112037,180202103,180202079,180202036,fulya.akdeniz, burcu.kir,ybecerikli}@kocaeli.edu.tr

Abstract. To cope with modern requirements, recognition of combined multi-digit numbers are necessary. State-of-the-art methods have used object detection models to recognize the individual digits and this way makes the automatic multi-digit detection/recognition techniques not effective, not certain and not stable. For this reason, it is necessary to apply a learning-based method/system to detect the multi- digits. In this study, YOLOv3-tiny and YOLOv5 based real-time multi-digit number detection/classification system has been designed. Images have been taken from the camera, the multi-digit number's region has been localized and finally, the multi-digit recognition process has been performed. In cases where more than one digit number is sent to the system, it is possible to find out how many digits this multi-digit number has by processing the probabilities of what each written number may be, separated by deep learning. MNIST and Street View House Number (SVHN) datasets have been used in the experimental study. The results have been analyzed comparatively and the training and test results have been given.

Keywords: Human-computer interaction · Deep learning · Multi-digit numbers · YOLOv3-tiny · YOLOv5

1 Introduction

Digitalization has become one of the most important developments of the time [1]. With digitalization, the issue of digitizing and scanning information in many fields [2] including using handwriting has begun to come to the fore. The advantages and convenience of number recognition systems also increase the attractiveness of these systems. Number detection, analysis and recognition are used effectively in many areas from vehicle license plate recognition to bank account control. However, there are fundamental difficulties in obtaining number detection and recognition solutions, such as determining the number of digits that can be experienced in the recognition of a printed number. Although it is quite easy for humans, it is a very difficult problem to perceive lines and curves on a floor as numbers. Number recognition is still not a fully resolved problem. The conversion of handwritten documents into digital recording copies that are more

M. Ben Ahmed et al. (Eds.): SCA 2022, LNNS 629, pp. 463–472, 2023.
https://doi.org/10.1007/978-3-031-26852-6_43

available through systems such as digital forms and databases is a major issue. The difficulty in handwriting recognition is due to a large number of different writing characters and differences from person to person, as well as the fact that the numbers are written in conjunction with each other. The human visual system is not affected by the size and direction differences of numbers, but in an automatic system, these concepts can cause problems in number recognition. Also, the image of a number written and imaged on paper will contain a lot of noise. For this reason, it is observed that the problem of recognizing handwritten numbers is a difficult and problem yet to be solved.

Multi-digit number recognition generally uses optical character recognition (OCR) methods [3–5]. The performance of OCR is correlated with the success of the detection methods. Multi-digit detection/classification from the photograph is still a difficult assignment due to the difference in the visual aspect of digits due to a wide range of designs, hues, intensity and blush. Furthermore, shadows, lighting, focus blurs, resolution of the camera lens distortion and motion etc. affect the detection of multi-digit numbers so easily [4].

The rest of this paper is organized as follows: Related works section explains different approaches recently used for multi-digit number detection. The method section gives us the models used in this paper, the essential objectives of this system architecture and the contribution of this paper to the literature. The experimental section shows the training and test phases. And finally, the Conclusion and Discussion section give the results of this paper.

2 Related Works

Albehadili et al. detected handwritten Arabic numerals using deep neural network methods and trained them on 70,000 MNIST data by assigning a learning rate of 0.01 and achieved a 98% success rate in their study [6]. Nagai et al. developed a new method for recognizing single Japanese historical cursive lines of text. The method reaches an accuracy rate of over 95% for texts consisting of 46 Hiragana characters, and over 84.08% from texts containing thousands of Kanji symbols [7]. Kim et al. performed edge-based segment image generation and license plate recognition for vehicle license plate recognition system. The results of the real-time license plate recognition method demonstrated that the rate of plate detection is 97.5% and the overall license plates character detection performance is 99.3% for an average of 1535 vehicles per day [8]. LeCun et al. created a handwritten number dataset (MNIST). This dataset contains 60,000 training and 10,000 test images [9]. Asif et al. worked on the number recognition algorithm up to 18 digits working in the field of scanning images and converting the scanned information into digital format and evaluated this method in the public SVHN dataset. Also, this system achieved over 96% accuracy in the detection of streets numbers. They found that they achieved 97.84% accuracy in the per-digit recognition task [10]. Salvador et al. developed a hybrid Hidden Markov Model (HMM) for handwritten texts. In the study, the structural part of the optical model is modeled with Markov chains and a Multilayer Sensor estimates the emission probabilities [11]. Bajaj et al. used three different types of properties in the Devanagari Digits classification: density properties, moment properties, and descriptive component properties. Their classification performs with 89.6%

accuracy for handwritten Devanagari numerals [12]. Hanmandlu et al. have presented classification of handwritten Hindi and English numerals in their work. The overall recognition rate was 95% for Hindi numerals and 98.4% for English numerals [13]. Khedher et al. have explained that the recognition of Characters is largely dependent on the attributes used. Various features of handwritten Arabic characters were selected. A recognition system was created according to the selected features. The system has been trained and tested with handwritten Arabic character samples. The recognition based on selected features gives an average accuracy of 88% for numbers and 70% for letters [14].

Research based on YOLOv5 models are so limited. Research mostly focuses on object detection. However, Annapareddy et al., analyzed character recognition, and classified handwritten medical texts and numbers. They used 363 manuscript images obtained from the University Teaching Hospital of Kigali in Rwanda. They developed the YOLOv5s model in their study. They obtained a mean prediction value of 98.43% for Single Class and a mean prediction value of 35.95% for Multi-Class [15]. Chandra et al. studied Optical Character Recognition and developed R-CNN, YOLOv5s and SVM methods in their work [16].

3 Methods

The main objectives of this proposed project can be summarized as follows:

- In the proposed study, multi-digit numbers in printed documents can be detected with high accuracy.
- With the deep learning model to be developed, multi-digit numbers can be recognized with high accuracy.
- With the proposed system, the recognition and overall computational complexity can be reduced.
- With the comprehensive experiments performed on two different data sets, it can be shown that the proposed method run effectively as the results obtained can be compared with the existing methods.

To achieve the above goals in this study,

- Taking images with the camera
- Detection of multi-digit numbers
- Finding the number of digits
- Classifying and estimating number of images with deep learning method on MNIST and SVHN datasets
- The results are analyzed comparatively and training and test results are given graphically.

The contribution of this study to the literature,

- We develop an end-to-end multidigit handwritten recognition system.
- We perform localization and identification of digits simultaneously in a multi-digit image using YOLOv3- tiny and YOLOv5s model.

The system overview is given in Fig. 1.

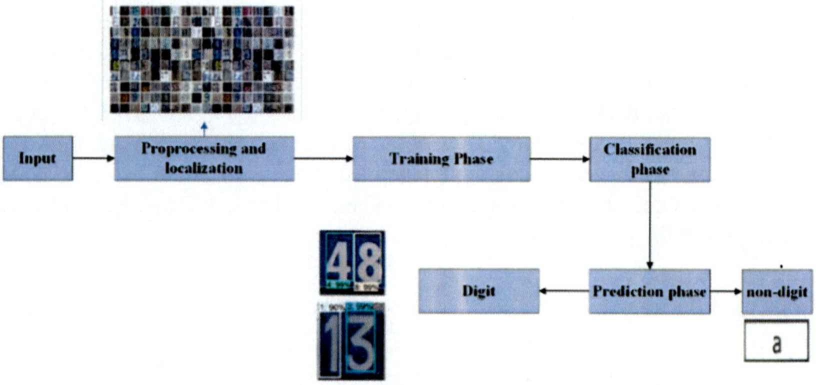

Fig. 1. System overview

3.1 Datasets

MNIST [9] and SVHN [17] datasets have used in the study. MNIST data are the adapted dataset for image processing and analysis. This dataset consists of handwritten number images. The SVHN dataset contains images of street numbers annotated with bounding boxes and labels. SVHN images consist of multi-digit numbers and sizes.

3.2 YOLOv3-Tiny

CNN models [18] have been extensively used in image detection problems. For all deep learning models, the YOLOv3 object detection algorithm [19] has more flexible and higher accuracy rate than many techniques in image processing. Recently, YOLOv3[20] has been used for detecting and localizing objects in a wide range of application fields including object detection, remote detection, medical domain and so on [21–24]. The YOLOv3 algorithm frames the objects or numbers it detects on the pictures with a bounding box. It divides the image given as input to the YOLOv3 algorithm into N × N boxes. All boxes have a confidence score. These boxes predict the object to be detected, therefore there may be more than one box at first. It uses the Non-Maximum Suppression (NMS) algorithm to solve this problem. The YOLOv3 detection process is given in Fig. 2 and Fig. 3.

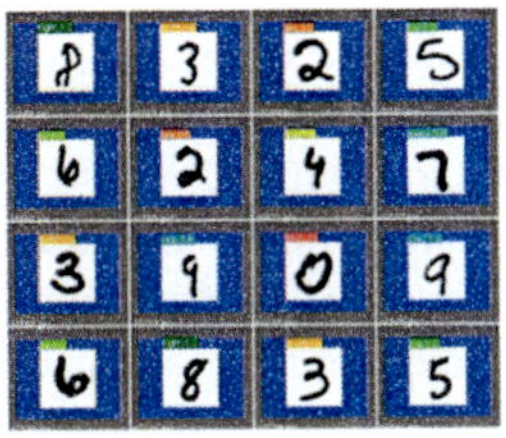

Fig. 2. YOLOv3-tiny on MNIST database

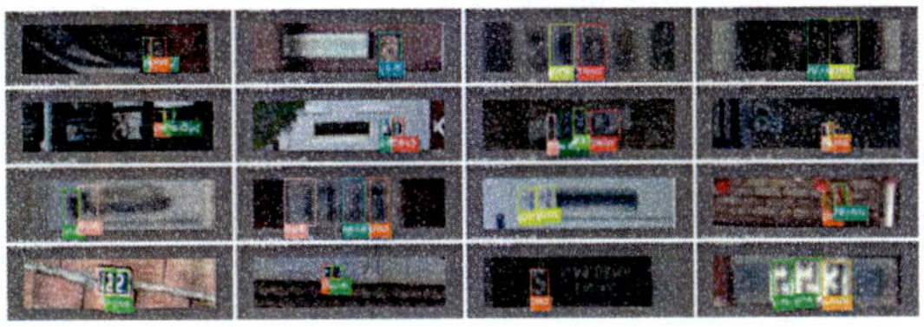

Fig. 3. YOLOv3-tiny on SVHN database

3.3 YOLOv5

YOLOv5 is the first YOLO model to run on the PyTorch framework, unlike other YOLO versions. Compared to previous versions of the YOLOv5 model, it is faster because the weight file size is 90% smaller. In addition, the detection accuracy of the network model is high compared to other versions. This greatly increases the effectiveness of the YOLOv5 model in real-time applications. Thanks to these advantages, it shows high performance in embedded systems. YOLOv5 has 4 different versions YOLOv5s, YOLOv5m, YOLOv5l and YOLOv5x. These architectures are different from each other in feature extraction, convolutional kernel of the network, size of the model, and model parameters [25,26]. The YOLOv5 detection process is given Fig. 4 and Fig. 5.

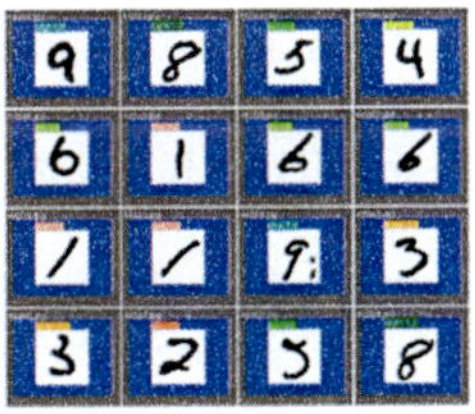

Fig. 4. YOLOv5 on MNIST database

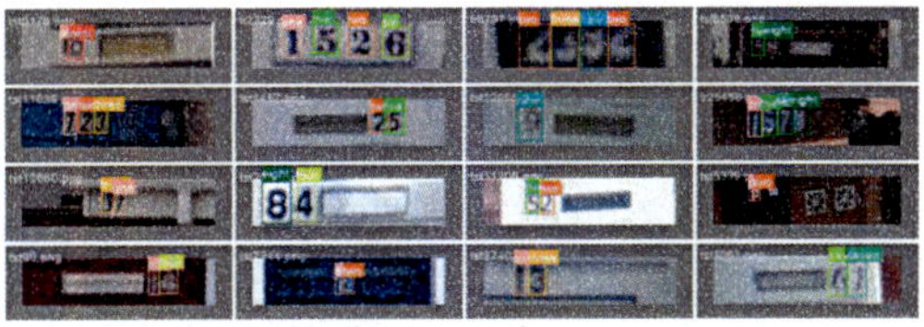

Fig. 5. YOLOv5 on SVHN database

3.4 Non-maximum Suppression (NMS) Algorithm

The NMS algorithm [27] is an algorithm employed in many computer vision, image processing and object detection. With this algorithm, an object is selected among many overlapping objects. The measure frequently used the overlap measure. Most object detection algorithms perform some kind of windowing. Thousands of windows are created directly on the property of the image. It assumes that only one of these windows contains objects and uses a classifier to obtain scores for each class. The NMS algorithm is used to select best of these boxes in accordance with the score value.

3.5 Optimization Algorithm

Stochastic Gradient Descent (SGD): SGD algorithm is a rapid and usable updating optimizer at every step. The parameters are renewed for each training sample. Because of these updates, the parameters have high variance values, causing the error function to fluctuate at different intensities, in this paper. In this optimizer, a few random samples are chosen for every update rather than the whole dataset. Although it is useful to use the entire dataset with less noise, it is inevitable to encounter problems if the dataset reaches very large sizes. Therefore, instead of the gradient sum of the cost function of all samples in the SGD algorithm, only one sample has the gradient of the cost function for each update [28, 29].

Adaptive Momentum (AdaM): AdaM algorithm is the first-order gradient-based optimization of stochastic objective functions based on adaptive low-order moment estimates. This algorithm's implementation is simple, quantitively forceful, requires little storage, in-variant from diagonal gradient scaling, and is suitable for large issues in terms of data or parameters. The algorithm is also suitable for non-stationary targets, very noisy and sparsely sloped issues [30, 31].

4 Experimental Analysis

The aim of this work is to detect individual digits in the multiple digits and classify them for automatic systems. This study is developed based on YOLOv3-tiny and YOLOv5 with high precision accuracy in locating and recognizing street door numbers and multiple digit numbers in the normal environment. The MNIST and SVHN datasets are used in our experiments. This paper selects 10000 train samples and 1500 test samples from both datasets. Thus, people can easily search for information based on door numbers or document recorded dates in the images. For training and testing part; We used Intel i7 – 11700F CPU with 2.5 GHz NVIDIA GeForce RTX 3070 which has 32 GB RAM for CPU and 8 GB for GPU.

Multiple digit number recognition in random images in automatic systems is admitted to be a difficult problem. In the literature, digit recognition processes are generally performed on mobile platforms. However, considering the low CPU performance and limited memory resources, the predicted performances cannot be obtained from the applications of deep learning architectures on the mobile platform. For this reason, as proposed in this study we are planning to find robust models for multi-digit systems and then we will use an embedded system with high CPU and GPU performance. Images have been obtained with a camera. The digit in the images taken from the camera is first preprocessed, segmented and detected to reduce the calculation steps. The aim here is to reduce the number of calculation operations by separating the multi-digit number into its digits and to obtain a robust multi-digit detect system. In addition, the system determines which number belongs to which number according to the localization information of the multi-digit numbers. In the detection process, the digits will be marked and the digit calculation of the number will be made. Table 1 gives Confusion matrix and F1 Curve for different datasets and optimization algorithm on YOLOv3-tiny and YOLOv5.

Table 1. Confusion matrix for different datasets and optimization algorithm and F1 Curve on YOLOv3-tiny and YOLOv5.

Model	Dataset/ Optimizier	Confusion matrix	F1 Curve
YOLOv3-tiny	MNIST/SGD	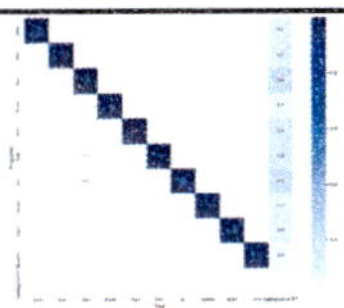	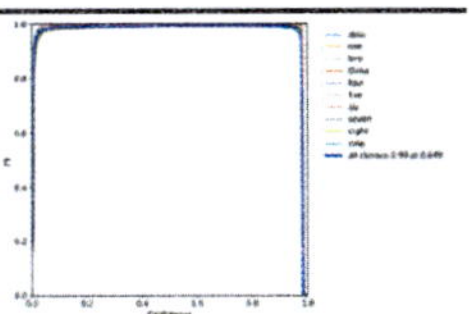
	MNIST/ADAM	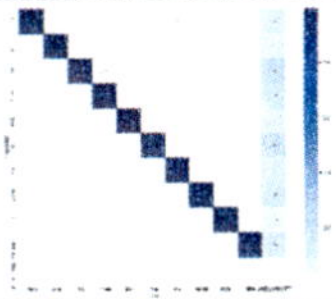	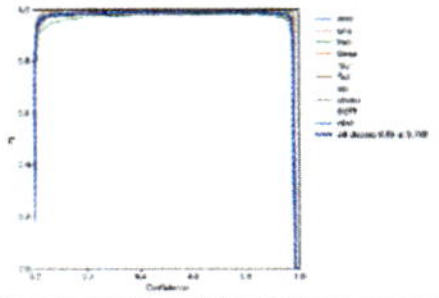
	SVHN/SGD	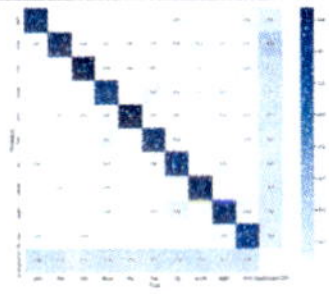	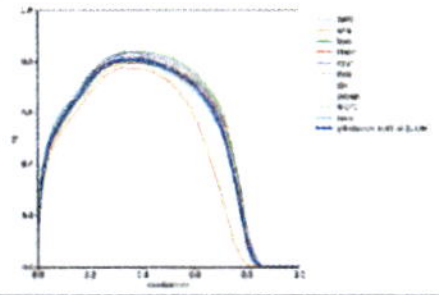
	SVHN/ADAM	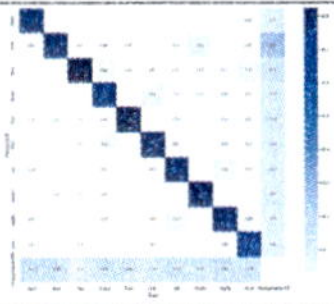	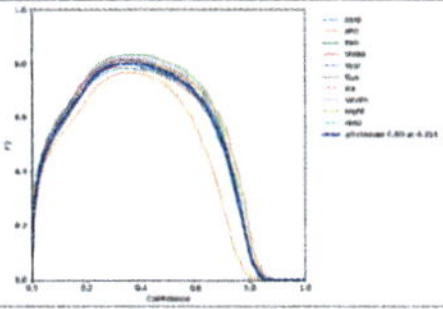
YOLOv5	MNIST/SGD	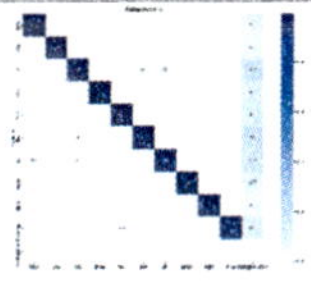	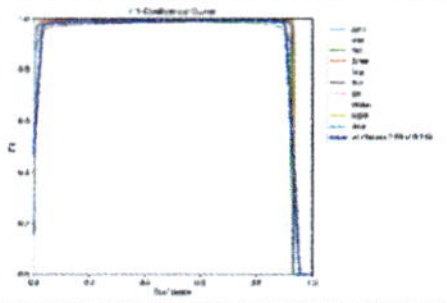
	MNIST/ADAM	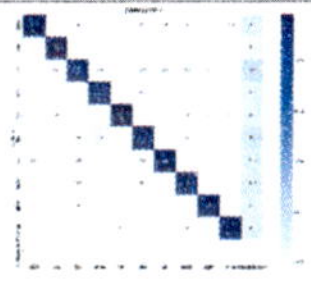	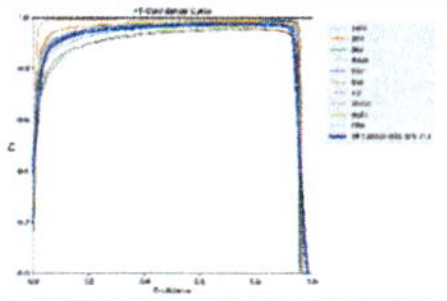
	SVHN/SGD	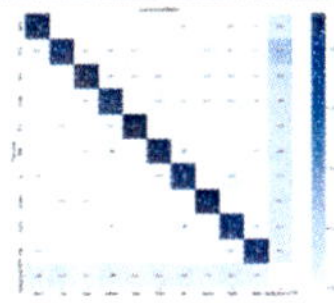	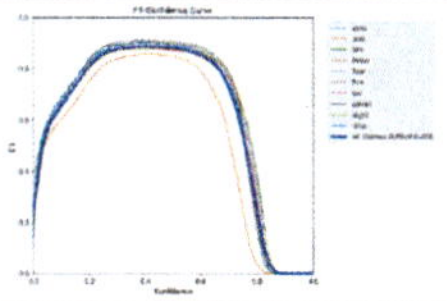
	SVHN/ADAM	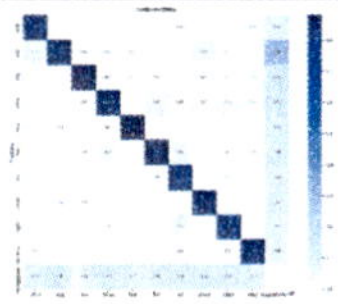	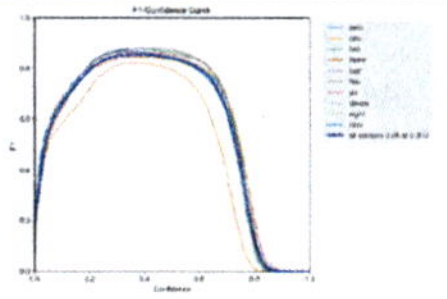

5 Conclusion and Discussion

In this study, multi-digit handwriting numbers are determined on the images obtained under daylight conditions, and these numbers are classified with a YOLOv3-tiny and YOLOv5 models on real-time systems with high performance. Also, this system locates and recognizes street door number and multiple digit numbers in the normal environment. These models have been tested on SVHN and MNIST datasets, respectively; it has been trained and tested with the AdaM and SGD optimizers. The comparison of different optimizers is given in Table 2.

Table 2. The comparison of different optimizers using MNIST and SVHN datasets

		Epoch	Train data	Validation data	Test data	F1-Score (%)	Recall (%)	Precision (%)
YOLOV3-tiny MNIST	AdaM	50	44800	11200	14000	99%	98.7%	99.3%
	SGD	50	44800	11200	14000	**99.3%**	99.3%	99.4%
YOLOV5 MNIST	AdaM	50	44800	11200	14000	97.6%	97.2%	98.1%
	SGD	50	44800	11200	14000	99.4%	99.4%	99.5%
YOLOV3-tiny SVHN	AdaM	50	29741	7435	9294	**81.9%**	77.3%	87.1%
	SGD	50	29741	7435	9294	80.7%	74.0%	88.8%
YOLOV5 SVHN	AdaM	50	29741	7435	9294	85.5%	81.6%	89.9%
	SGD	50	29741	7435	9294	**89.0%**	86.5%	91.6%

Experimental results show that the most successful result is the SGD optimizer for both deep learning models. YOLOv5 is better result than YOLOv3-tiny model. The strength of our paper is the employed YOLOv3 and YOLOv5, a real-time state-of-the-art object detection algorithm, for the first time in multi-digit number detection and recognition problem which is important to detect numbers such as house numbers and personal identification numbers.

Acknowledgements. The study was carried out in the Image Processing and Computer Vision Laboratory of the Computer Engineering Department of Kocaeli University, and supported by the TÜBİTAK 2209-A Project with project no:1919B012112745.

References

1. Akdeniz, F., Becerikli, Y.: Detection of copy-move forgery in audio signal with mel frequency and delta-mel frequency kepstrum coefficients. In: 2021 Innovations in Intelligent Systems and Applications Conference (ASYU), pp. 1–6. IEEE, October 2021
2. Akdeniz, F., Becerikli, Y.: Performance comparison of support vector machine, k-nearest-neighbor, artificial neural networks, and recurrent neural networks in gender recognition from voice signals. In: 2019 3rd International Symposium on Multidisciplinary Studies and Innovative Technologies (ISMSIT), pp. 1–4. IEEE, October 2019

3. Asif, M., Ahmad, M.B., Taj, I.A., Tahir, M.: A generalized multi-layer framework for video coding to select prediction parameters. IEEE Access **6**, 25277–25291 (2018)
4. Goodfellow, I.J., Bulatov, Y., Ibarz, J., Arnoud, S., Shet, V.: Multi-digit number recognition from street view imager using deep convolutional neural networks (2013). arXiv preprint, arXiv:1312.6082. Accessed 23 July 2020
5. Asif, M., Taj, I.A., Ziauddin, S.M., Ahmad, M.B., Raza, A.: An efficient inter prediction mode selection scheme for advanced video coding based on motion homogeneity and residual complexity. IEEJ Trans. Electr. Electron. Eng. **11**, 760–767 (2016)
6. Alwzwazy, H.A., et al.: Handwritten digit recognition using convolutional neural networks. Int. J. Innov. Res. Comput. Commun. Eng. **4**(2), 1101–110 (2016)
7. Nagai, A.: On the improvement of recognizing single-line strings of Japanese historical cursive. In: Proceedings of the 2019 International Conference on Document Analysis and Recognition, ICDAR 2019; Sydney, Australia, 20–25 September 2019; pp. 621–628 (2019)
8. Kim, J.-H., Noh, D.-S.: Vehicle license plate recognition system by edge-based segment image generation. J. Korea Contents Assoc. **12**(3), 9–16 (2012)
9. Lecun, Y.: The Mnist Database of handwritten digits. Erişim tarihi: 30.05.2022. http://yann.lecun.com/exdb/mnist/
10. Asif, M., et al.: Long Multi-digit Number Recognition from Images Empowered by Deep Convolutional Neural Networks. The Computer Journal (2021)
11. España-Boquera, S., Maria, J.C.B., Jorge, G.M., Francisco, Z.M.: Improving offline handwritten text recognition with hybrid HMM/ANN models. IEEE Trans. Pattern Anal. Mach. Intell. **33**(4) (2011)
12. Bajaj, R., Dey, L., Chaudhury, S.: Devnagari numeral recognition by combining decision of multiple connectionist classifiers. Sadhana **27**(part. 1), 59–72 (2002)
13. Hanmandlu, M., Ramana Murthy, O.V.: Fuzzy model based recognition of handwritten numerals. Pattern Recogn. **40**, 1840–1854 (2007)
14. Wang, Y., Huang, Y., Zheng, W., Zhou, Z., Liu, D., Lu, M.: Combining convolutional neural network and self-adaptive algorithm to defeat synthetic multi-digit text-based CAPTCHA. In: 2017 IEEE International Conference on Industrial Technology (ICIT), pp. 980–985. IEEE, March 2017
15. Annapareddy, N., et al.: Handwritten text and digit classification on rwandan perioperative flowsheets via YOLOv5. In 2022 Systems and Information Engineering Design Symposium (SIEDS), pp. 270–275. IEEE, April 2022
16. Chandra, A., Stefanus, R.: An end-to-end optical character recognition pipeline for Indonesian identity card. In 2021 9th International Conference on Information and Communication Technology (ICoICT), pp. 307–312. IEEE, August 2021
17. Netzer, Y., Wang, T., Coates, A., Bissacco, A., Wu, B., Ng, A.Y.: Reading Digits in Natural Images with Unsupervised Feature Learning NIPS Workshop on Deep Learning and Unsupervised Feature Learning (2011)
18. Savaş, B.K., Becerikli, Y.: Real time driver fatigue detection system based on multi-task ConNN. Ieee Access **8**, 12491–12498 (2020)
19. Zhao, L., Li, S.: Object detection algorithm based on improved YOLOv3. Electronics **9**(3), 537 (2020)
20. Redmon, J., Farhadi, A.: Yolov3: an incremental improvement (2018). arXiv preprint arXiv:1804.02767
21. Shao, Y., Zhang, X., Chu, H., Zhang, X., Zhang, D., Rao, Y.: AIR-YOLOv3: aerial infrared pedestrian detection via an improved YOLOv3 with network pruning. Appl. Sci. **12**(7), 3627 (2022)
22. Lv, N., Xiao, J., Qiao, Y.: Object detection algorithm for surface defects based on a novel YOLOv3 model. Processes **10**(4), 701 (2022)

23. Cheng, L., Li, J., Duan, P., Wang, M.: A small attentional YOLO model for landslide detection from satellite remote sensing images. Landslides **18**(8), 2751–2765 (2021). https://doi.org/10.1007/s10346-021-01694-6
24. Jiang, X., et al.: Automatic detection of coronary metallic stent struts based on YOLOv3 and R-FCN. Computational and Mathematical Methods in Medicine (2020)
25. Yan, B., Fan, P., Lei, X., Liu, Z., Yang, F.: A real-time apple targets detection method for picking robot based on improved YOLOv5. Remote Sens. **13**(9), 1619 (2021)
26. https://github.com/ultralytics/yolov5. Accessed 10 Aug 2022
27. Bottou, L.: Stochastic gradient descent tricks. Neural Networks: Tricks of the Trade, pp. 421–436. Springer, Heidelberg (2012)
28. Ünal, H.B., Vural, E., Savaş, B.K., Becerikli, Y.: Fruit recognition and classification with deep learning support on embedded system (fruitnet). In: 2020 Innovations in Intelligent Systems and Applications Conference (ASYU), pp. 1–5. IEEE, October 2020
29. Balci, T., Ogul, H.: Predicting Bank Return on Equity (ROE) using Neural Networks. In: 2021 IEEE 19th World Symposium on Applied Machine Intelligence and Informatics (SAMI), pp. 000279–000286. IEEE, January 2021
30. Kingma, D.P., Ba, J.: Adam: a method for stochastic optimization (2014). arXiv preprint arXiv:1412.6980
31. Mora, E.A.H., González-Huitrón, V., Rodríguez-Mata, A.E., Rangel, H.R.: Convolutional Neural Networks-based plant disease detection implemented on low-power consumption device. In: 2021 IEEE International Autumn Meeting on Power, Electronics and Computing (ROPEC), vol. 5, pp. 1–6. IEEE, November 2021

A Comparative Analysis of SVM, LSTM and CNN-RNN Models for the BBC News Classification

Yunus Karaman, Fulya Akdeniz, Burcu Kır Savaş(✉), and Yaşar Becerikli

Kocaeli University, Umuttepe Campus, 41001 Kocaeli, Turkey
{200202113,fulya.akdeniz,burcu.kir,ybecerikli}@kocaeli.edu.tr

Abstract. Developing technology, easier access to internet networks and the increase in the widespread use of the internet cause more text data to be produced than ever. Analyzing and classifying this data is one of the most difficult problems. Hence, automatic classification of text data is very important. In this study, the word embedding method, machine learning, deep learning and a hybrid model is proposed to classify news texts. GloVe method is used as a word embedding model and SVM (Support Vector Machine), LSTM (Long Short-Term Memory), CNN (Convolutional Neural Network) + RNN (Recurrent Neural Networks) models are used to classify data. Also, various optimization algorithms have been tested in deep learning models and the results of these algorithms are in a comparative manner. BBC-News dataset is used to classify news texts. The best accuracy results have been obtained as 98.0% in the SVM model, 97.0% in the LSTM model and 96.0% in the CNN + RNN model.

Keywords: Text document classification · Natural language processing · BBC news classification · GloVe · Hybrid model

1 Introduction

"Data is the new oil" is one of the most popular mottos highlighting the importance of data. With various communication tools such as social media data, websites, Wikipedia, news sites, and online newspapers, the size of the data produced is increasing day by day [1]. To distinguish these data, it is necessary to make classifications or categorize them. Most of the produced texts are in unstructured form. It is quite difficult to manage unstructured text manually. Hence, it is very important to put data into a structured form to be able to analyze them [2]. Natural language processing (NLP) analyzes large amounts of text data and one of the most significant tasks of NLP is to automatically categorize text data [3, 4]. Text classification suggestion systems are preferable in different areas such as sentiment analysis, information filtering, web search, information retrieval, news subject modeling, and theme extraction.

Considering the rates technology use by society, most people prefer web searches using information filtering. Because individuals getting these products frequently visit news sites, news classification is very important for the word to be disseminated in a

M. Ben Ahmed et al. (Eds.): SCA 2022, LNNS 629, pp. 473–483, 2023.
https://doi.org/10.1007/978-3-031-26852-6_44

timely manner and for the user to reach the desired news in a timely manner. In addition, classification process is vital for news sites to present the most accurate news to the user according to the relevant keywords. The automatic news classification system provides fast and accurate reporting of news, and recognition of different news according to the type of information. Automatic news classification provides fast access to news texts by analyzing the content of the news text and has critical importance in the processing of a large number of news texts simultaneously [5].

Researchers have employed various feature extraction algorithms and classification algorithms to categorize news texts. Dogru et al. [4] classified Turkish and English news texts. They used TTC-3600 and BBC-News datasets for Turkish Text classification and Doc2vec word embedding method in their studies. In the classification phase, they examined CNN, Naive Bayes (NB), Gaussian Naive Bayes (GNB), Random Forest (RF), and Support Vector Machine (SVM) methods. The best result for BBC news classification with the CNN algorithm is 96.41%. Shah et al. [2] compared Logistic Regression (LR), RF and K-nearest neighbor classification algorithms to classify BBC news items. The best accuracy in their study is 97.0% with the LR classifier using the TF-IDF Vectorizer feature. Sidiropoulos et al. [3] used BBC News and BBC Sports datasets to classify news texts in their study. They developed Fuzzy Set measures used to categorize news texts. Ahmed et al. [6] used the BBC News dataset to classify texts in their study. They analyzed LR, SVM, and K-Means algorithms in the classification phase. The accuracy of this study is between 93.0% and 96.0%. Mauni et al. [7] used the BBC News dataset to categorize the news in their study. They developed Deep Neural Network (DNN) in the classification phase.

2 Methodology

2.1 Dataset and Preprocessing

In the proposed study, the BBC News dataset is employed. This consists of 2225 articles and 935 words created between 2004 and 2005. This dataset includes a total of 5 topics: business, entertainment, politics, sports, and tech. The preprocessing steps in this study; stop word removal, lowercasing, remove punctuation, removing numbers, Tokenize and Lemmatization steps were applied [8–10].

Stop word Removal: Words such as "the", "a" and "is", which do not contribute much to the meaning in the texts, is removed from the text in the study. *Lowercasing*: This process converts the texts to the same form. *Remove Punctuation*: This operation removes punctuation marks in the text. *Removing numbers*: This process extracts numbers from the text. *Tokenize*: it is used to divide text into words. *Lemmatization*: used to remove common and meaningless words.

Word embedding methods are developed to express verbal expressions mathematically. Global Vectors for Word Representation (GloVe) word embedding methods were examined in the study. The GloVe method is known as an improved version of the Word2Vec and LSA algorithms. This method adapts probabilistic statistics to create a new objective function and try to find semantic relationships. The GloVe method is a new global logbilinear regression model that combines the advantages of two main models in the literature, global matrix decomposition and local context window methods. This method uses the word-word cooccurrence matrix statistical method instead of the entire matrix or individual context windows [11]. GloVe method is also effective in vector representation of words and interpretation of vectors and creates a vector representation of the word using the similarity between words. GloVe is based on two different approaches the Skip-gram and the Continuous Bag of Words Model (CBOW). The disadvantage of Skip-gram and Continuous Bag of Words Models are computation time and accuracy. On the other hand, Glove combines the approaches offered by the two models and obtains a better and more efficient method from these two models [12]. For all these reasons given, GloVe method is preferred as the word embedding method in the study. The J error function, $X_{i,j}$ weight function, gives the word size V.

$$J = \sum_{i,j}^{V} f(X_{i,j})(w_i^T + b_i + \overline{b_j} - \log X_{i,j})^2 \tag{1}$$

2.2 SVM

SVM is a supervised learning algorithm. With kernel functions SVM achieves high performance results in different data sets [13]. It is a machine learning algorithm that has linear separation in two-dimensional data and planar separation in three-dimensional data and hyperplane separation in multidimensional data [4].

2.3 CNN

The popularity of deep learning is increasing day by day. The CNN is a deep learning architecture and basically includes convolution layer, pooling layer and fully connected layers. Although the CNN model was used frequently in the field of image processing in the first years, it has been used a lot in text classification because it has shown high performance in the field of NLP in recent years [4, 14, 15]. In CNN-based text classification methods, the words in the text are converted into digital vectors. To analyze these digital vectors, these vectors must be taken as input in a digital way. Since CNN eliminates the feature extraction step in traditional machine learning algorithms, it is preferred more than machine learning algorithms in NLP studies [16, 17].

2.4 RNN

RNN model is a powerful model that can learn from sequential and time-series data such as signal data, text data, and each element in the input of the network is dependent on the previous output [18, 19]. The RNN model can detect word dependencies in a text data and store information [20]. Each word in the text is linked to feedback connections to the previous and following words. RNN method provides to consider the relationship between words.

2.5 LSTM

LSTM is an iterative neural network architecture used in the deep learning field. The LSTM architecture differs from a classical feedforward neural network architecture with its feedback connections. With Gate and Cell state structures, LSTM architecture determines which information is remembered and which information is forgotten. Also, this architecture is a robust and effective classifier that has become very extensive in recent years. The LSTM architecture is widely employed in various NLP studies [21]. LSTM method is better than the RNN method in terms of computation and its ability to process the information in the memory for a long time.

3 Experimental Analysis

This study aims to classify BBC news texts. In the preprocessing stage of the study, the data is organized using Stop word Removal, Lowercasing, Remove Punctuation, Removing numbers, Tokenize and Lemmatization methods and GloVe is used as a word embedding method to represent words in text analysis. In the classification phase, the results are analyzed on the BBC News dataset. Also, we have developed 3 different models machine learning, deep learning and a hybrid model. We calculated the metrics of accuracy, precision, Recall, F1-score to evaluate the classification performances of each model.

In this study, 3 different methods as SVM, LSTM, CNN + RNN models are developed and the results are given comparatively. In addition, different optimization algorithms have been implemented for each deep learning algorithm and graphics showing the results of these implementations are given in this paper.

In the first step of the study, we generate the SVM classifier and teste on the BBC-News texts. Precision, recall and F1- score of this model are given in Table 1. Also, the model accuracy is calculated to be 98.0%.

Table 1. The accuracy of SVM model.

Label	Precision	Recall	F1-score
0	95.0%	99.0%	97.0%
1	100%	96.0%	98.0%
2	100%	92.0%	96.0%
3	100%	100%	100%
4	96.0%	99.0%	97.0%

In the second step, we generate the LSTM method and classify the BBC-News texts dataset. The dataset is divided into 80% train and 20% test. The batch size is arranged as 32 and the epoch is determined as 200. Three different optimization algorithms as Adaptive Moment Estimation (Adam), Root mean square error probability (RMSprop) and Stochastic gradient descent (SGD) have been implemented in the LSTM method, and the performance results of these algorithms have been calculated.

Table 2 gives the results of Adam optimizer in the LSTM. The average accuracy of the study has been concluded as 97.0%.

Table 2. The results of Adam optimizer in the LSTM.

Label	Precision	Recall	F1-score
0	93.0%	99.0%	96.0%
1	97.0%	96.0%	97.0%
2	100%	92.0%	96.0%
3	99.0%	98.0%	99.0%
4	98.0%	98.0%	98.0%

Figure 1 shows training and validation accuracy graphics using the Adam optimizer in the LSTM. Training and validation loss graphics using the Adam optimizer in the LSTM are given in Figs. 1 and 2.

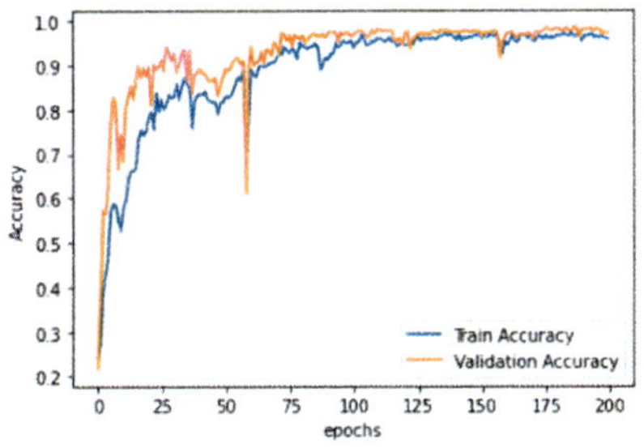

Fig. 1. Training and validation accuracy using the Adam optimizer in the LSTM.

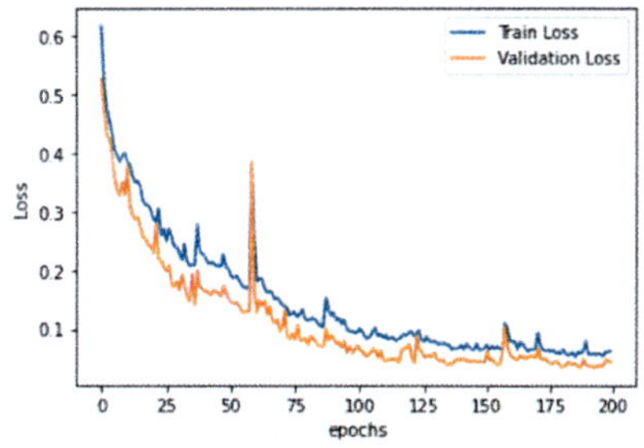

Fig. 2. Training and validation loss using the Adam optimizer in the LSTM.

Table 3 addresses the results of the RMSprop optimization method in the LSTM. The average accuracy of this model is calculated at 96.0%.

Figure 3 shows training and validation accuracy graphics using the RMSprop optimization method in the LSTM. Figure 4 Shows training and validation loss graphics using the RMSprop optimization method.

Table 4 gives the results of the SGD optimization algorithm in the LSTM. The average accuracy of this model is computed at 41.0%.

Figure 5 gives the training and validation accuracy graphics using SGD optimization algorithm in the LSTM and Fig. 6 shows training and validation losses graphics.

Table 3. The results of RMSprop optimization algorithm in the LSTM.

Label	Precision	Recall	F1-score
0	95.0%	95.0%	95.0%
1	95.0%	95.0%	95.0%
2	90.0%	97.0%	94.0%
3	99.0%	99.0%	99.0%
4	98.0%	94.0%	96.0%

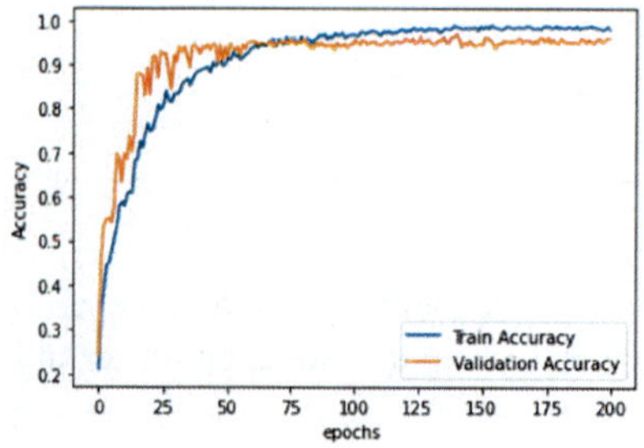

Fig. 3. Training and validation accuracy using the RMSprop optimization method in the LSTM.

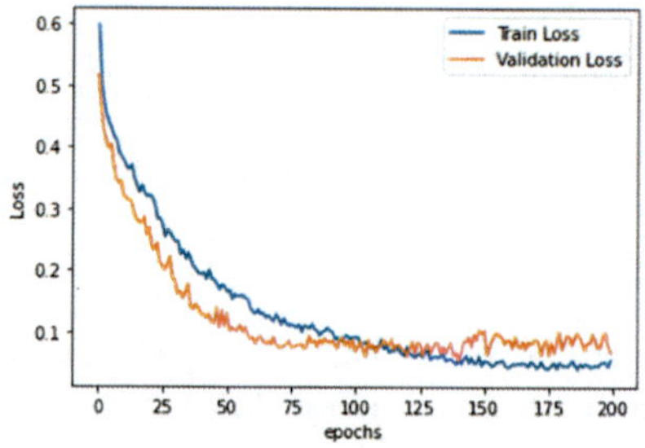

Fig. 4. Training and validation loss using the RMSprop optimization method in the LSTM.

Table 4. The results of SGD optimization algorithm in the LSTM.

Label	Precision	Recall	F1-score
0	0.0%	0.0%	0.0%
1	57.0%	5.0%	9.0%
2	60.0%	24.0%	34.0%
3	37.0%	94.0%	53.0%
4	47.0%	53.0%	50.0%

In the third and final step of this study, a hybrid model is developed. CNN and RNN models are merged together. The BBC news texts dataset is divided into 80% training phase, and 20% test phase. In the study, the batch size was taken as 32 and the results at 200 epochs were included. 3 different optimization algorithms are implemented and the impact of optimization algorithms on the performance is examined. Table 5 gives the results of the Adam optimization algorithm, which is the best result in the CNN + RNN hybrid algorithm. The average accuracy of the study is computed at 91.0%.

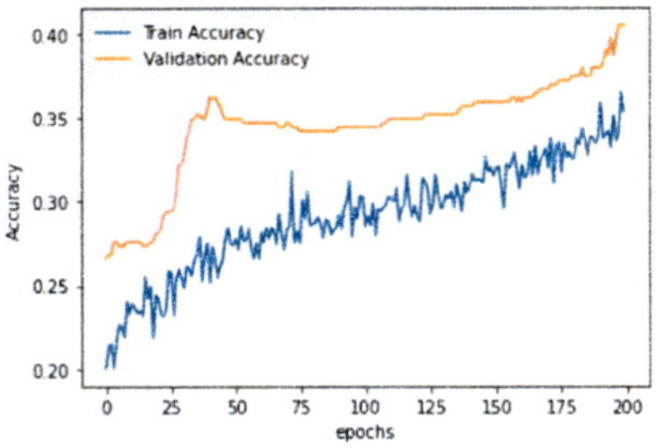

Fig. 5. Training and validation accuracy using the SGD optimization algorithm in the LSTM.

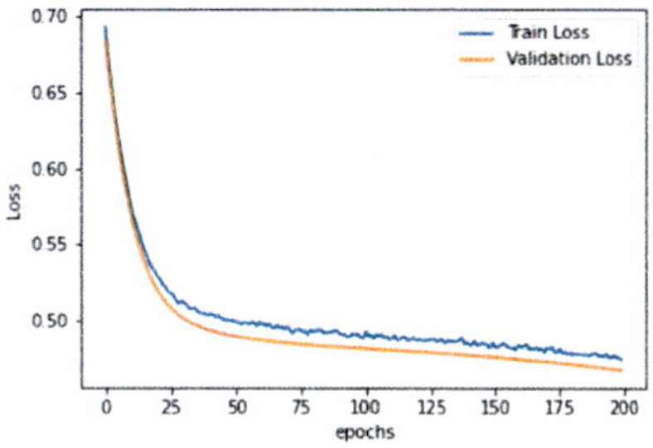

Fig. 6. Training and validation loss using the SGD optimization algorithm in the LSTM.

Table 5. The results of the Adam optimization algorithm.

Label	Precision	Recall	F1-score
0	93.0%	87.0%	90.0%
1	86.0%	91.0%	88.0%
2	80.0%	95.0%	87.0%
3	98.0%	96.0%	97.0%
4	90.0%	85.0%	87.0%

Figure 7 shows training and validation accuracy graphics using the Adam optimizer in the CNN + RNN. Figure 8 shows training and validation losses graphics using the Adam optimizer.

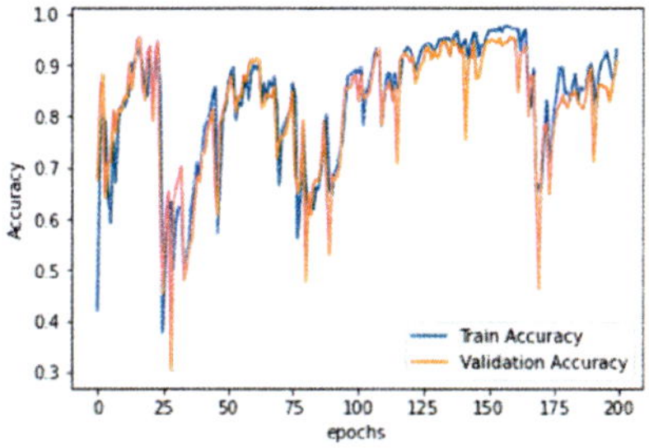

Fig. 7. Training and validation accuracy using the Adam optimization algorithm in the CNN + RNN.

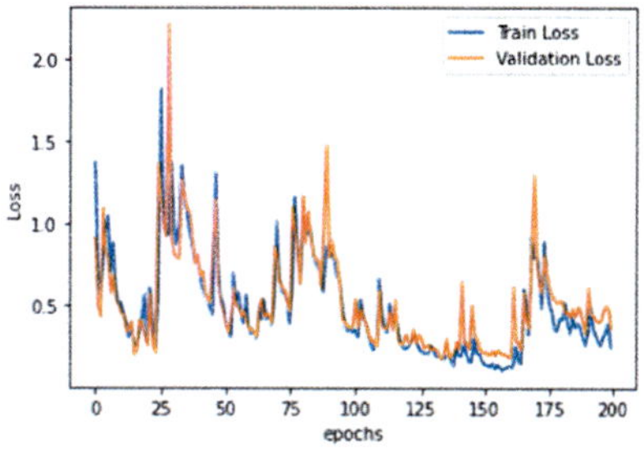

Fig. 8. Training and validation loss using the Adam optimization algorithm in the CNN + RNN.

Table 6 addresses the calculation of the RMSprop optimizer in the CNN + RNN model. The average accuracy of this model is calculated at 96.0%.

Table 6. The results of the RMSprop optimizer in the CNN + RNN.

Label	Precision	Recall	F1-score
0	98.0%	96.0%	97.0%
1	96.0%	94.0%	95.0%
2	93.0%	97.0%	95.0%
3	99.0%	99.0%	99.0%
4	94.0%	95.0%	95.0%

Figure 9 demonstrates the training and validation accuracy in the RMSprop optimization algorithm graphics in the CNN + RNN, and Fig. 10 demonstrates the training and validation losses graphics.

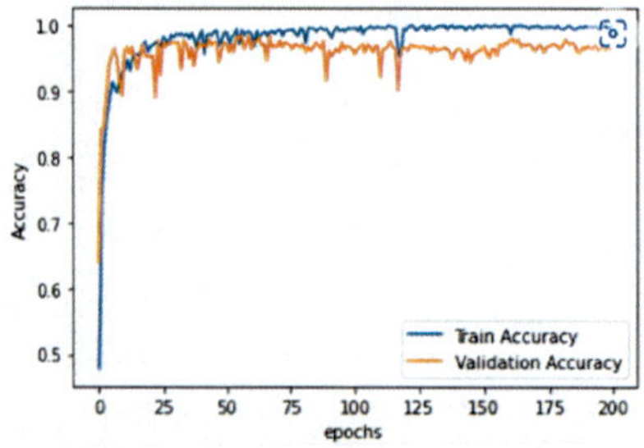

Fig. 9. Training and validation accuracy using the RMSprop optimizer in the CNN + RNN.

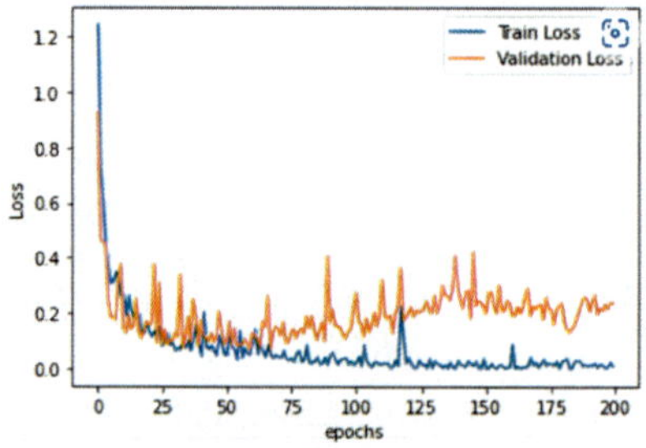

Fig. 10. Training and validation loss using the RMSprop optimizer in the CNN + RNN.

The results of the SGD optimization algorithm in the CNN + RNN are given in Table 7. The average accuracy of this model is calculated at 42.0%.

Table 7. The SGD optimization algorithm in the CNN + RNN.

Label	Precision	Recall	F1-score
0	100%	7.0%	13.0%
1	50.0%	12.0%	19.0%
2	29.0%	53.0%	37.0%
3	35.0%	79.0%	48.0%
4	76.0%	52.0%	62.0%

Figure 11 gives the training and validation accuracy in the SGD optimization algorithm in the CNN + RNN, and Fig. 12 gives the training and validation losses graphs.

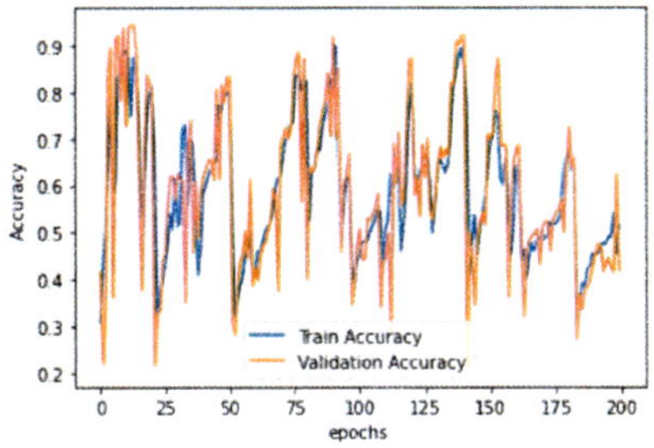

Fig. 11. Training and validation accuracy using the SGD optimization algorithm in the CNN + RNN.

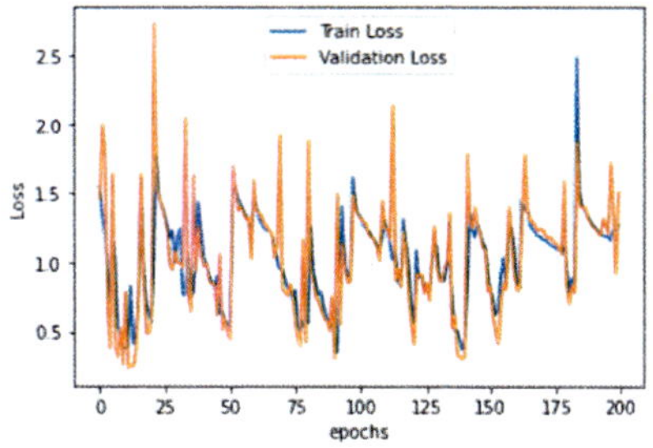

Fig. 12. Training and validation loss using the SGD optimization algorithm in the CNN + RNN.

4 Conclusions

In the proposed study, word vectors have extracted from BBC news texts with the GloVe model and have classified with SVM, LSTM and CNN + RNN models. In the deep learning models, different optimization algorithms have implemented and all the results have examined comparatively. The best classification result has obtained using the SVM model, with an average accuracy of 98.0%. The state-of-art of study is to analyze the performance results of machine learning, deep learning and a hybrid model comparatively.

Table 8. Related studies.

Literature	Database	Method	Performance Metric
[4]	BBC-News	Doc2Vec + CNN + (C-DS)	96.41%
[2]	BBC-News	TF-IDF Vectorizer, Logistic Regression	97.0%
[6]	BBC-News	Fuzzy vector, Sigmoidal SVM	96.0%
[22]	WebKB, BBC	TFDM, Naïve Bayes	96.7%
Proposed Method	BBC-News	GloVe, SVM Glove, LSTM GloVe, CNN-RNN	98.0% 97.0% 96.0%

A comparison of different studies is given in Table 8. This article outperforms many studies in the literature using the BBC-News dataset. This results proved the study to be robust.

Acknowledgements. This study is developed in the Kocaeli University, Computer Engineering Department, Image Processing and Computer Vision Laboratory and supported by the TÜBİTAK 2209-A Project with project no: 1919B012112844.

References

1. Urologin, S.: Sentiment analysis, visualization and classification of summarized news articles: a novel approach. Int. J. Adv. Comput. Sci. Appl. **9**(8) (2018)
2. Shah, K., Patel, H., Sanghvi, D., Shah, M.: A comparative analysis of logistic regression, random forest and KNN models for the text classification. Augmented Hum. Res. **5**(1), 1–16 (2020)
3. Sidiropoulos, G.K., Diamianos, N., Apostolidis, K.D., Papakostas, G.A.: Text classification using intuitionistic fuzzy set measures—an evaluation study. Information **13**(5), 235 (2022)
4. Dogru, H.B., Tilki, S., Jamil, A., Hameed, A.A.: Deep learning-based classification of news texts using doc2vec model. In: 2021 1st International Conference on Artificial Intelligence and Data Analytics (CAIDA), pp. 91–96. IEEE, April 2021
5. Hussain, A., Ali, G., Akhtar, F., Khand, Z.H., Ali, A.: Design and analysis of news category predictor. Eng. Technol. Appl. Sci. Res. **10**(5), 6380–6385 (2020)
6. Ahmed, R., Ahmad, T., Almutairi, F.M., Qahtani, A.M., Alsufyani, A., Almutiry, O.: Fuzzy semantic classification of multi-domain E-learning concept. Mobile Networks Appl. **26**(5), 2206–2215 (2021)
7. Mauni, H.Z., Hossain, T., Rab, R.: Classification of Underrepresented Text Data in an Imbalanced Dataset Using Deep Neural Network. In: 2020 IEEE Region 10 Symposium (TENSYMP), pp. 997–1000. IEEE, June 2020
8. Felber, T.: Constraint 2021: Machine learning models for COVID-19 fake news detection shared task (2021). arXiv preprint arXiv:2101.03717
9. Yahi, N., Belhadef, H.: Morphosyntactic preprocessing impact on document embedding: an empirical study on semantic similarity. In International Conference of Reliable Information and Communication Technology, pp. 118–126. Springer, Cham, September 2019
10. Chong, M., Specia, L., Mitkov, R.: Using natural language processing for automatic detection of plagiarism. In: Proceedings of the 4th International Plagiarism Conference (IPC-2010), June 2010
11. Pennington, J., Socher, R., Manning, C.D.: Glove: global vectors for word representation. In: Proceedings of the 2014 Conference on Empirical Methods in Natural Language Processing (EMNLP), pp. 1532–1543, October 2014
12. Sharma, Y., Agrawal, G., Jain, P., Kumar, T.: Vector representation of words for sentiment analysis using GloVe. In 2017 International Conference on Intelligent Communication and Computational Techniques (ICCT), pp. 279–284. IEEE, December 2017
13. Kayikcioglu, I., Akdeniz, F., Köse, C., Kayikcioglu, T.: Time-frequency approach to ECG classification of myocardial infarction. Comput. Electr. Eng. **84**, 106621 (2020)
14. Kır Savaş, B., Becerikli, Y.: Behavior-based driver fatigue detection system with deep belief network. Neural Computing and Applications, 1–13 (2022)
15. Savaş, B.K., Becerikli, Y.: Real time driver fatigue detection system based on multi-task ConNN. IEEE Access **8**, 12491–12498 (2020)
16. Wang, Z., & Qu, Z. (2017, October). Research on Web text classification algorithm based on improved CNN and SVM. In 2017 IEEE 17th International Conference on Communication Technology (ICCT) (pp. 1958–1961). IEEE
17. Ce, P., Tie, B.: An analysis method for interpretability of CNN text classification model. Future Internet **12**(12), 228 (2020)
18. Salehinejad, H., Sankar, S., Barfett, J., Colak, E., Valaee, S.: Recent advances in recurrent neural networks (2017). arXiv preprint arXiv:1801.01078
19. Akdeniz, F., Becerikli, Y.: Performance comparison of support vector machine, K-nearest-neighbor, artificial neural networks, and recurrent neural networks in gender recognition from voice signals. In: 2019 3rd International Symposium on Multidisciplinary Studies and Innovative Technologies (ISMSIT), pp. 1–4. IEEE, October 2019

20. Kim, S., Min, B.H., Kang, M., Demetriou, P.: Comparative Analysis of Deep learning-based News Topic Classification Models
21. Makarenkov, V., Guy, I., Hazon, N., Meisels, T., Shapira, B., Rokach, L.: Implicit dimension identification in user-generated text with LSTM networks. Inf. Process. Manage. **56**(5), 1880–1893 (2019)
22. Thirumoorthy, K., Muneeswaran, K.: Feature selection for text classification using machine learning approaches. Natl. Acad. Sci. Lett. **45**(1), 51–56 (2022)

ARCH Model and Nonlinear Autoregressive Neural Networks for Forecasting Financial Time Series

Mhamed El Merzguioui(✉), Younes Ait Taleb, and Mustapha El Jarroudi

Department of Mathematics, Faculty of Sciences and Techniques, Tangier, Morocco
mh.elmerzguioui@gmail.com

Abstract. A financial time series is most often based on representing financial prices (stock market indices). For this reason, the researchers focused on the development of stock market forecasting models because its wide range of financial applications and commercial benefits in order to maximize financial profit. In this paper, two approaches are proposed that ARCH modelling and non-linear autoregressive neural networks to forecast the daily prices of the NASDAQ 100 stock market index over a period from 02 January 2019 to 22 May 2020. The results of the simulation are obtained by the use of MATLAB software and Eviews.

Keywords: ARCH model · Nonlinear autoregressive neural networks · Forecasting

1 Introduction

The study of time series has a significant impact on several professional disciplines, particularly finance. Among all these areas, the prediction of stock market indices has been a hot topic of research and widely studied over the past two decades. As a result, researchers were motivated to adopt different approaches to develop accurate prediction models for the stock market. In general, the approaches used by researchers can be classified into two main classes:

The first is econometric models, which are statistical approaches such as linear regression, autoregression (AR) and moving autoregression average (ARMA), ARCH (Engle, 1982),

The second class uses artificial intelligence approaches to predict stock market indices. Examples of such approaches are artificial neural networks (NAR,..), support vector machines.

2 Methodology

The prediction of the daily prices of the NASDAQ 100 stock market index is a non-linear problem. This section represents a review of the ARCH model and NAR neural network model.

M. Ben Ahmed et al. (Eds.): SCA 2022, LNNS 629, pp. 484–498, 2023.
https://doi.org/10.1007/978-3-031-26852-6_45

2.1 ARCH Model

The process $(X_t)_{t \in \mathbb{Z}}$ satisfies a representation of a conditionally heteroscedastic autoregressive process of order p (Engle 1982), noted ARCH (p) if and only if:

$$X_t = \sigma_t \varepsilon_t, \sigma_t^2 = \alpha_0 + \sum_{j=1}^{p} \alpha_j X_{t-j}^2 \tag{1}$$

Here, $\alpha_0, \ldots, \alpha_p$ are parameter values in $\mathbb{R}$, the process ε_t is a white-noise independent of $\{X_{t-l}, l \geq 1\}$ for any $t \in \mathbb{Z}$.

Unlike ARMA models, ARCH models allow volatility, that is to say, variance to vary with time.

2.2 NAR Model

First, artificial neural networks are adjusted or formed for an input object to lead to a specific target. However, they are changed to make better based on a comparison of production (output) and target until the first matches up to the second.

The NAR network, which may be expressed in this format, can be used to forecast financial time series:

$$y_t = \varphi(y_{t-1} + y_{t-2} + \cdots + y_{t-d}) + \varepsilon_t \tag{2}$$

where φ is a non-linear function, y_t is the predicted value of the data series of y at a discrete-time set t, d previous values of the series and ε_t is the error at time t.

Network Architecture. The non-linear autoregressive neural network NAR has a multilayer architecture: input layer, hidden layer and output layer. Two transfer functions are used: a "TANGSIM" sigmoid in the hidden layer and a "PURELIN" linear function in the output layer.

Training Algorithm. The Levenberg-Marquard algorithm is based on the following equation:

$$w_{t+1} = w_t - \left[J(w_t)^t J(w_t) + \lambda I\right] J(w_t) E \tag{3}$$

where λ is a positive constant and J is the matrix of the first derivatives of the network errors concerning the weights, and the network error vector is denoted by E. The value of λ is arbitrarily set to 0.001.

Evaluation of the Performance. Two statistical criteria were used to evaluate the model's performance, namely the coefficient of determination R^2 and mean squared errors MSE, given by:

$$R^2 = 1 - \frac{\sum_{i=1}^{m} (y_i - o_i)^2}{\sum_{i=1}^{m} (y_i - r_i)^2} \tag{4}$$

$$MSE = \frac{1}{m} \sum_{i=1}^{m} (y_i - o_i)^2 \tag{5}$$

In which y represent the measured values, o the predict values, r the average of the measured values, and m is the number of values.

3 Results and Discussion

3.1 Prediction Using ARCH Model

The daily closing price data of 350 days of the NASDAQ 100 index is distributed in the following Fig. 1:

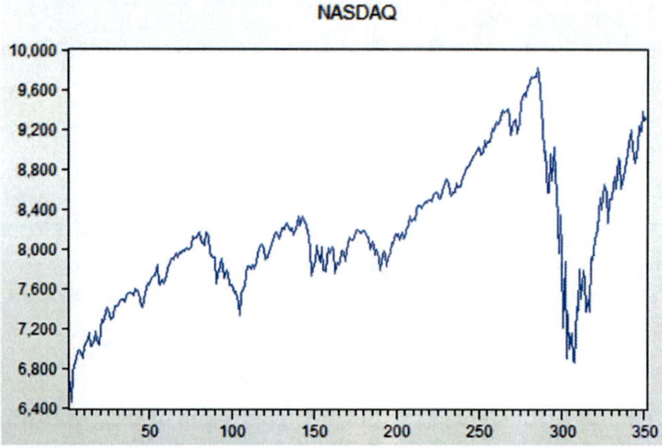

Fig. 1. The daily closing prices of the NASDAQ 100 index (Source: Yahoo Finance).

It seems that there is non-stationarity of the series confirmed statistically by the ADF unit root test (Augmented Dickey-Fuller, 1981) (Fig. 2):

		t-Statistic	Prob.*
Augmented Dickey-Fuller test statistic		0.825558	0.8893
Test critical values:	1% level	-2.571663	
	5% level	-1.941742	
	10% level	-1.616086	

Fig. 2. Result of the ADF test.

From this result, it can be seen that the value of p-value Augmented Dickey-Fuller is 0, 8893 greater than 0, 05, so the series is not stationary. To make the series stationary, we proceed to the logarithmic differentiation of first order using the formula (Lin, 2018):

$$R_t = log(\frac{P_t}{P_{t-1}}) \tag{6}$$

P_t : The closing price of the stock index at the time t.
P_{t-1} : The closing price of the stock index at the time $t - 1$.

By using the formula of stock market return in formula (6), the following is the procedure for converting data with EVIEWS software (Fig. 3):

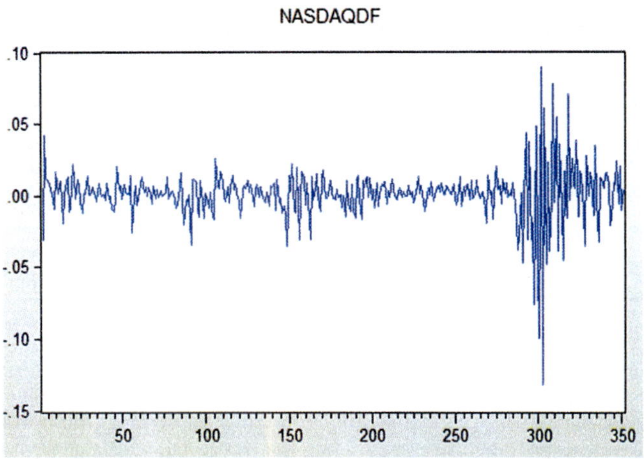

Fig. 3. The return NASDAQ 100.

Furthermore, the stationary of the return series is confirmed by the ADF unit root test, where the results are as follows (Fig. 4):

		t-Statistic	Prob.*
Augmented Dickey-Fuller test statistic		-7.259060	0.0000
Test critical values:	1% level	-2.571682	
	5% level	-1.941745	
	10% level	-1.616084	

Fig. 4. Result of the ADF test after differentiation.

Based on this result, there is no unit root because the value of p-value Augmented Dickey-Fuller is smaller than 0, 05, so the series is stationary. After that, the analysis of the correlogram and partial correlogram is used EVIEWS software, where the results are as follows (Fig. 5):

Autocorrelation	Partial Correlation		AC	PAC	Q-Stat	Prob
		1	-0.368	-0.368	47.696	0.000
		2	0.231	0.111	66.568	0.000
		3	0.008	0.145	66.590	0.000
		4	-0.056	-0.044	67.709	0.000
		5	0.138	0.086	74.534	0.000
		6	-0.258	-0.204	98.384	0.000
		7	0.355	0.228	143.59	0.000
		8	-0.306	-0.103	177.21	0.000
		9	0.309	0.189	211.75	0.000
		10	-0.121	-0.012	217.07	0.000
		11	-0.033	-0.071	217.46	0.000
		12	0.160	0.049	226.81	0.000
		13	-0.177	0.025	238.22	0.000
		14	0.230	0.046	257.54	0.000
		15	-0.169	0.056	268.04	0.000

Fig. 5. Correlogram and partial correlogram analysis.

From this figure, it is clear that there are significant values of the autocorrelation function and the partial autocorrelation function. So, to determine the best model, we will therefore examine all possible models in order to validate only one by the principle of parsimony; the appropriate model is ARIMA (1, 1, 9) (Fig. 6):

Variable	Coefficient	Std. Error	t-Statistic	Prob.
AR(1)	-0.302866	0.026546	-11.40928	0.0000
MA(9)	0.300629	0.036481	8.240694	0.0000
SIGMASQ	0.000279	8.66E-06	32.27965	0.0000

Fig. 6. ARIMA (1, 1, 9).

The next step is to exarmine the correlogram and partial correlogram of residues of ARIMA (1, 1, 9), where the results are as follows (Fig. 7):

Autocorrelation	Partial Correlation		AC	PAC	Q-Stat	Prob
		1	0.028	0.028	0.2852	
		2	0.121	0.120	5.4337	
		3	0.086	0.081	8.0582	0.005
		4	-0.016	-0.035	8.1507	0.017
		5	0.014	-0.006	8.2164	0.042
		6	-0.173	-0.178	18.888	0.001
		7	0.187	0.208	31.481	0.000
		8	-0.168	-0.156	41.629	0.000
		9	-0.020	-0.015	41.779	0.000
		10	-0.050	-0.067	42.684	0.000
		11	-0.029	0.026	42.995	0.000
		12	0.109	0.097	47.322	0.000
		13	-0.029	0.040	47.627	0.000
		14	0.190	0.085	60.815	0.000
		15	-0.020	-0.005	60.958	0.000

Fig. 7. Correlogram and partial correlogram of residues.

From this figure, there are significant values of residues autocorrelation, which means they are not a white noise. It can be confirmed by the Jarque-Bera test, where the result as follows (Fig. 8):

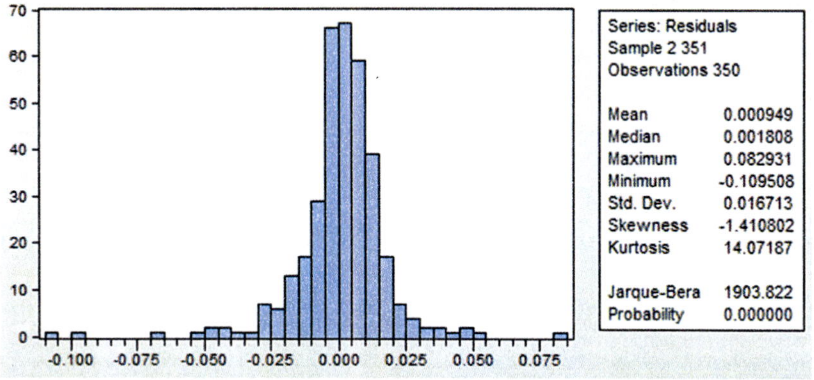

Fig. 8. Histogram of residues.

Thus, since the probability of the Jarque-Bera test is smaller than 0, 05, and then the residues are not normal. Hence, we reject the ARIMA specification in favour of heteroscedastic modelling (ARCH) – which is suitable for the study of time series with high variability/volatility (involving non-stationarity, non-normality or non-linearity) – to forecast future trends in NASDAQ 100 returns.

Study of the "NASDAQ" by ARCH Modelling. As part of the ARIMA family, ARCH modelling is likely to eliminate conditional heteroscedasticity. Note that we will estimate the ARCH model and proceed as follows (Fig. 9):

- Descriptive statistics and normality test of the series "NASDAQ".
- Study of the volatility of the "NASDAQ" series.
- Estimation of ARCH models and search for the optimal model.
- Forecasting.

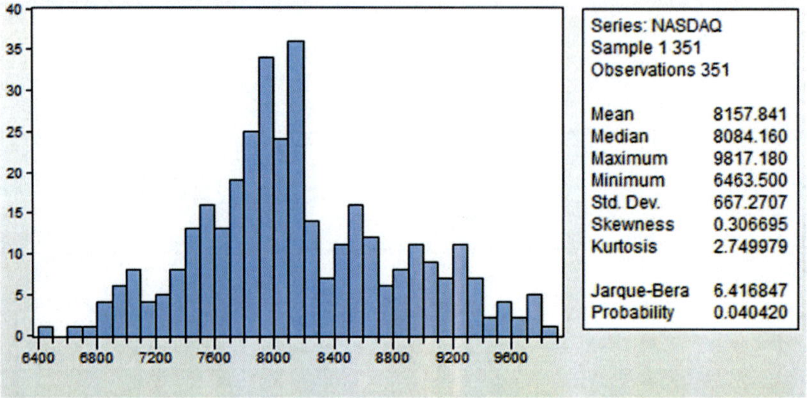

Fig. 9. Histogram of the raw series.

From this result, the Jarque-Bera test affirms the non-normality of the "NASDAQ" series (Fig. 10):

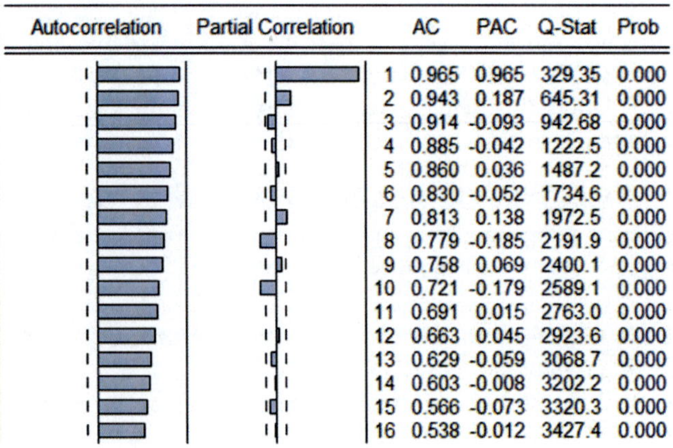

	AC	PAC	Q-Stat	Prob
1	0.965	0.965	329.35	0.000
2	0.943	0.187	645.31	0.000
3	0.914	-0.093	942.68	0.000
4	0.885	-0.042	1222.5	0.000
5	0.860	0.036	1487.2	0.000
6	0.830	-0.052	1734.6	0.000
7	0.813	0.138	1972.5	0.000
8	0.779	-0.185	2191.9	0.000
9	0.758	0.069	2400.1	0.000
10	0.721	-0.179	2589.1	0.000
11	0.691	0.015	2763.0	0.000
12	0.663	0.045	2923.6	0.000
13	0.629	-0.059	3068.7	0.000
14	0.603	-0.008	3202.2	0.000
15	0.566	-0.073	3320.3	0.000
16	0.538	-0.012	3427.4	0.000

Fig. 10. AC and PAC of the "NASDAQ" series.

According to the PAC, we can say that the series "NASDAQ" would follow a process of type AR (2). It also shows non-stationarity (slow decrease in terms of the AC); it would be much more variance. The next step is to study the volatility of the "NASDAQ" series, let us regress "DNASDAQ" to the square ("DNASDAQ2") on the same lagged series (this way of proceeding makes it possible to judge the autocorrelation of the variance of residues/heteroscedasticity (Fig. 11):

Variable	Coefficient	Std. Error	t-Statistic	Prob.
C	10134.97	3544.873	2.859051	0.0045
DNASDAQ2(-1)	0.520768	0.045829	11.36329	0.0000

Fig. 11. AC and PAC of the "NASDAQ" series.

The test results confirm the presence of heteroscedasticity of the errors, as the coefficient associated with DNASDAQ2 (−1) is statistically significant. We estimate an AR (1) from which we perform the ARCH presence test to judge the need for ARCH modeling (Fig. 12):

Variable	Coefficient	Std. Error	t-Statistic	Prob.
C	226.9710	95.55378	2.375322	0.0181
NASDAQ(-1)	0.973098	0.011679	83.31875	0.0000

Fig. 12. Estimation of the AR (1) model.

After that, the presence of the ARCH effect is tested and the result obtained is as follows (Figs. 13, 14, 15):

Heteroskedasticity Test: ARCH			
F-statistic	106.8933	Prob. F(1,347)	0.0000
Obs*R-squared	82.19062	Prob. Chi-Square(1)	0.0000

Fig. 13. Order ARCH test 1.

Variable	Coefficient	Std. Error	z-Statistic	Prob.
C	158.4756	40.41996	3.920726	0.0001
NASDAQ(-1)	0.983729	0.005173	190.1706	0.0000
		Variance Equation		
C	7661.704	458.2712	16.71871	0.0000
RESID(-1)^2	0.747933	0.104979	7.124627	0.0000

Fig. 14. Estimate of ARCH (1).

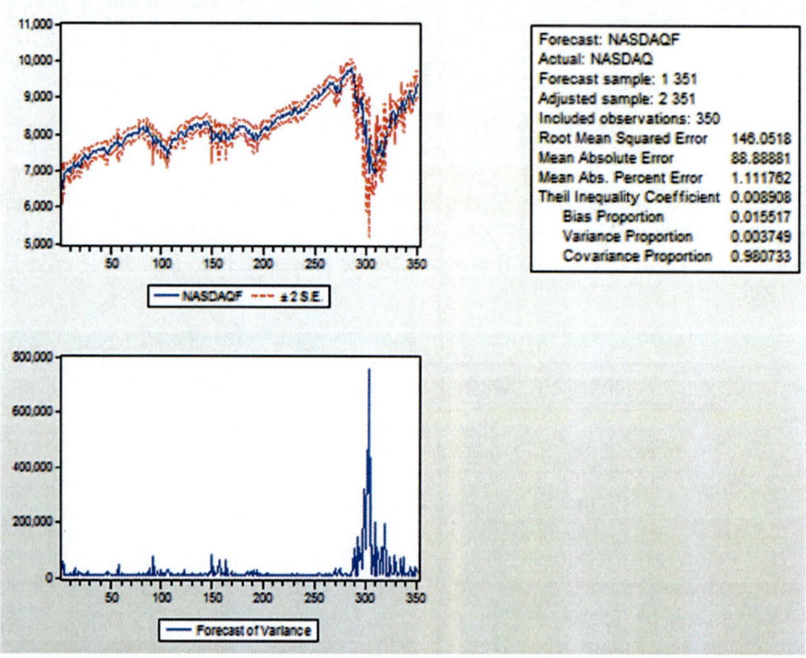

Fig. 15. Prediction of the variance.

The graph below represents the evolution of the real data of the raw and predicted series: the graphs show that the prediction is almost perfect (Fig. 16).

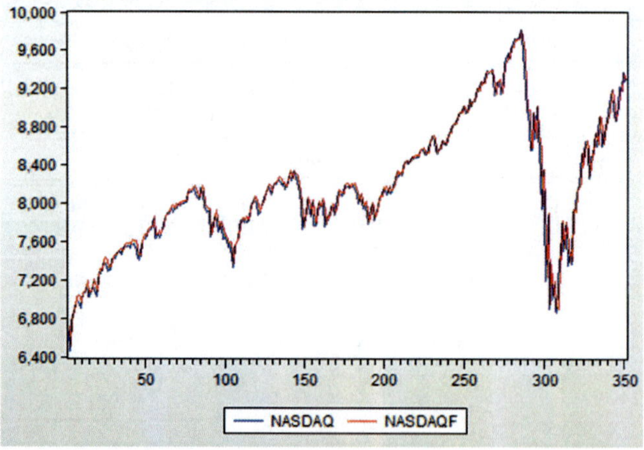

Fig. 16. Prediction of the variance.

3.2 Prediction Using NAR Model

In this section, the closing standard prices of the NASDAQ 100 index are taken into account from 7/8/2019 to 30/12/2019. The first 70 days are selected as a training dataset, and the last 29 days are used to test the predictive capacity of each of the three NAR networks.

The principle is to forecast the normalized closing price of the NASDAQ 100 index based on 70 past values. We choose $P_{t-1}, \ldots, P_{t-70}$ as inputs and we want to forecast the value of P_t.

The architecture of the three NAR networks to study their predictive capacity is (Figs. 17, 18, 19):

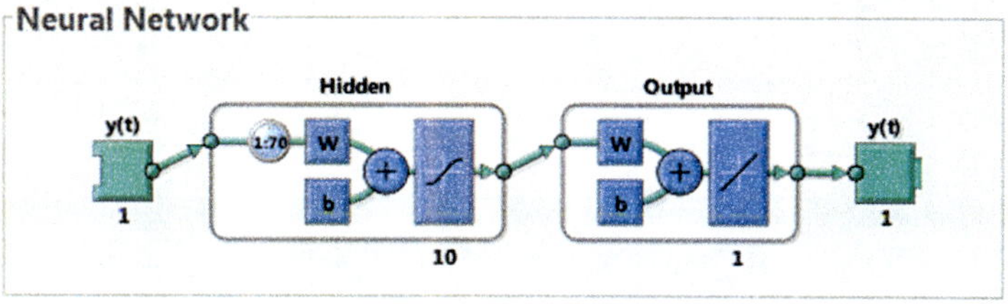

Fig. 17. NAR (70, 10, 1).

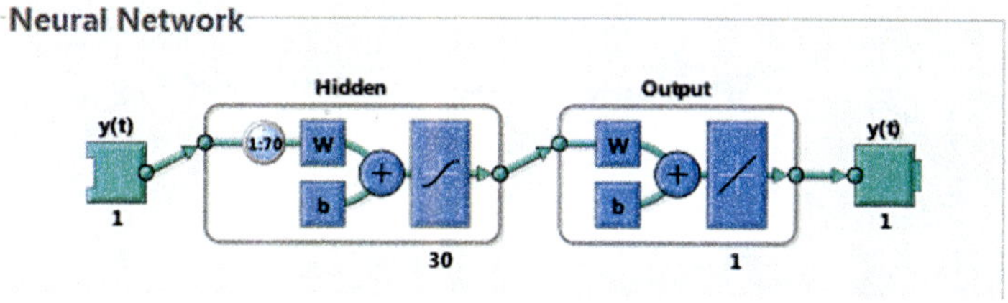

Fig. 18. NAR (70, 30, 1).

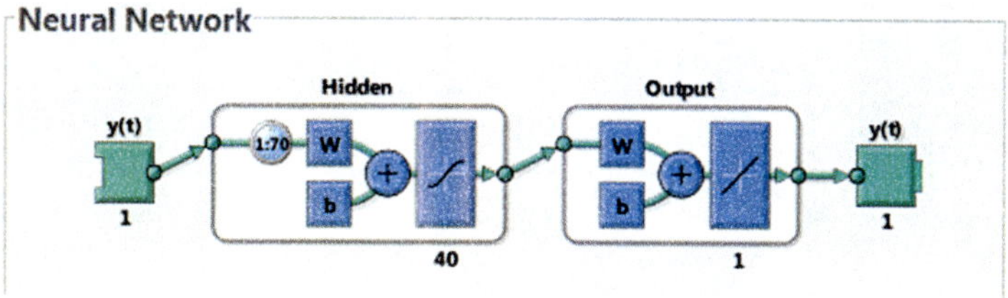

Fig. 19. NAR (70, 40, 1).

The MSE values for the three models are (Figs. 20, 21, 22):

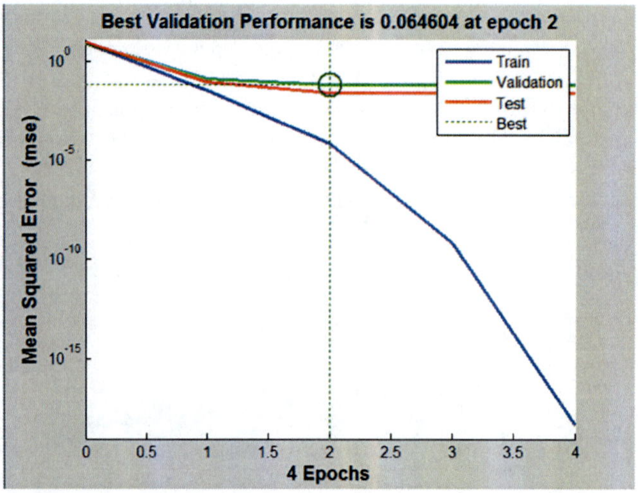

Fig. 20. MSE of NAR (70, 10, 1).

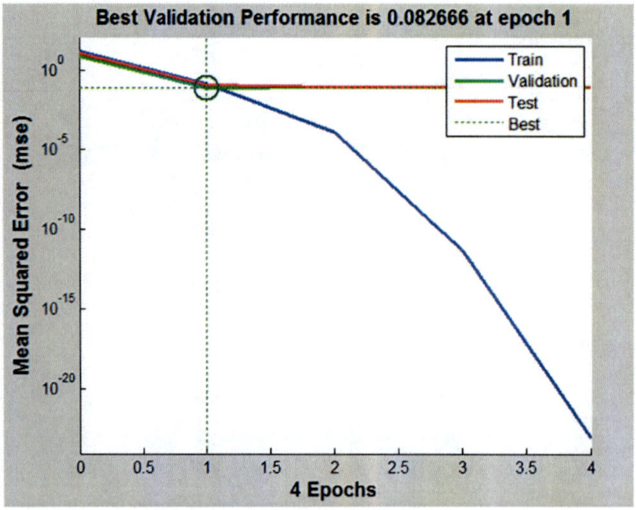

Fig. 21. MSE of NAR (70, 30, 1).

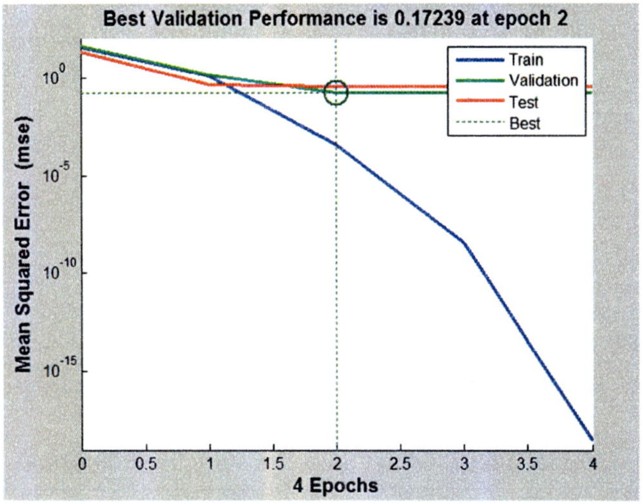

Fig. 22. MSE of NAR (70, 40, 1).

Based on these results, we note that NAR (70, 10, 1) is the best model of NAR networks that has a minimum error value of 0.064604 of *MSE*. Moreover, we can see the *MSE* and *R* values of this model in each part (training, test and validation) as following (Figs. 23, 24):

Results

	Target Values	MSE	R
Training:	70	7.00721e-5	9.99941e-1
Validation:	15	6.46035e-2	8.97336e-1
Testing:	15	2.58958e-2	9.76636e-1

Fig. 23. NAR (70, 10, 1) characteristics.

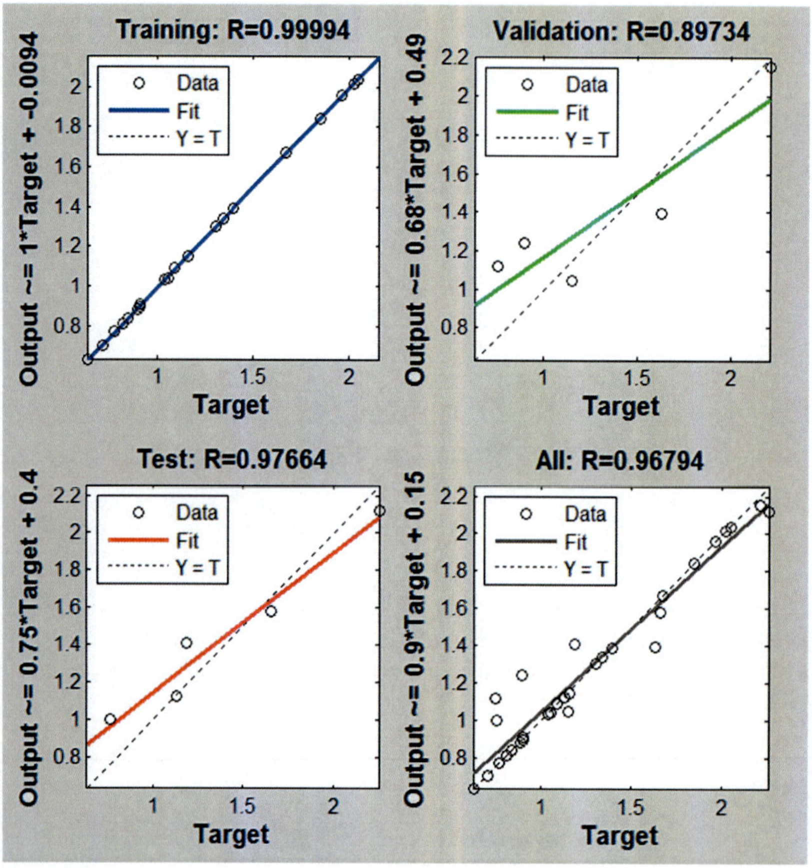

Fig. 24. *R* values of NAR (70,10,1).

A comparison between the outputs of the NAR (70, 10, 1) model and the targets is given in the following Fig. 25:

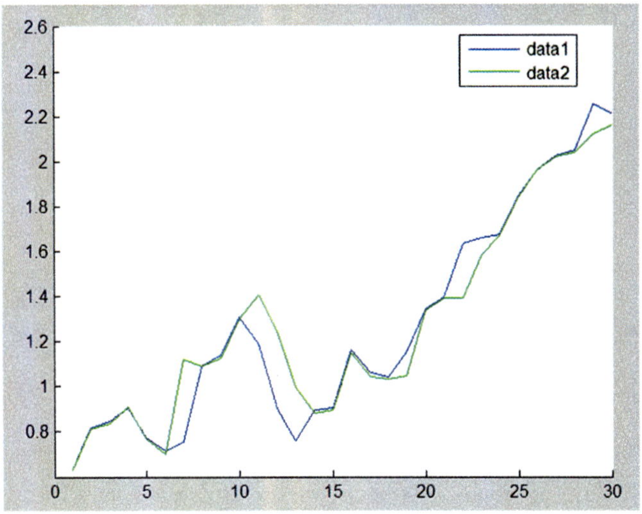

Fig. 25. Comparaison of outputs and targets.

4 Conclusion

On the NASDAQ 100 index from 02 January 2019 to 22 May 2020, it appears that the data is not stationary. For this reason, differentiation is needed by finding the value of stock returns from the NASDAQ 100 index data from 02 January 2019 to 22 May 2020. After differentiating by looking for return values, the next analysis can be done, namely looking for the ARIMA model. The results obtained after using the EVIEWS software are getting the ARIMA (1, 1, 9) model. The ARIMA (1, 1, 9) was chosen because it has the smallest AIC value. After that, it is necessary to study the "NASDAQ" series with ARCH modelling because of the presence of the ARCH effect and the results obtained after using the EVIEWS software are getting the ARCH (1) is the best model. Furthermore, for the second approach, the study shows that the NAR (70, 10, 1) is a powerful neural network architecture for forecasting the "NASDAQ" series.

References

1. Franke, J., Härdle, W.K., Hafner, C.M.: Statistics of Financial Markets. Library of Congress Control Number: 2007942157 (2007)
2. Bărbulescu, A., Băutu, E.: A hybrid approach for modelling financial time series. Int. Arab J. Inf. Technol. **9**(4), 327–335 (2012)
3. Bărbulescu, A., Pelican, E.: On the Sulina Precipitation Data Analysis Using the ARMA models and a Neural Network Technique. Recent Advances in Mathematical and Computational Methods in Science and Engineering, Part II (2008)
4. Lin, Z.: Modelling and forecasting the stock market volatility of SSE composite index using GARCH models. Futur. Gener. Comput. Syst. **79**, 960–972 (2018)
5. Lai, T.L., Xing, H.: Statistical models and methods for financial markets. Springer, New York (2008). https://doi.org/10.1007/978-0-387-77827-3

6. Brooks, C.: Introductory Econometrics for Finance. 2nd edn. Cambridge University Press, Cambridge (2008)
7. Sunarya, I.W.: Modelling and forecasting stock market volatility of nasdaq composite index. Econ. Acc. J. **2**(3), 181 (2019)
8. Pino, R., Parreno, J., Gomez, A., Priore, P.: Forecasting next-day price of electricity in the Spanish energy market using artificial neural networks. Eng. Appl. Artif. Intell. **21**(1), 53–62 (2008)

Mapping Human Development Indices in Moroccan Regions

Abroun Ikrame[1,3](✉), Azyat Abdelilah[2,3], Raissouni Naoufal[1,3], Ben Achhab Nizar[1,3], and Chahboun Asaad[1,3]

[1] RSGIS Lab, Remote Sensing GIS Unit, Tetouan, Morocco
abroun.ikrame9@gmail.com
[2] MIA Research Unit, Tetouan, Morocco
[3] ENSA, Tetouan, Morocco

Abstract. Human development (HD) is a concept about the thriving of the hu-man being. It's a strategy that focuses on the people and the opportunities avail-able to them. With the knowledge that economic growth does not necessarily bring about wealth for everyone. Human development is all about human free-doms: freedom to realize the full potential of every human life [1]. Increased income is merely a means to an objective. HD is a multidimensional concept that is measured using the human development index (HDI). The fundamental elements of HD are living a long and healthy life, being globally aware and knowledgeable, and having a fair standard of living. This index determines whether a country is developed, developing, or non-developed, as well as the degree to which the economy's impact on the standard of living. However, it has two main flaws: it isn't particularly useful at finer geographical scales and it does not provide a dynamic and less static account of progress or delays re-marked. In order to highlight regional disparities, we have focused in this work on the calculation of the indices namely Social Development Index (SDI), Afri-can Social Development Index (ASDI) and Multidimensional Local Develop-ment Index (MLDI) which are more suited to the sub-national level using the different calculation formulas mentioned in this paper. Finally, we have per-formed a cartographic analysis with comparison of the mentioned indices for the 12 regions of Morocco according to the new territorial division using GIS tools and data analysis techniques.

Keywords: Human development · Human development indices · Data collection · Data analysis · GIS

1 Introduction

Human progress is much more than only accumulating wealth, income, or economic growth. It must be human-centered. As a result, issues such as respect for human rights, the reduction of socioeconomic inequities and poverty, the promotion of equal opportunity for men and women, and environmental preservation are all relevant. This considers human resources not only as a means of growth, but more fundamentally, as

M. Ben Ahmed et al. (Eds.): SCA 2022, LNNS 629, pp. 499–510, 2023.
https://doi.org/10.1007/978-3-031-26852-6_46

an end of growth. According to the United Nations Development Program (UNDP), HD refers to a process of expanding human capacities and thus development opportunities that is available to everyone in society. To measure human development, UNDP established a composite Human Development Indicator (HDI) based on progress in three important dimensions of life: longevity, education, and a decent standard of living. While the HDI allows for comparison ranking between countries, it suffers from major shortcomings. To overcome these limitations, we will have used some indicators more adapted to the Moroccan context (SDI, ASDI, and MLDI) that are available at subnational scales and incorporate multiple characteristics of the HDI in this research. In our work, we try to deal with the following problem: how to measure the level of human, economic and social development at the regional level, taking into account the changes in regions, and how to determine inter-regional disparities and therefore, reveal persistent deficits.

2 State of Art

Recently, some works studied Human development and its different measurement methods. The measurement of human development is a rich field which has seen a veritable explosion of new and innovative international indices which evolve and develop over the years, the latter has recently seen remarkable improvements and great efforts by researchers at the level of included dimensions and method of meas- urement. Several articles have put forward different approaches, one of them is [2] which proposes a new National Sustainable Development Index (NSDI), based on the modification of the Human Development Index (HDI). The purpose of research was to improve the widely adopted HDI index by incorporating more comprehensive sus- tainability perspectives, so as to help policy makers to better analyze the sustainability-related issues facing their countries [2]. According to the findings, the term "sustainable development" refers to the coordination and configuration of economic, social, and environmental components of development, with a particular emphasis on balancing intra-generational welfare and maximizing total welfare across generations. The weights of 12 indicators were then determined using the entropy approach, and they proposed a new NSDI framework with 12 indicators from the dimensions of economics, resource environment, and society. [2] also measured the NSDIs of 163 nations around the world and compared it to the HDI and other well-known HDI modification indices to further validate the proposed index. The findings revealed that the NSDI is a valid and relatively complete index for assessing sustainable development that compensates for the inadequacies of other indices. Another article looks at how business intelligence (BI) technologies might help with sustainability manage- ment in organizations. The authors in [3] suggest that business intelligence (BI) can assist organizations in implementing and monitoring sustainable practices. Their main contribution in this study is to propose a conceptual model to aid in the integration of socio-environmental indicators into organizational sustainability strategies.

The methodological approach guiding [3] is an adapted version of grounded theory that begins with a genuinely inductive approach, i.e., one that starts from scratch without any theoretical model guiding data collection and analysis, but which inte- grates existing theoretical models in the last phase of data analysis [4]. As a result of this work, a BI conceptual model has been proposed that aims to support the processes of defining and monitoring socio-environmental indicators, as well as their integra- tion into the firm's business strategy and practices. This model is unique in that it aids in the integration of sustainability and social responsibility practices into corporate management systems that Uses BI tools. In the same context, [5] examines the literature on the role of Artificial Intelligence (AI) in the development of sustainable business models (SBMs). It presents a quantitative summary of the field's academic literature and analyzes the relationships between AI, quick developments in machine learn- ing, and sustainable development (SD). In [5] the authors conducted a systematic literature review (SLR) which adopts a replicable, scientific and transparent process [6], com-pared to traditional narrative reviews. This methodology is divided into two research steps: the first involves identifying, reading, and understanding significant documents; and the second involves a bibliometric examination of the identified articles [5]. After eliminating irrelevant and duplicate articles, all authors discussed their findings and prepared the SLR's numerous paragraphs. A total of 73 articles were included in the final list of works. The results of this work are organized as follows: the bibliometric aspects of the selected articles; the content of the articles; and the classification by secondary topic.

3 Human Development Frameworks

3.1 Human Development Framework

Generally, there are many different philosophical underpinnings in defining human development. The definition of human development and its aspects are very important as it will affect the choices of performance indicators. According to the Arab Human Development Report (2002) [7], the human development concept is broader than other people-oriented approaches to development. The human welfare approach looks at people as beneficiaries, but not as active participants in the process of development that shape their lives towards betterment. However, HD represents a multidimensional and holistic approach to development because it encompasses all aspects of well- being [8].

3.2 Social Development Framework

Social development can be defined in two ways. It can refer to improvement in the welfare and quality of life of individuals; or changes in societies which make development more equitable and inclusive for all members of a society [9]. These two definitions are based on the meaning of social which refers to people's welfare, and to relationships: between individuals and groups within a society [8].

4 Human Development Indices

4.1 Human Development Index

The HDI is an index that measures key dimensions of human development. The three key dimensions are: a long and healthy life – measured by life expectancy expressed by *IHealth*, access to education – measured by expected years of schooling of children at school-entry age and mean years of schooling of the adult population expressed by *IEducation*, and a decent standard of living measured by Gross National Income per capita adjusted for the price level of the country expressed by *IIncome* [10]. Once each of the individual indices have been calculated, they are aggregated to calculate the HDI. The HDI is calculated as the geometric mean (equally-weighted) of life expectancy, education, and GNI per capita, as follows in (1):

$$HDI = (IHealth * IEducation * IIncome)^{1/3} \tag{1}$$

4.2 Social Development Index

It is an index proposed by the [11] National Observatory for Human Development, based on two dimensions namely education and health. For the health dimension ex- pressed by Medical Index(MI), the indicator used is the rate of medical supervision which is measured at the provincial level. For the education dimension which is ex- pressed by School Index (SI), the indicator used is the success rate at the end of the college cycle. The value of the proposed SDI is the arithmetic mean of the indices SI and MI obtained for each of the two dimensions (2). The higher the SDI the more satisfactory the level of human development in the territory [11].

$$SDI = \frac{SI + MI}{2} \tag{2}$$

Its value can vary within a range of 80 to more than 125.

4.3 The Human Exclusion Index

Proposed by the High Commission for Planning (HCP)and developed in collaboration with the United Nations Economic Commission for Africa (ECA), it is also called the African Social Development Index (ASDI), this index makes it possible to diagnose socio-economic development national and local level, to assess and to identify social deficits in order to measure the impact of public resources on the populations who need them most. The dimensions thus retained for its calculation relate to infant mortality expressed by dExim, child malnutrition expressed by dExchm, illiteracy expressed by dExLr and unemployment among young people aged 15–24 expressed by dExYu, poverty measured at the national threshold expressed by dExnp and finally life expectancy at 60 years expressed by dExLe [12]. In order to assess the overall degree of human exclusion throughout the life cycle, we aggregate the levels of exclusion ineach of the six dimensions (3). Using a simple arithmetic sum, the overall level of exclusion can therefore be defined as [12]:

$$Hex^{v} = dEx^{im} + dEx^{chm} + dEx^{Lr} + dEx^{Yu} + dEx^{np} + dEx^{Le} \tag{3}$$

ASDI's value can range between 0 and 1, and It varies in the opposite direction, i.e. The lower the ASDI, the less the region is affected by human exclusion.

4.4 Multidimensional Local Development Index

The Multidimensional Local Development Index (MLDI) is a composite measure of accumulated progress in six key dimensions of development, education, health, employment, housing, living environment and social services. Its construction is based on the "RGPH 2014 Indicators" database (HCP, 2017), and on the indicator aggregation approach, recommended by the UNDP, for the calculation of human development indicators [13]. The health dimension is measured by the chances of survival beyond 70 and life expectancy at 60 expressed by (HI), education by the literacy rate(EI), the proportion of the population knowing Arabic and French and the proportion of the population having completed college, high school or higher, employment by unemployment rate and paid employment(EmI), the housing condition by the percentage of households living in non-basic housing, villas or Moroccan houses which is expressed by (ShI),Access to social services (SsI),and finally, the living environment expressed by (LeI). The dimensional indices are calculated for each of the six dimensions selected. Their simple arithmetic mean gives the MLDI (4) ,its value varies between 0 and 1 [13]:

$$MLDI = \sum_{1}^{6} \frac{HI + EI + EmI + ShI + SsI + LeI}{6} \tag{4}$$

5 Methods

5.1 Data Collection

Data collection is a systematic process of gathering observations or measurements, it allows you to gain first- hand knowledge and original insights into your research problem. Whatever philosophical standpoint the researcher is taking and whatever the data collection method the process will involve the generation of large amounts of data [14].

5.2 Data Sources

The data used in this project are those of four indices of human development for the regions of Morocco in 2014, we obtained this data from ONDH (The National Observatory for Human Development) on which we have based our calculations for the social development index (SDI) and the Multidimensional Local Development Index (MLDI). Also another data obtained from HCP which concerns the indexes ASDI and HDI. Basically external data source major steps are 4 [15]: Data source identification, Data extraction, Data transformation and Data loading.

Remark: Because of the lack in data we had to work with previously calculated and a little old indices such as ASDI, HDI and MLDI of the year 2014.

5.3 GIS Analysis

A Geographic Information System (GIS) is a system of computer software, hardware and data, personnel that make it possible to enter, manipulate, analyze, and pre-sent data, and the information that is tied to a location on the earth's surface. There are different definitions for Geographic Information System, each developed from a different perspective or disciplinary origin. Some focus on the map connection, some stress the database or the software tool kit and others emphasis applications such as decision support [16]. Nowadays, GIS offers tools allowing users to create interactive queries, and displaying the results of all these operations in different ways, presenting relationships, patterns, and developments in the form of maps, globes, reports, and diagrams [17]. We can perform two types of analysis with a GIS: spatial analysis and thematic analysis. The spatial analysis can be refined and made interactive, i.e., trans- formation, manipulation of maps, and applied simple mathematical facts. The spatial data can be derived from large databases providing detailed information and trends.

For example, multivariable or factor analysis allows changes in variables [18].

On the other hand, the thematic analysis consists to create a map for characteristics visual representation of a specific geographic entity. The illustration of these characteristics is realized by a variety of properties/attributes depending on interests. In a thematic map, we can use data from an existing source, such as census, socioeconomic or environment data. Thematic maps show the differences in density, em- ployment, income, age, and background of the population in a particular city, province, region, or country [19].

5.4 Study Area

Morocco is a North African country situated on the edge of continental Europe in the far northwest of Africa (Fig. 1). The Strait of Gibraltar is a 13-km (8.1-mile) waterway that connects Spain and Morocco. Morocco is bordered on the west by the North Atlantic Ocean and on the north by the western Mediterranean Sea. The Moroccan Plateau covers 2,310 km (1,440 miles) along the Atlantic coast from Tangier to Lagouira, and extends interior to the Saiss Plains near Fes and Tansift-Alhaouz near Marrakech. Due to its proximity to the Sahara Desert, the regions in the far southeast of the country are arid. Many regions as Figuig and Zagora have created palm tree oasis.

In 2015, Morocco adopted a new administrative division. It now has 12 Regions (Fig. 2), 77 Provinces and 1505 Municipalities according to HCP.

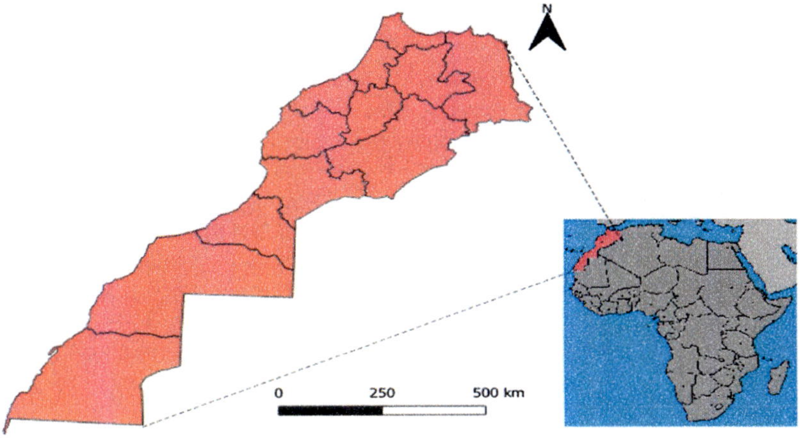

Fig. 1. The study area

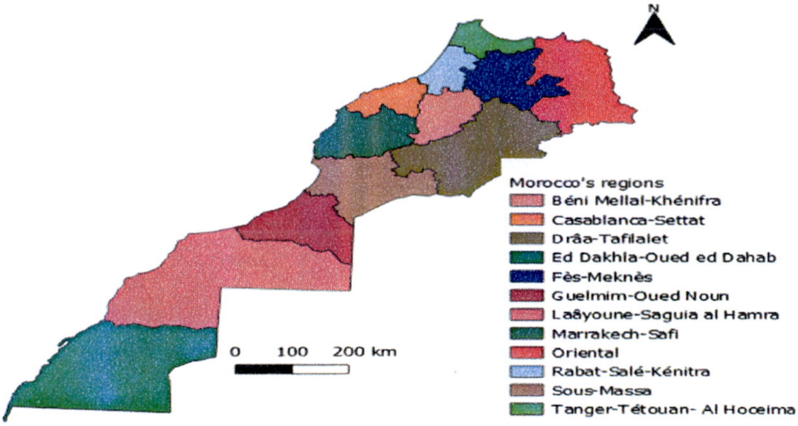

Fig. 2. The 12 regions of Morocco

6 Results and Discussion

In order to visualize our data, we have used Geo-graphic information system which is a computer system for capturing, storing, querying, analyzing, and displaying geospatial data. GIS is an indispensable tool in resource management, emergency planning, crime analysis, public health, land records management, precision farming, and many other fields. Geospatial data are spatially referenced and can be either vector or raster. Common GIS operations include data acquisition, data management, data query, vector data analysis, raster data analysis, and data display. An important trend is the integration of desktop GIS and web and mobile technologies, which has already led to the development of location-based services, collaborative web mapping, and volunteered geo-graphic information [20].

6.1 Social Development Index SDI by Region

In Fig. 3, we can see that the regions of Casa-Settat, Rabat-Salé-Kénitra, Oriental, and Fès-Meknes, as well as the areas of Béni Mellal-Khénifra and Marrakech-Safi, all have scores between 100 and 135 and their values are 135,133,116,116,110 and 105, respectively.

The regions of Eddakhla-Oued Eddaheb, which have a rating of 100, and Tanger-Tétouan-Al Hoceima, which have a value of 92, have an SDI that is comparable to the national average.

The four remaining regions of the Kingdom, namely Drâa-Tafilalet, Souss-Massa, Guelmim-Oued Noun, and Laayoune-Sakia El Hamra, join the poor performance class with values of 95 and require interventions with values of 80, 90, and 94, respectively.

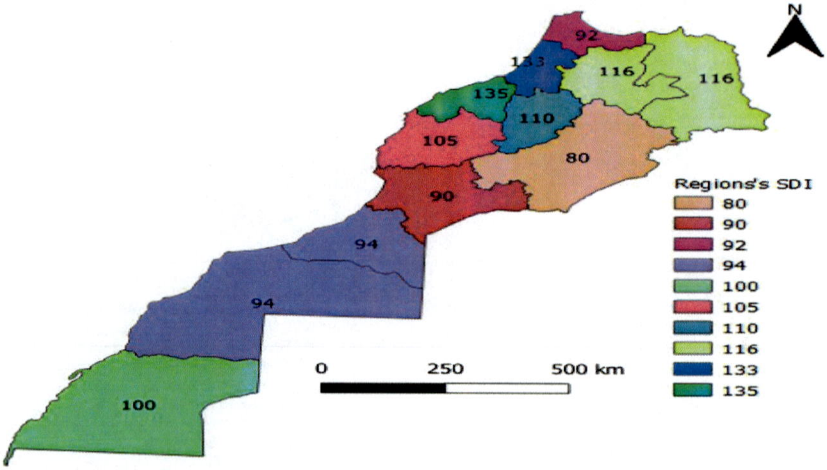

Fig. 3. SDI values for 2014

6.2 Human Exclusion Index ASDI by Region

Human exclusion is more common at the regional level in the regions of Fès-Meknes 1.00, the south 1.00, and Souss-Massa 1.00. In the regions of Béni-Mellal-Khénifra 0.96, Marrakech-Safi 0.98, and DarâaTafilalet 0.98, the human exclusion index is similar to the national average. Casablanca-Settat 0.67, Rabat-Salé-Kénitra 0.83, Ori- ental 0.74, and Tanger-Tetouan-Al Hoceima 0.85 are the regions least affected by human exclusion (Fig. 4).

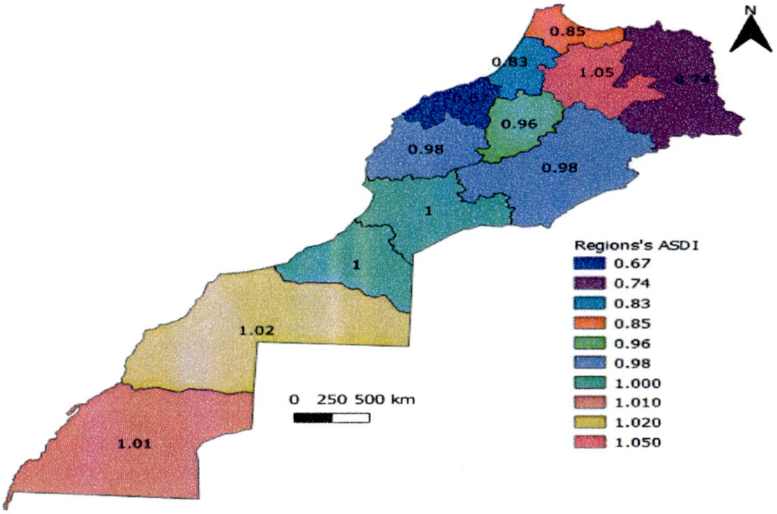

Fig. 4. ASDI values for 2014

6.3 Multidimensional Local Development Index MLDI by Region

Drâa-Tafilalet, Béni Mellal-Khénifra, Marrakech-Safi, and Tanger-Tetouan-Al Hoceima are the regions with development levels below the national average of the MLDI 0.700, as shown in (Fig. 5). The underdevelopment of these regions is attributable, first and foremost, to their mountain roads, which makes them difficult to access and infrastructure-poor. On the other hand, the more favored regions, include Laay- oune-Sakia El Hamra, Casablanca Settat, Eddakhla-Oued Eddahab, and Rabat-Salé- Kénitra. The southern regions, such as the Kenitra-Casablanca axis, so stand out from the rest of the regions in terms of economic and social performance. Guelmim- Oued Noun, Oriental, Souss-Massa and Fès-Meknes. Have similar levels of develop- ment to the national average.

6.4 Human Development Index HDI by Region

The Fig. 6 present the values of HDI in 2014 at na-tional level. In general, the HDI gives consistent results in the 12 regions. But we can deduce that the region Ed- dakhla-Oued Eddahab registers a value of 0.851 which is somewhat distant from other values, this is due to the remarkable progress made in the Eddakhla-Oued Edda-hab region in 2014. On the other hand, all the other regions have very similar values, we can conclude here, as we mentioned previously, that the HDI does not give very visible results in terms of finer geographical scales and therefore does not allow a clear comparison between regions.

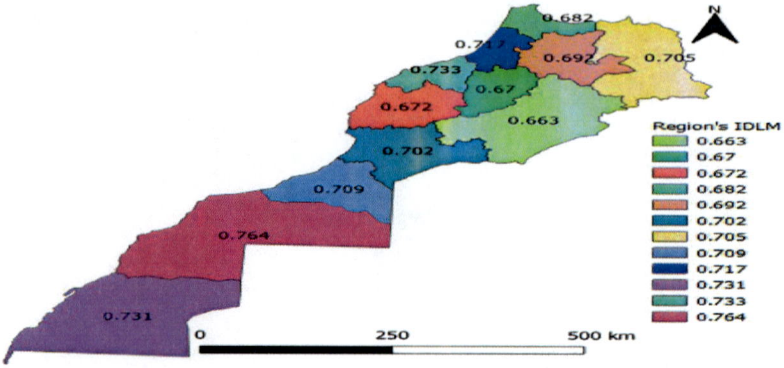

Fig. 5. MLDI values for 2014

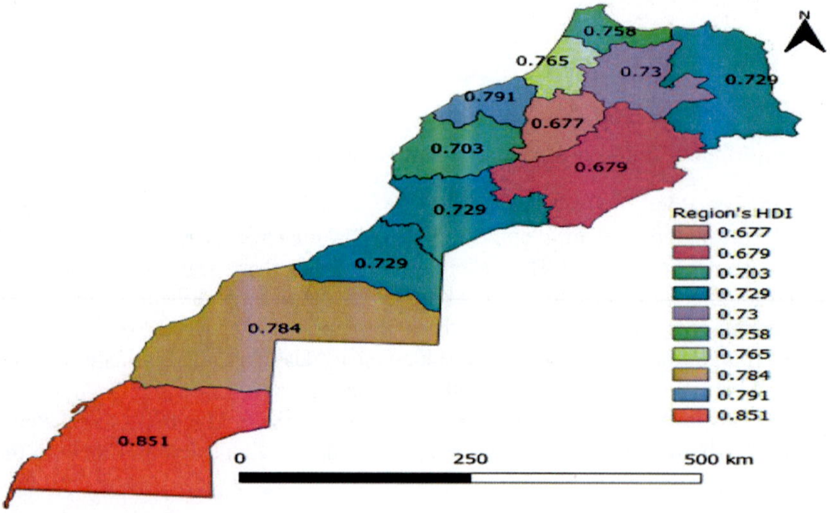

Fig. 6. MLDI values for 2014

6.5 Comparison of SDI, ASDI, HDI and MLDI Values of 2014

Based on the previously mentioned results on the maps and the values of the human development indices mentioned in this table (Table 1), we conclude that the region of Casablanca-Settat is considered to be the most developed in relation to most of the four indicators, as we find that its value is always the highest where SDI equals to 135, ASDI to 0.67, MLDI to 0.733 and HDI to 0.791.

According to the table (Table 1)., we have classified the regions into three categories, the good performance class, the average performance class and the least performed regions. The high performing regions for the four indices are Casablanca- Settat, Rabat-Salé-Kénitra, and Oriental. The average preforming regions as shows the table are Fès-Meknes, Marrakech-Safi and Béni Mellal-Khénifra. And the least preforming regions of Drâa-Tafilalet, Souss-Massa, Guelmim-Oued-Noun, Laayoune- Sakia-ElHamra, TangerTetouan-AlHoceima, Eddakhla-OuedEddaheb give minimum values for the four indices.

Table 1. Comparison of SDI, ASDI, MLDI and HDI values for 2014.

Region	Indices			
	SDI	ASDI	MLDI	HDI
TTA	92	0.85	0.682	0.758
Casa-Settat	135	0.67	0.733	0.791
Fes-Meknes	116	1.00	0.692	0.73
Rabat-Sale-Kénitra	133	0.83	0.682	0.758
Oriental	116	0.74	0.705	0.729
Marrakech-Safi	105	0.98	0.672	0.703
Souss-Massa	90	1.00	0.702	0.729
Béni-Melal-Khénifra	110	0.96	0.67	0.677
Darâa-Tafilalet	80	0.98	0.679	0.663
Eddakhla-OuedEddaheb	100	1.00	0.731	0.851
Laayoune-Sakia El Hamra	94	1.00	0.764	0.784
Guelmim-Oued Noun	94	1.00	0.709	0.729

7 Conclusion

The work presented in this paper has as a goal analyzing and comparison of the human development indicators according to regions and at finest territorial scales. The analysis carried out in this context makes it possible to identify the regions that show the most disparities in terms of social development measured by its fundamental pillars.

We collected the data of the four indices, tabulated and cleaned it. Then, analyses it with GIS. The results showed that the four indicators show that there are still regions that have large deficits in basic aspects of life, namely, education, health, un-employment, child malnutrition, personal income, etc, and this is evident in the south regions. Comparison with other regions such as Casablanca-Settat, Rabat-Salé, which are considered somewhat advanced in terms of health, education, youth employment. Nevertheless, the results can be used as a starting point for another study to be carried out and can be improved for a much more in-depth study.

Therefore, this study is open to adding other indicators of human development, which will enable us to conclude an in depth study on sustainable development.

References

1. Jahan, S.: Human Development Report Human Development for Everyone. In: By the United Nations Development Programme, 1 UN Plaza, New York, NY, p. 10017 (2016)
2. Jin, H., Qian, X., Chin, T., Zhang, H.: A global assessment of sustainable development based on modification of the human development index via the entropy method. Sustainability **12**(8), 1–20 (2020)
3. Pozzebon, M., Marlei, M.: Managing sustainability with the support of business intelligence: Integrating socio-environmental indicators and organisational context. J. Strategic Inform. Syst. **18**, 178–191 (2009)
4. Pandit, N.R.: The creation of theory: A recent application of the grounded theory method. The qualitative report, vol. 2, pp. 1–15 (1996)
5. Di Vaio, A., Palladino, R., Hassan, R., Escobar, O.R.: Artificial intelligence and business models in the sustainable development goals perspective: A systematic literature review. J. Bus. Res. **121**, 283–314 (2020).
6. Tranfield, D., Denyer, D., Smart, P.: Towards a methodology for developing evidence-informed management knowledge by means of systematic review. Br. J. Manag. **14**, 207–222 (2003)
7. bin Talal, H.: THE ARAB HUMAN DEVELOPMENT REPORT 2002: REVIEW AND REFORM. Arab Stud. Quart. **26**, 5−20 (2004)
8. Aziz, S.A., et al.: A critical analysis of development indices. Australian J.f Sustain. Bus. Society. **1** 37−53 (2015)ss
9. Davis, G.: A history of the social development network in The World Bank, 1973–2003. (2004)
10. Roser, M.: Human Development Index (HDI). Our World in Data (2014)SS
11. Disparités territoriales en matière de développement social au Maroc, Observatoire National du developpement Humain, Morocco (2014)
12. Le développement socio-économique régional ,Niveau et disparités, 2001–2017 HCP, Morocco (2017)
13. A. Union: African social developement index (ASDI): measuring human exclusion for structural transformation, vol. 68, pp. 1–78 (2016)
14. Cartographie du développement local multidimensionnel: Niveau et Déficit. ONDH
15. Mazumdar.: A survey on data storage and placement methodologies for cloud- big data ecosystem. J. Big Data **6**, pp. 1–37 (2019)
16. Wah, T.Y., Peng, N.H., Hok, C.S.: Building data warehouse. In: Proceedings of the 24th South East Asia Regional Computer Conference, vol. 15, pp. 51–56 (2007)
17. Ershad, Mr., Ali, E.: Geographic Information System (GIS): Definition, Development, Applications & Components, vol. 03 (2020)
18. Information Resources Management Association
19. Venkatramanan, C.R., Paramasivam, S.: Chapter 3 - An introduction to various spatial analysis techniques. In: GIS and Geostatistical Techniques for Groundwater Science, S. V. a. M. V. P. a. S. Y. Chung, Ed., Elsevier, pp. 23–30 (2019)
20. Erskine, M.A., Gregg, D.G., Karimi, J., Scott, J.E.: Business decision-making using geospatial data: a research framework and literature review. Axioms **3**, 10–30 (2014)
21. Chang, K.-T.: Geographic information system. In: International Encyclopedia of Geography: People, the Earth, Environment and Technology, pp. 1–0 (2016)

Applying Process Mining to Sensor Data in Smart Environment: A Comparative Study

Elkodssi Iman[1](✉), My Driss Laanaoui[2], and Hanae Sbai[1]

[1] Mathematics, Computer Science and Application Laboratory FST Mohammedia, Hassan II University of Casablanca, Casablanca, Morocco
elkodssi.iman@gmail.com, hanae.sbai@univh2c.ma

[2] Computer Systems Engineering Laboratory Cadi, Ayyad University of Marrakech, Marrakech, Morocco
d.laanaoui@uca.ma

Abstract. Business process management is a popular approach that has caught the attention of several researchers over the years. BPM has been extended to Business Process Intelligence in order to add more intelligence to managing business processes. Process Mining is one of the emerging areas of Business Process Intelligence, which aims to automatically discover, verify and improve business processes. Nowadays, the emergence of smart environment. The data that occurs in this environment are more varied and are kind of unstructured data. In this sense, applying process mining, to analyze processes in a smart context is a challenging task, since the process mining techniques relies on a structured event log. In this paper, we present a comparative study of existing approaches on applying process mining techniques in smart BPM, as a first step towards developing a new approach to discover and analyze business processes from sensor data perspective.

Keywords: Process Mining (PM) · Smart environments · Sensor log · Smart BPM

1 Introduction

In recent years, the emergence of a new concept of smart city has provided a futuristic vision, based on various issues resulted in its success. Thus, cities around the world are gradually moving from the smart city since then. In fact, Process Mining (PM) is becoming a widely used support tool, it helps users to construct processes quickly and objectively by analyzing business data, including event logs. In fact, process mining is an important bridge between data mining and business process modeling and analysis [1]. However, process mining is limited by the number of cases available for observation. Besides applying process mining techniques should examine data with different structures, namely structural, unstructured and semi-structured data. Today, we are faced with a more intelligent environment, in fact the smart city is a new concept of urban development. It is about improving the quality of life by making the city more adaptable and efficient, using new technologies that rely on an ecosystem of objects and services. The

M. Ben Ahmed et al. (Eds.): SCA 2022, LNNS 629, pp. 511–522, 2023.
https://doi.org/10.1007/978-3-031-26852-6_47

scope of this new way of managing cities includes: public infrastructure such as home automation. In addition, the field of transport as well as e-services and e-administration. [2] This has motivated the emergence of the smart Business Process Mining, this concept aims to modernize business process management by using process mining and modeling techniques to address automated business process discovery. So, the main issue will be to apply process mining to sensor logs, which record raw data generated from sensors, devices, that are connected to business processes. Many approaches [3–8] have been proposed in the literature to automatically discover business processes from sensor logs, by transforming these sensor logs into event logs. However, the majority focus on the problem of transforming unstructured logs into structured logs in order to generate the process model, all of these approaches don't identify the sensors and their data in the output process model.

This article proposes a comparative study of existing work concerning the management of business processes in an intelligent environment in terms of process mining. The organization of the rest of this article is as follows. Section 2 is dedicated to the basic concepts. Section 3 deals with the existing works related to the discovery of a process model in an intelligent environment in the context of Process Mining. The comparative study is given in Sect. 4. Finally, the last part, "conclusion and perspective" summarizes the work and the different points we have to develop in the future.

2 Background Review

2.1 Process Mining (PM)

Process Mining [9, 10] has emerged as a powerful family of Process Science techniques for analyzing and improving business processes. Although, Process Mining has managerial benefits, such as better decision making. The scientific literature has investigated it the mainly from a computer science standpoint and appears to have overlooked various possible applications. [11], While, the information in event logs was rarely used to analyze the underlying processes, Process mining aims at improving this by providing techniques and tools for discovering process, control, data, organizational, and social structures from event logs, i.e., the basic idea of process mining is to diagnose processes by mining event logs for knowledge [12]. From another point of view, the idea of Process Mining is to discover, monitor and improve real processes" [14], where "discover" presumably corresponds to the discovery type, "monitor" to the conformance type, and "improve" to the enhancement type. Another work [11], qualifies process mining as a set of data-driven techniques for diagnosing and enhancing business processes, by combining Machine Learning and Business Process Management (BPM). It leverages event data to identify process deviations and inefficiencies. PM thus bridges the gap between the model-driven approach and current data-driven BPM techniques. Process mining collects the digital traces of processes in order to analyze the different and sometimes excessive variants of a process flow. Process improvements based on this new transparency level create significant added value. To sum up, process mining is about three areas;

- **Process discovery**; which require an event log to generate a business process as output.

- **Process conformance**; requires two entries such as an event log and the related business process, to verify if it conforms to the given event log.
- **Process enhancement**; allows for enriching processes with new information discovered from event logs.

The event log is the main important element to apply process mining. It is a set of structured events. Each event is characterized by an ID, activity, resource time, and other information. The ID referred to an instance of a business process. An event log is a set of traces. A trace contains events with the same ID. Here's an example of an event log;

Table 1. Event log of E-Healthcare system PIKA ET AL., 2019.

Case ID	Activity	Type	Time	Resource	Age	language	Diagnosis
1	Register	Comp-Lete	12/01/2019 11:03	Ann	56	EN	Pancreatitis
1	Triage	Start	12/01/2019 14:55	Michael	56	EN	Pancreatitis
1	Blood Test	Start	12/01/2019 17:01	Robert	56	EN	Pancreatitis

The Table 1 shows a list of events, the Sequence (Register, Triage, Blood test) is a trace that has the same ID = 1

2.2 Smart BPM

Although widely used in recent years in public discourse and the media [16], there is no standardized and commonly accepted definition or terminology for a smart city [17]. The Smart City, also known as "intelligent environment". The word "smart" refers to an automated mechanism introduced to perform the desired activity within a given domain [18]. As a concept, smart-city is very popular in literature and is used throughout the world [19]. This concept is referred to, in literature, as the safe, secure, environmentally and efficient urban center for the future, with state-of-the-art infrastructure such as sensors, electronics and networks to drive economic growth sustainable and high quality of life [20]. The aim of a smart environment to provide people with automatic or semi-automatic services realizing the concept of ambient intelligence. Authors in the field of smart environments use, sometimes as synonyms, a variety of terms to refer to the state of the environment and the tasks humans perform in it. Such smart products are able to collect, process and produce information and they can make use of knowledge about themselves, their users and their context. The goal of a smart environment is to provide people with automatic or semi-automatic services. Managing business processes in a smart environment introduces the smart BPM. Since business process management is exactly concerned with coordinated action, there have been attempts to generalize the commonalities of the smart application scenarios, in terms of their dynamic adaptation and continuous learning towards smart business process management. The techniques that are related to smart business process management are: Smart Multi Process Management, Smart Process Model Management and Smart Process Instance Management.

2.3 Sensor Log

This concept is referred in the literature as a sequence of measurement values acquired from sensors deployed in the monitored space. [21]. In the context of raw sensor data events can be any kind of observations (e.g., a sensor value changed) with relevance to a certain process. Derived from e.g., networks of Wi-Fi-access points, or motion sensors in smart homes and smart factories. Event data like this does not necessarily have a unique identifier attached to it to identify who triggered the event [3]. A sensor database involves stored data and sensor data. Stored data includes the set of sensors that participate in the sensor database together with characteristics of the sensors (e.g., their location) or characteristics of the physical environment. These stored data are best represented as relations [22]. Some researchers agree that a distinctive feature of a "Sensor data" is a special case of data that can be represented as a data point in a high dimensional space and is produced by (multiple) [23].

3 Related Work

In this section we present related work on applying process mining to unstructured event logs with a focus on work on sensor logs. There are a many research works that deal with the discovery of activities at different levels. Particularly, approaches can be found for activity recognition from low-level events on different types of sensors where machine learning is used for data aggregation. Authors of [3] proposed frameworks to allow for discovering a process model from raw sensor data, which improves the quality of data aggregations, which machine learning is not able to do [24]. The framework consists of four steps: Event Correlation, Activity Discovery, Event Abstraction and Discovery of Processes. The work of [25] enriches existing business process models with communication activities. They add semantic information to an existing business process model. Their approach proves the feasibility of business process management techniques based on online CRM activities. They note that traditional process mining techniques produce incomplete process models because they only work on structured data, i.e. classified events, and ignore the semantics of communication activities. Another work [26] describes a method for leveraging informal communication to discover business processes. They extract event logs from emails and personal tasks. A small set of learning training is manually labeled. Supervised machine learning is then used to label the remaining activities. As well as, authors in [13] present an interesting approach to extracting an event log from personal emails. This approach uses k-means clustering of the subject and body of personal emails to assign emails to event classes. For each identified cluster, the user is then prompted to manually label of each cluster. Based on research inspiration on extracting events, a framework proposed in [4] to mine CRM data for process models. It can be interpreted as a generic sequence of steps namely: Extract Activities, Data selection, labelling activities and writing the event log. This method uses Latent Dirichlet Allocation (LDA), an unsupervised machine learning technique, to automatically detect and assign labels to activities.

Firstly, we will introduce criteria on which we are based in our comparative study, then we are going to present and discuss the comparative study results. A. Comparison criteria The works forming the comparative study are analyzed based on seven criteria which are defined in the following order:

- System type: the system that includes sensor devices, wearables, computer chips, interfaces, actuators, robots, and computers.
- Log input: are the data derived from different sources in a smart environment or another one which is populated with unstructured data.
- Nature Log input: on the ones hand, homogeneous data structures are the one in which the data elements have the same data type. All the data elements in the homogeneous belongs to the single data type. On the other hand, the heterogeneous data structures are the one in which the data elements don't belong to the same source data.
- Transformation Method: determines the method used to prepare event logs before applying process mining techniques. This criterion is used to evaluate the application of machine learning when preparing event logs before applying process mining techniques.
- Transformation Algorithm: indicates the Algorithm used to transformer raw unstructured data from the sensor logs, into event logs with a good level of structuring capable of coping with these traditional PM assumptions.
- Event log Category: it means the type of event log (Single Event Log, Multiple Event Log) used as input of process mining techniques. This will show how much the generated model represents all the execution cases of the process.
- Process Mining Area: it indicates the category of process mining techniques (discovery, conformance or enhancement) studied in the work. The objective is to know the most interesting area for research.
- Model resulting; show the kind of representation of the model discovered after applying process mining techniques.

In order to compare the different works, I, it is necessary to know the data sources used as input as shown in Table 1. So, data in smart environments can be generated using different sensors depending on the target environment. Location Sensor log data can be collected with smart homes to track the movement of residents around the home [3]. Additional ambient smart products, can provide other types of time series [6]. Another type of sensor used to generate unstructured contextual data such as online CRM [4], email systems [8]. Also in the field of associated systems produces the same type of data. More recent research uses the internet of Things (IOT) environment, which produces IoT data [41, 43], Raw IoT data [44] and history data [7, 42]. To transform the data of sensors into event. In order to apply the process mining techniques. There are approaches that use heterogeneous data structures like [4, 6, 44], the others are based only on homogeneous data structures.

Paper	System type	Log input	Nature Log input	Transformation Method	Process Mining Area	Transformation Algorithm	Event log Category
(A.Koschmider,2020) [2]	smart home	Location Sensor log	Homogeneous	Event Correlation Activity Discovery Event Abstraction	Discovery	--------	S.E.L
(Van Eck,2016) [3]	Smart Products	Time series	Heterogeneous	Sensor Data Segmentation Segment Labelling Multi-sensor Activity Creation	Conformance	centroid- based clustering techniques (k-means)	S.E.L
(Banziger,2018) [4]	CRM Online (emails, Phone Calls, tasks)	Unstructured text data	Heterogeneous	Extract Activities (Data structure Data selection) Pre-processing data Discover event classes and label activities(Number of topics, Labelling activities,Writing the event log)	Discovery	Term frequency-Inverse Document Frequency (Tf-Idf) Latent Dirichlet allocation (LDA)	S.E.L
(Francesco Piccialli,2019) [5]	Smart environment	Sensor data	Homogeneous	Unsupervised Learning	Enhancement	Mathematic al algorithm Hierarchical clustering algorithm	S.E.L

(continued)

(continued)

Paper	System type	Log input	Nature Log input	Transformation Method	Process Mining Area	Transformation Algorithm	Event log Category
(Álvaro Valencia-Parra1,2019) [6]	workstation aircraft	Complex data structure	Homogeneous	Domain-Specific Language(DSL)	Discovery	-------	S.E.L
(W.M.P van der Aalst, 2015) [7]	the redo log system	Historical Data (Redo Logs)	Homogeneous	Event Extraction Exploiting the Data Model Process Instance identification	Enhancement	---------	S.E.L

Concerning transformation method, several works like [3, 43] have opted for correlation method of log input introduced by data mining techniques. To cluster raw events into a set of unlabeled activity instances. Some other works like [8, 41, 44], choose to apply clustering method to group log input into clusters according to their process subject. However, other works use other methods to find the corresponding activity, for example [6] uses the segmentation method. LDA was introduced by [4]. It is a generative statistical model that allows to automatically detect latent topics of documents in a corpus, LDA also assigns a topic distribution to each document. The work of [5] employs Domain-Specific Language, which enables the specification of the recipes for extraction from sensors. Hierarchical clustering algorithms and k-means are the languages that dominate transformation phases despite the diversity of techniques used for discovering, conforming or enhancing processes.

Concerning Process extraction area: The process discovery area is the most area studied by many works such as [3, 4, 5, 8]. But few measures quality assessment of the processes discovered like [6, 42, 43] which focus on conformance checking, using existing models and traces extracted from process execution environment as inputs, to provide diagnostics about actual model. Another category of works like [7, 41, 44] study process extension area to generate an improved process model with high quality level. For all the input is a single event log. From the best of our knowledge, approaches do not use a collection of events logs to analyze sensors data. Against this background, we can deduce that applying process mining to sensor logs has benefits since it allows for connecting the business process intelligence field to smart environments. In fact, it helps to analyze sensors data from a process perspective. However, existing approaches focus only on how to transform sensor logs into event logs to apply process mining. However, work ignore the importance of the model discovered, in which sensors information is not yet highlighted by existing approaches. So, with the emergence of the smart BPM and internet of things, we propose a novel approach to discover sensor aware business process from sensor logs.

4 Our Approach

An overview of our initial framework can be presented in the following Fig. 1.

The main aim of our approach is to align sensor logs with process models by discovering a sensor aware process models, is composed of three following parts:

- Model driven Transformation component: takes as input a sensor log that is generated from smart object of a given smart environment, and generates as output an event log. It is based on Model Driven Architecture MDA, which allows for transforming each element of sensor log into event log, through a set of mapping rules.
- Semantic enrichment of event log component: takes as input an event logs and generates as output smart environment. Which allows to define an enrichment approach involving both the knowledge that we have built by extending the Business process management notation BPMN ontology which describes the structure of event logs by intelligent environment concepts and that of the business domain.

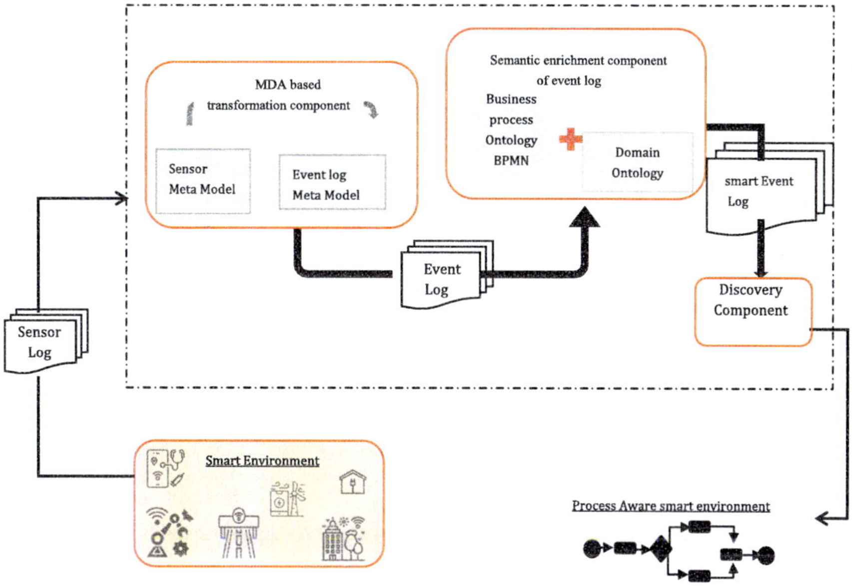

Fig. 1 An overview of our initial framework

- Discovery component: takes in input a smart event log and generates as output a process aware smart environment. It allows for applying a process mining technique such as fussy miner or heuristic miner, in order to discover the process, which presents the main behavior of smart devices captured by sensor log and transformed into event log.

5 Conclusion

Intelligent environments are composed of a wide range of connected devices. While providing a new set of opportunities, it also contributes to the complexity of business process management. Several research projects focus much more on transforming sensor data into structured event logs. In this article, we lead a comparative study of existing approaches for applying process mining in smart environment sensors logs into structured event logs, and do not discuss the main issue of discovering process aware sensor information. As future work, we will develop the mains components of our proposed framework.

References

1. Goncalves, J. C. d. A. R., Santoro, F.M., Baiao, F.A.: Business Process Mining from Group Stories. In: 13th International Conference on Computer Supported Cooperative Work in Design (2009)

2. Amina, S.R., Nassira, B.: La ville intelligente, une stratégie pour un développement urbain durable (2021)
3. Koschmider, A., Jansssen, D., Mannhardt, F.: Framework for process discovery from sensor data. In: 10th International Workshop on Enterprise Modeling and Information Systems Architectures EMISA (2020)
4. Banziger, R B., Basukoski, A., Chaussalet, T.: Discovering Business Processes in CRM Systems by Leveraging Unstructured Text Data. In: IEEE 20th International Conference on High Performance Computing and Communications (2018)
5. Valencia-Parra, A., Ramos-Gutierrez, B., Jesus Varela-Vaca, A., Teresa Gomes-Lopez, M., Bernal, A.G.: Enabling Process Mining in Aircraft Manufactures: Extracting Event Logs and Discovering Processes from Complex Data (2019)
6. van Eck, M.L., Sidorova, N., van der Aalst, W.M.P.: Enabling process mining on sensor data from smart products. In: IEEE Tenth International Conference on Research Challenges in Information Science (RCIS) (2016)
7. de Murillas, E.G.L., van der Aalst, W.M.P., Reijers, H.A.: Process mining on databases: Unearthing historical data from redo logs. In: Motahari-Nezhad, H.R., Recker, J., Weidlich, M. (eds.) 2015. LNCS, vol. 9253, pp. 367–385. Springer, Cham (2015). https://doi.org/10.1007/978-3-319-23063-4_25
8. Jlailaty, D., Grigori, D., Belhajjame, K.: A framework for mining process models from email logs (2016)
9. Van der Aalst, W.M., Van Dongen, B.F., Herbst, J., Maruster, L., Schimm, G., Weijters, A.J.: Workflow mining: a survey of issues and approaches. Data Knowl. Eng. **47**(2), 237–267 (2003)
10. Van der Aalst, W.M., Weijters, A.J.: Process Mining, Special Issue of Computers in Industry, Elsevier Science Publishers, Amsterdam (2004)
11. Zerbino, P., Stefanini, A., Aloini, D.: Process science in action: a literature review on process mining in business management. Technol. Forecast. Social Change (2021)
12. Van der Aalst, W.M., de Medeiros, A.K.A.: Process Mining and Security: Detecting Anomalous Process Executions and Checking Process Conformance. Electron. Notes Theor. Comput. Sci. **121**, 3–21 (2005)
13. Maita, A.R.C.: A systematic mapping study of process mining. Enterprise Inform. Syst. **12**(5), 505–549 (2017)
14. Van der Aalst, W.M.: Process-aware information systems: lessons to be learned from process mining. Trans. Petri. Nets Model Concurr. **2**, 1–26 (2009)
15. Wil van der Aalst: Process Mining. Springer Berlin Heidelberg, Berlin, Heidelberg (2016). https://doi.org/10.1007/978-3-662-49851-4
16. Stamos, A.: Good examples of smart green/resilient cities services: study, evaluation (2018)
17. U. N. C. o. S. a. T. f.: Development., (2016)
18. Ahad, M.A., Paiva, S., Tripathi, G., Feroz, N.: Enabling technologies and sustainable smart cities. Sustain. Cities Society **61**, 102301 (2020)
19. Nam, T., Pardo, T.A.: Conceptualizing smart city with dimensions of technology, people, and institutions . In: Proceedings of the 12th Annual International Digital Government Research Conference: Digital Government Innovation in Challenging Times (2011)
20. Caragliu, A., Bo, C.D., Nijkamp, P.: Smart cities in Europe. Research Memoranda ,University Amsterdam, Faculty of Economics, Business Administration and Econometrics. J. Urban Technol. **18**, 0048 (2009)
21. Leotta, F., Mecella, M., Sora, D.: Visual process maps: a visualization tool for discovering habits in smart homes. J. Ambient Intell. Humanized Comput. **11**(5), 1997–2025 (2019). https://doi.org/10.1007/s12652-019-01211-7

22. Bonnet, P., Gehrke, J., Seshadri, P.: Towards Sensor Database Systems. In: Tan, K.-L., Franklin, M.J., Lui, J.-S. (eds.) 2001. LNCS, vol. 1987, pp. 3–14. Springer, Heidelberg (2001). https://doi.org/10.1007/3-540-44498-X_1
23. Beddar-Wiesing, S., Bieshaar, M.: Multi-Sensor Data and Knowledge Fusion A Proposal for a Terminology Definition , University of Kassel, Germany (2020)
24. Koschmider, A., Mannhardt, F., Heuser, T.: On the Contextualization of Event-Activity Mappings. In: Daniel, F., Sheng, Q.Z., Motahari, H. (eds.) 2018. LNBIP, pp. 445–457. Springer, Cham (2019). https://doi.org/10.1007/978-3-030-11641-5_35
25. Jlailaty, D., Grigori, D., Belhajjame, K.: Multi-level clustering for extracting process-related information from email logs. In: International Conference on Research Challenges in Information Science (2017)
26. Brander, S., et al.: Refining process models through the analysis of informal work practice. In: 9th International Conference Business Process Management. BPM. Springer, Heidelberg (2011) https://doi.org/10.1007/978-3-642-23059-2_12
27. Avci, A., Bosch, S., Marin-Perianu, M., Marin-Perianu, R., Havinga, P.: Activity recognition using inertial sensing for healthcare, wellbeing and sports application. In: ARCS '10 - 23th International Conference on Architecture of Computing Systems (2010)
28. Yamada, M., Kimura, A., Naya, F., Sawada, H.: Change-point detection with feature selection in high-dimensional time-series data. In: Proceedings of the 23rd International Joint Conference on Artificial Intelligence, Beijing, China (2013)
29. Banos, O., Galvez, J.M., Damas, M., Pomares, H., Rojas, I.: Window size impact in human activity recognition (2014)
30. Cook, D.J., Krishnan, N.C., Rashidi, P.: Activity discovery and activity recognition: A new partnership (2013)
31. Lara, O.D., Labrador, M.A.: A survey on human activity recognition using wearable sensors, Commun. Surveys Tutorials, IEEE, **15**(3), 1192–1209 (2013)
32. Motahari-Nezhad, H.R., Saint-Paul, R., Casati, F.: Event correlation for process discovery from web service interaction log. VLDB J. **20**, 417–444 (2011). https://doi.org/10.1007/s00778-010-0203-9
33. Engel, R., van der Aalst, W.M.P., Zapletal, M., Pichler, C., Werthner, H.: Mining Inter-organizational Business Process Models from EDI Messages: A Case Study from the Automotive Sector. In: Ralyté, J., Franch, X., Brinkkemper, S., Wrycza, S. (eds.) CAiSE 2012. LNCS, vol. 7328, pp. 222–237. Springer, Heidelberg (2012). https://doi.org/10.1007/978-3-642-31095-9_15
34. Lee, I., Lee, K.: The Internet of Things (IoT): Applications, investments, and challenges for enterprises. Business Horizons **58**(4), 431–440 (2015)
35. Seiger, R., Zerbato, F., Burattin, A., Garcia-Banuelos, L.: Towards IoT-driven Process Event Log Generation for Conformance Checking in Smart. In: Proceedings of 2020 IEEE 24th International Enterprise (2020)
36. Folino, F., Guarascio, M., Pontieri, L.: Mining predictive process models out of low-level multidimensional logs. In: Jarke, M., Mylopoulos, J., Quix, C., Rolland, C., Manolopoulos, Y., Mouratidis, H., Horkoff, J. (eds.) 2014. LNCS, vol. 8484, pp. 533–547. Springer, Cham (2014). https://doi.org/10.1007/978-3-319-07881-6_36
37. Tax, N., Sidorova, N., Haakma, R., van der Aalst, W.M.P.: Event abstraction for process mining using supervised learning techniques. In: B, Y., Kapoor, S., Bhatia, R. (eds.) 2016. LNNS, vol. 15, pp. 251–269. Springer, Cham (2018). https://doi.org/10.1007/978-3-319-56994-9_18
38. Wanner, J., Herm, L.-V., Janiesch, C.: Countering the fear of black boxed in maintenance: Towards a smart colleague. In: Proceedings of the Pre-ICIS SIGDSA Symposium (2019)
39. van der Aa, H., Leopold, H., Reijers, H.A.: Efficient process conformance checking on the basis of uncertain event-to-activity mappings. IEEE Trans. on Knowl. Data Eng. **32**(5), 927-940 (2019)

40. Baier, T., Mendling, J., Weske, E.: Bridging abstraction layers in process mining. Inform. Syst. **46**, 123–139 (2014)
41. Piccialli, F., Casolla, G., Cuomo, S., Giampaolo, F., di Cola, V.S.: Decision making in iot environment through unsupervised learning. IEEE Intell. Syst. **35**(1), 27–35 (2019)
42. Gupta, M.: Nirikshan: Mining bug report history for discovering process maps, inefficiencies and inconsistencies, In Proceedings of the 7th India Software Engineering Conference. ACM (2014)
43. Seiger, R., Zerbato, F., Burattin, A., García-Bañuelos, L., Weber, B.: Towards IoT- driven Process Event Log Generation for Conformance Checking in Smart Factories. In: IEEE 24th International Enterprise Distributed Object Computing Workshop (2020)
44. Brzychczy, E., Trzcionkowska, A.: Process-oriented approach for analysis of sensor data from longwall monitoring system. In: Burduk, A., Chlebus, E., Nowakowski, T., Tubis, A. (eds.) ISPEM 2018. AISC, vol. 835, pp. 611–621. Springer, Cham (2019). https://doi.org/10.1007/978-3-319-97490-3_58
45. Hoornaar, T.J.: Extracting Real-life Workflow Models from Relational Data and Using These to Generate Field- based Usability Testing Scenarios at Philips Healthcare, Master's thesis, Eindhoven University of Technology (2017)
46. Rashidi, P., Cook, D.J., Holder, L.B., Schmitter-Edgecombe, M.: Discovering Activities to Recognize and Track in a Smart Environment. IEEE Trans. Knowl. Data Eng. **23**(4), 527–539 (2016)
47. Rautenburger, L., Liebl, A.: Process mining. In: Liermann, V., Stegmann, C. (eds.) The Digital Journey of Banking and Insurance, Volume II, pp. 259–275. Springer, Cham (2021). https://doi.org/10.1007/978-3-030-78829-2_15
48. Leotta, F., Mecella, M., Sora, D.: Visual process maps: a visualization tool for discovering habits in smart homes. J. Ambient Intell. Humanized Comput. **11**(5), 1997–2025 (2019). https://doi.org/10.1007/s12652-019-01211-7

Deep Background Matting

Saîd El Abdellaoui[1,2](✉) and Ilham Kachbal[1]

[1] LAPSSII, Technology High School, Cadi Ayyad University, B.P. 89, Safi, Morocco
Said.elabdellaoui@uca.ac.ma
[2] LRIT, Unité Associée au CNRST (URAC 29), Faculty of Sciences, Mohammed V University - Agdal, Rabat, Morocco

Abstract. Matting methods currently play a vital role in image processing and computer vision. In image matting, the most important step is picking specific elements from an image that will be embedded in a different background. Matting results can vary depending on an image's complexity, making it an ill-posed problem. In this paper, we present a deep learning method that solves the issue of poor performance when the foreground and background colors are similar, which the current methods suffer from. The encoder-decoder algorithm generates an alpha matte and foreground using the image itself, its soft segmentation and motion cues. These results are fed into a Generative Adversarial Network (GAN) in order to improve the matting process. Our proposed method achieves the performance levels of state-of-the-art algorithms, which makes it better than previous methods.

Keywords: Matting · Motion cues · Soft segmentation · GAN

1 Introduction

In order to create a high-quality matte, pre-processing is an essential step. This step classifies pure pixels, which are tough to classify using most pure mating methods. Many years of research have been carried out on the aspect of image matting; these studies found that image matting can be categorized into basic methods as it is shown in the Fig. 1.

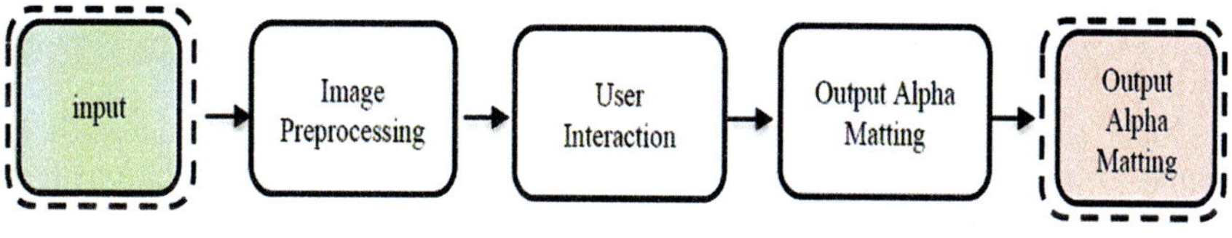

Fig. 1. Phase of image matting

M. Ben Ahmed et al. (Eds.): SCA 2022, LNNS 629, pp. 523–532, 2023.
https://doi.org/10.1007/978-3-031-26852-6_48

Image matting is the accurate extraction of a foreground object from an image. It has a fundamental role in image and video editing operations. Matting approaches, as currently practiced, do not translate well to typical everyday scenes. Matting problems have seven unknown values per pixel, compared to only three known values per pixel, which complicates the equation:

$$I_i = \alpha_i F_i + (1 + \alpha_i) B_i, \qquad \alpha_i \in [0, 1] \tag{1}$$

where F_i and B_i are the foreground and background colors of pixel i that are sequentially combined using α_i to represent its observed color I_i. The opacity parameter α takes a value between 0 and 1 with pixels having $\alpha = 1$ belonging to the foreground and those having $\alpha = 0$ belonging to the background. Image matting is a challenging task. There is often similarity in color between the foreground and background. Therefore, traditional color-based matting techniques fail to produce strong results. Furthermore, most natural images contain microscopic details like cells, fingers, or hair, making accurate extraction difficult.

2 Related Works

Existing mating methods are largely based on color to decide an alpha matte, however, positioning and other low-level features are also considered. There are two main categories of these methods: traditional matting methods and untraditional matting methods. The first one can be classified into propagation-based and sampling-based, while the second one is combinations of sampling- and propagation-based methods Fig. 2.

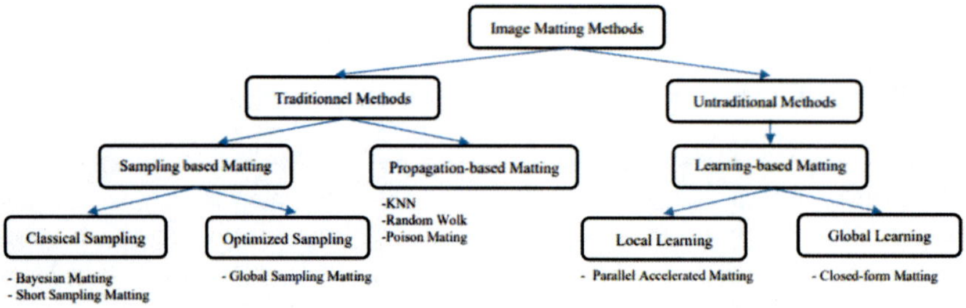

Fig. 2. Image matting methods & techniques

Using traditional methods, a trimap is usually required as input. Sampling-based approaches [1,2] determine which colors are competitors for a particular pixel by collecting a group of foreground and background pixels. A metric is then applied to determine the best combination of foreground and background. These pixels are considered to be "close" to the real foreground and background colors. After background and foreground colors have been resolved, alpha values will be calculated. Among sampled candidates, the metric nearly always includes a matting equation reconstruction error, potentially based on distances between samples or similarity between foreground and background samples.

These methods also may be divided into two major groups Classical sampling [3] and Optimized sampling [4].

In propagation-based methods [4,5], the alpha values are spread from known to unknown regions based on the affinities of neighbors.The technique is commonly used for scribble-based matting. The methods of propagation also include random walks, solving Poisson equations, and non-local propagation. Despite their ability to produce good mattes, propagation-based methods usually suffer from high processing difficulty and low memory efficiency.

Learning-based method [6–10]. Mixture of sampling-based and propagation-based matting. It is possible to divide these methods into two categories; local learning [11] and global learning [12]. Based on local learning, the algorithm is learned from the adjacent pixels of the pixel being estimated. The global learning based-methods fit the trimap-based mating better and learn the algorithm more efficiently with less distance between the marked pixels.

Recent research has demonstrated that deep learning methods provide a powerful way to map images to their alpha mattes. Yoon et Donggeun. [13] propose an architecture based on close-form mating to reconstruct alpha mattes [14] and KNN matting [15] take advantage of the meeting. Lutz et al. [8,23,24] implement generative adversarial networks based on encoder-decoder [7,25] architecture to help find the alpha matte. Cai et al. [5] test the assumptions of directly using the alpha matte values produced by a weak trimap and do the mating task separately, starting with trim adaptation and then estimating alpha values.

By using deep learning estimation, our paper propose to extract the foreground and alpha matting without the need for user interaction from inputs including the original image, soft segmentation, and motion cues of the subject (scribbles/trimap). Then, we use the generative adversarial network to refine composite creations on the one hand, and to improve the mating process on the other hand, because the results generated by this block may still have visible artifacts, especially when composing on a new background.

3 Our Approach

Our method employes a deep learning matting network generator G Fig. 3 that extracts alpha value and the foreground color for a given inputs and a discriminator network D that manipulates the training to generate realistic results [19]. In the first section of our network Fig. 3, we present a supervised network architecture based on an original image I, a motion cues M and soft-segmentation S [18]. In order to generate these last, for the soft-segmentation, morphological operations are used like erode and dilate to induce an automatic trimap and for the motion cues, we concatenate four of input images (I, I, I, I) converted to grayscale.

Our supervised network employs fully convolutional encoders [22] for I, S, and M. These encoders produces 256 deep features maps. Additionally, two separate encoders are used to enhance the mating process, which consists of 1*1 convolutional layers, ReLu, batch-Norm functions and up-sampling modules. To learn mapping from the shared representations, we use the ResNet bloc that combines the last features for each of the three pairs. Finally, these three features are passed to two residual blocks and decoders in order to generate foreground and alpha value respectively.

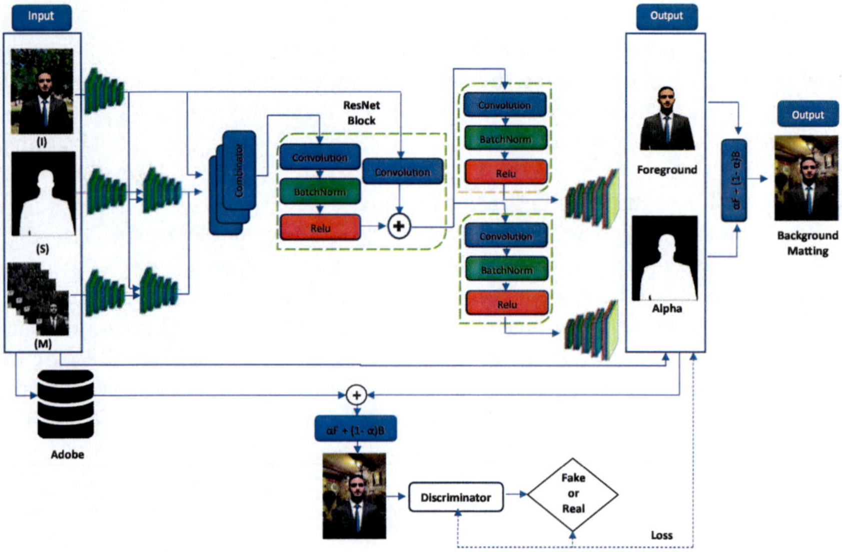

Fig. 3. Our approach network

When the foreground colors match the background colors, our supervised approach still fails in those regions, specially around the hairs, arms, or fingers. As a solution to this issue, we aim to use an adversarial network. Essentially, substantial errors in the estimated matte produce unrealistic composites over novel backgrounds. It is possible that a bad matte contains a chunk of the source background which when composited over a new background, will introduce huge visual artifacts. Therefore, we can train an adversarial discriminator to differentiate between fake composites and real images (that have already been captured) to improve the mating network.

In the second section Fig. 3, we describe a generative adversarial network (GAN) [8] in which the generator G attempts to produce a fake matte similar to Adobe's output, while the discriminator D determines if the compositing results are real or fake. Large datasets such as Adobe dataset [7] and COCO datasets [21] have greatly improved many computer vision tasks. While this may seem like a minor disadvantage, the fact remains that alpha data collection is significantly more challenging for image matting than for most other tasks. A very high quality alpha is also required, since algorithms should be able to capture fine differences in alpha for good results. Due to these difficulties, we rely on Adobe dataset [21]. From COCO dataset, each foreground object receives a randomly selected background image, which enables us to create a new unique image using the foreground, the provided alpha, as well as the background image.

Our generator G [8] presents a new encoder-decoder network that represents a RGB image using three-feature maps, the grayscale motion cue information using single features, and the soft segmentation using single features. In the encoder architecture, dilated convolution is used along with the ResNet 101 network. Rather than normal convolutions, the third and fourth blocks are convolutions at 2 and 4 rates, respectively. Moreover, the decoder contains a network of convolutional layers and skip connections. In the mating adversarial discriminator D, G is enforced to output more accurate results in order to minimize ζ_{GAN} the alpha prediction loss which is defined as:

$$\zeta_{GAN}(G, D) = logD(x) - log(1 - D(C(G(x))) \tag{2}$$

where the discriminator D aims to maximize the value of the function and the generator G wishes to minimize it. D(x) represents the probability of a sample is "real" and C(G(x)) represents a composition function where G tries to get alphas closer to the ground-truth alpha, while D tries to check if the compositing is fake or real. The above losses are combined to complete the objective of our network:

$$\zeta_{AlphaGAN}(G, D) = \zeta_{alpha}(G) + \zeta_{comp}(G) + \zeta_{GAN}(G, D) \tag{3}$$

To validate our mating approach, we compare our network with a residual encoder-decoder [20]. We find that when colors are similar, encoder-decoder networks generate holes whenever there is an excessive difference in color between them. In our case, when a segment of an individual matches the background color, our network focuses on the hard region to provide a more matte result. Eventually, we can summarize our contributions steps as Fig. 4:

A) We represent the image input that we feed to the network.
B) We provide a supervised network model that can detect the object we want to extract using soft-segmentation.
C) We calculate the alpha values from the image input with the help of motion cues and soft-segmentation.
D) We use alpha values for detecting and extracting the foreground.
E) We provide background matting by applying foreground extracted in other background.

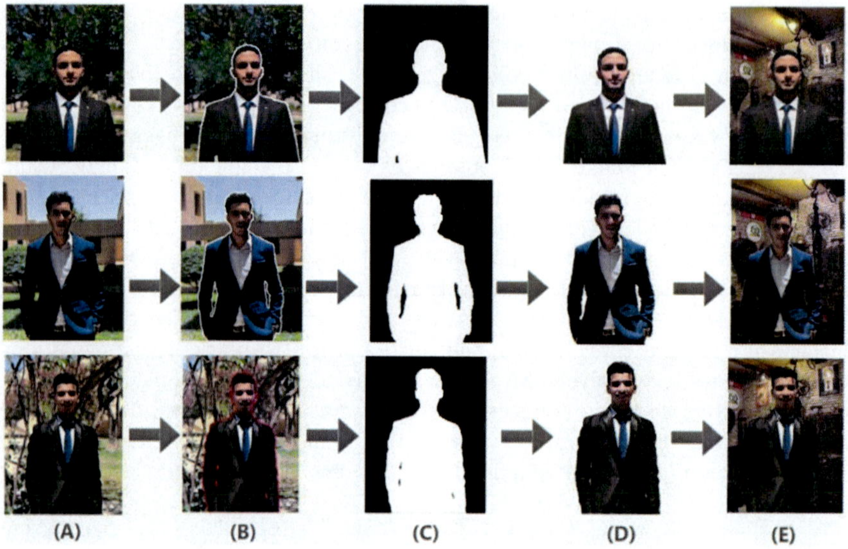

Fig. 4. Steps of our contribution

4 Discussion and Results

In order to test the efficiency of our approach further, we will compare it with other existing matting techniques like K-Nearest Neighbours (KNN), Random Walk (RW) and Closed-Form (CF) since our advantage is that we do not require the user to do anything. As opposed to these existing techniques that rely on hard constraints (trimap), our new approach generates a trimap automatically. Thus, our method offers a distinct advantage if the end product is a composition of the foreground in a new background with a refined quality especially when the color of the foreground matches the background. We provide performance results as confusion matrices [26] (see the Table 1) that depict the number of correct and incorrect classifications achieved. Columns of the matrix represent instances in predicted classes, and rows represent instances in actual (known) classes. There are many parameters that can be computed using the values of the true positive TP, true negative TN, false negative FN value, and false positive FP value of the trained model. The following are the five most commonly used parameters for algorithm evaluation: Sensitivity (S_{ens}), Specificity (S_{pec}), Precision (P_{rec}) and Accuracy (Acc). This parameters are defined successively as follows:

$$S_{ens} = \frac{TP}{TP + FN} \tag{4}$$

$$S_{pec} = \frac{TN}{TN + FP} \tag{5}$$

Fig. 5. Visual results of the proposed method

$$P_{rec} = \frac{TP}{TP + FP} \tag{6}$$

$$Acc = \frac{TP + TN}{TP + TN + FP + FN} \tag{7}$$

A higher number indicates a better result for the confusion matrix parameters in the Table 1. Comparing our algorithm to other methods, we find that ours has the highest average value. Our algorithm has a Sensitivity value of 97.3%, which means that the image has been classified correctly in all corners. Only 1.2% of unknown pixels have been classified incorrectly based on the Accuracy value of 98.80%. This error represents a so-called over classification and is not critical in our case since all detected regions of the image are typically used during automatic trimap.

Table 1. Evaluation parameters from confusion matrices

	Sensitivity	Specificity	Precision	Accuracy
KNN	92.22%	94.87%	83.15%	91.25%
RW	93.65%	95.64%	81.12%	82.66%
CF	73.53%	80.00%	56.17%	75.19%
Our	97.3%	97.88%	98.94%	98.49%

In Fig. 6 the input of K Nearest Neighbors, Random Walk and Closed-Form methods has very similar colors of the hair with the background. The best results of these methods has very obvious mistakes in those hard regions (red box green screen (a), (b), (c) and background matting (a), (b), (c)). Clearly, all variants of our approach have much better results (yellow box in green screen (d), alpha (d) and background matting (d)). The main reason is still the capability of our deep learning model to understand the complex context of images while the others like these methods or like a well-known application such as "remove.bg" cannot. To validate the performance of our approach, we tested on real data images taken by using a professional camera.

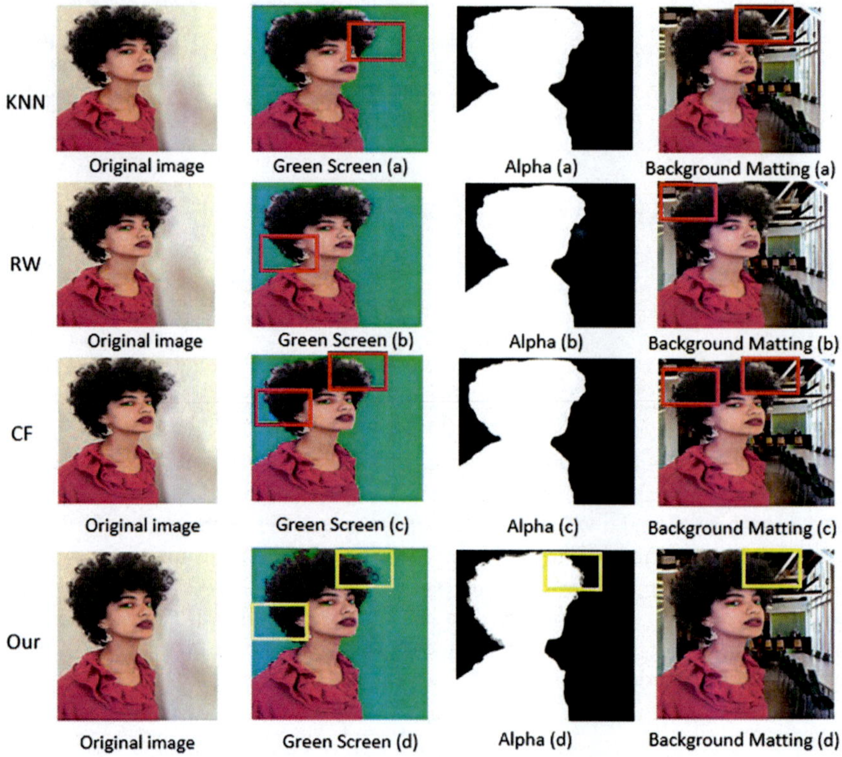

Fig. 6. Comparison of Random Walk matting and our approach

Figure 5 shows the result of our approach, we observe that we produce a high-quality output even with complicated backgrounds. Note that the hair details of girl in Fig. 5 are only recovered nicely by our method. Note that the composition examples of the foregrounds and new backgrounds are illustrated also in Fig. 5. We can see these compositions have high visual quality.

5 Conclusion

In this paper, a deep learning network has been proposed to solve the background matting problem where the task can be divided into two sub-tasks; the first one is foreground and alpha extraction, to achieve this last we have used a supervised network based on encoder-decoder network. The second one is alpha refinement using generative adversarial network. Furthermore, our proposed method shows high-quality performance establishing a new state-of the-art for background matting.

References

1. Aksoy, Y., Aydın, T.O., Pollefeys, M.: Designing effective inter-pixel information flow for natural image matting, arXiv:1707.05055 (2017)
2. Li, Y., Lu, H.: Natural image matting via guided contextual attention. In: Proceedings of the AAAI Conference on Artificial Intelligence, vol. 34, pp. 11450–11457 (2020)
3. Wei, T., et al.: Improved image matting via real-time user clicks and uncertainty estimation, arXiv preprint arXiv:2012.08323 (2020)
4. Yao, G., Yang, H., Wu, S.: A robust matting method combined with sparse-coded model. In: Hassanien, A.E., Xu, Y., Zhao, Z., Mohammed, S., Fan, Z. (eds.) BIIT 2021. LNDECT, vol. 107, pp. 119–130. Springer, Cham (2022). https://doi.org/10.1007/978-3-030-92632-8_12
5. Cai, S., et al.: Disentangled image matting. In: Proceedings of the IEEE/CVF International Conference on Computer Vision, pp. 8819–8828 (2019)
6. Zhang, Y., et al.: A late fusion CNN for digital matting. In: Proceedings of the IEEE/CVF Conference on Computer Vision and Pattern Recognition, pp. 7469–7478 (2019)
7. Xu, N., Price, B., Cohen, S., Huang, T.: Deep image matting. In: Proceedings of the IEEE Conference on Computer Vision and Pattern Recognition, pp. 2970–2979 (2017)
8. Lutz, S., Amplianitis, K., Smolic, A.: Alphagan: generative adversarial networks for natural image matting, CoRR, vol. abs/1807.10088 (2018)
9. Tang, J., Aksoy, Y., Oztireli, C., Gross, M., Aydin, T.O.: Learning based sampling for natural image matting. In: Proceedings of the IEEE Conference on Computer Vision and Pattern Recognition, pp. 3055–3063 (2019)
10. Hou, Q., Liu, F.: Context-aware image matting for simultaneous foreground and alpha estimation, CoRR, vol. abs/1909.09725 (2019)
11. Lee, P., Wu, Y.: Nonlocal matting. In: CVPR 2011, pp. 2193–2200. IEEE (2011)
12. Fuertes-Camacho, M.T., Graell-Martin, M., Fuentes-Loss, M., Balaguer-Fàbregas, M.C.: Integrating sustainability into higher education curricula through the project method, a global learning strategy. Sustainability **11**(3), 767 (2019)
13. Xu, N., Price, B., Cohen, S., Huang, T.: Deep image matting. Computer Vision and Pattern Recognition (2017). arxiv:1703.03872
14. Li, X., Li, J., Lu, H.: A survey on natural image matting with closed-form solutions. IEEE Access **7**, 136658–136675 (2019)
15. Chen, Q., Li, D., Tang, C.-K.: KNN matting. IEEE Trans. Pattern Anal. Mach. Intell. **35**, 2175–2188 (2013)

16. Chen, Q., Ge, T., Xu, Y., Zhang, Z., Yang, X., Gai, K.: Semantic human matting, CoRR, vol. abs/1809.01354 (2018)
17. Shen, X., Tao, X., Gao, H., Zhou, C., Jia, J.: Deep automatic portrait matting. In: Leibe, B., Matas, J., Sebe, N., Welling, M. (eds.) ECCV 2016. LNCS, vol. 9905, pp. 92–107. Springer, Cham (2016). https://doi.org/10.1007/978-3-319-46448-0_6
18. Yurtkulu, S.C., Sahin, Y.H., Unal, G.: Semantic segmentation with extended deeplabv3 architecture. In: 2019 27th Signal Processing and Communications Applications Conference (SIU), pp. 1–4. IEEE (2019)
19. El Abdellaoui, S., Kachbal, I.: Deep residual network for high-resolution background matting. Stud. Inf. Control **30**(3), 51–59 (2021)
20. He, K., Gkioxari, G., Dollar, P., Girshick, R.: Mask R-CNN. In: Proceedings of the IEEE International Conference on Computer Vision, pp. 2961–2969 (2017)
21. Sun, Y., Tang, C.-K., Tai, Y.-W.: Semantic image matting. In: Proceedings of the IEEE/CVF Conference on Computer Vision and Pattern Recognition (2021)
22. Abdellaoui, S.E., Kachbal, I.: Apparel E-commerce background matting. Int. J. Adv. Res. Eng. Technol. (IJARET) **12**(3), 421–429 (2021)
23. Sengupta, S., Jayaram, V., Curless, B., Seitz, S.M., Kemelmacher-Shlizerman, I.: Background matting: the world is your green screen. In: Proceedings of the IEEE/CVF Conference on Computer Vision and Pattern Recognition, pp. 2291–2300 (2020)
24. Xu, Y., Liu, B., Quan, Y., Ji, H.: Unsupervised deep background matting using deep matte prior. IEEE Trans. Circ. Syst. Video Technol. **32**(7), 4324–4337 (2021)
25. Cho, D., Tai, Y.-W., Kweon, I.: Natural image matting using deep convolutional neural networks. In: Leibe, B., Matas, J., Sebe, N., Welling, M. (eds.) ECCV 2016. LNCS, vol. 9906, pp. 626–643. Springer, Cham (2016). https://doi.org/10.1007/978-3-319-46475-6_39
26. Carrington, A.M., et al.: Deep ROC analysis and AUC as balanced average accuracy to improve model selection, understanding and interpretation. arXiv preprint arXiv:2103.11357 (2021)

Machine Learning Modeling to Estimate Used Car Prices

Mustapha Hankar(✉), Marouane Birjali, and Abderrahim Beni-Hssane

LAROSERI Laboratory, Computer Science Department, Faculty of Sciences, University of Chouaib Doukkali, El Jadida, Morocco
hankar.mustapha@gmail.com

Abstract. Due to the increased prices of new cars nowadays, many customers prefer to buy a used car. Several online websites are made to offer their third-party services to contact used car sellers with the potential buyers. However, many customers who log into these websites in order to buy or sell a used car don't have prior knowledge about its pricing value. The price of used cars often varies based on many features like production year, mileage, fuel type, fiscal power, mark, etc. In this paper, we designed a machine learning (ML) system that enables both potential buyers and sellers of used cars to estimate their car price before making a decision to buy or sell. In all tested ML models, gradient boosting regressor showed a high R-squared score (82%), a low root mean square error (RMSE), and a low mean absolute percentage error (MAPE). The system is finally deployed in a web application for a further usage of end users.

Keywords: Prediction · Estimation · Avito · Machine learning · Log transformation · Used car price · Regression · Assumptions · Streamlit

1 Introduction

The market of used car sales has been rising in recent years in Morocco. Many customers yet choose to buy a used car because of its affordable price compared to a new car. As a result, several local web platforms offer their third-party services for sellers and buyers of used cars such as Moteur[1] and Avito[2]. For many buyers and sellers of used cars in these platforms, it is very helpful to know the reasonable resale value of their car before making a decision to buy or sell. From the seller perspective, it would be difficult to estimate the price of his car before offering it to sell in the website. On the other side, the potential buyer does not have any prior knowledge about the real price of the car he is willing to buy. However, estimating the prices of used cars strongly depends on many factors such as mileage, year of production, fuel type, fiscal power, mark of the car, etc. Machine learning algorithms can be used effectively to estimate the resale value of used cars based on these factors [1, 2].

[1] https://www.moteur.ma/.
[2] https://www.avito.ma/.

M. Ben Ahmed et al. (Eds.): SCA 2022, LNNS 629, pp. 533–542, 2023.
https://doi.org/10.1007/978-3-031-26852-6_49

In this paper, we designed a system that implements a machine learning regression-based model to predict the price of used cars based on the features mentioned above. The system takes these features as inputs and then estimates the resale value of the car for a potential buyer (or a seller). An exploratory data analysis (EDA) is carried out to explore the main characteristics of the dataset beforehand, and to know which variables are important to the target variable. Afterwards, many supervised machine learning models are trained and tested on the collected dataset from Avito in order to select the best model for predicting the prices. In all tested models, gradient boosting regressor (GBR) outperformed other models with a high R-squared score, low root mean squared error (RMSE), and a low mean absolute percentage error (MAPE).

The key contributions in this paper are: (i) collecting a real dataset that contains the most important features about used cars from a local e-commerce website, (ii) using machine learning models to predict their price based on these features, and (iii) deploying the most accurate model in a web-based application for further usage of end users. The remainder of this paper is organized as follows. In the second section, we introduce and discuss some related work to our case study. The third section is dedicated to describe the methods and materials used to build the system. The final results are presented and discussed in Sect. 4. Finally, we conclude our work with some perspectives in Sect. 5.

2 Related Works

Monburinon et al. [3] conducted a comparative study on regression models for used car price prediction. The models were trained on a dataset collected from an e-commerce website. Experiment results showed that gradient boosted regression trees outperformed other models with mean absolute error (MSE) of 0.28, followed by random forest regression with MSE of 0.35 and multiple linear regression with MSE of 0.55 respectively.

In [4], Samruddhi et al. applied K-nearest neighbor (KNN) to predict the price of secondhand cars. The model is trained on a dataset of used cars from Kaggle website which is examined with different trained and test ratios. The proposed model reached an accuracy of 85% in predicting the price of secondhand cars. Narayana et al. [5] designed a predictive model to predict the cars selling price based on many factors. The model helps sellers and buyers to get an approximation about selling price of a used car and reduces the consumer risk in business. They implemented machine learning algorithms and regression techniques of statistics like linear, decision tree and random forest regressions in the prediction process.

Pal et al. [6] used random forest model on a Kaggle dataset to predict the used car prices. An exploratory data analysis process is performed to determine the impact of each predictor variable on the car price. Experimental results reached a training accuracy of 95.82%, and a testing accuracy of 83.63%. The model is capable to accurately predict the car prices using a feature selection method to choose the important features. In [7], Amik et al. developed a forecasting machine learning system that helps car buyers to estimate the price of a pre-owned car. The model is trained on a collected dataset from an e-commerce website for car selling. In all evaluated machine learning regression algorithms, XGBoost model was chosen with an accuracy (R^2 score) of 91%.

N. Kanwal and J. Sadaqat [8] used a multiple linear regression model to predict vehicle car price. They used a feature selection method to find the most relevant variables. The used dataset contains only selected variables to form the linear regression model. The result was impressive with an R^2 score of 98%. In [8], Van Thai et al. proposed a model for car prices prediction based on quantitative data analysis to build a knowledge-based system with qualitative data. The used data was collected from an e-commerce site with many non-structured, digital, non-numeric data fields. They developed a method for quantifying qualitative data based on ensemble machine learning algorithms. Experiments showed better results compared to other models.

3 Methodology

In this section, we present the overall workflow of the research methodology, materials, and technologies used to build a machine learning system that estimates the resale value of used cars. In Fig. 1, we illustrate the whole process of our work from collecting and preparing data to serving the model in a web-based application.

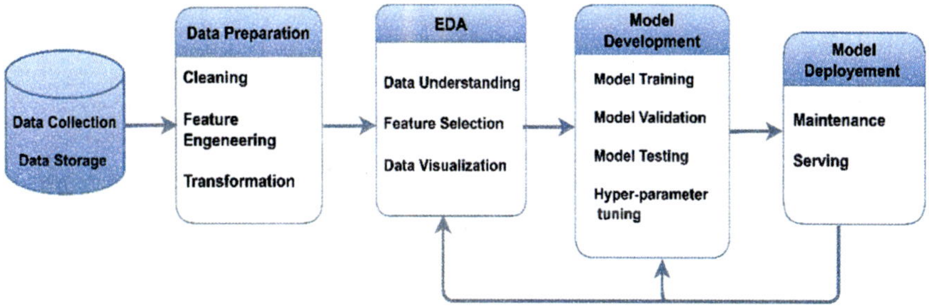

Fig. 1. The overall machine learning pipeline

3.1 Data Collection

The dataset provided for this study was collected from Avito, an ecommerce local website. For the data collection task, we coded a python script to crawl all offered-to-sell used car links in the website using a web scraping library called BeautifulSoup. We loop over the links to extract the most important car features such as price, mileage, year of production, fuel type, fiscal power, mark, model, city, sector, and type of the offer.

3.2 Data Preparation

After raw data has been collected and stored in a csv file, cleaning and preprocessing tasks are performed to prepare data for exploratory data analysis and model development. Irrelevant features like city, sector, and type are dropped from the dataset because they don't hold any useful information for the price of the car. Other features like mileage and fiscal power that contain missing values are imputed by their mean (or mode) to avoid information loss. Outliers within some numerical features are removed, because they negatively affect the mean, the standard deviation, and the normality of the feature distribution. The presence of outliers in a dataset may also decrease the performance of ML models [9]. After cleaning data, label encoding is performed to convert categorical features (fuel type, mark, model) to numerical features. Finally, standard scaling is applied on the whole dataset before training machine learning models.

3.3 Exploratory Data Analysis (EDA)

In every data science problem, exploratory data analysis is considered a crucial step to investigate and analyze data using statistical methods (mean, frequency, quantiles, etc.), and visualization charts (histograms, scatter plots, heatmaps, etc.). In the EDA phase, we try to discover useful insights and anomalies about data features and to assess prior assumptions about data beforehand [10, 11]. Tables 1 and 2 below show summary statistics about numerical and categorical features of the dataset.

Table 1. Summary statistics of numerical features

Features	Count	Mean	Std	Min	Max
Price	7765	96251.83	79292.98	4000	1580000
Production year	7765	2007.379	6.309803	1991	2018
Mileage	7765	128698.9	83606.91	4999	349999
Fiscal power	7765	6.523374	1.36627	4	20

Table 2. Summary statistics of categorical features

Features	Count	Unique	Top	Frequency
Fuel type	7765	2	Diesel	5609
Mark	7765	53	Renault	1082
Model	7765	461	Logan	309

Visualizing scatter plots of numerical variables (mileage, year of production, and fiscal power) against the price is useful to assess the relationship between these features and the price (Fig. 3), and plotting bar charts of categorical features shows the influence of these features on the price. Likewise, plotting a heatmap matrix between the input features offers an idea about the correlations among them (Fig. 4). Based on these visualizations, feature selection is performed to only maintain the important features for car price prediction. The selected features by importance are: year of production, fuel type, mark, model, fiscal power, and mileage. In the Fig. 2, we show the input features sorted by importance to gradient boosting estimator.

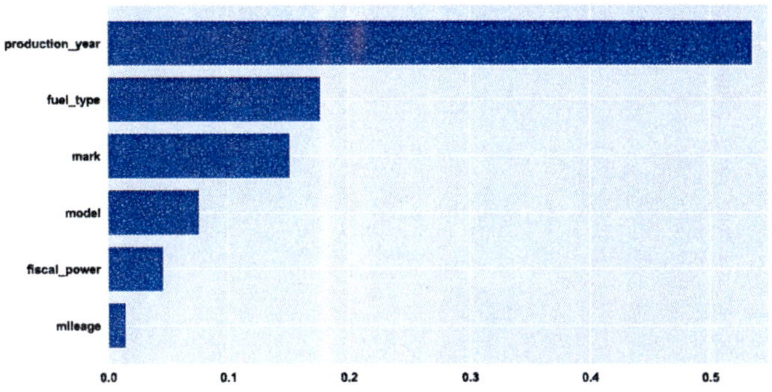

Fig. 2. Features importance to the car price.

The scatter plots in the Fig. 3 show that mileage, fiscal power, and production year have relationship with car prices.

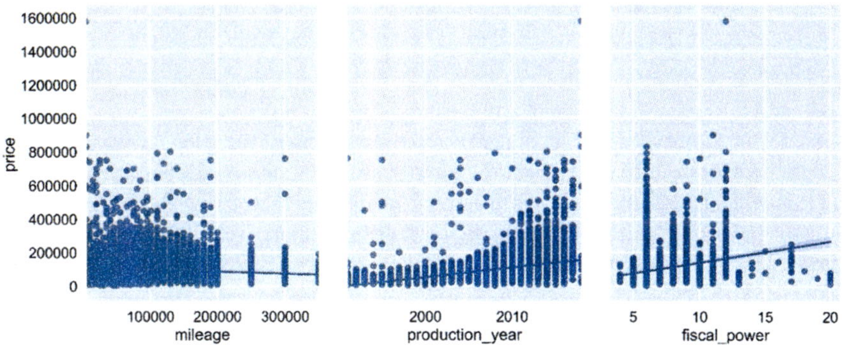

Fig. 3. Scatter plots between numerical variables and price.

The heatmap matrix in Fig. 4 shows that the correlations among predictor variables (mileage, production year, and fiscal power) are mostly weak, but we can notice a slight negative correlation between mileage and production year (−0.24). To avoid collinearity, mileage variable will be replaced by production year as the most correlated variable with the prices.

3.4 Model Development

Predicting used car prices is a regression problem in which we aim to predict a continuous value of an output variable based on many given values of the input variables. Regression analysis is a type of supervised machine learning techniques which are used to estimate a linear (also nonlinear) relationship between input variables and a response variable [12]. The general form of such a linear regression dependency is given by the Eq. (1):

$$y = \beta_0 + \beta_1 x_1 + \cdots + \beta_p x_p + \varepsilon \quad (1)$$

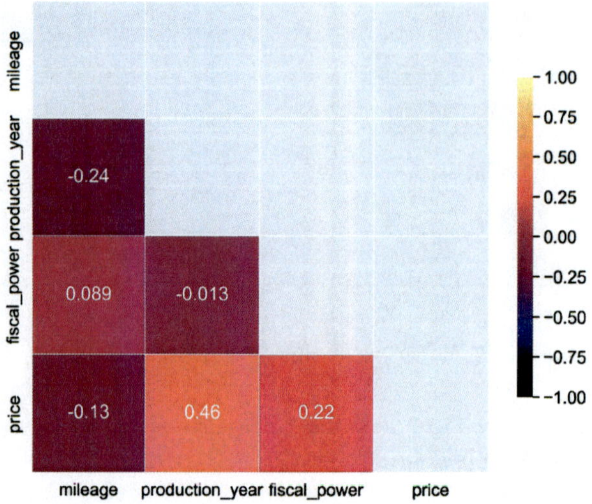

Fig. 4. Correlation matrix between numerical variables.

where y is dependent variable and $\beta_0, \beta_1, \beta_3, \ldots, \beta_p$ are regression coefficients and $x_1, x_2, x_3, \ldots, x_p$ are independent variables and the error term ε.

To select the best model to use as a price estimator in the proposed system, we have trained seven different regressors other than the baseline multiple linear regression (MLR) model. Namely, ridge regressor, k-nearest neighbors (KNN), random forest regressor (RFR), gradient bosting regressor (GBR), artificial neural networks (ANN), extreme gradient boosting (XGboost), and finally light gradient boosting machines (LightGBM). The dataset is separated randomly into training and validation sets. We trained all ML models on a portion of 80% of data, and the remaining 20% is left to test them. The regressors are then evaluated using three different metrics.

The coefficient of determination score in (2) measures the rate of variance in the target variable that is explained by the predictor variables [13]. The loss functions [14] in the Eqs. (3) and (4) measure respectively the root mean squared errors, and the mean absolute percentage errors between the predicted prices and the real prices.

$$R^2 = 1 - \frac{\sum \left(y_i - \hat{y}_i\right)^2}{\sum (y_i - \overline{y})^2} \tag{2}$$

$$\text{RMSE} = \sqrt{\frac{\sum_{i=1}^{N} \left(y_i - \hat{y}_i\right)^2}{N}} \tag{3}$$

$$\text{MAPE} = \frac{100}{N} \sum_{i=1}^{N} \left| \frac{y_i - \hat{y}_i}{y_i} \right| \tag{4}$$

where $\boldsymbol{y_i}$ are the observed values, $\hat{y}_i$ are predicted values, and $\overline{y}$ is the mean value of $\boldsymbol{y_i}$ observations. The range of R^2 score values varies between 0 and 1, and a higher value of R^2 score implies that the model is a best fit. Low values of RMSE and MAPE metrics mean that the model has a high predictive capability.

4 Results and Discussion

In the Table 3 below, we present the final evaluation results for all tested models in terms of R^2 score, RMSE error, and MAPE percentage error.

Table 3. Regression models evaluation results.

Model	R^2 score	RMSE	MAPE (%)
MLR	0.57	63933.52	31.66
KNN	0.70	51224.96	26.00
Ridge	0.58	52214.82	31.17
RFR	0.74	44939.79	21.97
GBR	**0.82**	**43839.92**	**19.69**
ANN	0.67	54957.98	27.83
XGboost	0.78	44565.45	20.29
LightGBM	0.66	48723.73	24.37

From the results above, we notice that GBR model outperformed other models with 0.82 in the coefficient of determination (R^2 score). This means that the model explains 82% of the variation in car price values. However, this result is not sufficient to evaluate the model as a good fit. We also observe that GBR model recorded a low value on minimizing RMSE errors (43839.92), and MAPE percentage errors (19.69%). The results still remain with no significance if regression assumptions [15] of the model are violated. One of these assumptions is to check if the residuals of the model are normally distributed.

The plot in Fig. 5 shows that GBR residuals follow a normal distribution across the x-axis. The other assumption to check is homoscedasticity, which means that the model residuals must have a constant variance across values of input variables. When homoscedasticity is violated (presence of heteroscedasticity), the results of the regression

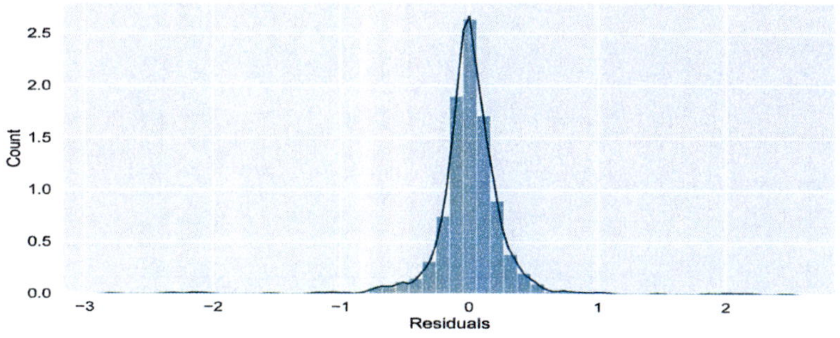

Fig. 5. Residuals distribution for GBR model

analysis become biased and hard to trust. Heteroscedasticity increases the variance of the regression coefficient estimates.

In Fig. 6, we show residual values versus fitted values scatter plot, and it is noticed that error terms are constantly spread across the values of independent variables. As a result, error terms remain unchanged with the values of car price. While car prices are not normally distributed and they are highly skewed in the right, we applied a natural log transformation for the dependent variable before training the models. The result was an obvious increase of R^2 score and RMSE errors are decayed even more.

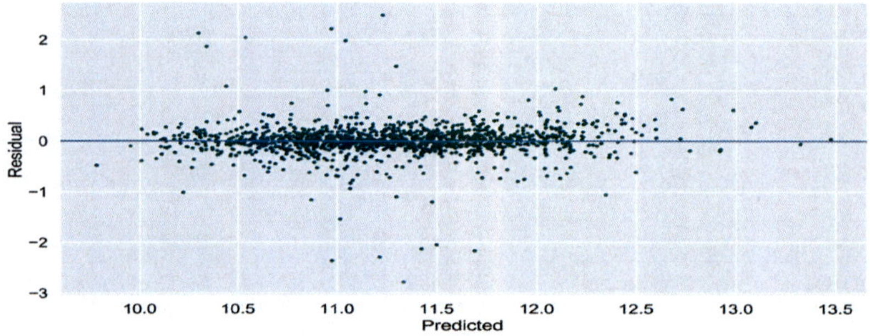

Fig. 6. Residuals vs predicted values of GBR model

Finally, we plot the real prices versus predicted prices in Fig. 7 to show the difference between the ground truth and the model predictions. The interpretation of these results state that the model learned from data to estimate a relationship between car features and its price. The selected model is then integrated in a web-based application using streamlit[3] framework to predict the price of used cars based on some given features as inputs. Finally, the system will be served for further usage to end users.

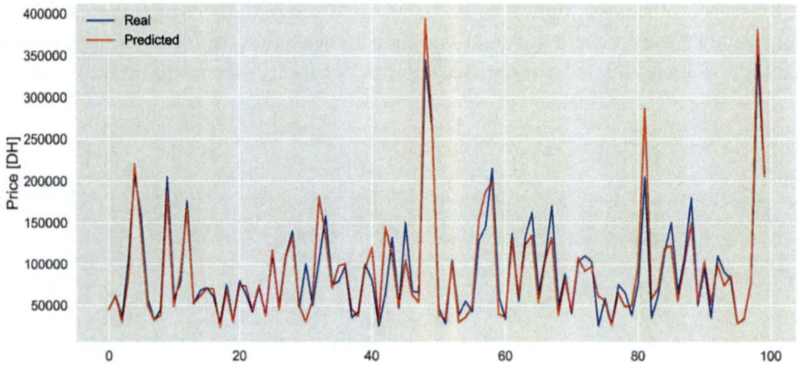

Fig. 7. Real prices vs predicted prices on test data

3 https://docs.streamlit.io/.

5 Conclusion

In this paper, we proposed a machine learning system that predicts the resale value of used cars in Morocco based on many car features like production year, fuel type, brand name, etc. The system takes these features as inputs to estimate an approximate price of a used car. The results showed that gradient boosting regressor model outperformed all tested models with a highest R^2 score, minimizing root mean squared errors, and mean absolute percentage errors. As a future work, we will improve the predictive power of the proposed system by training the machine learning model on a large-scale dataset, and by adding other important features to the price of used cars such as transmission type, horse power, number of cylinders, etc.

References

1. Gegic, E., Isakovic, B., Kečo, D., Mašetić, Z., Kevric, J.: Car price prediction using machine learning techniques. TEM J. **8**. 113–118 (2019). https://doi.org/10.18421/TEM81-16
2. Ganesh, M., Venkatasubbu, P.: Used cars price prediction using supervised learning techniques. Int. J. Eng. Adv. Technol. **9**, 216–223 (2019). https://doi.org/10.35940/ijeat.A1042.1291S319
3. Monburinon, N., Chertchom, P., Kaewkiriya, T., Rungpheung, S., Buya, S., Boonpou, P.: Prediction of prices for used car by using regression models. In: 2018 5th International Conference on Business and Industrial Research (ICBIR), pp. 115–119 (2018). https://doi.org/10.1109/ICBIR.2018.8391177
4. Samruddhi, K., Kumar, R.: Used car price prediction using k-nearest neighbor based model. Int. J. Innov. Res. Appl. Sci. Eng. **4**, 686–689 (2020). https://doi.org/10.29027/IJIRASE.v4.i3.2020.686-689
5. Narayana, C.V., Likhitha, C.L., Bademiya, S., Kusumanjali, K.: Machine learning techniques to predict the price of used cars: predictive analytics in retail business. Second International Conference on Electronics and Sustainable Communication Systems (ICESC) **2021**, 1680–1687 (2021). https://doi.org/10.1109/ICESC51422.2021.9532845
6. Pal, N., Arora, P., Kohli, P., Sundararaman, D., Palakurthy, S.S.: How much is my car worth? A methodology for predicting used cars' prices using random forest. In: Proceedings of the Future of Information and Communication Conference, Singapore, 5–6 April 2018, pp. 413–422 (2018)
7. Amik, F., Lanard, A., Ismat Tasin, A., Momen, S.: Application of machine learning techniques to predict the price of pre-owned cars in Bangladesh. Information **12,** 514 (2021). https://doi.org/10.3390/info12120514
8. Kanwal, N., Sadaqat, J.: Vehicle price prediction system using machine learning techniques. Int. J. Comput. Appl. **167**(9), 27–31 (2017)
9. Van Thai, D., Ngoc Son, L., Tien, P.V., Nhat Anh, N., Ngoc Anh, N.T.: Prediction car prices using quantify qualitative data and knowledge-based system. In: 2019 11th International Conference on Knowledge and Systems Engineering (KSE), pp. 1–5 (2019). https://doi.org/10.1109/KSE.2019.8919408
10. Wada, K.: Outliers in official statistics. Jpn. J. Stat. Data Sci. **3**(2), 669–691 (2020). https://doi.org/10.1007/s42081-020-00091-y
11. Komorowski, M., Marshall, D.C., Salciccioli, J.D., Crutain, Y.: Exploratory data analysis. In: Secondary Analysis of Electronic Health Records, pp. 185–203. Springer, Cham (2016). https://doi.org/10.1007/978-3-319-43742-2_15

12. Midway, S.: Principles of effective data visualization. Patterns. **1**, 100141 (2020). https://doi.org/10.1016/j.patter.2020.100141
13. Sarker, I.H.: Machine learning: algorithms, real-world applications and research directions. SN Comput. Sci. **2**(3), 1–21 (2021). https://doi.org/10.1007/s42979-021-00592-x
14. Chicco, D., Warrens, M.J., Jurman, G.: The coefficient of determination R-squared is more informative than SMAPE, MAE, MAPE, MSE and RMSE in regression analysis evaluation. Peer J. Comput. Sci. **7**, e623 (2021). https://doi.org/10.7717/peerj-cs.623
15. Hyndman, R.J., Koehler, A.B.: Another look at measures of forecast accuracy. Int. J. Forecast. **22**(4), 679–688 (2006). CiteSeerX 10.1.1.154.9771. https://doi.org/10.1016/j.ijforecast.2006.03.001
16. Osborne, J.W., Waters, E.: Four assumptions of multiple regression that researchers should always test. Pract. Assess. Res. Eval. **8**, Article 2 (2002). https://doi.org/10.7275/r222-hv23

Improvement Deep Leaning Network Performance by Increasing Its Complexity

Moulay Lhabib El Hadi(✉) and M'barek Nasri

MATSI EST, University Mohammed I, BP 717, 60000 Oujda, Morocco
mlhabibelhadi@gmail.com, nasrihome@gmail.com

Abstract. Performance evaluation is an important aspect of the machine learning process. However, it is a complex task. Therefore, it needs to be conducted carefully for the application of machine learning to image classification or other domains to be reliable. In this work, we have proposed a CNN (Convolutional Neural Network) classifier from the results of the current state of the art using a certain ML (Machine Learning) algorithm applied to the problems of classification of images into components of the results of several different models, going from the simplest to the most complicated possible. And at each stage, we analyze the faults by observing the behavior of each model and increasing the complexity of our classifier by integrating several parameters, such as: changing the learning algorithm, adding a layer, dropout, batch normalization, and global average pooling. After finding the architecture that suits us best, we get the maximum performance. The application of our classifier on the Cifar-10 and Cifar-100 datasets shows us the robustness of our approach since it achieves one of the best performances recorded on these datasets.

Keywords: Convolutional neural network · Learning algorithm · Dropout · Batch normalization · Global average pooling

1 Introduction

Several theoretical results show that there is a performance limit for any optimization algorithm that can be designed for a neural network [1, 2]. Theoretical results reveal that there are classes of insoluble problems, but it can be difficult to say whether a particular problem belongs to this class. Other results show that finding a solution for a network of a given size is insoluble, but in practice, we can easily find a solution using a larger network for which many parameters correspond to an acceptable solution. Also, in the context of learning a neural network, we don't usually worry about finding the exact minimum of a function, but only try to reduce its value enough to get a good test error. The theoretical analysis of whether an optimization algorithm can achieve this objective is extremely difficult [3].

In general, there are a lot of decisions to make when designing and configuring your deep learning models. Most of these decisions must be resolved empirically through trial and error and by evaluating them on real data. So it is extremely important to have a robust means of assessing the performance of your CNN models.

M. Ben Ahmed et al. (Eds.): SCA 2022, LNNS 629, pp. 543–553, 2023.
https://doi.org/10.1007/978-3-031-26852-6_50

In this work, the issue of image classification using a deep learning network is of particular interest. The results of different models range from being as simple to complex as they can be, starting with a simple CNN model that performs minimally and thinking about accelerating the performance of our model by components. We analyze the behavior of each model at each stage to spot any flaws, and then add more parameters to our classifier to make it more complex to get the best performance.

The remainder of the work is organized as follows: Deep learning is discussed in Sect. 2. The setup of learning parameters is described in general in Sect. 3. We outline our suggested strategy, experimental results, and discussion in Sect. 4. Conclusions and suggestions for further research are included in Sect. 5 as well.

2 Deep Learning

Deep learning is based on the concept of artificial neural networks and is designed to handle large amounts of data by adding layers to the network. As seen in Fig. 1, a deep learning model can extract characteristics from raw data by using multiple processing layers made up of various linear and non-linear transformations, and gradually learn about these characteristics through each layer with little assistance from a human. While in the case of machine learning, we had to manually define these functionalities [4, 5, 6, 7].

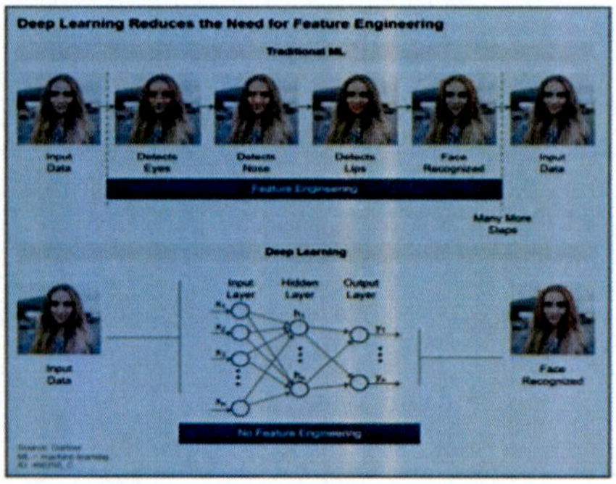

Fig. 1. Comparison between traditional ML and deep learning

Applications

Deep learning has made a go of the GO [8], learned to drive a car [9], diagnose cancer [10], and autism [11], and even become an artist [12].

3 Learning and Initiating Parameters

When working with deep neural networks, initializing the network with the right weights is very important if we want the network to work properly. We need to make sure the weights are in a reasonable range before we start training the network. For this, we will use Glorot initialization [17].

Consider a linear neuron:

$$y = w_1x_1 + w_2x_2 + \cdots + .w_nx_n + b \tag{1}$$

With x, the entry is made up of n components. This helps us to prevent the signal from exploding at a high value or disappearing at zero. This initialization process is known as Glorot initialization.

We want the variance to stay the same as we go through each layer. The variance of y:

$$\mathrm{var}(y) = \mathrm{var}(w_1x_1 + w_2x_2 + \cdots + .w_nx_n + b) \tag{2}$$

Calculate the variance of the terms in the parentheses on the right side of the above equation:

$$\mathrm{var}(w_ix_i) = E(x_i)^2\mathrm{var}(w_i) + E\Big(w_i)^2\mathrm{var}(x_i)\Big) + \mathrm{var}(w_i)\mathrm{var}(x_i) \tag{3}$$

Here, E () signifies the expectation of a given variable, which essentially represents the mean value.

We assumed that the inputs and weights were obtained from a Gaussian distribution of mean zero. Consequently, the term E () disappears and we obtain:

$$\mathrm{var}(w_ix_i) = \mathrm{var}(w_i)\mathrm{var}(x_i) \tag{4}$$

Note that b is a constant and has a zero variance, so it will disappear. Substitute in the original equation:

$$\mathrm{var}(y) = \mathrm{var}(w_1)\mathrm{var}(x_1) + \ldots + \mathrm{var}(w_n)\mathrm{var}(x_n) \tag{5}$$

Then, if we assume that the xi and wi are all independent and identically distributed, we can determine that the variance of y is:

$$\mathrm{var}(y) = n\ \mathrm{var}(w_i)\mathrm{var}(x_i) \tag{6}$$

So if we want the variance of y to be the same as x, then the term n should be equal to 1. Therefore:

$$n\mathrm{var}(w_i) = 1 \tag{7}$$

$$\mathrm{var}(w_i) = 1/n \tag{8}$$

We have arrived at the Glorot initialization formula. We must choose the weights from a Gaussian distribution with a zero mean and a variance of (1/n).

In the original article, the authors average the number of input and output neurons. So the formula becomes:

$$\mathrm{var}(w_i) = 1/n_{avg} \tag{9}$$

$$n_{avg} = \frac{n_{in} + n_{out}}{2} \tag{10}$$

The reason they do this is to preserve the signal during backpropagation too.

For the ReLu type activation functions, the following formula is used:

$$\mathrm{var}(\mathrm{w}) = \frac{2}{\mathrm{n}} \tag{11}$$

which makes sense. A ReLu unit [18] is zero for half of its inputs, so we have to double the size of the weight variance to keep the signal variance constant.

4 Contribution

In this part, we will present a comparative study of the different methods used to improve the performance of a model, either in time or inefficiency.

They are the most widely used datasets for machine learning research. For this, we have chosen the CIFAR-10 and CIFAR-100 datasets (https://www.cs.toronto.edu/~kriz/cifar.html). In the following table, we find the description of our datasets (Table 1).

Table 1. Description of CIFAR-10 and CIFAR-100 Datasets.

Dataset	Total of class	Total of color images	Total training set/class	Total test set/class
CIFAR-10	*10*	*60 000*	50 00	10 00
CIFAR-100	*100*	*60 000*	50 0	100

We started by training a classical convolutional neural network. The network has 1172410 trainable parameters and consists of 6 convolutional layers. A layer of Max Pooling is placed after each series of 2 convolutional layers, and finally an MLP (Multi-Layer Perceptron.) with 3 layers. For faster learning, ReLu will be the activation function as shown in Fig. 2.

We started with the classification of the CIFAR-10 dataset.

4.1 What Learning Algorithm to Train our Classifier?

We experimented with five of the best-known learning algorithms to choose the best. For this, we tested each of them on the same model described in Fig. 1. The weights are initialized with Glorot initialization and the biases are initialized to 0.1.

For this test, we used the cost function cross-entropy:

$$\mathrm{H}(\mathrm{p}, \mathrm{q}) = -\sum\nolimits_{\mathrm{i}} \mathrm{p}\log(\mathrm{q}) \tag{12}$$

With p the desired output and q the network output.

The following table summarizes the learning process after 20 epochs (Table 2).

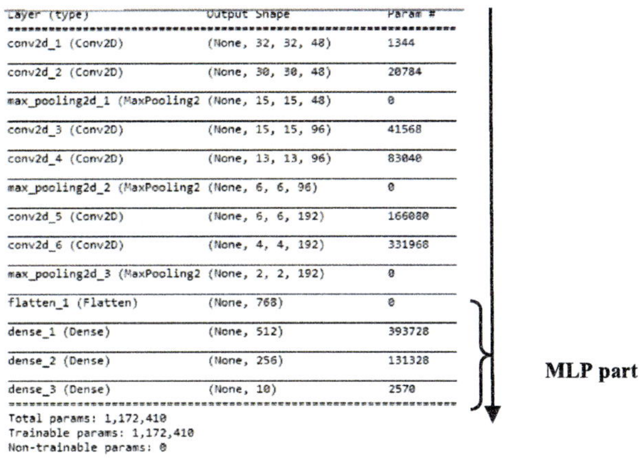

Layer (type)	Output Shape	Param #
conv2d_1 (Conv2D)	(None, 32, 32, 48)	1344
conv2d_2 (Conv2D)	(None, 30, 30, 48)	20784
max_pooling2d_1 (MaxPooling2	(None, 15, 15, 48)	0
conv2d_3 (Conv2D)	(None, 15, 15, 96)	41568
conv2d_4 (Conv2D)	(None, 13, 13, 96)	83040
max_pooling2d_2 (MaxPooling2	(None, 6, 6, 96)	0
conv2d_5 (Conv2D)	(None, 6, 6, 192)	166080
conv2d_6 (Conv2D)	(None, 4, 4, 192)	331968
max_pooling2d_3 (MaxPooling2	(None, 2, 2, 192)	0
flatten_1 (Flatten)	(None, 768)	0
dense_1 (Dense)	(None, 512)	393728
dense_2 (Dense)	(None, 256)	131328
dense_3 (Dense)	(None, 10)	2570

Total params: 1,172,410
Trainable params: 1,172,410
Non-trainable params: 0

Fig. 2. Illustration of the baseline model (Model 1)

Table 2. The Loss and execution time of each algorithm after 20 epochs

Algorithm	Loss	Execution time
SGD	2.03	2 h
ADAGRAD	2.05	2 h
MOMENTUM	1.25	2 h
ADAM	0.085	4 h 20 min

According to the table, the ADAM [15] algorithm, which proceeds with an adaptive gradient and a restricted accumulation of past gradients, progresses rapidly and minimizes the learning error from the start of the process. MOMENTUM [13] converges faster than SGD [16] and ADAGRAD [14]. The slow progression of ADAGRAD is explained by the fact that accumulating the square of the gradients in the denominator generates a drop in the learning rate but nevertheless remains faster than SGD.

Based on the table results. All the next experiments will be done with ADAM as a learning algorithm.

Now take the loss curve of the learning and testing phase.

In Fig. 3, We find that the error is well minimized at the beginning and reaches its lowest value after 7 epochs, then after the curve diverges completely; the test error begins to increase, while at the same time the learning error continues to decrease. It is over-learning. To correct this problem, our model 1 needs regularization. This deep learning has several strategies that solve this type of problem and we are interested in a particularly powerful technique called Dropout.

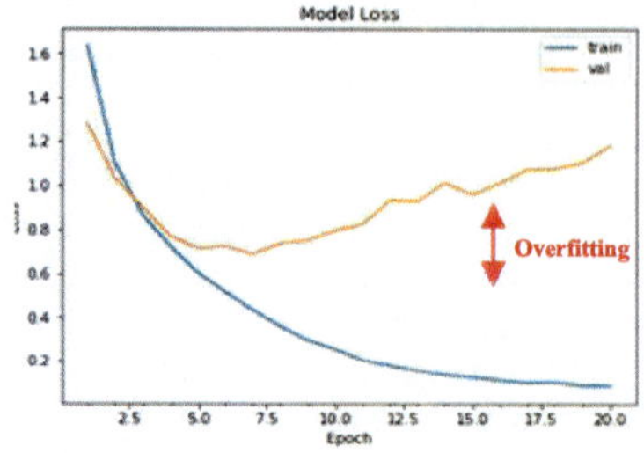

Fig. 3. The evolution of the test learning Loss with the ADAM algorithm over 20 epochs.

4.2 Dropout Effect

Dropout [19] is a technique where randomly selected neurons are ignored (temporarily) during learning. This means that their contribution to the activation of the succeeding neurons is temporarily suppressed during the propagation phase and all the weight updates are not applied to the neuron during the backpropagation phase.

During the test phase, all the neurons are kept, so we want the outputs of the neurons at the time of the test to be identical to their outputs at the time of learning (Fig. 4).

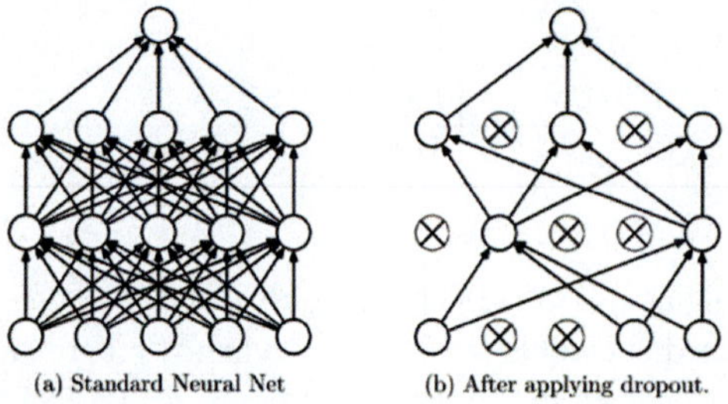

Fig. 4. Neural network model with dropout. (a): A standard neural network with 2 hidden layers. (b): application of Dropout to the network of (a).

Now that we know how dropout works, let's introduce it to our network and see how it impacts performance. Note: model 2 = model 1 + dropout.

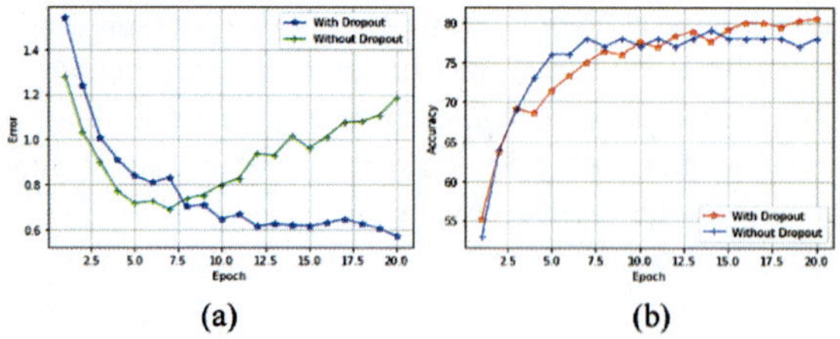

Fig. 5. (a) Evolution of the test loss with dropout and without dropout. (b) Evolution of test accuracy without dropout and with dropout

Table 3. Summary table of results previous after 20 epochs.

	Accuracy %		Execution time (s)
	Learning	Test	
Model 3	*93.43%*	*84.32%*	*3 h*
Model 2	*81.6%*	*80.56%*	*2 h 12 min*

From Fig. 5, and Table 3, the model with dropout performs better than the model without dropout, but even after 40 epochs, it is still unable to minimize learning error.

For this, we have several solutions:

- Let the learning last
- Increase the dropout probabilities: This strategy is not sure because increasing the probabilities of keeping neurons might not protect as effectively from overfit- ting.
- Change some network settings.

4.3 Batch Normalization Effect

Batch normalization [20] is a component that lies between the layers of the neural network and continuously takes the output of a particular layer and normalizes it before sending it to the next layer as input. This has the effect of stabilizing the learning process and considerably reducing the number of training periods necessary for learning deep networks.

We will introduce batch normalization in our network, and the results are as follows (Fig. 6)(Table 4):

Note: model 3 = Model 2 + Batch Normalization

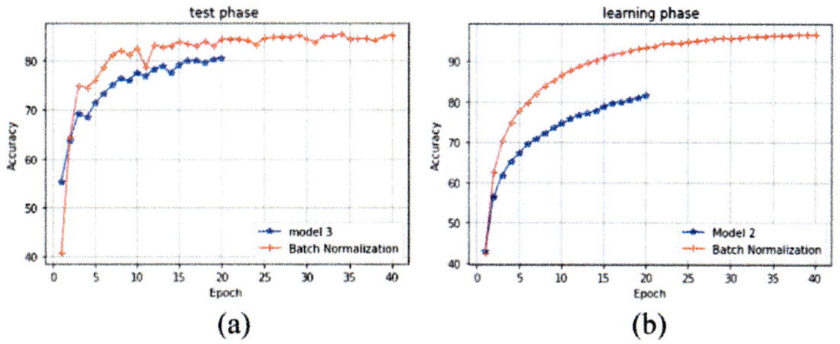

Fig. 6. Evolution of the precision curve in learning step (a) and in test phase (b) of models 3 and model 2

Table 4. A comparison between the accuracy of model 2 and model 3 in the learning phase and in the test phase after 20 epochs.

	Loss	Accuracy%	Execution time (s)
Model 1	*1.1832*	*78*	*6800*
Model 2	*0.5707*	*81*	*7222*

Figure 6. (a) shows that the allocation of a standardization batch allows us to achieve much better performance with more than 15% accuracy, stabilizes the learning process, and considerably reduces the number of training periods necessary for learning deep networks.

The same remark applies to the test phase (Fig. 6. (b)) such that normalizing the inputs by the mean and the standard deviation collected through learning is very effective and increases the accuracy of the model by 4%.

So the addition of a standardization batch increases the precision of learning and testing while increasing the training time. It normalizes the network input weights between 0 and 1.

4.4 Effect of Global Average Pooling

Global Average Pooling (GAP) [21] was proposed to replace the multi-layer perceptron part of a traditional convolutional neural network. The idea is to generate a feature map for each corresponding category. Instead of adding a perceptron after the feature maps, we take the average of each one and the result is injected into the soft-max function. An advantage of GAP over perceptron layers is that there are no parameters to train, so overfitting is avoided.

Model 4 = Model 3 + GAP.

Here is the behavior of our model after replacing the MLP part with a GAP (Fig. 7)(Table 5).

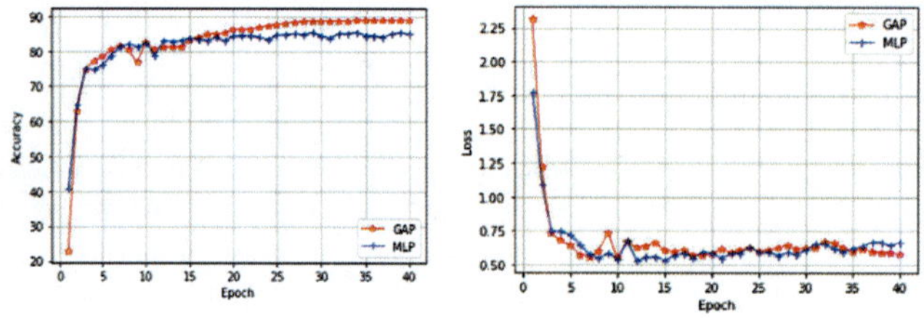

(a) (b)

Fig. 7. Evolution of the precision (a) and the loss (b) on the test set for model 4 and model 3

Table 5. Summary table of the previous results after 40 epochs.

	Loss	Accuracy
Model 4 (GAP)	0.58	88%
Model 3 (MLP)	0.66	85%

With 50% fewer parameters, the neural network with GAP greatly surpasses the model with unregulated MLP thanks to the ability of GAP to generalize better than MLP because there are no parameters to train. We can even distinguish in the figure overfitting in the curve of the MLP despite a convolutional part less inclined to overfitting by its nature and regularized despite everything with Dropout.

Now we have got our final model (Fig. 8). Since we intend to make the most of our network, we plan to allow learning to last a very long time and to make a backup of the parameters each time until there is no more improvement in performance on the test set, then restore the model to its state where it performed best.

Layer (type)	Output Shape	Param #
conv2d_1 (Conv2D)	(None, 32, 32, 48)	1344
batch_normalization_1 (Batch	(None, 32, 32, 48)	192
conv2d_2 (Conv2D)	(None, 30, 30, 48)	20784
batch_normalization_2 (Batch	(None, 30, 30, 48)	192
max_pooling2d_1 (MaxPooling2	(None, 15, 15, 48)	0
dropout_1 (Dropout)	(None, 15, 15, 48)	0
conv2d_3 (Conv2D)	(None, 15, 15, 96)	41568
batch_normalization_3 (Batch	(None, 15, 15, 96)	384
conv2d_4 (Conv2D)	(None, 13, 13, 96)	83040
batch_normalization_4 (Batch	(None, 13, 13, 96)	384
max_pooling2d_2 (MaxPooling2	(None, 6, 6, 96)	0
dropout_2 (Dropout)	(None, 6, 6, 96)	0
conv2d_5 (Conv2D)	(None, 6, 6, 192)	166080
batch_normalization_5 (Batch	(None, 6, 6, 192)	768
conv2d_6 (Conv2D)	(None, 4, 4, 192)	331968
batch_normalization_6 (Batch	(None, 4, 4, 192)	768
max_pooling2d_3 (MaxPooling2	(None, 2, 2, 192)	0
dropout_3 (Dropout)	(None, 2, 2, 192)	0
average_pooling2d_1 (Average	(None, 1, 1, 192)	0
flatten_1 (Flatten)	(None, 192)	0
dense_1 (Dense)	(None, 10)	1930

Total params: 649,402
Trainable params: 648,058
Non-trainable params: 1.344

Replacing MLP with GAP

Fig. 8. The final model

As the learning will last a long time, we can afford to drop the dropout probabilities to increase the generalization capacity of the network.

Figure 9 shows that after 200 epochs of learning, we reached the stability of the performance of our model After 100 epochs, the best-recorded performance is 91% accuracy with a minimization of the error of 0.37 which is one of the best-recorded performances on the cifar-10 base.

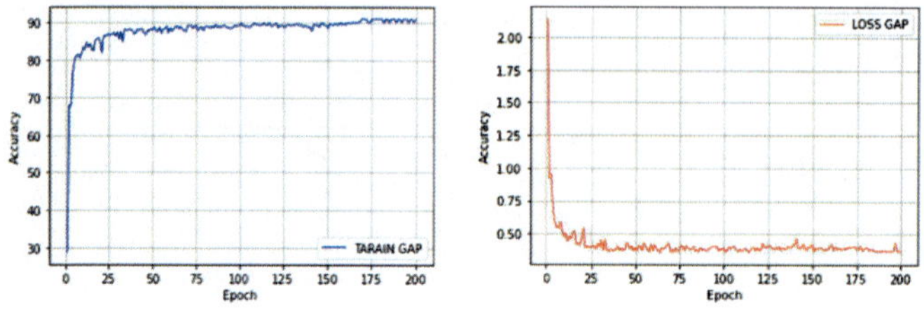

(a) (b)

Fig. 9. Evolution of the precision (a) and the loss (b) on the test set for final model.

Our reliance on CPU architecture in the CIFAR-10 dataset classification makes the execution time too expensive. To solve this problem, we used the GPU to test our final model on the CIFAR-100 dataset. The results are described in the following table.

Table 6. A summary table of classification results with the CIFAR-100 dataset

Dataset	Model	Epoch	Loss	Accuracy%	Total run time (sec)
CIFAR-100	Model 1	20	3.19	42.63	860
	Model 2	20	1.85	50.72	900
	Model 3 (MLP)	40	1.45	63.16	1 720
	Model 4 (GAP)	200	1.41	64.78	12 000

The results presented in Table 6 show that the results achieved have confirmed the effectiveness of our model, such that we succeeded in improving the performance of our model from 43% to 65%, and the previous comparison shows that several factors intervene to obtain a better classification, namely the number of epochs, the size of the dataset, and the depth of networks.

5 Conclusion

We have made several optimizations to a classical convolutional neural network model by proposing various parameters of the network architecture to understand the impact of each of them on the final result. After finding the architecture that works best for us, we decided to get the most out of the network by letting the learning last a long time until we obtained the maximum performance that our model can achieve, such that we succeeded in improving the performance of our model from 78% to 91% with the CIFAR-10 dataset and from 43% to 65% with the CIFAR-100, which are two of the best performances recorded based on this datasets. The comparison of the results has shown that the number of epochs, the size of the base, and the depth of networks are factors important for obtaining better results.

In our next work, we propose fine-tuning our model for better accuracy with the largest datasets.

References

1. Wolpert, D.H., Macready, W.G.: No free lunch theorems for optimization. IEEE Trans. Evol. Comput. **1**(1), 67–82 (1997)
2. Blum, A., Rivest, R.: Training a 3-node neural network is NP-complete. In: Proceedings of the1st International Conference on Neural Information Processing Systems, pp. 494–501. MIT Press (1988)
3. Goodfellow, I., Bengio, Y., Courville, A.: Deep Learning. MIT Press, Cambridge (2016)
4. Deng, L., Yu, D., et al.: Deep learning: methods and applications. Found. Trends R Sig. Process. flight **7**(3–4), 197–387 (2014)
5. Bengio, Y., Courville, A., Vincent, P.: Representation learning: a review and new perspectives. IEEE Trans. Pattern Anal. Mach. Intell. flight. **35**(8), 1798–1828 (2013)
6. Schmidhuber, J.: Deep learning in neural networks: an overview. Neural Netw. flight **61**, 85–117 (2015)
7. Schmidhuber, J.: Deep learning. Scholarpedia flight. **10**(11), 32832 (2015)
8. Gibney, E.: Google reveals secret test of AI bot to beat top go players. Nature **541**(7636), 142 (2017)
9. Bojarski, M., et al.: End to end learning for self-driving cars. arXiv preprint arXiv :1604.07316 (2016)
10. Esteva, A., et al.: Dermatologist level classification of skin cancer with deep neural networks. Nature **542**(7639), 115–118 (2017)
11. Hazlett, H.C.: Early brain development in infants at high risk for autism spectrum dis-order. In: Biological Psychiatry, vol. 73, pp. 115S–115S, Elsevier Science Inc 360 Park Ave South, New York, 10010–1710 USA (2013)
12. Gatys, L.A., Ecker, A.S. Bethge, M.: A neural algorithm of artistic style. arXiv preprint arXiv :1508.06576 (2015)
13. Qian, N.: On the momentum term in gradient descent learning algorithms. Neural Netw. flight. **12**(1), 145–151 (1999)
14. Duchi, J., Hazan, E., Singer, Y.: Adaptive subgradient methods for online learning and stochastic optimization. J. Mach. Learn. Res. flight. **12**, 2121–2159 (2011)
15. Kingma, D., Ba, J.: Adam: A method for stochastic optimization. arXiv pre-print arXiv: 1412.6980 (2014)
16. Spall, J.C. Introduction to Stochastic Search and Optimization: Estimation, Simulation, and Control. Hoboken, NJ: Wiley. pp. Sections 4.4, 6.6, and 7.5 (2003). ISBN 0-471-33052-3.
17. Glorot, X., Bengio, Y.: Understanding the difficulty of training deep feedforward neural networks. In: International Conference on Artificial Intelligence and Statistics (Aistats), flight. **9**, 249–256 (2010)
18. Nair, V., Hinton, G: Rectified linear units improve restricted boltzmann machines. In: Proceedings of the 27th International Conference on Machine Learning (ICML-10), pp. 807–814 (2010)
19. Srivastava, N., Hinton, G., Krizhevsky, A., Sutskever, I., Salakhutdinov, R.: Dropout: a simple way to prevent neural networks from overfitting. J. Mach. Learn. Res. flight. **15**(1), 1929–1958 (2014)
20. Ioffe, S., Szegedy, C.: Batch normalization: accelerating deep network training by reducing internal covariate shift. arXiv preprint arXiv: 1502.03167 (2015)
21. Lin, M., Chen, Q., Yan, S.: Network in network. arXiv preprint arXiv: 1312.4400 (2013)

Pedagogical Scenarisation for Virtual Environments of Training: Survey

Mohamed Fahim[1](✉) and Abdeslam Jakimi[2]

[1] ISISA, FS, Abdelmalek Essaadi University, Tetouan, Morocco
mfahim@uae.ac.ma
[2] FSTE, SEISE, Moulay Ismail University, Meknes, Morocco

Abstract. Virtual Environments for Training (VET), gained significant attention and popularity over the last years, thanks to the fast and remarkable technological progress. VET are based on virtual reality (VR) to offer users an outstanding experience. VET can not only enhance students to acquire knowledge and abstract concepts, but also develop specific skills by learning in an environment that offers high level of interaction. However, the success of a VET to reach its goal, do not depend only on the technological tools, but also on the pedagogical side. Thus, the trainer/teacher must create a pedagogical scenario to guide the learner through his experience in the VET. This paper aims to examine the existing literature to address currently used methods and approaches to model and create pedagogical scenarios for VET.

Keywords: Virtual reality · Education · Pedagogical scenario

1 Introduction

New technologies have radically changed the world in which we live. Education was not immune to this change. In fact, technologies have played an essential role in changing traditional practices of teaching and learning, furthermore they have made it possible to make many changes to the infrastructure of educational institutions today. One of those technologies is the virtual reality. It is regarded as a smart and digital technology that has many characteristics, for instance, it has the ability to allow users to live an immersive experience, carry out an activity in an artificially-generated environment, and interact with each other and with events. VR can be very helpful to create and deal with situations that are unavailable or unrealizable in real world, due to distance, time, cost, or safety factors. The VR's characteristics find a big interest in the field of education [1]. In fact, VR can provide limitless opportunities for innovation, learning, and high level of interaction between teachers and learners. Virtual Environment of Training (VET) which are based on VR, are the best application of VR in education. VET makes it possible to offer original and dynamic learning situations. However, the VET design is a complex task. The difficulties can be technical, cognitive, or both at the same time [2]. According to [3], the success of any VET depends on the combination of three points: technology, content, pedagogy. To fully create and describe a learning activity that will take place

M. Ben Ahmed et al. (Eds.): SCA 2022, LNNS 629, pp. 554–561, 2023.
https://doi.org/10.1007/978-3-031-26852-6_51

in a VET, designers take on the two sides: technology and content, and the teacher take on the pedagogy side. The best way to take into consideration the pedagogical side is to create a pedagogical scenario (PS). Researchers in the educational field generally agree to consider a pedagogical scenario as a specification of a future pedagogical activity as it will take place face-to-face or remotely using computer tools [4]. The scenario is then seen as a "guide" who will ensure the progress of the educational activity in the computer environment for which it was designed. Each pedagogical scenario is made up of a set of prerequisites, pedagogical activities and sequences, roles (generally: learner/trainer), and finally the resources used during the learning operation.

In this paper, we take a closer look at some approaches and tools that allows us to model pedagogical scenarios for VET, furthermore, we will focus on their architectures, and we will examine their strengths key, and limits.

The rest of the paper proceeds as follows: Sect. 2 describes existing projects and work for modeling pedagogical scenarios for VET. Section 3 is a comparison between different models allowing the creation of pedagogical scenarios, based on specific criteria. Finally Sect. 4 concludes the paper.

2 Literature Review

These last years, huge amount of educational researches on the use of VR has been published in the literature.

In fact VR has been extensively applied in education. Several contributions have been suggested in the literature using VR in the educational field. For instance, authors in [5–8], applied and discussed the utilization of VR for nurse education. Additional studies [9, 10], use VR to create training simulators that emulate a physical teaching laboratory.

Furthermore, authors in [11–13], proposed review studies of the different application of VR in the education field.

However, despite the richness and quality of those works, there is always a shortage of studies that takes into consideration the pedagogical side in the implementation phase of a VET.

Next subsections discuss the existed work in the literature, related to the modelization of pedagogical scenarios for VET.

2.1 POSEIDON

POSEIDON (PedagOgical ScEnario for vIrtual and informeD envirONment) [14] is a generic model of pedagogical scenarios for VET. It relies on a meta-model based on UML. POSEIDON makes it possible to describe an educational activity either as a "*black box*", it is the case with IMS-LD (Instructional Management Systems - Learning Design) [15], or in the form of an UML activity diagram. The actions referenced in this model are associated with a role, which can either correspond to a learner or to a trainer. These roles can be assumed by a human or by a virtual agent.

POSEIDON aims to be directly reusable in different virtual environments. It describes the information relating to the activity expected to be handled by the learners in the virtual environment (situations to identify, assistance to be provided, etc.).

The strength of this approach, alongside its genericity, lies in the fact that it covers different points of a real pedagogical scenario such as: educational objectives, prerequisites, educational activities, educational organizations and environments. However, this approach remains complex for people who are not computer scientists.

2.2 FORMID

FORMID (FORMation Interactive à Distance) [16] approach allows the trainer to create the scenario independently of the simulation. This approach allows both the reusability of simulations and scenarios, but also the possibility of using existing simulations.

The architecture implemented by FORMID is based on three components: the first one is FORMID-Author, which is used to create training sessions. The second is FORMID-Learner, which serves as an interface for and manages traces. The final component FORMID-Observer, which allows remote monitoring of learners by the trainer.

The pedagogical scenario used in the FORMID project is structured as follows:

- The initial situation: It is about placing the IEO (Interactive Educational Object) in the starting situation;
- The objective to be achieved: to check whether or not the learner has achieved the achievement objective assigned to him
- A set of behaviors to identify. These behaviors can represent:

- Resolution steps, such as mandatory waypoints;
- Situations to observe, which define what will be checked in addition to the mandatory passages. These situations can describe classic errors, deadlock situations, optional behaviors, etc.

The strength of this approach is that the scenarios created by FORMID are true pedagogical scenarios in the sense that they describe both the task to be carried out and the pedagogical specificities of this task.

2.3 GVT

GVT (Generic Virtual Training) [17] is a complete creation platform that facilitates the development of new VET. In other words, GVT offers a set of software that covers the life cycle of a VET, from its design to its exploitation by learners and trainers. GVT is intended for the creation of virtual environments for learning industrial procedures, first, in an individual mode, then in a collaborative mode in the second version of the architecture. In the industrial context of GVT, the maintenance procedures to be learned can be long, complicated and strictly scheduled.

The scenarios within the framework of GVT are expressed in LORA (Language for Object Relation Application) language [18]. LORA is a graphic language inspired by Grafcet (GRAphe Fonctionnel de Commande des étapes et Transitions). This graphic aspect of LORA was chosen with the aim of allowing the creation and editing of scenarios by experts who don't have enough knowledge in the IT field. This graphic language is also associated with an XML (eXtensible Markup Language) grammar. This format is used to export scenarios, with the aim of importing them into a tool for their modification or simply to have them executed by the GVT engine.

2.4 EMSAVE

EMSAVE [19] is a virtual environment designed to teach its users the basics of first aid. The simulation is based on a set of predefined scenarios, each putting the learner in front of a patient with need special care. Learners should carry out a combination of examinations and treatments following the appropriate first aid procedures. At each step, a set of actions is offered to the learner through a menu. If he chooses an appropriate action, the environment animates and evolves by applying the effects of the action. If not, the system provides the learner with a report summarizing the situation and giving clues to the correct action.

2.5 SELDON

SELDON (ScEnario and Learning situations adaptation through Dynamic OrchestratioN) [20, 21], is a model based on the centralized and indirect control of an emergent simulation from models of the script content. The environment is populated by autonomous virtual characters and the user has the freedom to act. SELDON depends on the TAILOR and DIRECTOR modules which constitute the main architecture of the SELDON system. The TAILOR module generates educational situations and a sequence of constraints according to the state of the user and the virtual environment, and the DIRECTOR module, which is an engine for dynamic adaptation of the scenario of the virtual environment, predict the evolution of the simulation, and adjust the state of the environment to guide its evolution towards the desired scenario.

2.6 POSVET

POSVET (PedagOgical Scenarisation for Virtual Environments for Training) [22] is directly reusable in different environments, the pedagogical scenario created using the POSVET model, relies on the MASCARET (Multi-Agent System for Collaborative Adaptive and Realistic Environment for Training) [23]: a virtual environment description language which contains all the necessary concepts. POSVET is a generic pedagogical model scenario for the VETs, i.e. applicable in multiple environments, regardless of the field or type of task to be performed,

POSVET is powerful but still have one big drawback: it is hard to handle by non-computer scientists.

2.7 VTS

The VTS (Virtual Training System) project [24], offers a training device for assembly operations in a virtual environment. The architecture of VTS is composed of three modules:

- Virtual Workspace: it is the virtual environment management infrastructure. This module handles object manipulation, collision detection, etc.
- Virtual Author: it is the component that allows the rapid creation of new exercises. The trainer through a process of authoring by doing allows creating the assembly procedure interactively.
- Virtual Mentor: identifies learner errors during learning and provides appropriate assistance.

The Virtual Workspace offers three training modes. The 3D animation mode: it allows the user to visualize the complete assembly through animations. The video mode: allows the visualization of the assembly by video clips. Finally, the interactive simulation mode allows the learner to carry out the assembly tasks through interaction with 3D objects.

In a pedagogical scenario produced by VTS, the task to be carried out by the learner is described by the list of assembly operations to be carried out as well as the instruction corresponding to each operation.

2.8 VR-Peas

VR-Peas (Virtual Reality PEdAgogical Scenarisation tool) [25], is intended primarily for teachers who don't have enough knowledge in computer science. VR-Peas allow teachers to express their pedagogical needs, to design and deploy their pedagogical situations. The VR-Peas architecture is composed of five main modules: Authentication module, VR-Peas module, VRLE (Virtual Reality Learning Environment), VRLE model and VR Data manager. The authentication module allows teachers to create accounts on the editor and be authenticated. Each teacher must enter his login and password. Moreover, The VR-Peas module is dedicated to the pedagogical scenarisation of educational situations by the teacher. This module is based on two sub-modules: the pedagogical management of scenarios module and the environment management module. Yet, the VR Data manager module provides a way to translate the elements of the model into concrete objects that can be interpreted by Unity software. Finally, The VRLE module represents the play mode allowing executing the scenario. This module is used by the teacher to test the result of his scenario via the generated VET.

The main strength of the VR-Peas is that offers the teacher not only the means to design his pedagogical scenario but also the means to operationalize it and generate his own VET.

3 Discussion

In this paper, we presented several models for the creation of pedagogical scenarios allowing the control of the user's behavior in the VET. By studying these different models of pedagogical scenarios, we have identified several parameters that seem to be significant to classify them:

- Genericity: The level of genericity of the description of the task to be performed in the environment. The majority of the models discussed before, with the exception of POSEIDON, POSVET and FORMID, are designed to create scenarios specific to a type of task. For example VTS is specific to assembly operations as well GVT is specific to the description of procedures. Finally, FORMID describes the activity to be carried out depending on environmental conditions.

- Lack of educational information: Some models do not provide any pedagogical information in the scenario. This is the case of EMSAVE for example. If during execution, a generic mechanism guides the learner towards the realization of the procedure, the pedagogical specificities of the exercise (classic errors, assists, etc.) are not taken into account in the scenario.
- Adaptability: Learning in a VET where the learner has the freedom of action, demands more adaptive techniques, to provide learners with tailored educational activities to enhance the quality of learning. In fact, the excellence of learners can be improved by recommending relevant educational resources and activities, based on each learner's ability and personal preferences.
- Reusability: The need to have a reusable model in order to avoid reinventing the process of creating a new scenario for each new environment. Thus, this characteristic therefore reduces implementation costs.
- Operationalizable: has the proposed approach been validated by a real case study?

The following table presents an assessment of the different solutions discussed in the second section, based on the criteria above (Table 1).

Table 1. Summary of the different approaches for the creation of pedagogical scenarios

Approach name	Generic	Adaptable	Operationalizable	Reusable	Propose an editor
POSVET [22]	✓	✓	×	✓	×
SELDON [21]	×	✓	✓	×	×
EMSAVE [19]	×	×	✓	×	×
GVT [17]	×	×	✓	✓	✓
FORMID [16]	✓	×	✓	×	✓
POSEIDON [14]	✓	×	✓	✓	✓
VTS [24]	×	×	✓	×	✓
VR-Peas [25]	×	×	✓	✓	✓

4 Conclusion

This current overview has summarized a brief state of the art of the models allowing the creation of pedagogical scenarios for VET. A pedagogical scenario presents a learning activity, created by a teacher/trainer in order to supervise the learning of his learners (before, during and after the activity).

Our future work focuses on the development of a model to generate pedagogical scenario for VET. This model must be generic in the sense that it is not specific to a particular field. Furthermore, it has to take into consideration the pedagogical information of the task or the activity to be carried out by the learner, finally, it must allow the development of these scenarios in an easy way and give the opportunity to non-computer scientist trainers to create their own scenarios effectively.

References

1. Oubahssi, L., Mahdi, O., Piau-Toffolon, C., Iksal, S.: A process of design and production of virtual reality learning environments. In: Auer, M.E., Tsiatsos, T. (eds.) The Challenges of the Digital Transformation in Education. Advances in Intelligent Systems and Computing, vol. 916, pp. 353–364. Springer, Cham (2020). https://doi.org/10.1007/978-3-030-11932-4_34
2. Carpentier, K., Lourdeaux, D.: Generation of learning situations according to the learner's profile within a virtual environment. In: Filipe, J., Fred, A. (eds.) Agents and Artificial Intelligence. Communications in Computer and Information Science, vol. 449, pp. 245–260. Springer, Heidelberg (2014). https://doi.org/10.1007/978-3-662-44440-5_15
3. Mishra, P., Koehler, M.J.: Technological pedagogical content knowledge: a framwork for teacher knowledge. Teach. Coll. Rec. **108**(6), 1017–1054 (2006)
4. Guillaume, D. : La scénarisation de l'évaluation des activités pédagogiques utilisant les Environnements Informatiques d'Apprentissage Humain. PhD thesis, Université de S voie. (2006)
5. Saab, M.M., Hegarty, J., Murphy, D., Landers, M.: Incorporating virtual reality in nurse education: a qualitative study of nursing students' perspectives. Nurse Educ. Today **105**, 1–7 (2021)
6. Bayram, S.B., Caliskan, N.: Effect of a game-based virtual reality phone application on tracheostomy care education for nursing students: a randomized controlled trial. Nurse Educ. Today. **79**, 25–31 (2019)
7. Mendez, K.J.W., et al.: Virtual and augmented reality: implications for the future of nursing education. Nurse Educ. Today **93**, 104531 (2020)
8. Plotzky C., et al.: Virtual reality simulations in nurse education: a systematic mapping review. Nurse Educ. Today **101**(9), 104868 (2021)
9. Shalaunda, M.R., Kent, J.C., Erica, D. M: The varied experience of undergraduate students learning chemistry in virtual reality laboratories. Comput. Educ. **175**, 104320 (2021)
10. Pascalis, T., Mario, W., Sulamith, F.: Remote lab meets virtual reality – enabling immersive access to high tech laboratories from afar. Procedia Manufact. **43**, 25–31 (2020)
11. Kamińska, D., et al.: Virtual reality and its applications in education: survey. Information **10**(10), 318 (2019)
12. Jaziar, R., Tim, A.M., Jennifer, F., Isabell, W.: A systematic review of immersive virtual reality applications for higher education: design elements, lessons learned, and research agenda. Comput. Educ. **147**, 103778 (2020)
13. Rafael, V-T., Sergio, T-O., Ramón, C-G., José, A.G-C.: Effects of virtual reality on learning outcomes in K-6 education: a meta-analysis. Educ. Res. Rev. **35**, 100434 (2022)
14. Marion, N., Querrec, R., Chevaillier, P.: Integrating knowledge from virtual reality environments to learning scenario models. a metamodeling approach. In: International Conference of Computer Supported Education, pp. 254–259 (2009)
15. Koper, R.., Olivier, B., et Anderson, T.: IMS Learning Design Information Model. IMS Global Learning Consortium (2003)
16. Guéraud, V., Lejeune, A., Adam, J.-M., Dubois, M., Mandran, N.: Supervising distant simulation-based practical work: environment and experimentation. In: The European Conference on Technology Enhanced Learning, pp. 602–608 (2009)

17. Gerbaud, S., Mollet, N., Ganier, F.: GVT : a platform to create virtual environments for procedural training. In: The 9th IEEE Virtual Reality Conference, pp. 225–232 (2008)
18. Mollet, N., Gerbaud, S., Arnaldi, B.: STORM: A Generic Interaction and Behavioral Model for 3D Objects and Humanoids in a Virtual Environment, pp. 95–100. Eurographics Symposium on Virtual Environments, Short Papers and Posters (2007)
19. Vidani, A.C., Chittaro, L., Carchietti, E.: Assessing nurses' acceptance of a serious game for emergency medical services. In: 2010 Second International Conference on Games and Virtual Worlds for Serious Applications, Braga, Portugal, pp. 101–108 (2010). https://doi.org/10.1109/VS-GAMES.2010.12
20. Barot, C., Lourdeaux, D., jean-marie, B., Amokrane, K., Lenne, D.: V3S : a virtual environment for risk-management training based on human-activity models. Presence : Teleoperators Virtual Environ. **22**(1), 1–19 (2013)
21. Barot, C., Lourdeaux, D., Lenne, D.: Using planning to predict and influence autonomous agents behaviour in a virtual environment for training. In: The International Conference on Cognitive Informatics and Cognitive Computing (ICCICC 2013), pp. 274–281 (2013)
22. Fahim, M., Jakimi, J., El bermi, L.: Pedagogical scenarization for virtual environments for Training: towards genericity, coherence and adaptivity. Int. J. Adv. Eng. Res. Sci. (IJAERS). **12**(3), 96–103 (2016)
23. Chevaillier, P.: Les systèmes multi-agents pour les environnements virtuels de formation. Laboratoire d'Informatique des SYstèmes Complexes. Habilitation à Diriger des Recherches (2006)
24. Brough, J., Schwartz, M., Gupta, S., Anand, D., Kavetsky, R., et Pettersen, R.: Towards the development of a virtual environment-based training system for mechanical assembly operations. Virtual Reality **11**(4), 189–206 (2007)
25. Mahdi, O.: Modèle et outil pour assister la scénarisation des activités pédagogiques orientées RV. PhD thesis, le mans université (2021)

Design of a Novel Patch Antenna with Dual-Band 27/38 GHz for 5G Applications

Anouar Es-saleh[1](✉), Mohammed Bendaoued[2], Soufian Lakrit[3], Rihab Roubhi[1], Mohamed Atounti[3], and Ahmed Faize[3]

[1] Mathematics and Information Systems Laboratory, FP of Nador, Mohammed First University, Oujda, Morocco
anouar_es-saleh@hotmail.com
[2] LMIET FST of Settat, University of Hassan 1st, Settat, Morocco
[3] Mathematics and Information Systems Laboratory, EST of Nador, Mohammed First University, Oujda, Morocco

Abstract. This paper present the design of an elliptical form patch antenna for 5G applications on a low-cost substrate Rogers RO4003C with permittivity of 3.55 of dimension $14 \times 6 \times 0.2\ mm^3$ in both bands (27 GHz and 38 GHz). The proposed antenna consists of the patch contains five ellipses with a slot of the same shape with a partial ground plane use to increase the bandwidth. The antenna shows a good matching at both bands. The results show two resonance frequencies that define two bandwidths defined by a return loss less than 10 dB and are respectively; at a frequency 26.75 GHz of 0.48 GHz, and at a frequency 38 GHz of 0.428 GHz. Also, the obtained gain has a good value and it's very remarkable according to the small size of the structure, with a peak value of 5.731 dB at 26.75 GHz and a peak value of 4.75 dBi at 38 GHz. The study of this antenna has been performed by using Computer Simulation Technology-Microwave Studio (CST).

Keywords: 5G · Ellipse antenna · Mmwave · Patch antenna

1 Introduction

To meet the demand for increasing data rates and channel capacity, mobile phone networks have been built in stages, starting with the first generation (1G) and ending with the fourth generation (4G). However, due to an increase in the number of mobile users, mobile traffic is increasing year after year in global situations, and this trend is predicted to continue in the foreseeable future. As a result, the present 4G mobile network's available data rate and channel capacity will be insufficient to accommodate future mobile communication technology advances in the long run [1]. Besides to the advances in the field of wireless communications have a significant impact on economic and social development in the recent years; for that fifth-generation (5G) technology has risen to the top of the 2020 generation's priority list. 5G is a new technology that offers both evolutionary and revolutionary services aims to triple data transmission rates compared to the fourth generation (4G). It is the next generation of technology that will enable

M. Ben Ahmed et al. (Eds.): SCA 2022, LNNS 629, pp. 562–571, 2023.
https://doi.org/10.1007/978-3-031-26852-6_52

ultra-high data rates, extremely low latency, increased capacity, and high service quality. It's worth noting that 5G technology will open up new ways to overcome traditional development barriers [2].

5G data ranges from 5 to 50 Gbit/s and require the use of antennas with circular polarization and MIMO (Multiple Input, Multiple Output) to achieve the expected speed. The higher frequency spectrum for 5G technology includes frequency bands such as 24.25 GHz–27.5 GHz, 26.5 GHz–27.5 GHz, 37.0–40.0 GHz, and 37.0–43.5 GHz [3].

The mmWave spectrum between 28 GHz (27.5–29.5 GHz) and 38 GHz (37–38.6 GHz) is being heavily considered for next-generation 5G wireless systems, which would allow for significant performance improvements over fourth-generation systems [4, 5].

There are various types of antennas that meet the requirement of having a high gain for 5G applications, such as corrugated antennas, cavity-backed antennas, and SIW aperture antennas. Recently, much attention has been focused on the design of multi-functional antennas, which have a combination of characteristics in one single structure for different applications. Recently, Microstrip antennas are promising candidates for fifth-generation wireless communications because of their small size, light weight, low cost, ease of fabrication and integration and their capability of integration with other systems. Higher gain and better radiation characteristics are the primary requirements for fifth-generation antennas [6, 7].

The purpose of this work is to create a novel patch antenna array and optimize its performance for 5G wireless applications at 27 GHz and 38 GHz. The main objective is to achieve broad bandwidth, standard gain, and high efficiency for 27 GHz and 38 GHz 5G applications band in a very compact size.

The proposed antenna designed using substrate having physical dimensions of 14 $\times$ 6 $\times$ 0.2 mm^3. All the antenna structures are designed, simulated, and analyzed using the CST Microwave Studio electromagnetic 3D simulator toll. Sufficient impedance bandwidth preferred unidirectional stable radiation patterns, satisfactory gain, radiation efficiency as well as a miniaturized size make the proposed antenna a suitable candidate for future 5G applications.

2 Antenna Design

Although several shapes of microwave antenna (f > 1 GHz) are used, a microstrip antenna are the most commonly encountered [8]. In this work we used an elliptical form patch antenna.

To design the proposed antenna we used CST Microwave Studio, the radiation element are depicted on a substrate of Rogers RO4003C with a permittivity of 3.55 with dimension 14 $\times$ 6 $\times$ 0.2 mm^3 and a partial ground plane with dimension 12.5$\times$6 mm^2, the radiation element is feeding by a microstrip line with 50 Ω.

The optimized design parameters of the proposed antenna are listed in the Table 1., and the design structure is given in Fig. 1.

The patch contains 5 ellipses, each ellipse has an area of $\pi \times R1 \times R2 = 5.10$ mm^2 and elliptical slot contains 5 ellipses, each ellipse has an area of $\pi \times R3 \times R4 = 1.34$ mm^2.

Table 1. The design parameters of proposed antenna

Parameter	Value (mm)
L_G	12.5
$W_G = W_{SUB}$	6
$H_F = H_G$	0.035
L_{SUB}	14
X_{sc}	6
T_{SUB}	0.2
L_F	7.6
W_F	0.54
R_1	0.5
R_2	3.25
R_3	0.33
R_4	1.3

The Slots are used to increase the bandwidth and miniaturize the patch antenna, compared to the basic element, the slot loaded modified patch radiator has a significant improvement in term of reflection coefficient, operating bandwidth and gain.

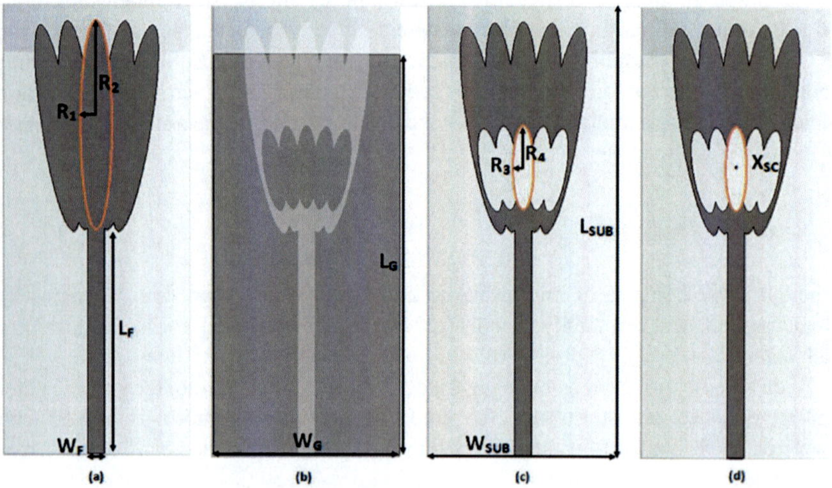

Fig. 1. Geometry of proposed antenna (a) basic element without slots, (b) partial ground, (c) proposed antenna with slots and (d) X_{sc} position of slot

2.1 Antenna Parameters Analysis

Before fabricating an antenna, it is necessary to investigate the effects of major design parameters on the resonance performance of the proposed antenna.

The area of the elliptical slot and their position, and ground plane length has a significant impact on the performance of the proposed antenna.

2.1.1 Effect of Ground Plane Length on the Performance of the Proposed Antenna

A parametric analysis has been carried out to investigate the impact of the ground plane length on the resonant characteristics of the proposed antenna. The simulation results (reflection coefficient) depending upon the variations of the ground plane length from ($L_G = 6$ mm to $L_G = 12.5$ mm) are shown in Fig. 2.

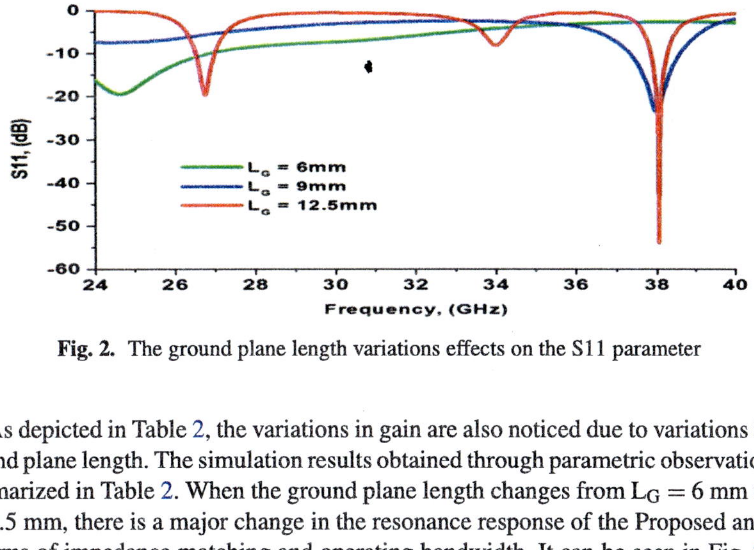

Fig. 2. The ground plane length variations effects on the S11 parameter

As depicted in Table 2, the variations in gain are also noticed due to variations in the ground plane length. The simulation results obtained through parametric observation are summarized in Table 2. When the ground plane length changes from $L_G = 6$ mm to $L_G = 12.5$ mm, there is a major change in the resonance response of the Proposed antenna in terms of impedance matching and operating bandwidth. It can be seen in Fig. 2 and Table 2 that if the ground plane length proposed $L_G = 12.5$ mm, the reflection coefficient decreases to – 19.817dB and –53.983 at 26.75 GHz and 38.048 GHz respectively with – 10 dB impedance bandwidth of about 0.48 GHz (26.51 GHz to 27 GHz) and 0.428 GHz (37.84 GHz to 38.26 GHz) at 26.75 GHz and 38.048 GHz respectively.

2.1.2 Effect of Slot Radius on the Performance of the Proposed Antenna

A parametric analysis has been carried out to investigate the impact of the slot radius on the resonant characteristics of the proposed antenna.

The simulation results (reflection coefficient) depending upon the variations of small elliptical slot radius from $R_3 = 0.26$ mm to $R_3 = 0.34$ mm shown in Fig. 3, we observe that the best result is when $R_3 = 0.33$ mm.

Table 2. Simulated results of the antenna by varying the dimensions of the ground plane length

Length of ground plane LG (mm)	Frequency (GHz)	Bandwidth (GHz)	−10 dB bandwidth (GHz)	Gain (dB)	S11 (dB)
6	24.6	4.118	22.77–26.89	5.215	−19.077
9	38	1.327	37.31–38.63	5.94	−22.177
12.5	26.75	0.48	26.51–27	5.731	−19.817
	38.048	0.428	37.84–38.26	4.75	−53.983

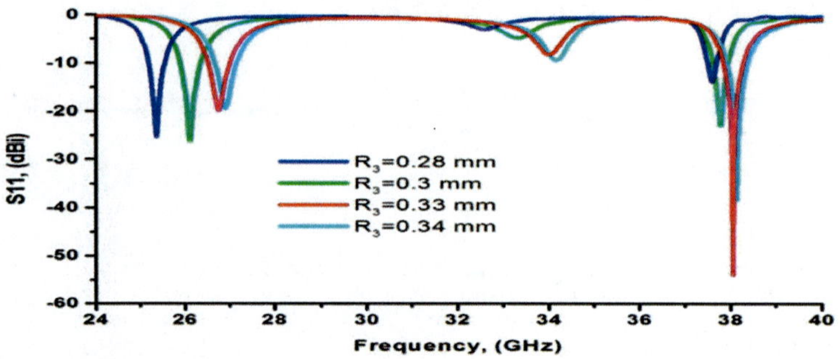

Fig. 3. The slot small radius variations effects on the S11 parameter

The simulation results (reflection coefficient) depending upon the variations of large elliptical slot radius from $R_4 = 0.8$ mm to $R_4 = 1.5$ mm shown in Fig. 4, we observe that the best result is when $R_4 = 1.3$ mm.

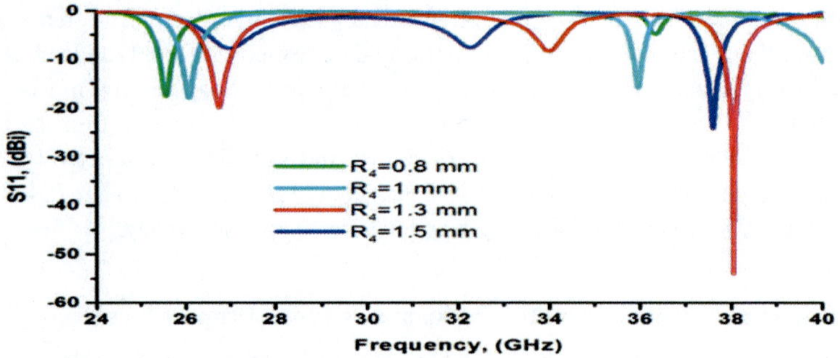

Fig. 4. The slot large radius variations effects on the S11 parameter

2.1.3 Effect of Slot Position on the Performance of the Proposed Antenna

A parametric analysis has been carried out to investigate the impact of the slot position on the resonant characteristics of the proposed antenna. The simulation results (reflection coefficient) depending upon the variations of position elliptical slot (position of X_{SC}) from $X_{center} = 5.5$ mm to $X_{center} = 7$ mm shown in Fig. 5, we observe that the best result is when $X_{center} = 6$ mm.

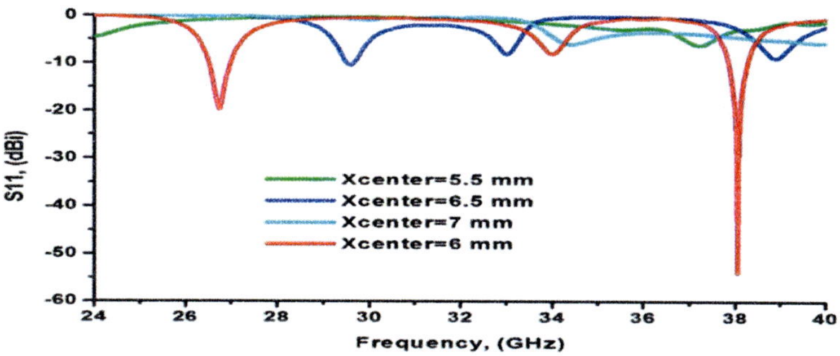

Fig. 5. The slot position variations effects on the S11 parameter

2.2 Results and Discussion

2.2.1 Reflection Coefficient

The comparison of the variations of reflection coefficient vs frequencies for the basic patch antenna with and without the presence of elliptical slot and with partial ground plane (GP) and full ground plane are depicted in Fig. 6.

The simulation results show that the reflection coefficients of the patch antenna with slot and with partial ground plane present the value of return loss – 19.817 dB at 26.75 GHz with a bandwidth of 0.48 GHz, from 26.51 GHz to 27 GHz, and the value of return loss – 53.983dB at 38.048 GHz with a bandwidth of 0.428 GHz, from 37.84 GHz to 38.26 GHz, completely covering the two 5G frequency bands from 26.51 GHz to 27 GHz and from 37.84 GHz to 38.26 GHz.

On the other hand we seen a single pick of the return loss of −15.26 dB at the frequency 30.15 GHz for the antenna without slot and full ground plane; also we observe a bad value when we used a partial ground plane and a patch without slot.

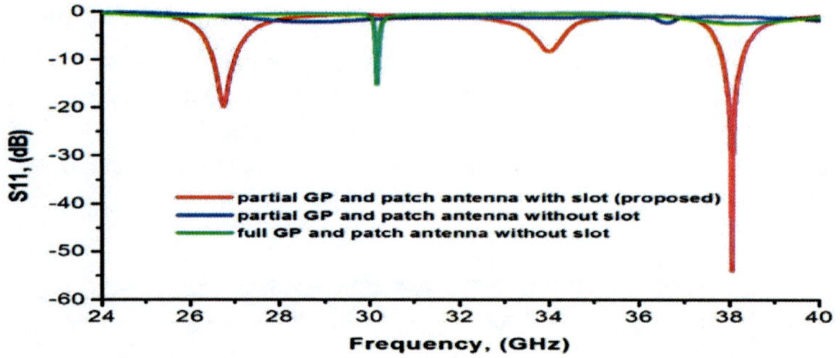

Fig. 6. Reflection coefficient vs frequencies comparison of proposed antenna with and without slots in the presence of a partial ground plane and the other full ground plane

2.2.2 Gain

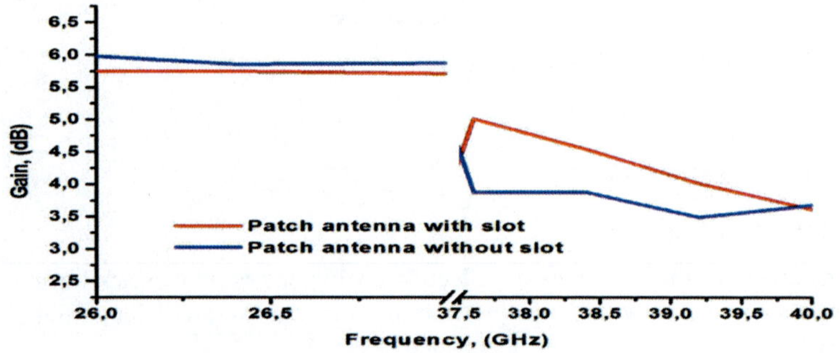

Fig. 7. Gain vs frequencies comparison of proposed antenna with and without slots

The comparison of the variations of gain vs frequencies for the basic patch antenna with and without the presence of elliptical slot are depicted in Fig. 7.

In the patch antenna with slot, it can be observed that the proposed work exhibits an average gain of 5.731 dB around of 28 GHz and a gain of 4.75 dB around of 38 GHz,

2.2.3 Radiation Efficiency

Figure 8 illustrates the efficiency of the radiation for the basic patch antenna with and without the presence of elliptical slot, note that the efficiency is 98.26% at a frequency of 27 GHz, and it is valued at 96.82% at a frequency 38 GHz.

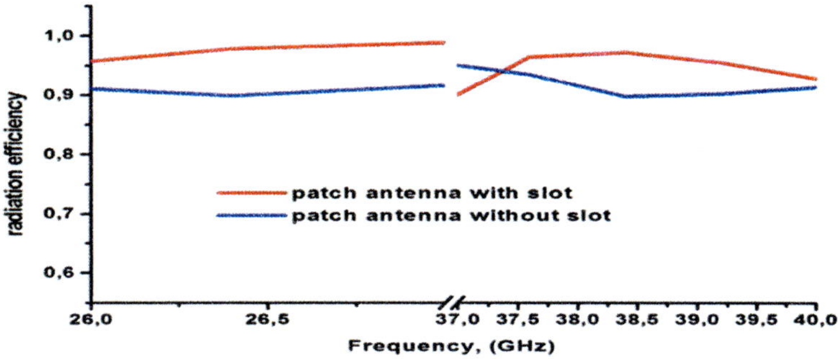

Fig. 8. Radiation efficiency vs frequencies comparison of proposed antenna with and without slots

2.2.4 Radiation Pattern

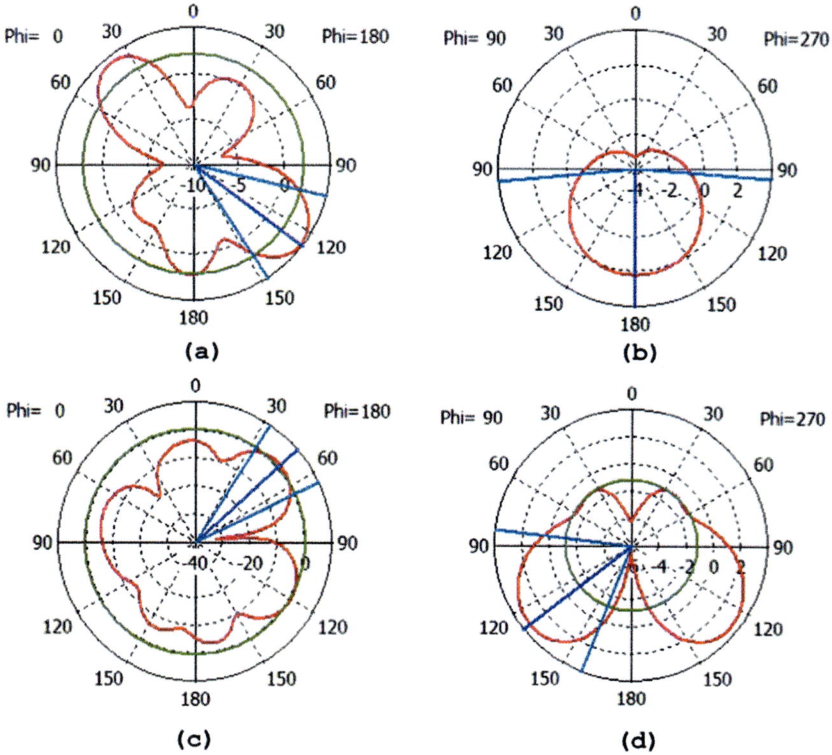

Fig. 9. Radiation pattern of proposed antenna: (a) E-plane at 26.7 GHz,(b) H-plane at 26.7 GHz, (c) E-plane at 38 GHz, (d) H-plane at 38 GHz

Figure 9 illustrates the radiation pattern of the proposed antenna. For both 27 and 38 GHz, it can be observed at 38 that the designed antenna exhibits a radiation pattern in E-Plane (XZ-plane) Fig. 9(c) almost omnidirectional radiation patterns and bidirectional radiation pattern in H-Plane (YZ-plane) Fig. 9(d). At 26.7 GHz we can see that the radiation pattern of the proposed antenna for H-plane Fig. 9(b) remains directional and for E-plane Fig. 9(a) remains bidirectional. The designed antenna exhibits stable radiation patterns over the operating band which is a primary requested for 5G wireless communication applications.

2.3 Comparison of the Proposed Antenna with Other Existing Antenna Designs

Table 3. Comparison of performance of designed antenna with other existing antenna

Reference papers	Performance measure					
	Size in (mm^2) L_{SUB} × W_{SUB}	Center frequency	Patch gain (dB)	Impedance bandwidth (GHz)	Level of reflection coefficient (S11)	Radiation efficiency
[1]	10.591 × 13.802	38.1	4.43	3.7	−34	97.94%
[9]	15 × 15	29.7	5.42	6.4	−35.8	85%
[10]	1.3 × 1.2	28	3.75	3.34	−41	–
		38	5.06	1.395	−18	–
[11]	30 × 35	28	8.3	4.1	−41.5	79%
[12]	55 × 110	27.946	7.18	–	−27.84	91.24%
		37.83	9.24		−18.35	89.63%
Proposed work	6 × 14	26.75	5.731	0.48	−19.817	98.26%
		38.048	4.75	0.428	−53.983	96.82%

The proposed antenna maintains good stability among the characteristics parameters (gain, bandwidth, return loss) in a dual band 28 GHz and 38 GHz compact dimension. The geometrical configuration of the proposed antenna is novel compared to other structures reported in the literature for next generation 5G applications.

3 Conclusion

An ellipse-shaped compact patch antenna operates in dual bands (26.51–27 GHz) and (37.84–38.26 GHz), having bandwidth of 480 MHz and 428 MHz respectively for 5G applications. Compared to the basic patch antenna, the modified slot loaded patch radiator exhibits a significant improvement in the reflection coefficient, the operating bandwidth and in the gain. The proposed antenna shows a stable radiation patterns with a peaks gain of 5.73 dBi at 26.7 GHz and 4.75 dBi at 38 GHz and maximum radiation efficiency of 98.26% for the first band and 96.82% for the second band. Hence, the proposed antenna is suitable for 5G wireless communication systems.

References

1. Bilal, A., Sudipta, D., El Faylali, H., et al.: Analysis, design and fabrication of a square slot loaded (SSL) millimeter-wave patch antenna array for 5G applications. J Cir. Syst. Comput. **30**(05), 2150086 (2021)
2. Kumar, S., Dixit, A.S., Malekar, R.R., et al.: Fifth generation antennas: A comprehensive review of design and performance enhancement techniques. IEEE Access **8**, 163568–163593 (2020)
3. Pant, M., Malviya, L., Choudhary, V.: Gain and bandwidth enhancement of 28 GHz tapered feed antenna array. In: 2020 11th International Conference on Computing, Communication and Networking Technologies (ICCCNT), p. 1–4. IEEE (2020). Author, F.: Contribution title. In: 9th International Proceedings on Proceedings, pp. 1–2. Publisher, Location (2010)
4. Deckmyn, T/, Cauwe, M., Ginste, Dries V., et al.: Dual-band (28, 38) GHz coupled quarter-mode substrate-integrated waveguide antenna array for next-generation wireless systems. IEEE Trans. Antennas Propag. **67**(4), 2405–2412 (2019)
5. Ali, W., Das, S., Medkour, H., et al.: Planar dual-band 27/39 GHz millimeter-wave MIMO antenna for 5G applications. Microsyst. Technol. **27**(1), 283–292 (2021)
6. Goudarzi, A., Honari, M.M., Mirzavand, R.: Resonant cavity antennas for 5G communication systems: A review. Electronics **9**(7), 1080 (2020)
7. Amjadi, S.M., Sarabandi, K.: A low-profile, high-gain, and full-band subarray of cavity-backed slot antenna. IEEE Trans. Antennas Propagat. **65**(7), 3456–3464 (2017)
8. Cormier, G.: GELE5222 Chapitre 9 :Antennes microruban. Universite de Moncton Hiver (2012)
9. Hussain, N., Awan, W.A., Ali, W., et al.: Compact wideband patch antenna and its MIMO configuration for 28 GHz applications. AEU-Int. J. Electron. Commun. **132**, 153612 (2021)
10. Ahmad, W., Khan, W.T.: Small form factor dual band (28/38 GHz) PIFA antenna for 5G applications. In: 2017 IEEE MTT-S International Conference on Microwaves for Intelligent Mobility (ICMIM), pp. 21–24. IEEE (2017)
11. Khalid, M., Iffat Naqvi, S., Hussain, N., et al.: 4-Port MIMO antenna with defected ground structure for 5G millimeter wave applications. Electronics **9**(1), 71 (2020)
12. Marzouk, H.M., Ahmed, M.I., Shaalan, A.H.A.: Novel dual-band 28/38 GHz MIMO antennas for 5G mobile applications. Progress Electromagn. Res. C **93**, 103–117 (2019)

Using Machine Learning with Pyspark for Solving a Big Data Problem: Searching for Particles in High Energy Physics with Big Mass (HEPMASS)

Mourad Azhari[1](✉), Khalid Ahaji[1], Abdallah Abarda[2], Badia Ettaki[3], and Jamal Zerouaoui[4]

[1] Center of Guidance and Planning, Rabat, Morocco
azharimourad@yahoo.fr

[2] Laboratoire de Modélisation Mathématiques et de Calculs Economiques, FSJES, Université Hassan 1er, Settat, Morocco

[3] Laboratory of Research in Computer Science, Data Sciences and Knowledge Engineering, Department of Data, Content and Knowledge Engineering, School of Information Sciences, Rabat, Morocco

[4] Laboratory of Engineering Sciences and Modeling, Faculty of Sciences, Ibn Tofail University, Kenitra, Morocco

Abstract. The search for exotic particles is one of the most challenging topics for physicists. This work aims to resolve the Big Data problem in the exotic particle area using the Apache Spark environment with the MLlib library. For this purpose, we analyze and discuss the performance of the following methods: Logistic Regression (LR), Decision Tree (DT), Random Forest (RF), and Gradient Boosted Tree (GBT). The results are very encouraging and prove that these proposed methods run efficiently with an Accuracy exceeding (83%). While the Gradient Boosted Tree (GBT) reaches a high AUC score 85%, better than Decision Tree (DT) and Random Forest. In this manuscript, we attempt to solve a big data problem using the "HEPMASS" dataset. Our data includes 10.5 million examples collected from the UCI website for the experimental phase.

Keywords: Exotic particles · Pyspark · Machine learning (ML) · Logistic regression (LR) · Decision tree (DT) · Random forest (RF) · Gradient boosted tree (GBT) · Area under the curve (AUC) · Accuracy · Computation time (CT)

1 Introduction

Exotic Particle detection in high-energy constitutes a big challenge for the High-energy Physics area. Machine learning methods are a powerful tool that can improve the performance of Exotic Particles identification. These methods are proposed to resolve a binary classification problem (signal and background

M. Ben Ahmed et al. (Eds.): SCA 2022, LNNS 629, pp. 572–579, 2023.
https://doi.org/10.1007/978-3-031-26852-6_53

events) in the context of the "HEPMASS" dataset. Various studies related to Exotic Particle search have been advanced in the Particle physics field. Hence, ATLAS, detector physicists studied the predictions of the Standard Model. With this in mind, Higgs Boson Machine Learning Challenge has been tenuous in 2014 [1]. Many algorithms have been used to study basic algorithms such as Decision Tree, Naive Bayes, K-Nearest Neighbors, etc., advanced algorithms: KMeans Clustering, Support Vector Machine, etc., and ensemble methods such as Bagging, Random Forest, and Boosting. The Accuracy score obtained with Gradient Boosting Classifier achieved 84% followed by Support Vector Machines (SVM) with linear kernels 76%. Chen and Tong introduced a regularized version of the Gradient Boosting method with an efficient implementation [2]. Baldi et al. have implemented the Deep Networks classifier on "Higgs", "Susy" and "HEPMASS" datasets [3–5]. Alves advanced Stacking Machine Learning classifiers, multivariate statistical analysis approach, to detect Higgs Bosons at the LHC, this method outperforms the Boosted Decision Trees and Deep Neural Network applied to Particle Physics [6].

This work tries to solve the Exotic Particle classification problem with four Machine Learning methods using Pyspark environment: Logistic Regression (LR), Decision Tree (DT), Random Forest (RF), and Gradient Boosted Tree (GBT). Therefore, we evaluate the performance of these ML methods using three metrics: Accuracy, AUC, and Computation Time (CT) metrics. In the experimentation phase, we use Big dataset such as the "HEPMASS" with 10.5 million examples. This manuscript is presented in the following structure: The first section will try to provide a brief relevant work to Exotic particle detection, the second section will present the Spark environment with proposed methods. While the third one will describe the "HEPMASS" dataset and will present the evaluation metrics. The fourth section will explain the modeling process of the proposed methods and summarizes our experimental results and their consequent analysis.

2 Proposed Methods

This paper aims to applied different algorithms, with Pyspark Environment, to resolve the classification problems of Exotic Particle discovery: Logistic Regression (LR), Decision Tree (DT), Random Forest (RF), and Gradient Boosted Tree (GBT).

2.1 Pyspark Environment

Apache Spark is a powerful tool reserved for Big Data. It includes a Machine Learning library (MLlib) designed to ML algorithms. Spark works well in memory, supports many programming languages (java, scala, python, SQL, R). In particular, It runs very well with python. Also, Spark manipulates successfully Big data set as such a "HEPMASS" dataset with the Pyspark API [7,8].

Apache Spark works rapidly, dynamic, In-Memory Computation speed swift, and reusable [9].

2.2 Logistic Regression Method

Logistic Regression (LR) is a classification method with powerful predictors. In the context of binary discrimination, LR is designed to predict and explain a categorical response with two modalities (positive, negative) depending on predictors. It focuses only on the ratio of the probabilities. In practice, we use the spark.ml logistic regression module to predict a binary outcome [10].

2.3 Decision Tree Method

Decision Tree Method is a supervised Machine Learning classifier used to solving regression or classification problems [11]. It is a tree where each node designs an attribute, each branch represents a decision rule and each leaf represents an outcome. We can build a decision tree for classification problems applying many algorithms as ID3 and CART [12].

2.4 Random Forest Method

The Random Forest is a Machine Learning method that consists of multiple decision trees to build the final decision using the majority voting [13–16]. The principal benefit of RF is that it decreases the risk of over-fitting, offers a high level of accuracy and runs efficiently in large databases [17].

2.5 Gradient Boosted Tree Method

The Boosting method consists of converting weak learners into strong learners to minimize a loss function. The main idea of Boosting is to join a new classifier to the ensemble sequentially. Gradient Boosted tree (GBT) is an ensemble method that uses an ensemble of decision trees to predict a target label. The spark.ml implementation supports GBTs for binary classification [18,19].

3 Dataset and Evaluation Metrics

3.1 Description of "HEPMASS" Dataset

In this manuscript, we use the "HEPMASS" dataset, collected from UCI site web. We use 28 features as independent variables and a dependent variable with two classes (signal and background events) :

- 28 Particle features: 'f0', 'f1', 'f2', 'f3', 'f4', 'f5','f6', 'f7', 'f8', 'f9', 'f10', 'f11', 'f12', 'f13', 'f14', 'f15', 'f16', 'f17', 'f18', 'f19', 'f20', 'f21', 'f22', 'f23', 'f24', 'f25', 'f26', 'mass'
- dependent variable: # label

The dependent variable is the class label (1 for signal, 0 for background). It consists of 50% (5 250124 examples) of the signal and 50% of background events (5 249876 examples). We thus find that the distribution of the "HEPMASS" dataset is balanced.

3.2 Split Data Set

In the purpose to get a true estimation of the algorithm performances, "HEPMASS" dataset is partitioned into a training set (66.67%) and test set (33.33%) (see Table 1).

Table 1. "HEPMASS" dataset description

Description	Examples	Features	Training data	Testing data	Signal (s=1)	Background (b=0)
Quantity	10500000	28	7000000	3500000	5250124	5249876

3.3 Evaluation Metrics

In this work, we use three metrics to evaluate our algorithms:

- Accuracy: proportion of the total number of predictions that were correct. It is calculed using confusion matrix of actual and predicted class (Table 2):

Table 2. Confusion Matrix of actual and the predicted class

Actual	Predicted	
	Positive Class	Negative Class
Positive Class	True Positive (TP)	False Negative (FN)
Negative Class	False Postive (FP)	True Negative (TN)

$$Accuracy = \frac{TP + TN}{TP + TN + FP + FN} \tag{1}$$

- Receiver Operating Characteristic Curve-Area Under the curve ((ROC-AUC);
- Computation Time (CT).

Table 3. Performance metrics with "HEPMASS" dataset

Algorithms	Accuracy	AUC	CT (s)
LR	0.836396	0.836395	407.599
RF	0.821090	0.821069	530.517
DT	0.83391	0.83394	348.971
GBT	0.83391	0.85310	3656.89

Table 4. Performance metrics with "SUSY" dataset

Algorithms	Accuracy	AUC	CT (s)
LR	0.8576	0.7884	2148.734
RF	0.7658	0.774	654.944
DT	0.7416	0.7546	627.392
GBT	0.7862	0.793	5468.769

Table 5. Performance metrics with "HIGGS" dataset

Algorithms	Accuracy	AUC
LR	0.6421	0.6843
RF	0.6764	0.6715
DT	0.6357	0.642
GBT	0.7062	0.705

4 Experimental Results and Analysis

4.1 Experimental Results

Table 3 Present the performance of input features with the "HEPMASS" dataset using proposed algorithms. In the same way, the Table 4 and Table 5 give the proposed metrics with "SUSY" and "HIGGS" datasets [20,21]. These results are analyzed, interpreted and discussed in the next section, using accuracy and CT metrics. We use AUC metric as validation method.

4.2 Analysis and Discussion

In this part, we compare different score algorithms on "HEPMASS" dataset using Accuracy, AUC, and Computation Time metrics. We compare also these results to the related dataset as "Higgs" and "SUSY" dataset [20,21]. Generally, using AUC and Accuracy scores, all Machine Learning methods proposed run efficiently. Applying the Accuracy metric, Fig. 1 shows that the GBT, DT, and LR algorithms run well with 83% and better than RF classifier with 83%. On the other hand, using AUC metric, we show clearly that the GBT method achieved

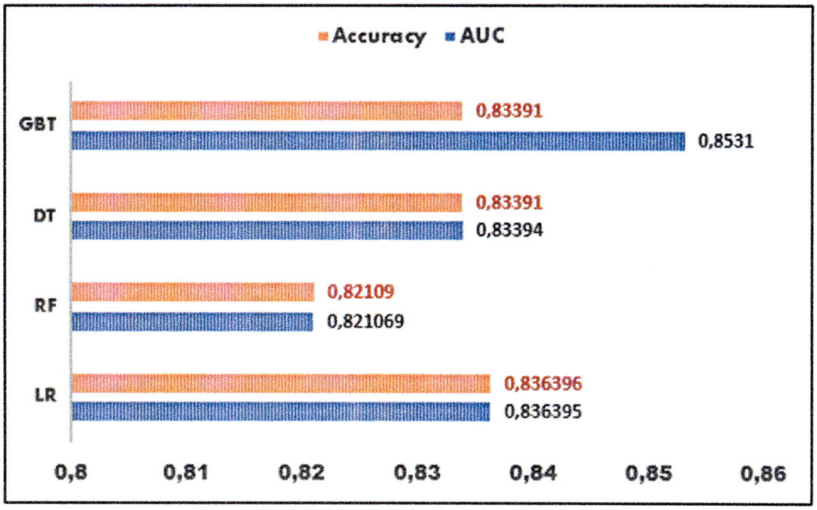

Fig. 1. AUC and Accuracy scores with "HEPMASS" dataset

a higher AUC score with 85% better than LR and DT methods with 83%, that better than RF with 82% (See Fig. 1). We remark also that the AUC and Accuracy metrics are equal for all classifiers (AUC $\approx$ Accuracy $\approx$ 83%) . Hence, we can conclude that the AUC metric is a validation method for the accuracy metric. These results are also confirmed by related work, particularly, in the HIGGS and SUSSY datasets [20,21]. Referring to existing work, these results are closes to the score of Higgs challenge 2014 (83%) and very closes those achieved applying the Deep Learning method [3]. On the other hand, these results are similar to AUC score registered on "SUSY" dataset using LR model (85%) (See Table 2) and better than Accuracy score obtained on "Higgs" dataset, collected from UCI site, with GBT method (70%). Concerning the Computation Time, Fig. 2 shows that the DT classifier runs fastly (348 seconds), whereas the GBT method trained very slowly in 3656 seconds. While DT was learned in 407 seconds and RF in 530 seconds (see Fig. 2).

In the same way, the results obtained with "SUSY" and "Higgs" datasets confirm that the GBT algorithm was learned very slowly. This latency can explained by the high number of learners used in the training phase. Thus, generally, a slow-classifier works efficiently with a good accuracy metric. Moreover, AUC metric approves its effectiveness.

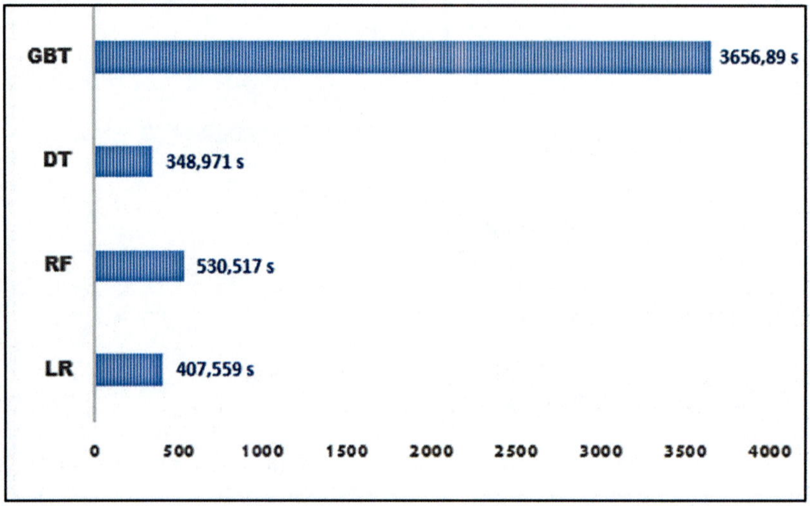

Fig. 2. Computation Time metric with "HEPMASS" dataset

5 Conclusion

In the domain of Searching for Exotic Particles with the "HEPMASS" dataset, The results prove we that the Pyspark environment runs well with big data as the "HEPMASS" dataset and all proposed algorithms work efficiently and registered a score varies within 82% and 85%. Furthermore, we have found that the GBT classifier achieved a higher score ($AUC = 85\%$) followed by LR, DT with an accuracy score of (83%) and RF with 82%. According to the computation time, the GBT method works very slowly With 3656 s.

In brief, the algorithms presented in this paper are well adapted to solve big data problems, especially in the context of "HEPMASS" and "SUSSY" datasets. The GBT algorithm proves its effectiveness despite its latency in the training phase.

References

1. Adam-Bourdarios, C., Cowan, G., Germain, C., Guyon, I., Kégl, B., Rousseau, D.: The Higgs boson machine learning challenge. In: NIPS 2014 Workshop on High-energy Physics and Machine Learning, 19–55 (2015). https://hal.inria.fr/hal-01208587/document
2. Chen, T., He, T.: Higgs boson discovery with boosted trees. In: NIPS 2014 workshop on high-energy physics and machine learning, 69–80 (2015). https://proceedings.mlr.press/v42/chen14.pdf
3. Baldi, P., Cranmer, K., Faucett, T., Sadowski, P., Whiteson, D.: Parameterized neural networks for high-energy physics. Eur. Phys. J. C, 76(5), 1–7 (2016). https://link.springer.com/content/pdf/10.1140/epjc/s10052-016-4099-4.pdf

4. Sadowski, P.J., Whiteson, D., Baldi, P.: Searching for Higgs boson decay modes with deep learning. Adv. Neural. Inf. Process. Syst. **27**, 2393–2401 (2014)
5. Sadowski, P., Collado, J., Whiteson, D., Baldi, P.: Deep learning, dark knowledge, and dark matter. In: NIPS 2014 Workshop on High-energy Physics and Machine Learning, PMLR, 81–87 (2015)
6. Alves, A.: Stacking machine learning classifiers to identify Higgs bosons at the LHC. J. Instrum. **12**(05), T05005 (2017)
7. Meng, X.: MLlib: machine learning in apache spark. J. Mach. Learn. Res. **17**(1), 1235–1241 (2016)
8. Assefi, M., Behravesh, E., Liu, G., Tafti, A.P.: Big data machine learning using apache spark MLlib. In: 2017 IEEE International Conference on Big Data, Big Data), IEEE, 3492–3498 (2017)
9. Armbrust, M., et al.: Spark SQL: relational data processing in spark. In: Proceedings of the 2015 ACM SIGMOD international conference on management of data, 1383–1394 (2015)
10. Peng, H., Liang, D., Choi, C.: Evaluating parallel logistic regression models. In: 2013 IEEE International Conference on Big Data, IEEE, 119–126 (2013)
11. Loh, W.Y.: Classification and regression trees. Wiley Interdisc. Rev. Data Min. Knowl. Disc. **1**(1), 14–23 (2011)
12. Genuer, R., Poggi, J.M.: Arbres CART et Forêts aléatoires, Importance et sélection de variables (2017)
13. Breiman, L.: Random forests. Mach. Learn. **45**(1), 5–32 (2001)
14. Azhari, M., Alaoui, A., Achraoui, Z., Ettaki, B., Zerouaoui, J.: Adaptation of the random forest method. In: Proceedings of the 4th International Conference on Smart City Applications-SCA, 19 (2019)
15. Azhari, M., Alaoui, A., Abarda, A., Ettaki, B., Zerouaoui, J.: Using ensemble methods to solve the problem of pulsar search. In: Farhaoui, Y. (ed.) BDNT 2019. LNNS, vol. 81, pp. 183–189. Springer, Cham (2020). https://doi.org/10.1007/978-3-030-23672-4_14
16. Azhari, M., Alaoui, A., Abarda, A., Ettaki, B., Zerouaoui, J.: A comparison of random forest methods for solving the problem of pulsar search. In: Ben Ahmed, M., Boudhir, A.A., Santos, D., El Aroussi, M., Karas, İR. (eds.) SCA 2019. LNITI, pp. 796–807. Springer, Cham (2020). https://doi.org/10.1007/978-3-030-37629-1_57
17. Liaw, A., Wiener, M.: Classification and regression by RandomForest. R News **2**(3), 18–22 (2002)
18. Ganjisaffar, Y., Caruana, R., Lopes, C.V.: Bagging gradient-boosted trees for high precision, low variance ranking models. In: Proceedings of the 34th International ACM SIGIR Conference on Research and Development in Information Retrieval, pp. 85–94 (2011)
19. Roe, B.P., Yang, H.J., Zhu, J.: Boosted decision trees, a powerful event classifier, In: Statistical problems in particle physics, astrophysics and cosmology, 139–142 (2006)
20. Azhari, M., Abarda, A., Ettaki, B., Zerouaoui, J., Dakkon, M.: Using machine learning with PySpark and MLib for solving a binary classification problem: case of searching for exotic particles. In: Recent Advances in Intuitionistic Fuzzy Logic Systems and Mathematics, Studies in Fuzziness and Soft Computing **395**, 109–118 (2021)
21. Azhari, M., Abarda, A., Ettaki, B., Zerouaoui, J., Dakkon, M.: Higgs Boson discovery using machine learning methods with Pyspark. Procedia Comput. Sci. **170**, 1141–1146 (2020)

Color Image Encryption Using Improved Vigenère Method Followed by a Permutation

Abdellah Abid(✉), Mariem Jarjar, Abdelhamid Benazzi, and Abdellatif Jarjar

MATSI Laboratory Mohamed First University, Oujda, Morocco
abdellahab90@gmail.com

Abstract. In this paper we will develop a new method of color image encryption, based on a deeply improved Vigenère scheme, using two large substitution matrices designed from the most used chaotic maps in cryptography. After vectorizing the image, we begin by computing an initial value to modify the first pixel of the image, then we apply the new Vigenère technique followed by a chaotic permutation to increase the complexity of the algorithm. To overcome the differential, a diffusion function was established between the encrypted pixel and the adjacent clear pixel. Simulations performed on a large number of images of different sizes and formats ensure that our approach is not exposed to known attacks.

$$Notation\{G_t = Z/tZ^{ring}$$

Keywords: Vigenère table · Chaotic map · Chao permutation · Encryption function · Diffusion function

1 Introduction

In the world of mathematics, the chaos theory describes the behavior of dynamical systems, whose state changes as a function of time, and is very sensitive to the initial condition value. Many researchers are attempting to exploit the properties of chaos theory to develop new encryption techniques or to improve existing methods, such as Hill [1, 2], Cesar, Vigenère, [3, 4, 5] Feistel [6] .

M. Ben Ahmed et al. (Eds.): SCA 2022, LNNS 629, pp. 580–590, 2023.
https://doi.org/10.1007/978-3-031-26852-6_54

1.1 Vigenère Classical Technique

The classical Vigenère method is based on a static matrix (V) defined by the following algorithm.

$$algorithm\ 1\begin{cases} Fist\ Row \\ For\ i = 1\ to\ 26 \\ V(1, i) = i \\ Nexti \\ folloying\ Rows \\ For\ i = 2\ to\ 26 \\ For\ j = 1\ to\ 26 \\ V(i, j) = V(i-1, mod(j+1, 26)) \\ Next\ j, i \end{cases}$$

Let (P): plain text, (C): cypher text; (K): Encryption key, (V) Vigenere matrix and (l): length of clear text. So

$$\begin{cases} C_i = V(P_i, K_i) = (P_i + K_i) \quad mod\ 26 \\ P_i = V(C_i, K_i) = (P_i - K_i) \quad mod\ 26 \end{cases} \text{for each i in } [1 \ldots l] \tag{1}$$

1.2 Our Contribution

Our contribution in this paper is the creation of new chaotic substitution matrices of size (256,256) replacing the static s-box of Vigenère. A strong diffusion function will be implemented in our approach.

2 The Proposed Method

Based on chaos [7], our method is based on the two most used chaotic maps in cryptography. At the end of this work, a detailed analysis of the performance of our methodology will be discussed and compared with other referencing systems.

2.1 Chaotic Sequences Design

The Logistics Map:
The logistic map [8] is a recurrent sequence described by a simple polynomial of second degree defined by the following equation

$$\begin{cases} u_0 \in]0, 5\ 1[, \quad \mu \in [3, 75\quad 4] \\ u_{n+1} = \mu u_n(1 - u_n) \end{cases} \tag{2}$$

Skew Tente Map

In mathematics, the skew tent map [9] with parameter μ is a recurrent sequence v_n defined by

$$\begin{cases} v_0 \in]0\ \ 1[\quad \mu \in]0,5\ \ 1[\\ v_{n+1} = \begin{cases} \frac{v_n}{\mu} & if\ \ 0 \prec v_n \prec \mu \\ \frac{1-v_n}{1-\mu} & if\ \ \mu \prec v_n \end{cases} \end{cases} \tag{3}$$

Chaotic Vector Construction

Our work requires the construction of four chaotic vectors ($K1$), (K2), (K3) and (K4), with a coefficient of (G_{256}), and a binary vector (VB).

$$Algorithm\ 2 \begin{cases} for\ \ i = 1\ to\ 3\,\mathrm{nm} \\ K1(i) = mod\left(E\left(u(i)*10^{11}\right), 254\right) + 1 \\ K2(i) = mod\left(E\left(v(i)*10^{11}\right), 253\right) + 2 \\ K3(i) = mod\left(E\left(\frac{3*u(i)+2*v(i)}{5}*10^{11}\right), 254\right) + 1 \\ K4(i) = mod\left(E\left(\frac{4*u(i)+3*v(i)}{7}*10^{11}\right), 253\right) + 2 \\ Next\ i \end{cases}$$

$$Algorithm\,3 \begin{cases} for\,i = 1\,to\,3\,\mathrm{nm} \\ if\ u(i) \geq v(i)\ then \\ VB(i) = 0\,else\,VB(i) = 1 \\ end\ if \\ Next\,i \end{cases}$$

Global Permutation Design

Sorting the vector (VB) in a large decreasing order generates the global permutation (GP). This method is illustrated by the following algorithm

$$Algorithm\,4 \begin{cases} k = 1 \\ For\,i = 3\,\mathrm{nm}\,to\,1 \\ If\ VB(i) = 0\,Then\,GP(i) = k \\ k = k + 1 \\ end\ if \\ Next\,i \end{cases} \begin{cases} For\,i = 3\,\mathrm{nm}\,to\,1 \\ If\ VB(i) = 1\,Then\,GP(i) = k \\ k = k + 1 \\ end\ if \\ Next\,i \end{cases}$$

Example:

VB	0	0	1	1	1	0	1	0	1	1	0	1	0	1	1	0	0	1	1	0
sort	8	7	9	10	11	6	12	5	13	14	4	15	3	16	17	2	1	18	19	0

$$GP = \begin{pmatrix} 0 & 1 & 2 & 3 & 4 & 5 & 6 & 7 & 8 & 9 & 10 & 11 & 12 & 13 & 14 & 15 & 16 & 17 & 18 & 19 \\ 8 & 7 & 9 & 10 & 11 & 6 & 12 & 5 & 13 & 14 & 4 & 15 & 3 & 16 & 17 & 2 & 1 & 18 & 19 & 0 \end{pmatrix}$$

Fig. 1. Permutation in G_{20}

2.2 Encryption Process

Original Image Vectoring

After the three (RGB) color channels extraction and their conversion into size vectors $(\boldsymbol{Vr})$, $(\boldsymbol{Vg})$, $(\boldsymbol{Vb})$ $(1, \boldsymbol{nm})$ each, a concatenation is established to generate a vector $\boldsymbol{X}(\boldsymbol{x}_1, \boldsymbol{x}_2, \ldots\ldots.., \boldsymbol{x}_{3\boldsymbol{nm}})$ of size $(1, 3\,\text{nm})$.

Initialization Value Calculation

The initialization value (IV) must be recalculated to change the value of the starting pixel.

$$Algorithm\,5 \left\{ \begin{array}{c} for\; i = 2\; to\; 3\,\text{nm} \\ IV = IV \oplus X(i) \oplus K1(i) \\ Next\; i \end{array} \right.$$

Advanced Vigenère Methods

This technique requires the establishment of two substitution matrices $(VG1)$ and $(VG2)$ of size (256, 256), through the process described by the following steps.

- a permutation $(P1)$ obtained by descending ordering the first $256\,values$ of the sequence $(k1)$
- a permutation $(P2)$ obtained by increasing the ordering the first $256\,values$ of the sequence $(k1)$,
- a permutation $(P3)$ obtained by increasing the ordering the last $25\,values$ of the sequence $(k3)$,
- The construction of $(VG1)$ is given by the following algorithm

$$algorithm\,6 \left\{ \begin{array}{c} Fist\; Row \\ For\; i = 1\; to\; 256 \\ VG1(1, i) = P1(i) \\ VG1(2, i) = P2(i) \\ Next\; i \end{array} \right. \left\{ \begin{array}{c} Next\; lines \\ For\; i = 3\; to\; 256 \\ For\; j = 1\; to\; 256 \\ VG1(i, j) = VG1(i-1, VG1(i-2, j)) \\ next\; j, i \end{array} \right.$$

- The construction of $(VG2)$ is given by the following algorithm

$$algorithm\,7\left\{\begin{array}{c} Fist\,Row \\ For\,i = 1\,to\,256 \\ VG2(1,i) = P3(i) \\ Next\,i \end{array}\right\} \left\{\begin{array}{c} Next\,lines \\ For\,i = 2\,to\,256 \\ For\,j = 1\,to\,256 \\ VG2(i,j) = VG2(i-1, mod(j + K4(i), 256)) \\ next\,j, i \end{array}\right.$$

Example:

	(VG1)	0	1	2	3	4	5	6	7
P1	0	3	5	0	6	2	7	1	4
P2	1	2	7	1	4	3	5	0	6
	2	4	3	5	0	6	2	7	4
	3	2	7	1	4	3	5	0	6
	4	0	6	2	7	1	4	3	5
	5	7	5	2	4	6	1	0	3
	6	1	3	2	6	0	4	7	5
	7	2	0	4	7	6	1	3	5

	(VG2)	0	1	2	3	4	5	6	7
P3	0	0	4	5	7	1	2	6	3
	1	6	2	1	7	3	0	4	5
	2	7	1	2	4	5	6	3	0
	3	5	2	3	0	6	7	1	4
	4	0	4	1	3	2	6	7	5
	5	6	3	4	7	0	5	1	2
	6	4	7	5	1	6	0	3	2
	7	5	1	7	2	3	4	6	0

Fig. 2. Vigenère tables

Classical Vigenère Function Expression. The classical Vigenère's replacement function is giving by the following formula

$$Y(i) = V(K, X(i)) \tag{4}$$

(K) key duplicated to the size of the text to be encrypted.

New Vigenère Mathematica Expression. The following equation illustrates the effective expression of the $Y(i)$ image of the pixel through the new advanced Vigenère (AV) method.

$$AV(X(i)) = Y(i) = VG1(K1(i), VG2(K2(i), X(i) \oplus K3(i))) \tag{5}$$

Improved Vigenère Spread Function Expression. The encryption process using the function (AV) is illustrated by the following figure.

$$\forall\, i > 1 \quad \Phi(X(i)) = VG2(K4(i), VG1(K3(i), X(i) \oplus K2(i))) \tag{6}$$

Φ is the diffusion function.

The Encryption Algorithm.

The encryption process is defined by the following algorithm,

$$We\,note : AV\left(X'(i)\right) = Y(i)$$

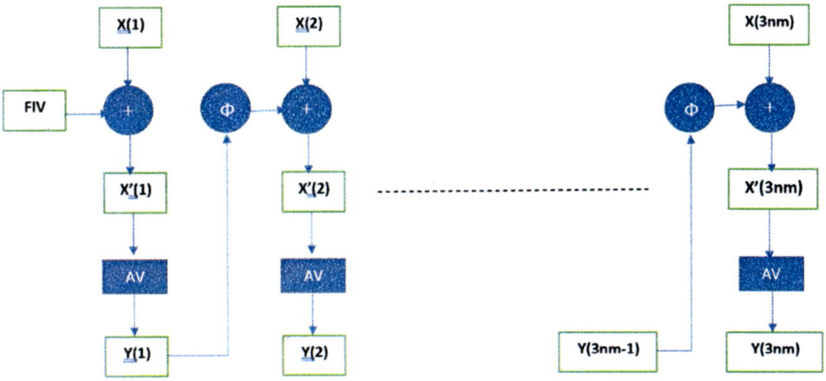

Fig. 3. Encryption process

$$Algorithm\,8\begin{cases} X'(1) = IV \oplus X(1) \\ Y(1) = AV\big(X'(1)\big) \\ For\,i = 2\,to\,3\,\text{nm} \\ \alpha = \Phi(Y(i-1)) \\ X'(i) = \alpha \oplus X(i) \\ Y(i) = AV\big(X'(i)\big) \\ Next\,i \end{cases}$$

The output vector (Y) will be subjected to global permutation (GP) to possibly suppress any correlation.

$$Algorithm\,9\begin{cases} For\,i = 1\,to\,3\,\text{nm} \\ Z(i) = GP(Y(i)) \\ Next\,i \end{cases}$$

The vector (Z) constitutes the image encrypted by our algorithm.

2.3 Decryption of Encrypted Images

Decryption Matrix Structure

Each row of the encrypted S-box is a permutation in (G_{256}), so the decryption matrix will consist of reverse permutations. For this reason, two decrypted $S-Box$ generations are given by the following algorithm

$$Algorithm\,10\begin{cases} for\,i = 1\,to\,256 \\ for\,j = 1\,to\,256 \\ GV1(i, VG1(i,j)) = j \\ GV2(i, VG2(i,j)) = j \\ Next\,j, i \end{cases}$$

Example:

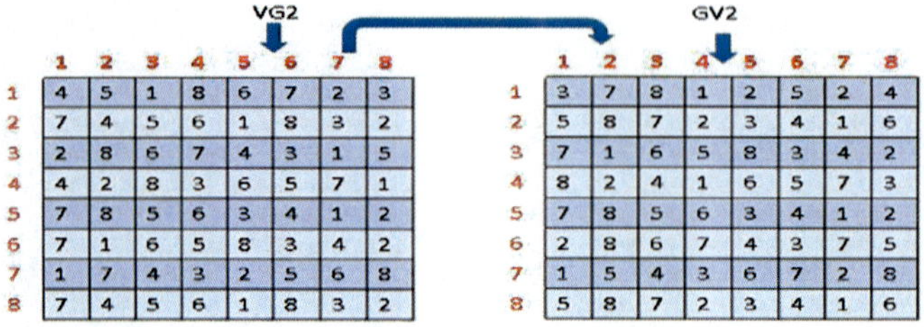

VG2

	1	2	3	4	5	6	7	8
1	4	5	1	8	6	7	2	3
2	7	4	5	6	1	8	3	2
3	2	8	6	7	4	3	1	5
4	4	2	8	3	6	5	7	1
5	7	8	5	6	3	4	1	2
6	7	1	6	5	8	3	4	2
7	1	7	4	3	2	5	6	8
8	7	4	5	6	1	8	3	2

GV2

	1	2	3	4	5	6	7	8
1	3	7	8	1	2	5	2	4
2	5	8	7	2	3	4	1	6
3	7	1	6	5	8	3	4	2
4	8	2	4	1	6	5	7	3
5	7	8	5	6	3	4	1	2
6	2	8	6	7	4	3	7	5
7	1	5	4	3	6	7	2	8
8	5	8	7	2	3	4	1	6

Fig. 4. Vigenère inverse table

Decryption Function

By following the same logic of Vigenère's traditional technique, we obtain

$$\begin{cases} if\ z = VG(y, x) \\ \quad Then \\ x = GV(y, z) \end{cases} \tag{7}$$

Decrypt the Encrypted Image

Our algorithm is a symmetric encryption system, so the same key will be used in the decryption process.

Reverse Permutation. The inverse permutation (PG) of (GP) is given by the following algorithm

$$Algorithm\,11 \begin{cases} For\, i = 1\, to\, 3\,\text{nm} \\ PG(GP(i)) = i \\ \quad Next\, i \end{cases}$$

After vectorization of the image encrypted in vector (Z), an intervention of the permutation (GP) to recover the vector (Y). This operation is determined by the following algorithm

$$Algorithm\,12 \begin{cases} For\, i = 1\, to\, 3\,\text{nm} \\ Y(i) = PG(Z(i)) \\ \quad Next\, i \end{cases}$$

The Reciprocal of the Lap Function.

$$X(i) = GV2(K2(i), GV1(K1(i), Y(i))) \oplus K3(i) \oplus \Phi(Y(i-1)) \tag{8}$$

A recalculation of the initialization value will make it possible to retrieve the exact value of pixel $.X(1)$

3 Examples and Simulations

All experiments are performed Python programming language running under ubuntu 20.04, on a basic i5 personal computer, 8 GB RAM, and 500 GB hard disk.

3.1 Key-Space Analysis

The global size of our key is well above $2^{180} \gg 2^{110}$, which is enough to avoid any brutal attacks.

3.2 Secret Key's Sensitivity Analysis

The sensitivity to the private key of our system can be seen by the following diagram.

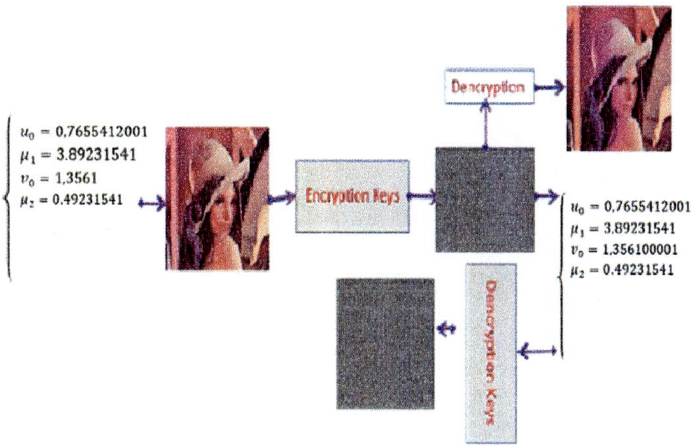

Fig. 5. Encryption key sensitivity

3.3 Statistics Attack Security

Entropy Analysis
The entropy of an image of size (n, m) is given by the equation below

$$H(MC) = \frac{1}{t}\sum_{i=1}^{t} -p(i)\log_2(p(i)) \tag{9}$$

$p(i)$ is the probability of occurrence of level (i) in the original image.

Correlation Analysis
The correlation of an image of size (n, m) is given by the equation below

$$r = \frac{\operatorname{cov}(x, y)}{\sqrt{V(x)}\sqrt{V(y)}} \tag{10}$$

We noticed that the entropy of all images tested by our algorithm is close to 8.

All the correlation measures of the images tested by our system are very close to zero. This can protect our methods from statistical attacks.

Histogram Analysis
All images tested by our algorithm have a uniformly distributed histogram.

Table 1. Original images and cipher images

name	Original image	size	Cipher image
Img1	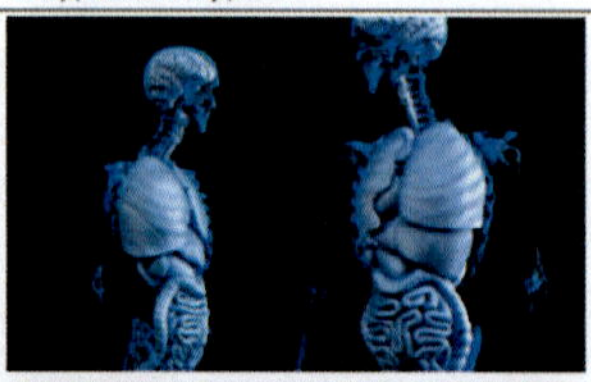	243x411	
Img 2	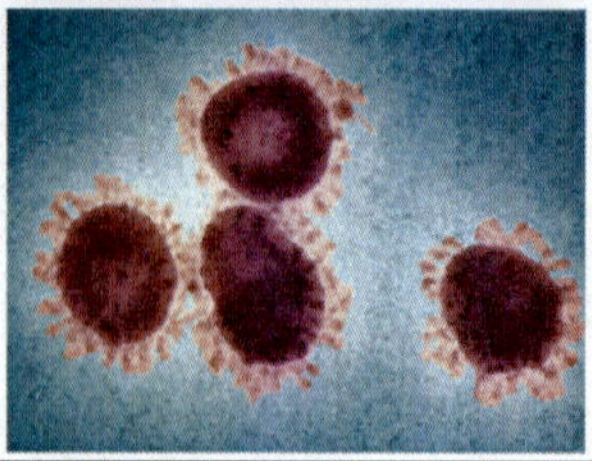	807x593	

Table 2. Correlation and Entropy of some tested images

Name	Vertical correlation			Horizontal correlation			Diagonal correlation			Entropy		
	Red	Green	Blue	Red	Green	Blue	Red	Green	Blue	Red	Green	Blue
Img1	−0.0042	−0.0008	−0.0029	−0.0004	−0.0060	−0.0025	−0.0056	−0.0009	−0.0057	7.9978	7.9981	7.9980
Img2	−0.0009	−0.0006	0.0031	−0.0005	0.0001	−0.0005	0.0016	−0.0004	−0.0002	7.9995	7.9995	7.9995

Table 3. Encrypted image histogram

name	Original image	Cipher image
Img1	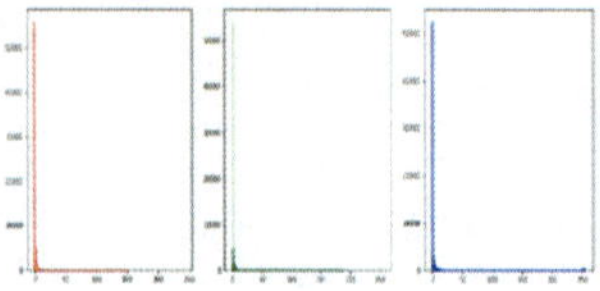	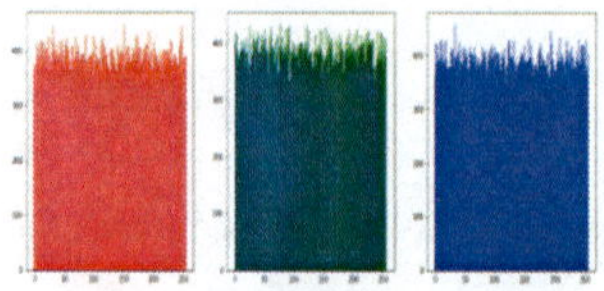
Img2	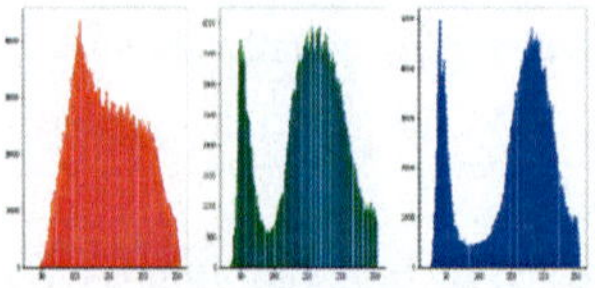	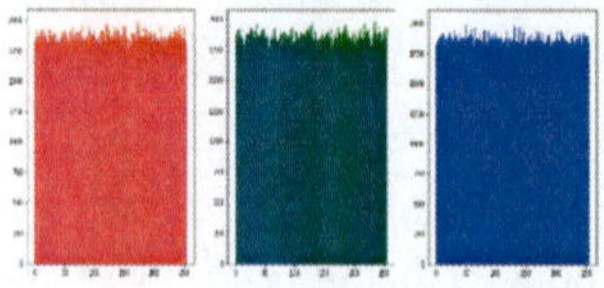

3.4 Differential Analysis

The NPCR Constant

It is determined by the equation below

$$NPCR = \left(\frac{1}{nm} \sum_{i,j=1}^{nm} D(i,j) \right) *100 \tag{11}$$

$$With \quad D(i,j) = \begin{cases} 1 & if \quad C_1(i,j) \neq C_2(i,j) \\ 0 & if \quad C_1(i,j) = C_2(i,j) \end{cases}$$

The UACI Constant

The *UACI* mathematical analysis of an image is given by the next equation

$$\text{UACI} = \left(\frac{1}{nm} \sum_{i,j=1}^{nm} Abs(C_1(i,j) - C_2(i,j)) \right) *100 \tag{12}$$

Avalanche Effect

The mathematical expression of this avalanche effect is given by

$$\text{AE} = \left(\frac{\sum_i bit\ change}{\sum_i bit\ total} \right) *100 \tag{13}$$

Table 4. Differential parameters

Name	NPCR	UACI	EA
Img1	99.9994%	33.5355%	50.1794%
Img2	99.9997%	33.6323%	50.2036%

3.5 Performance Time

The following figure shows the encryption and decryption time.

4 Math Security

Our system is protected against brute force attacks because of the large size of the key space, moreover the chaotic character of the system increases the difficulty to unlock all encrypted images using statistical attacks. as well as, the high sensitivity to the initial values of the chaotic maps and the diffusion function confirmed the robustness of our system.

Table 5. performance time

Name	Encryption time	Decryption time
Img1	0.7534	0.7621
Img2	3.7341	3.7453

5 Conclusion

Chaos theory has taken a place in cryptography because of the high sensitivity to the initial condition of chaotic systems. Our method was based on this character to build two chaotic substitution matrices of Vigenère. Initially we calculated an initial value to start the encryption process, as well as we use a diffusion and confusion function to protect our system against differential attacks. The simulation results can ensure that our system is not exposed to the different known attacks.

References

1. Acharya, B., et al.: Image encryption using advanced hill cipher algorithm. ACEEE Int. J. Signal Image Process. **1**(1) (2010)
2. Putrie, V.M., et al.: Super encryption using transposition-hill cipher for digital color image. In: 2018 International Seminar on Research of Information Technology and Intelligent Systems (ISRITI) (2018)
3. Triandi, B., et al.: Improve security algorithm cryptography vigenere cipher using chaos functions In: 2018 6th International Conference on Cyber and IT Service Management (CITSM) (2018)
4. Tan, C.M.S., et al.: A hybrid encryption and decryption algorithm using caesar and vigenere cipher. In: Journal of Physics Conference Series, Asian Conference on Intelligent Computing and Data Sciences (ACIDS), vol. 1997 , 24–25 May 2021, Perlis, Malaysia (2021)
5. Jarjar, A.: Vigenere Followed by Multiple Bit-Level Permutations to Encrypt a Color Image. Preprints, 2021040647 (2021) (doi: https://doi.org/10.20944/preprints202104.0647.v1)
6. Zhang , X., et al.: image encryption method based on the feistel network and dynamic DNA encoding. IEEE Photonics J. **10**(4) (2018)
7. Suneja, K., Dua, S., Dua, M.: A review of chaos based image encryption. In: 2019 3rd International Conference on Computing Methodologies and Communication (ICCMC), pp. 693–698 (2019). doi: https://doi.org/10.1109/ICCMC.2019.8819860
8. Luo, Y., Yu, J., Lai, W., Liu, L.: A novel chaotic image encryption algorithm based on improved baker map and logistic map. Multimedia Tools and Applications **78**(15), 22023–22043 (2019). https://doi.org/10.1007/s11042-019-7453-3
9. Ahmad, J., Khan, M.A., Ahmed, F., Khan, J.S.: A novel image encryption scheme based on orthogonal matrix, skew tent map, and XOR operation. Neural Comput. Appl. **30**(12), 3847–3857 (2017). https://doi.org/10.1007/s00521-017-2970-3

Design of the UTEA 1 Experimental Rocket with a Greater Stable Vertical Range for the Study of Atmospheric Variables

Edy Ambia Vasquez[1](✉), Carolina Soto Carrion[1], Wilber Jiménez Mendoza[2], and Isabel Milagros Jiménez Soto[3]

[1] Technological University of the Andes, Abancay, Apurímac 03001, Peru
eam-iav@utea.edu.pe
[2] Micaela Bastidas National University of Apurimac, Abancay, Apurímac 03001, Peru
wjimenez@unamba.edu.pe
[3] University of Applied Sciences, San Isidro, Lima 15073, Peru
u201511196@upc.edu.pe

Abstract. The purpose was to evaluate an experimental design of 3 factors at 2 levels: type of warhead (conical with blunt tip and parabolic), type of fin (trapezoidal and delta), number of fins (3, 4) and their interactions for the construction of the experimental rocket UTEA 1, which has a stable flight and reaches a higher altitude. The rocket was tested on a scale of 1:5 assigning a rating from 1 to 10: 1 very bad, 5 fair, 10 very good. There was no substantial difference between the tests, but one of the tests with the conical nose cone with blunt tip, trapezoidal fin and 4 fins, went well, that is why the rocket with conical nose conical blunt tip, trapezoidal fin and 4 fins, with a rocket length of 150 cm (fuselage 120 cm and warhead 30 cm), total rocket mass 2000 g (structural mass 1475 g, propellant 410 g, payload 115 g). It was determined that the stability and vertical range of the rocket does not depend on the design of the fins, number of fins and the shape of the warhead because the stability of the rocket is related to the center of pressure and the center of gravity, but the fins act as stabilizer elements and their effectiveness depends on their size, shape, number of fins. Likewise, the increased vertical range of the rocket depends on its power and amount of propellant and not on the type of warhead, type of fins, and number of fins.

Keywords: Rocket · Fins · Warheads · Fuselage · Propeller

1 Introduction

The purpose of the research has been to determine that the stability and vertical range of the rocket depends on the design of the fins, number of fins and the shape of the fuselage warhead. Sounding rockets are used to monitor atmospheric variables such as pressure, temperature and relative humidity [1]. In Europe the pioneers of experimental rocketry formed amateur rocketry societies, in America they had their beginnings in the universities of the United States becoming popular as a hobby in the late 1950s, follow

M. Ben Ahmed et al. (Eds.): SCA 2022, LNNS 629, pp. 591–608, 2023.
https://doi.org/10.1007/978-3-031-26852-6_55

ing the launch of Sputnik, in the mid-70s it spread worldwide. In Brazil and Argentina carried out many studies and designs of experimental rocketry [2]. Currently weather balloons and radiosondes are used for the acquisition of atmospheric data, these data are taken in the troposphere which is the atmospheric layer that is in contact with the surface of the earth and is about 10 km high and this is where all atmospheric phenomena occur [3], however, the impact is not natural. One solution is the use of sounding rockets to measure these complex atmospheric variables [4].

1.1 Related Word

In 1895 a Peruvian engineer invented the liquid-fueled engine and the first modern rocket propulsion system in 1900. Pedro Paulet helped man board the Moon, wrote Wernher von Braun, chief designer of the Apollo 11 Saturn V rocket, in his book His- tory of Rocketry and Space Travel [5].

Soviet physicist Konstantin Tsiolkovsky (Izhevskoye, 1857 - Kaluga, 1935), a Russian research scientist in aeronautics and astronautics was a pioneer in space and rocket research, in the development and use of wind tunnels for aerodynamic studies [6]. That is why Konstant ın Tsiolkovski is considered the father of cosmonautics. (The Economist America, 2014) says, that he designed a ship to 'retropropulsion' for interplanetary travel guided by the designs and the prototype called "Autobolido" that in 1895 had designed Pedro Paulet. The Peruvian engineer Pedro Paulet, (Arequipa, 1874 Buenos Aires 1945), considered as a precursor of the propulsion airship and pioneer of the space age. The journalist [7] says that he was inspired by the books of the French writer Jules Verne entitled From the Earth to the Moon (1865) and Around the Moon (1869), which helped to forge that futuristic and advanced spirit. The first liquid propellant rocket engine was built by the Peruvian P. E. Paulet Mostajo, in 1896, using hydrogen perox ide and gasoline [8]. He designed his own aerospace spacecraft, and his formulas and experiments served as a reference for the main designers of the U.S. space agency NASA. With his efforts, in 1912, the American professor Robert Goddard and the German scientist Hermann Julius Oberth in 1923, perfected their experimental engines on the basis of Paulet's initial conception [9].

German engineer Wernher von Braun (Wyrzysk 1912-Virginia 1977), member of Hitler's Schutzstaffel (SS); designed the V2 rocket, precursor of space rockets. He surrendered to the Americans at the end of World War II, and offered the army unused V2 rockets, tons of documents and a hundred of his best engineers and scientists [10]. In his book World History of Astronautics (which he wrote together with Fred Ordway), Wernher von Braun, former director of NASA's Marshall Space Center, and director of the Saturn V program, which took man to the moon, said: 'Peter Paulet, in those years (1900), being in Paris, experimented with his small engine weighing two and a half kilos and achieved a hundred kilograms of force.

For this fact, Paulet must be considered as the pioneer of the liquid fuel propulsion engine". Moreover, in his "History of Rocketry and Space Travel", von Braun acknowledges that, with his efforts, Paulet helped man to reach the moon [11].

Design and implementation of the UTEA rocket 1 Thermodynamics is the part of physics that studies heat transfers, energy conversion and the ability of systems to produce work [12].

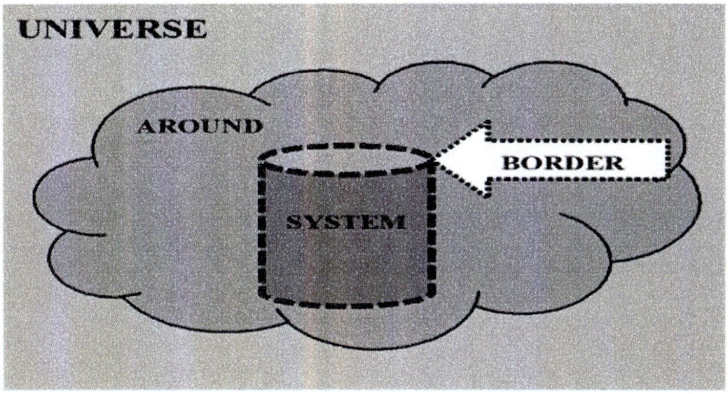

Fig. 1. Components of a thermodynamic system

Aerodynamics is a branch of physics dedicated to fluid mechanic's study. That is, it specializes in studying the principles and laws that gover the motion of a body immersed in air to analyze its interactions.

In the case of a rocket moving through the Earth's atmosphere, it generates a resultant force on the body called aerodynamic force (See Figs. 1 and 2).

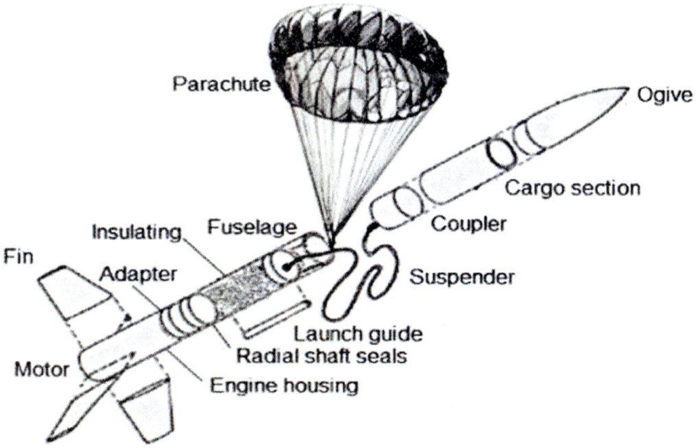

Fig. 2. Parts of a solid rocket propellant

The functions of the parts of a rocket are: The fuselage is the structural part of the rocket to which the other parts such as the wings, and the various systems and equipment are attached.

The fins serve to provide stability and maintain the direction of flight of the rocket. The warhead has an aerodynamic shape (See Fig. 3) so that the rocket achieves greater displacement when passing through the air and there are the following types of warheads:

The motor housing has the function of containing the rocket motor the adapter has the objective of coupling the diameter of the fuselage to the motor diameter. The retainer ring is the stopper of the rocket motor so that the thrust of the rocket motor is transmitted to the fuselage. The guide tube is used to attach the model to the launch ramp to help guide the model during takeoff. The suspensor is intended to hold the warhead and the fuselage together. The rocket motor is the rocket booster the parachute is used to return the rocket to the surface slowly without suffering significant damage the insulator functions as a piston and thermal insulator" between the motor ejection and the model's recovery system.

A. Type of warheads

- Parabolic warhead
- Conical warhead
- Spherical warhead

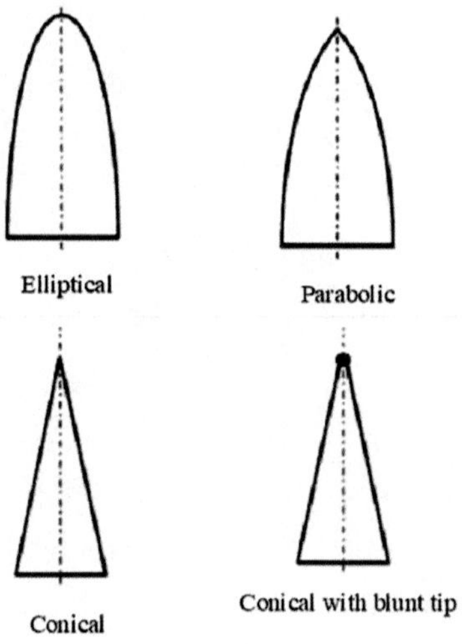

Fig. 3. Type of warheads

B. Type of fins

There are 4 basic fin types: trapezoidal fins, square fins, square fins, delta fins and triangular fins (See Fig. 4).

C. Stabilizers

Designing a new rocket is basically a fin design problem. First, an initial rocket design was determined to fit requirements such as the desired size of the body tube and any other desired special features. Second, the center of gravity (C.G.) of the design was calculated using a technique, Third, the position of the center of pressure C.P the design was calculated. Fourth, the result was compared with the center of

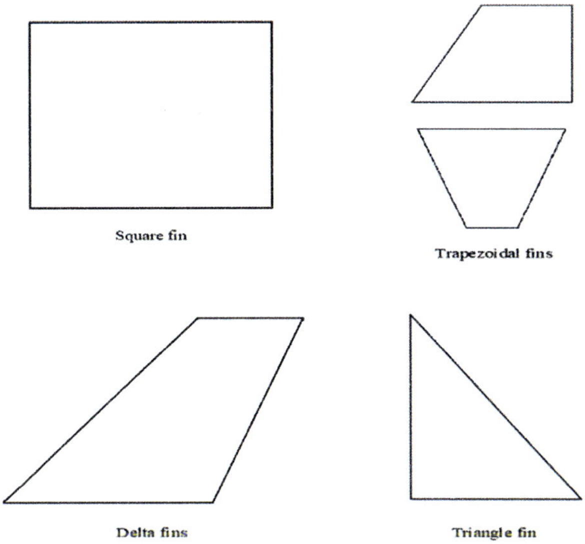

Fig. 4. Fin type

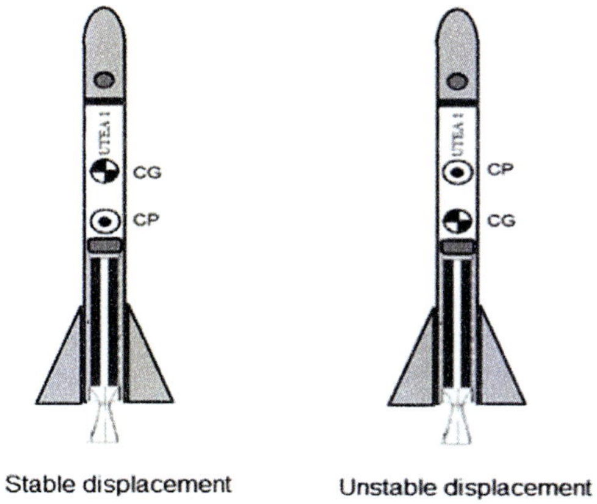

Fig. 5. Center of gravity and pressure

gravity to see if the design is stable and has adequate static margin. Again, a good value for the static margin is the largest diameter of the rocket body [13].

The flight stability of the rocket is a function of the location of gravity center relative to the pressure center. That is, the center of gravity must be above of the center pressure and generally tends to the center of the rocket structure to have a stable flight and achieved by incorporating fins at the base of the rocket so as to increase the area of contact with the air and consequently decrease the center of pressure (See Fig. 5).

One of the most important requirements in the design of the rocket has been to determine the CG and CP to ensure stability throughout the flight.

D. Rocket engine

Rocket engines are internal combustion, autonomous systems that do not need external gases to generate the combustion reaction because all the fuel components are carried inside the rocket. Among the chemical type engines are classified by the physical state of the fuel:

- Liquid propellant.
- Solid propellant.
- Gaseous propellant.
- Hybrid propellant.

E. Propellant grain

A propellant grain has a cylindrical configuration that can be of the cartridge type (separate from the motor) and of the type adhered to the casing. The geometric shape that the grain adopts is obtained by molding or extraction.

F. Solid propellant

Once ignition of the solid propellant has started, it cannot be turned off by generating the combustion gases to provide the necessary thrust for rocket liftoff.

One of the most widely used propellants by rocketry enthusiasts is KNO3+ sugar because it is more accessible. The propellant used in experimental amateur rocket engines can be a simple composition, being the combination of two main constituents fuel and oxidizer. Such is scarce of the propellants with sugar bases [14].

G. Vertical motion of a rocket

In the vertical motion of a rocket there are two stages: The stage when the rocket is launched with fuel and the stage at the instant the fuel is exhausted, until it reaches the apogee of the rocket.

If a rocket at instant (t), has a mass (m), which carries a velocity v respect to the Earth, which is the inertial reference frame. At instant (t + At), a mass (Au) of fuel is ejected with a constant velocity - or relative to the rocket, as a consequence the velocity of the remaining mass (m-Au) of the rocket increases by v + Av.

H. Factorial design for the rocket flight tests

The simplest factorial designs are 2-factor: n levels of the factor f, arranged in a factorial design. Observations can be described by the linear or first-order statistical model [15, 16].

$$\gamma_{ijk} = \mu + \tau_i + \beta_j + (\tau\beta)_{ij} + \varepsilon_{ijk}$$

where:

$i = 1, 2, \ldots.a$
$I = 1, 2, \ldots.a$
$j = 1, 2, \ldots.b$
$k = 1, 2, \ldots.n$

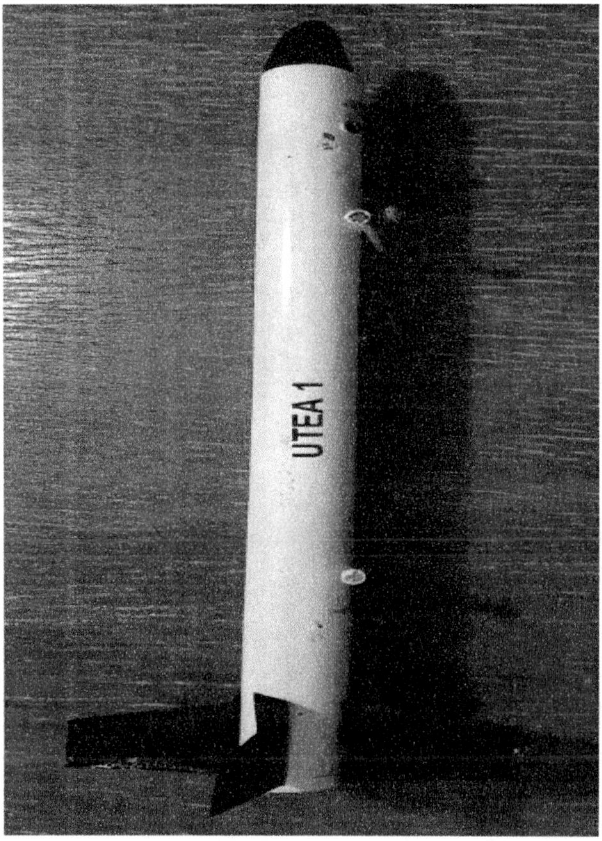

Fig. 6. UTEA 1 rocket at 1:5 scale

μ = is the media general effect
τ i = is the effect of del i-n level factor line A
β j = is the effect del j-n level factor column
B ($\tau\beta$)ij = is the interaction effect between t y b
EijK = is the random error component.

The factorial design of E = nf +2, of 2 levels and 3 factors, in the fermentation process corresponds to 8 experimental tests that are each one of the vertices of a cube and two tests replicated in the center to identify the most efficient variables of the process and serves to explain the possible existence of a curvature (Fig. 7).

It has been determined as hypothesis: The stability and vertical range of the experimental rocket UTEA 1 depends on the design of the fins, number of fins and the shape of the fuselage warhead. Operationalization of variables.

Independent variables
X1 = Type of fuselage nose cone (TO)
X2 = Number of fins (NA)

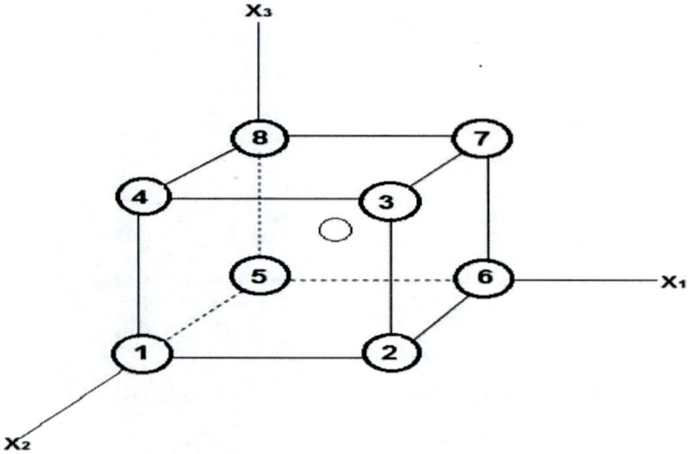

Fig. 7. Factorial design F = 3 replicated in the center

X3 = Rocket fin type (TA)
Dependent variable
Y = Aerodynamic stability of the experimental rocket UTEA1 Parameters
Room temperature
Atmospheric pressure: 513 mm Hg Rocket length: 150 cm (30 cm) 1:5 Total rocket mass: 1:5 of 2000 g (400 g)

I. Characterization of the sample Sample length

The aerodynamic analyses were performed at 1:5 scale of the research unit of 150 cm length. That is, as a basis for the development of all tests was 30 cm length of the rocket (See Fig. 8 and Fig. 6).

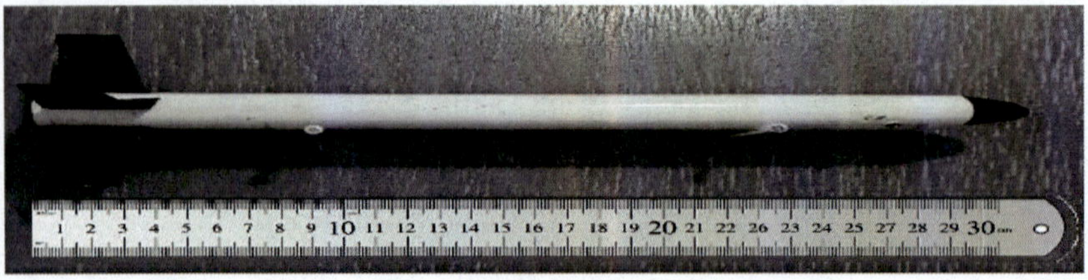

Fig. 8. 1:5 scale rocket

The samples were subjected to the treatment of an aerodynamic tunnel (See Fig. 9) and water rocket launches were used to determine the 3 factors and their levels assigning values for each rocket flight whose scale was 1 very bad, 5 regular and 10 very good.

The 2k full factorial design was used, this model at 2 levels (lower, upper), in each factor must predict all the points of the design.

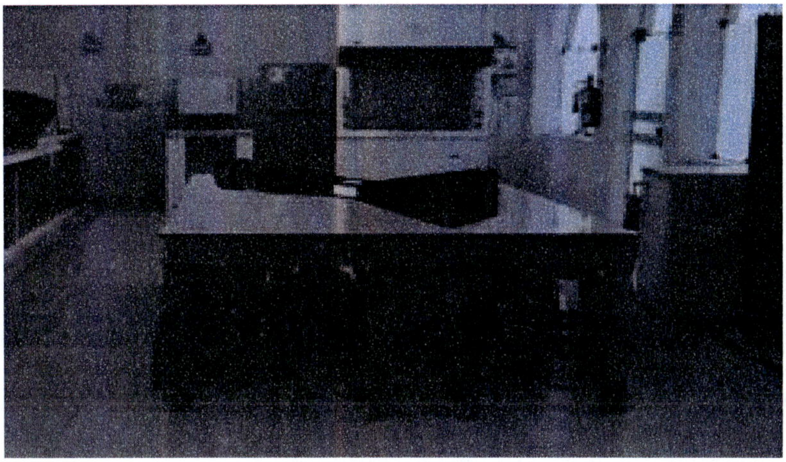

Fig. 9. Wind tunnel (design)

Tests for the application of the factorial design

It can be adequately developed through an empirical mathematical model defined by:

$$Y = f(x, u, z, h).$$

For the simulation we have considered the full factorial design at 2 levels with 3 variables defined by the equation

$$E = nf$$

where: E = Total number of tests nf = Factorial part (full factorial design) 2 represents the two levels of each factor f are the factors

$$E = 2f = 23 = 8$$

The total number of experimental trials is 8 trials which corresponds to the vertices of a cube.

J. Tests for the full factorial design Summary of the design

Factors: 3 Base design 3;8

Runs: 8 Replicates: 1

Blocks: 1 Center points: 0

We proceeded to select the 3 most significant variables and their corresponding levels. See Table 1

X1 = Type of fuselage nose cone: (TO)

X2 = Number of fins: (NA)

X3 = Type of fins of the rocket (TA)

Table 1. Variables and levels for the experimental design.

Variables		Levels	
		Lower n (-)	Higher n (+)
X1	Fuselage warhead type (TO)	A (conical with blunt tip)	B (parabolic)
X2	Rocket fin type (TA)	C (trapezoidal)	D (Delta)
X3	Number of fins	E (3)	F (4)

A coded matrix and natural scale of the independent variables was proposed for the UTEA 1 rocket flight tests. And the rocket flight stability was considered as a response variable (Tables 2 and 3).

Table 2. Natural scale matrix of independent variables.

Variables		Levels
	lower n (−)	Higher n (+)
Fuselage warhead type (TO)	−	+
Rocket Fin Type: (TA)	−	+
Number of fins: (NA)	−	+

Table 3. Coded array of independent variables.

Variables		Levels
	Lower Z_1	Higher Z_3
Fuselage warhead type (TO)	A	B
Rocket Fin Type: (TA)	C	D
Number of fins: (NA)	E	F

K. Dimensions and mass of UTEA 1 rocket Fuselage

It was determined based on the ratio of fuselage length and fuselage diameter which should be between 5 and 25 [17]. In this case the fuselage diameter is 21/8 in. which is equivalent to 6.0325 cm. Then, the length of the fuselage is 120 cm since the diameter is 6 cm × 20 = 120 cm (See Fig. 10).

a) Warhead: The length of the warhead is determined based on the percentage that generally ranges between 15% and 30% of the length of the fuselage. In this case,

Fig. 10. Fuselage with fins

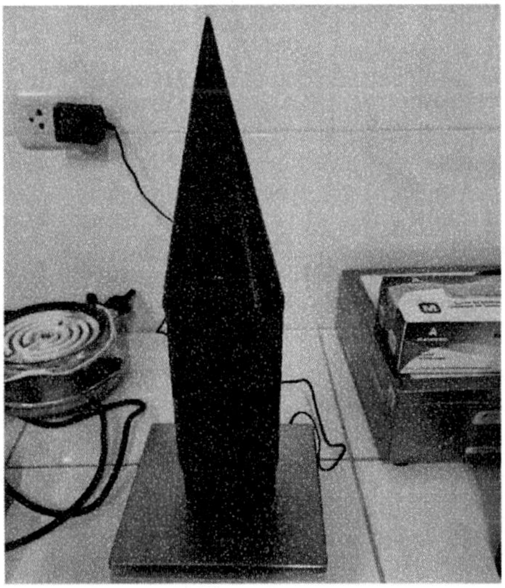

Fig. 11. Ogive

the length of the warhead is 30 cm since 25% of the length of the fuselage was taken, which is 120 cm (Fig. 11).

b) Fins: The fins to be used in the tests are trapezoidal and delta type. The fuselage will have 3 and 4 fins. Likewise, the dimensions of the fins were taken according to the diameter ratio of the fuselage (Fig. 12).

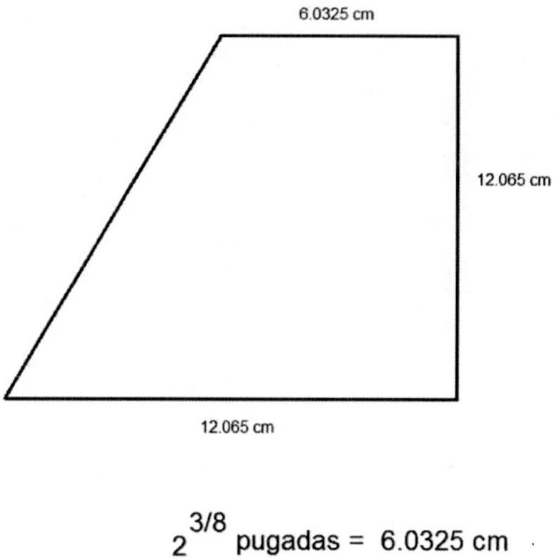

Fig. 12. Fins

c) Mass of the rocket: The total mass of the rocket is 2000 g: the propellant is 410 g, the payload is 115 g. So, the structural mass is 1475 g. that would be 73.75 (Fig. 13).

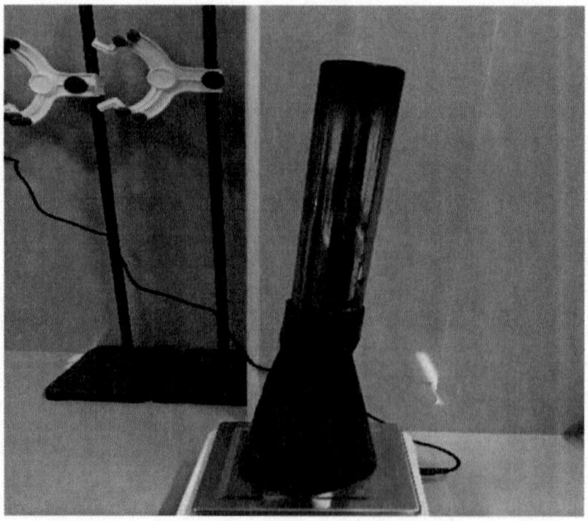

Fig. 13. Engine mass

d) Center of pressure: Aerodynamic forces exert pressures on all the components of the rocket that are exposed to air resistance and are generally: the warhead cone, fuselage, fins and nozzle [13].

$$Dcp = \frac{do * Ao + df * Af + da * Aa + dto * Ato}{At}$$

where dCP distance from a reference line to CP of the vehicle, do= distance from the reference line to the geometrical center of the warhead, df= distance between the reference line and the geometric center of the fuselage, da = distance from the reference line to the geometrical center of the fins, dto = distance from the reference line to the geometrical center of the area exposed to the nozzle airflow,

Ao = exposed area of the warhead,

Af = exposed area of the fuselage,

Aa = exposed area of the fins,

Ato = exposed area of the nozzle, and nozzle, and At = sum of the areas of the mentioned components.

e) Center of Gravity: The center of gravity of a rocket is the point which all the weight of the rocket appears to be concentrated. That is, there is as much weight distributed in front of the rocket's center of gravity as there is behind it [13]. That is, it is the point where all the weight of the rocket is concentrated and all the aerodynamic forces exerted by all the components of the rocket with the mass corresponding to each of these are concentrated on its geometric center:

Warhead or Fuselage f

Recovery system sr Propulsion system sp Payload pay pl

Face insulator pay aispl

Determination of the CG distance on the rocket.

Dcg = (do * mo + df * mf + da * ma + dsr * mr + dpl * mpl + dsp * msp + daispl * maispl)/mt.

Source: NASA. Determining Center of Gravity-cg

f) Rocket stability: Rocket stability consists of maintaining the flight trajectory at all times. That is to say, that it does not lose the desired direction in the presence of an external disturbance; and to guarantee this, center of pressure must be considered (CP) remains below the center of gravity (CG). For your model rockets has safe and predictable flights, they must be stable. A model rocket will be stable only if its center of pressure is behind its center of gravity. Before flying any model rocket, you must make sure that it is stable. Obviously, this boils down to finding the locations of the rocket's center of gravity (C.G) and center of pressure (C.P) [13]. Then the rocket stability margin would be expressed by the distance from the center of gravity minus the distance from the center of pressure:

$$\text{ME} = \text{dCG}{-}{-}\text{dCP}$$

g) Parachute: The surface of the parachute must take into account the gravity, the chosen descent speed, the mass of the rocket without propellant and the drag coefficient (see Fig. 14).

To obtain the surface area of the parachute by the equation to determine (S), variables such as gravity, the chosen speed of descent, the mass of the rocket without propellant and the drag coefficient must be taken into account [18].

Standard gravity = 9.80665 m/s^2 Descent velocity = 7 m/s

Total mass of the rocket = 2000 g

Drag coefficient = 0.75 typical model rocket

The density of air at normal atmospheric pressure and at 15 °C, is 1,225 kg/m^3, although it decreases with humidity, temperature, and altitude, and therefore influences the kinetic energy of the wind (Fig. 15 and Table 4).

$$S = \frac{2 * g * m}{\rho * Cd * v^2}$$

$$s = \frac{2 * 9.81\frac{m}{s^2} * 2kg}{1.225\frac{kg}{m^3} * 0.75 * 7^2(\frac{m}{s})^2} = 0.87163$$

$$R = \frac{\sqrt{0.87163}}{3.141592} = 0.52673$$

$$\mathrm{D} = 0.52673 \times 2 = 1.05346$$

Fig. 14. Parachute diameter

Fig. 15. Rocket prototype with 1:5 scale rocket

2 Results

Table 4. Effects of variables and rocket stability.

N°	Warhead Type	Fin type	Number of fins	Rocket mass	Rocket length	Result
	TO	TA	NA	g	cm	y
1	A	C	3	2000	150	7
2	B	C	3	2000	150	4
3	A	D	3	2000	150	5
4	B	D	3	2000	150	3
5	A	C	4	2000	150	10
6	B	C	4	2000	150	9
7	A	D	4	2000	150	8
8	B	D	4	2000	150	4

2.1 Statistical Analysis for Response Variable

Using the data in Chart 1, the effects of the variables under study on the stability of the UTEA 1 rocket were determined.

Factorial regression: Results vs. warhead type; Fin type; Number of fins; Number of fins (Tables 5 and 6).

Table 5. Estimated effects and coefficients for rocket stability.

Finished	Effect	Coefficient	Effect of coefficient	Value T	Value p
Constant		6.250	*	*	*
Warhead type	−2.500	−1.250	*	*	*
Fin type	−2.500	−1.250	*	*	*
Number of fins	3.000	1.500	*	*	*
Warhead type * fin type	−0.500	−0.250	*	*	*
Warhead type * number of fins	0.000	0.000	*	*	*
Fin type * number of Fins	−1.000	−0.500	*	*	*
Warhead type * fin type * number of fins	−1.000	−0.500	*	*	*

Table 6. Model summary.

S	R- square	R-square (tight)	R-square (predetermined)
*	100.00%	*	*

Asterisks represent missing values that cannot be calculated because the model is saturated.

Analysis of variance for response variable was used. To determine the actual significance of the effects of the variables, analysis of variance was performed.

2.2 Mathematical Model

Once the results of Table 1 have been obtained, the mathematical model representing the aerodynamic flight process of the UTEA1 rocket will be determined, with a factorial design at two levels and three factors (Table 7).

Regression equation in uncoded units

$Y = 6.250 - 1.250\ A - 1.250\ B + 1.500\ C - 0.2500\ A * B + 0.000000\ A * C - 0.5000\ B * C - 0.5000\ A * B * C$

The mathematical model does not represent the regression equation to achieve a higher stable range of the experimental rocket UTEA 1 because the variables nor their interactions are significant. That is, they are not significant in achieving stable vertical range (Fig. 16).

Table 7. Estimated coefficients for stability.

Terms	Coefficient
Constant	6.250
Warhead type	−1.250
Fin type	−1.250
Number of fins	1.500
Warhead type * fin type	−0.2500
Warhead type * number of fins	0.000000
Fin type * number of fins	−0.5000
Warhead type * fin type * number of fins	−0.5000

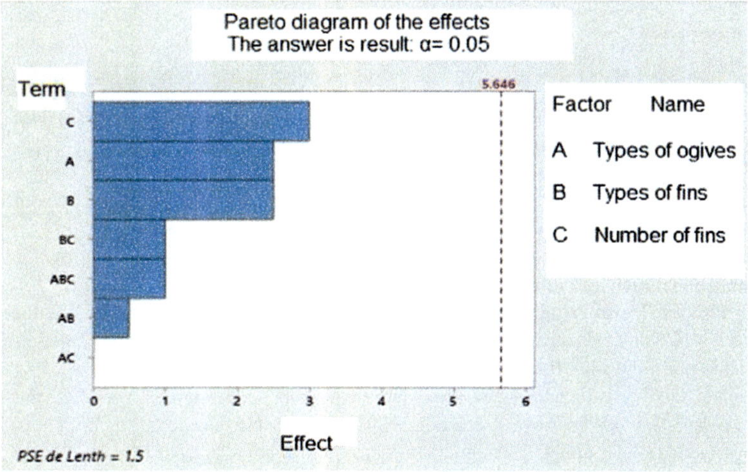

Fig. 16. Pareto chart

The tests performed with the factors and levels according to the experimental design in Chart 1, are not significant. That is, the interaction between the factors and levels are practically equal, therefore, there is not much difference in the aerodynamic stability between these factors according to the Pareto diagram, giving us a clear reference that none of the three factors nor their interactions are significant.

NOTE * The specified residual type cannot be plotted, because MSE = 0, the degrees of freedom for error = 0.

2.3 Conclusions

The stability and vertical range of the experimental rocket UTEA 1, does not strictly depend on the fins design, number of fins and shape of the fuselage warhead. The stability of the rocket is directly related between the center of pressure and the center of gravity of the rocket. The fins act as stabilizing components and their effectiveness depends on their size, shape and number of fins. The vertical range of the rocket flight depends on its power and the amount of propellant in the rocket motor. There was not much difference between the tests with the factors and levels of the experimental design and the prototype of the experimental rocket UTEA 1 with 4 trapezoidal type fins and blunt-tipped conical warhead cone has been designed for a test with a good performance.

References

1. Galindo Herrera, B.Á.C.A., Freddy A., Calderón Rojas, J.O.: Structural and aerodynamic design of a sounding rocket to be used of atmospheric variables acquisition. (2016). https://repository.unilibre.edu.co/bitstream/handle/10901/9471/Diseñocohetesonda%2CrevistaEGSIN.pdf?sequence=1&isAllowed=y
2. Correal, H., Castaño, M.A., Rincon, D.A., Aguillon, O.A.: Design and construction of an amateur rocket type G. 16
3. Giraldo, E.G.I., Parra, J.: Operational study of radiosonde procedures in Latin America. TECNOESUFA **23**(1), 36–42 (2015)
4. Jaramillo, J.C.D., Valencia, J., Torres, J., Gaviria, E., Quintero, C.: Rocket Design to Propulsion to a Target, Quindio's University, pp. 1–6 (2009)
5. Collave, Y.: Pedro Paulet, the Peruvian whose designs inspired the engineers of Apollo 11. https://elcomercio.pe/tecnologia/ciencias/pedro-paulet-peruano-cuyos-disenos-inspiraron-ingenieros-apolo-11-ecpm-noticia-657333-noticia/?ref=ecr
6. Arlazorov, M.S.: Konstantin Tsiolkovsky (2019). https://www.britannica.com/biography/Konstantin-Eduardovich-Tsiolkovsky
7. Gonzales, D.: Pedro Paulet, the Peruvian genius who was ahead of his time and founded the space ag. El Comer (2021). https://elcomercio.pe/eldominical/ciencia/pedro-paulet-el-genio-peruano-que-fundo-la-era-espacial-noticia/
8. Ordoñez, G.: Design, manufacture and experimentation of a jet engine for research purposes, no. 66 (2014). file:///E:/ProyectocoheteUTEA1/proyectocohete/ordoñez_cg.pdf
9. The American Economist: The legacy of Pedro Paulet, The Father of Modern Astronautics, 69 years after his death (2019). https://www.eleconomistaamerica.pe/sociedad-eAm-
10. Agency, A.: Wernher von Braun, the german force behind apollo 11, Her. https://www.elheraldo.co/ciencia/wernher-von-braun-la-fuerza-alemana-detras-del-apolo-11-651238
11. Paulet, S.M.: Pedro Paulet: Peruvian space pioneer. Sci. Technol. (5) (2002)
12. Ambía, E.: Edico looking for Sophia (2019). https://edico3004.blogspot.com/2017/10/sistemas-concepto-desistema.html
13. Barrowman, J.: Calculating the center of pressure. Techical informticon report 33 (1988)
14. Nakka, R.: Solid propellant rocket motor theory (1997). http://www.nakka-rocketry.net/
15. Montgomery, D.: Designs and analysis of experiments Iberoamerican (1991)
16. Cochran, G., Cox, W.: Disseños experimentale"s. Trillas (2001)
17. Fleeman, E.L.: Tactical Missile Design. American Institute of Aeronautics and Astronautics, Michigan (2001)
18. Riveros, L.A., Rodriguez, F.A.: Design and construction of an amateur rocket controlled by actuating a vector thrust nozzle (2010)

Comparison of Deep Convolutional Neural Networks and Histogram of Oriented Gradients Based Feature Extraction in Concrete Bridge Crack Images

Hajar Zoubir(✉), Mustapha Rguig, Mohammed Elaroussi, and Rachid Saadane

Centre des Etudes Doctorales de l'Ecole Hassania des Travaux Publics, Casablanca, Morocco
zoubir.hajar.cedoc@ehtp.ac.ma

Abstract. Cracks in reinforced concrete bridges adversely affect the durability and the safety of the structure. In an attempt to automate their recognition in bridge elements to minimize the risks associated with the conventional methods based on the subjective judgment of bridge inspectors, vision-based techniques (i.e. Image Processing, Machine Learning, and Convolutional Neural Networks) have been widely investigated to detect cracks in concrete images. This work compares the performance of the Histogram of Oriented Gradients (HOG) feature descriptor and a pretrained Deep Convolutional Neural Network (DCNN) in extracting features from a concrete crack dataset. The dataset contains over 13,000 uncracked and cracked images with different patterns of cracks and includes challenging concrete surface conditions that make feature extraction more complex. The extracted features are fed to a Support Vector Machine classifier to perform the classification task. The experimental results show that the DCNN based approach achieves better performance (98.31% accuracy, 99% precision, 97% recall, and 98% F1-Score) compared to the HOG-based method, demonstrating that features extracted by the pretrained deep network outperform the handcrafted features of the HOG descriptor.

Keywords: Bridge concrete cracks · Feature extraction · Histogram of oriented gradients · Deep convolutional neural network · Transfer learning · Support vector machine

1 Introduction

Reinforced concrete bridges are subject to several deterioration mechanisms leading to severe damages in their structural elements. Cracks are one of the most common defects encountered in these bridges and their presence is hazardous to the durability of the bridge and its structural safety. Hence, a timely and accurate detection of these defects is paramount to ensure functionality and safety requirements of the bridge.

Visual inspections are regularly carried out by trained inspectors to examine bridge elements and evaluate their physical condition based on inspection manuals and established standards and procedures [1]. Manual processing of bridge images captured while

M. Ben Ahmed et al. (Eds.): SCA 2022, LNNS 629, pp. 609–616, 2023.
https://doi.org/10.1007/978-3-031-26852-6_56

conducting visual inspection is tedious, error prone and relies on the experience of inspectors. To address these limitations, vision-based techniques have been investigated by researchers to automate crack recognition in concrete surface images. Considering cracks as edges that represent a sharp change in pixel intensity, edge detectors have been widely used to detect the pixels belonging to cracks in concrete surface images [2, 3]. Particularly, spatial edge detectors (e.g. Prewitt, Sobel, and Canny) are generally based on filters that compute the image gradients and intensify the high gradient areas that represent edges [2]. Hand crafted features like the Histogram of Oriented Gradients (i.e. HOG) [4] are used to detect cracks in concrete surfaces [5].

Although these techniques provide relevant features that are usually fed to Machine Learning classifiers (e.g. Support Vector Machine [6], Nearest Neighbor [7]) for image classification, their performance is limited by the complexity of concrete surface conditions (e.g. light, roughness and humidity) [8], requiring multiple necessary preprocessing and enhancement operations to perform the recognition task.

To overcome the mentioned limitations, Deep Convolutional Neural Networks (i.e. DCNNS) have been employed to automate crack recognition in concrete images. These feedforwrad models contain several convolutional layers that automatically learn features from a training dataset by optimizing a loss function through backpropagation. Dorafshan et al. [9] compared the performance of edge detectors and the AlexNet [10] model to classify cracks in the SDNET dataset [11]. Their study showed the superiority of the AlexNet model over the edge detectors. Xu et al. [12] constructed a crack dataset and proposed an end-to-end crack detection model based on a Convolutional Neural Network. Since these networks require massive datasets to be trained, Transfer Learning techniques have been employed to transfer knowledge from large source domains to specific target domains [13]. In the context of concrete crack recognition, these learning schemes employ pretrained state-of-the art networks on the large benchmark ImageNet [14] dataset to classify images of a small scale concrete crack dataset [8, 15, 16]. Two common settings are generally used: feature extracting where all the pretrained network parameters are frozen and features from the training set are extracted and fed to a classifier and fine tuning where parts of the network are retrained to update their parameters through backpropagation [17].

We compare in this work the performance of the pretrained network VGG16 [18] on the ImageNet Dataset and the HOG feature descriptor in extracting features from a concrete crack dataset. We feed the extracted features to a Support Vector Machine (i.e. SVM) classifier to perform the classification task. The evaluation of the two methods is conducted based on performance metrics defined later in the paper.

The rest of the paper is organized as follows: the Sect. 2 briefly explains the HOG descriptor and the architecture of the VGG16 model. The Sect. 3 presents the dataset studied in this paper and the experimental setup followed to recognize cracks in images using the two classification methods. Results are presented and discussed in Sect. 4 and conclusions are provided in Sect. 5.

2 Background

This section presents a brief explanation of the HOG features descriptor and the architecture of the VGG16 model used in this paper.

2.1 Histogram of Oriented Gradients

The Histogram of Oriented Gradients (i.e. HOG) [4] is a feature descriptor used to describe edge gradient distribution and orientation.

The vertical and horizontal gradients of an image are calculated by applying convolution filters that are commonly used to find edges in an image. The magnitude and the direction of the gradients are computed following Eqs. (1) and (2) respectively:

$$G = \sqrt{g_x^2 + g_y^2} \tag{1}$$

$$\theta = \arctan\frac{g_y}{g_x} \tag{2}$$

where:

g_x: represents the gradient in the x direction.

g_y: represents the gradient in the y direction.

The image is divided into small cells and the HOG of each cell is calculated. This histogram is split into nine bins corresponding to angles from 0–180° and shows the frequency distribution of the gradient orientations. The contribution of each pixel within the cell is weighted by its gradient magnitude. The cells are combined into blocks and the HOG features of the image are generated by combining all the resulting block features. Figure 1 presents the HOG features of four concrete images with cracks from the dataset constructed by the authors in [19].

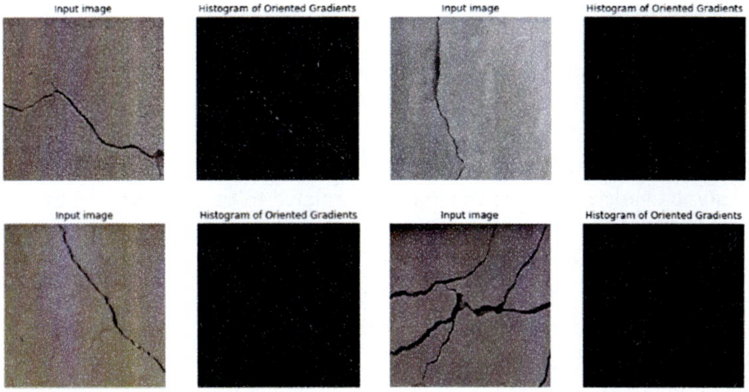

Fig. 1. HOG features of concrete crack images from [19].

2.2 VGG16 Network

VGG16 is a state-of-the art Deep Convolutional Neural Network proposed in 2014 by the Visual Geometry Group [18]. The model contains 13 convolutional layers to extract features, 5 max pooling layers to reduce the number of parameters to learn and

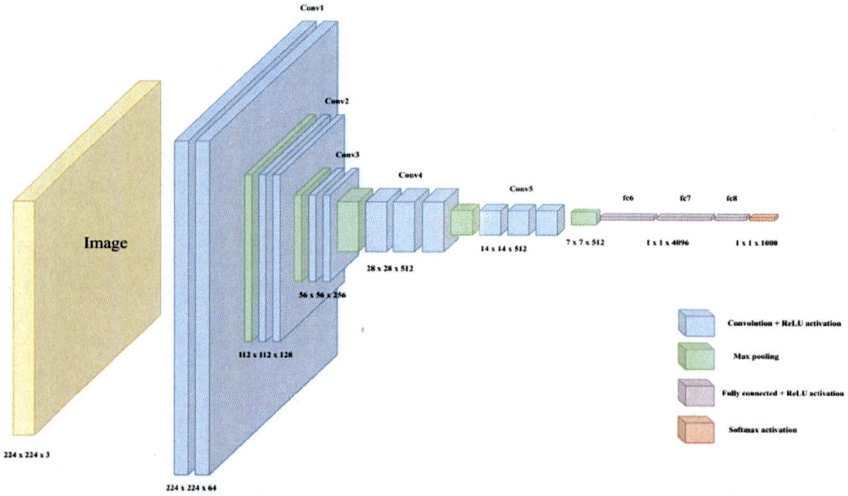

Fig. 2. Architecture of the VGG16 model.

3 fully connected layers that map the flattened features to the Softmax layer where class probabilities are calculated. The Rectified Linear Unit (ReLU) activation function is used to increase the nonlinearity of the model.

The architecture of the model is presented in Fig. 2.

3 Experimental Setup

The dataset studied in this paper was created by combining two concrete crack datasets: the dataset constructed by the authors in [19] and the dataset shared by Xu et al. in [12]. The resulting dataset contains 5360 and 7648 cracked and uncracked (background) images, respectively. These images include different representations of cracks in concrete and some challenging surface conditions related to the lighting and the presence of markings, stains and some minor defects (e.g. scaling). Figure 3 presents sample examples from the dataset.

The dataset is randomly split into training (70%) and testing (30%) sets.

The HOG parameters were set as follows: orientations = 9; pixels per cell = (8, 8) and cells per block = (3, 3).

The pretrained VGG16 used in this paper contains the weights of the imageNet dataset that consists of millions of images belonging to thousands of classes. The weights before the fully connected layers are maintained and used to extract features from the dataset studied in the present work.

The pretrained VGG16 and the HOG are used to extract the features from the training dataset. The extracted features are fed to an SVM classifier to perform the classification task on the test set. Figure 4 shows the two classification methods explored in this paper.

The performance of the two methods is evaluated based on the following metrics:

$$\text{Accuracy} = \frac{\text{TP} + \text{TN}}{\text{TP} + \text{TN} + \text{FP} + \text{FN}} \tag{3}$$

$$\text{Precision} = \frac{\text{TP}}{\text{TP} + \text{FP}} \tag{4}$$

$$\text{Recall} = \frac{\text{TP}}{\text{TP} + \text{FN}} \tag{5}$$

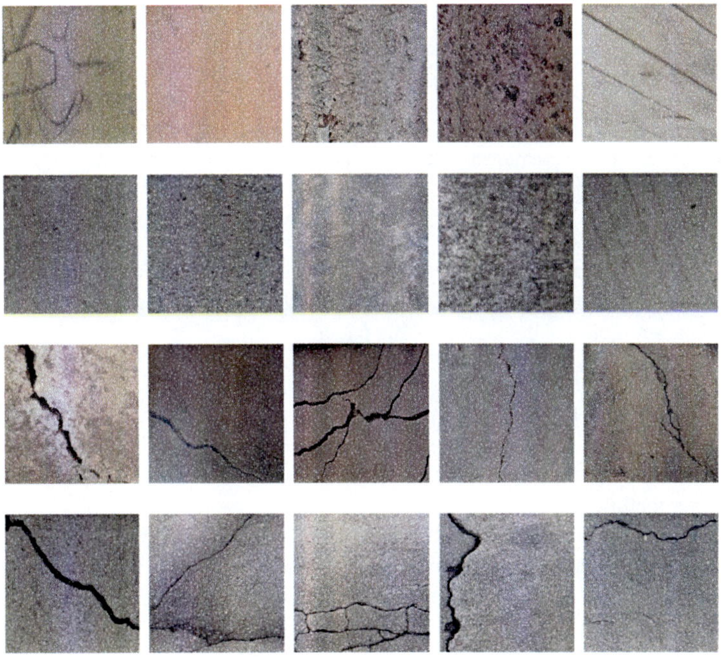

Fig. 3. Sample images of the studied dataset (row 1: uncracked images from [19]; row 2: uncracked images from [12]; row 3: cracked images from [19]; row 4: cracked images from [12])

$$\text{F1} - \text{Score} = \frac{2}{\frac{1}{\text{Precision}} + \frac{1}{\text{Recall}}} \tag{6}$$

where:

TP (True Positives): refer to the number of cracked images that are correctly classified as cracked;

TN (True Negatives): refer to the number of non-cracked images that are correctly classified as non-cracked;

FP (False Positives): refer to the number of non-cracked images that are incorrectly identified as cracked;

FN (False Negatives): refer to the number of cracked images that are incorrectly identified as non-cracked.

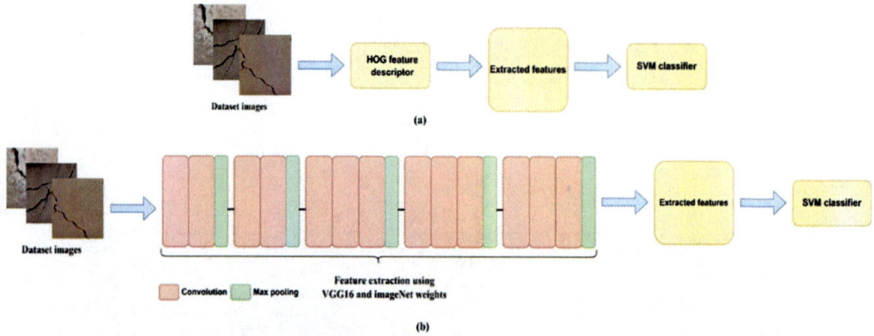

Fig. 4. Crack classification methods: (a) feature extraction with the HOG feature descriptor; feature extraction with the pretrained VGG16 model on the imageNet dataset (b).

All the experiments in this paper were conducted in Google Colaboratory with the 12 GB NVIDIA Tesla K80 GPU provided by the platform. The Scikit-learn library [20], the Scikit-image Python package [21] and the Pytorch library [22] are used to implement the classification methods discussed in this paper.

4 Results and Discussions

Table 1 displays the accuracy, precision, recall and F1-Score results of the two classification methods.

The DCNN based method achieved a high classification accuracy of 98.31% over 84.63% yielded by the HOG based method. Precision, recall and F1 Score results are also higher in the DCNN based approach and their values are comparable between the two classes (e.g. 99% and 98% F1-Scores for Background and Cracks respectively), which demonstrates that the features extracted by the pretrained network are robust to distinguish and recognize the two classes of the dataset. In the HOG based approach, besides the low Precision, Recall and F1-Score results achieved by the classification method, a notable difference can be seen between the two classes results (e.g. 87% and 81% F1-Scores for Background and Cracks respectively) meaning that the handcrafted features of the HOG descriptor are not capable of accurately separating and recognizing

Table 1. Accuracy, precision, recall and F1-Scores of the two classification methods

Classification method	Accuracy	Precision		Recall		F1 score	
		Background	Cracks	Background	Cracks	Background	Cracks
HOG + SVM	84.63%	86%	83%	89%	78%	87%	81%
Pretrained VGG16 + SVM	98.31%	98%	99%	99%	97%	99%	98%

crack and background images. This can be attributed to the challenging surface condition of concrete in a real world inspection covered in the studied dataset. Images that contain markings, minor defects and concrete joints are likely to be mistaken for cracks since they represent edges in the image and then are likely extracted as features by the HOG descriptor to represent cracks. Figure 5 shows examples of HOG features of background images with markings and concrete joints.

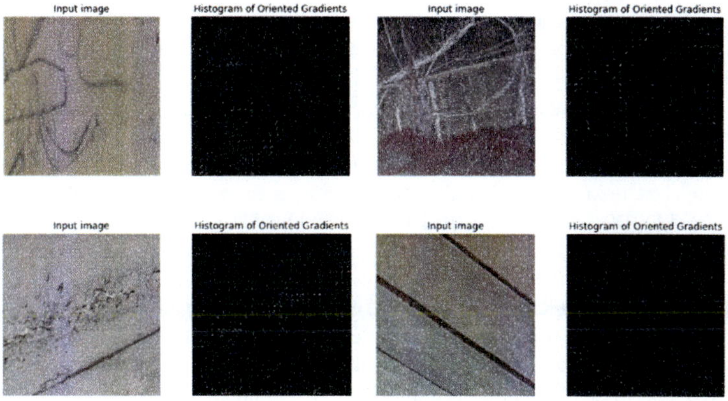

Fig. 5. Examples of HOG features of background images with markings (row 1) and concrete joints (row 2)

5 Conclusions

This work compared the performance of the Histogram of Oriented Gradients (HOG) feature descriptor and the pretrained VGG16 model on the imageNet dataset in extracting features from crack images with challenging and complex surface conditions. The extracted features were fed to a Support Vector Machine classifier to perform the classification task. The experimental results demonstrate that the DCNN based approached achieved high accuracy (98.31%), precision (99%), recall (97%) and F1 score (98%) outperforming the HOG based approach. These results showcase the superiority of the extracted features by the DCNN over the handcrafted features of the HOG descriptor in the context of a challenging dataset that covers the different representations of concrete surfaces in the bridge inspection real world.

References

1. Omar, T., Nehdi, M.: Condition assessment of reinforced concrete bridges: current practice and research challenges. Infrastructures **3**(3), 36 (2018). https://doi.org/10.3390/infrastructures3030036
2. Dorafshan, S., Thomas, R.J., Maguire, M.: "Benchmarking image processing algorithms for unmanned aerial system-assisted crack detection in concrete structures. Infrastructures **4**(2), 19 (2019). https://doi.org/10.3390/infrastructures4020019
3. Abdel-Qader, I., Abudayyeh, O., Kelly, M.E.: Analysis of edge-detection techniques for crack identification in bridges. J. Comput. Civ. Eng. **17**(4), 255–263 (2003). https://doi.org/10.1061/(ASCE)0887-3801(2003)17:4(255)

4. Dalal, N., Triggs, B.: Histograms of oriented gradients for human detection. In: 2005 IEEE Computer Society Conference on Computer Vision and Pattern Recognition (CVPR 2005), vol. 1, pp. 886–893, June 2005. https://doi.org/10.1109/CVPR.2005.177
5. Meng, L., Wang, Z., Fujikawa, Y., Oyanagi, S.: Detecting cracks on a concrete surface using histogram of oriented gradients. In: 2015 International Conference on Advanced Mechatronic Systems (ICAMechS), pp. 103–107, August 2015. https://doi.org/10.1109/ICAMechS.2015.7287137
6. Cortes, C., Vapnik, V.: Support-vector networks. Mach. Learn. **20**(3), 273–297 (1995). https://doi.org/10.1007/BF00994018
7. Breiman, L.: Random forests. Mach. Learn. **45**(1), 5–32 (2001). https://doi.org/10.1023/A:1010933404324
8. da Silva, W.R.L., de Lucena, D.S.: Concrete cracks detection based on deep learning image classification. In: Proceedings, vol. 2, no. 8 (2018). https://doi.org/10.3390/ICEM18-05387
9. Dorafshan, S., Thomas, R.J., Maguire, M.: Comparison of deep convolutional neural networks and edge detectors for image-based crack detection in concrete. Constr. Build. Mater. **186**, 1031–1045 (2018). https://doi.org/10.1016/j.conbuildmat.2018.08.011
10. Krizhevsky, A., Sutskever, I., Hinton, G.E.: ImageNet classification with deep convolutional neural networks, p. 9 (2012)
11. Dorafshan, S., Thomas, R.J., Maguire, M.: SDNET2018: an annotated image dataset for non-contact concrete crack detection using deep convolutional neural networks. Data Brief **21**, 1664–1668 (2018). https://doi.org/10.1016/j.dib.2018.11.015
12. Xu, H., Su, X., Wang, Y., Cai, H., Cui, K., Chen, X.: Automatic bridge crack detection using a convolutional neural network. Appl. Sci, **9**(14), 2867 (2019)
13. Zhuang, F., et al.: A comprehensive survey on transfer learning. Proc. IEEE **109**(1), 43–76 (2021). https://doi.org/10.1109/JPROC.2020.3004555
14. Deng, J., Dong, W., Socher, R., Li, L.J., Li, K., Fei-Fei, L.: ImageNet: a large-scale hierarchical image database. In: 2009 IEEE Conference on Computer Vision and Pattern Recognition, pp. 248–255, June 2009. https://doi.org/10.1109/CVPR.2009.5206848.
15. Su, C., Wang, W.: Concrete cracks detection using convolutional NeuralNetwork based on transfer learning. Math. Prob. Eng. **2020**, e7240129 (2020). https://doi.org/10.1155/2020/7240129
16. Yang, Q., Shi, W., Chen, J., Lin, W.: Deep convolution neural network-based transfer learning method for civil infrastructure crack detection. Autom. Constr. **116**, 103199 (2020). https://doi.org/10.1016/j.autcon.2020.103199
17. Gao, Y., Mosalam, K.M.: Deep transfer learning for image-based structural damage recognition: deep transfer learning for image-based structural damage recognition. Comput.-Aided Civil Infrastruct. Eng. **33**(9), 748–768 (2018). https://doi.org/10.1111/mice.12363
18. Simonyan, K., Zisserman, A.: Very deep convolutional networks for large-scale image recognition. arXiv:1409.1556 [cs] (2015). http://arxiv.org/abs/1409.1556. Accessed 16 Apr 2021
19. Zoubir, H., Rguig, M., Elaroussi, M.: Crack recognition automation in concrete bridges using Deep Convolutional Neural Networks. MATEC Web Conf. **349**, 03014 (2021). https://doi.org/10.1051/matecconf/202134903014
20. Pedregosa, F., et al.: Scikit-learn: machine learning in Python. J. Mach. Learn. Res. **12**(85), 2825–2830 (2011)
21. van der Walt, S., et al.: Scikit-image: image processing in Python. PeerJ **2**, e453 (2014). https://doi.org/10.7717/peerj.453
22. Paszke, A.., et al.: Automatic differentiation in PyTorch (2017). https://openreview.net/forum?id=BJJsrmfCZ. Accessed 13 Mar 2022

The Study of the Unsupervised Classification Method Using the K-means Algorithm by a Proposition of a Simple Initialization Technique

Rahma Ouchani(✉), Mohammed Merzougui, and M'barek Nasri

Applied Mathematics, Signal Processing and Computer Science Laboratory (MATSI), Higher Institute of Technology (ESTO), Mohammed First University (UMP), BV Mohammed VI B.P. 524, Oujda 60000, Morocco

rahmaouchani01@gmail.com

Abstract. Image segmentation is a fundamental step in image processing, The existing segmentation techniques are numerous, but they are generally grouped into three main approaches which are the contour approach, the region approach and the cooperative approach. The segmentation by region approach aims to group the pixels with the same characteristics into homogeneous regions. It is characterized by the measure of uniformity of the regions constructed into the image. Generally, we distinguish three families of algorithms for the region approach: classification methods, region growth methods, and methods that divide or merge regions according to the chosen criterion. Our study is based on the unsupervised classification method using the K-means algorithm. This is the most known and used clustering algorithm, because of its simplicity of implementation. We know that this algorithm still suffers from several drawbacks such as The number of classes must be fixed at the beginning, The result depends on initial draw or starting values of the centers of the classes, so a main problem in the initialization phase. For this we propose to improve this behind by a simple initialization technique based on the optimization of the distance between the objects inside each class to obtain a set of compact and clearly separated clusters.

Keywords: Image segmentation · Classification methods · K-means algorithm

1 Introduction

Image segmentation is one of the most important techniques of image processing, is a fundamental step in the field of image processing we can mention as examples, object recognition, computer vision, medical imaging, and image tracking and analysis, etc. Image segmentation is a low-level image processing task whose main objective is to partition the digital image into different segments or regions containing similar characteristics in terms of color, intensity or texture so that the image can be easily understood for analysis. There are various image segmentation techniques, are generally grouped into three main approaches which

M. Ben Ahmed et al. (Eds.): SCA 2022, LNNS 629, pp. 617–625, 2023.
https://doi.org/10.1007/978-3-031-26852-6_57

are the contour approach, the region approach and the cooperative approach. We have tried to propose a classification of these methods according to the following scheme:

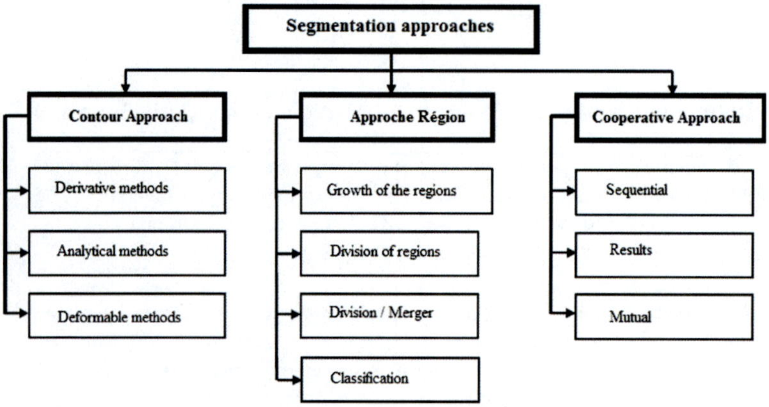

Fig. 1. Approaches to image segmentation.

1.1 Contour Approach

In the contour approach, we consider that the primitives to be extracted are the contrast lines separating regions of different and relatively homogeneous gray levels, or regions of different texture. In practice, it is a question of recognizing the transition zones and locating the boundary between the regions as well as possible. There are several methods such as derivative methods, analytical methods and deformable methods.

1.2 Region Approach

Image segmentation into homogeneous regions is based on the intrinsic properties of the regions. The choice of these properties determines what is called segmentation criteria. These criteria can be the gray level value [1], the color [2], the texture [3], or a combination of several information [4]. From an algorithmic point of view, segmentation consists in assigning to each pixel of the image a label of belonging to a given region. It is defined mathematically by Zucker in the following way [Zucker, 1976] : Segmenting an image I into N regions is equivalent to partitioning it into N subsets R_1, R_2.........R_n such that:

1. I = $\cup_i$ Ri
2. R_i consists of ∀i related pixels.
3. P(R_i) = true i.
4. P($R_i \cup R_j$) = false for all i, j , with R_i and R_j adjacent in I.

The first condition indicates that each pixel of the image must belong to a region R_i and the union of all regions forms the whole image. The second condition is related to the structure of the regions, it defines a region as a set of pixels that must be related. The third condition expresses that each region must respect a uniformity predicate. The last condition implies that the non-fulfillment of this same predicate for the meeting of two adjacent regions. We distinguish four types of methods, Growth of regions [5], Division of regions, Division/Merge and Segmentation by classification. The method used in this article is the Classification Segmentation:

This method consists in grouping and classifying the pixels of an image into classes according to their properties. Each point in the image is associated with a vector of attributes. The classification is then performed on these attribute vectors in order to achieve a limited number of homogeneous regions within the image. It is a procedure in which similar pixels of an image are identified and grouped into a single class. There are two main types of classification:

• Supervised Classification: It is based on the knowledge of each class defined by a probabilistic approach. It is based on the learning of discriminating properties on a sample of already classified data.

• Unsupervised Classification: Automatic separation of the image into clusters without any a priori knowledge of the classes. It is based on a measure of distance between the attribute vectors. The most common algorithms for this category are K-means, Isodata, and Fuzzy c-means.

1.3 Cooperative Approach

Cooperative region-contour segmentation combines the two segmentation techniques, region-based segmentation and contour-based segmentation, taking advantage of the complementary nature of the information on these two approaches. Thus, segmentation by region-contour cooperation can be expressed as a mutual aid between these two concepts are used to improve the final segmentation result.

2 Related Work

Many works have been done in the field of image segmentation using different methods, such as threshold value [6] based segmentation method, edge based segmentation method and region based segmentation method (classification method). The K-means algorithm is one of the simplest and most effective clustering algorithms, many researches implemented to improve the initialization of the center. Some of the existing recent works that show better results are cited.

- Simon Tongbram and all proposed a new method uses an improved subtractive algorithm [7] which is based on the distance relationship between data points and cluster center and provides a more accurate center than the conventional subtractive clustering method. and the centroids of the modified SC algorithm

are used in the k-means algorithm to segment the image. experimental results validate a good performance of the method. - Alan Jose and all proposed a new method to generate the center of the cluster by reducing the mean square error of the final cluster with very little increase in execution time.
- Dhanachandra Nameirakpam and Khumanthem Manglem Singh proposed to segment the image using k-clustering algorithm, using subtractive cluster [8] to generate the initial centroid value. At the same time, partial contrast stretching is used to improve the quality of the original image and median filter is used to improve the segmented image. The proposed clustering algorithm has better segmentation.
- K. A. Abdul Nazeerand all proposed the k-means algorithm which combines a systematic method consisting of two approaches. The first approach is to find the initial centroid and another to assign the data point to the clusters. The proposed algorithm reduced the time complexity without sacrificing the cluster accuracy.

In this paper, we propose an image segmentation algorithm based on k-means algorithm with a simple centroıdes initialization technique, the experimental results validate the superiority of the proposed method and its performance over the methods.

3 K-means Clustering

Clustering method is one of the unsupervised learning methods in which a set of essential features is separated into uniform groups; K-means clustering technique is a widely used approach that has been applied to solve large-scale image segmentation tasks. There are different types of clustering: hierarchical clustering, Fuzzy C-means clustering, K-means clustering. The K-means method is one of the most commonly used clustering techniques for various applications [9]. The K-means clustering algorithm [10, 11] is widely used due to its simple process and fast computation. The image is divided into clusters, i.e., "k" clusters of data. The centroid of a cluster is represented by a point such that the sum of the distances between all other data points in the cluster and this point is the smallest. The choice of initial cluster centers is very important because it prevents the clustering algorithm from producing incorrect decisions. The most common initialization procedure chooses the initial cluster centers randomly from the input data [12]. K-means algorithm is an algorithm based on the initial centroids of clusters, it is used to separate similar data into clusters.

The operation of the algorithm is:

Chooses k data values as initial cluster centers, then finds the distance between each cluster center and each data value and assigns it to the nearest cluster, updates the averages of each cluster, repeat this process until the criterion does not match. K-means clustering aims to divide the data into k clusters in which each data value belongs to the cluster with the closest mean [13].

1. Initialize number of cluster k and centre.
2. For each pixel of an image, calculate the Euclidean distance d, between the center and each pixel of an image using the relation given below.

$$d = \|p(x, y) - c_k\| \tag{1}$$

3. Assign all the pixels to the nearest centre based on distance d.
4. After all pixels have been assigned, recalculate new position of the centre using the relation given below.

$$c_k = \frac{1}{k} \sum_{y \in c_k} \sum_{x \in c_k} p(x, y) \tag{2}$$

5. Repeat the process until it satisfies the tolerance or error value.
6. Reshape the cluster pixels into image.

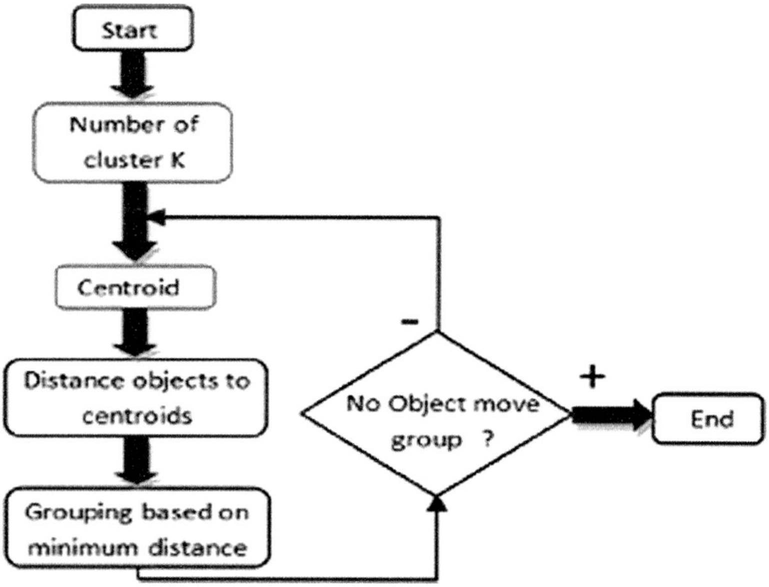

Fig. 2. K-means algorithm process

Advantages of the Algorithm:

- The k-means algorithm is very popular due to the fact that it is very easy to understand and implement.

- Its conceptual simplicity and speed
- Applicable to large data, and also to any type of data (even textual), by choosing a good notion of distance.

Disadvantages of the Algorithm:

- The number of classes must be fixed at the beginning,
- The result depends on the initial drawing of the class centers,
- The clusters are built in relation to non-existent objects (the middles).

4 Proposed Algorithm

However, the main limitation of this method is the dependence of the results on the values (initial centers). To each initialization corresponds a different solution (local optimum) which can in some cases be very far from the optimal solution (global optimum). A naive solution to this problem consists in launching the algorithm several times with different initialization and retain the best grouping found. The use of this solution remains limited because of its cost and because a better partition can be found in a single run.

In this paper we present a simple initialization technique of the k-means with the aim of maximizing the separability and compactness of the groups.This technique consists in determining according to the value of k given a spectrum of image color starts from the value of the brightest pixel to the darkest in order to build a set of groups K of color well determined to choose the initialization centroïdes, not in a totally random way but more exactly, in order to reduce the number of iterations and the numerous repetitions of the algorithm launch.

1. Initialize number of cluster k
2. creation of the distance matrix of the color spectrum.
3. choice of the centroïdes from the created spectrum (k groups)
4. For each pixel of an image, calculate the Euclidean distance d, between the center and each pixel of an image using the relation given below.

$$d = \|p(x, y) - c_k\| \tag{3}$$

5. Assign all the pixels to the nearest centre based on distance d.
6. After all pixels have been assigned, recalculate new position of the centre using the relation given below.

$$c_k = \frac{1}{k} \sum_{y \in c_k} \sum_{x \in c_k} p(x, y) \tag{4}$$

7. Repeat the process until it satisfies the tolerance or error value.
8. Reshape the cluster pixels into image.

5 Experimental Results

We segmented the images using k-means clustering algorithm, with the new technique to generate the initial centroid. We can conclude that the proposed clustering algorithm has better segmentation. In this method the number of distance calculations as well as the time reduced and the accuracy also improved. If the number of pixels in the image is increased, the execution time also increases.

The results that are obtained by using k-means clustering shown in below figures.

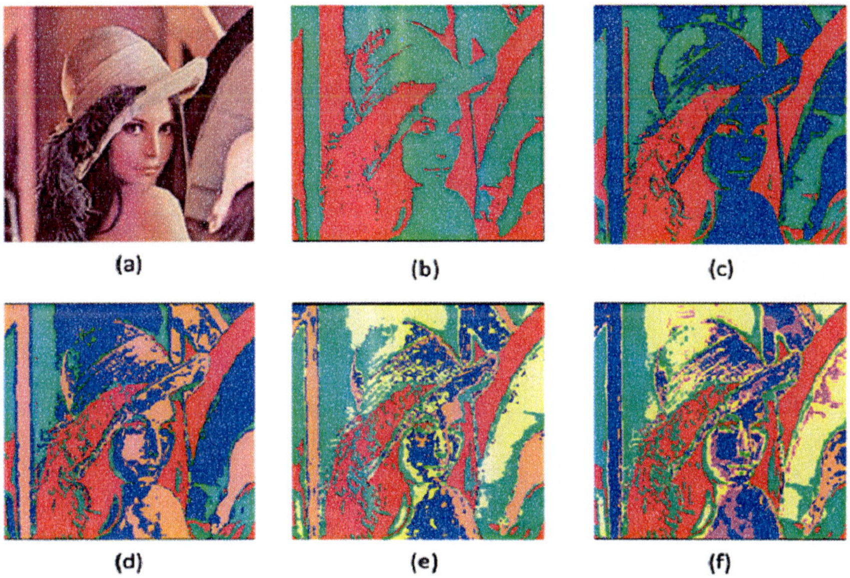

Fig. 3. Image segmentation results of real images: (a) original image, (b) k = 2, (c) k = 3, (d): k = 4, (e): k = 5, (f): k = 6.

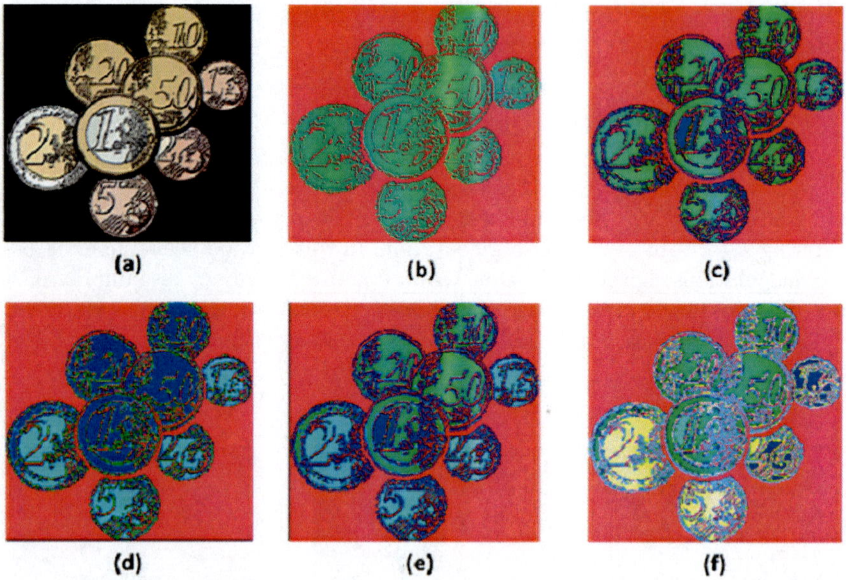

Fig. 4. Image segmentation results of real images: (a) original image, (b) k = 2, (c) k = 3, (d): k = 4, (e): k = 5, (f): k = 6.

6 Conclusion

This paper proposes a new image segmentation method based on k-means clustering methods using a new initialization technique. This technique consists in determining according to the value of k given a spectrum of image color starts from the value of the brightest pixel to the darkest in order to build a set of groups K of color well determined to choose the initialization centroïdes.We can conclude that we have successfully implemented the K-means clustering by this new technique which gives good results.

References

1. Sezgin, M., Sankur, B.: Survey over image thresholding techniques and quantitative performance evaluation. J. Electron. Imag. **13**(1), 146–165 (2004)
2. Lucchese L., Mitra, S.K.: Color image segmentation: A state-of-the-art survey. In: Proceedings of the Indian National Science Academy (INSA-A), Vol. 67-A(2), pp. 207–221 (2001)
3. Materka, A., Strzelecki, M.: Texture Analysis Methods - A review. Technical University of Lodz, Institute of Electronics, Bruxelles (1998)
4. Tremeau, A., Borel, N.: A region growing and merging algorith to color segmentation. Pattern Recogn. **30**(7), 1191–1204 (1997)

5. Li, C., Kao, C.Y., Gore, J.C., Ding, Z.: Minimization of region-scalable fitting energy for image segmentation. IEEE Trans. Image Process. **17**(10), 1940–1949 (2008)
6. Alamr, S.S., Kalyankar, N.V., Khamitkar, S.D.: Image segmentation by using threshold techniques. Comput. Sci. **2**(5), 83–86 (2010)
7. Tian, L., Han, L., Yue, J.: Research on Image Segmentation based on Clustering Algorithm. In: Int. J. Signal Process. Image Process. Pattern Recogn. **9**, 1–12 (2016)
8. Tongbram, S., Shimray, B.A., Singh, L.S.: Segmentation of image based on k-means and modified subtractive clustering. Indonesian J. Electr. Eng. Computer Sci., **22**(3), 1396 1403 ISSN: 2502–4752 (2021). https://doi.org/10.11591/ijeecs.v22.i3.pp1396-1403
9. Dhanachandra, N., Manglem, K., Jina Chanu, Y.: Procedia Computer Science - December (2015). https://doi.org/10.1016/j.procs.2015.06.090
10. Celebi, M.E., Kingravi, H.A., Vela, P.A.: A comparative study of efficient initialization methods for the k-means clustering algorithm. Expert Syst. Appl. **40**(1), 200–210 (2013). https://doi.org/10.1016/j.eswa.2012.07.021
11. Alhawarat, M., Hegazi, M.: Revisiting k-means and topic modeling, a comparison study to cluster arabic documents. IEEE Access **6**, 42740–42749 (2018). https://doi.org/10.1109/ACCESS.2018.2852648
12. Panda, S.: Color Image Segmentation Using K-means Clustering and Thresholding Technique. In: IJESC (2015)
13. Shinde, S., Tidke, B.: Improved K-means Algorithm for Searching Research Papers. Int. J. Comput. Sci. Commun. Netw. **4**, 197–202 (2014)

Interest of Remote Sensing and GIS in the Study of Water Erosion: Case of the Zat Watershed (High Atlas, Morocco)

Jihad Bouaida(✉), Omar Witam, and Monsif Ibnoussina

Geosciences Semlalia Laboratory, Geology Department, Faculty of Sciences Semlalia, Cadi Ayyad University, Marrakech, Morocco
jihad.bouaida@ced.uca.ma

Abstract. The Zat watershed is a left bank tributary of the Tensift watershed. It is located to the south-east of the Marrakech city. This basin presents all the factors that can generate the risk of water erosion, namely, the accidental topography, the limited vegetation cover, as well as a lithology favourable to erosion.

The aim of this study is to analyse the main factors that can lead to water erosion and to identify vulnerable areas in the Zat watershed. The study is carried out by applying the revised universal soil loss equation. The introduction of geographic information systems and remote sensing is essential to obtain significant results.

The results obtained from the RUSLE model show that the climatic aggressiveness, the topographic factor, the erodibility and the vegetation cover factor, present significant values which means the strong presence of erosion at the level of the Zat watershed. Indeed, the average value obtained for soil losses is 347 t/ha/year, with a standard deviation of 341 t/ha/year.

Keywords: RUSLE · Water erosion · Watershed · Zat · GIS · Remote sensing

1 Introduction

Water erosion is a challenge that poses serious socio-economic and environmental problems in many regions both nationally and internationally. It is the main cause of soil degradation in Morocco [1–3].

Characterised as a global phenomenon [4], water erosion is the interaction of several factors, both natural and anthropogenic, such as climate change, population increase, lack of vegetation cover, deforestation or overgrazing, lithological nature of the soil, etc. [4–8].

Water erosion is a major factor of soil degradation in Morocco. About 40% of Moroccan land is affected by erosion. The consequences of this phenomenon are very costly for the Moroccan economy [8–10].

The Universal Soil Loss Equation (RUSLE), provides a simple and comprehensive framework for the evaluation of the factors generating soil erosion. Rainfall rate (R),

M. Ben Ahmed et al. (Eds.): SCA 2022, LNNS 629, pp. 626–632, 2023.
https://doi.org/10.1007/978-3-031-26852-6_58

topography (LS), erodibility (K), vegetation cover (C) and erosion control practices (P) are the main factors evaluated under the empirical RUSLE model [11–15].

The main objective of this study is the qualitative and quantitative estimation of water erosion in the Zat watershed in order to map soil losses in the watershed.

2 Materials and Methods

2.1 Study Area

The watershed of Zat is part of the northern slope of the High Atlas of Marrakech, it belongs to the hydraulic system of the Tensift watershed. The Zat wadi originates at an altitude of 3920 m and has a surface area of 577 km^2 and a perimeter of 167 km. It is bounded by the Haouz plain to the North, the axial zone to the South, the R'Dat watershed to the East and the Ourika watershed to the West [16].

The Zat watershed is characterised by a dense and complex hydrographic network and an elongated shape that has a direct impact on the flow of the basin. The upstream part of the basin extends over the axial zone of the High Atlas chain, characterised by the outcrop of magmatic and metamorphic rocks of the terminal Precambrian, while the downstream part of the basin extends over the northern sub-atlasic zone where the post-Hercynian cover outcrops [17, 18] (Fig. 1).

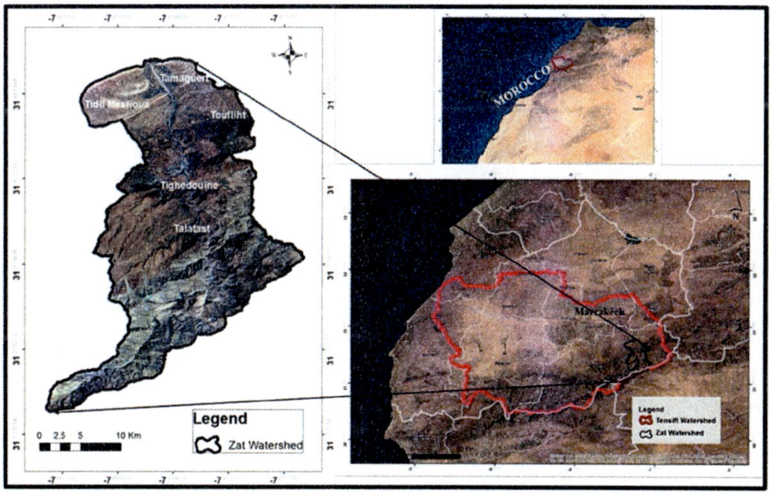

Fig. 1. Geographical situation of the Zat basin

2.2 Methodology

The Revised Universal Soil Loss Equation is adequate to quantify soil loss due to water erosion. It is a product function of five factors controlling water erosion; climatic aggressiveness, soil erodibility, slope steepness and length, vegetation cover and erosion control measures [13–19].

$$A = R.K.LS.C.P$$

A is the annual soil loss rate in t/ha/year.

Climatic Aggressiveness (R)
The equation of Wischmeier and Smith (1978), shows that the estimation of the R-factor is based on the knowledge of the kinetic energy and the average intensity of each storm drop over a period of 30 min. Other authors [20–22] have developed equations involving monthly and annual rainfall to determine the R-factor:

$$Log\, R = 1.74 * Log\, \Sigma (Pi^2/P) + 1.29$$

where Pi is monthly precipitation and P is annual precipitation.

This factor is calculated using data from different rainfall stations in and around the Zat watershed. It was applied to a twenty (20) year rainfall series. The database used was retrieved from the Tensift, Marrakech water basin agency (ABHT).

Soil Erodibility (K)
Expressed in t.h/ha, the soil erodibility factor is used to determine the resistance of all soil types to erosion based on their texture, presence of organic matter, permeability and depth [23].

$$K = [2.1 * 10 - 4 * (12 \; - \; MO)\, M \; * 1.14 \; + 3.25 * (S - 2) \; + 2.5 * (P - 3)]/100$$

where, M is (% fine sand + % silt) (100 - % clay); MO is the percentage of organic matter; S is the structure code; P is the permeability class

The K-factor is estimated at the level of the Zat watershed, using the lithological map, field missions and a bibliographic synthesis of further studies, in order to know the friability rate of the substrate of this watershed.

Topographical Factor (LS)
Topographic factors represent the effect of topography on erosion [24]. The length and inclination of slopes contribute to the generation and transport of sediments in a given area. The factor is calculated from a digital terrain model (DTM-30 m) by overlaying maps of slope length (L in m) and slope (S in %).

$$LS = Power\left(\left[flow\, acc\right] * resolution/22.1,\; 0.4\right) * Power\left(Sin\left(\left[Slope\right] * 0.01745\right)/0.09,\; 1.4\right) * 1.4$$

Land Cover Factor (C)
The C factor expresses the influence of different land use methods on the erosion rate, mainly related to the vegetation cover rate. It varies with time and nature. This factor shows the positive effect of vegetation cover on the stability of soil particles [25]. The values of the C-factor vary from 0.001 for well-covered soils to 1 for bare soils. These

values were determined using high-resolution Landsat 8 OLI satellite images. Image processing was used with ArcGIS software using supervised classification [26].

Anti-erosion Practices Factor

The anti-erosion practices factor presents the effect of practices to change the profile, slope, direction of runoff flow and reduce erosion [27]. It is determined from the land use map and the P-factor classification table [13] (Fig. 2).

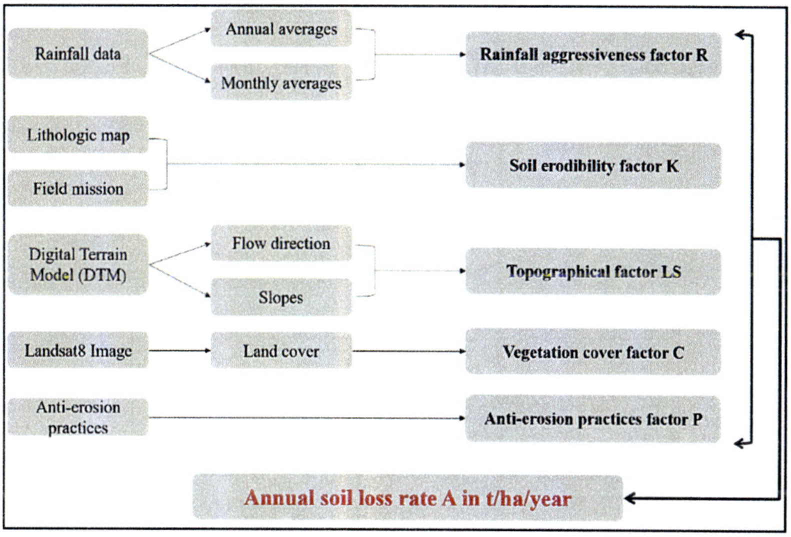

Fig. 2. Organizational chart of the methodology

3 Results and Discussion

In order to estimate the rate of soil loss in the Zat watershed, the empirical model of the Universal Soil Loss Equation was used. The introduction of Geographic Information Systems and remote sensing was essential for this study.

Annual Soil Erosion Rate

After the analysis and study of all the factors of the RUSLE model, we present the main results obtained.

The values of climatic erosivity vary between 50 MJ.mm/ha.h.yr downstream, and 98 MJ.mm/ha.h.yr upstream of the watershed. In comparison with other watersheds, the Zat watershed has lower erosivity values than the Ourika watershed [8], and the Oum Er-Rbia watershed [15]. However, these values are higher than those obtained in the N'Fis watershed [24].

Soil erodibility in the Zat varies between a value of 0.1 for resistant geological formations and 0.5 t. ha.h/ha.MJ.mm for friable formations. Overall, land vulnerable to water erosion has low values of the erodibility factor [28]. The topographic factor has high values in the upstream part of the basin, while the downstream part is characterised by values below 5.

For the vegetation cover factor, values between 0.05 and 0.4 correspond to land that is more or less protected by vegetation, which accounts for only 20% of the total watershed area. The value 1 refers to bare ground. The lack or absence of vegetation cover increases the vulnerability of the catchment to water erosion.

The values of the anti-erosion practice factor depend on the slopes of the area. These values vary between 0 and 1, with the maximum value assigned to land where no erosion control practices are used.

The different factors of the RUSLE model (R, K, LS, C, P) were combined in a GIS environment to generate the soil loss map of the Zat watershed. The resulting map shows the spatial distribution of the erosion rate in the catchment. The rate varies between areas of the catchment depending on the influence of various factors in the RUSLE model. As a result, the areas vulnerable to erosion are concentrated in the upper part of the basin. The average soil loss is 347 t/ha/year, and the standard deviation is 341 t/ha/year. About 65% of the total area of the Zat watershed is influenced by water erosion (Fig. 3).

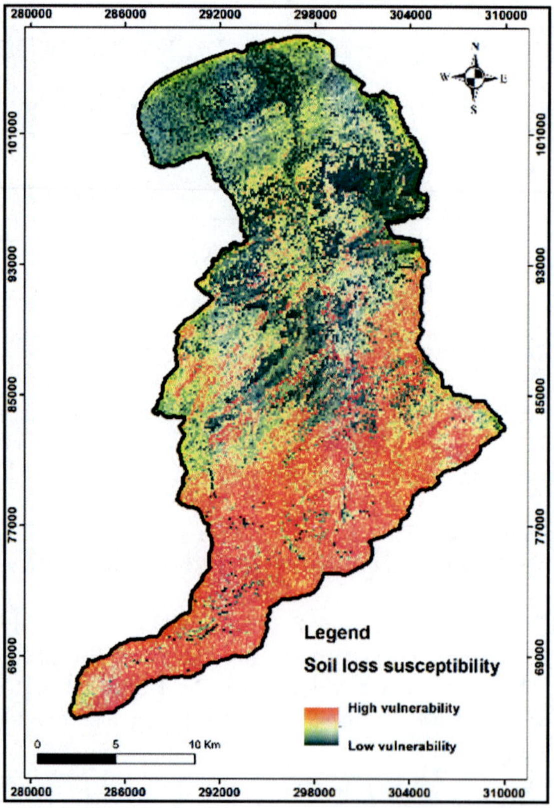

Fig. 3. Soil loss susceptibility map of Zat watershed

4 Conclusion

The estimation of the erosion phenomenon in the Zat watershed was carried out with the help of the Universal Soil Loss Model, which introduces the use of geographic information systems. The combination of all RUSLE factors in the context of GIS resulted in a soil loss map of the catchment. The basin has a high vulnerability to erosion in its upstream part, as the upstream part of the basin is specified by high altitudes, rugged topography, limited vegetation cover and lack of erosion control practices.

The soil loss map of the Zat watershed can be considered as a tool to support decision making in development and management of erosion control practices.

Acknowledgment. The authors would like to thank the Department of Geology of the Faculty of Sciences of Semlalia and the Geosciences Semlalia Laboratory for facilitating the field missions and providing favourable conditions for scientific research. We also thank the Tensift Hydraulic Basin Agency (ABHT) for their welcome and for providing us with rainfall data.

References

1. Mostephaoui, T., Merdas, S., Sakaa, B., Hanafi, M.T.: Cartographie des risques d'érosion hydrique par l'application de l'équation universelle de pertes en sol à l'aide d'un Système d'Information Géographique dans le bassin versant d'El Hamel (Boussaâda). Journal Algérien des Régions Arides **11** (2013)
2. Cheggour, A.: Mesures de l'érosion hydrique à différentes échelles spatiales dans un bassin versant montagneux semi-aride et spatialisation par des S.I.G. Application au bassin versant de la Rhéraya, Haut Atlas, Maroc. Marrakech (2008)
3. Ait Mlouk, M., Algouti, A.B., Algouti, A.H., Ourhzif, Z.: Assessment of river bank erosion in semi-arid climate regions using remote sensing and GIS data: a case study of Rdat River, Marrakech, Morocco, Estudios Geologicos **74**(2) (2018). https://doi.org/10.3989/egeol.43217.493
4. Meliho, M., Khattabi, A., Mhammdi, N., Hongming, Z.: Cartographie Des Risques De L'erosion Hydrique Par L'equation Universelle Revisee Des Pertes En Sols, La Teledetection Et Les Sig Dans Le Bassin Versant De L'ourika (Haut Atlas, Maroc). Eur. Sci. J. ESJ **12**(32), 277 (2016). https://doi.org/10.19044/esj.2016.v12n32p277
5. le Bissonnais, Y., Thorette, J., Bardet, C., Daroussin, J.: L'érosion hydrique des sols en France. Soil erosion View project State of French soils View project. Rapport INRA, IFEN **106** (2002). https://www.researchgate.net/publication/266884294
6. Das, B., Paul, A., Bordoloi, R., Tripathi, O.P., Pandey, P.K.: Soil erosion risk assessment of hilly terrain through integrated approach of RUSLE and geospatial technology: a case study of Tirap District, Arunachal Pradesh. Model. Earth Syst. Environ. **4**(1), 373–381 (2018). https://doi.org/10.1007/s40808-018-0435-z
7. Ganasri, B.P., Ramesh, H.: Assessment of soil erosion by RUSLE model using remote sensing and GIS - A case study of Nethravathi Basin. Geosci. Front. **7**(6), 953–961 (2016). https://doi.org/10.1016/j.gsf.2015.10.007
8. Ayt Ougougdal, H., Khebiza, M.Y., Messouli, M., Bounoua, L., Karmaoui, A.: Delineation of vulnerable areas to water erosion in a mountain region using SDR-InVEST model: a case study of the Ourika watershed, Morocco. Sci Afr **10** (2020). https://doi.org/10.1016/j.sciaf.2020.e00646

9. el Hafid, D., Akdim, B.: Quantification De L'erosion Hydrique En Utilisant Le Modèle Rusle Et Déposition Intégrée Dans Un Sig. Cas Du Bassin Versant De L'oued Isly (Maroc Oriental). Eur. Sci. J. ESJ **14**(5), 373 (2018). https://doi.org/10.19044/esj.2018.v14n5p373
10. Markhi, A., Laftouhi, N., Grusson, Y., Soulaimani, A.: Assessment of potential soil erosion and sediment yield in the semi-arid N′fis basin (High Atlas, Morocco) using the SWAT model. Acta Geophys. **67**(1), 263–272 (2019). https://doi.org/10.1007/s11600-019-00251-z
11. Phinzi, K., Ngetar, N.S.: The assessment of water-borne erosion at catchment level using GIS-based RUSLE and remote sensing: A review. Int. Soil Water Conserv. Res. **7**(1), 27–46 (2019). https://doi.org/10.1016/j.iswcr.2018.12.002
12. Mohammed, S., et al.: Estimation of soil erosion risk in southern part of Syria by using RUSLE integrating geo informatics approach. Remote Sens. Appl.: Soc. Environ. **20** (2020). https://doi.org/10.1016/j.rsase.2020.100375
13. Wischmeier, W., Smith, D.: Predicting Rainfall Erosion Losses a Guide to Conservation Planning. Department of Agriculture, Science and Education Administration, vol. 537, p. 67 (1978)
14. Sadiki, A., Bouhlassa, S., Auajjar, J., Faleh, A., Macaire, J.-J.: Utilisation d'un SIG pour l'évaluation et la cartographie des risques d'érosion par l'Equation universelle des pertes en sol dans le Rif oriental (Maroc): cas du bassin versant de l'oued Boussouab. Bulletin de l'Institut Scientifique, Rabat, section Sciences de la Terre **26**, 69–79 (2004)
15. Yjjou, M., Bouabid, R., El Hmaidi, A., Essahlaoui, A., El Abassi, M.: Modélisation de l'érosion hydrique via les SIG et l'équation universelle des pertes en sol au niveau du bassin versant de l'Oum Er-Rbia. Int. J. Eng. Sci. (IJES)**3**(8), 83–91 (2014)
16. Bouaida, J., Witam, O., Ibnoussina, M., Delmaki, A.E.F., Benkirane, M.: Contribution of remote sensing and GIS to analysis of the risk of flooding in the Zat basin (High Atlas-Morocco). Nat. Hazards **108**(2), 1835–1851 (2021). https://doi.org/10.1007/s11069-021-04758-x
17. Riad, S., Mania, J., Bouchaou, L.: Variabilité hydroclimatique dans les bassins-versants du Haut Atlas de Marrakech (Maroc). Sécheresse (Montrouge) **17**(3), 443–446 (2006). https://doi.org/10.1684/sec.2006.0005
18. el Arabi, E.H., Diez, J.B., Broutin, J., Essamoud, R.: First palynological characterization of the Middle Triassic, implications for the first Tethysian rifting phase in Morocco. Comptes Rendus - Geoscience **338**(9), 641–649 (2006). https://doi.org/10.1016/j.crte.2006.04.001
19. Smith, D.D., Wischmeier, W.H.: Factors affecting sheet and rill erosion. Trans. Am. Geophys. Union **38**(6), 889–896 (1957)
20. Kalman, R.: Le facteur climatique de l'érosion dans le bassin de Sebou. Projet Sebou, Rapp. inédit (1967)
21. Arnoldus, H.M.J.: Methodologie used to determine the maximum potentiel average soil loss due to sheet and rill erosion in Morrocco. Bulletin F.A.O, Chichester (1980)

Accelerating NLP for Technologically Underserved Languages: A Corpus for Moroccan Dialect

Hajar Zaidani[1](✉), Maryeme Zaim[1], Abderrahim Maizat[1], Mohammed Ouzzif[1], and Charif Mahmoudi[2]

[1] RITM Laboratory ESTC, CED ENSEM, Hassan II University Casablanca, Casablanca, Morocco
zaidanihajar@gmail.com, maryeme.zaim@ensem.ac.ma

[2] Laboratoire d'Algorithmique Complexité et Logique France, Creteil, France
charif.mahmoudi@lacl.fr

Abstract. This paper describes our approach to define a dataset to accelerate the development of Natural Language Processing (NLP) techniques for dialects spoken by technologically underserved populations. We designed and implemented a multimodal solution to collect data, mapping and annotating both the textual representation and the voice recording of utterances targeting specific samples of the native speakers. Our project is inspired from the Moroccan strategy, implemented through Law 55.19, which aims to simplify administrative procedures and formalities in order to strengthen the degree of trust between the public administration and users (citizens).

Moreover, this paper describes the format that we defined for these datasets and the results of our experimentation with the Moroccan dialect. The availability of this dataset will be helpful to researchers due to the lack of Moroccan dialect datasets for NLP fields.

Keywords: Moroccan dialect · Natural language processing · Voice assistant · Corpus · Data collection

1 Introduction

Speakers of the Arabic language worldwide has exceeded 300 million, including people of North Africa and the Gulf countries. It varies between Standard Arabic, which has fixed and unchanging rules, and dialects, which differ from one country to another and from one region to another. People in the same country, living in a particular region, may not understand the dialects of other regions. These dialects were born out of the influence of classical Arabic.

Although Arabic is difficult, it is among the most widely spoken languages in the world. Different forms of Arabic dialects exist throughout the Arab world. This dialect diversity makes it hard for people from different Arab countries to understand each other. It should be noted that Arabic dialect was considered a partially under-resourced

M. Ben Ahmed et al. (Eds.): SCA 2022, LNNS 629, pp. 633–642, 2023.
https://doi.org/10.1007/978-3-031-26852-6_59

language. As a result, we suffer from a lack of dialect data [1]. Therefore, since the effectiveness of NLP tasks - such as sentiment analysis - depends on the availability of representative resources, there is a strong need for such resources in these dialects [2]. For instance, Moroccan dialectal Arabic remains under investigated, which makes it isolated and difficult to understand by the majority of the other Arab countries. Moreover, Moroccan Darija was hardly ever published before the beginning of the Internet and social media. [3]. Therefore, some studies have focused on North African dialects such as Moroccan, Algerian, and Tunisian [4].

In this paper, we present an approach to create a corpus of Moroccan dialect. We aim to collect two types of data from a Facebook Chatbot. This corpus will be beneficial in accelerating NLP by being an open resource to developers. It can be used in many applications, such as voice assistant or virtual assistant because artificial intelligence is the research trend of this decade. It may also serve in translation applications, where text and voice can be used at the same time. Finally, this corpus can be of use in GPS applications, which help users find their way without the need to look at the screen the whole time. This paper is organized as follows:

Section 2 presents a background review of the efforts made in Arabic NLP, Moroccan dialect, and how to link them to digitalization. In Sect. 3, we provide more details about dataset and multimodal corpus building. Section 4 outlines experimentation with Moroccan Dialect, while the process of collecting data, preprocessing and analyzing is represented in Sect. 5. Section 6 discusses how we could expand this approach, using NLP, and proposes a system architecture. In the last Section, a conclusion is provided, and areas of future research are suggested.

2 Background

2.1 Voice Assistant

Today, there are many applications with voice user interfaces, which we use in smartphones, TVs, smart homes and other products. They allow users a vocal interaction with the machine in order to perform a variety of actions. The Voice User Interface (VUI) used as a web navigator can perform such tasks as researching information, ordering products, making purchases, managing agendas and appointments, or checking the weather.

Moreover, the VUI is capable of controlling home equipment by voice to perform very specific commands, such as turning on the light, closing the window curtains, and starting the heater. Applications with VUI are more practical and quicker than using the keyboard. It is especially good as a hands-free solution for users whether in the kitchen, in the car, or while practicing sports. These applications use NLP techniques to treat our voice commands. It is also worth noting that Arabic, which is now recognized as the fourth most used language on the Internet, has witnessed an increase in natural language processing research in the last ten years.

The recent research in Arabic natural language processing NLP focuses on new learning techniques. Supervised ML algorithms are widely used in NLP because they are easy to implement and give high accuracy, from SVM, NAIVE BAYES CLASSIFICATION (NB), DECISION TREES, KNN [5] to text categorization such like spam, legitimate messages, translation, named entity recognition, sentiment analysis, or speech recognition [6]. In addition, deep learning algorithms are new research orientations in the domain of NLP such as deep neural networks (DNN), convolutional neural networks (CNN), long and short memory (LSTM), recurrent neural networks (RNN) [7]. In recent years, natural language processing has also developed for Arabic dialects.

2.2 Moroccan Dialect

Moroccan Arabic or Darija is the main language in Morocco as it is the language of daily communication, spoken in almost all parts of the country. This language, although technically Arabic, is strongly influenced by other languages such as French, Spanish, due to the colonial occupation [8]. Darija uses many Arabic words but vowel changes and differences in pronunciation make the words different from Modern Standard Arabic (MSA). For this reason, it is classified as one of the most difficult Arabic dialects.

3 Multimodal Corpus Building

Arabic dialect studies have developed rapidly in recent years since people have started using dialect in social media [9]. Hence, many researchers tend to create corpus using social media, like Twitter, Facebook, YouTube, etc.... Twitter is more popular in the Arabian Gulf area, whereas Facebook is more popular in North Africa [10]. For all of the reasons we mentioned before, we chose Facebook as a source for building our Moroccan dialects corpus.

There are two goals of our corpus. First, we aim to build a dataset for dialect of a particular under-resourced Moroccan Darija [1]. Second, we turn to make this corpus specific and useful. Consequently, we choose Moroccan administration as a topic in order to avoid multiple dialects and just focus on Moroccan dialect in general.

The corpus is created by exploiting a Chatbot on Facebook to interact with users, asking them specific questions, looking for answers in different formats of data: voice and text. Therefore, to reach a huge number of users, we create a video advertisement, containing why we are working on this project, and how everyone can help us and interact with the Chatbot to answer questions. This video was published in different groups on Facebook so that we can collect big data in order to apply algorithms of deep learning.

4 Experimentation with Moroccan Dialect

Resources and tools for Arabic dialects, especially Moroccan dialect are still lacking [4] even though Moroccan Dialectal Arabic (Darija) is a language with 21 million native speakers [3]. That is why we have decided to choose Moroccan dialect to conduct research in this new field, and create our own dataset. Apple's Siri, Amazon's Alexa, Microsoft's Cortana and Google Assistant are the most popular voice assistants [11], all of which use English or another language, but do not understand Arabic. Salma is the only voice assistant that can understand it. Unfortunately, no voice assistant can understand Moroccan dialect. This is the second reason why we have chosen to work on Moroccan dialect. The use of voice assistants facilitates communication between public administration and citizens. It is also an option that brings together artificial intelligence techniques allowing richer service experiences that increasingly resemble human experiences. The use of these technologies lowers the burden on the organization and its human agents [12].

This corpus can also be used in other domains not only in public administration. Part of this corpus is stable and not modifiable. Each user can exploit it and develop it according to the desired domain, such as voice assistants for healthcare or education. For example, Amazon launched Alexa virtual assistant features to help users establish routines while at home and provide advice, information and guidance on COVID-19. Moreover, an AI Chatbot-based virtual assistant called LTKA-Bot is proposed. It is a virtual assistant to provide services regarding course activities such as recording session data and services related to attendance, task assignment and scoring management [13].

5 Results

In this section, we describe how we collected the dataset, and present the results of the study.

5.1 Data Collection

First, we chose Facebook as a platform to collect the data of our corpus due to its popularity, and the simple use of its functions. We created a Facebook page, with images that explain our aim. After that, on this page, we create a Chatbot.

Chatbot is a recent technology used to conduct interaction data and services online. Instead of contacting a real human agent, Chatbot allows conversation via text or text to speech. Chatbot is a software application, used especially as mobile messaging applications. The recent developments in artificial intelligence and machine learning boost the development of Chatbot; this is why giant companies, like Facebook and Google, were racing to integrate Chatbot in their applications. In order to exploit this new and strong technology, we gave our Chatbot three questions to ask users. The questions were related to Moroccan Administration. The answers should be provided in text format, in Moroccan dialect with Arabic script, and as a voice message.

Conversations between users and Chatbot generated a lot of data, with which it was difficult to deal. Therefore, we were obliged to store this data in an organized database in order to facilitate its exploiting later. We chose an open-source platform, which allowed us to automate storing data from Facebook. It enabled us to make links between Chatbot on Facebook and a database to store conversations. Later, we could download this database with different formats, such as Excel. This diagram shows the entire data collection process described above:

The downloaded Excel file contains eight columns: username, date and time, answer to question one, answer to question two, answer to question three, link to audio file one, link to audio file two, link to audio file three, in MP3 format. As we mentioned before, we made a short video to explain the aim of this project, and how users can interact with Chatbot. To reach a big number, we shared it on different social media, like LinkedIn, Facebook, and WhatsApp.

5.2 Data Preprocessing

In order to follow the crucial steps of data cleaning, to choose an appropriate algorithm, we proceeded to a quick manual data cleaning of the corpus so as to prepare it for the next step. First, we eliminated the date and hour column as well as the empty rows and the non-significant answers. Based on the name field, we distinguished between women and men's answers, before eliminating them to maintain user confidentiality. Secondly, we checked the consistency between the answers in text format and the voice recordings. After that, we filled the empty fields with the audio contents sent by the same user. Finally, we replaced the French letters with Arabic letters. The result was an Excel sheet with three columns, each of which contains the user's response. The figure below shows the format of the Excel file (Fig. 1):

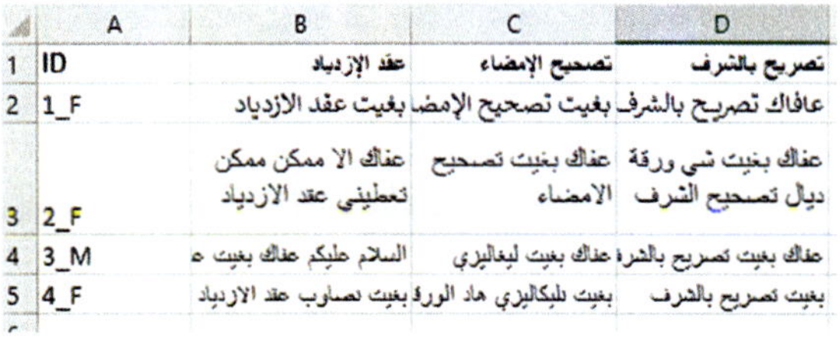

	A	B	C	D
1	ID	عقد الإزدياد	تصحيح الإمضاء	تصريح بالشرف
2	1_F	بغيت عقد الازدياد	بغيت تصحيح الإمض	عافاك تصريح بالشرف
3	2_F	عفاك الا ممكن ممكن تعطيني عقد الازدياد	عفاك بغيت تصحيح الامضاء	عفاك بغيت شي ورقة ديال تصحيح الشرف
4	3_M	السلام عليكم عفاك بغيت ع	عفاك بغيت ليغاليزي	عفاك بغيت تصريح بالشر
5	4_F	بغيت نصاوب عقد الازدياد	بغيت نليكاليزي هاد الور	بغيت تصريح بالشرف

Fig. 1. Dataset before preprocessing

In order to prepare our dataset for automatic cleaning, we grouped the 3 columns in a single one named user response, and then we added a new column, which allowed us to label each row of the previous column. The cleaning of an Arabic corpus is different from a Latin language. Our goal was to obtain a coherent corpus, without special characters. To achieve this goal, we used the Python language, with the libraries: pandas, numpy, string, and re. This code cleaned the corpus, removing punctuation, special characters and unifying the writing of some letters like (Table 1):

Table 1. Unifying special letters

Letters before preprocessing	Letters after preprocessing
ڭ، گ	ك
آ، أ، إ	ا

We obtained a dataset with a new format, as shown in Fig. 2 below:

5.3 Data Analysis

In this section, we present and describe some statistics related to the collected dataset. This dataset contains two types of data in Moroccan dialect, text and voice. Both of

5	4	السلام عليكم عفاك بغيت عقد الازدياد	1	انثى
6	5	بغيت نصاوب عقد الازدياد ديالي	1	انثى
7	6	بغيت ناخد عفاك عقد الازدياد	1	انثى
8	7	بغيت لخلوق	1	ذكر

Fig. 2. Dataset after preprocessing

them are answers to three specific questions, related to how to ask for a service in a Moroccan administration. Collected data was useful with a percentage of 73%, after deleting non-significant answers and non-complete answers. The figure below illustrates this usefulness (Fig. 3 and Table 2).

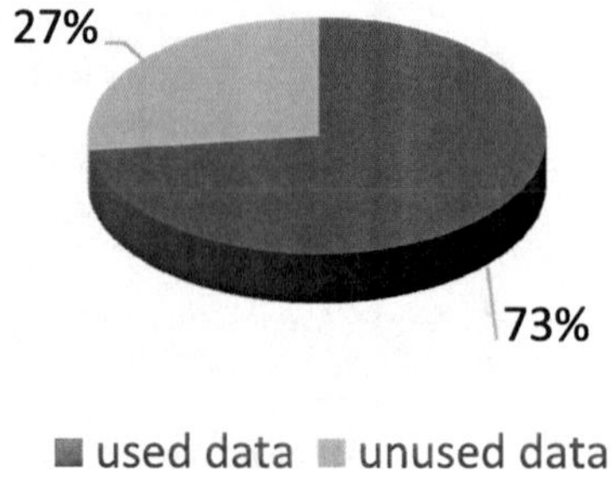

Fig. 3. Usefulness of the data collected

Table 2. Some forms of politeness

"..... السلام عليكم عافاك الله يخليك بغيت"
"...ممكن الله يخليك ناخد"
" ...لله يرحم الوالدين غنصاوب "

During their interactions with Chatbot, users use greeting formulas and forms of politeness before asking services. Text and audio recordings include answers from both sexes, of which women represent the largest part. The figure below shows that 77% of respondents are women, while the percentage of men who participated is a low 23% (Fig. 4).

The participants were divided into three age ranges. The first one, comprising 67% of participants, is between 20 and 25. The second category, which makes up 23% of participants, is between 26 and 30. The third one, including no more than 10% of participants, ranges between 31 and 35. Figure 5 below shows these results:

6 Discussion

In our daily life, the use of deep learning, Natural Language Processing (NLP), and speech recognition has become widespread. These new technologies are used in different

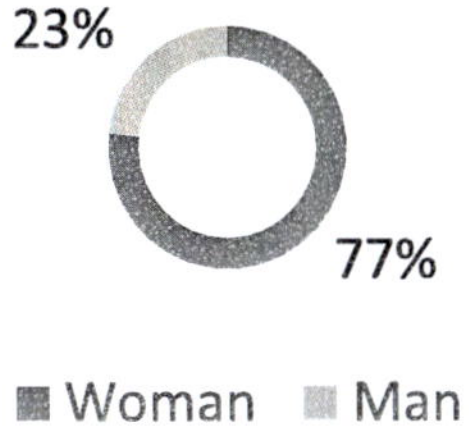

Fig. 4. Gender of the users

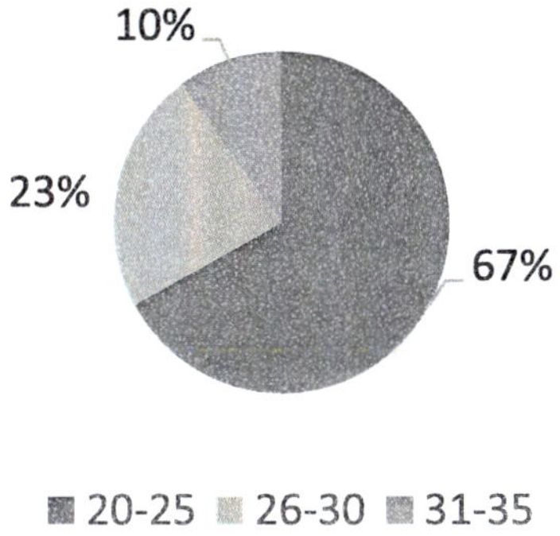

Fig. 5. Age range graph

fields, ranging from finance to healthcare. There is a great need to combine deep learning, NLP and speech recognition to create a real-world application. To achieve this goal, many steps need to be taken [14, 15]. First, we need to specify the suitable AI technology to be used; namely, either deep learning or machine learning.

Deep Learning is generally used for complex tasks that until recently could only be accomplished by humans. This learning system enabled the development of Automatic Speech Recognition (ASR) systems. Two types of neural networks have marked a technological breakthrough in the field of speech processing: "deep" networks and recurrent networks [16]. In addition, in the last few years, deep learning has been producing superior results on various NLP tasks. For instance, it enables multi-level automatic feature representation learning.

Recent NLP research is now increasingly focusing on the use of new deep learning methods. The important step following the selection of the appropriate technology is collecting data. Using a Chatbot on Facebook, we built our corpus. This Chatbot reacts with users, who answer questions. The corpus contains text and speech data, we can use it to create an application by exploiting deep learning technology in NLP, and speech recognition. How could we develop collecting data, and do it better?

Our approach deals with a specific topic, requests for services from Moroccan administrations, using special questions, which the user answers in text and voice format, in Moroccan dialect. Therefore, in order to develop this approach, and to use it in other services besides public administration, we need to ask the user more questions, so that we can expand our corpus. Our method is based on the question-answer principle, which is a crucial task in natural language processing. The question answer system is essentially based on a Chatbot, allowing users to give answers to any questions. As we know, the same question can be phrased in different ways or by using different words.

For example, all these requests have the same meaning (Table 3):

Table 3. Phrases with the same meaning

بغيت عقد الإزدياد	**birth certificate**
عطيني الخلوق	
شهادة الميلاد عافاك	

The major revolution took place when natural language processing was used to solve the task of answering questions.

For this reason, our corpus will be beneficial as such corpus is not yet available. It will boost research in NLP, and speech recognition, based on question answering systems. To establish communication between the human and the machine, there are several technical studies to put in place a conversational interface that allows this direct interaction. It is a complex system that integrates the fields of deep learning, speech recognition, and natural language processing. The objective is to conduct this theoretical study, by developing an application based on this technology. The figure below illustrates the proposed system architecture of this application (Fig. 6):

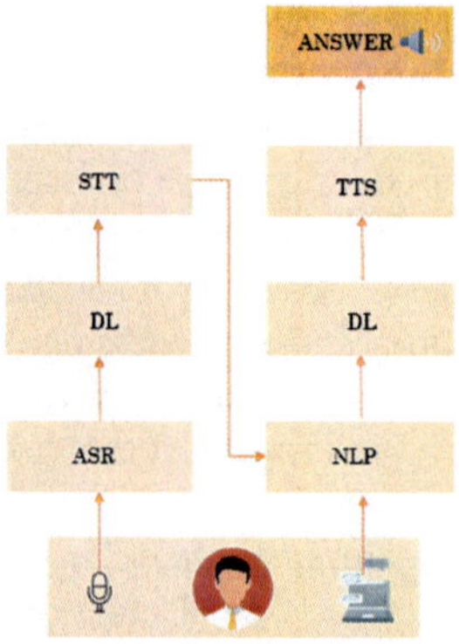

Fig. 6. Proposed system architecture

The user asks a question, in different forms, either voice or text. If the question is in text form, Natural Language Processing (NLP) is used to derive meaning from the transcribed text. Deep learning has been applied to understand language because of its ability to generalize context and accurately respond to language tasks. Last, text-to-speech (TTS) is used for the artificial production of human speech from text. So, when you hear the system's reply to your question, it sounds naturally human. If the question is in voice format, we apply automatic speech recognition (ASR) processes. The audio signal is transcribed into text (STT) using deep learning technology this time. Once the speech recognition step has converted the spoken question into text, we need to understand the language, intent and context by following the same steps while dealing with questions in text format.

7 Conclusion and Future Works

We presented a method of data collection for underserved languages, in particular a corpus for the Moroccan dialect given the lack of corpora and research in this field. We chose the Moroccan administration as our domain since the majority of Moroccans use the dialect in public administrations. The proposed approach was based on the creation of a Facebook Chatbot, which interacted with users to collect data in an implicit way. First, we created a Facebook page containing a Chatbot. Through this, users were asked questions about some of the services they request from the Moroccan administration. Then, thanks to an API, we were able to collect the text-format data from the Facebook page, in an excel file. We also downloaded the audio files. Finally, we obtained a dataset, to which we could apply machine learning and deep learning algorithms, after pre-processing and removal of erroneous data.

We have chosen this specific structure of our dataset in order to create a robust application. Another goal behind our choice is to benefit from the development of speech recognition and other natural language processing techniques, which will help us in our future research, using the proposed system architecture.

References

1. Hamed, I., Denisov, P., Li, C.Y., Elmahdy, M., Abdennadher, S., Thang Vu, N.: Investigations on speech recognition systems for low-resource dialectal Arabic–English code-switching speech. Comput. Speech Lang. **72**, 101278 (2021)
2. Alowisheq, A., et al.: MARSA: multi-domain Arabic resources for sentiment analysis. IEEE Access **9**, 142718–142728 (2021)
3. Labied, M., Belangour, A.: Moroccan dialect "Darija" automatic speech recognition: a survey. In: 2021 IEEE 2nd International Conference on Pattern Recognition and Machine Learning (2021)
4. Alsudais, A., Alotaibi, W., Alomary, F.: Similarities between Arabic dialects: Investigating geographical proximity. Inf. Process. Manag. **59**(1), 102770 (2022)
5. Marie-Sainte, S.L., Alalyani, N., Alotaibi, S., Ghouzali, S., Abunadi, I.: Arabic natural language processing and machine learning-based systems. IEEE Access **7**, 7011–7020 (2019)
6. Antoun, W., Baly, F., Hajj, H.: Arabert: transformer-based model for Arabic language understanding, arXiv preprint arXiv:2003.00104 (2021)
7. Al-Ayyoub, M., Nuseir, A., Alsmearat, K., Jararweh, Y., Gupta, B.: Deep learning for Arabic NLP: a survey. J. Comput. Sci. **26**, 522–531 (2017)
8. Allak, A., Benelallam, I., Habbouza, H., Amallah, M.: Toward an end-to-end voice to sign recognition for dialect moroccan language. In: Saidi, R., El Bhiri, B., Maleh, Y., Mosallam, A., Essaaidi, M. (eds.) Advanced Technologies for Humanity, pp. 253–262. Springer, Cham (2022). https://doi.org/10.1007/978-3-030-94188-8_24

9. Alshutayri, A., Atwell, E.: Exploring twitter as a source of an Arabic dialect corpus. Int. J. Comput. Linguist. **8**, 37–44 (2017)
10. Alshutayri, A., Atwell, E.: A social media corpus of Arabic dialect text. Computer-Mediated Communication and Social Media Corpora, Clermont-Ferrand, Presses Universitaires Blaise Pascal (2019)
11. Hoy, M.B.: Alexa, siri, cortana, and more: an introduction to voice assistants. Med. Ref. Serv. Q. **37**, 81–88 (2018)
12. Acco Tives Leão, H., Canedo, E.D.: Best practices and methodologies to promote the digitization of public services citizen-driven: a systematic literature review. Information **9**(8), 197 (2018)
13. Terzopoulos, G., Satratzemi, M.: Voice assistants and artificial intelligence in education. In: Proceedings of the 9th Balkan Conference on Informatics, pp. 1–6 (2019)
14. Nassif, A.B., Shahin, I., Attili, I., Azzeh, M., Shaalan, K.: Speech recognition using deep neural networks: a systematic review. IEEE Access **7**, 19143–19165 (2019)
15. Young, T., Hazarika, D., Poria, S., Cambria, E.: Recent trends in deep learning based natural language processing. IEEE Comput. Intell. Mag. **13**(3), 55–75 (2018)
16. Alam, M., Samad, M.D., Vidyaratne, L., Glandon, A., Iftekharuddin, K.M.: Survey on deep neural networks in speech and vision systems. Neurocomputing **417**, 302–321 (2020)
17. Nagajyothi, D., Siddaiah, P.: Speech recognition using convolutional neural networks. Int. J. Eng. Technol. 133–137 (2018)
18. Otter, D.W., Medina, J.R., Kalita, J.K.: A survey of the usages of deep learning for natural language processing. IEEE Trans. Neural Netw. Learn Syst. **32**(2), 604–624 (2020)
19. Zhang, L., Wang, S., Liu, B.: Deep learning for sentiment analysis: a survey. Wiley Interdiscip. Rev. Data Min. Knowl. Discov. **8**, e1253 (2018)

A Comparative Analysis of Top Ten Real Estate Advertising Web Sites in Morocco

Saloua Bensiali[1,2](✉), Zainab Azough[2], Lahoussaine Baamal[1,2], and Ayoub El Adraoui[2]

[1] Department of Applied Statistics and Computer Science, Hassan II Institute of Agronomy and Veterinary Medicine, Rabat, Morocco
s.bensiali@iav.ac.ma

[2] Network of Excellence on Land Governance in Africa, North Africa Node, Hassan II Institute of Agronomy and Veterinary Medicine, Rabat, Morocco

Abstract. Among many platforms used in the real estate market, advertising websites play a relevant role in real estate offers. In this paper, we work on a comparative study of the ten popular real estate portals in Morocco. To achieve this aim, we identify the most determining criteria for assessing the role of a property website in the real estate market. Then, we carry out a comparative analysis of these real estate portals by assessing the defined criteria based on their structure and content. Our work helps to distinguish the two most competitive portals compared to the others. This comparative study enables the positioning of real estate portals in Morocco. It helps suggesting relevant Internet Data Sources (IDS) to analyze real estate offers market in Morocco.

Keywords: Real estate portals · Real estate advertising · Property websites · Real estate market · Real estate offers · Benchmarking · Morocco

1 Introduction

The real estate market consists of all transactions carried out on real estate. There are mainly two types of operations, namely sales (or purchases) and rentals. These operations involve two parties. In the case of a purchase, there are buyers on one hand and the sellers or their representatives, called intermediaries, on the other. In the case of rental, the operation is made between the tenants on one hand and the owners or brokers as their representatives, also called intermediaries, on the other. In summary, as stated by G. Maier & S. Herath [1], in economic terms, real estate market can be seen as the market where the supply of and the demand for real estate meet and where real estate is traded.

Real estate market analysis is based on several aspects. Indeed, it is mainly a question of observing various parameters related to the offers and demands available on the market. The offer is represented by the supply of real estate placed on the market by sellers, owners or intermediaries. Demand is raised by buyers or tenants available to buy or rent real estate. Offers and demands are materialized by the transactions made between the market parties, hence the interest of examining the details related to these

M. Ben Ahmed et al. (Eds.): SCA 2022, LNNS 629, pp. 643–654, 2023.
https://doi.org/10.1007/978-3-031-26852-6_60

transactions. In addition, there are also other essential issues, such as the conditions for granting real estate loans and the evolution of real estate taxation, that are strongly correlated with the dynamics of this market.

In this article, we focus on one single facet of real estate market analysis, which is the analysis of real estate offers. In fact, there are four best known ways to list real estate offers. First, advertisements for sale or rental can be observed on the premises of the properties in the form of advertising billboards. This approach is especially valid if we are aiming to collect offers of new properties made by the projects of real estate developers, as their advertisements are more adapted and more visually accessible. However, locating these ads for large-scale collection requires big human efforts and considerable time. Secondly, announcements related to real estate offers are disseminated through various media channels. The actions of collecting and digitizing these advertisements require also a lot of time, especially if we are targeting a long period. Thirdly, real estate agencies have a share of the stock of properties available on the market as they play the role of intermediaries in sales and rental transactions. These agencies own a part of the information related to the stock but would not share access to this data in general. Finally, real estate ads are increasingly published on social media and websites dedicated exclusively or partially to the real estate market. These offers are freely accessible, digital, public, and available for longer periods of time. Thus, the Internet plays an important role in the housing market, as a source of information for potential buyers and for price and demand forecasting as demonstrated in [2, 3]. Hence our interest in Internet Data Sources (IDS) for real estate offers.

Moreover, concerning the relevance of IDS for real estate market statistics, M. E. Beręsewicz in [4] asserts that it is possible to use IDSs to produce estimates of the offer price per m2 with an acceptable level of absolute relative bias. Also, M. E. Beręsewicz shows that IDSs are especially suitable for providing statistics about the secondary real estate market. Therefore, we chose to deal with IDSs in the second-hand real estate market, particularly for residential properties located in urban areas in Morocco. According to our knowledge, there is no similar published work of research analysing IDS for local real estate market in Morocco to date.

Given that online real estate advertisements can be found on many Moroccan internet resources, we initiated this study by selecting the most popular real estate internet websites, in order to use them as a base of our comparative study. Then, we identified criteria that are widely used in real estate market to conduct the benchmarking study. In particular, the main contribution of this study is that it categorizes the IDS of Moroccan real estate ads to give an overview on their relevance and reliability in the secondary real estate market.

The article is structured as follows: In Sect. 2, we review the literature in order to identify the most determining criteria to distinguish real estate offers. Then, in Sect. 3, we reveal our study concerning the benchmark of online real estate resources in Morocco. Finally, in Sect. 4, we present the summary of the benchmarking and make a detailed analysis of the most relevant internet resources for real estate offers. The article ends with some conclusions and indicates some future work.

2 Potential Criteria in Real Estate Market

The real estate market is managed by the same principles everywhere in the world. These principles are then projected on each local real estate market to derive specific analyses depending on the context of each study region. Thus, we conduct a literature review to identify the key factors that affect the real estate market. Then we define the criteria on which our benchmarking will be based.

2.1 Literature Review

Property pricing is the most important dimension in real estate market. Indeed, the objective value of a property is its estimated market value. While this appraisal value is thought to be primarily systematic, it does contain subjective components, over and above individual differences among appraisers, that can affect valuation [5]. Thus, as stated by K.M. Lusht in [6], "In this kind of market, price-setting is a probabilistic exercise, with some degree of inefficient pricing inevitable. The extent and magnitude of the mispricing is an empirical question". Besides, the property's listing price may be viewed as the seller's best guess of the property's Fair Market Value (FMV), which comes before the price negotiation process [5].

G.Maier & S.Herath in [1] testify that the real estate market is typically segmented into various submarkets along different dimensions. And they indicate that the most important dimensions are the type of real estate and its characteristics in space and time. The question of what factors affect real estate pricing has been discussed in the literature for a very long time [5-6-7-8]. K.M. Lusht in [6] draws attention to the data problems, like sampling period of time and number of observations, when making real estate appraisal based on historical data. Specifically, from [5, 7, 8] we will cite in the following some particular factors, among others, related to the residential property type: the size of the household in square feet of living space, the condition of the household and building quality, and other significant features like financing arrangements. Also, local knowledge (such as proximity to shopping areas and parks) may be relevant. Indeed, as demonstrated in [7, 8], the location, that gives a distance from the city center, is the most important influential factor on the property value, in many countries around the world.

The potential impact of location in the appraisal of the property value is explained by several features: distance to work, proximity to health institutions and other important facilities, proximity of public transport, good traffic connection, parking facilities, quiet neighborhood, view, orientation, tidiness of neighborhood, and green environment [9]. Besides, we emphasize residential property orientation, which is another factor equally decisive for the appraisal of the property value as verified in [10]. For instance, according to interviewed Slovenian experts in [8], non-optimal orientation, which means receiving small amounts of sunlight during the day in the residential properties, decreases the residential property value.

2.2 Criteria Selection for Benchmarking Real Estate Internet Data Sources Content

In consideration of all aforementioned factors, and by referring to our study goal, we identify seven criteria, described below, that we will use later in our comparative study.

Listed in Table 1, we use them in order to conduct a comparative analysis of the content of inspected real estate offers IDS.

Table 1. Selective criteria for real estate IDS benchmark

Criterion code	Criterion name	Criterion description
a	Price	Property price offered by the seller and mentioned in the ad within the IDS
b	Features	It indicates the presence of important property features in the ads (size, number of rooms, bathrooms, floors, etc.) within the IDS
c	Conditions	It mentions the presence of property age or conditions in the ads within the IDS
d	Geolocation	It indicates the availability of location information for properties placed in advertisement, within the IDS,
e	Orientation	It shows availability, within the IDS, of orientation information of properties that are placed in advertisements
f	Ads number	Total number of Moroccan real estate ads in the IDS in question
g	Ads lifetime	It serves to approximate the duration for which announcements stay in the IDS. It is expressed in number of months

3 Multicriteria Evalaution and Benchmarking of Moroccan Real Estate Portals

At first, we have refined a list of the websites used in Morocco to disseminate the real estate offers. Then, we examined the popularity of these resources. Finally, we did a benchmark study on the selected internet data sources according to the real estate pre-established criteria.

3.1 Inventory of IDSs on Moroccan Real Estate Market

We are targeting internet data sources that publish real estate ads throughout Morocco, knowing that there is no other kind of study in the literature about the use of online services by real estate market participants in Morocco. Therefore, we don't recognize the right information about what, or how many, web services brokers use to make real estate appraisal.

Thus, we started the selection of IDSs for real estate by choosing a random web-panel sample using search engines. We refined this sample, by rejecting non-relevant portals, non-available websites, and the websites that are dedicated to the advertisements of a specific Moroccan region or to the offers of a particular agency. Around thirty websites were checked to select the most suitable for our study. Those websites belong to different categories. For instance, there are portals dedicated just to real estate announcements, while others publish ads of many types. Also, those portals are either dedicated to advertisers from Morocco or covering various countries of the world including Morocco. In total we opted for ten resources. Table 2 lists their URLs, specifies their category whether it specializes in real estate or not, and shows if they are dedicated to Moroccan ads or not. Screenshots of the first three websites in that list are shown in Fig. 1.

Table 2. Selection of IDS for Moroccan real estate announcements by categories

IDS No.	IDS Name	IDS URL	Real estate portal (exclusively)	Moroccan ads only
1	Marocannonces	marocannonces.com		✓
2	Mubawab	mubawab.com	✓	
3	Avito	avito.ma		✓
4	Louervendreaumaroc	louervendreaumaroc.com	✓	✓
5	Sarouty	sarouty.ma	✓	✓
6	Repimmo	repimmo.com	✓	✓
7	Immolist	immolist.ma	✓	✓
8	1000-annonces	1000-annonces.com	✓	✓
9	Snapimmo	snapimmo.ma	✓	✓
10	Darna	darna.ma	✓	✓

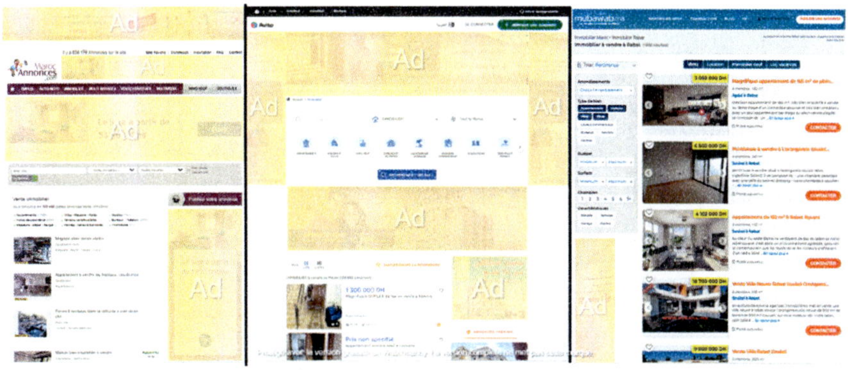

Fig. 1. Some Moroccan real estate market portals

3.2 Comparison Between Moroccan Real Estate IDS in Terms of Popularity

In the next step, we examine the popularity of these real estate advertisement web services in Morocco. Therefore, knowing that language is an important enabler of ecommerce adoption [12], we identify the ads language within the IDS. In fact, Arabic (Ar) is the first official and written language in Morocco, followed by French (Fr) considered as an international language that serves as Morocco's primary language in commerce and economics. As Table 3 shows, all the selected resources contain advertisements written either only in French, or by advertiser's choice from both languages. However, they differ in terms of display strategy; for example, while No. 1 and No. 3 merge both languages in the display, No. 2 displays all ads in one language according to the website visitor's choice.

The popularity is then measured according to two approaches. On one hand, we used data analysis from Alexa.com, which gives, among others, the traffic rank of the websites all over the world and in a given country. On the other hand, we used Google Trends service[1] which provides a relative metric based on queries made by Google Search users. In fact, this service takes a particular query's search volume and divides it by the total number of searches completed in the given time period and geographic region; the resulting ratio is then scaled on a range of 0–100. Table 3 shows the Alexa rank and the average corresponding Google Trends index for each IDS URL over twelve months (03/2021 - 02/2022) for both the real estate category and all search categories. Both measures are related to the country of Morocco.

Table 3. Measure of popularity of Moroccan real estate IDS

IDS No.	Language		Alexa rank	Google trends index mean	
	Fr	Ar		Real estate category	All categories
1	✓	✓	98	-	15,81
2	✓	✓	363	6.75	19,23
3	✓	✓	37	33,42	68,11
4	✓		13746	-	-
5	✓	✓	1646	5.52	14,36
6	✓		-	-	-
7	✓	✓	-	-	-
8	✓		13278	-	-
9	✓	✓	-	-	-
10	✓		-	-	-

By observing the Table 3 we notice the following:

1 https://trends.google.com/.

- Google Trends shows the search trend in the real estate category for only three resources (No. 2, No. 3 & No. 5).
- In the case of all categories one other resource is added by Google Trends (No. 1), next to the other three. This resource is not specialized on real estate but publishes ads in other categories. So, it seems that the popularity of this portal in the real estate category is negligible by Google Trends.
- Resources that rank highest by Alexa rank in Morocco (No. 3 & No. 1) are bilingual. This observation promotes the role of language in favour of popularity of digital advertising resources.
- In particular, resource No. 3 ranks first in both approaches. This portal is popular in Morocco, not only for real estate ads but for all types of ads, which these results confirm. Concerning its Google Trends index, it is superior to the indices of other sites more than three times (for all categories) and five times (for real estate category). Regarding the Alexa rank of this website, it is included in the top 50 websites, in terms of traffic rank in Morocco (rank 37). Additionally, according to another Alexa indicator, which is site's overlap score, among all our selected resources, No. 3 overlaps just with No. 1 and No. 2, with about third of its audience, which goes with Google Trends results, neglecting resource No. 5 which is also well ranked by Google Trends.

Even if there are resources that are not considered popular by Alexa or Google Trends, there may be better approaches to measure the popularity in our case. Because Alexa rank is considered against all sites that are not necessarily interested in ads of real estate. Besides, in the Google Trends approach, users tend to enter search terms such as apartments for sale, apartments to buy rather than write specific IDS names.

M. E. Beręsewicz in [11], suggests also the measure of the number of advertisements by web resource, as a complementary indicator to judge about its popularity, and most of web portals provide this information. In fact, within our study, we consider this information, as mentioned before, as a criterion of comparison between IDSs in addition to the other determining factors. Thus, we will be able to confirm the popularity and reliability of all the consulted sources afterwards through those indicators.

3.3 Multicriteria Evaluation of Moroccan Real Estate Market IDS

Preliminary IDS analysis has so far marked the following IDS No. 1, No. 2, No. 3 & No. 5. The next objective is to refine the analysis further according to the content verification criteria already identified for the real estate market. Our goal is to verify, in terms of content and structure, the importance of the IDS which have been distinguished, and also to review among the other less classified sites, according to our analysis of Sect. 3.2, how can we evaluate them in terms of structure and content.

While browsing the portals of our study, we noticed that there is a lot of noise inside some ones, in terms of the presence of web banner ads, especially N.1 and N.4 hold a significant amount of this noise. Also, there are portals that present the real estate advertisements in a single batch without separating between sales and rentals categories. The Table 4 shows the difference between the portals in terms of these last two parameters.

By browsing Table 4, we notice the next:

Table 4. Comparison between Moroccan real estate IDS in terms of structure

IDS No.	Separates sales from rentals	Contains Ads noise
1	✓	✓
2	✓	
3		✓
4	✓	✓
5	✓	✓
6	✓	✓
7	✓	
8	✓	✓
9		
10	✓	

- Among the well-ranked websites so far, the only website that doesn't separate sales and rentals is No. 3, and the only website that doesn't have high noise levels is No. 2.
- Compared to other Web services, the most structured sites, i.e., which are without advertising noise and have a clear separation between the types of targeted real estate operations are: No. 2, No. 7 and No. 10.

The comparison between portals in terms of content according to criteria defined above (Table 1), is made in the month of February 2022. It is specified in Table 5, without forgetting to mention that even if the portals of our study offer the verified criteria, there are ads that do not reveal some of these features, by choice of the advertisers.

Table 5. Benchmark of Moroccan real estate IDS

IDS No.	a	b	c	d	e	f	g (months)
1	✓	✓				342560	6
2	✓	✓	✓	✓		136232	6
3	✓	✓	✓	✓	✓	120484	12
4	✓	✓				57412	168
5	✓	✓		✓		13359	6
6	✓	✓	✓			10000	12
7	✓	✓	✓			6929	12
8	✓	✓				5907	72
9	✓	✓	✓			4062	60
10	✓			✓		1930	Ads date not available

Ads number by portal is a key indicator about its popularity, but we should be aware that this number may be biased because multiple advertisements may refer to the same property, or may include false or outdated ads [11]. Ads number by portal is represented in Fig. 2. It can be seen that the volume of real estate ads in Moroccan IDSs is mainly composed of the No. 1, No. 2, No. 3 and No. 4; The No. 1 is dominant of approximately 50% of the ads volume.

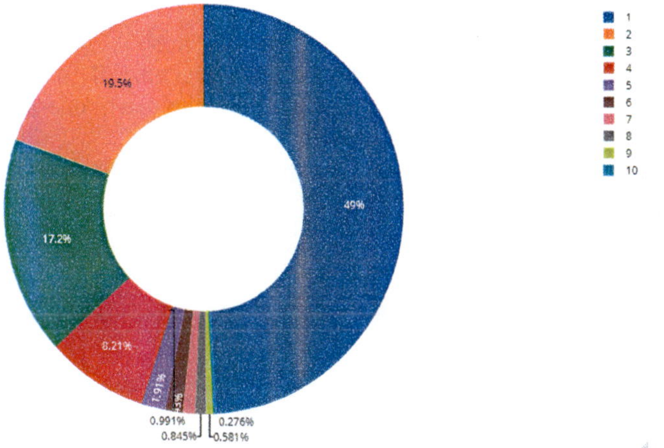

Fig. 2. Real estate ads number per portal

Ads lifetime is an important indicator of novelty and timely relevance of real estate ads, the shorter the ad lifetime, the more the IDS is up-to-date. The shortest lifetime is 6 months and it is present in IDSs No. 1, No. 2 and No. 5.

From further criteria in the Table 5, now we can refine the proposal of the well ranked sites:

- No. 1 even if it is an up-to-date portal with a huge amount of real estate ads, but it is noisy, and doesn't give property condition, geolocation and orientation.
- No. 2 is an up-to-date portal that meets all the predefined real estate criteria except orientation.
- No. 3, meets all the predefined real estate criteria.
- No. 5 is an up-to-date portal that gives geolocation, but doesn't give property condition and orientation, and offers a smaller number of real estate ads.

Among the other less classified sites, referring to Table 5, we notice the following:

- No. 4 is a less up-to-date portal, that leaves out in its real estate ads, the property condition, geolocation and orientation.
- No. 6 & No. 7 are both up-to-date portals, that offer the majority of criteria with the exception of geolocation and orientation.

- No. 8 & No. 9 are both less up-to-date portals, that do not offer geolocation and orientation.
- No.10 offers just the price and geolocation, and it does not advise the ads date.

From the results of the comparative analyses, we observe that No. 2 and No. 3 websites stand out from the others in many aspects: first in the trends, then in the determined criteria of the benchmark. Therefore, we figure out their importance in the real estate market landscape in Morocco.

Knowing that No. 2 is dedicated to real estate ads, while No. 3 is more versatile to other types of ads, both of them are distinguished from other IDSs by a close and significant total ads number. Yet, they differ in structure, property orientation feature and ads lifetime. Indeed, No. 2 is more structured than No. 3, because the first gives real estate ads without noise and with only one language at a time, while the last gives a listing of real estate ads merging between Arabic and French besides noise from recommended ads. In particular, No. 3 is the only portal that starts to take into consideration orientation until now.

Figure 3 shows the distribution of the number of ads in the two portals No. 2 and No. 3 by the main cities of Morocco (Agadir, Casablanca, Fes, Marrakech, Rabat, Tangier).

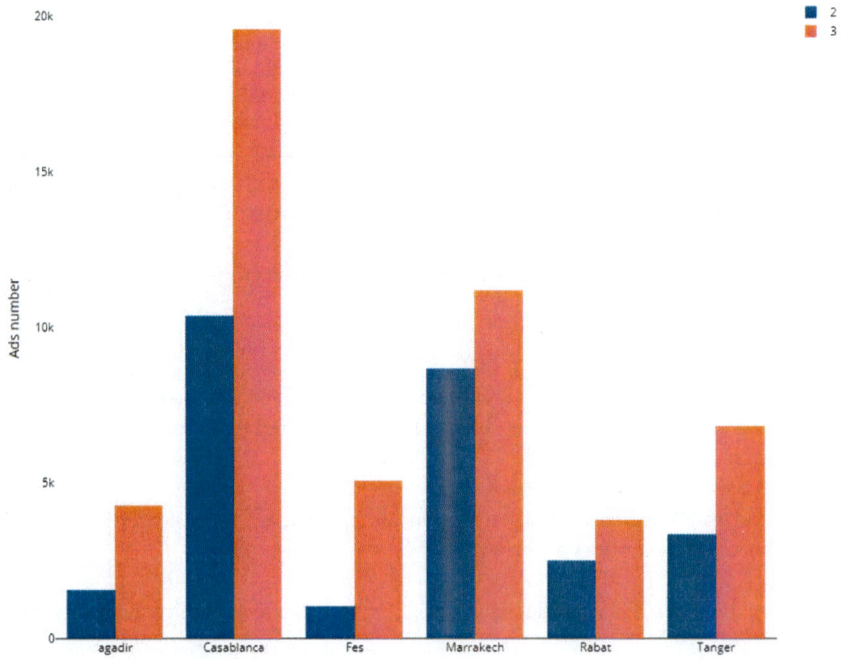

Fig. 3. Real estate ads number per Moroccan cities inside the portals No. 2 and No. 3

From Fig. 3, we observe the subsequent:

- For Rabat and Marrakech, the volume of advertisements is almost similar in the two portals.
- For Tangier and Casablanca, the volume of advertisements inside the portal No. 3 is double than that of No. 2.
- For Fes and Agadir, the volume of real estate advertisements in N°3 is more dominant (more than double) than that in N°2.

4 Discussion

The most important criterion in the offer: the price, is almost present in all the studied IDS. This is from the basic needs for any client searching for offers, so it is a core criterion. However, this value does not generally match exactly the market value of the property, because it is often amplified compared to the final transaction value which comes after negociation. Thus, it is necessary to be aware of this consideration when analyzing and making interpretations from these collected data.

Property features are also practically present in all the studied IDS, and constitute important information to document for enriched and more appealing offers.

Property conditions are less present than the other criteria, considering it complementary information to give to the client. It may be given in further steps of negotiation, but it is still present in 50% of the studied IDS.

Geolocation is an important and core criterion to display in an offer. Its inclusion in Moroccan real estate IDS has emerged and it is expanding among them. In fact, embedded maps are gradually being included in these portals.

Orientation is an advanced criterion mostly valued among real estate appraisers. It is important in defining residential well-being and comfort, but it is still poorly documented in Moroccan real estate IDS.

5 Conclusions and Future Work

In this paper, the authors made a study of important features of internet data sources in the context of real estate offers. Based on those features, they made a progressive multi-criteria evaluation of the top ten real estate advertising websites in Morocco. They have then obtained as a result of their benchmarking study that there are two main internet data sources to consider for the analysis of real estate market in Morocco. As future work, we plan to consider those portals for data collection in the context of Moroccan real estate market analysis.

References

1. Maier, G., Herath, S.: Real Estate Market Efficiency. A Survey of Literature. SRE - Discussion Papers (2009)
2. Lee, K.O., Mori, M.: Do conspicuous consumers pay higher housing premiums? Spatial and temporal variation in the United States. Real Estate Econ. **44**(3), 726–763 (2016)
3. Wu, J., Deng, Y.: Intercity information diffusion and price discovery in housing markets: evidence from Google searches. J. Real Estate Finan. Econ. **50**(3), 289–306 (2015)

4. Beręsewicz, M.E.: On representativeness of internet data sources for real estate market in Poland. AJS **44**(2), 45–57 (2015)
5. Northcraft, G.B., Neale, M.A.: Experts, amateurs, and real estate: an anchoring-and-adjustment perspective on property pricing decisions. Organ. Behav. Hum. Decis. Process. **39**(1), 84–97 (1987)
6. Lusht, K.M.: The real estate pricing puzzle. Real Estate Econ. **16**(2), 95–104 (1988)
7. Dolaeva, Z.N., Kidakoeva, M.M.: Research of residential real estate pricing. Sci. Europe **8**(8–1), 85–87 (2016)
8. Ferlan, N., Bastic, M., Psunder, I.: Influential factors on the market value of residential properties. Eng. Econ. **28**(2), 135–144 (2017)
9. Appraisal Institute: The Appraisal of Real Estate, 14th edn. Appraisal Institute, Chicago (2013)
10. Ma, A., Narwold, A.: Which way is up? Orientation and residential property values. J. Sustain. Real Estate **11**(1), 40–59 (2019)
11. Beręsewicz, M.E.: Internet data sources for real estate market statistics. Doctoral dissertation, Poznan University of Economics (2016)
12. Lawrence, J.E., Tar, U.A.: Barriers to e-commerce in developing countries. Inf. Soc. Justice J. **3**(1), 23–35 (2010)

Robust Vector Control of Synchronous Reluctance Motor Using Space Vector Modulation Algorithm

Yassine Zahraoui(✉), Mohamed Moutchou, and Souad Tayane

Higher National School of Arts and Crafts (ENSAM), Electrical Engineering Department, Complex Cyber-Physical Systems Laboratory, Hassan II University, Casablanca, Morocco
yassine.zahraoui1-etu@etu.univh2c.ma

Abstract. This paper aims to develop a robust vector control (VC) of a three-phase synchronous reluctance motor (SynRM). The inverter is controlled by a space vector modulation (SVM) algorithm. The objective is to guarantee a fast response on one hand and to improve the dynamic performance on the other hand. SVM algorithm replaces the conventional pulse width modulation (PWM) in order to obtain a smooth control sequence of the inverter switches. The goal is to generate an output voltage vector that approaches as much as possible the reference voltage vector. Simulation results show that the proposed vector control with the proposed numerical space vector algorithm provides good performance dynamic characteristics for different operating ranges. The robustness of the overall system is tested under load torque disturbance, reversal, low, and high-speed ranges in order to show the system stability. All simulations have been realized in MATLAB/Simulink.

Keywords: Field-oriented control · Synchronous reluctance motor · Space vector modulation · Robust control · Fast dynamic response

1 Introduction

The synchronous reluctance machine (SynRM) has received much attention for many applications in industry, especially for electrical vehicles, in recent years due to its simple structure and low manufacturing cost. The SynRM is also functionally robust, it is relatively inexpensive as it has no associated permanent magnets which is also an advantage for high-temperature applications [1]. Therefore it has no demagnetization problem or associated losses. SynRM has a clear advantage compared to an induction machine because it has a natural salience, particularly for the sensorless control. In addition, there are no rotor losses, which allows a higher mass torque than that of an induction machine [2]. These different advantages seem to give it chances for new developments [3,4].

M. Ben Ahmed et al. (Eds.): SCA 2022, LNNS 629, pp. 655–665, 2023.
https://doi.org/10.1007/978-3-031-26852-6_61

However, SynRM also has significant drawbacks. The salience of the rotor causes high ripples on the electromagnetic torque, the origin of this salience is the torque itself. That can result in vibrations and acoustic noise. The power factor of this type of machine is generally low, leading to over-sizing of the inverter [5]. In addition, it is very sensitive to magnetic saturation, which has a strong impact on the developed average torque [6].

The three-phase inverter is normally controlled by the conventional pulse width modulation (PWM), but this technique does not reject the inverter output voltage harmonics, the space vector modulation (SVM) algorithm reduces these harmonics and guarantees a fixed switching frequency. This technique still has one major lab disadvantage, it increases the switching loss per period and reduces the opening-closing switches cycle, so it requires better switches performance than the PWM [7]. This point is concerning hardware implementation at the first level [8].

2 Synchronous Reluctance Motor Modelling

In this part, the classical model based on the Park transformation is presented. To do this, the following assumptions are taken into consideration [9]:

- Magnetic materials are isotropic and non-saturated, (there is no need to distinguish apparent inductance from incremental inductance);
- The phenomenon of hysteresis as well as the iron losses are neglected;
- The inductance varies in a sinusoidal way (hypothesis of the first harmonic);
- Capacitive coupling between the windings is ignored.

In the absence of a zero-sequence current component, the electrical equations in the *d*-*q* reference frame are [10]:

$$\begin{bmatrix} v_{ds} \\ v_{qs} \end{bmatrix} = R_s \begin{bmatrix} i_{ds} \\ i_{qs} \end{bmatrix} + \begin{bmatrix} L_d & \\ & L_q \end{bmatrix} \frac{d}{dt} \begin{bmatrix} i_{ds} \\ i_{qs} \end{bmatrix} + p\omega_m \begin{bmatrix} 0 & -L_q \\ L_d & 0 \end{bmatrix} \begin{bmatrix} i_{ds} \\ i_{qs} \end{bmatrix} \tag{1}$$

The electromagnetic torque is expressed as:

$$T_{em} = p(L_d - L_q) i_{ds} i_{qs} \tag{2}$$

L_d and L_q are the inductance of the direct and the quadrature axis. In the frame related to the rotor, the total flux through the windings d and q is expressed by:

$$\begin{cases} \phi_{ds} = L_d i_{ds} \\ \phi_{qs} = L_q i_{qs} \end{cases} \tag{3}$$

hence

$$\begin{bmatrix} v_{ds} \\ v_{qs} \end{bmatrix} = R_s \begin{bmatrix} i_{ds} \\ i_{qs} \end{bmatrix} + \frac{d}{dt} \begin{bmatrix} \phi_{ds} \\ \phi_{qs} \end{bmatrix} + p\omega_m \begin{bmatrix} 0 & -1 \\ 1 & 0 \end{bmatrix} \begin{bmatrix} \phi_{ds} \\ \phi_{qs} \end{bmatrix} \tag{4}$$

and

$$\phi_s = \sqrt{\phi_{ds}^2 + \phi_{qs}^2} \tag{5}$$

Figure 1 shows the vector diagram of the synchronous reluctance motor in the steady-state. The rated parameters of the machine used in the simulation are listed in Table 1.

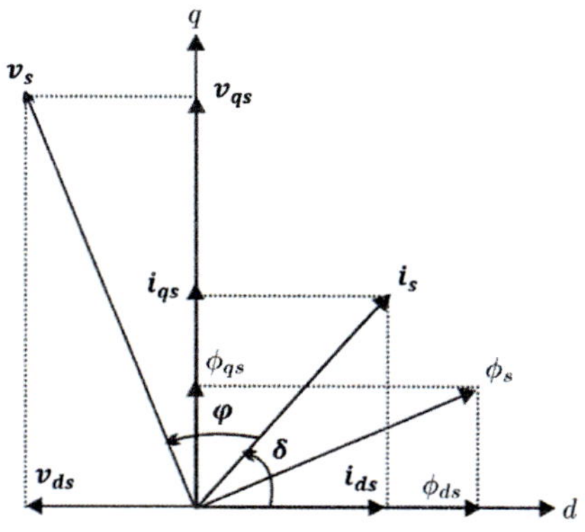

Fig. 1. SynRM steady-state vector diagram

Table 1. Rated power and parameters of the used SynRM

Rated power	1.1 kW
Rated speed	1500 tr/min
Rated voltage	220/380 V
Rated current	3.2 A
Rated torque	7 N.m
Frequency	50 Hz
Pole pair	2
Stator resistance	6.2 Ω
Apparent stator inductance L_d	0.34 H
Apparent stator inductance L_q	0.105 H
Inertia moment	0.008 N.m.s^2
Viscous friction coefficient	0.0001 N.m.s/rad

3 SynRM Vector Control

The vector-controlled synchronous reluctance motor is a very accepted method for high-performance drive system response [11]. This technique is essentially based on decoupling the flux and the torque in order to produce stator current components. As a result of this condition, the q-axis component of stator current is set to zero while the d-axis reaches the nominal value of the magnetizing current [12,13]:

$$\begin{cases} i_{ds} = i_d \\ i_{qs} = 0 \end{cases} \tag{6}$$

3.1 Current and Speed Loop Regulation

In this section, we will analyze the classical controllers for all loops. The proportional-integral (PI) controller uses a proportional action on the measure and an integral action on the error in order to cancel the steady-state error. This structure is used for the current loops, an IP structure is used for the speed loop. The synthesis of these controllers is performed by the pole compensation method [14].

Stator Current Regulation: The block diagram of the current control loop by a PI controller is illustrated by Fig. 2.

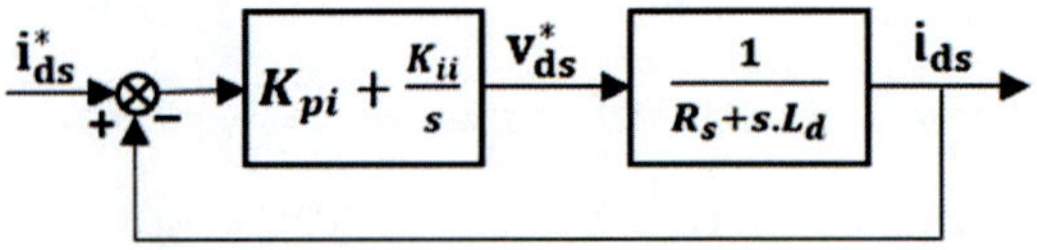

Fig. 2. Stator current regulation loop

The open-loop transfer function is given by:

$$G_{oi}(s) = K_{pi}(s + \frac{K_{ii}}{K_{pi}})\frac{1}{s}\frac{\frac{1}{L_d}}{s + \frac{R_s}{L_d}} \tag{7}$$

By offsetting the dominant pole, the equivalence condition is set as follows:

$$\frac{K_{ii}}{K_{pi}} = \frac{R_s}{L_d} \tag{8}$$

The transfer function in the closed-loop, after calculation and development, is:

$$G_{fi}(s) = \frac{1}{\tau_i s + 1} \tag{9}$$

with

$$\tau_i = \frac{L_d}{K_{pi}}$$

where τ_i ($\tau_i = 0.001\,\mathrm{s}$) is the current loop response time, the same controllers coefficient values are adopted for the two current loops. We can calculate the controller parameters from Eq. (8), by simple identification, then we get:

$$\begin{cases} K_{pi} = \frac{L_d}{\tau_i} \\ K_{ii} = K_{pi}\frac{R_s}{L_d} \end{cases} \tag{10}$$

Speed Regulation: In order to achieve good performance for both the track (response compared to the reference) and to the speed control (response

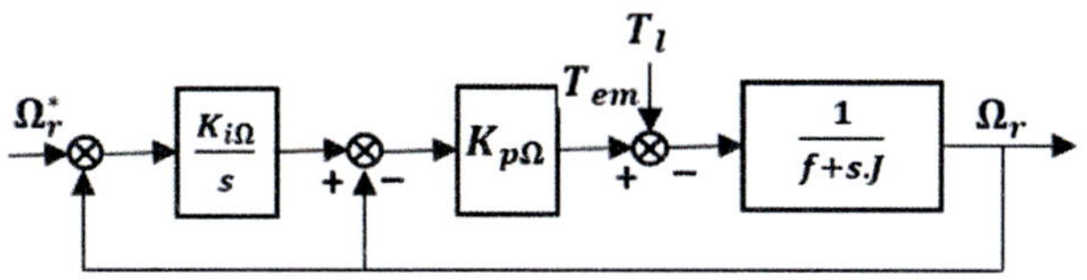

Fig. 3. Speed regulation loop

compared to the disturbance), we consider the functional diagram of the speed control loop shown schematically in Fig. 3 [15]. According to the block diagram of the speed regulation, we have:

$$\Omega(s) = \frac{1}{Js + f}(C_e - C_r) \tag{11}$$

The transfer function of the closed-loop, after calculation, is expressed by:

$$G_{f\Omega}(s) = \frac{1}{\frac{J}{K_{i\Omega}K_{p\Omega}}s^2 + \frac{K_{p\Omega}+f}{K_{i\Omega}K_{p\Omega}}s + 1} \tag{12}$$

The transfer function (12) can be identified to a second-order system in the form:

$$F(s) = \frac{1}{\frac{1}{\omega_n^2}s^2 + \frac{2\zeta}{\omega_n}s + 1} \tag{13}$$

This implies the identities:

$$\begin{cases} \frac{J}{K_{i\Omega}K_{p\Omega}} = \frac{1}{\omega_n^2} \\ \frac{K_{p\Omega}+f}{K_{i\Omega}K_{p\Omega}} = \frac{2\zeta}{\omega_n} \end{cases} \tag{14}$$

If we choose $\zeta = 1$ we will get a relationship between ω_n and the wanted speed response time τ_Ω, which allows us to freely set the system dynamic. This relation is written as: $\omega_n \tau_\Omega = 4.75$. Having properly selected the damping coefficient $\zeta = 1$ and response time $\tau_\Omega = 0.1s$ and next ω_n, we can calculate the controller parameters from Eq. (14), by simple identification, then we get:

$$\begin{cases} K_{p\Omega} = 2J\zeta\omega_n - f \\ K_{i\Omega} = J\frac{\omega_n^2}{K_{p\Omega}} \end{cases} \tag{15}$$

4 Space Vector Modulation Algorithm

Space vector modulation (SVM) is an algorithm for the control of pulse width modulation (PWM). It is used for the creation of alternating current (AC) waveforms; most commonly to drive three-phase AC-powered motors at varying speeds from DC using multiple amplifiers [16–18]. The SVM technique is following the next principles:

- Sampling the reference signal at regular intervals T called modulation period;
- Standardization of all the switches states of the same half-bridge in the center and at both ends of the period.

The desired three sinusoidal voltages to the output are represented by a single vector called the reference voltage vector. We approximate this vector during each modulation interval by acting on the complementary switches K_1 and $K_1^{'}$, K_2 and $K_2^{'}$, K_3 and $K_3^{'}$ shown in Fig. 4.

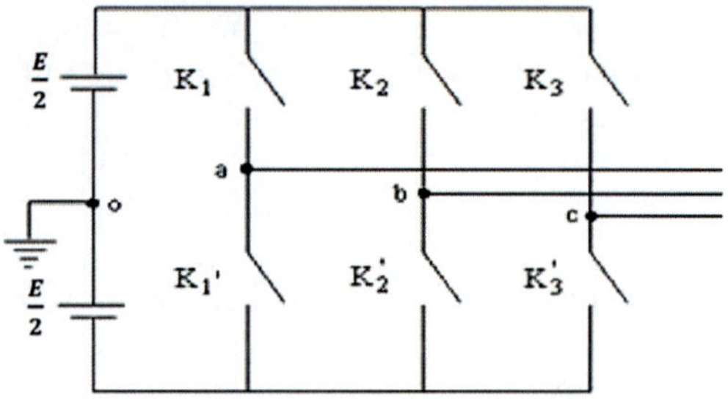

Fig. 4. Two-level three-phase inverter simplified scheme

Voltage vectors are represented by the following equation:

$$\mathbf{V}_i = \begin{cases} \sqrt{\frac{2}{3}} E e^{j(i-1)\frac{\pi}{3}} \\ 0 \end{cases} \tag{16}$$

The ends of these six vectors define the summits of a regular hexagon since two successive vectors are at an angle of $\frac{2\pi}{3}$, Each pair of vectors $\mathbf{V}_i$ and $\mathbf{V}_{i+1}$ define the limits of one of the hexagon six sectors, as demonstrated in Fig. 5.

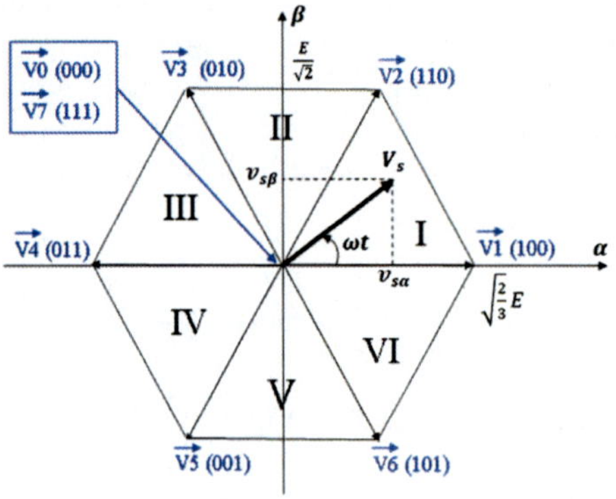

Fig. 5. The switching polygon representation

Note that in sector 6, the notation $\mathbf{V}_{i+1}$ corresponds to the vector $\mathbf{V}_1$. The vector $\mathbf{V}_s$ has an amplitude equal to $\frac{r\sqrt{\frac{3}{2}}E}{2}$ and it is rotating in the direction counterclockwise at an angular velocity equal to the angular frequency ω_r of the desired voltages. At every moment, the vector $\mathbf{V}_s$ can be expressed as a linear combination of two vectors $\mathbf{V}_i$ and $\mathbf{V}_{i+1}$ which are adjacent to it. Therefore, an approximation of $\mathbf{V}_s$ can be generated using a combination within a very short time of two switch states corresponding to the vectors $\mathbf{V}_i$ and $\mathbf{V}_{i+1}$. Inverter and voltage vector different topologies are clearly shown in Fig. 6.

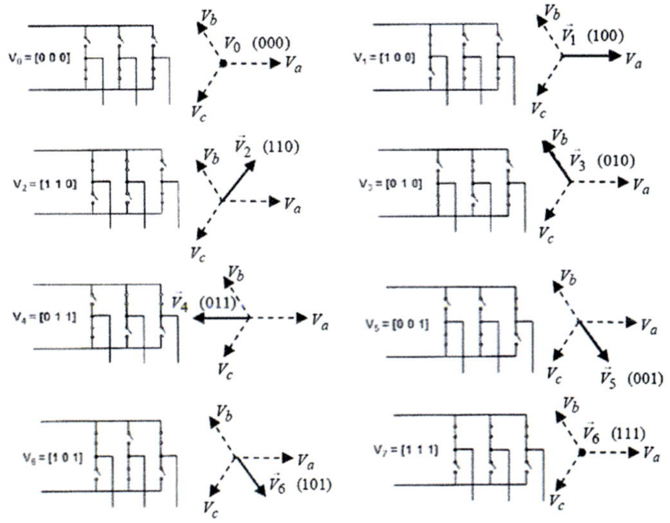

Fig. 6. Inverter and voltage vector different topologies

5 Simulation Results

Figures 7, 8, and 9 represent the simulation results under a load disturbance, while Fig. 10 represents the simulation obtained from a no-load operating at all. The reference of the stator flux is set to 1 Wb. The speed error plot in each figure represents the difference between the reference and the measured speed.

5.1 Operating with Load Torque

In Fig. 7, a load torque of 5 N.m is applied at $t = 0.7$ s and removed at $t = 1.7$ s. This can be clearly noticed from the torque response of Fig. 7 (c).

5.2 Operating with Reverse Speed

In this case, a test of control performance is realized by applying reverse rotational speed from +100 rad/s to −100 rad/s at $t = 1$ s (Fig. 8).

5.3 Operating at the High-Speed Region

In this case, the speed is carried out for +80 rad/s and then for a higher speed of +160 rad/s (Fig. 9).

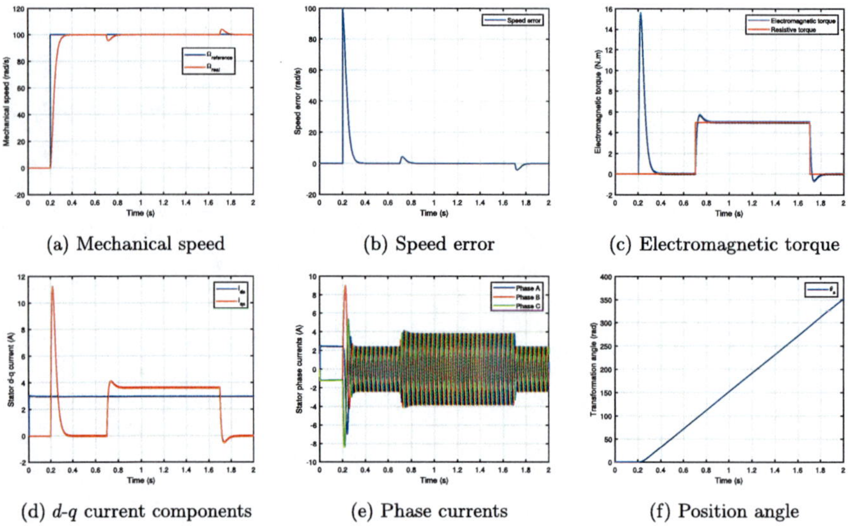

(a) Mechanical speed (b) Speed error (c) Electromagnetic torque

(d) *d*-*q* current components (e) Phase currents (f) Position angle

Fig. 7. Test with load torque

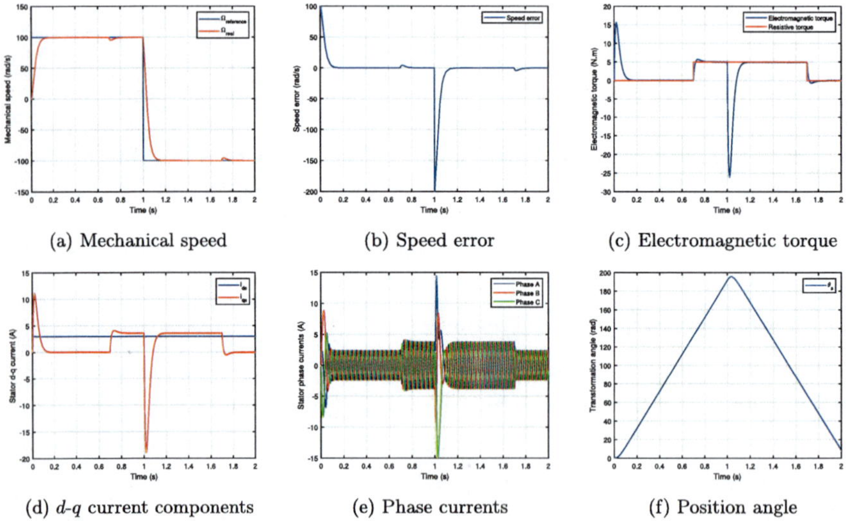

(a) Mechanical speed (b) Speed error (c) Electromagnetic torque

(d) *d*-*q* current components (e) Phase currents (f) Position angle

Fig. 8. Test with reverse speed

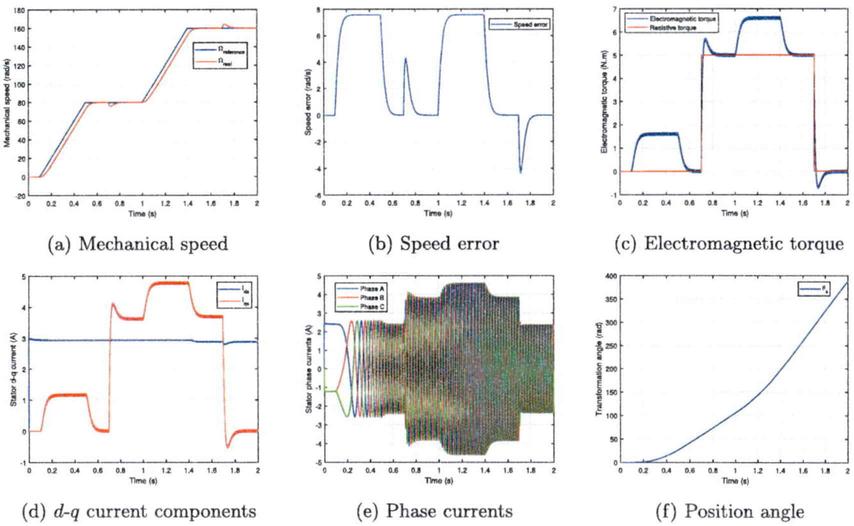

(a) Mechanical speed (b) Speed error (c) Electromagnetic torque

(d) *d*-*q* current components (e) Phase currents (f) Position angle

Fig. 9. Test at high-speed region

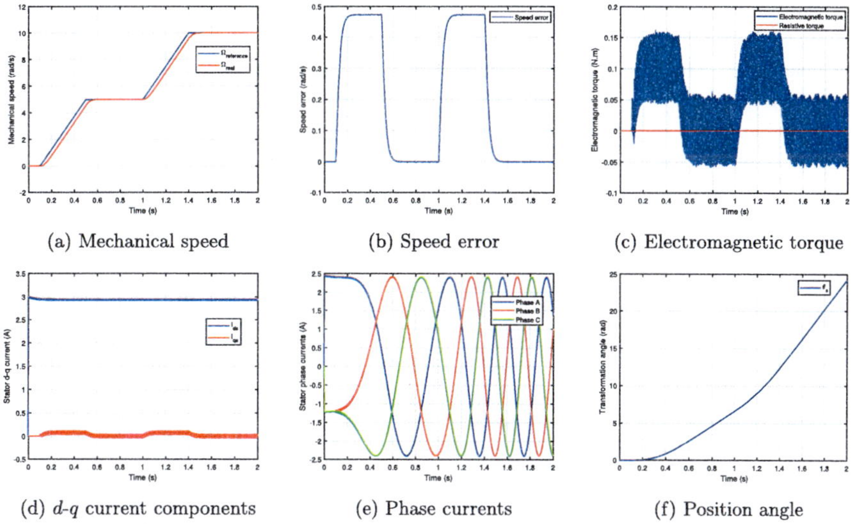

(a) Mechanical speed (b) Speed error (c) Electromagnetic torque

(d) *d*-*q* current components (e) Phase currents (f) Position angle

Fig. 10. Test at low-speed region

5.4 Operating at the Low-Speed Region

In this case, the speed is carried out for a very low speed of +5 rad/s and then for +10 rad/s (Fig. 10).

5.5 Results Analysis

The results shown in Fig. 7 demonstrate the good tracking of the reference speed. The speed regulation loop rejects quickly the disturbance of the applied load. The response time is less than 0.2 s, it represents the time the system needs to catch the reference. Once the steady-state is reached, the system no longer needs torque. The speed regulator cuts the torque demand, it is never zero because it should overcome the friction of the machine which is represented by the friction torque (resistive torque). Figure 8 shows how the system can deal with an instant reverse speed. The control law maintains high performance even when reversing the direction of rotation. Strong torque is noticed at the time of the transition of the reference, this high value of approximately −8 N.m is a reaction of the vector control to keep the fast pursuit. In Figs. 9 and 10, good performance of high and low-speed tracking is achieved with less static error.

6 Conclusion

In this research, the performance of the vector control applied to the synchronous reluctance motor for the rotor speed track has been designed and simulated. The obtained simulation results have confirmed the efficiency and the precision of this proposed control during the sudden load torque, reversal, high and low-speed regions. The space vector modulation gives better control and robustness, the key is the numerical algorithm that allows a good generation of inverter input voltage. This research can be developed into a direct vector version, this type of control requires a good determination of flux and phase module. But it ensures a proper decoupling between the flux and torque, regardless of the machine's operating point.

References

1. Wang, S., et al.: Design and performance analysis of a novel synchronous reluctance machine. Int. J. Appl. Electromagnet Mech **63**(2), 249–265 (2020)
2. Zahraoui, Y., Akherraz, M., Fahassa, C., Elbadaoui, S.: Robust control of sensorless sliding mode controlled induction motor drive facing a large scale rotor resistance variation. In: Proceedings of the 4th International Conference on Smart City Applications, ser. SCA 2019, pp. 1–6. Association for Computing Machinery, New York, October 2019
3. Heidari, H., et al.: A review of synchronous reluctance motor-drive advancements. Sustainability **13**(2), 729 (2021)
4. Zhang, Z.: Advanced non-permanent-magnet reluctance machines for traction applications: a review. In: 2021 IEEE 12th Energy Conversion Congress Exposition - Asia (ECCE-Asia), pp. 2052–2058, May 2021
5. Dinh, B.M., Tan, D.T., Vuong, D.Q.: Electromagnetic design of synchronous reluctances motors for electric traction vehicle. In: Sattler, K.-U., Nguyen, D.C., Vu, N.P., Long, B.T., Puta, H. (eds.) Advances in Engineering Research and Application. LNNS, pp. 373–378. Springer, Cham (2021)

6. Tawfiq, K.B., Ibrahim, M.N., El-Kholy, E.E., Sergeant, P.: Performance improvement of synchronous reluctance machines-a review research. IEEE Trans. Magn. **57**(10), 1–11 (2021)
7. Peng, T., Xu, Y., Tao, H., Chen, Z., Fan, X.: A SVPWM-based fault-tolerance control method for three-level inverter under different working conditions. In: 2021 33rd Chinese Control and Decision Conference (CCDC), pp. 6390–6395, May 2021
8. Zahraoui, Y., Akherraz, M., Fahassa, C., Elbadaoui, S.: Induction motor harmonic reduction using space vector modulation algorithm. Bul. Electric. Eng. Inf. **9**(2), 452–465 (2020)
9. Ferrari, S., Pellegrino, G.: FEA-augmented design equations for synchronous reluctance machines. In: 2018 IEEE Energy Conversion Congress and Exposition (ECCE), pp. 5395–5402, September 2018
10. Lenin, N.C., Sanjeevikumar, P., Iqbal, A., Mbohwa, C.: Linear synchronous reluctance motor—a comprehensive review. In: Konkani, A., Bera, R., Paul, S. (eds.) Advances in Systems, Control and Automation. LNEE, vol. 442, pp. 45–70. Springer, Singapore (2018). https://doi.org/10.1007/978-981-10-4762-6_5
11. Ismaeel, S.M., Allam, S.M., Rashad, E.M.: Current vector control techniques of five-phase synchronous reluctance motor drive systems. In: 2019 21st International Middle East Power Systems Conference (MEPCON), pp. 1180–1185, December 2019
12. Babetto, C., Bacco, G., Bianchi, N.: Synchronous reluctance machine optimization for high-speed applications. IEEE Trans. Energy Convers. **33**(3), 1266–1273 (2018). Conference Name: IEEE Transactions on Energy Conversion
13. Zahraoui, Y., Bennassar, A., Akherraz, M., Essalmi, A.: Indirect vector control of induction motor using an extended kalman observer and fuzzy logic controllers. In: 2015 3rd International Renewable and Sustainable Energy Conference (IRSEC), pp. 1–6, December 2015
14. Zahraoui, Y., Fahassa, C., Akherraz, M., Bennassar, A.: Sensorless vector control of induction motor using an EKF and SVPWM algorithm. In: 2016 5th International Conference on Multimedia Computing and Systems (ICMCS), pp. 588–593, September 2016
15. Li, C., Wang, G., Zhang, G., Zhao, N., Xu, D.: Review of parameter identification and sensorless control methods for synchronous reluctance machines. Chin. J. Electric. Eng. **6**(2), 7–18 (2020). Conference Name: Chinese Journal of Electrical Engineering
16. Boztas, G., Aydogmus, O., Guldemir, H.: Design and implementation of a high-efficiency low-voltage synchronous reluctance motor. Electr. Eng. **104**, 717–725 (2021)
17. Vajsz, T., Számel, L., Handler, A.: Analysis of direct torque control with space vector modulation for synchronous reluctance motor motion control applications and a comparison with other torque-control algorithms. Power Electron. Drives **4**(39) (2019)
18. Vajsz, T., Szamel, L., Handler, A.: An investigation of direct torque control and hysteresis current vector control for motion control synchronous reluctance motor applications. Power Electron. Drives **4**, 115–124 (2019)

Machine Learning Applications for Consumer Behavior Prediction

Nouri Hicham(✉) and Sabri Karim

Research Laboratory on New Economy and Development (LARNED), Faculty of Legal Economic and Social Sciences AIN SEBAA, Hassan II University of Casablanca, Casablanca, Morocco
nhicham191@gmail.com

Abstract. Customers are growing increasingly interested in the degree of service that businesses can provide. Because the services provided by different providers are not significantly different, there is greater competition among firms to maintain and improve the quality of their services. This research study investigates how machine learning can help businesses learn more about their clients and provide them with more personalized services. It also acts as a pattern on how firms can improve the effectiveness of their marketing efforts to acquire new clients and establish long-term connections with existing customers.

In marketing, the prediction of consumer behavior is frequently used to schematize service offerings and develop targeted marketing programs. We have implemented six different machine-learning algorithms to improve further our ability to forecast consumer behavior. We have presented six machine learning models to improve performance, including Random Forest, Gradient Boosting, Logistic Regression, LightGBM, XgBoost, and Decision Tree, to achieve better results. Gradient Boosting, according to this study's findings, is the best overall model for performance and efficiency.

Keywords: Consumer behavior · Machine learning · Prediction

1 Introduction

Marketing history can be defined as developing tactics, concepts, and best practices for maximizing a given type of activity [1]. This tale has a long history further than most people think. For millennia, people have traded goods and services. Thus, the ancients began to use marketing strategies to do business. As a result, the complexity and variety of tactics used in today's world of rapid advancement and exponential digitization have expanded considerably. Marketers have access to e-commerce analytics, customer relationship management databases, artificial intelligence, mobile and Internet of Things analytics, and enormous amounts of unstructured data [1, 2]. There is no longer a shortage of tools or knowledge that prevents marketing plans from being developed and implemented. Dealing with the complexity of big data and technologies for optimizing marketing efforts is a significant challenge.

M. Ben Ahmed et al. (Eds.): SCA 2022, LNNS 629, pp. 666–675, 2023.
https://doi.org/10.1007/978-3-031-26852-6_62

Before this, there was nothing remotely similar to it anywhere else in the realm of digital marketing research and technology [3]. When it comes to marketing research, artificial intelligence (AI) is quickly becoming a valuable tool for companies because of its ability to perform human-like activities, such as learning on their own or making judgments based on big datasets [4]. Studying how technological improvements affect a firm's ability to manage consumer needs and deliver solutions and products [5] is a current focus for researchers. An extensive body of work has already been done on how various technological advancements impact marketing. Recent calls for study have encouraged increased exploration into AI-related themes and their roles in marketing. However, research on the confluence of consumer behavior and artificial intelligence (AI) has recently gained growing attention [6].

According to the latest research, machine learning (ML) and deep learning (DL) are becoming increasingly crucial in marketing. AI-based platforms, like Microsoft Cognitive, Google Assistant, IBM Watson, and Amazon Lex are being employed by many businesses to improve their productivity and profitability. Google, Microsoft, IBM, and Amazon are just a few companies using AI-based platforms to improve their performance. This strategy promotes customer involvement across all advertising networks while enhancing market forecasting and automating capabilities. Because of their predicted expansion from $10.1 billion in 2018 to $126 billion by 2025, DL and ML have been labeled the most important corporate technologies [7]. The Society for Industrial and Organizational Psychology (SIOP) has named machine learning the number one workplace advancement on their list of the top business advancements [8].

As a result of this study, companies will be able to understand their customers better and provide them with a fantastic experience by using machine learning to boost marketing efforts and forecast consumer behavior. This provides a beautiful illustration of how firms can improve their marketing efforts by utilizing what they already do well.

2 Related Work and Conceptual Background

Business and marketing have relied more and more on machine learning (ML) in the recent decade [9, 10]. Human-like behaviors such as optimizing marketing campaigns, making strategic judgments based on enormous information, or learning autonomously are becoming extremely valuable to enterprises [11]. When it comes to sectors traditionally assumed to necessitate human-level cognitive growth, machine learning, in especially, has made significant progress [12]. ML, the central paradigm of contemporary AI research, can be credited with a large portion of this accomplishment [13].

The success of machine learning in marketing has been replicated in research, with significant contributions made recently, notably from 2017 onward, demonstrating the effectiveness of machine learning in marketing. When machine learning first gained attention in the 1980s, researchers concentrated on robots and expert systems [14], and the field has continued to grow ever since. Because of three essential factors: the accessibility of processing capacity, the rise of Big Data, and the development of machine learning techniques and technological enablers, it has regained popularity amongst marketing practitioners and scholars in recent years [15, 16].

For organizations to achieve their objectives and recruit new consumers or retain existing customers [10], various strategies and machine learning techniques are employed. A critical step in optimizing marketing campaigns and ensuring that the correct customers are served is the ability to anticipate behavior of customers [19]. With the help of a collection of machine learning techniques [17], this unsupervised categorization method aims to group unlabeled data with ordinary qualities. Many of the standard machine learning models that are used to construct clustering, such as support vector machines, artificial neural networks (ANNs), and random forests are still considered "black boxes." Gradient boosting machines, which are more advanced techniques, have lately been accessible [20–22].

The amount of machine learning methods that were presented to assess customer behavior resulted from this achievement. The machine learning technologies Support vector machines (SVM), Decision Trees (DT), and Random Forests (RF) [18] are reliable and straightforward to grasp when it comes to forecasting client behavior. According to the evaluation metrics Accuracy, Recall, Precision, and F1-score, in [26], the findings of the random forest are more accurate than the results of many other machine learning techniques. On the other hand, the authors in [27] devise a hybrid prediction model for user purchasing behavior that incorporates SVM and logistic regression (LR) approaches. They undertake empirical research to see how effective the model performs. According to the findings, the fusion model outperforms the single model in prediction accuracy.

3 Methodology

In this section, we would go over the machine learning techniques that were used in this research. We should start by taking a look at the overall architecture of our investigation.

3.1 Global Architecture

There are four key architectural steps depicted in Fig. 1.

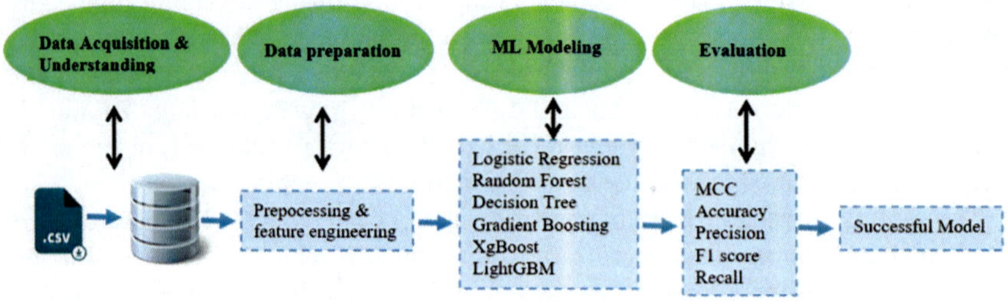

Fig. 1. Global architecture

Step 1: Data Acquisition & Understanding

Two thousand two hundred forty clients were included in this study's trial, containing 28 variables related to marketing data from the international food company, iFood. Which today has several hundred thousand current customers and feeds nearly one million people annually. They sell a variety of exotic fruits, meat, sweet products, wine, and fresh fish. Those can also be categorized into normal and gold items. Customers can purchase and buy things from the company via the company's website, catalogs, and physical storefronts. It is shown in Fig. 2 that there is a correlation between each feature of the customer dataset and the variables listed in Table 1.

Table 1. Variables description in dataset

Variable	Type	Description
ID	Integer	Customer's unique identifier
Year_Birth	Integer	Customer's birth year
Education	Object	Customer's education level
Marital_Status	Object	Customer's marital status
Income	Object	Customer's yearly household income
Kidhome	Integer	Number of children in customer's household
Teenhome	Integer	Number of teenagers in customer's household
Dt_Customer	Object	Date of customer's enrollment with the company
Recency	Integer	Number of days since customer's last purchase
MntWines	Integer	Amount spent on wine in the last 2 years
MntFruits	Integer	Amount spent on fruits in the last 2 years
MntMeatProducts	Integer	Amount spent on meat in the last 2 years
MntFishProducts	Integer	Amount spent on fish in the last 2 years
MntSweetProducts	Integer	Amount spent on sweets in the last 2 years
MntGoldProds	Integer	Amount spent on gold in the last 2 years
NumDealsPurchass	Integer	Number of purchases made with a discount
NumWebPurchases	Integer	Number of purchases made through the company's web site
NumCatalogPurchass	Integer	Number of purchases made using a catalogue
NumStorePurchases	Integer	Number of purchases made directly in stores
NumWebVisitsMonh	Integer	Number of visits to company's web site in the last month
AcceptedCmp3	Integer	1 if customer accepted the offer in the 3rd campaign, 0 otherwise
AcceptedCmp4	Integer	1 if customer accepted the offer in the 4th campaign, 0 otherwise
AcceptedCmp5	Integer	1 if customer accepted the offer in the 5th campaign, 0 otherwise
AcceptedCmp1	Integer	1 if customer accepted the offer in the 1st campaign, 0 otherwise
AcceptedCmp2	Integer	1 if customer accepted the offer in the 2nd campaign, 0 otherwise
Response	Integer	1 if customer accepted the offer in the last campaign, 0 otherwise
Complain	Integer	1 if customer complained in the last 2 years, 0 otherwise
Country	Object	Customer's location

Step 2: Data Preparation

Data selection, data transformation and data preprocessing, are all part of the preparation phase for a machine learning techniques [23, 24].

a - data selection: It is necessary to select a subset of all the available data to begin working with it at this level. Consider the data that is now available and the data that is missing and the data that may be deleted.

b - Data preprocessing: We are transforming raw data into a form that can be easily comprehended by a machine. Raw data mining is crucial, considering we can't do much with raw data. Make absolutely sure that our data is of excellent quality before using machine learning algorithms.

c - Data transformation: Feature engineering, often referred to as data transformation, describes processes for converting previously processed data. This step consumes a lot

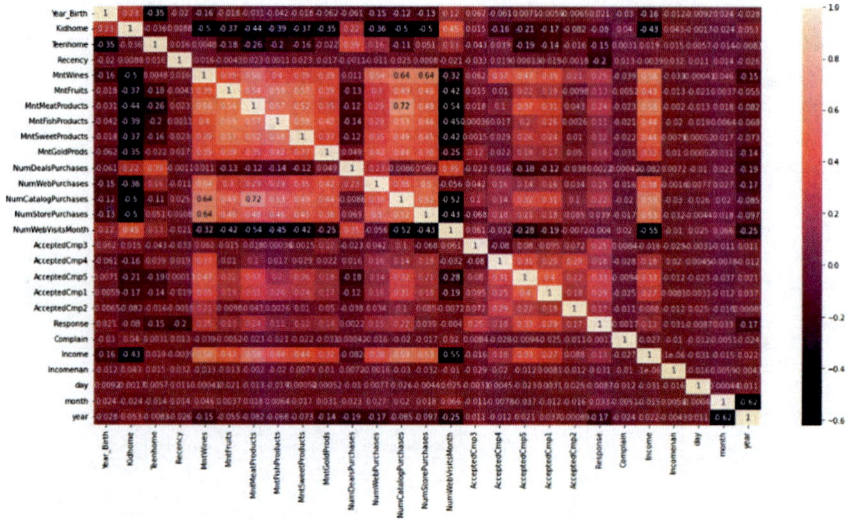

Fig. 2. Correlation between each features

of time to develop features using given data. However, they could significantly increase the effectiveness of algorithms for machine learning.

Step 3: Machine Learning Modeling

Section 3.2 will detail some of the most powerful machine learning algorithms. Using artificial intelligence (AI) to improve advertising campaigns has been a great success. Here, we will go into deeper depth on applying these algorithms to analyze our dataset.

Machine learning has become an integral part of our daily lives. It is inevitable that we will be introduced to artificial intelligence daily. For several reasons, establishing metrics that provide accurate and reliable assessments is essential because of machine learning models. We will use Precision, F1 score, Accuracy, Recall, and Matthew's correlation coefficient (MCC) to evaluate our model's performance.

It is clear that accuracy is most significant parameter for evaluating the performance of machine learning programs [25].

$$\text{Accuracy} = (\mathrm{T_N} + \mathrm{T_P} + \mathrm{F_N} + \mathrm{F_P})^{-1} \times (\mathrm{T_N} + \mathrm{T_P}). \tag{1}$$

where $\mathrm{T_N}$, $\mathrm{F_N}$, $\mathrm{T_P}$ and $\mathrm{F_P}$ are respectively the number of true negatives, false negatives, true positives, and false positives.

We need more specific metrics to improve the performance of our models. The F1 score is calculated by taking the harmonic mean of recall and precision:

$$\text{Precision} = (\mathrm{T_P} + \mathrm{F_P})^{-1} \times \mathrm{T_P}. \tag{2}$$

$$\text{Recall} = (\mathrm{T_P} + \mathrm{F_N})^{-1} \times \mathrm{T_P} \tag{3}$$

$$\text{F1 score} = 2 \times (\text{Recall} + \text{Precision})^{-1} \times (\text{Recall} \times \text{Precision}). \tag{4}$$

In additional of this, The MCC stands for Matthew's correlation coefficient. To evaluate models statistically, it is utilized. For a 2 × 2 contingency table, chi-square statistics can be used to calculate the change between forecasted and actual values.

It is calculated in the following way:

$$\mathrm{MCC} = (\mathrm{TP} * \mathrm{TN} - \mathrm{FP} * \mathrm{FN})\sqrt{(TP + FP)(TP + FN)(TN + FP)(TN + FN)} \quad (5)$$

3.2 The Used Machine Learning (ML) Models

Machine Learning methods that were employed in our work are discussed in this section, and they are classified as follows:

Logistic Regression (LR)
When it comes to classification challenges, Logistic regression (LR) [1] is among the most extensively used approaches; the sigmoid function is employed to convert a linear model into a logistic model in this technique. The LR method links a categorical variable to one or more independent variables grouped. The following is the equation for it:

$$\frac{1}{1 + e^{-Y}} \quad (6)$$

For any given collection of n unknown variables, Y denotes a row of vectors, including the elements κ which are not reliant on each other and their weights λ:

$$\mathrm{Y} = \lambda_0 + \lambda_1\kappa_1 + \lambda_2\kappa_2 + \ldots \lambda_n\kappa_n, \ \text{with } n \in \mathbb{R} \quad (7)$$

When it comes to classification, the average is used for regression, and most votes are used for classification. One of the essential characteristics of the Random Forest is that it can manage data sets that contain variables that are both categorized and continuous, such as those seen in regression and classification. When used correctly, it results in productive results for categorization difficulties [2].

Decision Tree (DT)
In mathematics, a DT is a type of self-learning that makes use of time series forecasting in the tree structure that has been learned through supervised practice to identify the best alternative. This includes automated clustering based on a training data set and data type predicting for data that has not previously been classified. Each attribute is assigned a value by the rank widget in [21], dependent on the attribute's relation to the class.

Gradient Boosting (GB)
When performing supervised tasks such as classification or regression, GB is a method for combining several weak learners to construct a solid interactive ensemble [23]. There are numerous real-world circumstances where this cutting-edge machine-learning technology provides cutting-edge results. Because of its adaptability, it has gained a good

reputation; it manages diverse tasks and data by default, and it has a large number of open-source solutions to choose from.

Extreme Gradient Boosting (XGBoost)

Extensive research has shown that XGBoost is one of the most effective supervised learning methods, and it is one of the techniques used to apply this technique. This method can solve regression and classification problems, among other things. A key reason why data scientists use XGBoost is the speed with which it does out-of-core computations.

LightGBM

One of Microsoft Research Asia's Gradient-Boosted Decision Trees frameworks (GBDT) was LightGBM, released in 2017. Its purpose is to improve computational efficiency to make it possible to conduct more efficiently significant data prediction issues [25]. Like other boosting strategies, grouping weak learners together to create a powerful one is the goal of GBDT. The decision tree utilized in the GBDT approach can only be a regression tree in this case since each tree learns the conclusions and residuals of all preceding trees. The LightGBM algorithm has been proposed for use at this time. The forecast accuracy is not affected, but the performance of the prediction and the quantity of memory consumed are significantly affected [24].

4 Experimental Results

The findings of our investigations examining the performance of the most commonly used diverse analytical techniques, which fall into distinct categories of machine learning and are used to forecast consumer behavior, are presented in this section of the paper. There are a variety of machine learning techniques available, including Logistic Regression,Gradient Boosting,Random Forest, XgBoost, Decision Tree, and LightGBM.

The accuracy of the different models' performance is the first thing we look at, so we start there. It is the ability to appropriately discriminate between believable and non-credible scenarios, referred to as accuracy. Figure 3 demonstrates the accuracy of many algorithms used to conduct our research and analyze the data. We can see from the accuracy test results that Gradient Boosting produces the best result, which is approximately 89.12%. Both XgBoost and LightGBM can be advised, with accuracy rates of 88.90% and 88.39%, respectively. Finally, with an accuracy rate of 70.65%, Logistic Regression can be recommended. Figure 3 summarizes the accuracy findings, which include the specifics.

MCC, recall, F1 score, and precision are the additional metrics employed in our study to evaluate the machine learning methods. Table 2 summarizes the findings of our research.

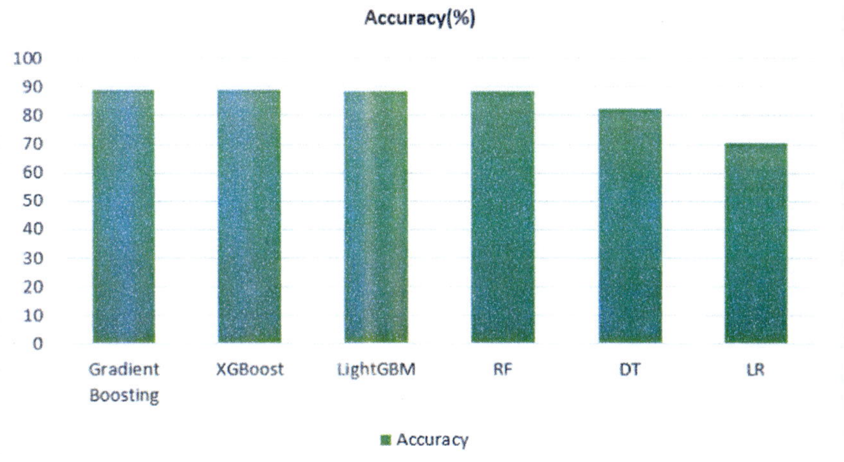

Fig. 3. Accuracy of ML technics

Table 2. Performance evaluation of different models

	Gradient Boosting	XGBoost	LightGBM	RF	DT	LR
Accuracy	0.8912	0.8890	0.8839	0.8828	0.8225	0.7065
Precision	0.7077	0.6961	0.6700	0.6999	0.4213	0.2998
Recall	0.4815	0.4778	0.4627	0.3880	0.4181	0.6907
F1 score	0.5693	0.5624	0.5438	4975	0.4148	0.4164
MCC	0.5242	0.5155	0.4928	0.4627	0.3137	0.3021

5 Conclusion and Future Work

As a result of our study paper, companies can learn more about their customers, deliver a better customer experience, increase customer retention in order to enhance earnings by developing a deep connection with them. Logistic Regression, Gradient Boosting, XgBoost, Decision Tree, and LightGBM are some machine learning techniques that we have presented. The Gradient Boosting model had the best accuracy, recall, precision, F1 score, and MCC scores of 89.12%, 48.15%, 70.77%, 56.93%, and 53.42%, as determined by the results of this investigation. Deep learning and hybrid models can be added to the mix of research. The performance can be assessed using other metrics. Several datasets from various domains can be used to verify the model.

References

1. Palmatier, R.W., Crecelius, A.T.: The "first principles" of marketing strategy. AMS Rev. **9**(1–2), 5–26 (2019). https://doi.org/10.1007/s13162-019-00134-y
2. Chintagunta, P., Hanssens, D.M., Hauser, J.R.: Marketing science and big data. Mark. Sci. **35**(3), 341–342 (2016). https://doi.org/10.1287/mksc.2016.0996
3. Crittenden, W.F., Biel, I.K., Lovely, W.A.: Embracing digitalization: student learning and new technologies. J. Mark. Educ. **41**(1), 5–14 (2019). https://doi.org/10.1177/0273475318820895
4. Huang, M.-H., Rust, R.T.: A strategic framework for artificial intelligence in marketing. J. Acad. Mark. Sci. **49**(1), 30–50 (2020). https://doi.org/10.1007/s11747-020-00749-9
5. Kumar, V., Rajan, B., Venkatesan, R., Lecinski, J.: Understanding the role of artificial intelligence in personalized engagement marketing. Calif. Manag. Rev. **61**(4), 135–155 (2019). https://doi.org/10.1177/0008125619859317
6. Davenport, T., Guha, A., Grewal, D., Bressgott, T.: How artificial intelligence will change the future of marketing. J. Acad. Mark. Sci. **48**(1), 24–42 (2019). https://doi.org/10.1007/s11747-019-00696-0
7. Tractica: the global revenue of AI software will reach 126 billion US dollars in 2025, focusing on consumer, financial services and other applications | GSAS. https://www.gsas.edu.hk/tractica-the-global-revenue-of-ai-software-will-reach-126-billion-us-dollars-in-2025-focusing-on-consumer-financial-services-and-other-applications/. Accessed 10 Feb 2022
8. Featured Articles, SIOP. https://www.siop.org/Research-Publications/Items-of-Interest/ArtMID/19366/ArticleID/3361/Top-10-Workplace-Trends-for-2020. Accessed 11 Feb 2022
9. van Giffen, B., Herhausen, D., Fahse, T.: Overcoming the pitfalls and perils of algorithms: a classification of machine learning biases and mitigation methods. J. Bus. Res. **144**, 93–106 (2022). https://doi.org/10.1016/j.jbusres.2022.01.076
10. Vemuri, V.K.: The AI Advantage: How to Put the Artificial Intelligence Revolution to Work: by Thomas H. Davenport, 248 p. MIT Press, Cambridge (2018). ISBN 9780262538008. J. Inf. Technol. Case Appl. Res. **22**(1), 70–72 (2020). https://doi.org/10.1080/15228053.2020.1756084
11. Bock, D.E., Wolter, J.S., Ferrell, O.C.: Artificial intelligence: disrupting what we know about services. J. Serv. Mark. **34**(3), 317–334 (2020). https://doi.org/10.1108/JSM-01-2019-0047
12. Gregoriades, A., Pampaka, M., Herodotou, H., Christodoulou, E.: Supporting digital content marketing and messaging through topic modelling and decision trees. Expert Syst. Appl. **184**, 115546 (2021). https://doi.org/10.1016/j.eswa.2021.115546
13. Dwivedi, Y.K., et al.: Setting the future of digital and social media marketing research: perspectives and research propositions. Int. J. Inf. Manag. **59**, 102168 (2021). https://doi.org/10.1016/j.ijinfomgt.2020.102168
14. Potrawa, T., Tetereva, A.: How much is the view from the window worth? Machine learning-driven hedonic pricing model of the real estate market. J. Bus. Res. **144**, 50–65 (2022). https://doi.org/10.1016/j.jbusres.2022.01.027
15. Chan, C.C.H., Hwang, Y.-R., Wu, H.-C.: Marketing segmentation using the particle swarm optimization algorithm: a case study. J. Ambient. Intell. Humaniz. Comput. **7**(6), 855–863 (2016). https://doi.org/10.1007/s12652-016-0389-9
16. Lemmens, A., Gupta, S.: Managing Churn to Maximize Profits. SSRN Electron. J. (2017). https://doi.org/10.2139/ssrn.2964906
17. Martínez, A., Schmuck, C., Pereverzyev, S., Pirker, C., Haltmeier, M.: A machine learning framework for customer purchase prediction in the non-contractual setting. Eur. J. Oper. Res. **281**(3), 588–596 (2020). https://doi.org/10.1016/j.ejor.2018.04.034

18. Hu, X., Yang, Y., Chen, L., Zhu, S.: Research on a prediction model of online shopping behavior based on deep forest algorithm. In: 2020 3rd International Conference on Artificial Intelligence and Big Data (ICAIBD), Chengdu, China, pp. 137–141 (2020). https://doi.org/10.1109/ICAIBD49809.2020.9137436
19. Berti-Equille, L.: Learn2Clean: optimizing the sequence of tasks for web data preparation. In: The World Wide Web Conference on - WWW 2019, San Francisco, CA, USA, pp. 2580–2586 (2019). https://doi.org/10.1145/3308558.3313602
20. Frye, M., Schmitt, R.H.: Structured Data Preparation Pipeline for Machine Learning-Applications in Production, p. 6 (2020)
21. Ağbulut, Ü., Gürel, A.E., Biçen, Y.: Prediction of daily global solar radiation using different machine learning algorithms: evaluation and comparison. Renew. Sustain. Energy Rev. **135**, 110114 (2021). https://doi.org/10.1016/j.rser.2020.110114
22. Tangwannawit, S., Tangwannawit, P.: An optimization clustering and classification based on artificial intelligence approach for internet of things in agriculture. IAES Int. J. Artif. Intell. IJ-AI **11**(1), 201 (2022). https://doi.org/10.11591/ijai.v11.i1.pp201-209
23. Katsov, I.: Introduction to algorithmic marketing: artficial intelligence for marketing operations. Ilya Katsov, Sunnyvale, Calivornia (2018)
24. Liang, W., Luo, S., Zhao, G., Wu, H.: Predicting hard rock pillar stability using GBDT, XGBoost, and LightGBM algorithms. Mathematics **8**(5), 765 (2020). https://doi.org/10.3390/math8050765
25. Ju, Y., Sun, G., Chen, Q., Zhang, M., Zhu, H., Rehman, M.U.: A model combining convolutional neural network and LightGBM algorithm for Ultra-short-term wind power forecasting. IEEE Access **7**, 28309–28318 (2019). https://doi.org/10.1109/ACCESS.2019.2901920
26. Valecha, H., Varma, A., Khare, I., Sachdeva, A., Goyal, M.: Prediction of consumer behaviour using random forest algorithm. In: 2018 5th IEEE Uttar Pradesh Section International Conference on Electrical, Electronics and Computer Engineering (UPCON), Gorakhpur, pp. 1–6 (2018). https://doi.org/10.1109/UPCON.2018.8597070
27. Hu, X., Yang, Y., Zhu, S., Chen, L.: Research on a hybrid prediction model for purchase behavior based on logistic regression and support vector machine. In: 2020 3rd International Conference on Artificial Intelligence and Big Data (ICAIBD), Chengdu, China, pp. 200–204 (2020). https://doi.org/10.1109/ICAIBD49809.2020.9137484

LSTM Deep Learning Method for Radiation Short and Long-Term Prediction

Mohamed Khala[1](✉), Houda Abouzid[2], Sara Teidj[3], Omar Eloutassi[1], and Choukri Messaoudi[1]

[1] Optoelectronics and Applied Energy Techniques, Faculty of Science and Technology, Moulay Ismail University of Meknes, Meknes, Morocco
mo.khala@edu.umi.ac.ma
[2] Computer Science and New Technology Trends, Abdelmalek Essaadi University, Tetouan, Morocco
[3] Department of Physics, Faculty of Science and Technology, Moulay Ismail University of Meknes, Meknes, Morocco

Abstract. The renewable energy market, like conventional energy, is often unstable. Solar energy in particular presents disturbances linked to the spatiotemporal fluctuations that solar radiation undergoes. To remedy this problem, it is essential to predict its fluctuations that have exhausting consequences for integration into the electricity network and reduce the dependence on fossil fuels essential for socio-economic development. The main objective of this study is to propose a prognostication model based on artificial intelligence to predict with great accuracy the fluctuations of global solar radiation (GSR). Indeed, we adopted a deep dynamic recurrent memory artificial neural network (ANN) model which can manipulate the dynamics of atmospheric factors and is capable to save the history of these characteristics. For this purpose, we developed two long-short term memory (LSTM) models for short and long term forecasting of global solar radiation time series. The first model is a monovariate LSTM, the second one is a multivariate LSTM, after selecting the input variables (predictors) by correlation study, the models are trained by dividing the data set into two subsections, the first subsection is used for model training and the second subsection is reserved for testing. The indices and results established in this work demonstrate the robustness attributed to our method which can provide power grid managers at Er-Rachidia, MOROCCO (Latitude: 31°55′53″N; Longitude: 4°25′35″ W; Altitude: 1039 m) credible predictions to ensure better supervision of the energy system and to adapt solar energy to the electricity production system, especially multivariate LSTM model which proved more accuracy compared to monovariate LSTM.

Keywords: Global solar radiation · Prediction · Deep artificial neural network · Long short time memory

1 Introduction

The world faced severe crises related to energy issues. To ease these crises, it is essential to make a transition to renewable energies. Nevertheless, the integration of this type of energy in the Grid is not always obvious and often unsuitable and destabilizes the Grid.

M. Ben Ahmed et al. (Eds.): SCA 2022, LNNS 629, pp. 676–696, 2023.
https://doi.org/10.1007/978-3-031-26852-6_63

The deployment of renewable energy sources and their storage reinforces the idea of the Smart Grid to solve the problem of interruption of renewable energy, stabilize the frequency of grid integration and prevent back-feeding to avoid the reversal of the electric flow [1]. Since December 2015, Morocco has set a challenge to accelerate its energy transition toward a smart grid by increasing the installed power from renewable energies to up 52% by 2030 [2].

Current scientific research dedicated to the integration of renewable energies is directed towards the exploitation and application of new technologies to accurately predict solar energy. Artificial intelligence with its rapidly growing in several areas [3–5] remains a very relevant choice, especially with the appearance of new information technologies and the availability of a large database. Indeed, several recent researches based on Machine Learning (ML) algorithms have been adopted with the aim of the prediction and optimization of energy systems.

In this context, Voyant et al. [6] listed several journals associated with ML processes that aim to predict global solar radiation. Several methods of predictors demonstrate that it is the ANNs that are generally used to establish a set of predictors ($> 70\%$ of cases). Some articles have reviewed various hybrid approaches involving decision tree-based ML methods and non-linear learning i.e. Deep Learning (DL) methods such as Long- and Short-Term Memory neural networks (LSTM) and Convolution Neural Networks (CNN) [7–9]. Other papers have focused on LSTM and/or Gated Recurrent Unit (GRU) memory neural networks to highlight and use time series of weather data that helps in optimizing the power of output of photovoltaic parks [10–12]. LSTM networks have shown great importance in terms of storing and transporting information over long distances and capturing long-term dependencies. Recently Jallal et al. [13, 14] proposed two DL algorithms, able to handle nonlinearities and dynamic behaviour of meteorological data and also able to generate accurate real-time predictions of RSG data, the first algorithm achieves a coefficient of correlation $R = 99.38\%$ and the second one achieves a coefficient of correlation $R = 99.13\%$.

For this purpose, we are interested in predicting the time series of global solar radiation with less complexity. We are motivated by the research mentioned above, the optimal parameters for the proposed model will be examined and designated. Another strong point of the suggested technique is demonstrated by the considerable prediction data which reaches 217 days.

2 Methodology

2.1 LSTM Theory

The standard Recurrent Neural Networks (RNN) are challenged with the complications of the disappearance of gradient i.e. the vanishing gradient problem, which does not allow to change parameters (weight matrices) considering the previous entries [15]. However, with the appearance of LSTM type networks, answers have been attributed to these problematics [3]. An LSTM network, or recurrent short- and long-term memory network or even more explicitly, an intelligent memory cell with feedback connections. In practice, this type of network is frequently used, and it offers solutions to the difficulties of gradient disappearance.

In this section, we discuss the LSTM theory. Unlike the recurrent unit, which simply calculates a weighted sum of the input signals and applies a nonlinear function, each LSTM unit maintains one memory at a time. An LSTM unit consists of three gates, an input gate that can accept or reject the upgrade of the LSTM system, another output that checks if the state of the cell is transferred out of this system, and a for-gotten gate with internal memory providing the ability to perform a cell state reset. An LSTM unit thus can decide whether to keep or forget the existing memory via the introduced gates. Intuitively, if the LSTM unit detects an important feature of an input sequence at an early stage, it easily carries that information over a long distance, thereby capturing potential long-distance dependencies. A comparison between simple RNN and LSTM units is shown in Fig. 1.

LSTMs are the most interesting RNN models used for DL to solve time-series data prediction problems [9, 10].

The parameters of an LSTM network shown in Fig. 1 are calculated from the following formulas:

The LSTM network input gate i_t decides how much current information should be transmitted according to the following equation:

$$i_t = \sigma\left(x_t U^i + h_{t-1} W^i\right) \tag{1}$$

The forget gate f_t decides which information to forget from the previous state as follows:

$$f_t = \sigma\left(xU^f + h_{t-1} W^f\right) \tag{2}$$

The output gate O_t defines the internal state information that should be transmitted using the following equation:

$$O_t = \sigma\left(x_t U^o + h_{t-1} W^o\right)$$

The cell state also called the internal memory C_t is updated in two steps. The first one consists of the information about the state of the cell

$$\widehat{C}_t = \tanh\left(xU^g + h_{t-1} W^g\right) \tag{3}$$

The U and W correspond to each of the weight matrices applied to the inputs of the associated operation. After calculating the cell state Ct, the internal memory C_t is updated from $\widehat{C}_t$ as follows:

$$C_t = \sigma\left(f^t * C_{t-1} + i_t * \widehat{C}_t\right) \tag{4}$$

where * is an element-by-element multiplication operation. The state of a cell C_t, is used to get the state of the next time step using the following formula:

$$h_t = \tanh(C_t) * O_t \tag{5}$$

where x_t, h_{t-1} and C_{t-1} are respectively the input to the hidden layer, the previous hidden state of the previous time step, and the previous state of the cell of the previous time step.

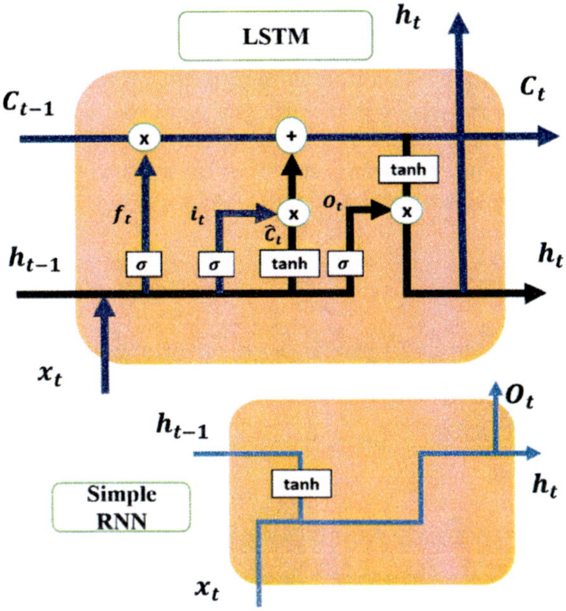

Fig. 1. Comparison between simple RNN and LSTM cells

Activation Function

There are several types of activation functions, the choice of the activation function depends on the application, whether or not it is necessary to have binary outputs. The mentioned activation functions are the Sigmoid (σ), and Tangent-hyperbolic (tanh) functions [14]:

Sigmoid function:

$$\sigma(x) = \frac{1}{1 + e^{-x}} \tag{6}$$

Hyperbolic tangent function;

$$\tanh(x) = \frac{e^{x} - e^{-x}}{e^{x} + e^{-x}} \tag{7}$$

3 Database Collection and Preprocessing

The PROPRE.MA project initiated by Cadi Ayyad University, MOROCCO and sponsored by the Research Institute for Solar Energy and Renewable Energies (IRESEN), the data used during the design of this RSG prediction system are drawn from the meteorological station installed within this project in the Faculty of Sciences and Techniques of Er-Rachidia. The data collected between 01/01/2016 and 01/10/2020 are analyzed and processed to obtain information that may be useful for learning our predicting system.

3.1 Data Collection

The weather data measured by the sensors and instruments of the weather station are sent in real-time by wireless connection in an orderly manner to the weather station console. The Weather link data management software installed on our PC retrieves the recorded weather data at a frequency of 30 min, and allows us to visualize them in the form of bulletins or graphs over several periods of time as shown in Fig. 2.

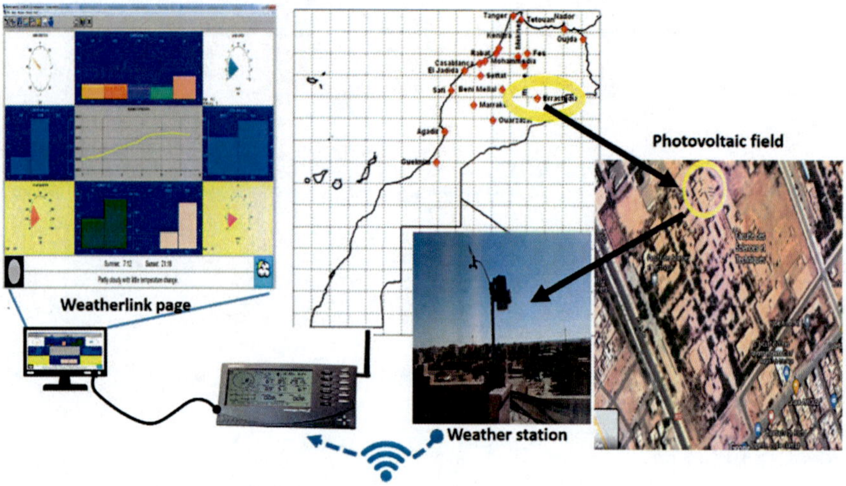

Fig. 2. Data collection process

3.2 Data Cleaning

Before analyzing the collected data and adjusting the prediction models, it is essential to process and purify its content. As a matter, missing data can mislead the learning process or be contributing to misinterpretation of the data. To avoid any such issues, we replace the missing values with the average of measured values of the previous and the following days at the same time step, provided that the total number of missing values on the same day does not exceed 12-time steps (6 h).

3.3 Selection of Predictors by a Correlation Study

Statistical studies are proposed to master the database which allows the analysis of the relationship between the elements that contribute to this database.

Figures Fig. 3 and Fig. 4 show Pearson's (R) and Phik's (φk) correlation indicators respectively [16]. It can be considered that the three variables: THWindex, Solar Energy, and UVindex can build the best predictors for the prediction of global solar radiation since the last three fields are strongly correlated and interactive with the field from GSR.

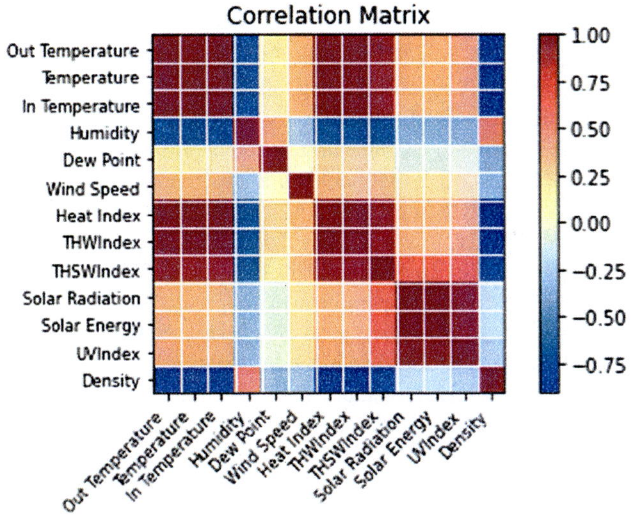

Fig. 3. Pearson's correlation of the global solar radiation parameters.

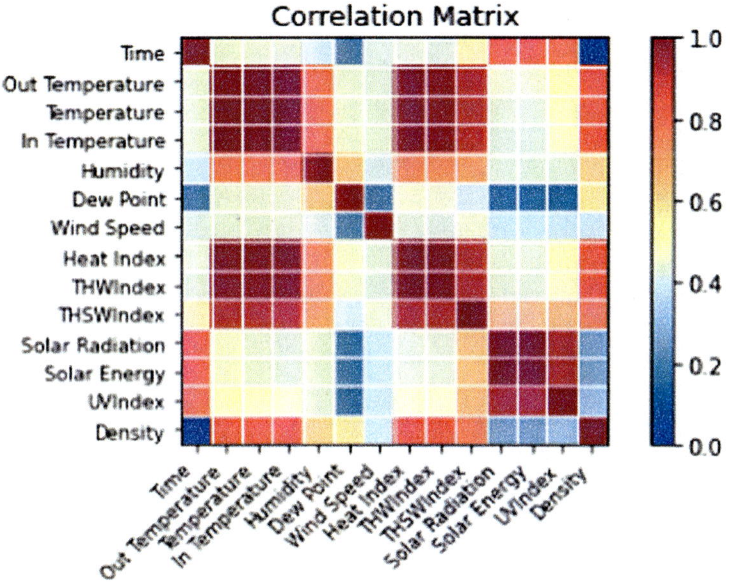

Fig. 4. Phik (φk) correlation matrix of the global solar radiation parameters.

3.4 Model Architecture

The architecture of the model is, in general, a set of an input layer, which receives the collected data, an output layer, which produces information, calculated in the form of predictions and a hidden layer used to connect the input layer, and the output layer using processing unit neurons [17]. However, recent established studies seek to improve the performance of the models by increasing the number of hidden layers and the number of nodes [10], the optimization of the model is done randomly at the beginning and then validated later by error estimators and the correlation coefficient [17, 18].

In this study, all the basic LSTM models are structured in the same form of the Adam learning algorithm [19] to train the LSTM models. Each model receives as input a sequence of successive input vectors of size $N - 1$ with N the number of base observations to predict the elements y(t) [20]. In the case of multivariate LSTM, these vectors include subsets of the vectors for the $P = 3$ parameters (the predictors selected in section A). However, oscillating windows of univariate LSTM contain the RSG parameter only. The samples of the resulting input data will be in the form of the 2D vectors of dimension $[n, P] = [200, 1]$ for monovariate LSTM, and in the form of the 3D vectors $[N, n, P] = [N, 200, 3]$ in case of multivariate LSTM. Each sample of the output data is a 1-dimensional vector.

3.5 Optimization Parameters Choice

Before proceeding to train each neural network model it is essential to normalize the database for scaling high contrast data for harmonized and easier computation afterwards. Indeed, we converted the characteristic values to a range of (0, 1) by the MinMaxScaler() transformer from the formula:

$$X_n = \frac{X - X_{min}}{X_{max} - X_{min}} \tag{8}$$

where X_n is the normalized value, X is the actual value, X_{max} is the maximum value among all the values of the variable linked to X, and X_{min} is the minimum value among all the values of the variable associated with X.

Many tests have been performed to collect the appropriate hidden layers numbers and the number of neurons in each layer based on the values of Mean Squared Error (MSE) and Mean Absolute Error (MAE) calculated by the formulas mentioned in the research papers [7, 13]. MSE is practically essential in Machine Learning, it is often the error metric that is frequently used (loss function).

3.6 Classification Type

Deep neural networks such as LSTM can simulate single-variable problems and problems with multiple input variables. This research work aims to develop univariate LSTM models and multivariate LSTM models that can accurately predict short-term GSR at Er-Rachidia. Then, compare the results obtained by each approach. In the study conducted in this reference [21], the authors developed and compared two mentioned LSTM classes to predict global horizontal irradiation. However, they combined several variables to build multivariate models. Moreover, among these combinations, the model that includes the temperature with the global horizontal irradiation as an input has the highest performance.

Monovariate LSTM Model
In this class, we consider that the GSR varies only in time. Consequently, the input tensor contains one feature as a variable. Nevertheless, we win in computation time and optimizing sensors. But, we lose the diversity of the variables that can advantageously characterize the problem of prediction to solve.

Multivariate LSTM Model
The time series forecasting method using multiple input variables for LSTM networks offers a great advantage in handling complex problems that are difficult to handle by conventional methods. In this class, we established a sense of the relationship of all variables with the GSR using the correlation matrix and then used the selected variables to develop a multivariate LSTM model. Therefore, the input tensor shapes for LSTM models this time is in the form (N, 200, 4).

3.7 Number of Hidden Layers Optimization

Two steps were defined to select the hidden number layers:

- **Step1**: following several tests, we fixed, while comparing the error obtained in each test, the number of neurons to 50 in the model with a single hidden layer.
- **Step2**: we test the model performance by increasing the number of hidden layers by one for each test. The results are summarized in Table 1 and Fig. 5.

Table 1. Hidden layers number variation versus calculated MSE and MAE.

Learning algorithm «Adam»			
Number of epochs	Number of hidden layers	MSE	MAE
30	1	0.0234	0.060
	2	0.0107	0.057
	3	**0.0022**	**0.0269**
	4	0.0023	0.0272
	5	0.0024	0.0271

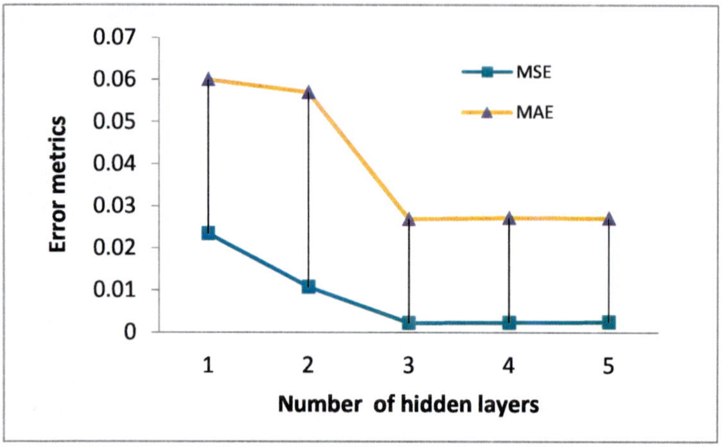

Fig. 5. Hidden layers number variation diagram versus the error metrics.

The model showed a convergence near a minimum error of MSE = 0.0022 and MAE = 0.0269 from the third layer, beyond this layer the model learning became almost stationary. Based on the above figure, the number of 3 hidden layers turned out to be the best one for the prediction model adopted in this study. In what follows, we will fix the number of layers in 3 and seek the number of optimal neurons for each layer.

3.8 Number of Neurons Optimization

By the same process as earlier, we carried out error tries, which narrowed the range of tests on the number of neurons in the three layers in (20, 60) neurons. Indeed, after several tests it has been observed that beyond 60 neurons the error committed by the model has become almost stable, similarly, below 20 neurons the performance of the model no longer improves. Consequently, the tests are carried out according to the combinations of Table 2 and the calculation results are grouped in Table 3.

The results show that on one hand, the increase in the number of neurons in the hidden layers beyond (50-50-50) neurons do not improve the learning rate of the model according to the values of the MSE, but also the MAE drops slightly, on the other hand, the reduction in the number of neurons in the three layers below (50-30-30) neurons increases the learning error of the model. Though, we scanned our tests between these extremities until the most adequate configuration (50-40-30) neurons which corresponds to an MSE = 0.0017 and MAE = 0.0180.

Table 2. Combination interval for neuron number in the hidden layer.

Hidden layer 1	Hidden layer 2	Hidden layer 3
(20-60)	(20-60)	(20-60)

Table 3. MSE and MAE values for different hidden neurons number.

Number of neurons in a hidden layer			MSE	MAE
Layer 1	Layer 2	Layer 3		
60	60	60	0,0022	0,0261
55	55	55	0,0022	0,0267
50	50	50	0,0022	0,0269
45	45	45	0,0023	0,0281
40	40	40	0,0023	0,0287
55	30	20	0,0028	0,0330
50	**40**	**30**	**0,0017**	**0,0180**
50	30	30	0,0025	0,0298
50	45	30	0,0029	0,0338
50	40	20	0,0028	0,0331

Therefore, an LSTM recursive neural network with deep machine learning by three hidden layers and 50, 40, and 30 neurons respectively in each layer was found to be the best implementation for predicting the semi-hourly global solar radiation at the Er-Rachidia site with a minimum of MSE. Suffice to say that the evolution of MSE and MAE during model training adheres to model convergence towards minimum error as shown in Fig. 6.

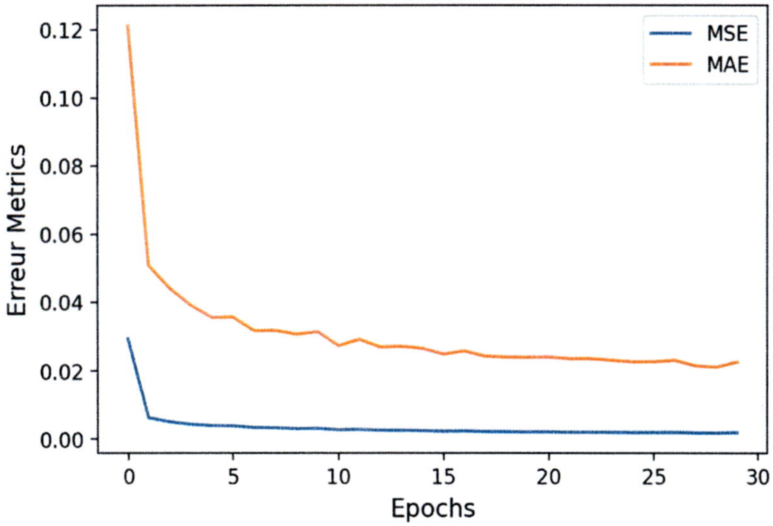

Fig. 6. MSE and MAE evolution versus epoch's number.

4 Realization of Predictions and Discussions

Two variants of the LSTM network (LSTM-multivariate and LSTM-monovariate) were experimented with within this part to predict RSG at the Er-Rachidia level as accurately as possible with learning rate (Learning rate = 0.001) and introduce one (Dropout = 0.1) after the first hidden layer to avoid the risk of overfitting [22]. The performance of the two varieties is evaluated according to the MSE and the MAE. The batch size is given as 100 and 48 for monovariate LSTM and multivariate LSTM respectively.

The raw data run through the model in the form of oscillating windows [10] as input and following the flowchart in Fig. 7 to estimate the output RSG. The first step in machine learning is data acquisition. Then, the acquired data was shared by various interested parties and summarized into useful information. The steps included in this process are data selection, purification and preprocessing, also data scaling. The data is divided into two disjoint subsets, a training set and another reserved for testing and evaluating the model according to different appropriate statistical indicators. The correlation coefficient R is chosen to assess the performance of the models in this study. R is calculated by the following formula:

$$R = \sqrt{\frac{\sum_{i=1}^{N}(x_i - y_i)^2}{\sum_{i=1}^{N}(x_i - \overline{y})^2}} \quad (9)$$

where, x_i is the observed value of semi-hourly RSG, y_i is the estimated value of semi-hourly RSG, $\overline{y}$ is the average of observations, and N is the total number of observations.

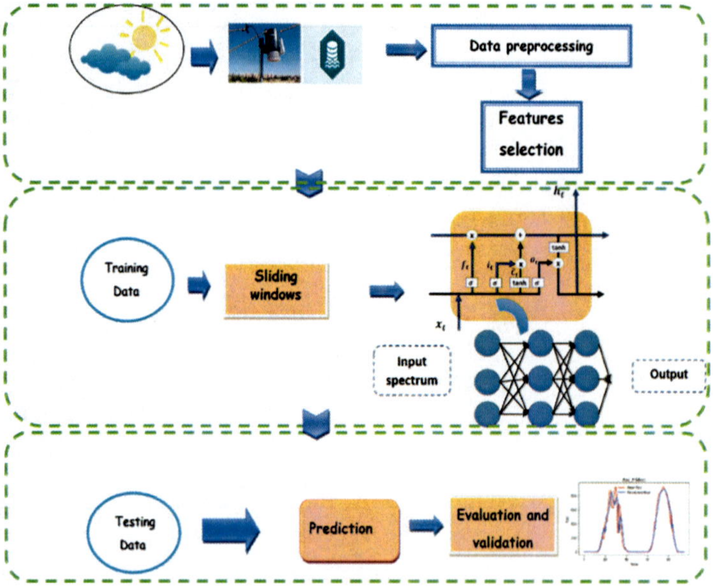

Fig. 7. DATA processing diagram and learning process.

4.1 First Experience

Over the period from 01/10/2016 until 13/03/2017 two simulations were carried out by the two methods monovariate LSTM and multivariate LSTM by the data shown in figure Fig. 8. Over this period, the two ANNs are tested by the totality of the data (training data and test data): the training data to control and visualize the training phase and the test data are processed in installments: 20%, 30%, and 40% of the database to study the stability of the models.

Based on the coefficient of determination R and the error metrics, both methods showed high accuracy, with prediction quality stability up to 40% as shown in Table 4. Emphasizing that the multivariate LSTM model converges faster than monovariate LSTM. A comparison between the estimated values in different forms and the measured values for the two methods is presented in Fig. 9 and Fig. 10.

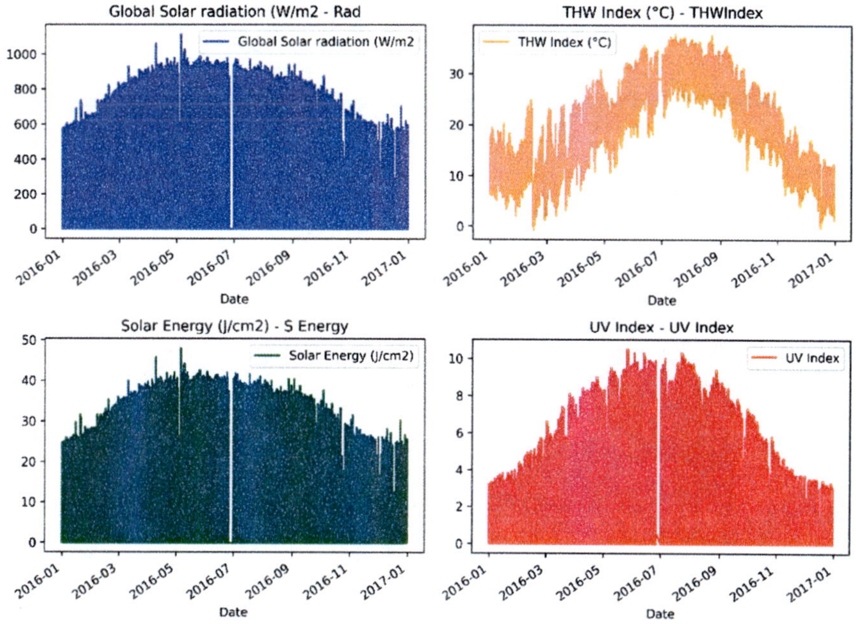

Fig. 8. Selected parameters for building ANN models for 2016

The four parameters shown in Fig. 8 are very important. Indeed, the actors of alternative energies especially solar energy need to track the amount of solar radiation for solar energy production purposes. However, the THWindex field uses humidity, temperature and wind measurements to calculate an apparent temperature, and the UV index assigns a number between 0 and 16 to the current UV intensity, which will help to analyze changing UV radiation levels.

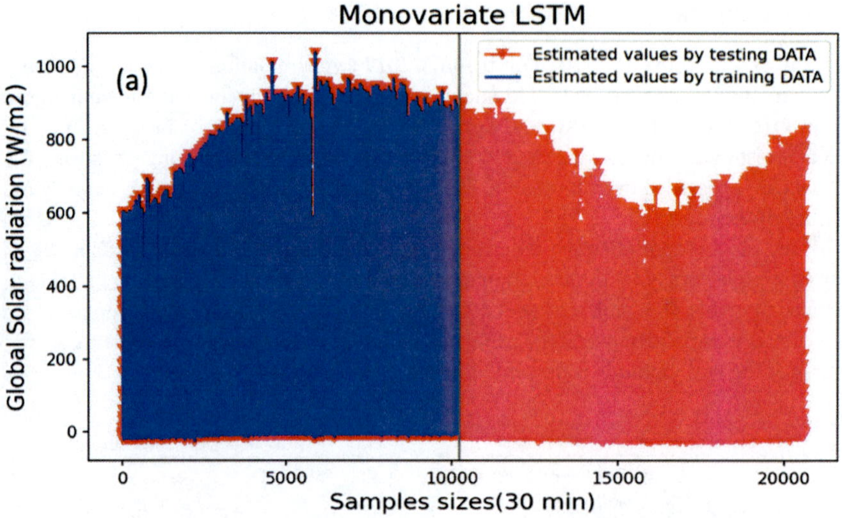

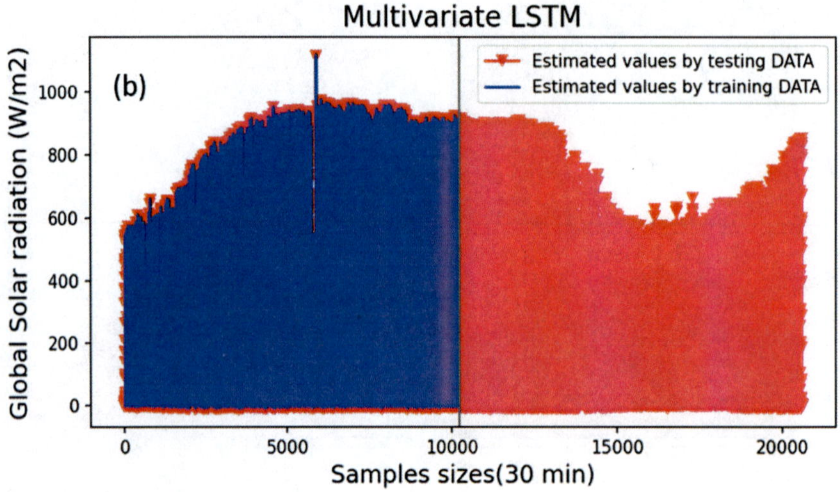

Fig. 9. Estimated GSR in the training and testing phases—(a) by univariate LSTM, (b) by multivariate LSTM.

Considering the convergence of the loss function as a function of epochs and the values of the significant correlation coefficients obtained at the end of the tests (Table 4), can guarantee that the learning rate and the other optimization parameters are selected appropriately. The maximum correlation coefficient obtained for this first simulation is R = 98.97% (Fig. 12).

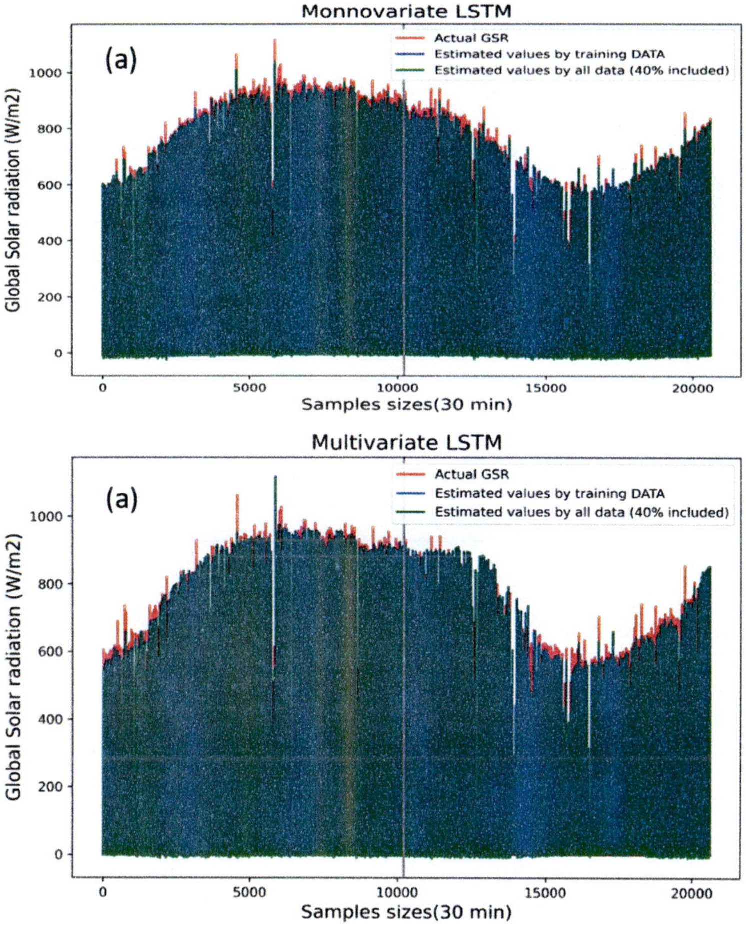

Fig. 10. Estimated and measured GSR in the training and testing phase—(a) by univariate LSTM, (b) by multivariate LSTM.

4.2 Error Metrics Synthesis of the Developed Models

To prove the stability of our models, we have calculated the MSE, MAE, and R for the two approaches developed in this work during the training and the testing phase.

Table 4. Model performance in terms of MSE, MAE, and R indicators during the training and testing phases.

	Test size	Train			Test		
		MSE	MAE	R	MSE	MAE	R
LSTM_MONO	20%	0.0014	0.0152	0.9908	0.0010	0.0125	0.9800
	30%	0.0018	0.0221	0.9895	0.0010	0.0160	0.9890
	40%	0.0017	0.0219	0.9897	0.0019	0.0222	0.9886
LSTM_MULTI	20%	0.0010	0.0153	0.9929	0.0010	0.0139	0.9803
	30%	0.0011	0.0152	0.9865	0.0011	0.0153	0.9897
	40%	0.0011	0.0162	0.9932	0.0017	0.0207	0.9827

Based on the mean square error and the mean absolute error values given in Table 4, the scores obtained for MSE and MAE indicate that the two models examined showed great stability, especially the LSTM_MULTI model which committed an error of MSE $= 0.0017\text{W/m}^2$ at 40% test size compared to LSTM_MONO model which reached an error of MSE $= 0.0019\text{W/m}^2$. The results obtained in the training and testing phase mean that our model is correctly calibrated, also it's able to overcome learning problems such as vanishing gradient.

Validation Loss Function

For the validation task, we choose an MSE loss function to validate the predictive model. Therefore, a graph is also created in Fig. 11 showing two-line plots, the top showing the MSE for the training dataset over epochs, and the bottom showing the MSE for the testing dataset over epochs. The graph shows that the training process has indeed converged. But we can see that MSE shows an irregular evolution in the test phase, although the dynamics of MSE do not seem to be too affected compared to the training phase, given that the nature of GSR sometimes shows large outliers.

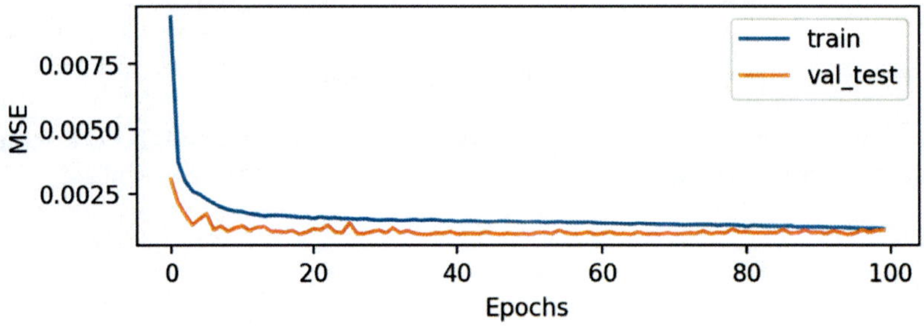

Fig. 11. Evolution of the loss function versus the number of epochs for the training and testing phase.

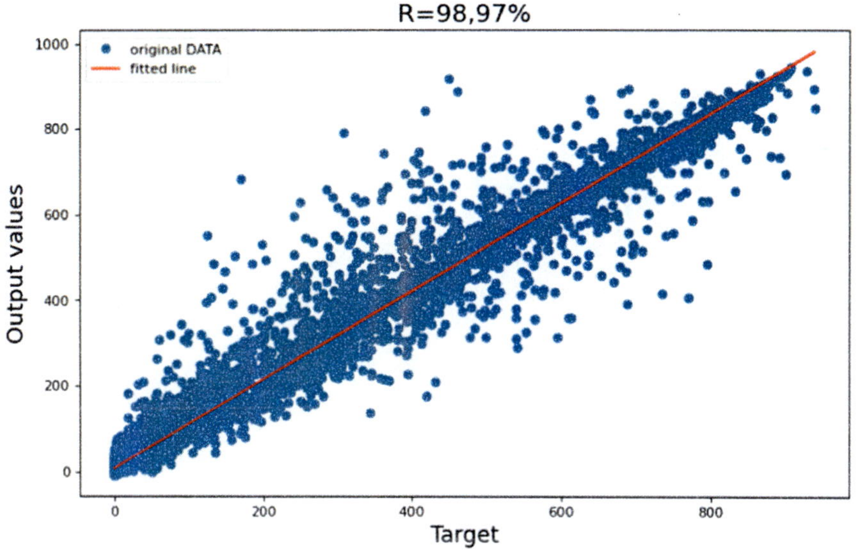

Fig. 12. Scattering diagram for testing data obtained in the first experience.

4.3 Second Experience

The second experiment is performed by monovariate LSTM based on the RSG time series between 01/10/2016 and 01/10/2020 (Fig. 13). However, this period is tested by 20% of all data chosen between 2019-08-05 and 2020-10-01 of the GSR time series. Model convergence is established at 30 epochs. Figure Fig. 14 shows that predicted SRG versus actual GSR wish is very accurate. The predictions of the RSG time series show high accuracy and a strong correlation between the measured values and the estimated values with a correlation coefficient of R = 98.83%, (Fig. 15).

The performance demonstrated by our model in the two experiments examined shows that it could be applied to optimize the operation of solar energy-based electricity services by predicting solar radiation. Several studies have recently been carried out internationally to simulate solar radiation [14], the technique proposed in this study marks a high accuracy in terms of correlation coefficient compared to the majority of studies shown in Table 5.

The main common point between the studies that have marked a great correlation between the estimated and the real values of GSR based on the coefficient of determination R is the integration of dynamic and recurrent ANN models such as LSTM models that can handle the complexity of prediction problems. The prime keys to success in this work are: The data analysis examined based on the correlation study which allows selecting the most strongly correlated features that contribute to learning models with less error and the adoption of a dynamic neural network LSTM to deal with the dynamics of GSR which is influenced by the weather conditions changes.

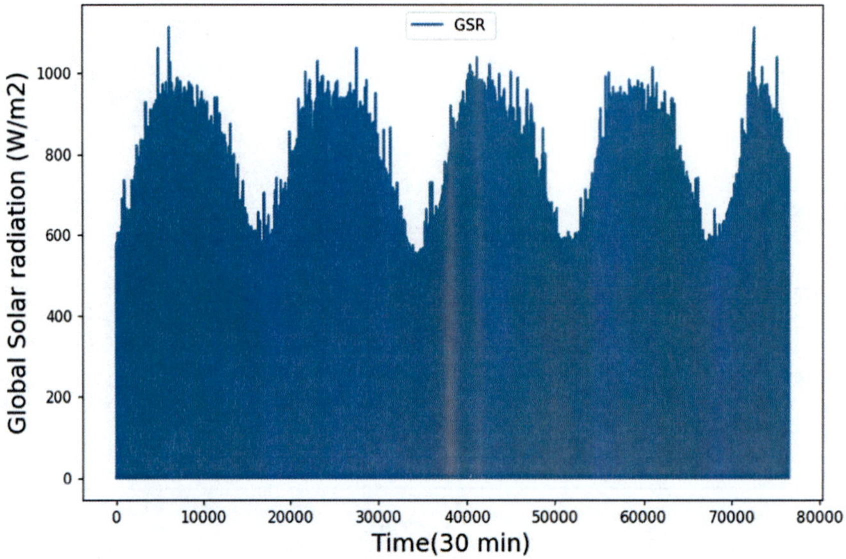

Fig. 13. Global solar radiation time series measurement between 2016 and 2020.

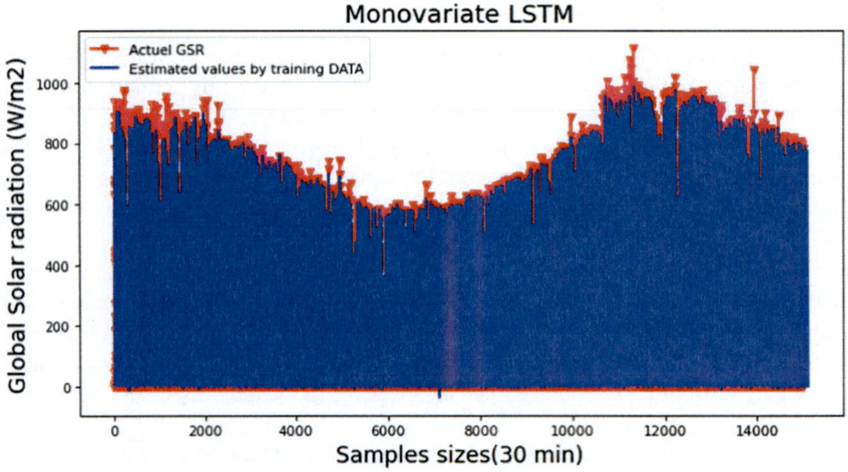

Fig. 14. Estimated GSR and actual GSR for the second experience

For a prediction dedicated to the short applications [30] of the GSR time series by our model, a zoom on 2 days (Fig. 16) and 10 days (Fig. 17) of prediction. The figures show the difference between real GSR and GSR predicted by our approach. For this purpose, the developed prediction model allows the power grid managers to optimize the operation of energy systems, and to generate simulated data of the GSR when missing values of this parameter are detected.

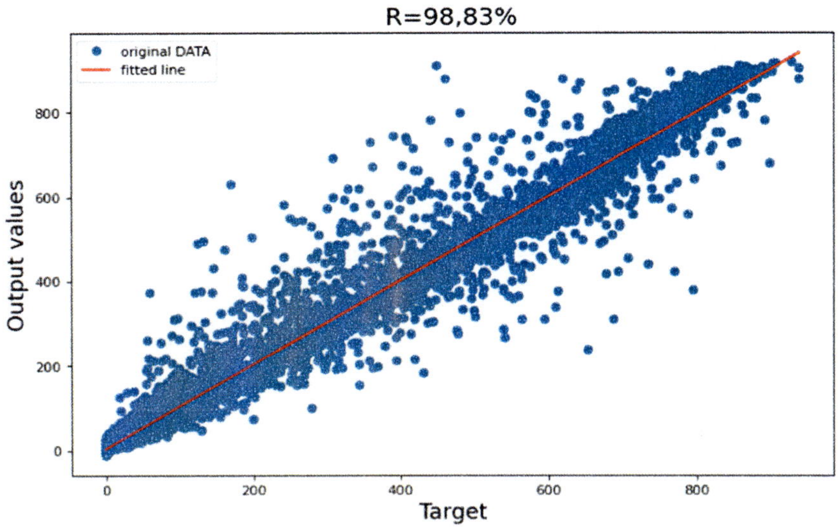

Fig. 15. Scattering diagram for testing data obtained in the second experience.

Table 5. Some recent studies on the prediction of GSR time series and their correlation coefficient.

Reference	Applied technique	R (%)
[13] (2020)	ANN	99.13
[23] (2018)	SVR-PLS	93.3
[24] (2018)	Empirical model	96.7
[24] (2018)	Empirical model	82.5
[25] (2019)	DENFIS	94.2
[25] (2019)	DENFIS	97.9
[26] (2020)	PVGIS	97.1
[26] (2020)	HILIOSAT	97.2
[26] (2020)	HILIOSAT	97.6
[27] (2020)	ANN	98.49
[28] (2020)	ANN	97.93
[29] (2021)	ANN (SR and Wind)	96.93–99.97
Present work (2022)	**Proposed model (ANN/LSTM)**	**98.97**

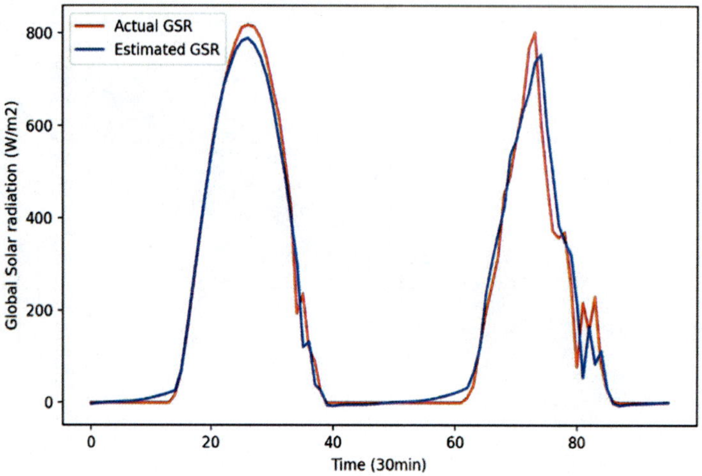

Fig. 16. Predicted GSR and actual GSR for two days.

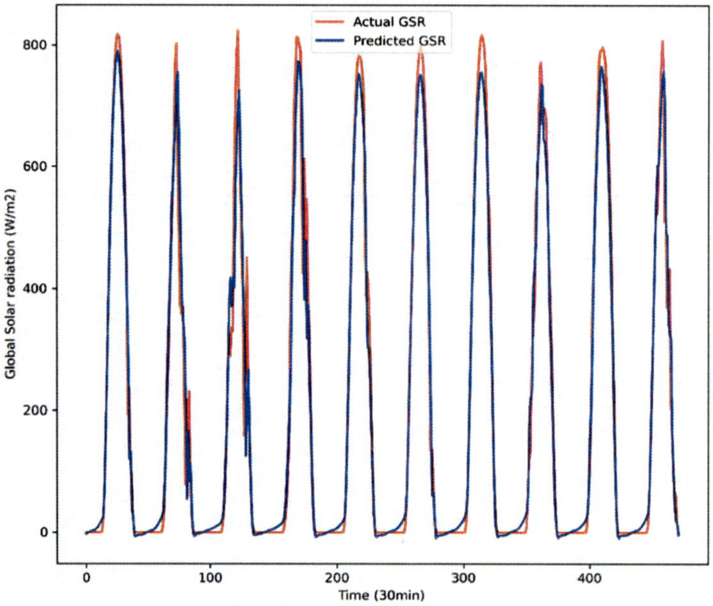

Fig. 17. Predicted GSR and actual GSR for ten days.

5 Conclusion

The study we conducted in this article was meant to improve the solar irradiation prediction tools using Deep Learning methods. The aim is to innovate in terms of new technologies and then invest in smart systems to respond to a crucial need which is the integration of solar energy into the grid, in a fast dynamic and appropriate way. Afterwards studying statistically, the database and selecting strongly correlated fields to the RSG parameter, we adopted a model based on LSTM neural networks to identify the half-hourly RSG changes at Er-Rachidia, Morocco. We also controlled the dynamism over time of this parameter during model training with a well-optimized calibration. The pertinent results obtained show the robustness of our method. We reached a strong Pearson correlation at R = 98.97% given by the multivariate LSTM model.

References

1. Tan, K.M., Babo, T.S., Ramachandaramurthy, V.K., Kasinathan, P., Solanki, S.G., Raveendran, S.K.: Empowering smart grid: a comprehensive review of energy storage technology and application with renewable energy integration. J. Energy Storage **39**, 102591 (2021). https://doi.org/10.1016/j.est.2021.102591
2. Le Ministère de l'Energie, des Mines et de l'Environnement. https://www.mem.gov.ma/Pages/secteur.aspx?e=2. Accessed 5 Sept 2021
3. Cheng, P.M., Tejura, T.K., Tran, K.N., Whang, G.: Detection of high-grade small bowel obstruction on conventional radiography with convolutional neural networks. Abdom. Radiol. **43**(5), 1120–1127 (2017). https://doi.org/10.1007/s00261-017-1294-1
4. Schmidhuber, J.: Deep Learning in neural networks: an overview. Neural Netw. **61**, 85–117 (2015). https://doi.org/10.1016/j.neunet.2014.09.003
5. Goldberg, Y.: A primer on neural network models for natural language processing. J. Artif. Intell. Res. **57**, 345–420 (2016). https://doi.org/10.1613/JAIR.4992
6. Voyant, C., et al.: Machine learning methods for solar radiation forecasting: a review. Renewable Energy **105**, 569–582 (2017). https://doi.org/10.1016/j.renene.2016.12.095
7. Huang, X., et al.: Hybrid deep neural model for hourly solar irradiance forecasting. Renewable Energy **171**, 1041–1060 (2021). https://doi.org/10.1016/j.renene.2021.02.161
8. Citakoglu, H.: Comparison of artificial intelligence techniques via empirical equations for prediction of solar radiation. Comput. Electron. Agric. **118**, 28–37 (2015). https://doi.org/10.1016/j.compag.2015.08.020
9. Al-Hajj, R., Assi, A., Fouad, M., Mabrouk, E.: A hybrid lstm-based genetic programming approach for short-term prediction of global solar radiation using weather data. Processes **9**(7), 1187 (2021). https://doi.org/10.3390/pr9071187
10. Aslam, M., Lee, J.M., Kim, H.S., Lee, S.J., Hong, S.: Deep learning models for long-term solar radiation forecasting considering microgrid installation: a comparative study. Energies **13**(1), 147 (2019). https://doi.org/10.3390/en13010147
11. Qing, X., Niu, Y.: Hourly day-ahead solar irradiance prediction using weather forecasts by LSTM. Energy **148**, 461–468 (2018). https://doi.org/10.1016/j.energy.2018.01.177
12. Alzahrani, A., Shamsi, P., Dagli, C., Ferdowsi, M.: Solar irradiance forecasting using deep neural networks. Procedia Comput. Sci. **114**, 304–313 (2017). https://doi.org/10.1016/j.procs.2017.09.045
13. Jallal, M.A., El Yassini, A., Chabaa, S., Zeroual, A., Ibnyaich, S.: AI data driven approach-based endogenous inputs for global solar radiation forecasting. Ing. des Syst. d'Inf. **25**(1), 27–34 (2020). https://doi.org/10.18280/isi.250104
14. Jallal, M.A., El Yassini, A., Chabaa, S., Zeroual, A., Ibnyaich, S.: A deep learning algorithm for solar radiation time series forecasting: a case study of el kelaa des sraghna city. Rev. d'Intelligence Artif. **34**(5), 563–569 (2020). https://doi.org/10.18280/ria.340505
15. Bengio, Y., Simard, P., Frasconi, P.: Learning long-term dependencies with gradient descent is difficult. IEEE Trans. Neural Netw. **5**(2), 157–166 (1994)

16. Phik (ϕk). https://towardsdatascience.com/phik-k-get-familiar-with-the-latest-correlation-coefficient-9ba0032b37e7. Accessed 28 Jan 2022
17. Premalatha, N., Valan Arasu, A.: Prediction of solar radiation for solar systems by using ANN models with different back propagation algorithms. J. Appl. Res. Technol. **14**(3), 206–214 (2016). https://doi.org/10.1016/j.jart.2016.05.001
18. Apaydin, H., Feizi, H., Sattari, M.T., Colak, M.S., Shamshirband, S., Chau, K.-W.: Comparative analysis of recurrent neural network architectures for reservoir inflow forecasting. Water **12**(5), 1500 (2020). https://doi.org/10.3390/W12051500
19. Adam. https://towardsdatascience.com/adam-latest-trends-in-deep-learning-optimization-6be9a291375c. Accessed 3 Oct 2021
20. Cui, Z., Ke, R., Wang, Y.: Deep Bidirectional and Unidirectional LSTM Recurrent Neural Network for Network-Wide Traffic Speed Prediction. IEEE (2018)
21. Mandal, A.K., Sen, R., Goswami, S., Chakraborty, B.: Comparative study of univariate and multivariate long short-term memory for very short-term forecasting of global horizontal irradiance. Symmetry **13**(8), 1544 (2021). https://doi.org/10.3390/sym13081544
22. Mele, B., Altarelli, G.: Lepton spectra as a measure of b quark polarization at LEP. Phys. Lett. B **299**(3–4), 345–350 (1993). https://doi.org/10.1016/0370-2693(93)90272-J
23. García-Hinde, O., et al.: Evaluation of dimensionality reduction methods applied to numerical weather models for solar radiation forecasting. Eng. Appl. Artif. Intell. **69**, 157–167 (2018). https://doi.org/10.1016/J.ENGAPPAI.2017.12.003
24. Bailek, N., et al.: A new empirical model for forecasting the diffuse solar radiation over Sahara in the Algerian Big South. Renew. Energy **117**, 530–537 (2018). https://doi.org/10.1016/j.renene.2017.10.081
25. Kisi, O., Heddam, S., Yaseen, Z.M.: The implementation of univariable scheme-based air temperature for solar radiation prediction: new development of dynamic evolving neural-fuzzy inference system model. Appl. Energy **241**, 184–195 (2019). https://doi.org/10.1016/j.apenergy.2019.03.089
26. Ener Rusen, S., Konuralp, A.: Quality control of diffuse solar radiation component with satellite-based estimation methods. Renew. Energy **145**, 1772–1779 (2020). https://doi.org/10.1016/J.RENENE.2019.07.085
27. Guti, P.A., et al.: Evolutionary artificial neural networks for accurate solar radiation prediction. Energy **210**, 118374 (2020). https://doi.org/10.1016/j.energy.2020.118374
28. Üstün, İ, Üneş, F., Mert, İ, Karakuş, C.: A comparative study of estimating solar radiation using machine learning approaches: DL, SMGRT, and ANFIS. Energy Sour. Part A **44**(4), 10322–10345 (2020). https://doi.org/10.1080/15567036.2020.1781301
29. Shboul, B., et al.: Original article A new ANN model for hourly solar radiation and wind speed prediction: a case study over the north & south of the Arabian Peninsula. Sustain. Energy Technol. Assessments **46**, 101248 (2021). https://doi.org/10.1016/j.seta.2021.101248
30. Mishra, M., Byomakesh, P., Nayak, J., Naik, B., Kumar, S.: Deep learning and wavelet transform integrated approach for short-term solar PV power prediction. Measurement **166**, 108250 (2020). https://doi.org/10.1016/j.measurement.2020.108250

COVID-19 Contact-Tracing Technology, Users' Privacy and Security

Anasse Hanafi(✉), Mohammed Bouhorma, and Lotfi Elaachak

LIST Laboratory, Abdelmalek Essaâdi University, Tangier, Morocco
anasse.hanafi@etu.uae.ac.ma

Abstract. Nowadays, the entire world is struggling to adapt and survive the SARS-CoV-2/COVID-19 pandemic, the new mutations in the Coronavirus disease is causing damage and disruption across the world. Taking preventive measures to control the spreading of the virus, including lockdowns, curfews, social distancing, masks, vaccination are not enough to stop the virus. Many countries have sought to support their contact tracers with the use of digital contact tracing apps to manage and control the spread of the virus. Using the new technologies to adapt the prevention measures furthermore enhancing the existing ones, will definitely be more efficient. There are many contact tracing apps that have already been launched and used since 2020. There has been a lot of speculations about the confidentiality and security aspects of these apps and their possible violation of data protection principles. In this paper we propose a system of contact tracing, we explain how this system treats sensible information to preserve the user's identity and protect their personal information.

Keywords: Contact tracing · Privacy · Security · Data protection

1 Introduction

As COVID-19 continues to claim lives and shake the global economy, governments are urgently seeking new and innovative tools to address the crisis. Digital solutions based on geolocation data are emerging to help authorities monitor and contain the spread of the virus. Some are powered by mobile Call Data Records (CDRs), i.e., data produced by telecommunications service providers regarding cell phone calls or other telecommunications transactions, which provide valuable information on population movements. As network operators serve large portions of the population across entire nations, the movements of millions of people at fine spatial and temporal scales can be measured in near real time. The resulting information and trends are invaluable for governments seeking to track the COVID-19 outbreak, alert vulnerable communities, and understand the impact of measures like physical distancing and lockdowns.

Mobile apps delivering health advice are already a significant part of the healthcare ecosystem and have proven to be effective for prevention [1], early diagnosis (e.g., symptom checkers) and connecting users with health care providers. Local health services and emergency units. Today, new applications aimed at consumers are emerging as part

M. Ben Ahmed et al. (Eds.): SCA 2022, LNNS 629, pp. 697–705, 2023.
https://doi.org/10.1007/978-3-031-26852-6_64

of the monitoring of the pandemic. Today, new applications aimed at consumers are emerging as part of the monitoring of the pandemic. These apps are the result of partnerships between technology companies, universities, clinicians and governments, who are ultimately responsible for their funding, further development and implementation. Although they do not necessarily capture the entire population (for example, the elderly who may not have a smartphone or may not be familiar with it), these applications provide public authorities with an additional tool to monitor and contain the spread of the virus.

2 State of the Art

As the Scientific Council reminded, new technologies are a useful tool for tracing the chains of contamination, especially when the circulation of the virus is high. These applications provide public authorities with an additional tool to monitor and contain the spread of the virus. The most cited applications include:

TraceTogether, developed by the Singapore Government Technology Agency (GovTech) in conjunction with the Ministry of Health [2], this app uses Bluetooth to track people who have been exposed to the virus. This information is used to identify close contacts based on the proximity and duration of an encounter between two users. It then alerts those who come into contact with someone who has tested positive or who is at high risk of carrying the coronavirus. Once an individual is confirmed positive or suspected of being infected, they can choose to allow hospitals, the Ministry of Health and third parties access to app data to help identify contacts. Singapore aims to make the underlying data exchange protocol open-source, while preserving user privacy.

C-19 COVID Symptom Tracker: [3] The aim of this application developed in the United Kingdom as part of a partnership between doctors and scientists from King's College London, a health data science company (company spun off of King's College), and Guy's & St Thomas' Hospitals National Institute for Health Research Center, is to slow the outbreak of COVID-19 by helping researchers identify: i) the rate of spread of the viruses in different areas; (ii) high risk areas in the UK; and iii) those most at risk, by improving understanding of the symptoms related to the underlying disease state. According to the researchers, the data from the study can reveal critical information about the symptoms and course of the infection in different people. They can also help researchers understand why some infected people develop more severe or even fatal symptoms, while others show only mild symptoms.

The Japanese application against COVID-19, called COCOA (for "Contact Confirming Application") [4], was designed to alert users who have been in close contact (within a radius of 1 m for more than 15 min) with a person who's the COVID-19 test came back positive. This application was originally developed as early as March 2020 by volunteers from Microsoft Japan Co. In early May, the Ministry of Health took control of the development by applying the company Persol Process and Technology Co., which itself sub-handled application maintenance work at information technology company MTI Ltd.

Wiqaytna [5], a corona virus exposure notification mobile application has been named by the Ministry of Health in Morocco, This mobile solution is launched as part of a national awareness campaign, under the theme "Stay vigilant, protect-us mutually", the objective of which is to encourage citizens to continue to adopt preventive measures limiting the spread of this virus, namely frequent washing of hands with soap and water or a hydro-alcoholic solution, wearing a mask and physical distancing, and to which is added the use of the "Wiqaytna" application, available from June 1, 2020 for download on Google Play, App store, and on www.wiqaytna.ma. The application is part of the strengthening of the existing system for the management of contact cases, set up by the Ministry of Health [6], which has demonstrated its effectiveness, and which remains the basic system. Wiqaytna, based on Bluetooth technology and whose use is voluntary, notifies its users in the event of prolonged physical proximity to another user who tests positive for Coronavirus in the last 21 days following contact. Following this notification, the teams of the Ministry of Health carry out an assessment of the risk of exposure, and come into contact with the notified case if necessary. It should be noted that this application is produced jointly by the Ministry of Health and the Ministry of the Interior in collaboration with the Digital Development Agency (ADD) [7], the National Telecommunications Regulatory Agency (ANRT), and with the gracious, voluntary and voluntary contribution of Moroccan companies that are experts in their fields. It is authorized by the National Commission for the Control of Personal Data Protection (CNDP) and available in open source.

3 Related Work

Pan-European Privacy-Preserving Proximity Tracing (PEPP-PT) [8] is an open-source protocol designed to facilitate digital contact tracing of infected individual. More than 130 scientists, technologists and experts from eight European countries – including France, Germany and Italy – took part in a non-profit initiative that led to the development of a free application to analyze signals Bluetooth between cell phones to detect users who have been in close proximity to each other. The app temporarily stores encrypted data locally, and if users subsequently test positive for COVID-19, it can alert anyone who has been near the infected person in previous days.

3.1 Privacy and Security Risk

We describe how PEPP-PT Protocol works in order to study the associated security and privacy risks. The developed application works on 4 steps, installation, normal operation, handling infected patients and contact tracing.

When the application is installed on a smartphone, a unique identifier is generated and sent to the server. This identifier is used to derive ephemeral identifiers which will be transmitted to the phone to be broadcast by Bluetooth. at the server level, the ephemeral identifiers are generated by encrypting the identifier sent by the mobile application by a periodically renewed global broadcast key. At the level of the second step, in the event of contact between two applications, they exchange their ephemeral identifiers. In the event that a user has been affected by the virus, the application uploads to the server the list of ephemeral identifiers that it has recovered from other applications. at the server level, a decryption of ephemeral identifiers is made to recover the unique identifier generated by the mobile application to calculate the risk of infection and send a notification.

This architecture raises the following privacy concerns: as the server application generates the ephemeral identifiers, the server can, at any time, assemble the past and future ephemeral identities of any user, by decrypting their unique identifier. This ability positions the backend server in a position of control with high potential for function creep. Feature creep is a highly relevant discussion in the current pandemic, as a contact tracing tool, can easily be turned into a surveillance instrument with far-reaching human rights implications.

4 Method

To combat the spread of the corona virus, we propose an anonymous tracking system to preserve the identity of users. The system is based on the client-server architecture, a mobile application (client) allowing the collection of tracing information, and the notification of the user in the event of contact with an individual who tested positive. A second application (server) installed on web server stores contact cases between users (anonymously) and handle the processing of requests sent from mobile applications.

When a user installs and launches the application on their phone, a registration form is displayed, allowing them to enter personal information such as first name, last name and mail address. This information is useful for contacting the user in the event of contact with another user who tested positive and will be encrypted and stored (with additional information, like the status of the covid infection) in a database. Only authorized persons will be able to access this data (for example health personnel).

Once the user has entered his personal information, the mobile application generates a universal unique identifier (UUID) that does not reveal any personal information about the user. This identifier is saved in a local storage accessible at any time and only by the mobile application. The purpose of generating this identifier is the encryption and decryption of the tracking data that will be sent to the server.

After generating the UUID, a process of discovering nearby devices using Bluetooth low energy is initiated. This process allows in case of contact with a person using the application on his phone to send this data to the server. In parallel, the mobile application sends requests to the server with a configurable time interval. These requests contain the name of the Bluetooth device and launch processing at the server application level, this processing makes it possible to find out if the user who sends his Bluetooth device name has been in contact with a person who has been tested positive.

4.1 The Proposed Algorithms

The proposed system is made up of two parts, a first part allows the collection of tracing information and sending this data to the server, the second part makes it possible to identify if a user has been in contact with a person who has tested positive.

The first algorithm shows the actions that are executed at the mobile application level, it is the process of discovering nearby devices

Function Browsing and advertising
For each interval of 10 seconds:
Start scanning for nearby devices using Bluetooth low energy for 3 seconds
If a device is detected then:
Get the detected device's name
Encrypt that name with the UUID stored locally
Send both UUID and encrypted name to the server
End if
End for
End of function

The second algorithm shows the actions that are executed at the server application level, it is the process of detecting and notifying users if they have been in contact with a person who tested positive. Note that this process is initiated by the mobile application by sending the Bluetooth device's name

Function check infection (device name):
For each stored tracking object
Retrieve the UUID
Encrypt "device name" with the UUID
Get the infection status of the UUID
If the encrypted device name is the same as the one stored in the database and the infection status is positive then:
send notification to alert the user
End if
End for
End of function

5 Results and Discussion

The proposed system contains two-parts, the mobile application and the backend-side (Fig. 1). The mobile application is used to trace nearby contacts, using Bluetooth low energy to get the device name of other users (Fig. 2), as well as store the user's UUID on a local storage and inform the user of any possible exposure (Fig. 3). For the backend part, the principal role is the storage of the tracking object, the combination of the generated UUID and the encrypted Bluetooth device name using that UUID (Fig. 4).

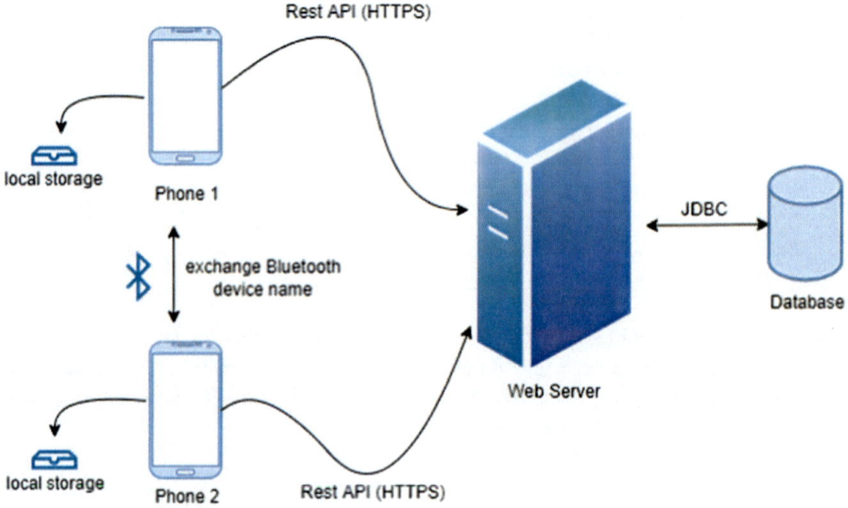

Fig. 1. Overview of the system architecture

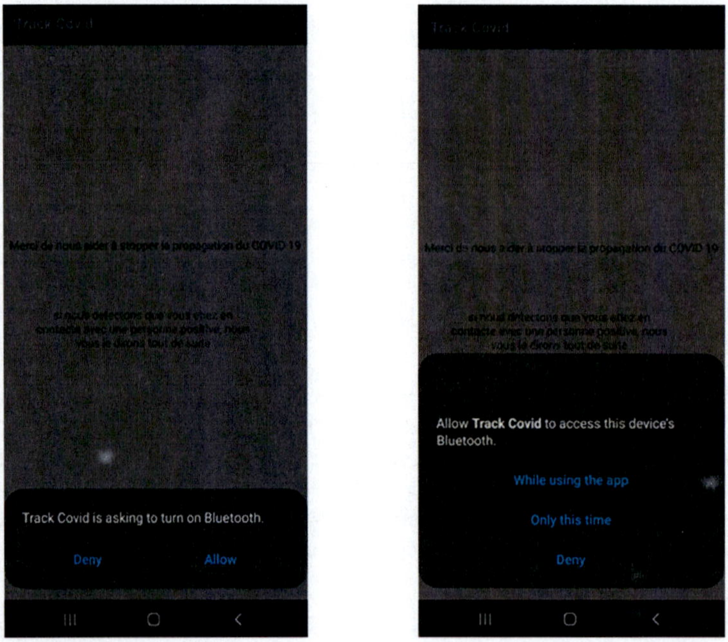

Fig. 2. Mobile application: request for Bluetooth activation, and authorization to access Bluetooth by the application

Fig. 3. Notification of a user who has been in contact with a person who tested positive

id	tracking_time	track_id	tracked_encoded_device_name
1	2022-03-23 14:15:13.866000	73c95a85-b27b-4eb1-b03c-66abe12b3cd3	MzreXFpGuYBGhzd+/WO98mOfnPs7ZHvCUIKcU73hczl=
2	2022-03-23 07:17:14.774193	d50de910-30e7-4426-bed1-b7f6ad348fe5	BU3CHzY11kq+DZVCr9lc2g==
3	2022-03-23 10:16:22.999117	5b75d425-9206-48d2-b7b3-c77f1ba778fe	Lq/xejTo18rrMlxjmABBQcy3FrLWoaplJu/e6sitbw8=
4	2022-03-23 15:07:53.486000	9a41febe-6ef4-412f-9d2c-7ba7b4ed2837	/pzQjB7e1vX5k09sNMPVkFF7Vquf11LvXZ9qHGXUZns=
5	2022-03-23 08:24:29.542613	833e0c60-60f5-4ee4-b301-d0601057e345	BjaNJwhfMyF50V1MD50/pJ40gtUeLLt48hv6vBfr8So=
6	2022-03-23 15:19:00.997078	9e3d5e43-623e-4d86-baa3-9f80b5931bf3	1CsBjNxyeVYUCFC0UjK2gQ==
7	2022-03-23 16:22:26.806827	699dcc20-8e60-4741-978f-b64367b72bf4	BU3CHzY11kq+DZVCr9lc2g==
8	2022-03-23 06:34:01.047141	5b47aa7a-9720-4515-8a27-802de12561ae	OM/BGjJo59rtMlrmmABVQcy7FrLHoaplJu/e5sihhwD=

Fig. 4. Overview of tracking data

Figure 5 shows the database model structure used to store UUIDs and tracking objects. As we can see, there is no relation between the two entities, the user's personal information is encrypted to preserve the anonymity of the user. Even if the encrypted device names are decoded using the UUIDs using the right algorithm by the server, the result does not reveal any personal information about the user.

We will shed the light on the most important part of the proposed system, which is the security and the data confidentiality:

- No GPS localization are shared with the server
- No sharing data with advertising networks or analytic frameworks
- Communications between mobile application and the server is made through HTTPS protocol

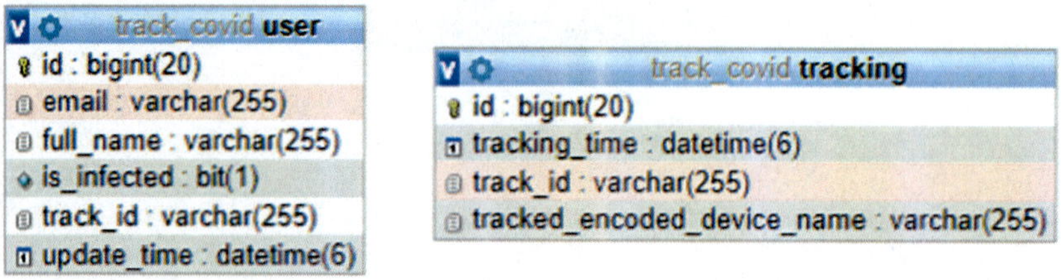

Fig. 5. Overview of the database model used for user management and anonymous tracking.

- The creep function is not possible because at any time no information can be extracted from the recorded data

Note that users' personal information is not mandatory and is not used when tracking. The purpose of recording this data is to help health personnel contact people who have been in contact with positive cases.

6 Conclusion

Digital technologies offer powerful tools to governments in their efforts to control the COVID-19 pandemic, but it is important to recognize their privacy and data protection implications. Contact tracing apps should be implemented transparently, in consultation with key stakeholders, with strong privacy protections built into design criteria, and as part of open-source projects (where applicable).

The elements below must be taken into account before the development of an IT solution for tracking and tracing during this pandemic period:

- The legal basis for using these technologies, which varies depending on the type of data collected (for example, personal, sensitive, pseudonymized, anonymized, aggregated, structured or unstructured data).
- The proportionate nature of the use of these technologies and the resulting data collection, the way in which data is stored, processed and shared, and the recipients of this data (including default security and protection protocols of privacy implemented).
- The time frame during which more invasive technologies that collect personal data can be used to combat the crisis. Data should only be retained for as long as necessary to fulfill the specific purpose for which it was collected.

Acknowledgments. This project is subsidized by the MENFPESRS and the CNRST as part of the program to support scientific and technological research related to "COVID-19" (2020). Also, we acknowledge financial support for this research from CNRST.

References

1. Ortega-García, J.A., Ruiz-Marín, M., Cárceles-Álvarez, A., Campillo I López, F., Claudio, L.: Social distancing at health care centers early in the pandemic helps to protect population from COVID-19. Environ. Res. **189**, 109957 (2020). https://doi.org/10.1016/j.envres.2020.109957
2. Bay, J., et al.: BlueTrace: a privacy-preserving protocol for community-driven contact tracing across borders. Government Technology Agency-Singapore, Technical report (2020)
3. Department of Health and Social Care: Next phase of NHS coronavirus (COVID-19) app announced (2020). https://www.gov.uk/government/news/next-phase-of-nhs-coronavirus-covid-19-app-announced. Accessed 17 Jan 2022
4. Ministry of Health Labour and Welfare: COVID-19 Contact-Confirming Application (2021). https://www.mhlw.go.jp/stf/seisakunitsuite/bunya/cocoa_00138.html
5. ADD.GOV Homepage. https://www.add.gov.ma/lancement-de-wiqaytna-application-mobile-de-notification-dexposition-au-coronavirus-responsable-de-la-maladie-covid-19. Accessed 22 Mar 2022
6. Moroccan Ministry of health Homepage. Note d'Orientations Générales pour le développement du Digital au Maroc à horizon 2025 (2020). https://add.gov.ma/storage/pdf/Avril_NOG_ADD_fr_SITE_VF.pdf. Accessed 24 Dec 2021
7. ADD Homepage. https://www.add.gov.ma/presentation-de-ladd. Accessed 13 Jan 2022
8. Clarke, L.: PEPP-PT vs DP-3T: The coronavirus contact tracing privacy debate kicks up another gear. https://tech.newstatesman.com/security/pepp-pt-vs-dp-3t-the-coronavirus-contact-tracing-privacy-debate-kicks-up-another-gear. Accessed 06 May 2020

Application of AI and IoT in the Containment of the Covid19 Pandemic

Ikram Ben Abdel Ouahab(✉), Lotfi Elaachak, Fatiha Elouaai, and Mohammed Bouhorma

Computer Science, Systems and Telecommunication Laboratory (LIST), University Abdelmalek Essaadi, FSTT, Tangier, Morocco
{ibenabdelouahab,lelaachak,mbouhorma}@uae.ac.ma

Abstract. Over 3 years, we are living under various restrictions due to the covid'19 pandemic. First, was the lockdown, then authorities allow people to go outside for a very limited duration. Second, the obligation of wearing a facemask to protect ourselves and others. Third, after being vaccinated with three shots, everyone has a vaccination passport or some call it a Health pass. This pass allows people to go whatever they want and to travel outside the country. We build a complete application able to manage and monitor: 1/ Social Distance, 2/ Facemask detection, and 3/ check the vaccination passport. We use deep learning models to do classification and detection. The resulting application is very flexible and user-friendly. Also, it could be used anywhere; schools, universities, public spaces, private buildings, etc. We tested our application with various devices including Raspberry Pi, wireless camera, and laptop.

Keywords: Deep learning · Covid19 · Object detection · IoT · Raspberry

1 Introduction

In two decades high-tech, digital technology, and the Internet have given us a feeling of omnipotence, a bit like the unstoppable superheroes in movies who regularly save the world. However, at the time of the coronavirus we are helpless, we understand that superheroes do not exist, that techno cannot do everything, and for some, this is a kind of disillusionment. That being, there are still many cases of research and innovation that may help us manage the crisis, or even get out of it. Some examples are given below.

The Medical research Mendel has assembled a task force of more than 50 AI scientists, physicians, and clinical experts to treat COVID-19 with clinical artificial intelligence. The goal is to answer key questions and corroborate findings faster and more accurately by absorbing insights from the medical literature and cross-referencing them with data from electronic medical records (EMRs) of COVID-19 patients.

Researchers from West Virginia University's Rockefeller Institute for Neuroscience worked with smart ring company Oura on a system that can help slow the spread of COVID-19 [1]. The project plans to use an AI-based forecasting system, which will predict symptoms of fever, cough, or fatigue, similar to those of the virus. The system

M. Ben Ahmed et al. (Eds.): SCA 2022, LNNS 629, pp. 706–715, 2023.
https://doi.org/10.1007/978-3-031-26852-6_65

can even produce symptoms 24 h before they occur. The smart ring is also being used by researchers to monitor any temperature changes and differentiate whether the changes are from COVID-19 or stress or anxiety. In addition, there is also the Apple Watch [2], which can measure blood oxygen saturation levels to help monitor the health status and respiratory stress of patients with Covid-19 who are maintained at home.

Apple and Google are developing together a mobile application for tracking Covid-19 patients and alerting people who have potentially been in close contact with them [3]. Thanks to Bluetooth, the app will be able to notify a person who has been in contact with a carrier of the coronavirus, the objective being to differentiate between people who have just crossed paths from afar (i.e., at a distance normally without risk of contamination) and those who have been nearby for a certain period.

In this paper, we developed a full application of smart surveillance using Raspberry Pi, where we can monitor social distance, facemask detection, and Read QR of vaccination passports. The usage of our application is not limited to a given environment, but it can be used anywhere and anytime. Also, we can take advantage of it with raspberry, with a laptop, or other devices.

The remaining of this paper is structured as follows. Section 2, we state some related projects using Raspberry pi in the context of covid19, found in the literature. Then in Sect. 3, we present the materials and methods used in the proposed solution. After that, Sect. 4 we give a detailed description of the proposed application. In Sect. 5, the demonstration of the solution then we conclude and give some perspectives.

2 Related Works

We found many works using Raspberry pi and IoT as prevention in the era of covid19 in various domains. In this section, we'll be discussing some of these interesting projects.

In [4], researchers focus on asymptotic patients that are extremely risky. Without knowing they're infected, they may spread the infection to others, additionally, asymptotic patients may be having extremely serious lung damage. So, earlier prediction and monitoring of asymptotic patients are required to help protect their life and prevent the virus from spreading. The proposed solution is based on a raspberry pi with sensors to detect asymptotic patients.

Furthermore, more and more research deals with X-ray images with machine learning techniques in the medical field. In [5], CT scans and chest X-ray images are used for covid19 classification. They proposed a raspberry-based solution to diagnose covid19 from CT scans and W-ray images. The authors used a deep learning model to train on data and make the covid classification task reach high accuracy of 99.3% for X-ray chest images.

Otherwise, a device prototype was proposed in [6] based on raspberry pi, camera, and sensors. The prototype checks the body temperature and facemasks. The raspberry pi was used as the central element of the solution, then the result is displayed on a screen connected directly to raspberry and camera. Facemask detection is based on mouth and nose detection. We tested a similar application for facemask detection using Haar Cascade models in previous work [7]. Facemask detection based on nose and mouth detection isn't effective. In the previous experimentation, we demonstrate that this kind of model can be bypassed by covering the mouth and nose using a hand for example.

Other researchers developed an independent facemask vending machine in the form of a kiosk [8]. The core of their prototype is a raspberry pi and a camera. So, when the user wants to buy a facemask goes to the kiosk and gets it facemask without any person-to-person contact. For payment, a QR code is scanned to choose the number of masks and to process payment.

3 Materials and Methods

In the implementation of our project, we use several technologies and knowledge. As hardware, we used the latest version of Raspberry Pi 4, with Intel Neural Compute Stick NCS2, the raspberry camera, a laptop, and one NVIDIA Quadro T1000 to train our models (Fig. 1).

Fig. 1. Materials and laboratory prototype

- Raspberry pi [9] is a cheap and tiny computer used in different applications and domains. Its recent versions (RPi4) have a more and more powerful processor and a higher RAM (8Go). Another advantage is its small size and weight, which allow us to put it anywhere.
- Intel Neural Compute Stick 2 (NCS2) [10] is used with OpenVino Intel's distributions toolkit. It's a simple USB plug-in device composed of VPU and used with a set of software tools to manage deep learning inference. It's a hardware accelerator used usually with powerless devices such as the raspberry pi.
- Raspberry Pi camera rev 1.3 is also a tiny and cheap device that is used directly into the CSI connector on the raspberry board. It's a low-quality camera used only in the test prototype. Practically, any camera can do the job. We used also the laptop camera without any problem.
- A computer is also needed to do tests. We used also an NVIDIA Quadro T1000 GPU to train, test, and evaluate deep learning models. The GPU is also used to test the resulting application performances.

On the other side, we used pre-trained models, such as Yolov3, MobileNet2. As a framework, we worked with TensorFlow, Keras API, and OpenCV for image processing using Python Language. Then to build the Graphical User Interface (GUI), we adopt Tkinter, a powerful Python API for GUI.

3.1 YOLO for Social Distance

You Only Look Once (YOLO) is a state-of-the-art, real-time object detection system. It is extremely fast and accurate. The YOLOv3 model has several advantages over classifier-based systems. It looks at the whole image at test time so its predictions are informed by the global context in the image. It also makes predictions with a single network evaluation unlike systems like R-CNN which require thousands for a single image. This makes it extremely fast, more than 1000x faster than R-CNN and 100x faster than Fast R-CNN. Especially version 3 of YOLO uses a few tricks to improve training and increase performance, including multi-scale predictions, a better backbone classifier, and more [11].

YOLOv3 pretrained model was trained using the COCO database provided in [12]. YOLOv3 uses a new network-Darknet-53that has 53 convolutional layers, which is deeper than the older version YOLOv2 and it also has residuals or shortcut connections. Based on the literature, the actual Darknet-53 is been more powerful than Darknet -19 and more efficient than ResNet-101 or ResNet-152. So, we performed a real-time social distancing tool using the YOLOv3 pretrained model [13].

3.2 Fine-Tuning MobileNetV2 for Facemask Detection

Fine-tuning is a transfer learning method, that takes a convolutional neural network pretrained model and fine-tune it with new data to obtain a brand-new classifier that fit out needs. This is exactly what we did to perform facemask classifier. To this end, we take MobileNetV2 pretrained model, and we add five more layers, then we train the resulting architecture on with_mask/without_mask database. We presented the detailed process of this model in [7]. The facemask model gives more than 99% of accuracy, recall and F1-score.

4 The Proposed Application

We build a GUI including 3 main functionalities: Social Distance, Facemask, and Vaccination Passport check. Firstly, only authorized persons can access the application by an authentication code. Once accessed, the user will choose the application he wants to use from the three above. At the same time, we display the real-time frame from the used camera. When one application is chosen the result appears directly on the screen. Let's see in detail what happens with each application individually.

4.1 Monitoring Social Distance

According to the CDC (Centers for Disease Control and Prevention), we should stay at least 6 feet away from other people, indoors in public. Social distance is very important to prevent the risk of contamination by the covid'19. Especially, people with a high risk of getting very sick should pay more attention and don't get close to anybody. High-risk people are more likely to get severely ill from covid19 [14], including Having cancer, chronic kidney disease, chronic liver disease, chronic lung diseases, dementia or other neurological conditions, diabetes, heart conditions, HIV infection, immunocompromised state, mental health conditions, overweight and obesity, pregnancy, tuberculosis, and others. In order, to protect ourselves and the high risked people in society, everyone should be responsible for leaving a social distance and not getting close to each other in public and private spaces.

Monitoring social distance passes by several steps using any camera. First, we need to detect the presence of a person in a frame and get its exact location. We use a pre-trained model: MobileNet-SSD300 by Google. Single Shot object detection (SSD), uses a single shot to recognize multiple objects, including a person, in the image. Second, we have more than one person detected with their locations in the image, we calculate the Euclidean distance between each two respectively, in real-time.

In a simple scenario, when the camera captures 2 individuals, P1 and P2. MobileNet-SSD model detects their presence and gets their locations. Locations are the center of the detected box, so $P1(x_{p1}, y_{p1})$ are coordinates of the center of P1. And, the same thing with $P2(x_{p2}, y_{p2})$. Then, we calculate the distance between two points P1 and P2 using the classical formula of Euclidean Distance (Fig. 2).

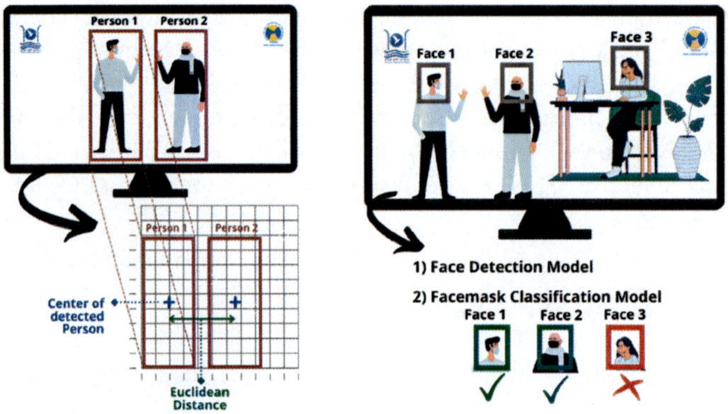

Fig. 2. The process of social distance and facemask detection

In a more complicated scenario, more than 2 individuals are detected. So here, we make sure that all the permutations of distances are calculated. In other words, if we have 5 detected persons, we need to calculate the distance between each 2 of the 5, which means 10 possible combinations. After getting, the distance between detected individuals. We can decide if they are respecting social distance or not. If the social distance is not respected, a loud vocal alert is generated. Also, a log file recording information's about the non-respect of social distance is locally saved. Find out the implementation of the social distance detection approach, using Raspberry Pi, in [13, 15, 16].

4.2 Facemask Detection

To respond to the needs of new changes brought on by the Covid-19 outbreak. Facemasks must be worn to protect both the wearer and those around them. In this section, we show the use of deep neural networks to recognize a person's face in real-time and determine if they are wearing a facemask. If this is not the case, we will issue a caution alert. The facemask detection application can be used anywhere on university campuses, including corridors, classrooms, libraries, and administration an example.

In addition to staying away from others by respecting social distance, we need also to wear a facemask. It is recommended to wear the most protective mask. Nowadays, facemasks are not limited to medical uses, but they extend to the fashion world. We find more and more creative facemasks, and we worry about the effectiveness in terms of protection from the coronavirus [17]. Also, it's hard to keep wearing a correct facemask for long periods. That's why we find some weird manners of wearing facemasks; under the nose, under the mouth, etc.

We build a real-time facemask detector [7], it's able to detect the person's face and know if he's wearing a mask or not. The facemask detector model contains two main parts: First, a pretrained face detection model. Second, a facemask classifier using finetuned CNN model. This application was tested previously on a Raspberry Pi and Intel NCS2 using satisfactory results.

4.3 Checking the Vaccination Passport

All over the world, countries create what we call the vaccination passport. It has many others names but refers to the same thing (covid certificate, covid pass, vaccination proof, green pass …). It's an administrative document containing personal information (name, birthday, ID card, vaccine type, etc.). In addition, there is a QR code. This code lets authorities verify if the vaccination passport is fake or not.

In Morocco, the vaccination passport has two QR codes. The first is called the National QR code, and the second is an international QR code. This feature allows Moroccans to travel abroad with a valid vaccination passport. Every country developed its official mobile application to read the QR code in the vaccination passport. For example, we found in Play Store: "TousAntiCovid", "TAC Verif" and many others. These two applications could read the International QR code perfectly. In Morocco, we have "Checkvax.ma", a mobile application able to read national QR codes.

In general, these QR reader applications have pros and cons. One disadvantage is the limitation on mobile users so the obligation of having Android to use the previous applications. Another disadvantage is the privacy of user data. This is a huge topic that has been discussed by several community parties. Moving to advantages, one is the offline option. Going behind the scenes, we understand that a QR reader is none other than a decoder of the QR code.

To decode the international QR code from vaccination passport, we use these techniques:

- The base 45 decoder is very famous in cybersecurity. It's often used to convert binary data to text format, with 45 characters defined.
- Zlib is a compressed format of data
- Cbor is a format data signed in COSE format (CBOR Object Signing and Encryption).

Once we get the encrypted data from a QR code, we use a base45 decoder to get compressed data. We decompress it using zlib to get a cose message that needs to be decoded. Finally, we load the cbor object which contains pieces of personal information from the original QR code of the vaccination passport, in JSON format.

In our application, to not make it limited to local use, we will be using only the international QR code. In this manner, our proposed solution could be used in any corner of the globe without any problem. In addition, we don't save any personal data respecting the privacy of users. We just make sure that the scanned QR code is real and has all the information. So, once we detect the good structure of these data and their presence, we consider the pass as real. Otherwise, if the QR code was not able to read from the application that's mean it's neither fake nor not in a corresponding format. For example, trying to scan the national QR code with this application will not succeed. Because our proposed application works only with the International QR code (Fig. 3).

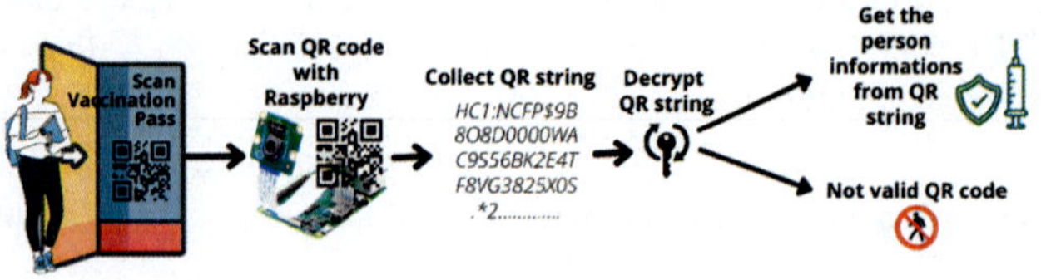

Fig. 3. The process of checking vaccination passport in entries

Today, we are testing this application in our department. One advantage is that we can check the presence of students directly while they are scanning their QR vaccination passports in the entry door. So, we verify the vaccinal passport and get the full names of our students at the same time. Regarding data privacy, students should note their presence by giving their full names anywhere. Also, this information will not be used anywhere, it's only for the teacher and administration.

5 Experimentation

Our application was developed using Python. To secure the application we make a login page using an authentication code, as we named password. The idea behind this is that only an approved person can access the app via the code delivered from the administration. For example, we declare a list of codes to be functional including capital letters, numbers, and special characters. So, the application will not be accessible to the public.

After getting access to the application, a real-time camera is shown on the screen, this is a live stream not doing any function. Then we put a ComboBox to allow the user to choose between available functionalities at a time. Our available options are Live, Facemask, Social Distance, and QR vaccination code. Once the user chooses an option from the previous list, the result will be given directly in the frame (Fig. 4).

For the facemask option, our solution highlight faces with rectangles in the frame using two different colors. Red refers to not wearing a mask, and Green refers to wearing a mask correctly. Many faces could be detected and highlighted simultaneously.

For the social distance option, the full body of a detected person will be highlighted with rectangles for each person. Under the image, we display the number of detected persons and when more than 1 person is detected, we give also the distance found between each two of them.

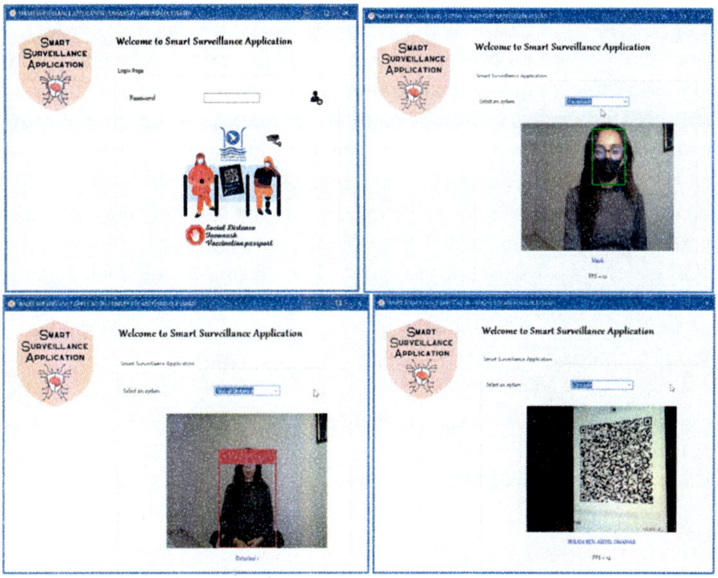

Fig. 4. Smart surveillance application: Demo

For the QR vaccination code option, when the user scans a random QR code an error message appears under the image. But when the right code is scanned, we display only the name of the person under the image. The QR code could be scanned from a paper or a smartphone without any problem. We tested many International QR vaccination codes of friends, family, and colleagues from Morocco and Spain, they work properly by extracting and displaying the user's full name.

In addition to the application, we build a laboratory prototype based on raspberry pi. It composed of a Raspberry Pi 4 8Go, a Raspberry Camera Rev1.3, Intel NCS2, a tablet as monitor, and a mini keyboard. Basically, our prototype is very small and can be used anywhere to use the smart surveillance application inside to ensure flexibility.

Wrapping all together, we used IoT devices and the power of AI in image processing to build the Smart surveillance application pack (application and prototype). At this time, we're testing this solution in LIST (Computer science, systems and telecommunication laboratory) at University Abdelmalek Essaadi, Tangier. We received satisfaction from the application users. Also, we noted some recommendations to improve the quality of this application.

6 Conclusion and Perspectives

Technology alone hasn't done any miracles yet, and it may never do. But, with the resources mobilized, financial and human, and the ability of some firms to pool their efforts, it will certainly help to take a step forward, especially during these delicate phases. The fastest-growing pandemic Covid-19 leads to change in our lifestyle. The danger of the Covid-19 is that it spreads from one person to another through contact or within proximity of an infected person. So, social distance and facemask have been forced upon us to protect ourselves and limit the propagation of the new virus.

Overall, this study offered a smart monitoring application that is not only suited for universities but can also be applied to another environment. The social distance application employs advanced deep learning models as part of this work. A Facemask Detector that can monitor and notify people if they are not wearing their mask was also successfully developed and tested. Finally, the vaccination passport reader based on decoding QR codes was also part of the application. Once implemented in a classroom the proposed solution can then verify the students' presentations and generate a log file with all of the necessary information during the exam period. In addition, real-time warnings were built to ensure prompt and effective intervention.

Acknowledgements. This project is subsidized by the MENFPESRS and the CNRST as part of the program to support scientific and technological research related to "COVID-19" (2020). Also, we acknowledge the financial support for this research from CNRST.

References

1. Researchers use Oura smart rings to predict onset of COVID-19 symptoms I Engadget. https://www.engadget.com/wvu-oura-smart-ring-coronavirus-study-162058734.html. Accessed 04 Apr 2022
2. Phelan, D.: Apple Watch Can Detect Covid-19 Before Symptoms Arise, New Study Shows. https://www.forbes.com/sites/davidphelan/2021/01/17/apple-watch-can-detect-covid-19-before-symptoms-arise-new-study-shows/. Accessed 04 Apr 2022
3. Apple releases new COVID-19 app and website based on CDC guidance. https://www.apple.com/newsroom/2020/03/apple-releases-new-covid-19-app-and-website-based-on-CDC-guidance/. Accessed 04 Apr 2022
4. Nv, R.K., Arun, M., Baraneetharan, E., Stanly Jaya Prakash, J., Kanchana, A., Prabu, S.: Detection and monitoring of the asymptotic COVID-19 patients using IoT devices and sensors. Int. J. Pervas. Comput. Commun. (2020). https://doi.org/10.1108/IJPCC-08-2020-0107

5. Hosny, K.M., Darwish, M.M., Li, K., Salah, A.: COVID-19 diagnosis from CT scans and chest X-ray images using low-cost Raspberry Pi. PLoS ONE **16**, e0250688 (2021). https://doi.org/10.1371/journal.pone.0250688
6. Barabas, J., Zalman, R., Kochlan, M.: Automated evaluation of COVID-19 risk factors coupled with real-time, indoor, personal localization data for potential disease identification, prevention and smart quarantining. In: 2020 43rd International Conference on Telecommunications and Signal Processing (TSP), pp. 645–648 (2020). https://doi.org/10.1109/TSP49548.2020.9163461
7. ben Abdel Ouahab, I., Elaachak, L., Bouhorma, M., Alluhaidan, Y.A.: Real-time facemask detector using deep learning and raspberry pi. In: 2021 International Conference on Digital Age Technological Advances for Sustainable Development (ICDATA), pp. 23–30 (2021). https://doi.org/10.1109/ICDATA52997.2021.00014
8. Chappidi, S., Nag, K., Shukla, D., Jindal, S.K.: New COVID-19 normal: An experimental prototype of smart face mask vending machine—An indispensable kiosk. In: Sikdar, B., Prasad Maity, S., Samanta, J., and Roy, A. (eds.) Proceedings of the 3rd International Conference on Communication, Devices and Computing, pp. 125–133. Springer, Singapore (2022). https://doi.org/10.1007/978-981-16-9154-6_13
9. Teach, Learn, and Make with Raspberry Pi – Raspberry Pi. https://www.raspberrypi.org/
10. Intel® Neural Compute Stick 2. https://www.intel.com/content/www/us/en/develop/hardware/neural-compute-stick.html, last Accessed 25 Sept 2020
11. Redmon, J., Farhadi, A.: YOLOv3: An Incremental Improvement. arXiv:1804.02767 [cs] (2018)
12. Lin, T.-Y., et al.: Microsoft COCO: Common objects in context. In: Fleet, D., Pajdla, T., Schiele, B., Tuytelaars, T. (eds.) ECCV 2014. LNCS, vol. 8693, pp. 740–755. Springer, Cham (2014). https://doi.org/10.1007/978-3-319-10602-1_48
13. Ben abdel ouahab, I., Elaachak, L., Elouaai, F., Bouhorma, M.: A smart surveillance prototype ensures the respect of social distance during COVID19. In: Ben Ahmed, M., Rakıp Karaş, İ, Santos, D., Sergeyeva, O., Boudhir, A.A. (eds.) SCA 2020. LNNS, vol. 183, pp. 1197–1209. Springer, Cham (2021). https://doi.org/10.1007/978-3-030-66840-2_91
14. CDC: COVID-19 and Your Health. https://www.cdc.gov/coronavirus/2019-ncov/need-extra-precautions/people-with-medical-conditions.html. Accessed 31 Jan 2022
15. Ben Abdel Ouahab, I.: SocialDistanceSSDRaspberry (2022)
16. Ben Abdel Ouahab, I.: SocialDistanceYOLORaspberry (2022)
17. Kwan, R.Y.C., Lee, P.H., Cheung, D.S.K., Lam, S.C.: Face mask wearing behaviors, depressive symptoms, and health beliefs among older people during the COVID-19 pandemic. Front. Med. **8** (2021)

CNN Model for Change Detection of Argania Deforestation from Sentinel-2 Remote Sensing Imagery

Soufiane Idbraim(✉), Zakaria Mimouni, Mohamed Ben Salah, and Mohamed Reda Dahbi

IRF-SIC Laboratory, Ibn Zohr University, Agadir, Morocco
s.idbraim@uiz.ac.ma

Abstract. Convolutional neural networks (CNN) have been used increasingly in several remote sensing applications. Although studies have used deep learning algorithms for a variety of remote sensing tasks in recent years, they are still relatively unexplored for deforestation mapping, which is an essential step in the process of the conservation of ecosystems and their biodiversity.

This study aims to map deforestation of Argania forest, an UNESCO Biosphere reserve in the Souss-Massa region in Morocco between 2015 and 2020. We worked on a CNN-based model for change detection. We used Sentinel-2 images and implemented a method involving training a CNN classifier with two different patch sizes in a post-classification framework to compare the classification maps and deduce the change map. The tests showed a significant change between the two maps, and the change detection method works relatively well given the lack of annotated data, with a classification accuracy that has reached 89%.

Keywords: Remote sensing image · Change detection · CNN · Deforestation

1 Introduction

The observation of deforestation and its evolution on a regional and continental scale, particularly since the 1980s, has progressively led various actors (national and international) to seek to implement so-called sustainable forest management [1]. Remote sensing is almost unanimously suggested as a tool for measuring deforestation. The importance thus given to remote sensing has led space agencies and scientific research institutes to become involved in the development of remote sensing-based methods for change detection and their implementation in national remote sensing monitoring systems. However, processing remote sensing images involves various processing procedures before reaching change detection map, and it strongly depends on the applied method.

Deep learning has only taken off in remote sensing since 2014 (with 3 published papers), followed by a notable rapid rise in interest toward deep learning applied to

M. Ben Ahmed et al. (Eds.): SCA 2022, LNNS 629, pp. 716–725, 2023.
https://doi.org/10.1007/978-3-031-26852-6_66

remote sensing [2]. Prior to the development of DL, remote sensing methods focused on traditional classifier such as Support Vector Machine (SVM) and Random Forest (RF) for image classification and other tasks (change detection,... etc.) [3]. SVM has received a lot of attention due to its ability to handle high dimensionality data and work well with limited training samples, while RF has gained popularity due to its ease of use (relative insensitivity to classification parameters) and generally high accuracy. In recent years, the advent of deep learning has led to a renewed interest in neural networks. Remote sensing has shifted its focus to deep learning, and deep learning algorithms have achieved significant success in many remote sensing imagery analysis tasks, including Land Use (LU) and Land Cover (LC) classification, scene classification, object detection and change detection [4, 5].

A wide body of remote sensing research has suggested that DL-based change detection approaches are superior to the traditional in terms of feature extraction [6, 7]. Due to the powerful modeling and learning capabilities. Still, there is no go-to method for all Remote Sensing applications utilizing DL attempting to solve the Change Detection task. A multitude of techniques have been proposed and applied in the literature. We distinguish two major categories. First, the post classification comparison methods as in [5, 8–10] consist of two classifiers, which can usually be converted as classification tasks into multiple labels such as Forest, Cropland, Urban, Water, Bare soil, Grassland ... etc. and trained in a joint or independent way. It provides a classification map for each period data and the change map with change directions can be obtained by comparing classification maps. The second category is the post-comparison analysis methods as in [11–14]. In these methods, multi-temporal satellite imagery is supplied to a deep neural network which is trained in an end-to-end manner to detect changes and produce the final change map allowing a direct comparison of spectral, spatial and textural features of the input images. In the first category, namely post classification comparison, errors in the classification step are propagated further down the process into the final change map. This drawback can still be mitigated by improving the classification accuracy. This can be achieved by using CNN architectures which have proven their effectiveness in optical images among others [15].

An example of deforestation related to Morocco, concerns the Argania (Argania spinosa), an endemic tree, which constitutes the second forest reserve of the country. The Argania tree represents an irreplaceable element in the ecological balance, a major contributor to biodiversity and a necessity ensuring socio-economic welfare in the region. Yet, the situation of the Argania forest is far from ideal. A series of threats urges us to take serious measures for its conservation. The surface of these forests has been reduced in recent decades, mainly by transformation to intensive irrigated crops, overgrazing and urban development. Although not as frequent, fires are also a serious threat to Argania forests.

In this study, in order to quantify the degradation of this forest resource, a novel design of a deep learning model for Argania change detection is proposed, taking into account the constraints related to the processing of satellite imagery and problems related to learning algorithms. We used multi-temporal satellite images to develop dynamic maps documenting temporal changes of Argania surfaces.

2 Material and Methods

2.1 Study Site and Data

The study area is located in the region of Souss, it is in the southwest of Morocco. Geologically, it is the alluvial basin of the Oued Souss, separated from the Sahara by the Anti-Atlas Mountains. The natural vegetation in the Souss is savanna dominated by the Argan (Argania spinosa), a local endemic tree found nowhere else, part of the area is now a UNESCO Biosphere reserve to protect this unique habitat [16].

We have acquired our dataset using Sentinel-2, thanks to its free, open data policy and convenient revisit rate, it is a great and consistent source of optical imagery of the widely areas of Argania presence.

Sentinel-2 is an Earth observation mission of the Copernicus program that systematically acquires high spatial resolution (10 m to 60 m) optical images of land and coastal waters. The mission is a constellation of two satellites, Sentinel-2A and Sentinel-2B. The mission supports a wide range of services and applications such as agricultural monitoring, emergency management, land cover classification or water quality or water quality [17].

The collected data consist of two images captured on 07/12/2015 and 01/30/2020. The images have four multispectral bands (Red, Green, Blue and Near Infrared) with a spatial resolution of 10 m. The studied site was subset into the "Admine" forest and its surrounding regions (Fig. 1) with spatial extents of 1300 × 1200 pixels (about 156 hectares of surface).

In this study, we have selected the "Admine" forest as the main center of deforestation developed between Agadir and Biougra. Agadir's international airport, urban sprawl, and the surrounding area appropriated for greenhouse cultivation, are the driving forces for this spatial distribution of deforestation of the Argan [18].

Fig. 1. Location of the studied site of Admine forest in the true color composite Sentinel-2 image taken on 07/12/2015.

2.2 Technical Approach

Flowchart

The DL-based change detection frameworks can be one of the following types: single-stream, dual-stream, or Multi-Model Integrated [2]. Since the change detection task in our project is based on two data periods, i.e., two inputs, the dual-stream structure is the one chosen. There are three variants in this category: the Siamese structure, the transfer learning-based structure and the post-classification structure.

The literature review [2] led us to the choice of the post-classification structure (Fig. 2) which consists of two classifiers, which can usually be converted into classification tasks and trained jointly or independently. This structure provides a classification map for the data of each period and the change map with the change directions can be obtained by comparing the classification maps.

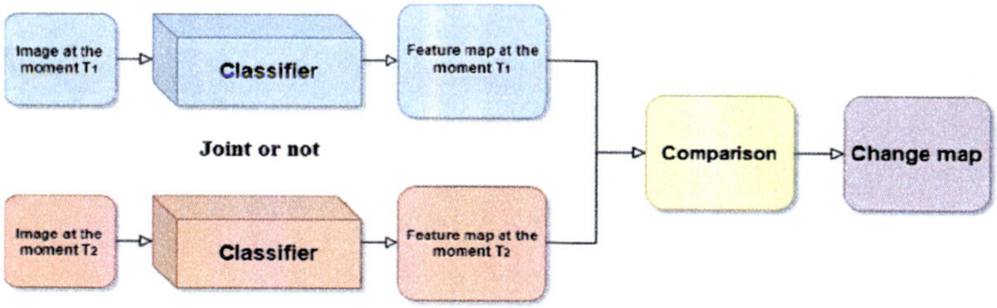

Fig. 2. Post-classification change detection structure

Data Structure

We labeled our dataset manually, since we are working on a patch-based classification, other labeling techniques, most often used for a pixel-based classification, did not work for our case. We use a binary labeling system in which the surfaces that represent the argan tree are positive and the other surfaces are then negative.

We chose a patch size of 32 in order to be able to eliminate, at least reduce the amount of noise and thus have clean patches. Adding to this being able to detect small changes since that Argania changes are really tiny in short periods.

One of the problems we encountered was the extreme lack of annotated data, so the increase of the number of training samples was a necessity to provide sufficient data to train the model. For simplicity, we assume that we do not need to consider what is beyond the boundaries of the image, we only use flips and 90-degree rotations, other angles rotations may result missing sections of the images and thus the need to interpolate some information.

The final dataset consists of 52,251 (4)-band images. We split our dataset into train, validation, and test while ensuring that the distribution of classes is maintained, which means that the percentage of Argania and not-Argania images is kept the same in the three groups.

DL Modeling Approach

To build our CNN, we have to go through many experiments, because we have to refine the parameters with respect to the dataset. Some of these parameters include the size of the filters, the number of filters in each convolution layer, the number of convolution layers, etc.

We tried different filter sizes, finally we opted for 3x3 filters because of their accuracy results. Each convolution layer is followed by a ReLu activation layer (Fig. 3). The number of convolution layers was decided due to the limited patch size, adding more layers could lead to loss of information.

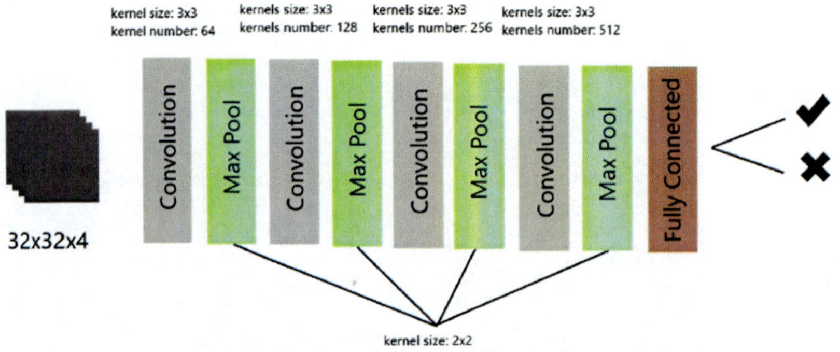

Fig. 3. The architecture of the proposed model

Training Parameters

Training our model requires to specify a number of parameters, such as choosing the appropriate optimizer and specifying an initial learning rate, and splitting dataset into training, validation and test sets. Our settings are:

- The used optimizer is Adam which combines the moment of SGD (Stochastic Gradient Descent) and the adaptive nature of RMSProp (Root Mean Squared Propagation). The moment gives the optimizer a short-term memory.
- We used the Categorical Cross Entropy as a loss function. It compares the predicted label and the actual label and calculates the loss.
- Choosing an appropriate learning rate is also quite important, When the learning rate is too small, training is not only slower, but may become permanently stuck with a high training error. And when the learning rate is too large, oscillations result.
- When training a model, it is often recommended to reduce the learning rate as the training progresses. This is why we use learning rate schedules, these are rules we set up to apply a decreasing effect on the learning rate.

– Regarding the dataset split, we use the following percentages: Training (70%), Validation (10%) and test (20%), and we introduce new whole images for testing because it makes more sense to test the model on an image with full composition.

To build and train the models, we used the Keras [19] python library, a high-level wrapper for the well-known Tensorflow library [19].

3 Results

3.1 Classifier Preparation

We trained two classifiers on two different datasets:

– Model A is trained on the dataset where the patch size is equal to 32.
– Model B is trained on a dataset with a patch size of 128.

Model A is trained for 50 epochs, while model B is trained for 15 epochs because it starts to overfit quite early, so we stop training it early to keep its best state. (Fig. 4) presents plot of the accuracy and loss on training and validation for models A and B.

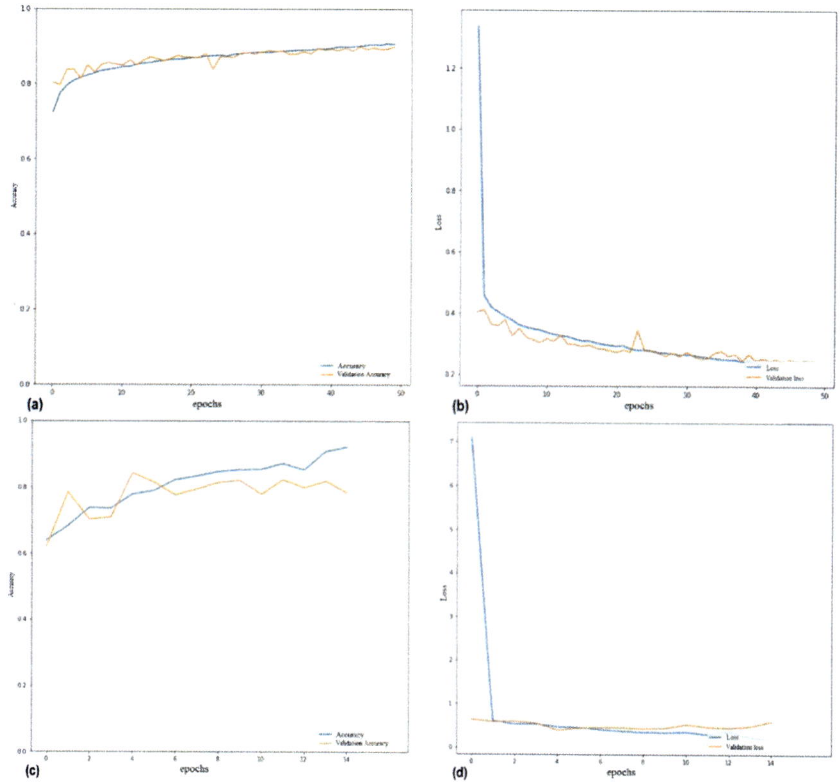

Fig. 4. Accuracy and loss during the training of Classifier A (a-b) and Classifier B (c-d)

Although Model B achieve respectable performance in theory, in practice, due to the large patch size, model B had a number of problems, including misclassification errors. So, by reducing the size of patch, we not only reduce the occurrences of this problem, but we also increase efficiency of our model when used to detect changes. Table 1 summarizes the completed results.

Table 1. Results of the training of the two classifiers

	Classifier A	Classifier B
Accuracy	90.50%	92.09%
Validation Accuracy	89.82%	78.22%
Test Accuracy	89.93%	86.30%

3.2 Change Detection

We have acquired two images of the same area at two different periods (07/12/2015 and 01/30/2020). We used the classifier A to obtain the classification maps of the two images (Fig. 5).

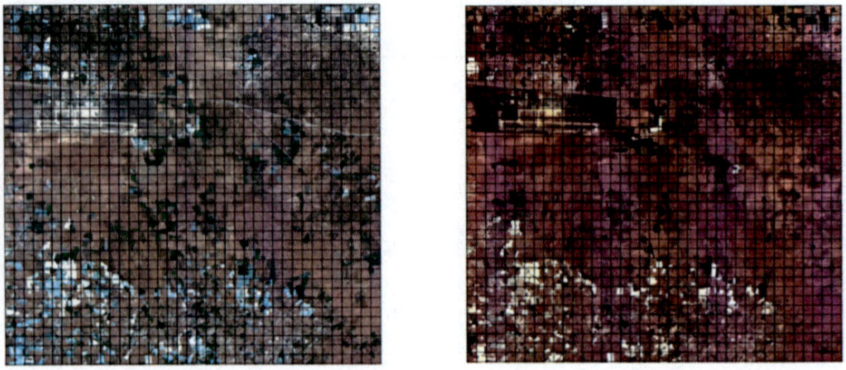

(a) (b)

Fig. 5. (a) An image of date 01/30/2020; (b) an image of date 07/12/2015

Change detection then consists of simply comparing the before and after class of each patch. Since our classification scheme consists of 2 classes: Argania and Not Argania, the change detection results would be expressed as follows:

- Was Argania, now not Argania (change).
- Was Argania, still Argania (no change).
- Was not Argania, now, is Argania (change).

– Was not Argania, still not Argania (no change).

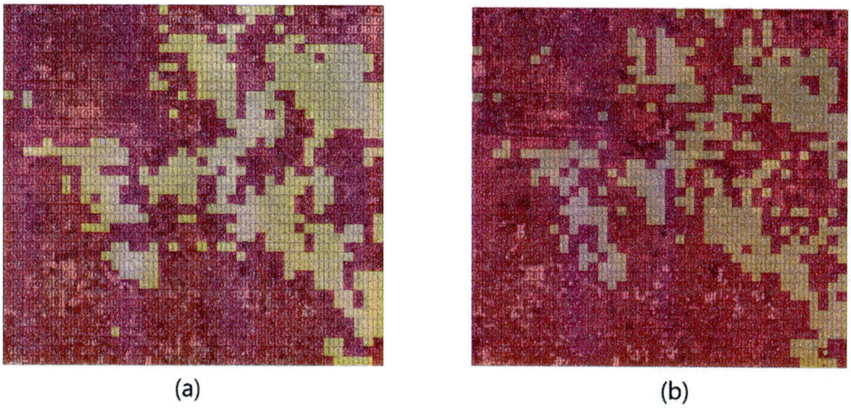

Fig. 6. The classification maps obtained from classifier A: (a) image of date 01/30/2020; (b) image with date 07/12/2015.

All that remains is to compare the two maps, obtaining the map of change is done by making the difference between the two maps (Table 2).

Table 2. Change detection (Argan deforestation) between 2015 and 2020.

Duration	Total studied area (Km^2)	Change		No change	
		Km^2	%	Km^2	%
	151.76	3.89	2.56	147.87	97.44

4 Discussion

The two classification maps in Fig. 6 show that there is a significant change between the two maps, this change may reflect both the losses of the Argania area (deforestation) and gains of the Argania area (plantation of Argania). Given that no Argania tree plantation operations have been carried out in the "Admine" forest over the last ten years [20], we can only consider that the rate of change obtained 2.56% concerns deforestation which is alarming for a small five-year period.

It is worth mentioning that even though this method is relatively simple and efficient for multi-spectral classification, it is highly dependent on the performance of the model, because if the classification maps are not accurate, the comparison between the two maps will also be inaccurate, so the best course of action is to improve the classifier accuracy as much as possible.

Despite their advantages, DL-based change detection algorithms are still not as accessible or easy to use. In addition to the fact that they require specific hardware for training, they require a relatively large number of samples. So, improving the classifier can be done by adding other sources of data. But the problem that arises is that we need really old images to notice a difference, since the changes in the Argania area are unnoticeable in the last few years and the launch of the Sentinel 2 was only 7 years ago, so a 5-year period is not really enough to introduce a noticeable difference. Thus, in the continuation of the work of this study, the use of multisource and multi-resolution imagery, especially Spot (10m) and LandSat TM-ETM (30m) images is conceivable, since we have images of these satellites for the years: 1988, 1992, 2002, 2007 and 2011, such a comparison will certainly lead to remarkable results of change.

Following this path will certainly be confronted to some problems related to the different parameters of the sensors used by the other satellites, this makes it a necessity to adapt our model to deal with all forms of data fed to it, essentially, we will explore the fusion of images, which can be done in multiple levels either by integrating a features fusion block inside our model to benefit from the features extracting capabilities of CNNs, or using multiple classifiers each trained on a specific sensor's imagery and fusing the decisions of the classifiers to determine a final decision using certain aggregation methods.

5 Conclusions

In this study, we have proposed a CNN-based model for remote sensing changes detection, and more particularly to detect the deforestation of Argania area. We started by acquiring Sentinel-2 data and enriching it with data augmentation to provide more samples for our classifier. We then trained a patch-based CNN classifier; we explored two different patch sizes to study their effect on classification accuracy, as small patches are more likely to be homogeneous. Finally, we used our classifier in a post-classification change detection structure to study Argania changes over an area of 151.76 km2 for between 2015 and 2020.The results were satisfactory but improvements can be made. Future work will focus on adapting the new patch sizes, which was time consuming as manual labeling requires a lot of human intervention, another perspective to consider is the inclusion of different remote sensing data sources and fusing them to achieve even better performance of our classifier and ultimately our change detection task.

Acknowledgement. This work was supported by the Ministry of Higher Education, Scientific Research and Innovation, the Digital Development Agency (DDA) and the CNRST of Morocco (ALKHAWARIZMI/2020/29).

References

1. Karsenty, A., Pirard, R.: Changement climatique: faut-il récompenser la déforestation évitée? Nat. Sci. Soc. **15**, 357–369 (2007)
2. Shi, W., Zhang, M., Zhang, R., Chen, S., Zhan, Z.: Change detection based on artificial intelligence: State-of-the-art and challenges. Remote Sens. **12**(1688) (2020)

3. Hussain, M., Chen, D.M., Cheng, A., Wei, H., Stanley, D.: Change detection from remotely sensed images: From pixel-based to object-based approaches. ISPRS J. Photogramm. Remote Sens. **80**, 91–106 (2013)
4. Cao, C., Dragicevic, S., Li, S.: Land-use change detection with convolutional neural network methods. Environments **6**(25) (2019)
5. Chen, Y., Ming, D., Lv, X.: Superpixel based land cover classification of VHR satellite image combining multi-scale CNN and scale parameter estimation. Earth Sci. Inf. **12**(3), 341–363 (2019)
6. Zhang, W., Lu, X.: The spectral-spatial joint learning for change detection in multispectral imagery. Remote Sens. (11), 240 (2019)
7. Fang, B., Pan, L., Kou, R.: Dual learning-based siamese framework for change detection using bi-temporal VHR optical remote sensing images. Remote Sens. (11), 1292 (2019)
8. Ienco, D., Gaetano, R., Dupaquier, C., Maurel, P.: Land cover classification via multitemporal spatial data by deep recurrent neural networks. IEEE Geosci. Remote Sens. Lett. **14**(10), 1685–1689 (2017)
9. Gallwey, J., Robiati, C., Coggan, J., Vogt, D, Eyre, M.: A Sentinel-2 based multispectral convolutional neural network for detecting artisanal small-scale mining in Ghana: Applying deep learning to shallow mining. Remote Sens. Environ. 248 (2020)
10. Xie, G., Niculescu, S.: Mapping and monitoring of land cover/land use (LCLU) changes in the Crozon Peninsula (Brittany, France) from 2007 to 2018 by machine learning algorithms (support vector machine, random forest, and convolutional neural network) and by postclassification comparison (PCC). Remote Sens. **13**(3899) (2021)
11. Peng, D., Zhang, Y., Guan, H.: End-to-end change detection for high resolution satellite images using improved UNet++. Remote Sens. **11**(1382) (2019)
12. Wiratama, W., Lee, J., Sim, D.: Change detection on multi-spectral images based on feature-level U-Net. IEEE Access **8**, 12279–12289 (2020)
13. Zhan, Y., Fu, K., Yan, M., Sun, X., Wang, H., Qiu, X.: Change detection based on deep siamese convolutional network for optical aerial images. IEEE Geosci. Remote Sens. Lett. **14**(10), 1845–1849 (2017)
14. Larabi, M.E.A., Chaib, S., Bakhti, K., Hasni, K., Bouhlala, M.E.A.: High-resolution optical remote sensing imagery change detection through deep transfer learning. J. Appl. Rem. Sens. **13**(4) (2019)
15. Paul, J., Shankar, B.U., Bhattacharyya, B., Datta, A.K.: Unsupervised change detection in remote sensing images using CNN based transfer learning. In: Singh, M., Tyagi, V., Gupta, P.K., Flusser, J., Ören, T., Sonawane, V.R. (eds.) ICACDS 2021. CCIS, vol. 1440, pp. 463–474. Springer, Cham (2021). https://doi.org/10.1007/978-3-030-81462-5_42
16. UNESCO. https://en.unesco.org/biosphere/arab-states/arganeraie. Last updated October 2018
17. European Space Agency. https://www.esa.int/Applications/Observing_the_Earth/. Accessed 01 Mar 2022
18. El Fasskaoui, B.: Fonctions, défis et enjeux de la gestion et du développement durables dans la Réserve de Biosphère de l'Arganeraie (Maroc). Études Caribéennes (12) (2009)
19. Chollet, F.K.: https://github.com/fchollet/keras. Accessed 01 Mar 2022
20. Département des Eaux et Forêts: bilan d'activités 2016–2017–2018, http://www.eauxetforets.gov.ma/Pages/Publications.aspx. Accessed 01 Mar 2022

Bowstring Bridge: Suspension Removal and Base for Progressive Collapse Analysis

Hanaa Maimouni[1(✉)], B. Kissi[2], and H. Khatib[2]

[1] Faculty of Sciences Ben M'sick, University Hassan II, Cdt Driss El Harti, B.P 7955 Casablanca, Sidi Othmane, Morocco
hanaa.maimouni-etu@etu.univh2c.ma
[2] ENSAM CASABLANCA, 150 Bd du Nil, Casablanca, Morocco

Abstract. The following study presents one of the main challenges in structural engineering, which is the progressive collapse of steel bridges. This phenomenon can be triggered by multiple factors such as extreme or abnormal loading, a local lack of resistance, or a simple human error during design or construction. If not handled correctly, progressive collapse can lead to catastrophic disproportionate consequences. This paper is an initiation towards enhancing progressive collapse resistance of bridges; it will highlight major aspects related to structural failures: history and previous research, methods, and approaches to design bridges against progressive collapse. Finally, a case study on a bowstring bridge using Euro codes, the Alternative Load Path (ALP) method, and Numerical linear static analysis. The aim is to evaluate the steel structure's behavior; particularly the variation of normal stress and displacement in different suspension removal case scenarios. The study leads to noticing the importance of proper spacing of suspension lines in a bridge, understanding the variation of stress and displacement in different scenarios, and finally the limitations of the static linear analysis and the requirement of further analysis such as quasi-static analysis, and non-linear dynamic procedure in the time domain.

Keywords: Steel bridges · Bridge response · Progressive collapse analysis · Stress behavior · Displacement behavior · Linear analysis · Static analysis · Numerical analysis

1 Introduction

The progressive collapse in structural engineering is the spread of an initial local failure of a structure resulting in the partial or total collapse of the latter. In other words, it can be defined as the dynamic mechanism during which the structure vibrates until it finds a new equilibrium position or collapses. The progressive collapse has played an important role in many catastrophic events such as the collapse of the Alfred P. Murrah Federal Building (Oklahoma City, 1995) and the World Trade Center towers (New York, 2001).

M. Ben Ahmed et al. (Eds.): SCA 2022, LNNS 629, pp. 726–738, 2023.
https://doi.org/10.1007/978-3-031-26852-6_67

Although many research efforts have contributed to the important progress made until this moment, low probability events are still neglected in the ordinary design; admissible failure probabilities cannot be specified for low probability/high consequence risks [1]. In addition, limited attention was paid to the progressive collapse of bridges in particular [2].

The following paper presents some of the notorious structural collapses in history, the methods used to design against progressive collapse, and a case study of a bowstring bridge in which the alternative load path method is initiated by removing suspension lines in different case scenarios to evaluate the variation of stress and displacement in the linear static domain.

2 Structural Failures

The records of Structural Engineering witnessed some major catastrophic structural failures that made engineers and scientists work harder to sharpen their craft and come up with optimized solutions to design, analyze, test, build, and inspect buildings and bridges. Scheer's book [3] discussed most of the structural failure cases, presenting buildings, bridges, and other structures. Some of the notorious events of progressive collapse are:

Buildings

- Ronan Point (London, United Kingdom, 1968)
- Hyatt Regency Walkway (Kansas, United States of America, 1981)
- A.P. Murrah Federal Building (Oklahoma, United States of America, 1995)
- Sampoong superstore (Seoul, South Korea, 1995)
- World Trade Center (New York, United States of America, 2001)

Bridges

- Tacoma Narrows bridge (Washington, United States of America, 1940)
- Viadotto Cannavino (Celico, Italy, 1972)
- Tasman Bridge (Hobart, Australia, 1975)
- Haeng-Ju Grand Bridge (Seoul, South Korea, 1992)

The occurrence of these events put light once and again on the importance of scientific research and innovation in building durable and robust structures, especially those of high consequence class.

3 State of the Art

Researchers Jose M. Adam, Fulvio Parisi, Juan Sagaseta, and Xinzheng Lu published a very interesting bibliometric analysis [4] to demonstrate the growing interest related to progressive collapse and robustness research. They worked on the Scopus Database; particularly the search area of Engineering by setting 50 years between 1968 and 2017, using the words 'progressive collapse' and 'structure' in the title, keywords, or abstract of the document. They also specified the type of the latter as reviews or articles. Some of the findings were that the Peak of scientific production related to this topic was mainly after the catastrophic collapse of the World Trade Center in 2001.

The rising number of papers is a clear indicator that due to the major structural failures mentioned previously, and thanks to the advancement in materials, design, and analysis approaches; which automatically set higher expectations of engineering structures. Codes and guidelines began to consider robustness and anti-progressive collapse measures in their content.

The leading countries by author's origin in this topic according to the same study were the United States of America and China, followed by the United Kingdom. Whereas the top three publishers were Elsevier, ASCE (American Society of Civil Engineers), and Science Press. The journals that were mostly concerned were Engineering Structures, Journal of performance of constructed facilities, and Journal of constructional steel.

The first time the topic of the progressive collapse was mentioned goes back to Gutteridge [5] in 1942, decades after that the same topic was interesting to systematic research and thus was presented in conferences and events. Later on, interesting methods, approaches, and case studies were proposed, among them the works of Yokel et al. [6], Leyendecker and Ellingwood [7], and Gross and McGuire [8]. As mentioned earlier, research on this particular topic intensified after the collapse of the WTC in 2001: workshops, funded projects, publications, journals, and events, among many other actions that were taken to tackle the issue of robust and resilient buildings.

The early publications were limited to the study of particular structures with very specific parameters and initial damage. However, later on, researchers such as Lalkovski and Starossek worked on comprehensive approaches to design against progressive collapse [9]. Researchers studied different aspects related to progressive collapse such as prescriptive rules, design methods, probabilistic procedures, reliability, and the distinction between direct and non-direct methods, robustness, numerical methods, and experimental testing. Still, many publications focused on buildings and thus literature presents limited studies and data concerning the progressive collapse of bridges [10–12].

4 Codes and Guidelines

The following table contains the main standards and guidelines related to progressive collapse analysis (Table 1):

Table 1. International codes and guidelines related to progressive collapse

Country	Codes and guidelines
UK	The Building (Fifth Amendment) Regulations 1970 and The building regulations 2010
Europe	EN 1991-1-7: Eurocode 1 – Actions on structures – part 1–7: General actions – Accidental actions
USA	Design of buildings to resist progressive collapse DoD (UFC 4-023-03) International Building Code (IBC 2009) Minimum design loads for buildings and other structures (ASCE/SEI 7-98, ASCE/SEI 7-02) Minimum design loads and associated criteria for buildings and other structures (ASCE/SEI 7-16) Alternative path analysis and design guidelines for progressive collapse resistance (GSA2013)
China	Code for anti-collapse design of building structures (CECS 392:2014)
Canada	Canadian Commission on Building and Fire Codes (NBCC 1995)
Australia	National construction code (NCC 2016)

5 Design Methods Against Progressive Collapse

Before choosing the proper design methods against progressive collapse, it is essential to mention that designers must go through design requirements and design objectives first. During this phase, designers will look if a design against collapse is necessary based on the exposure and the significance of the possible consequences. Once design requirements are set, the objectives come into play; in which we answer many questions such as: what are the abnormal loads to consider? What are the load combinations? What is the safety factor? How much damage is allowed?

Design requirements and design objectives are necessary before adopting a design approach. In fact, this latter depends on the type of structure, its alignment in space, and the design objectives. Design methods can be classified into two categories: direct methods and indirect methods; indirect methods are related to design prescriptive rules, whilst direct methods are more performance-based and refined approaches.

Direct methods can be either preventing local failure or assuming and limiting the local failure.

Preventing local failure aims to strengthen the key elements by designing these items with high safety against local failure; that is achieved by increasing the local resistance, assuming accidental loads such as blast in the design phase, and installing protective measures such as barriers, control of public access, or else. Assuming local failure is the method mostly used; the main direct methods that have been studied in literature are the Alternative load path and Segmentation (or compartmentalization).

The Alternative load Path (ALP) is deterministic and explicit, it is based on element removal and assessing the ability of the structure to redistribute the loads. It presents many refinement levels chosen according to the classification of risk and consequence.

Segmentation or compartmentalization is concerned with ensuring structural insulation and limiting the zone of propagation of collapse concerning the elimination of continuity at the boundaries or near the edges of compartments by inserting plastic hinges or other structural special elements.

On the other hand, indirect methods are the prescriptive rules found in codes, guidelines, and literature. The most common ones are providing ductility, enabling catenary action, and setting up tying force.

As for numerical methods [13–15], the four types that have been used for structural analysis are finite element method, discrete element method, applied element method [16], and cohesive element method.

6 Application: Bow String Bridge

As an initiation to progressive collapse design measures, the following case study highlights the impact of suspension lines' removal on the structural behavior and suggests further refinement levels as well as extensive analysis to perform in future work.

6.1 Geometry

The bridge's dimensions are inspired and based on the original design of the actual Tramway Bowstring Bridge built-in 2021 for the tramway T3 and T4 lines in Oulad Ziane Casablanca, Morocco. The structure consists of a composite bowstring bridge with a total span of 51, 18 m formed by a steel deck with beams connected to the arch and its bracings via suspension lines. Its deck supports two sidewalks of 2 m each and a 7 m tramway platform. Two abutments in both ends support the deck. The arch is inclined at 10°.

The structural elements of the bridge are all-welded reconstituted profiles as shown in the following table, except the suspension lines which are a circular filled section having a diameter of 10 cm (Table 2).

Table 2. Sections of main structural elements

Element	section	b	h_w	b_1	b_2	t_w	t_f	t_{f2}
Main girder	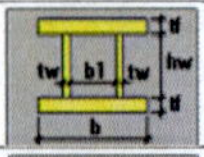	60	65	45	-	2,5	5	-
Beams	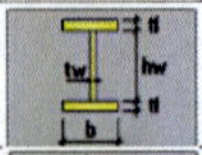	40	63	-	-	3	6	-
Arch	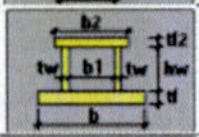	65	65	59	70	3	5	5
Arch bracing	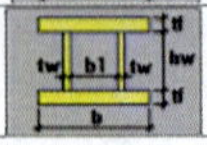	60	33	50	-	2,5	2,5	-

6.2 Materials and Standards

The materials used for this particular bridge are steel and concrete with the following grades and classification:

- Steel for beams: S355NL
- Steel for suspension lines: 460N
- Concrete for deck: 40/50

The standards used are:

- Actions on structure EN 1991 (EC1)
- Design of infrastructures for the movement of tram-train or light rail type equipment IN3128 (EF 1 C)
- Design of steel structures EN 1993 (EC3)

This study focuses only on steel parts of the bridge.

6.3 Methods and Equations

Design against progressive collapse has been interesting to research for so long; many methods and recommendations are found in the literature. Alternative Load Path (ALP) has been particularly studied in the past for several buildings assessments, and it was employed in a few applications of bridges. This method is useful in Cable-Stayed bridges to evaluate the removal of different cables and structural response and thus taking the necessary measures.

In this application, ALP is used for Bow String Bridge to evaluate if it is appropriate for this type of bridge along with a numerical elastic linear analysis. Different cases of suspension lines removal are suggested to compare the results of stress and displacement. Since the analysis is in the linear elastic domain, it is evident to assess the application of Hooke's general Law, particularly the following equation:

$$\sigma = C \times \varepsilon \tag{1}$$

where
C is the elastic Factor (Pa (N/m^2))
ε is the strain (unitless)
σ is the normal stress (Pa (N/m^2))
Given normal stress is expressed for each structural element as follows:

$$\sigma_x = \frac{N}{A} + \frac{M_z w}{I_z} + \frac{M_y v}{I_y} \tag{2}$$

where
N is the normal effort (N)
A is the section's air (cm^2)
M_z is the bending moment along the axis (Oz) (N.m)
M_y is the bending moment along the axis (Oy) (N.m)

I_y is the moment of inertia along the axis (Oy) (cm^4)
I_z is the moment of inertia along the axis (Oz) (cm^4)
v, w are respectively the corresponding distances of neutral fiber on (Oy) and (Oz) (mm)

Moreover, the strain is related to displacement (U) as mentioned in the equations below:

Lagrange strain tensor

$$\underline{\underline{E}} = \underline{\underline{\varepsilon}} + \frac{1}{2}\underline{\underline{H^T}}.\underline{\underline{H}} \tag{3}$$

where $\underline{\underline{\varepsilon}}$ is the infinitesimal strain tensor

$$\underline{\underline{\varepsilon}} = \frac{1}{2}\left(\underline{\underline{H}} + \underline{\underline{H^T}}\right) \tag{4}$$

And $\underline{\underline{H}}$ is the gradient displacement tensor

$$\underline{\underline{H}} = \overrightarrow{Grad}\left(\vec{U}(\vec{x}, t)\right) \tag{5}$$

The previously mentioned method will permit us to first define which scenario is safe and which one is critical, and second to verify if the displacement increases when the stress increases as well.

6.4 Loads and Combinations

The considered loads in this application are the total weight (G1), superstructure (G2), Tramway moving loads (Qt), sidewalk traffic loads (Qs), wind (W), and temperature (T) loads. The linear combinations used in the numerical model are presented below (Table 3).

Table 3. Linear load combinations

Limit state	Combination
ELU fundamental 1	1,35 G + 1,45 Qt + 1,6 Qs + 0,9T + 0,9 W
ELU fundamental 2	1,35 G + 1,16 Qt + 1,5 Qs + 0,9 T + 0,9 W
ELU fundamental 3	1,35 G + 1,16 Qt + 1,2 Qs + 1,5 T + 0,9 W
ELU fundamental 4	1,35 G + 1,16 Qt + 1,2 Qs + 0,9 T + 1,5 W
ELS rare 1	G + Qt + 0,8 Qs + 0,6 T + 0,6 W
ELS rare 2	G + 0,8 Qt + 0,6 T + W
ELS rare 3	G + 0,8 Qt + T + 0,6 W
ELS frequent 1	G + 0,6 Qt
ELS frequent 2	G + 0,5 W
ELS frequent 3	G + 0,6 T + 0,5 W

6.5 3D Modelling

The bridge is modeled in 3D as shown in the following figure. The structure contains several rigid connections from the center of the deck to the supports in the four ends, and additional ones connecting the arc and the deck (Fig. 1).

Fig. 1. 3D model of bow string bridge

6.6 Global Structural Behavior

Before diving into the suspension lines removal scenarios, it is essential to understand the general structural response of the bridge in its original state. The study focuses only on the stress-displacement variation, thus these parameters are the only ones to be extracted from the global analysis (Fig. 2).

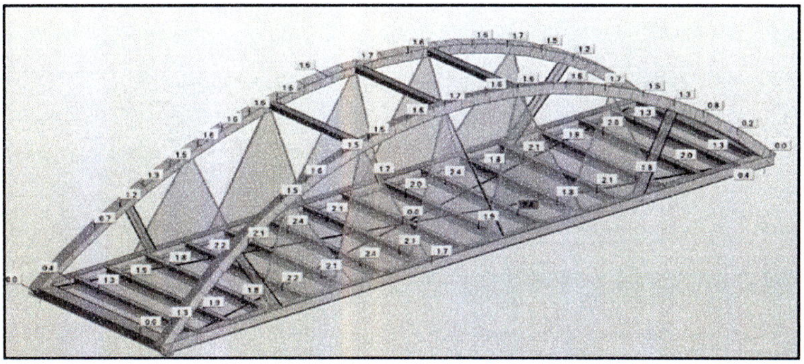

Fig. 2. Numerical model showing the different values of displacement in the original state

The numerical model shows the max value of displacement and normal stress, the following table presents these maximums (Table 4).

Table 4. Stress and displacement max values in the original state

Parameter	Combination	Max value
Normal stress (MPa)	ELU	238,11
Displacement (cm)	ELS	2,4

6.7 Suspension Removal Case Scenarios

After having established the original values of displacement and normal stress, which will be taken as a reference, three case scenarios will be evaluated as illustrated below.

- **Scenario 1: The removal of the two first suspension lines**

The first scenario concerns the removal of the first couple of suspensions where the arrow is pointing; the model, as well as the results, are shown in the figure and table below (Fig. 3) (Table 5).

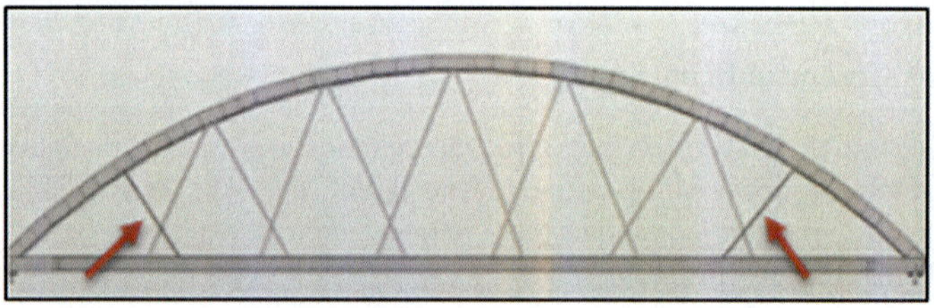

Fig. 3. Illustration of case scenario 1

Table 5. Stress and displacement max values in scenario 1

Parameter	Combination	Max value
Normal stress (MPa)	ELU	298,31
Displacement (cm)	ELS	3,4

- **Scenario 2: The removal of central suspensions**

The second scenario concerns the removal of central suspensions as illustrated in the following figure; the results are presented in the table below (Fig. 4) (Table 6).

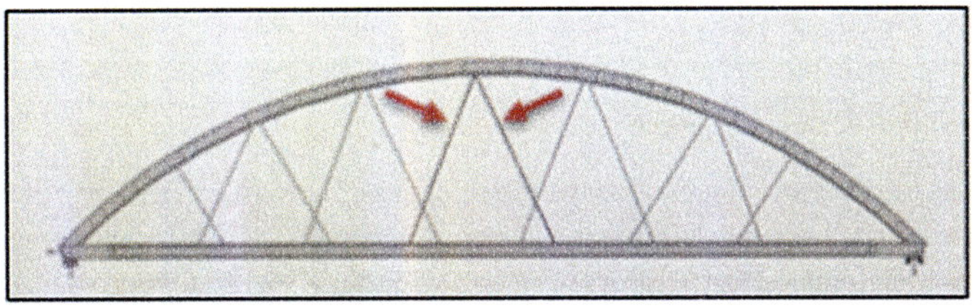

Fig. 4. Illustration of case scenario 2

Table 6. Stress and displacement max values in scenario 2

Parameter	Combination	Max value
Normal stress (MPa)	ELU	240,56
Displacement (cm)	ELS	3,2

- **Scenario 3: The removal of several adjacent suspensions**

Finally, we will remove several adjacent suspensions; the results, as well as the model, are the following (Fig. 5) (Table 7):

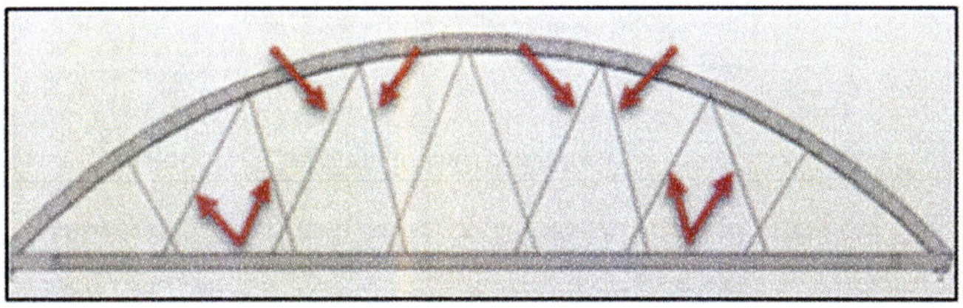

Fig. 5. Illustration of case scenario 3

Table 7. Stress and displacement max values in scenario 3

Parameter	Combination	Max value
Normal stress (MPa)	ELU	365,67
Displacement (cm)	ELS	5,2

7 Results Comparison

7.1 Displacement

The previous results show clearly that the displacement increases in the case of removing more than two adjacent suspensions; the structure's ability to redistribute loads is better when we have suspensions well-spaced along the deck allowing a good continuity.

7.2 Stress

The data collected in the previous scenarios were used to draw a graph illustrating the variation of normal stress values in each of the original state and the three case scenarios (Fig. 6).

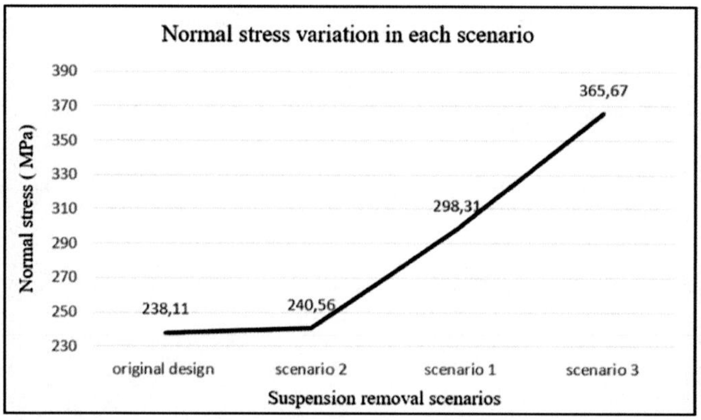

Fig. 6. Stress variation depending on suspension removal scenarios

The graph demonstrates that the normal stresses reach their maximum value in scenario 3, which also emphasizes the importance of intermediate suspension lines for a better overall structural response. Scenario 2 on the other hand is the safest one and the closest to the original state.

7.3 Stress-Displacement Graph and Discussion

After noticing both the variation in displacement and stress, the data is used to draw the stress-displacement graph to assess the linearity of the previous parameters (Fig. 7).

Note that:

O1: original state
S1: scenario 1 of suspension removal
S2: scenario 2 of suspension removal
S3: scenario 3 of suspension removal

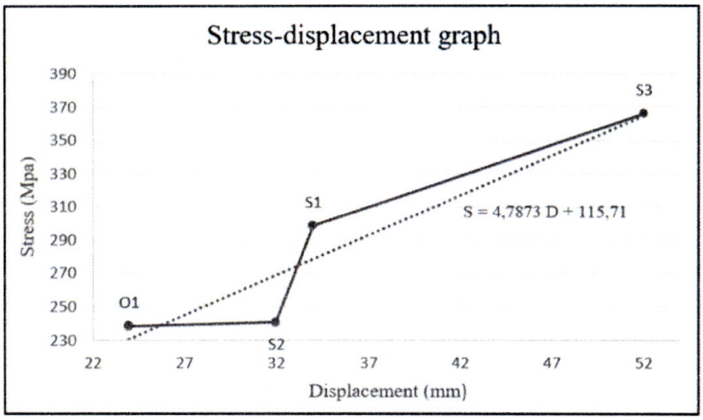

Fig. 7. The stress-displacement graph

The graph validates the results of stress and displacement in the sense that scenario 3 is the most critical one whilst scenario 2 is the safest case. Moreover, the linearity between the stress and displacement is applicable by interpolations.

The study provides interesting initial results through displacement and stress measures that demonstrate how the strength and spacing of suspension lines are vital for a bowstring bridge to work effectively. Another important parameter to study is the robustness, which is known to influence the behavior of a structure when exposed to high risks of failure [17–19].

The ALP method is a good start to determine the critical scenario of suspension removal leading to the highest risk of progressive collapse, and thus design a bridge against this phenomenon. In our case, the results showed that removing several adjacent suspensions is the most critical scenario compared to removing only the central suspensions or the ones on both ends of the deck. The justification of these results is in terms of the influence of supports, and rigid connections on the structural response as well as the arch bracing, and the design profiles.

8 Conclusion

The current study is appropriate initiation to simulate the issue of progressive collapse of bridges by evaluating the impact of suspension removal scenarios on the structure's overall behavior. Moreover, this analysis is particularly helpful to suggest preliminary measures to increase the progressive collapse resistance of the bridge. The latter can be achieved either by installing protection measures to avoid the failure of the suspensions in the first place or by considering the actual loss scenarios. In the second case, some of the propositions could be increasing the local resistance in the areas connecting the suspensions to the deck, adding stiffeners in the girders, or reinforcing the materials taking into consideration different loss scenarios.

Although these results present good insights on building a durable and collapse-resistant bridge, It is still not thorough to eliminate the risk of progressive collapse. Combining static and dynamic nonlinear methods is by far the most recommended when dealing with progressive collapse in bridges. Thus, this work must be extended by performing a quasi-static analysis, dynamic non-linear analysis in the time domain, referring to the probabilistic theory of reliability, studying other parameters other than displacement and stress, and performing the same analysis in different numerical models and refinement levels.

References

1. Starossek, U.: Progressive collapse of bridges – aspects of analysis and design. In: Invited Lecture, International Symposium on Sea-Crossing Long-Span Bridges, Mokpo, Korea, February, pp. 15–17 (2006)
2. Das, R., Pandey, A.D., Soumya, Mahesh, M.J., Saini, P., Anvesh, A.:: Progressive collapse of a cable-stayed bridge. In: 12th International Conference on Vibration Problems, ICOVP 2015. Procedia Eng. **144**, 132–139 (2016)
3. Scheer, J.: Versagen von Bauwerken: Band 2: Hochbauten und Sonderkonstruktionen. Earnst and John, vol. 2, Berlin, Germany (2001)
4. Adam, J.M., Parisi, F., Sagaseta, J., Lu, X.: Research and practice on progressive collapse and robustness of building structures in the 21st century. Eng. Struct. **173**, 122–149 (2018)
5. Gutteridge, H.: Proneness to damage of plant through enemy action. Proc. Inst. Mech. Eng. **147**(1), 99–109 (1942)
6. Yokel, F.Y., Wright, R.N., Stone, W.C.: Progressive collapse: US office building in Moscow. J. Perform. Constr. Facil. **3**(1), 57–75 (1989)
7. Leyendecker, E.V., Ellingwood, B.R.: Design methods for reducing the risk of progressive collapse in buildings. Build. Sci. Ser. **98** (1977)
8. Gross, J.L., McGuire: progressive collapse resistant design. J. Struct. Eng. **109**(1), 1–15 (1983)
9. Starossek, U.: Progressive Collapse of Structures, 2nd edn.. Institution of Civil Engineers (ICE) (2018). ISBN 978-0-7277-6168-2
10. Milic, I., Ivankovic, A.M., Syrkov, A.: Skokandic, D: Bridge failures, forensic structural engineering and recommendations for design of robust structures. Gradjevinar **73**(7), 717–736 (2021)
11. Saydam, D., Frangopol, D.M: Time-dependent performance indicators of damaged bridge superstructures. Eng. Struct. **33**(9), 2458–2471 (2011)
12. Starossek, U.: Avoiding disproportionate collapse of major bridges. Struct. Eng. Int. J. Assoc. Bridge Struct. Eng. **19**(3), 289–297 (2009)
13. Adam, J.M., Parisi, F., Delatte, N.J., Carper, K.L.: Using numerical models for the analysis of structural failures. In: IF CRASC'17. Milan (2017)
14. Kunnath, S.K., Bao, Y., El-Tawil, S.: Advances in computational simulation of gravity induced disproportionate collapse of RC frame buildings. J. Struct. Eng. **144**(2), 1–18 (2018)
15. El-Tawil, S., Li, H., Kunnath, S.: Computational simulation of gravity-induced progressive collapse of steel-frame buildings: current trends and future needs. J. Struct. Eng. **140**(8),1–12 (2014)
16. Domaneschi, M., Pellecchia, C., De Iuliis. Cimellaro, G.P., Morgese, M., Khalil, A.A., Ansari, F.: Collapse analysis of the Polcevera viaduct by the applied element method, Eng. Struct. **214**, 110659 (2020)
17. Fan, B.-H., Wang, S.-G., Chen, B.-C : Robustness assessment framework for through tied-arch bridge considering tie-bar failure. J. Bridge Eng. **27**(5), 04022028 (2022)
18. Praxedes, C., Yuan, X.-x., He, X.-H-C.: A novel robustness index for progressive collapse analysis of structures considering the full risk spectrum of damage evolution. Struct. Infrastruct. Eng. **18**(3), 376–394 (2022)
19. Bontempi, F.: Elementary concepts of structural robustness of bridges and viaducts. J. Civ. Struct. Heal. Monit. **9**(5), 703–717 (2019). https://doi.org/10.1007/s13349-019-00362-7

Split Ring Resonator (SRR) Incorporated Patch Array Antenna with Enhanced Gain and Bandwidth for 5.8 GHz WLAN Applications

Mohammed Bendaoued[1] (✉), Anouar Es-Saleh[2], Soufian Lakrit[3], Arnab De[4], Sudipta Das[5], and Ahmed Faize[3]

[1] LMIET FST of Settat, University of Hassan 1st, Settat, Morocco
mohammed.bendoued@gmail.com
[2] Mathematics and Information Systems Laboratory, FP of Nador, Mohammed First University, Oujda, Morocco
[3] Mathematics and Information Systems Laboratory, EST of Nador, Mohammed First University, Oujda, Morocco
[4] K.K. Group of Institutions, Dhanbad, Jharkhand, India
[5] Department of Electronics and Communication Engineering, IMPS College of Engineering and Technology, Malda, West Bengal, India

Abstract. The main objective of this research work is to present a new study on the design of a microstrip antenna array using a new technique based on integration of the circular split ring resonator (SRR) for the purpose of gain and bandwidth enhancement in order to use it for WLAN and Wi-Fi communication applications at 5.8 GHz. The simple antenna array resonates at 5.8 GHz with a gain of 6.48 dBi and an impedance bandwidth of 400 MHz (5.59–5.99 GHz). But, the same array structure offers significant improvements in its gain characteristics and enhancement in operating bandwidth due to the incorporation of circular shaped split ring resonators (SRR). The −10 dB impedance bandwidth of the SRR incorporated proposed antenna is 404 MHz covering 5.60 to 6.04 GHz. The peak gain of the antenna shows an improved value of 7.78 dBi at 5.8 GHz with an enhancement of 1.30 dBi. The electrical dimension of the patch is $0.23\lambda \times 0.3\lambda$. The CST Microwave studio has been utilized for the design and detailed simulation analysis of the suggested two element array antenna.

Keywords: Microstrip antenna · WLAN · Circular split ring resonator · SRR · Gain enhancement

1 Introduction

The rapid development of modern wireless technologies has stimulated the growth of research for microstrip antennas. These printed radiating patch elements are compact, lightweight, and low cost. The low-gain issues of the microstrip patch antennas can be resolved by grouping them as an array of patches. The antenna arrays with high gain,

M. Ben Ahmed et al. (Eds.): SCA 2022, LNNS 629, pp. 739–746, 2023.
https://doi.org/10.1007/978-3-031-26852-6_68

broad-bandwidth and efficient radiation characteristics are highly desirable for electrical/electronic equipment to be used in modern wireless communication systems. The electronic equipment integrated with a well-designed microstrip array antenna are highly reliable, cost-effective, and efficient [1–3]. This type of radiating element is becoming increasingly popular in many wireless systems such as RFID, medical applications, satellite communications, radar and wireless communications [4, 5]. However, there is electromagnetic mutual coupling between radiating patches of the designed array antenna due to the presence of surface waves, which leads to significant degradation of antenna parameters such as radiation characteristics, bandwidth, gain, beam forming errors, etc. [6, 7]. The reduction of mutual coupling between closely spaced antenna elements is quite a difficult task. However, various strategies have been applied to reduce the same. Some of the most common methods include placing the patches at different orientations around the circuit board [8, 9], implementation of field defective structures [6, 7], insertion of slots [9], electromagnetic band gap structures [9–11], parasitic elements between the antennas [12, 13], and inclination of patch elements along the elevation [14, 15]. However, it becomes very difficult to maintain the desired radiation properties of the antenna by offering a competent amount of mutual coupling reduction among the patch elements. In recent times, the electromagnetic research community has implemented the antenna structures based on the new artificial structure called meta-materials, which has proved to be an advantageous approach to these limitations. Inclusions of metamaterials, such as, MSRR [16], square SRR [17], rectangular SRR [23, 25], circular SRR [18, 24], etc. are directly used as loading elements for size reduction, mutual coupling reduction, enhancement of bandwidth and gain. Meta-materials are the composite structures of periodically arranged split ring resonators (SRR) to generate negative permeability and permittivity. Therefore, these materials have a negative refractive index and are therefore also known as double negative materials (DNG) or negative index materials (NIM). Meta-materials can suppress electromagnetic wave propagation in certain frequency bands [19]. Therefore, this type of material can reduce the mutual coupling between antenna array elements due to the suppression of wave propagation. This motivated the authors of this work to design a two-element patch array antenna loaded with metamaterial in the form of circular SRR for characteristic improvement of the antenna. In particular, we have aimed to enhance the impedance bandwidth and gain of the designed array antenna to be implemented for high speed 5.8 GHz WLAN applications. The advantages of the suggested array antenna include low cost, compact dimension, simple geometry, improved gain, and entire frequency band coverage for 5.8 GHz WLAN.

2 Geometry of the Antenna Array

The idea in this study is to design an antenna array based on a radiating base element that is narrow band, and then while integrating well optimized SRR cells, this antenna array can be made suitable for demanding modern wireless communication standards such as 5.8 GHz WLAN. Initially, the antenna element was validated at the 5.8 GHz centre frequency, then a 2-element antenna array was designed to operate in a narrow band centered on 5.8 GHz. Figure 1 shows the geometry of the proposed two element antenna array operating at 5.8 GHz with optimized dimensions. The basic elements are rectangular in shape as symbolized by Wp (width of the patch) and Lp (length of the patch) connected by a 50 Ω microstrip feed line. The corresponding reflection coefficient of this simple array antenna configuration is presented in Fig. 2. It can be observed that the designed simple array antenna resonates at 5.8 GHz with S11 of −32 dB. It offers good impedance matching over the operating band ranging from 5.59 GHz to 5.97 GHz with −10 dB impedance bandwidth of 400 MH.

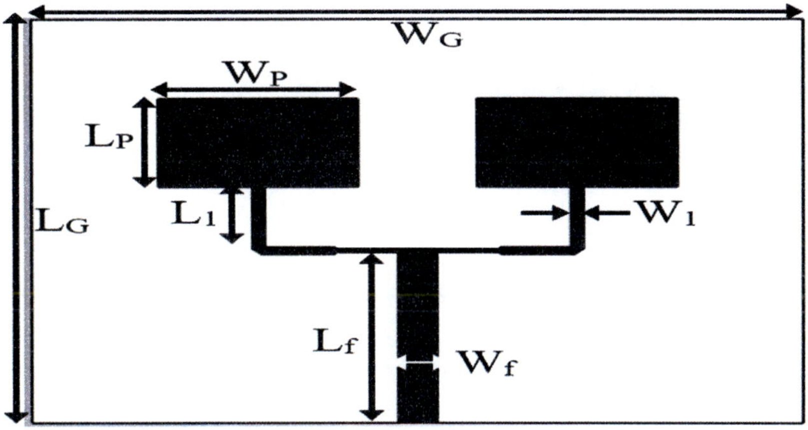

Fig. 1. Geometry of antenna array

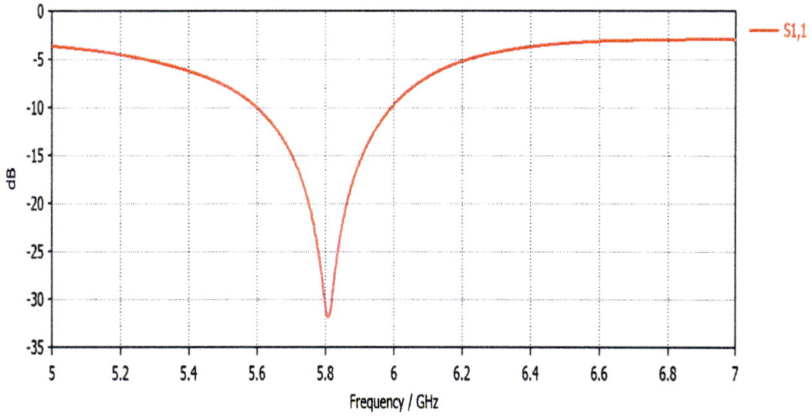

Fig. 2. Reflection coefficient S_{11} vs frequency of the proposed array

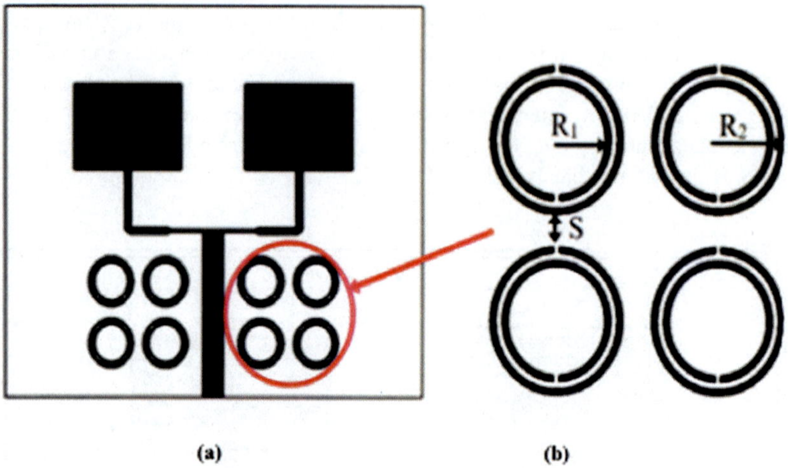

Fig. 3. (a) Proposed antenna array loaded with CSRR (b) Enlarged view of CSRR

Then, since the patch antenna can be modelled by two radiating slots separated by a transmission line that has the same length as the patch antenna, the SRR cell was coupled at the antenna width. A series of optimization and parametric studies have been carried out on the dimensions of the resonator and its position with respect to the radiating element. The proposed antenna array was obtained by introducing these circular SRR resonators as shown in Fig. 3. The optimal structural dimensions of the radiating patches, feed line and SRR are shown in the following Table 1.

Table 1. Optimized parameters of the proposed array antenna

Parameter	R1	R2	S	W_G	W_P	W_f	W_1	L_G	L_P	L_1	L_f
Value (mm)	2.4	3	1.4	60.74	15.74	3	1	50.61	11.11	7.5	21.35

3 Results and Discussion

In this section, the simulated results of the proposed antenna array loaded with SRR and without the presence of SRR are presented and discussed. The variations in S_{11} parameter and gain characteristics are observed due to the incorporation of split ring resonators within the geometry proposed antenna (Fig. 4). The comparisons of simulated S_{11} parameters and gains for these two cases (with SRR and without SRR) are depicted in Fig. 5 and 6, respectively. It is observed that the constructed simple array antenna resonates at 5.8 GHz with a bandwidth of 400 MHz. However, when the array antenna is loaded with circular SRRs, the operating bandwidth is slightly improved by 4 MHz but the gain of the antenna is improved very significantly by an amount of 1.30 dBi leading to a maximum peak gain of 7.78 dBi at 5.8 GHz operating frequency. The simulated radiation patterns of the proposed array antenna are also examined in E and H planes of observations and presented in Fig. 7. The radiation patterns presented for both planes confirm the directive nature of the designed antenna.

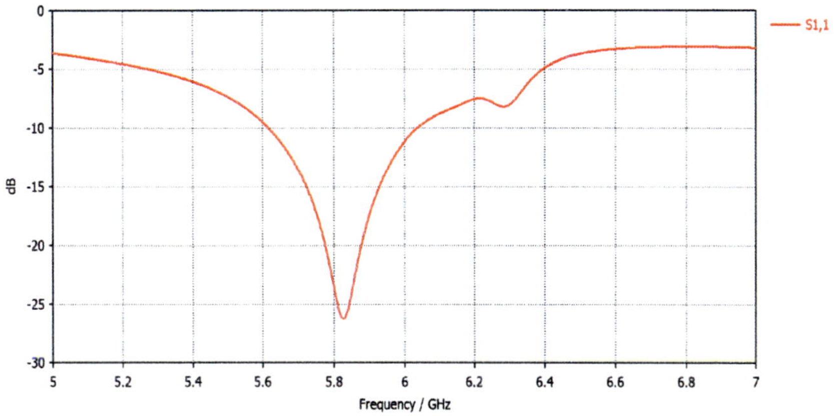

Fig. 4. Reflection coefficient vs frequency of array with SRR

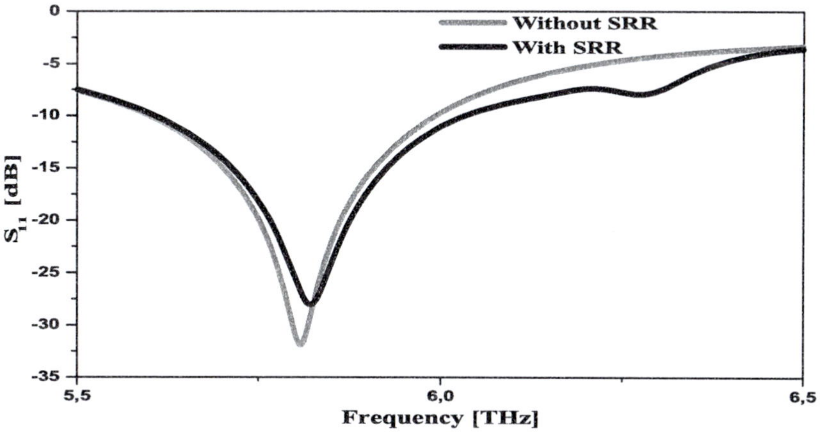

Fig. 5. S_{11} comparison of array antenna with and without SRR

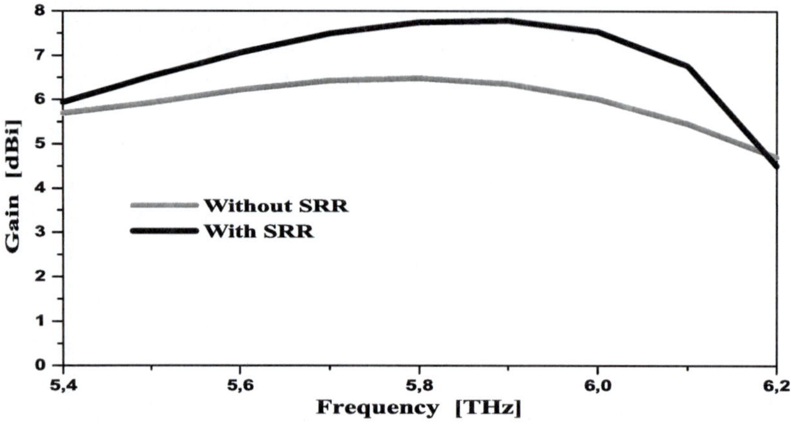

Fig. 6. Gain comparison of array antenna with and without SRRs

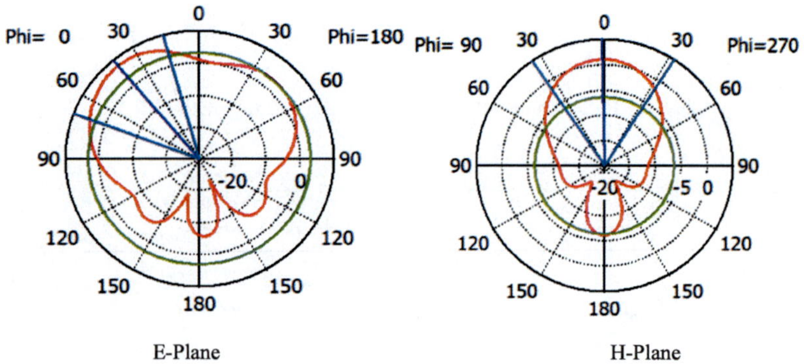

Fig. 7. Radiation patterns of proposed array antenna with SRRs at 5.8 GHz

Table 2. Comparison of the proposed array performance with other compact array antennas

Ref	Size (mm^2)	Relative permittivity	Resonance frequency (GHz)	Bandwidth (MHz)	Gain (dBi)
[20]	52.5 × 67.1	4.3	5.8	425	5.7
[21]	50 × 36.16	4.4	4.8	300	N
[22]	54 × 50	2.65	5.15/6.75	110/310	6.1/8.2
Proposed work	60.74 × 50.61	4.4	5.8	400	7.78

The performance comparison analysis of the proposed array antenna with other WLAN 5.8 GHz band antennas is summarized in Table 2; the proposed design is highly attractive due to its much better performance in terms of gain and impedance bandwidth.

4 Conclusion

In this paper, an SRR loaded microstrip patch antenna array has been presented. The effect of circular SRR loading on the designed array antenna has been investigated for the characteristics improvement of the proposed array antenna. The suggested array antenna offers bandwidth enhancement with significant improvement in its gain value due to the presence of SRR structures. The peak gain of the antenna has been increased by an amount of 1.30 dBi at the operating resonant frequency of 5.8 GHz. The advantage of this study is that the performance of the array antenna is improved without any increase in its size and cost. The proposed light weighted, low coast array antenna can be used for 5.8 GHz WLAN applications.

References

1. El Alami, A., Bennani, S.D.: Study, design and optimization of a new structure of patch antenna linear array for RFID applications. WSEAS Trans. Commun. **14**, 26–32 (2015)
2. Lakrit, S., Ammor, H., Terhzaz, J., Tribak, A.: Design of a new high-gain multiband and wideband rectangular patch antenna for C, X and Ku band applications. Walailak J. Sci. Technol. **14**(4) (2017)
3. Karli, R., Ammor, H.: Rectangular patch antenna for dual-band RFID and WLAN applications. Wireless Pers. Commun. **83**(2), 995–1007 (2015)
4. El Alami, A., Bennani, S.D., Slimani, A.: Optimization of the radiation performances of square shaped patch antenna for RFID reader. WSEAS Trans. Inf. Sci. Appl. **12**, 194–199 (2015)
5. Bendali, A., Fellahi, A.E., Srifi, M.N., Bri, S., Habibi, M.: A novel adaptive array antenna for a RFID applications. International Symposium on Advanced Electrical and Communication Technologies (ISAECT) **2018**, 1–5 (2018). https://doi.org/10.1109/ISAECT.2018.8618767
6. Singh, H., Sneha, H.L., Jha, R.M.: Mutual coupling in phased arrays: a review. Hindawi Int. J. Antennas Propagat. **2013**, 348123 (2013)
7. Dandekar, K.R., Ling, H., Xu, G.: Effect of mutual coupling on direction finding in smart antenna applications. Electron. Lett. **36**(22), 1889–1891 (2000)
8. Ying, Z., Zhang, D.: Study of the mutual coupling, correlation efficiency of two PIFA antennas on a small ground plane. In: Proceedings of IEEE Antennas Propagation Society, vol. 3B, pp. 305–308, Washington, DC, July 2005
9. Wong, K.L., Chou, J.H., Su, S.W., Su, C.M.: Isolation between GSM/DCS and WLAN antennas in a PDA phone. Microw. Opt. Technol. Lett. **45**(4), 347–352 (2005)
10. Ou Yang, J., Yang, F., Wang, Z.W.: Reducing mutual coupling of closely spaced microstrip MIMO antennas for WLAN application. IEEE Antennas Wirel. Propagat. Lett. **10**, 310–312 (2011)
11. Ebadi, S., Semnani, A.: Mutual coupling reduction in waveguide-slot-array antennas using electromagnetic bandgap (EBG) structures. IEEE Antennas Propag. Mag. **56**(3), 68–79 (2014)
12. Assimonis, S.D., Yioultsis, T.V., Antonopoulos, C.S.: Design and optimization of uniplanar EBG structures for low profile antenna applications and mutual coupling reduction. IEEE Trans. Antennas Propag. **60**(10), 4944–4949 (2012)

13. Yang, F., Rahmat Samii, Y.: Microstrip antennas integrated with electromagnetic band-gap (EBG) structures: a low mutual coupling design for array applications. IEEE Trans. Antennas Propagat. **51**(10), 2936–2946 (2003)
14. Minz, L., Garg, R.: Reduction of mutual coupling between closely spaced PIFAs. Electron. Lett. **46**(6), 392–394 (2010)
15. Ibraheam, M., Krauss, A., Irteza, S., Matthias, A.H.: Reduction of mutual coupling in compact antenna arrays using element tilting. In: Proceedings of Microwave Conference (GeMIC), pp. 1–4, Aachen, Germany, December 2014
16. Selvi, N.T., Pandeeswari, R., Selvan, P.T.: Multiband metamaterial-inspired antenna using split ring resonator. Progr. Electromag. Res. C **81**, 41–52 (2018)
17. Can, S., Soysal, D.: Estimating the frequency and bandwidth of square-split ring resonator (S-SRR) designs via support vector machines (SVM). Communications Faculty of Sciences University of Ankara Series A2-A3 Physical Sciences and Engineering, c. 63, sayı. 2, ss, pp. 127–141, Ara. (2021). https://doi.org/10.33769/aupse.992350
18. John, C., Rajesh, L.: Design and implementation of microstrip circular patch antenna for 5G applications. In: 2nd International Conference on Electrical, Communication, and Computer Engineering (ICECCE) (2020)
19. Pozar, D.M.: Microwave Engineering. John Wiley & Sons, New York (2008)
20. Arora, C., Pattnaik, S.S., Baral, R.N.: SRR Inspired Microstrip patch antenna array. Progr. Electromagn. Res. C **58**, 89–96 (2015). https://doi.org/10.2528/PIERC15052501
21. Sekharbabu, B., Reddy, K.N., Madhu, N.: Rectangular Microstrip patch antenna array with corporate feed network for WLAN applications. Int. J. Pure Appl. Math. **118**(20), 3769–3776 (2018)
22. Gao, X.J., Cai, T., Zhu, L.: Enhancement of gain and directivity for microstrip antenna using negative permeability metamaterial. AEU-Int. J. Electr. Commun. **70**(7), 880–885 (2016)
23. Lakrit, S., Das, S., El Alami, A., Barad, D., Mohapatra, S.: A compact UWB monopole patch antenna with reconfigurable Band-notched characteristics for Wi-MAX and WLAN applications. AEU-Int. J. Electr. Commun. **105**, 106–115 (2019)
24. Lakrit, S., Das, S., Ghosh, S., Madhav, B.T.P.: Compact UWB flexible elliptical CPW-fed antenna with triple notch bands for wireless communications. Int. J. RF Microwave Comput. Aided Eng. **30**(7), e22201 (2020)
25. Lakrit, S., Das, S., Madhav, B.T.P., Babu, K.V.: An octagonal star shaped flexible UWB antenna with band-notched characteristics for WLAN applications. J. Instrum. **15**(02), P02021 (2020)

State of Charge Estimation of Lithium-ion Batteries Using Artificial Intelligence Based on Entropy and Enthalpy Variation

Hind Elouazzani[1(✉)], Ibtissam Elhassani[1], Mohammed Ouazzani-Jamil[2], and Tawfik Masrour[1]

[1] Artificial Intelligence for Engineering Sciences Team (IASI), Laboratoty Mathematical Modeling, Simulation and Smart Systems(L2M3S), Ecole Nationale Supérieure d'Arts et Metiers (ENSAM)-Meknes, Moulay Ismail University of Meknes, Meknes, Morocco
hi.elouazzani@edu.umi.ac.ma

[2] Laboratoire Systèmes et Environnements Durables (LSED), Faculté des Sciences de l'Ingénieur (FSI), Université Privée de Fès (UPF), Fez, Morocco

Abstract. Lithium-ion batteries are widely employed in industrial, consumer, and military applications. Nowadays, lithium-ion batteries' technology is gaining the interest of both research and industry fields to ensure better performance of LIB by investing mostly in monitoring the internal and hidden states.

The state of charge is one of the most critical hidden states that reflects the battery's performance because it describes the percentage of energy left in the battery, but currently, there is no direct way to measure it. Therefore, achieving a high-accuracy state of charge estimation remains a challenging problem to solve.

This paper presents a new approach for SOC assessment based on data-driven techniques but using thermodynamic measurements, especially entropy and enthalpy, to describe the battery since they give accurate information about the battery performance during charge, discharge, and cycle aging.

Keywords: Lithium ion battery · State of charge (SOC) · Entropy/Enthalpy profiles

1 Introduction

Lithium-ion batteries are a type of rechargeable battery that were originally designed for portable electronics use. However, in recent decades, they have become the leading battery type for electric vehicle (EV) and hybrid electrical vehicle (HEV) manufacturers due to their high voltages (nominal voltage is around 3.7 V), high energy efficiency, low self-discharge rates, no memory effect, long lifespan recyclable, and safety [14,26].

M. Ben Ahmed et al. (Eds.): SCA 2022, LNNS 629, pp. 747–756, 2023.
https://doi.org/10.1007/978-3-031-26852-6_69

The state of charge is one of the critical indicators of battery performance because it influences its longevity, reliability, safety and security, but unfortunately, there is no direct way to measure it. To overcome this problem, a larger number of various models and algorithms have been proposed and developed to estimate the battery SOC based on measurable parameters such as current, voltage, and temperature. However, some experiences have shown that the battery thermodynamics, especially entropy and enthalpy profiles, can better simulate the battery's behavior because of their high sensitivity to the battery's structural changes during charge, discharge, and aging. Therefore, this paper presents a new method for SOC estimation using the battery's entropy and enthalpy variation as inputs to a set of machine learning algorithms in order to build a robust model able to estimate the battery state of charge with good accuracy and low error.

The remainder of this paper is composed as follows. Section 2 introduces the most commonly used methods in the literature for SOC assessment and highlights their advantages and limits. Section 3 provides more details about our method and the generation of the dataset used. The methodology followed and the models built are described in Section 4. Section 5 shows the results obtained. Finally, Sect. 6 summarises the main concluding remarks.

2 Background and Related Work

2.1 Definition of State of Charge

Battery State-of-Charge is defined as the remaining capacity (Q_{rem}) as a percentage of the maximum available capacity (Q_{max}) [12].

$$SOC = \frac{Q_{rem}}{Q_{max}} \cdot 100\% \tag{1}$$

This equation is usually used to express SOC, but actually, it has other different definitions [19] and can be expressed in other different ways as well; for lithium-ion batteries (LIB), the state of charge is directly connected to the physical quantity of lithium present in the negative or positive electrode particles, or equivalently, SOC measures the amount of charge stored in the electrode calculated in terms of the lithium average concentration. There is currently no sensor capable of directly measuring this concentration [1], making correct estimation of SOC for LIB extremely difficult. Furthermore, Q_{max} does not accurately reflect the true battery capacity claimed by manufacturers because it varies over time due to ambient temperature, battery aging, and internal complicated chemical interactions [9]. Possible physical flaws in a cell's architecture and fabrication, on the other hand, are also significant mechanical factors [10].

On account of these incalculable aspects, high-accuracy SOC estimation remains a challenging problem to solve. A large number of algorithms and methods have been proposed to address the problem, of which we will highlight the most frequently employed ones in the upcoming section.

2.2 SOC Estimation Methods

Based on the previous studies concerning SOC estimation in the literature, these methods can be classified into these four main categories:

Looking-Up Table Based Method. Also known as the direct measurement method, it use physical battery properties such as voltage and impedance to estimate SOC, usually known as the open circuit voltage method and impedance methods successively [18].

Coulomb Counting. This highly frequent method, also known as the Ampere-hour method, is usually used as a reference to validate the accuracy of other estimation methodologies. It measures the discharging/charging current of a battery and integrates it over time with a given initial value SOC(0) (2).

$$SOC(t) = SOC(0) + \frac{1}{Q_{max}} \int_0^t i_{meas}(\tau) d\tau \tag{2}$$

SOC(0) is the initial state of charge, i is the battery measured current and Q_{max} is the maximum available capacity.

Model-Based Filtering Methods. This approach consists of two major parts: the first part is describing the battery behavior using a series of mathematical equations to develop the battery model, and the second part consists of using this model to develop computer programs that simulate the behavior in a realistic environment. The most commonly used battery models are Equivalent Circuit Battery models which simulate the dynamic characteristics of the battery using electrical circuit components such as a voltage source, capacitors, and resistors. Researchers have proposed several models to reproduce the battery behavior, such as the Rint model, the RC model, the Thevenin model, and the PNGV model [5,16]. Mostly, these models are employed to construct a state-space model that is used by adaptive filters [8] such as kalman filter (KF) [15], extended Kalman filter (EKF) [6] and unscented Kalman filter (UKF) [24] to obtain the optimal estimate of battery SOC. The idea behind this method is to provide the same input to the reel system and its model and compare the measured and predicted outputs. The obtained error is used to update the estimated model's state.

Data Driven Methods. Also known as machine learning methods or black-box models, these techniques are used to obtain the nonlinear relationship between the battery SOC and the measured parameters including voltage, current, and temperature using a large amount of empirical data [23].

The frequently employed machine learning methods for SOC estimation are recurrent neural network [2], Feed-forward neural network, support vector machine, and Hybrid method [3,11,13].

2.3 Comparing SOC Estimators

To better understand the differences between the methods discussed previously, Table 1 summarizes the major advantages and drawbacks of each approach.

Table 1. Summary of SOC estimators benefits and limits

Method	Benefits	Limits
Open circuit voltage	* Simple * Inexpensive * Independent of battery models	* Requires long rest time (current= 0) * Affected by battery aging * Sensitive to temperature
Coulomb counting	* Simple * Inexpensive * Independent of battery models	* Capacity measured under Lab conditions and affected by battery aging * Sensitive to initial SOC * Depends on current sensor precision
Model based filtering	* Real-time applications * Reliable, dynamic and accurate * Insensitive to the initial SOC and measurement noise	* Depends on sensor accuracy and battery models * Model sensitive to battery aging * Modeling the battery needs extensive knowledge of internal phenomena with long time for the experiments * Increasing the state variables increases the time for SOC estimation
Data driven	* Higher accuracy * No need of specific mathematical model * No need of extensive knowledge	* Accuracy related to the training dataset * High computational cost * Time consuming

2.4 Batterie Entropymetry

Several studies have been devoted to characterization of the materials of the battery's electrodes using thermodynamic measurements, especially the entropy and enthalpy changes. It has been found that the temperature and structural disorders in electrodes affect strongly the battery's entropy and enthalpy functions. Therefore, the entropy and enthalpy profiles show obvious changes taking place in the material's structure that can barely be detected by the OCV profiles [25,27].

So, due to the power of thermodynamic measurements in revealing interesting features of batteries, R. Yazami has applied the entropy-metry to five major battery issues, such as chemistry recognition, assessment of the state of charge (SOC), state of health (SOH), state of security (SOS), and battery fast charging, but this paper will focus just on the state of charge estimation using data collected by R. Yazami during his experiences.

3 Experimental and Problematic

3.1 Experimental

In general, there are two steps to the experience: The first is the measuring of discharge capacity with Arbin equipment under defined discharge current, voltage limitations, and discharge capacity conditions. The cells are then introduced to BA 2000 equipment, which measures electrochemical thermodynamics (ETM). Cells are discharged intermittently by 5% under a specific rate and for a long enough period to produce a defined increment at each SOC. After a period of time, the temperature is gradually reduced (5 °C per step) until a new stable OCV is reached [26].

Using Eqs. 3 and 4, $\Delta S\, and\, \Delta H$ data are collected and computed from temperature dependence and open-circuit voltage (OCV) at various SOCs are collected and computed.

$$\Delta S = nF \frac{\delta OCV(SOC)}{\delta T} \tag{3}$$

$$\begin{aligned} \Delta H &= -nF\,[OCV(SOC) - T\frac{\delta OCV(SOC)}{\delta T}] \\ &= -nF\,OCV(SOC) + T\Delta S \end{aligned} \tag{4}$$

3.2 Problematic

The goal of this study is to exploit some data-driven algorithms for the state of charge assessment of lithium-ion batteries (LIBs) by receiving as input entropy and enthalpy change collected in different battery aging states as expressed by the battery cycle.

Concerning the dataset used for training and testing the models, we are inspired by thermodynamic measurements collected by R. Yazami during his experience on lithium-ion batteries [26] and we generated more synthetic samples respecting the statistical properties of the original data using data augmentation tools.

4 Methodology

4.1 Pre-processing

This step aims at organizing, cleaning, preparing, and visualizing the data as shown in Figa. 1 and 2

4.2 Modeling

Since we are facing a regression problem, we built multiple machine learning models to create a good mixture of simple linear and nonlinear algorithms, train

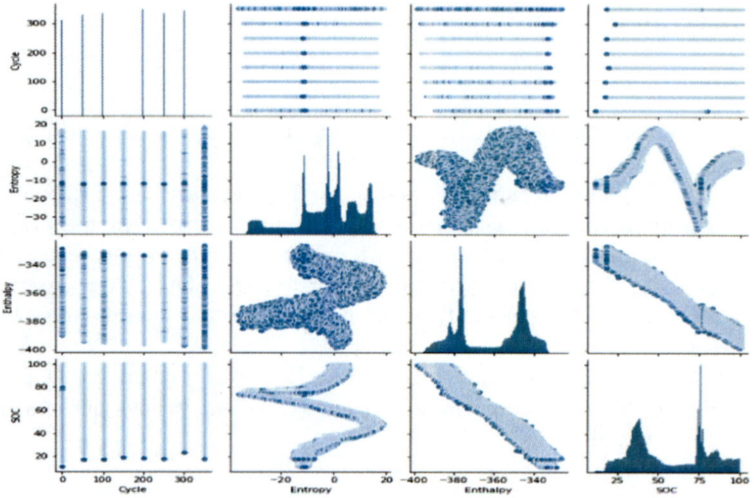

Fig. 1. Plot pairwise relationships in the dataset

them on 70% of data and evaluate them on the rest of 30% unseen data to compare their performance.

The models created are Linear Regression (LR) [22], RIDGE Regression [7], LASSO Regression [21], Elastic Net Regression (ELN), K Nearest Neighbors (KNN) [28], Support Vector Regression (SVR) [4], Decision Tree Regressor (DT) [17] and Neural Network (NN) [20].

5 Results and Discussion

After training the different estimators and adjusting their hyper-parameters to obtain the best model for each estimator, we evaluated each model and compared their performance to find out which one is the most efficient and provides the most accurate result according to three criteria:

- **Training Time:** The training of all models is done on a machine with processor Intel(R) Xeon(R) CPU E5-2680 2.4 GHz and memory of 32768 MB RAM.
- **Mean square error (MSE):** is the average of the square of the distances between observed values and predicted ones.
- **R-squared:** (R^2 or the coefficient of determination) measures how close the obtained values of the forecast are to the true values (the goodness of fit).

Before discussing the results of training and testing the models, Fig. 3 shows a good analogy between the true SOC (measured by R. Yazami during his experiences) and the SOC estimated by all the models, which confirms the high ability of entropy and enthalpy to reveal hidden and complicated battery features.

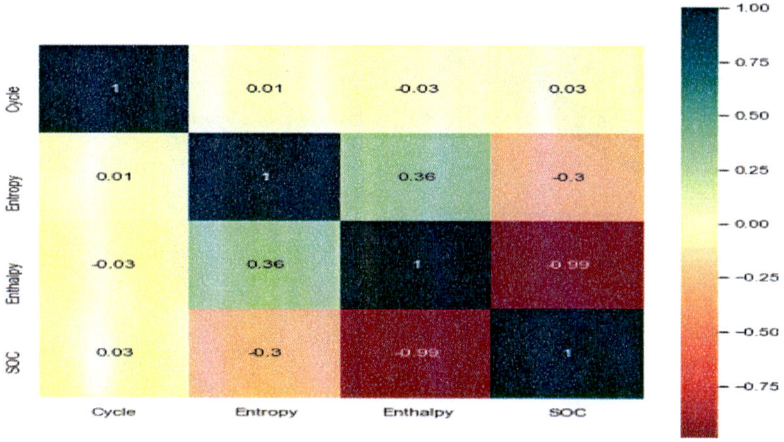

Fig. 2. Correlation matrix

Table 2. Comparison of models' performance.

Model	Training time (sec)	MSE	R-square
Linear regression	1.086	5.697	0.98957
SGD	17900.5	5.697	0.98957
RIDGE	0.201	5.68	0.98957
LASSO	0.300615	5.705	0.9895
Elastic Net	0.2769	5.699	0.9895
K Neighbors	2237.93	0.5438	0.9990
Decision Tree	27.47	0.924	0.9983
SVR	874.76	4.875	0.991053
Neural Network	261.069	5.223	0.9904

Table 2 reports the results obtained after running the models according to the selected criteria to make the comparison between their performances easier. So, the table shows that for the current problem, the models provide virtually comparable R-squared results, but they differ in terms of training time and error; K neighbors had the lowest mean squared prediction error (0.5438) followed by decision tree (0.924), despite being slower. Thus, the choice between the two algorithms is not trivial; it all depends on the purpose of the analysis and the size of the data. If the purpose of the analysis is to have the lowest error regardless how long the model takes for training, the knn is the best choice, but in the case of a huge dataset and a high cost of model training, the decision tree is more likely to be the superior selection.

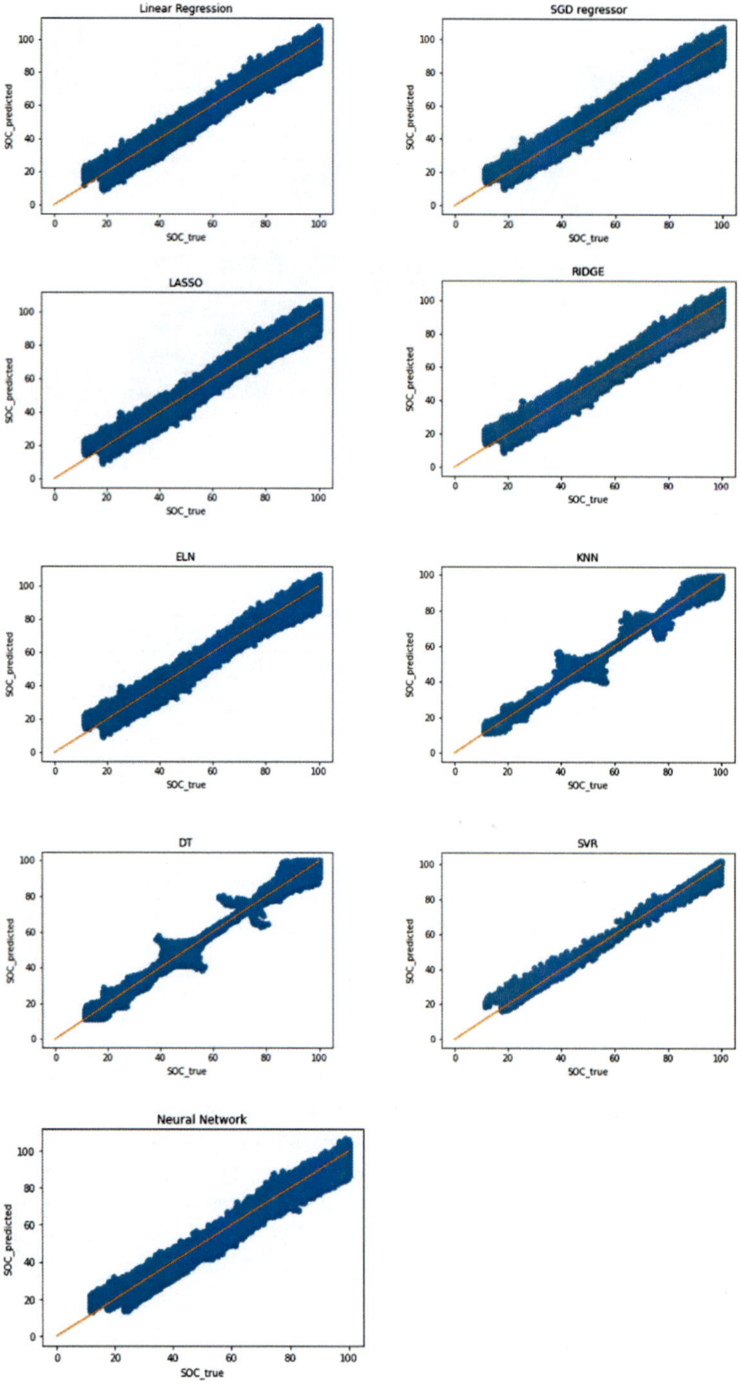

Fig. 3. SOC observed VS SOC predicted by each model.

6 Conclusion

For the SOC assessment of LIB, this paper proposed a collection of machine learning models with thermodynamic inputs. The proposed method, on the one hand, benefits from data-driven methods' high accuracy, simplicity, and low cost; on the other hand, it benefits from the quality of data used because the entropy and enthalpy profiles are highly sensitive to battery structural changes and better simulate its behavior.

Models training and testing were carried out on data generated from R. Yazami's electrochemical thermodynamic measurements, which resulted in a satisfactory SOC estimation. A comparison of the proposed models revealed that the decision tree showed the best results in terms of low MSE (0.54) and good R^2 (99%), it was successful in capturing the non-linear correlation between the battery cycle, entropy change, and enthalpy change as inputs and the SOC as an output.

References

1. Attanayaka, A., Karunadasa, J.P., Hemapala, K.: Estimation of state of charge for lithium-ion batteries - a review. AIMS Energy **7**(2), 186–210 (2019)
2. Bian, C., Yang, S., Liu, J., Zio, E.: Robust state-of-charge estimation of li-ion batteries based on multichannel convolutional and bidirectional recurrent neural networks. Appl. Soft Comput. 108401 (2022)
3. Chang, W.-Y.: The state of charge estimating methods for battery: a review. Int. Scholar. Res. Not. **2013** (2013)
4. Drucker, H., Burges, C.J.C., Kaufman, L., Smola, A., Vapnik, V., et al.: Support vector regression machines. In: Advances in Neural Information Processing Systems, vol. 9, pp. 155–161 (1997)
5. He, H., Xiong, R., Fan, J.: Evaluation of lithium-ion battery equivalent circuit models for state of charge estimation by an experimental approach. Energies **4**(4), 582–598 (2011)
6. He, Y., Li, Q., Zheng, X., Liu, X.: Equivalent hysteresis model based SOC estimation with variable parameters considering temperature. J. Power Electron. **21**(3), 590–602 (2021)
7. Hoerl, A.E., Kennard, R.W.: Ridge regression: biased estimation for nonorthogonal problems. Technometrics **12**(1), 55–67 (1970)
8. Hossain, M., Haque, M.E., Arif, M.T.: Kalman filtering techniques for the online model parameters and state of charge estimation of the li-ion batteries: a comparative analysis. J. Energy Storage **51**, 104174 (2022)
9. How, D.N.T., Hannan, M.A., Lipu, M.S.H., Ker, P.J.: State of charge estimation for lithium-ion batteries using model-based and data-driven methods: a review. IEEE Access **7**, 136116–136136 (2019)
10. Ilott, A.J., Mohammadi, M., Schauerman, C.M., Ganter, M.J., Jerschow, A.: Rechargeable lithium-ion cell state of charge and defect detection by in-situ inside-out magnetic resonance imaging. Nat. Commun. **9**(1), 1–7 (2018)
11. Ipek, E., Yilmaz, M.: A novel method for SOC estimation of li-ion batteries using a hybrid machine learning technique. Turk. J. Electr. Eng. Comput. Sci. **29**(1), 18–31 (2021)

12. Jiang, J., Zhang, C.: Fundamentals and Applications of Lithium-Ion Batteries in Electric Drive Vehicles. Wiley, New York (2015)
13. Lipu, M.S.H., et al.: Data-driven state of charge estimation of lithium-ion batteries: algorithms, implementation factors, limitations and future trends. J. Clean. Prod. 124110 (2020)
14. Languang, L., Han, X., Li, J., Hua, J., Ouyang, M.: A review on the key issues for lithium-ion battery management in electric vehicles. J. Power Sources **226**, 272–288 (2013)
15. Mastali, M., Vazquez-Arenas, J., Fraser, R., Fowler, M., Afshar, S., Stevens, M.: Battery state of the charge estimation using Kalman filtering. J. Power Sources **239**, 294–307 (2013)
16. Meng, J., Luo, G., Ricco, M., Swierczynski, M., Stroe, D.-I., Teodorescu, R.: Overview of lithium-ion battery modeling methods for state-of-charge estimation in electrical vehicles. Appl. Sci. **8**(5), 659 (2018)
17. Myles, A.J., Feudale, R.N., Liu, Y., Woody, N.A., Brown, S.D.: An introduction to decision tree modeling. J. Chemometr. J. Chemometr. Soc. **18**(6), 275–285 (2004)
18. Rodrigues, S., Munichandraiah, N., Shukla, A.K.: A review of state-of-charge indication of batteries by means of ac impedance measurements. J. Power Sources **87**(1–2), 12–20 (2000)
19. Sauer, D.U., Bopp, G., Jossen, A., Garche, J., Rothert, M., Wollny, M.: State of charge–what do we really speak about. In: The 21st International Telecommunications Energy Conference. pp. 6–9 (1999)
20. Specht, D.F., et al.: A general regression neural network. IEEE Trans. Neural Netw. **2**(6), 568–576 (1991)
21. Tibshirani, R.: Regression shrinkage and selection via the lasso. J. Roy. Stat. Soc.: Ser. B (Methodol.) **58**(1), 267–288 (1996)
22. Weisberg, S.: Applied Linear Regression, vol. 528. Wiley, New York (2005)
23. Yang, F., Song, X., Fan, X., Tsui, K.-L.: State-of-charge estimation of lithium-ion batteries via long short-term memory network. IEEE Access **7**, 53792–53799 (2019)
24. Yang, F., Zhang, S., Li, W., Miao, Q.: State-of-charge estimation of lithium-ion batteries using LSTM and UKF. Energy **201**, 117664 (2020)
25. Yazami, R.: Thermodynamics of electrode materials for lithium-ion batteries. In: Lithium Ion Rechargeable Batteries (2009)
26. Yazami, R.: Method an system for assessing a state of charge/discharge (SOC/SOD) for an electrochemical cell, 25 February 2021. US Patent App. 16/770,560
27. Yazami, R., Maher, K.: Thermodynamics of lithium-ion batteries. In: Lithium-Ion Batteries, pp. 567–604. Elsevier (2014)
28. Zhang, Z.: Introduction to machine learning: k-nearest neighbors. Ann. Transl. Med. **4**(11) (2016)

Covid-19: Impact of the Lockdown on the Behavior and on Waste Management, Case Study Ajdir, Morocco

Ahmed Ait Errouhi[1](✉), Jihane Gharib[2], Yassir Bouroumine[3], and Mahdi Raji[4]

[1] Laboratory of Advanced Systems Engineering, National School of Applied Sciences of Kénitra, Ibn Tofail University, 14000 Kénitra, Morocco
ahmed.aiterrouhi@uit.ac.ma
[2] Laboratory of Engineering, National School of Applied Sciences of Kénitra, Ibn Tofail University, 14000 Kénitra, Morocco
[3] Laboratory of Applied Geophysics, Geotechnics; Engineering Geology and Environment, Mohammadia School of Engineering, Mohammed V University, 10090 Rabat, Morocco
[4] Teorif Group Pizzorno, Al Hoceima, Morocco

Abstract. Since the containment measures placed in several countries to deal with Covid-19 pandemic, air and noise pollution has been significantly reduced, but what about soil pollution and greenhouse gas emissions from waste management? Covid-19 has given a break to the earth by immobilizing a very large part of the world economy, industrial activity, and transport and by having an important modification on the Moroccan consumer behaviour. This demographic evolution and change of consumption do not infect the resources, but are a source of degradation and pollution of the different environmental components. Among these sources, we find the production of household and similar waste. Our objective is to rate the impact of this pandemic on waste production, in Ajdir landfill, El Hoceima, which reflects the activity of the Moroccan citizen (between March 20 and April 27, 2020), and the initiatives that have been taken and launched to solve the problems at the level of each province through the establishment of a provincial master plan for the management of household and similar waste.

Keywords: Covid-19 · AHP · Environment · Waste management · Landfill

1 Introduction

1.1 A Subsection Sample

A land that was perishing because of intense human activity, as a result of rapid urbanization and population growth, the amount of waste produced worldwide each year is expected to rise to 3.4 billion tones over the next three decades, up from 2.01 billion tons in 2016, which not only dries up natural resources but also generates pollution that is putting the future of the planet in distress [1].

The new coronavirus has had an unprecedented impact in most countries of the world. The virus has affected almost every country in the world, has spread to more

M. Ben Ahmed et al. (Eds.): SCA 2022, LNNS 629, pp. 757–768, 2023.
https://doi.org/10.1007/978-3-031-26852-6_70

than 180 million people and has caused approximately 4 million deaths [2]. Ironically, Covid-19 gave the earth a break. Nature is back in full force. Pollution has been greatly reduced [3]. The fauna and flora are no longer under attack and are now benefiting from what humans have taken away from them. Henceforth, human beings are confined to their homes, thus involuntarily giving back to nature what belongs to it [4].

In Morocco, waste reflects the human activity of production, consumption and use. The waste we discard is the result of our lifestyles and consumption patterns [5, 6]. The model of the consumer society pushes us to produce waste, since we must constantly consume goods to keep the economy going. The more we consume, the more waste we generate [7, 8].

The impact of the economic crisis linked to Covid-19 is being felt more and more. After the supermarket rushes of the early days, there is now a retreat at home while waiting for the pandemic to spread [9]. In response to quarantine policies, established in most countries, Morocco has restricted travel, and Moroccan consumers have rethought their sense of priorities. Consumers are strongly influenced by the environment in which they live [10]. The crisis we are experiencing today has resulted in a significant change in Moroccan consumer behaviour [11].

Nowadays, and in front of different problems that humanity knows because of the degradation of the environment, it has become essential to consider the environment and to recommend the necessary means for its protection in a global context based on a sustainable development. Indeed, and according to the official results communicated, the planet knows an overexploitation of its resources and suffers from the various sectors of anthropic activity due to a demographic explosion of the population and which aims at the continuous improvement of its way of life and well-being [12].

This demographic evolution and change of consumption do not infect the resources, but are a source of degradation and pollution of the different environmental components. Among these sources, we find the production of household and similar waste.

The waste sector is experiencing a strong growth in production, which is obvious and normal, in view of population consumption's change and the ever-increasing population growth. While waste management is becoming more and more complicated due to several criteria including:

- Financial cost;
- Location of landfills;
- Environmental constraints;
- Anthropic constraints;

Al Hoceima region contains a controlled landfill that currently receives and assimilates waste from eight municipalities. Each year, this landfill picks up to 36,000 tons of waste. This quantity continues to extend over the years. The waste collected is directly covered in bins and racks, without any sorting at the entrance or at the source [12]. However, and since its commissioning in 2008, this landfill is considered a source of olfactory and visual nuisance and continuous complaints from the local population due to bad odors and degradation of the road network. This calls into question the choice of the landfill site and whether it meets the technical requirements, and whether only these criteria should be addressed in the case of such a project. The purpose of our study is to appraise the impact of this change on Moroccan consumer behaviour during this period of containment on waste generation and management, and to contribute to the development of prosperous scientific methodologies and tools to assist in the decision making process for landfill site's choice by using appropriate criteria and by integrating different stakeholders opinions of the project in order to evaluate the choice of the current site and to define, if necessary, the most appropriate site in the commune [12].

2 Change in Consumption of the Local Population

Before identifying trends in household food consumption requires several approaches. The objective of this article is to have a comparative analysis of the impact of the evolution and the change in behavior of the Moroccan consumer during two distinctive periods before and during confinement showing the emergence. When we discuss the evolution of consumption, we are still interested in the explanations of behavior which are very often influenced by various standards, sometimes very exceptional.

Based on the results of the tonnage produced by the municipalities for the same period - from March 01 to April 27 - for the last 5 years, we can see an −11% drop in waste production - from March 20 to April 27 - between 2019 and 2020, while for the other years, we see an annual change of around +4%.

This decrease in waste is mainly due to a change in the population's lifestyle and its consumption by the decrease in the ratio of waste production per inhabitant. Indeed, for the same period for the previous years, the daily average waste production ratio for each inhabitant was 0.7 kg/hab/day, while for the same period of the year 2020, this ratio reached 0.62 kg/hab/day, i.e. an average drop of −0.08 kg per day for each inhabitant (Table 1), (nomenclature: B.L = before lockdown; D.L = during lockdown; Pop = population).

Table 1. Daily Average waste production ratio

Year	Population	Daily average (kg/d)		Ratio	
		B.L	D.L	B.L	D.L
2015	149848	93361	100389	0,62	0,67
2016	151096	88393	91081	0,59	0,60
2017	152370	98494	102048	0,65	0,67
2018	153672	105323	107921	0,69	0,70
2019	155001	107563	108757	0,69	0,70
2020	156357	105500	96595	0,67	0,62

2.1 Reduction in the Amount of Leachate Produced

For the volume of leachate produced, and given the decrease in waste produced, the volume produced will also experience a decrease. To this end, and to accurately assess the volume of non-produced leachate, we will use two approaches:

Theoretical approach: based on estimation of the leachate volume by theoretical assumptions and the volume of waste;

The leachate production estimate of 474 tons was made as follows (Table 2):

Table 2. Estimation of leachate volume

Waste quantity/day	t	–474
Conversion rate	t/m^3	0.8
Waste volume/day	m^3/d	–592,925
Leachate rate	%	50%
Leachate volume	m^3/d	–296.4625

For the estimate of the leachate produced, the pluviometry was not considered. Indeed, during the lockdown period we saw a partial absence of rains, the intensity of which is negligible compared to the area of the landfill. Based on this theoretical approach, the volume of leachate produced is estimated at 296 m^3.

Experimental approach: Based on the comparison of the leachate level measurements in the existing storage basins for the same period between 2019 and 2020.

For the good management of the site and the control of the leachate production, daily measurements are carried out at the level of the storage basins to control the production and the remaining storage margin. Based on the available data provided by the company responsible for operating the site, during the same period, the stock was 9,900 m^3 of leachate in 2019, and 9,465 m^3 in 2020, i.e. a decrease in the quantity of leachate stored of 435 m^3 of leachate.

2.2 Optimization of the Locker Area Used

The management of solid waste and its burial requires developed, impermeable and draining surfaces. And the higher the daily amount of waste produced, the more land is needed, while taking into consideration the expected waste height and compaction rate to ensure dome stability. Based on the amount of waste that was not produced during the confinement period and for a compaction rate of 0.8 t/m^3 and a waste height of 10 m, the locker area that was not exploited is estimated at 59 m^2. This area in reality can reach 150 m^2 if we consider the operating mode which takes place in layers of 3 m for a dumping quay 100 to 300 m in length (Table 3).

Table 3. Optimization of the locker area used

Volume reduced	m^3	−593
Landfill height	m	10
Field's surface	m^2	−59

2.3 Reduction of Green House Gases Emissions

Solid waste management is a source of greenhouse gases, both operationally and technically. Indeed, burying waste is a job that requires the use of different machines to allow the spreading, compaction and covering of the waste. While the waste, itself, by the fermentation process is a source of various Green House Gases (GHG). Apart from the fermentation process, the diesel consumption of the trucks has decreased by more than 10% during the period of this confinement, that is to say a reduction in the greenhouse gases emitted. Regarding waste, and the lack of reliable data on GHG production, we can say that a clear decrease in GHG emissions from waste has taken place given the significant decrease in waste produced during this period.

2.4 Evaluation of the Cost of Environment Degradation Gained

In order to better stimulate the impact of this pandemic on the environment, it is necessary to calculate the cost of environmental degradation gained. To this end, a calculation was made of the degradation cost gained for each environmental component of the landfill according to the following assumptions (Table 4):

Table 4. Cost of environmental degradation gained

Cost	Estimation method adopted
Reduction of leachate produced	Cost of leachate treatment
Locker optimization	Development cost m^2
Air pollution	Diesel consumption

Based on the calculation assumptions adopted, the results are as follow: (nomenclature: Qty = quantity; U = unity; DHs = Moroccan currency; CDG = Cost of degradation gained).

Table 5. Cost of environmental degradation gained

Designation	Quantity	Unity	Unity/DHs	Cost of degradation gained
Quantity of waste	474	T	120	56 880
Leachate treatment cost	435	m^3	260	113 100
Locker optimization	59	m^2	400	22 531
Fuel consumption	906.5	1	8.5	8475
TOTAL	-	-	-	200 986 DHs

3 Evaluation of the Current Site of the Intermunicipal Landfill of Al Hoceima

3.1 Geographic Information System, GIS

The assessment of a new waste disposal site is a sophisticated process as it calls for considerable expertise in diverse areas such as social, environmental, technical and economical. Methods of evaluating new landfill sites take into consideration guidelines such as distance to roads, habitation, infrastructure elements and the propensity of soil to leach contaminants [13].

GIS has been found to play a meaningful role in the domain of sitting of waste disposal sites. The potential advantage of a GIS-based approach for sitting gets from the fact that it not only diminishes time and amount of site selection, but also affords a digital data bank for long-term monitoring of the site. GIS may also play a decisive role in maintaining account data to ease collection operation and furnish customer service, studying optimal locations for transfer stations, planning routes for vehicles transporting waste to transfer stations and from transfer stations to landfills, as well as long-term monitoring of landfills (Fig. 2) [13].

3.2 Analytic Hierarchy Process, AHP

The Analytic Hierarchy Process is an approach that aids decision makers in clarifying and designing complex decisions [13, 14]. However, AHP is a decision-aid and a mathematical tool authorizing decision maker to quantify intangible criteria and evaluating choices in a various degree, hierarchical format of objectives with regard to criteria, sub-criteria, and alternatives. Pairwise comparisons are used to collect the weights of seriousness for the decision criteria, and the relative performance measures of the alternatives in terms of each decision criterion. If the comparisons are not fully steady, then AHP will provide a scheme for enhancing consistency [14].

The analysis uses the AHP approach to identify options or alternatives for suitable waste disposal. The AHP helps to break down the problem into small parts with the aim of assisting decision makers in preference assessment [13, 14]. First, the problem is constructed into a structured hierarchy. The top of the hierarchy represents the goal level; the next level is that of criteria and sub-criteria, and the lowest level denotes options or alternatives. Pairwise comparisons permit the analyst to concentrate only on one element at a time: "how strongly important is one criterion related to another with regards to the goal" [15].

Once the hierarchical model is realized for all the levels, we will put in place a matrix of analysis in pairs to evaluate the importance of each one. Each criterion must be identified, with its definition and intensity. It is necessary in fact that the sum of the weights of all the sub-criteria of the same criteria, is equal to 1. Then, we will compare the two branches of the same level two by two, by assigning an encrypted note (on a predefined scale, from 1 to 9) to the way we feel the difference between the two criteria. We begin by weighing between them the criteria with the highest ranks. Thus, we have this whole scale of values to fill a matrix of judgment [15].

Table 6. Criteria rating scale

Criteria Wi	9	...	1	...	9	Criteria Wj
When: Wi to Wj = 1, they are of equal importance						
Wi to Wj = 3, Wi is moderately favored						
Wi to Wj = 5, Wi is strongly favored						
Wi to Wj = 7, Wi is clearly dominant						
Wi to Wj = 9, Wi is super dominant						

Based on the results of the surveys and discussions with the stakeholders and the analysis and interpretation of the data sheets, we started to calculate criteria and sub-criteria weights using AHP multi-criteria analysis method (Table 6). To do this, we used an application that we developed for this purpose and available on android (Fig. 1).

Table 7. Calculation of the criteria weight

Criteria	Eco	Socio-cultural	Environment	Weight
Environment	3	7	1	0.65
Economy	1	5	0,33	0.28
Socio-cultural	0,2	1	0,14	0.07

As the table shows, the weight of the environmental criterion is 0.65, followed by the economic criterion with a weight of 0.28 and finally the socio-cultural criterion with a weight of 0.1. These results reveal the importance that stakeholders attach to the environment through its various components.

Several criteria and exclusion sub-criteria must be retained and adopted for the choice of the site of the landfill. Indeed, the choice of the site is conditioned by the respect of the distances, notably the lithology of the ground, to protect the physical, human and natural environment. To this end, it was essential to define all the criteria to be adopted for the case of our research based primarily on the literature and engineering studies in the field, and the adaptation of each criterion and condition to our case. Indeed, the exclusion criteria cannot be generalized to the whole territory and must be adapted considering the conditions, the framework and the characteristics of the study area. For example, the criterion of slopes for the case of Al Hoceima, we adopted 10%, given that the average slopes of the Province exceed 5% because of geographical and geomorphological situation. While for other regions, this criterion should not exceed 5%. In general, the choice of exclusion criteria for each criterion must absolutely be adapted taking into consideration the characteristics and specificities of the study area for the environmental components as for the basic infrastructure, Table7 illustrates the processed data:

Table 8. Exclusion criteria

Criteria	Distance
Wind direction	Downward of residential areas considering to prominent wind direction
Historical site	>500 m of historical site
Roads	500 m of main roads
Littoral	>2 km
Forests	>500 m
Agglomeration	>500 m
Lithology	Must not be on permeable lithologic layer
Slope	<10%
Hydrology	>600 m
Geology	>1 km from a geological risk area

The extraction of suitable sites for the establishment of a landfill is subject to the application and respect of several exclusion criteria as mentioned on the Table 7. In the case of Ajdir municipality, all the exclusion criteria were applied to each component of the database environment. In fact, we proceeded to the buffer option in the GIS software, applying the minimum distances that must be respected for each data in order to define the free areas that respect each criterion treated. Once we had the free spaces for each criterion, we overlaid all the data in a single map. The objective of this operation is the determination of all the free areas that respect all the required criteria.

After achieving the exclusion zones for each criterion. We superposed the various maps to be able to designate the most suitable sites. By comparing the exclusion criteria suggested to implant a landfill in the zone of our study. The results of this step are displayed above (Fig. 2). This step based on GIS allowed us to select 3 favorable sites (Fig. 2), that can shelter a landfill site in the municipality of Ajdir.

The calculation of the sub-criteria weight was done in the same way and allowed us to classify all the sub-criteria by weight. Based on the results of using the exclusion criteria for the current landfill of Ajdir in GIS database, it was found that several criteria were not respected in the selection of this site. The exclusion criteria that have not been met are:

- Distance from dwellings;
- Hydrographic network;
- Forestry and arboriculture

On the basis of the calculation of the criteria and sub-criteria weights, we were able to have a classification of the sub-criteria to be taken into consideration, by their importance while integrating the stakeholders of such a project. The following Table 8 presents this classification:

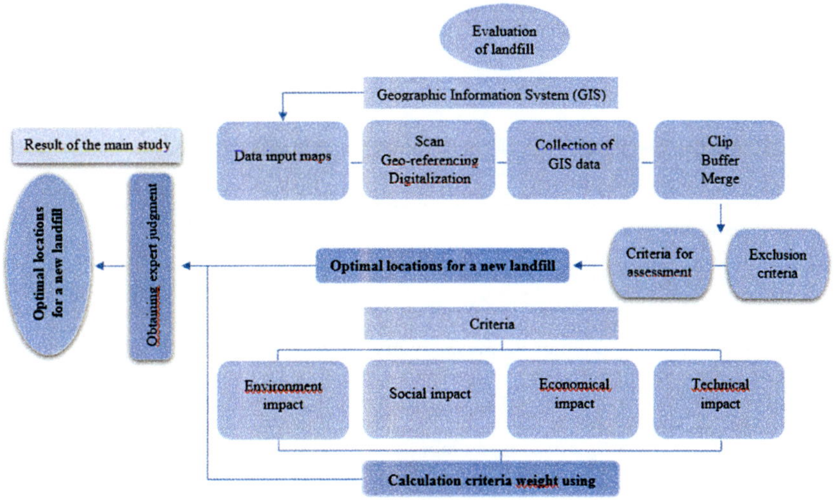

Fig. 1. Graph representing the methodology adopted to assess the current discharge and the developed application.

As the table shows, the coastal sub-criterion is the most important with a weight of 25.7%, given the economic and tourist role of this component for the Province. This is followed by geology with a weight of 18.4%, land status with 17.6%, lithology with 10.2% and distance from the waste barycenter with a weight of 6.1%.

In the light of GIS and AHP exclusion results, it can be seen that the criteria that were not met in the exclusion do not appear in the top 5 sub-criteria that are important to the stakeholders. To this end, it can be said that the current site of the Ajdir, although it does not meet the established exclusion criteria, it does meet the main sub criteria of the stakeholders.

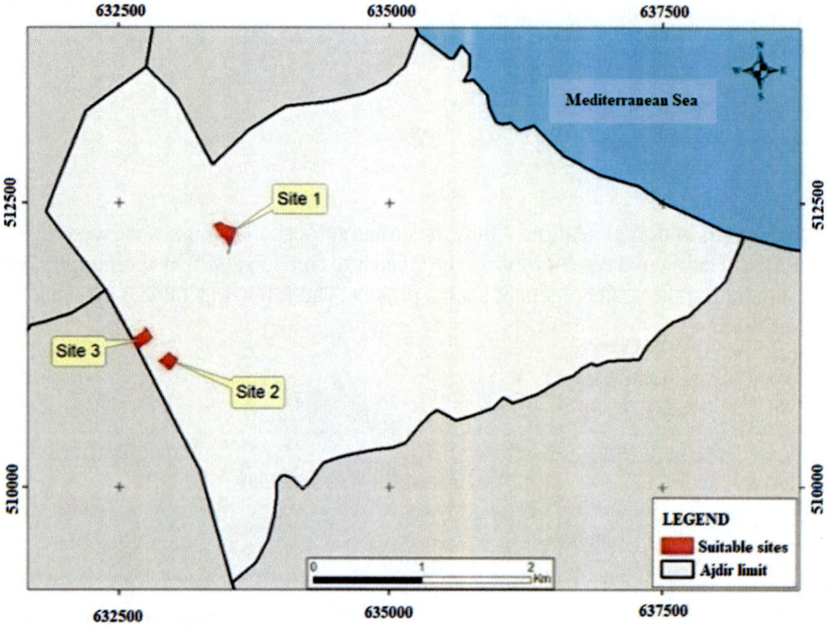

Fig. 2. Sites selected after assessment of buffer zones

Table 9. Summary of AHP results

Criteria		Sub-cietria	Weight	Rank
Environment	65%	Hydrology	5.6%	6
		Geology	18.4%	2
		Lithology	10.2%	4
		Forest	3.1%	8
		Slope	1.9%	11
		Coastline	25.7%	1
Economy	28%	Distance	6.1%	5
		Price of the land	2.9%	9
		Distances to roads	1.2%	12
		Land	17.6%	3
Socio-cultural	7%	Agglomeration	4.7%	7
		Wind direction	2.0%	10
		Historical sites	0.5%	13

4 Conclusion

The crisis caused by Covid-19 has created a particularly difficult climate. The pandemic has changed the experience of humans in general. We can expect large scale behavior change for quite a while. Indeed, consumer behavior has been forced to change instantly on a large scale. Anxieties over the availability of goods encouraged panic buying of bulk items. The financial ambiguity and the prospect of a serious, long-term recession create a difficult context that has impacted the visions, perceptions and behaviors of consumers. These felt concerns have had a respite effect and a positive effect on improving the environment and particularly in waste management, which has seen a remarkable decline during confinement. In this work, we also addressed a main feature which is the suitability of the current landfill site and the determination of the problematic related to the evaluation of this current site using GIS and AHP method used as an effective decision and support tool for planners and decision makers based on environmental, socio-economic criteria as well as their respective sub-criteria. The suitable sites obtained to implant a landfill site have distinct characteristics, advantages and disadvantages in other areas. Meanwhile, to classify the sites obtained between favorable and least favorable, the sites were compared using the AHP multi-criteria analysis method and GIS. Before using this contrast, it was necessary to make a comparison between environmental, economic and socio-cultural criteria. To have a comparison basis, and to select the criterion that must be favored in choosing the suitable site. The stakeholders; decision-makers, elected officials, experts, researchers and the local population; were consulted to scale environment, economy and socio- cultural criteria and compare them. In this sense a mobile application was developed to represent and facilitate the calculations of the criteria and sub-criteria weights. It turned out that, despite the efforts made to find better solutions, the current site still raises the concern of stakeholders.

References

1. WorldBank. Récupéré sur (September, 2018). https://www.banquemondiale.org/fr/news/press-release/2018/09/20/global-waste-to-grow-by-70-percent-by-2050-unless-urgent-action-is-taken-world-bank-report
2. World Health Organization, Récupéré sur (2021) https://www.who.int/emergencies/diseases/novel-coronavirus-2019
3. Sharma, P., Dhar, A.: Effect of hydrogen supplementation on engine performance and emissions. Int. J. Hydrogen Energy **43**(15), 7570–7580 (2018). https://doi.org/10.1016/j.ijhydene.2018.02.181
4. Boumahrou, L. [Environnement] Le COVID-19 a donné un répit à la planète, (April, 2020), https://www.ecoactu.ma/le-covid-19-repit-planete/
5. Saadat S., Rawtani D., Hussain C.M., Environmental perspective of COVID 19, Science of the Total Environment 728, (2020), https://doi.org/10.1016/j.scitotenv.2020.138870
6. ARCplus, Accessed (April 2020), https://www.acrplus.org/en/municipal-waste-management-covid-19
7. World Health Organization, Accessed (April 2020), https://www.who.int/publications/i/item/WHO-2019-nCoV-IPC-WASH-2020.4
8. Mourad, M.: Recycling, Recovering and Preventing "food waste": Competing Solutions for Food Systems Sustainability in the United States and France, vol. 126, pp. 461–477 (2016), https://doi.org/10.1016/j.jclepro.2016.03.084

9. Messner, W.: The institutional and cultural context of cross-national variation in COVID-19 Outbreaks. (2020). MedRxiv. https://www.medrxiv.org/content/https://doi.org/10.1101/2020.03.30.20047589v1
10. Dubois, B.M.V.: Récupéré sur (May 2020). https://www.universalis.fr/encyclopedie/consommation-comportement-du-consommateur/
11. Lefebure, A.: Récupéré (April 2020). https://telquel.ma/2020/04/07/comment-le-confinement-a-change-les-habitudes-de-consommation-des-marocains_1678189
12. Bouroumine, Y., Bahi, L., Ouadif, L., Elhachimi, D., Ait Errouhi A.: Sitting MSW landfill combining GIS and analytic hierarchy process (AHP), case study: ajdir, Morocco, Int. J. Adv. Res. Eng. Technol. **11**(5), 318–328, (2020). https://ssrn.com/abstract=3628393
13. Ait, E.A., Bahi, L., Ouadif, L., Akhssas, A., Bouroumine, Y., Bahi, A.: Evaluation of landfill site choice using AHP and GIS case study: Oum Azza. Morocco, MATEC Web of Conferences **149**, 02047 (2018). https://doi.org/10.1051/matecconf/201814902047
14. Elhamdouni D., Arioua A., Elhmaidi A., et al. Geomatics tools and AHP method use for a suitable communal landfill site: Case study of Khenifra region – Morocco. J. Mater. Environ. Sci. **8**(10), 3612–3624 (2017). https://doi.org/10.19044/esj.2018.v14n20p61
15. Ait Errouhi, A., Gharib, J., Bouroumine, Y., Bahi, A.: The choice of optimal sites for controlled landfill using the analytical hierarchy process and the geographic information system: a case of Oum Azza in Morocco. In: The International Conference on Innovation, Modern Applied Science & Environmental Studies, E3S Web Conference, vol. 234 (2021). https://doi.org/10.1051/e3sconf/202123400066

Development and Performances Comparison of Memory Architectures for Multiprocessor System on Chip at the SystemC/TLM

Kaoutar Aamali[1](✉), Abdelhakim Alali[2], Mohamed Sadik[1], and Zineb El Hariti[1]

[1] National High School for Electricity and Mechanics, Hassan II University of Casablanca, Casablanca, Morocco
kaoutar.aamali@ensem.ac.ma

[2] Faculty of Sciences Ben M'Sick, Hassan II University of Casablanca, Casablanca, Morocco

Abstract. Performance estimation of multiprocessor system-on-chip at a high level of abstraction is essential to address today's growing electronic design challenges and the complexity of software applications. As well, minimizing the data transfer time, between the memory and the processing unit, remains a major problem to ensure real-time, especially with complex algorithms that use more data. We present in this paper, a development description of the mono processor open-access platform, the Library of Power and Temperature at the Transaction-Level-Modeling (LIBTLMPWT) based on a Microblaze soft-core processor, built using SystemC at Transaction-Level Modeling (TLM) with shared memory architecture, towards the Multiprocessor System on Chip platform (MPSoC). The paper describes and compares two memory architectures: shared memory architecture and distributed data memory architecture for MPSoC. Thus, the paper proposes a description and comparison of the performance estimation of the two architectures for platform speed, temperature, and power consumption. While based on different benchmarks of the game of life application.

Keywords: Shared memory · MPSoC · SystemC · Performance

1 Introduction

This Systems on a Chip (SoC) are designed to meet the processing power of applications, and by dint of the complexity of embedded systems and especially the software applications [15]. Multiprocessor systems-on-a-chip (MPSoC) (see Fig. 1) integrates all necessary components for an application [16]. By this way can join more flexibility and programmability to these devices. MPSoC have become more used in recent years with the evolution of technology in various fields such as multimedia, artificial intelligence, medicine… these devices allow good development and very sophisticated products at a relatively low cost [6, 17].

In the previous articles, we have dealt with the importance of the choice of the level of abstraction for the co-simulation of systems on chips [1], and we proposed a descriptive study at these different levels. The use of a register transfer level (RTL) is characterized

M. Ben Ahmed et al. (Eds.): SCA 2022, LNNS 629, pp. 769–779, 2023.
https://doi.org/10.1007/978-3-031-26852-6_71

by good precision. But on the other hand, a slow simulation speed, which leads us to choose a faster and higher level of abstraction like Transaction-Level Modeling (TLM) concept [5]. Thus to guarantee the best performance, and as discussed in our previous studies [2] which showed that the memory architecture has a direct impact on the speed of the system.

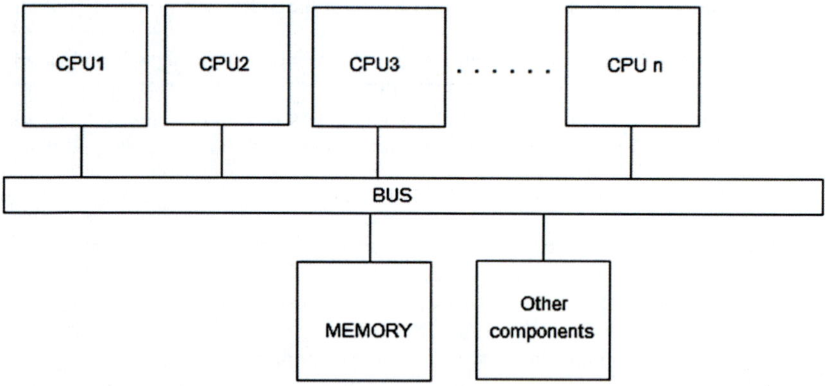

Fig. 1. Multiprocessor system on chip architecture (MPSoC).

In this paper, we will show the development and performance estimation of the Library of Power and Temperature at the Transaction-Level-Modeling (LIBTLMPWT). Section 2 presents research based on systems on a chip in different levels of abstraction. In Sect. 3, we will de-script the LIBTLMPWT platform, then the power and temperature model co-simulation [18]. In Sect. 4, we will present and detail the development of two memory architectures based on the LIBTLMPWT platform. Section 5 further explains the new memory architecture and we present a performance estimation in terms of power consumption and speed up. Ultimately, we make some conclusions and perspectives in Sect. 6.

2 Related Work

The PNX8500 chip is considered one of the first System-On-a-Chip (SOC) solutions from the Nexperia-DVP platform. Which is a multimedia system on chip solution for digital TV [19]. Philips Viper Nexperia architecture is based on two on-chip CPUs with Direct Memory Access (DMA), three buses. Two buses for each processor and one for the external memory. Since the 1990s the world discovered the launch of the first supercomputer[3], this machine had 64 arithmetic logic units and was controlled by four control units. Each CPUs had cache memory with 8 kb, and they are connected by bus of high speed. Then we saw the development of multiprocessors system on chip, which follows the complexity of the applications [20].

The optimization of SoC is due firstly to the optimization of the architecture and SoC memory design [10]. Among the studies carried out at the academic or industrial level, many papers treated the improvement of cache memory architecture. Which allows the improvement in the speed of the system. Loi et al. presented a new architecture for a multi-core system with shared L2 cache, this architecture presented advantages in terms of inter-core communication [13], which is faster, and shows a bandwidth of 28 GB/s, which presents 89% of the maximum channel. On the other hand, Loghi et al. shows a comparative energy and performance study that take into account the consistency of the cache in the MPSoCs [12].

The majority of performance studies based on shared memory architecture of MPSoC, Nogueira et al. presented solution on MPSoC platforms for scheduling, mapping and data placement with multi-bank shared memory [14]. So as to reduce the total energy consumption for data communication at the time of applications execution. A method proposed by Lee et al. allows transferring a large amount of data to Processing Elements (PEs) close to memory, instead of shared memory [11]. Which has improved bandwidth and energy consumption that has been reduced by 23.9% on average (use of the H.264 High Definition (HD) decoder). El Hariti et al. implemented in LIBTLM-PWT platform models describing the static and dynamic power, this platform is based on a mono processor on-chip at systemC with shared memory architecture, at the high level of abstraction (TLM), the 65nm and 22nm technologies were used in this study. The simulation results run different benchmarks set on the game of life application and clearly prove that the use of new technology leads to an increase in static energy.

On the other hand Chen et al. based on Distributed Shared Memory (DSM) [7], the approach in principle divides the memory into a private part and a shared part. The use of different experiments parts by the author (FFT, matrix multiplication, and H.264), has shown an improvement in performance up to 37.89%. The comparison of the performances carried out by Garibotti et al. [8] and Aamali et al. [2] between the various types of memories architectures and the performances of the systems. Proof a reduction of up to 50% of the power dissipation specified in the Distributed Shared Memory (DSM) compared to the shared memory, which despite its frequent use among researchers but it remains limited especially in speed.

This paper presents the development of the LIBTLMPWT; the platform has allowed us to carry out a study and estimate of the performance of multiprocessor systems-on-chip such as speed, energy consumption, and temperature.

3 Platform Overview

3.1 SytemeC and Transaction Level Modeling

SystemC language is built in standard C++. To move on to design and verification, classes have been created, and the C++ language extended with a set of libraries. The language is given by Open SystemC Initiative (OSCI) and affirmed as an IEEE standard in 2005. Recently, SystemC has become a frequently used language for system-level modeling, signal modeling, or performances modeling. SystemC is a library that provides the components necessary to create an algorithm model, with which we can simulate, validate and optimize the system [21].

RTL level used by VHDL and Verilog remains effective and offers a very easy synthesis. However, with the complexity of the systems the simulation at this level becomes more difficult for designers. The transaction level model (TLM) is a high level of abstraction where the structure and interconnection of registers are hidden in modules. TLM is an abstraction level modeling above RTL, whose main purpose is to speed up the simulation, the concept of this level is based on sending transactions between different classes, groups of OSCI proposed the first standard for transaction-level modeling in 2005. The version 2.0 standard was published in 2008, with an update 2.0.1 published in 2009. TLM has recently been recognized for improving modeling efficiency, platform building, and performance analysis. It is ideal for the system-level design that is increasingly recommended in the design as well as in the verification and debugging communities.

3.2 LIBTLMPWT

The Hardware/Software co-simulation in this case study (see Fig. 2) based on an open-source Library of Power and Temperature at the Transaction-Level- Modeling (LIBTLM-PWT). The hardware side present a model of system on chip mono processor of type MicroBlaze, instruction shared memory (RAM), data memory (SRAM), an interrupt controller, screen controller VGA, and a bus to connect these elements.

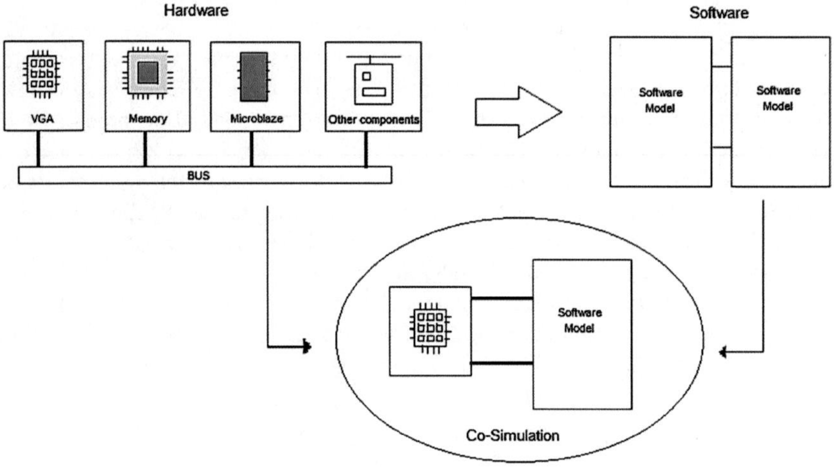

Fig. 2. The Hardware/Software Co-Simulation.

In order to guarantee a high simulation speed, the platform used the approach of Transaction-Level Modeling (TLM) that is based on the transmission of the transactions, implemented with the standard SystemC. The SystemC/TLM modeling has been successful in architectural exploration, verification, and performance analysis. In order to estimate the performance of the system LIBTLMPWT, the ATMI model is a principle model how calculated the temperature of each component based on its power consumption. Which allows the obtaining of the static and dynamic power, on the other hand, the implementation model expressed by the Eq. 1. Is based on a linear model, which slightly lacks accuracy, in our case, the short – circuit power of the short-circuit current which flows when the Metal Oxide Semiconductor PMOS and NMOS transistors are active at the same time is neglected in this approach.

$$P = P\mathit{static} + P\mathit{dynamic} \quad (1)$$

Pstatic: Static power is determined as the power consumed by the device when it is in rest mode. For MOSFET technology, the static power is related to the length of the transistor channel as well as the thickness of t-he gate.

Pdynamic: The following power is generally consumed when the inputs are active. The capacitors charge and discharge and the power increase correspondingly.

To improve simulation speed, the LIBTLMPWT platform uses an interesting concept it is a Direct Memory Interface (DMI) technique [16, 17], by accelerating access to the memory. Therefore, instead of sending transactions, it suffices to provide the initiator with a pointer to the memory array. Despite the importance of the platform, it uses a single processor and a simple architecture. Our task is to develop a more complex architecture that is based on a multiprocessor architecture system and a distributed shared memory architecture and integrate a model of performance estimation.

4 Memory Architectures Developed for MPSOC

The most important factors on which it is possible to act to improve the performance of a parallel architecture implemented, are mainly the choice of the processor, the means of communication, and the memories [4]. To study the memory architecture of MPSoCs using the LIBTLMPWT platform, the first step consists of the creation of the hardware part based on all hardware components, this depends on the needs of our application. The hardware platform consists of several processors in our case we opted for two processors as well as the peripherals connected to these processors through a Bus (SimSoC). For our platform, we use two processors (MicroBlaze) around which several basic components are associated. The first architecture created is a shared memory architecture (see

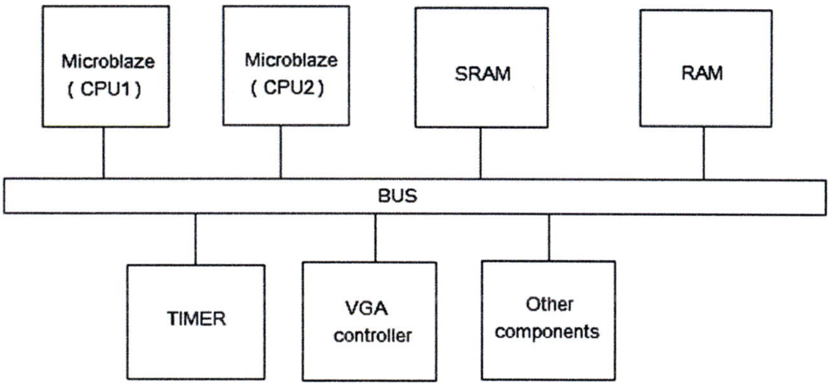

Fig. 3. MPSoC shared memory architecture of LIBTLMPWT

Fig. 3), the system is based on two processors, memory partitioned into data memory (SRAM) and instruction memory (RAM), a Timer temperature sensor, an interrupt controller, a VGA screen controller. Access and communication over Bus is managed by the Direct Memory Interface mechanism, which accelerates memory access by providing the initiator with a pointer to the memory.

Among the problems recognized during development, it was software programming because we had to solve several problems. As distributing software to different multiprocessors, and solving access memory issues. Efficient access is determined by access latency. The latter depends in part on the physical distance, from the memory bank to which one wishes to access, on the other hand on the crossing of the interfaces, in our case the bus interface, and therefore on the bandwidth of the communication links which believes the access latency. For This first architecture, Unified Memory Access (UMA) was the Hierarchical shared memory used, where processors share a single memory (data and instructions) and the access time to this memory is identical. Since the problem with this type of architecture is that the memory only supports one access at a time, the second architecture proposed is an architecture with distributed data memory, the principle used for this memory architecture is to distribute memory space. Each processor had its own data memory (SRAM) (see Fig. 4) with Non-Uniform Memory Access (NUMA) [9], For the instructions, both processors share memory instruction (RAM). The computing elements in this type of architecture are connected to several physically distributed memory banks. The communication is done by passing transactions. The processor uses the Direct Memory Interface mechanism.

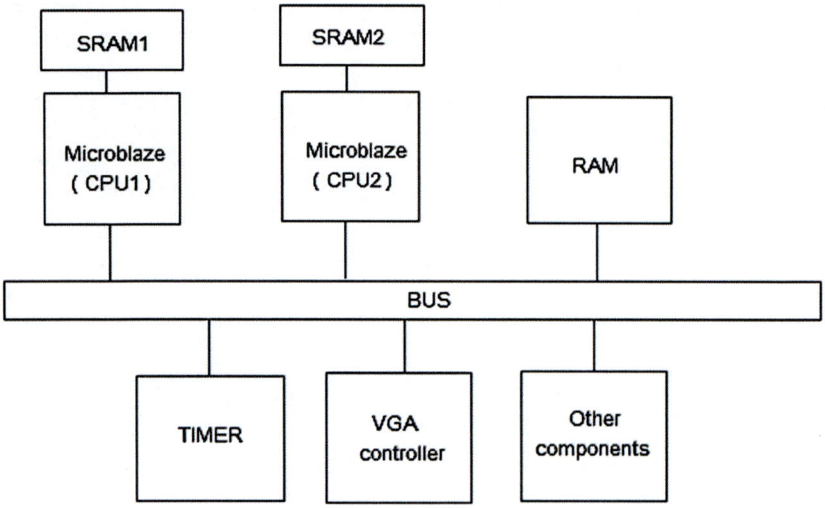

Fig. 4. MPSoC distributed data memory architecture of LIBTLMPWT

5 Results and Performance Analysis

In this section, we perform a simulation to evaluate the proposed memory architectures in terms of speedup and energy consumption, applying different benchmarks of the game of life application in our platform. The performance evolution on the experiments multiprocessor architecture is built on 22 nm technology nodes, using the initial parameters: T = 26.2 °C, Vdd = 1 V, and f = 100 MHz.

5.1 Power Consumption

The parallel architecture proposed is based on two MicroBlazes processing several instructions at the same time. The first MicroBlaze executes the game of life. As long as, the second MicroBlaze executes the lifepolling software on the target platform. The use of the Analytical Model of Temperature (ATMI) in Microprocessors component allowed the calculation of the temperature of each component at the platform based on its power consumption (see Fig. 5).

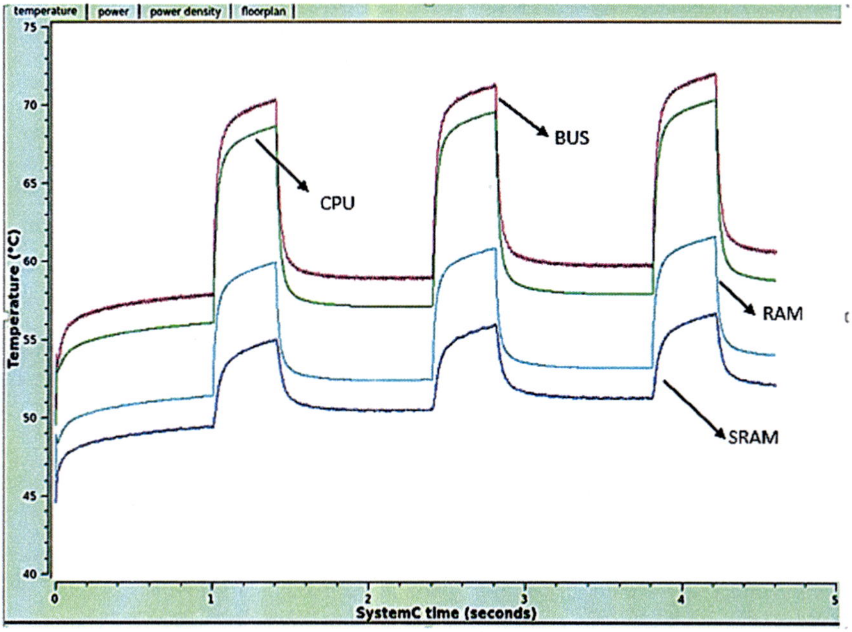

Fig. 5. Temperature of MPSoC LIBTLMPWT for the game of life application

As already mentioned, the temperature calculation is done using the ATMI component, which is packaged as a C library, called from SystemC. In order to calculate the power, the initial temperature is fixed and the ATMI is called several times. Moreover, during the calculation, the temperature sensor defines a callback method that triggers an interruption when the temperature reaches certain thresholds.

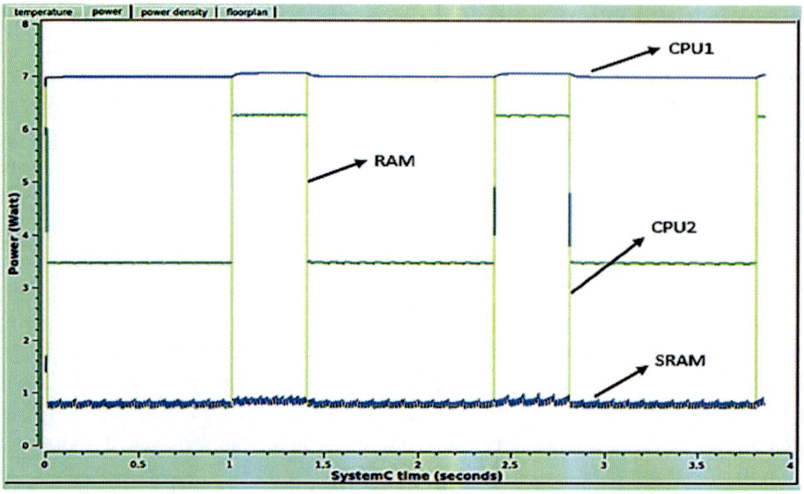

Fig. 6. Power consumption of f shared memory architecture of multiprocessor LIBTLMPWT platform for the Game of life and Life polling application.

Table 1. Power consumption of main components of shared memory architecture of Multiprocessors LIBTLMPWT platform during the execution of Game of life and Lifepolling.

Component	CPU1	CPU2	VGA	BUS	TIMER	INTC1	INTC2	SRAM	RAM
Power consumption (W)	3.90	7.01	2.51	8.55	0.34	0.52	0.52	0.825	4.85

Table 2. Power consumption of main components of distributed data memory architecture ofmultiprocessors LIBTLMPWT platform during the execution of Game of life and Lifepolling.

Component	CPU1	CPU2	VGA	BUS	TIMER	INTC1	INTC2	SRAM1	SRAM2	RAM
Power consumption (W)	3.90	7.01	2.51	8.55	0.34	0.52	0.52	0.70	0.825	4.85

The following two figures present an observation of the power consumption of the shared memory architecture (see Fig. 6), and the distributed data memory architecture result (see Fig. 7), on the other hand, as can be seen in Tables 1 and 2, which specifies the consumption of each component as a function of time for the two developed design.

The power consumption of the life polling benchmark is higher than the life application for both architectures. Thus, the power of the bus dominates the components with a value of 8.55 W. Comparing total power consumption of the two architectures, the distribution of data memory was led by a slight increase in power consumption. 30.33 W vs 29.63 W for shared memory architecture, which is expected, due to the adding of other hardware components.

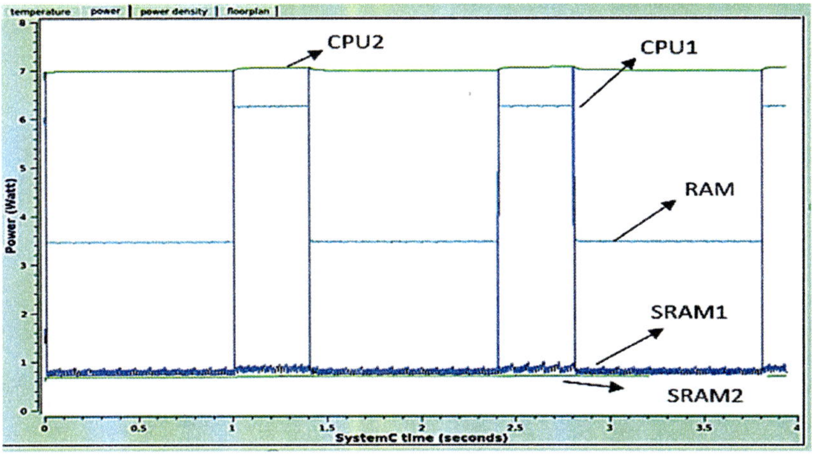

Fig. 7. Power consumption of Distributed data memory of multiprocessors LIBTLMPWT platform for the Game of life and Life polling application

5.2 Speedup

Open access platform simulation is one of the most usual simulation techniques for MPSoCs systems. Nevertheless, the complexity of the systems leads to latency in the simulation speed. However, we proposed a comparison of the speedup evolution between the developed memory architectures. The distribution of the data memory allowable us to reduce the distance between the processor and the memory. Our developed system is purely parallel; each processor executes its own software, using an identifier between each CPU and its executed application. Which means that is no competition between the two CPUs since each processor has its own data memory.

The simulation results shows an improvement in the distributed memory system speed, the shared memory system shows a longer response time of 40 s compared to the distributed memory architecture. The results will be clearly noticeable with the increase in the number of processors, which will mark the speed of distributed data memory architecture.

6 Conclusion and Perspectives

We presented the development of two memory architectures, one with shared memory and the second with a distributed data memory for a Multiprocessor System on Chip (MPSoC) in a co-simulation platform LIBTLMPWT, at a high level of abstraction (TLM) based on Xilinx MicroBlaze processors. In order to predict the results before moving to the conception of the system. The result of the development and performances estimation based on the game of life benchmark and SystemC simulator showed that at the level of energy consumption the distribution of data memory was led by a slight increase in power consumption in the order of 2.30% compared to the shared memory architecture. On the other hand, we have observed an optimization in terms of system speedup but which remains relatively not significant given the use of only two processors. Our perspective is to increase the number of processors in order to re-study the speed of the platform and optimize the memory architecture while distributing the instruction memory.

References

1. Aamali, K., Alali, A., Sadik, M., El Hariti, Z.: A review of the different levels of abstraction for systems-on-chip (SoC). E3S Web Conf. **229**, 01025 (2021). https://doi.org/10.1051/e3sconf/202122901025
2. Aamali, K., Sadik, M., Alali, A., Hariti, Z.E.: Literary study of memory architectures for multiprocessor system on chip (MPSoC). JARDCS **11**, 1067–1071 (2019). https://doi.org/10.5373/JARDCS/V11SP11/20193138
3. Ackland, B., et al.: A Single-chip, 1.6-billion, 16-b MAC/s multiprocessor DSP. J. Solid State Circuit. **35**, 412–424 (2000)
4. Aguilar, M.A., Leupers, R., Ascheid, G., Murillo, L.G.: Automatic parallelization and accelerator offloading for embedded applications on heterogeneous MPSoCs. In: Proceedings of the 53rd Annual Design Automation Conference, pp 1–6. ACM, Austin Texas (2016)
5. Alali, A., Assayad, I., Sadik, M.: Modeling and simulation of multiprocessor systems MPSoC by SystemC/TLM2 (2014). arXiv:14080982
6. Ali, H., et al.: A survey on system level energy optimisation for MPSoCs in IoT and consumer electronics. Comput. Sci. Rev. **41**, 100416 (2021). https://doi.org/10.1016/j.cosrev.2021.100416
7. Chen, X., Lu, Z., Jantsch, A., Chen, S., Chen, S., Gu, H.: Reducing virtual to physical address translation overhead in distributed shared memory based multi core network on chips according to data property. Comput. Electr. Eng. **39**, 596–612 (2013). https://doi.org/10.1016/j.compeleceng.2012.04.009
8. Garibotti, R., Butko, A., Ost, L., Gamatie, A., Sassatelli, G., Adeniyi-Jones, C.: efficient embedded software migration towards clusterized distributed memory architectures. IEEE Trans. Comput. **65**, 2645–2651 (2016). https://doi.org/10.1109/TC.2015.2485202
9. Lameter, C.: An overview of non uniform memory access. Commun ACM **56**, 54–59 (2013). https://doi.org/10.1145/2500468.2500477
10. Lee, K.-B., Chang, T.-S.: SoC memory system design. In: Lin Y-LS (ed) Essential Issues in SOC Design. Springer Netherlands, Dordrecht, pp 73–118 (2007). https://doi.org/10.1007/1-4020-5352-5_4
11. Lee, S.-H., Yoon, Y.-C., Hwang, S.-Y.: Communication-aware task assignment algorithm for MPSoC using shared memory. J. Syst. Archit. **56**(7), 233–241 (2010). https://doi.org/10.1016/j.sysarc.2010.03.001
12. Loghi, M., Poncino, M., Benini, L.: Cache coherence tradeoffs in shared-memory MPSoCs. ACM Trans. Embed. Comput. Syst. **5**, 383–407 (2006). https://doi.org/10.1145/1151074.1151081
13. Loi I, Benini L: A Multi banked multi ported non blocking shared L2 cache for MPSoC platforms, 6p. (2014)
14. Nogueira, B., Andrade, E., Tavares, E.: Power-aware scheduling of real-time applications onto MPSoC platforms with multi-bank shared memory. Microprocessors Microsyst. **67**, 93–102 (2019). https://doi.org/10.1016/j.micpro.2019.03.004
15. Ranjan Panda, P., Dutt, N.D.: Memory architectures for embedded systems-on-chip. In: Sahni, S., Prasanna, V.K., Shukla, U. (eds.) High Performance Computing — HiPC 2002. LNCS, vol. 2552, pp. 647–662. Springer, Heidelberg (2002). https://doi.org/10.1007/3-540-36265-7_61

16. Wolf, W.: The future of multiprocessor systems-on-chips. In: Proceedings in the 41st annual conference on Design automation, p. 681. ACM Press, San Diego, CA, USA (2004)
17. Wolf, W., Jerraya, A.A., Martin, G.: Multiprocessor System-on-chip (MPSoC) technology. IEEE Trans. Comput. Aid. Des. Integr. Circuits Syst. **27**, 1701–1713 (2008). https://doi.org/10.1109/TCAD.2008.923415
18. Xie, Y., Hung, W.-l: Temperature-aware task allocation and scheduling for embedded multiprocessor systems-on-chip (MPSoC) Design. J. VLSI Signal Process. Syst. Signal Image Video Technol. **45**(3), 177–189 (2006). https://doi.org/10.1007/s11265-006-9760-y
19. Sander, I., Jantsch, A., Tenhunen, H.: The platform as an interface in a soc design curriculum. In: Ionescu, A.M., Declercq, M., Kayal, M., Leblebici, Y. (eds.) Microelectronics Education, pp. 33–37. Springer Netherlands, Dordrecht (2004). https://doi.org/10.1007/978-1-4020-2651-5_6
20. IXDP2805_2855_Prod_Brief.qxd. 8
21. SystemC TLM. https://accellera.org/community/systemc/about-systemc-tlm. Accessed 15 Feb 2022

Corporate Finance and Governance: Role of Financial Policy in Governance of Banks Listed on the Stock Exchange Market

Abdelhamid Boulaksili[1] (✉), Mhamed Hamiche[2], Ouail El imrani[2], Abdelrhani Bentahar[3], Abdelfattah Lahiala[2], Najoua Chaouche[2], and Yousra El Hajel[4]

[1] Faculty of Legal, Economic and Social Sciences of Tetouan, Abdelmalek Essaâdi University, Tetouan, Morocco
boulaksili@gmail.com
[2] Abdelmalek Essaâdi University, Tetouan, Morocco
[3] Mohammed I University, Oujda, Morocco
[4] Mohammed V University, Rabat, Morocco

Abstract. The performance and accountability of boards of directors and effectiveness of governance mechanisms continue to be a matter of concern. Unlike the Corporate governance for conventional financial institutions, the corporate governance for participatory financial institutions has a more specific targets, in the sense that it has to ensure that the running of the institution is in accordance with the principles of participative finance.

Therefore, participatory financial institutions require an additional body in its corporate governance framework, i.e., a *Shari'ah* governance council, which is a vital function that guides participatory financial institutions on Shariah compliance. So, the financing policy is still supposed to be an action variable indirectly used by participatory banks' shareholders to capture value.

Certainly, the governance of participatory is a topic that is nowadays of growing interest in various fields of economics and management (especially in the field of participative finance). Thus, corporate governance is often considered - in a restrictive way - as a component of corporate finance.

In this spirit, this study aims to investigate empirically the relationship between financial policy and corporate governance in participatory banks. The financing policy is still supposed to be an action variable indirectly used by participatory banks' shareholders to capture value. This would allow them to reduce impairment losses caused by monitoring or control costs. The econometric tests conducted on the listed participatory banks on Bahrain Bourse have allowed appreciating if shareholders exercise has disciplinary control over managerial decisions by constraining the managers in their choice of financing.

Keywords: Corporate finance · Corporate governance · Financing policy · Governance mechanisms · Participatory banks listed

M. Ben Ahmed et al. (Eds.): SCA 2022, LNNS 629, pp. 780–792, 2023.
https://doi.org/10.1007/978-3-031-26852-6_72

1 Introduction

The subprime mortgage crisis of 2008 has heightened interest in the relationship between the governance and performance of banks. The Corporate Governance (CG) is then listed among the major themes called upon to mobilize the banking firms with a view to transforming, in the long term, their mode of organization and management. According to Albert (1994, p. 11) it is more than that, "it is a real revolution even within capitalism" [Our own translation].

In addition, the theoretical reflection on the theme of governance is characterized by the absence of a grid of structured reading which gives "a strange feeling of scientific weightlessness" (Pastré 1994, p. 15) [Our own translation]. Later, Charreaux (1997, p. 493) notes that "the theory of the Corporate Governance remains to build". Despite, the CG is at the heart of the understanding of several economic and sociologic contemporary mechanisms. Various authors have proposed definitions that appear to be different[1]. Additionally, (Mallin 2018) highlighted that CG theories were influenced by different aspects including; finance, economics, accounting, law and organizational behavior.

Certainly, the governance of participatory banks is a topic that is nowadays of growing interest in various fields of economics and management (especially in the field of participative finance). Thus, corporate governance is often considered - in a restrictive way - as a component of corporate finance (Charreaux and Schatt 2005). This association of governance with finance is essentially due to the article considered as the founder of the field: "Theory of the Firm: Managerial Behavior, Agency Costs and Ownership Structure", published by Jensen and Meckling in 1976.

Indeed, various authors have proposed definitions that appear to be different. The narrower definitions are limited to the study of the relations between shareholders and managers in the firms listed in the stock markets. Also, the definition which M. Albert (1994) retains is considered brutal: "Do not give to the managers of the firms a single goal, which is of maximizing profit and dividends" (Albert 1994) [Our own translation]. By contrast, the broadest definitions take in their fields of study all of the stakeholders for the whole of firms, whatever their size, whether they are listed or not. Pastré (1994, p. 18) has defined it as the: "all the rules on functioning and control that govern, in a given historical and geographical framework, the life of businesses" [Our own translation].

Subsequently, Blair (2002, p. 273) offers a more precise definition according to which: "governance systems, broadly defined, set the ground rules that determine who have what control rights under what circumstances, who receives what shares of the wealth created, and who bears what associated risks". The conceptions of power, control, management and responsibility constitute a central variable, implicit or explicit, of most of the definitions, but the emphasis is on one or the other of these dimensions and their mode of articulation varies with the authors.

Furthermore, the governance system of participatory banks has principles based on participative finance that establishes restrictions and limitations to it, which implies alignment with these restrictions, both for the members of the Board of Directors and the *Shari'ah* Board, as well as for the mangers and all employees, but also to their

[1] The review of the literature related to this theme reveals the profusion of definitions, from the wider to the narrower, with extremely varied properties.

interactions with the community around them. In this regard, the CG can be equated with the set of guiding standards that are consistent with the principles of *Shari'ah* and that address the multiple roles of Participatory Banking and Financial Institutions (PBFIs), seeking to ensure accountability and proper distribution of power among the various stakeholders and avoid conflicts of interest (Boulaksili et al. 2020, p. 12).

All the elements that we have exposed allow us to certify that: "the concept of PBFIs governance refers directly to the influence of common and specific mechanisms on value creation". Since managers play a main role in the determination of the general policy of the firm, the central objective of the governance of PBFIs is then, using the incentive levers and mechanisms for monitoring, align their behavior on the criterion of the maximization of shareholder value. As well as, the CG implies that the delimitation of the discretionary space of the leaders is the result of the interaction of a set of mechanisms. The latter are not confined only to internal and external mechanisms. In effect, all the means used to protect investors not to lose their funds and receive the salary expectations are included under the generic concept of the CG.

From there, the objective assigned to this research is to better understand the role of financing policy as a mechanism of control of the action of the managers. This is done in order to ensure a type of management centered on the maximization of the market value of the assets of the shareholders. The financial structure is supposed to be an action variable used indirectly by the shareholders of Participatory Banking Institutions (PBIs) to appropriate the value.

This would enable them to reduce the loss of the value generated by the costs of monitoring or control. Of this fact, our main ambition is to provide elements of a response to the following question: what is the role of financial policy, through its contribution to the improvement of shareholders' wealth, in the governance of Participatory Banking Institutions?

In order to achieve this goal, we will use the method hypothetico-deductive approach. From the prior observation of the phenomenon, a system of questions and a review of the literature will allow us to establish a theoretical prebuilt. Thus, the answer to the latter, we will establish a hypothetical model.

This model will be empirically tested to validate the research hypotheses using the Generalized Least Squares (GLS) method. This model will be the subject of empirical tests applied to the PBIs listed on the stock exchange to validate the assumptions of the search via the Generalized Least Squares method (GLS).

2 Review of the Literature

We can identify two types of approach to CG. The principal/agent theory, very largely dominant at the present time, is the one that presents the limits the more serious, because it finds its theoretical relevance in a very reductive representation of the firm, especially in favoring a vision strictly contractual of the firm and by focusing the analysis only on the shareholders.

By contrast, the second approach, based on a cognitive vision of the firm, opens up real prospects for the renewal of the analysis of the CG. It allows particularly going toward a representation of the question of the Governance extended to all stakeholders.

These approaches are often presented by their supporters as the principle's models underpinning CG. This has resulted in the debate on CG being polarized into two perspectives, the shareholding approach perspective with an emphasis on principal/agent approach and stakeholder approach perspective emphasizing a wider range of interested parties (Jensen 2001), with each perspective claiming superiority over the other. However, Meckling (1976) associates the concept of CG with three main approaches: agency approach, stakeholder approach and stewardship approach.

Through a succinct review of the literature, we will try to show the term of the second axis. How have mechanisms of discipline evolved and control of managers, and therefore the vision of researchers, in facing the problem of governance. Because the theories of the firm and theories of governance are intimately related (Charreaux 2002), we explain how our theoretical positioning means rethinking in depth the mechanisms of government business. That is to say, the first element will be the subject of exposure of different mechanisms of governance. The second element will be devoted to the exhibition of the role of the financial structure in the corporate governance of PBFIs.

3 The Experimental Plan and Development of Assumptions

By referring to the theoretical constructs presented in the general framework of reference, we propose to test empirically the relationship between the investment opportunities of banks and the choice of their financing policies.

Based on the article of (Collins et al. 1994), we will estimate two models. Therefore, the model of the multiple regressions is written as follows:

$$Y_i = \beta_0 + \beta_1 X_1 + \beta_2 X_2 + \beta_3 X_3 + \beta_4 X_4 + \varepsilon; \text{ (With: i = 1,2).}$$

The variables: Y_i, X_1 and X_2 will be explained.

Thus, the equation of the multiple regressions is written: $\widehat{Y} = b_0 + b_1 X_1 + b_2 X_2 + b_3 X_3 + b_4 X_4$

- b_0, b_1, b_2, b_3 *and* b_4 *are the estimates of* $\beta_0, \beta_1, \beta_2, \beta_3$ *and* β_4;
- $\hat{Y}$ *corresponds to the estimated value of the dependent variable.*

3.1 The Variables of the Study of the Relationship Between Investment Opportunities and the Policy of Debt

It distinguishes between two types of variables: dependent and independent variables.

- **The variables of indebtedness of Participatory Banking Institutions (dependent variables)**

It has chosen four variables "proxy" for the funding policy of the Participatory Banking Institutions.

$$Y_1 = \frac{Total\ debts}{Total\ assets}$$

$$Y_2 = \frac{Long - term\ liabilities}{Total\ assets}$$

$$Y_3 = \frac{Deposits}{Shareholders'\ equity}$$

$$Y_4 = \frac{Deposits}{Shareholders'\ equity}$$

- **The variables of growth opportunities of Participatory Banking Institutions (independent variables)**

$$X_1 = Tobins'\ Q = \frac{Market\ value\ of\ shares}{Accounting\ value\ of\ shares}$$

- X_1 : Represents a variable "proxy" of the opportunity for investment at the level of the banks. This ratio should be high for institutions with more of growth opportunities. It should be noted that the market value of the shares reflects the property of investors in the value of the available assets and in that of the growth opportunities. The book value of the shares reflects in its share the exclusive property of investors in the assets available.
- X_2 : Return On Operating Equity defined as:

$$X_2 = ROOE = \frac{Operating\ Profit}{Average\ Total\ Equity}$$

- X_3 : Return On Average Assets defined as:

$$X_3 = ROOA = \frac{Net\ Income}{Average\ Total\ Assets}$$

- X_4 : Constitutes a control variable "proxy" of the size of the firm. This variable allows to give judgments concerning the relationship between the X_1 variable and the choice of financial policies, all other things are equal.

The following Table 1 summarizes the variables in our hypothetical model.

Table 1. Description of variables

Variables	Codifications	Unit
Indicators of the financial structure of PBIs		
Y_1	Total debt ratio	%
Y_2	Long-term debt ratio	%
Y_3	Deposit ratio	%
Y_4	Reserves ratio	%
Indicators of shareholder wealth and the performance of PBIs		
X_1	Tobin's Q: Financial performance	Number
X_2	Operational efficiency: ROAE	%
X_3	ROAA	%
Control variable		
X_4	Log of the Total Asset of the PBIs, a proxy for size	Number

3.2 The Commitments of the Liabilities of PBIs as a Mechanism of Control and Discipline of the Managers

The formulation of the assumptions of research represents the culmination of the conceptual reflection. The objective of this second point is to introduce the system of assumptions of our research. As well, we will present in the first place the basic assumption of our research. By the following, we will present the under assumptions that flow from it.

- **The Basic Hypothesis**

In order to be competitive, PBIs must cope with the several problems which are mainly the activism of their management team. We have seen that there are several modalities, both internal and external, which aim at the discipline of the managers. However, the problem that arises is that these mechanisms, despite their importance, do not arrive to curb definitively the opportunism of the managers. For this reason and for others, the PBIs must develop their system of control and discipline. Beyond, in response to our research problem, our basic assumption is thus formulated:

1. H1: the financing policy constitutes an indirect mechanism of the governance of Participatory Banking Institutions.

- **The Derived Hypothesis**

The derived assumptions derive from the basic assumption. So, these assumptions will appear what are the policies of funding that contribute to the discipline of the managers.

– The opportunity of investment and growth of Participatory Banking Institutions

In the perspective of the costs of bankruptcies, the presence of growth opportunities should limit the use of the debt since the value of these assets is virtually zero in the case of liquidation. Concerning the conflict of interests between the shareholders and managers, the choice of a policy of external funding is also influenced by this variable. The existence of growth opportunities promotes the alignment of interests and the financing by the own capital is then a wise choice (Jung et al. 1996).

By contrast, in case of their absence, the debt is the appropriate vehicle because it disciplines managers. Indeed, the growth of the assets reflects the needs of funding resulting from the financial policy of the companies. The more this growth is strong, the more the firm is faced with the problem of funding. In the design of the theory of agency and according to Myers (1977), the costs of agency between shareholders and creditors will be much higher than the value of the opportunities for future growth is more important in relation to the current value of the assets in place. Then, we must expect a negative relationship between the level of debt and growth opportunities.

However, this negative correlation is due in large part to firms which produce their own funds. It is for this reason that Rajan and others argue for an alternative explanation: the enterprises would have a tendency to issue equity when the market value of the shares is high. Hovakimian et al. (2001) noted that this behavior can be explained by a decrease in the optimal ratio due to the modification of the portfolio of assets of the company in favor of growth opportunities. In referring to theoretical considerations and to the results found by the majority of the empirical work, therefore, we put forward the following hypothesis:

2. H2: The variable investment opportunity is negatively correlated with the commitments of the Participatory Banking Institutions (liabilities in the balance sheet).

To measure the investment opportunities, we have chosen therefore the ratio X_1 detailed above. In fact, the choice of our ratio is justified by the interests of comparability with the majority of studies and by the inadequacy of the information relative to alternative measures.

– The size of Participatory Banking Institutions

Several explanations argue in favor of the existence of a positive relationship between the size and the decision of the debt. In effect, the probability of bankruptcy is reduced in the companies of significant size since the diversification of activities reduces the volatility of the cash flows and therefore the probability of bankruptcy (Titman and Wessels 1988, Rajan and Zingales 1995). It follows that the firms of large size must be more indebted. Ferri and Jones (1979) note that the large firms which are generally listed on the stock exchange have more facility for access to the capital markets to obtain the financial resources they need. The positive relationship between the size and the debt is confirmed in a number of studies.

An approach based on the asymmetries of information can lead to assume a negative correlation between the size and the decisions of indebtedness. It may be, as the note Rajan and Zingales (1995) that the size used to measure reverse of the information obtained by the external investors. However, they do not observe that the large companies emit more sensitive titles to asymmetries of information. In addition, the negative relationship between the size and the indebtedness remains specific to Germany. Kremp et al. (1999) also note the importance of bank financing for businesses of small sizes in Germany.

The characteristics of the right of the bankruptcy and the system of the main bank (Hausbank) German offers an explanation to the German exception which is not based on the asymmetries of information with all external investors. Other researches find weak relationships (Kester 1986 and Ozkan 2001) or non-significant (Remmers et al. 1974; Kremp et al. 1999) and are reluctant to conclude. In the context of our study, we will assume that:

3. H3: size is positively correlated with the commitments of the financial institutions and Participatory Banking Institutions.

However, most of previous studies have defined bank size as the nature log of book value of total assets. In this study, we also define bank size by taking Log of total banks' assets as follows.

The previous existing studies have recognized mutually positive as well as negative effects of size on bank performance. For example, Akhtar et al. (2011) reported that bank size has an insignificant impact on performance, whereas, Smirlock (1985) reported a significant and positive impact of size on the performance of banks.

3.3 Finding and Analysis of Results

The objective of this last point has two parts: As for the first part, we are going to present the banks studied. For the second part, we will try to analyze and interpret the obtained results.

- **The identification of studied banks**

The analysis that we are going to perform focuses on a sample of 11 PBIs listed on the stock exchange market[2]. These banks have been selected primarily according to their ability to disseminate the main information to conduct our study and by their homogeneity with respect to their modes of organizational governance. In effect, the financial indicators of the banks in question will be detailed during the analysis and the interpretation of results.

[2] It is: al Baraka Banking Group, Al Salam Bank-Bahrain, Bahrain Islamic Bank, Bank ABC Islamic, Ithmaar Bank, Khaleeji Commercial Bank, Qatar Islamic Bank, RHB Islamic Bank Berhad, CIMB Group, Al Rajhi Banking & Investment Corporation, Hong Leong Islamic Bank Berhad.

- **The analysis and interpretation of the results**

To achieve our study, we must bring together a wide range of indicators. Following the consolidation of the necessary information, we have calculated the values of the independent and dependent variables of our model of research. Therefore, on the basis of the calculations made[3], it obtains the following data in order to estimate our models.

So, we are going to present in what follows the results obtained by the relationship between the funding policy of PBIs and investment opportunities. Thus, based on the theoretical models exposed in the general framework of reference, we will try to conclude also on what theory this relationship can be explained[4].

– The variables of the Model Y1 (Tables 2 and 3):

Table 2. Multiple regression results

Model		Unstandardized Coefficients		Standardized Coefficients			Correlations			Collinearity Statistics	
		B	Std. Error	Beta	t	Sig.	Zero-order	Partial	Part	Tolerance	VIF
1	(Constant)	,071	,276		,258	,797					
	X2_ROAE	1,078	,368	,405	2,926	,004	,204	,275	,253	,391	2,559
	X3_ROAA	-10,847	3,205	-,419	-3,385	,001	-,132	-,314	-,293	,490	2,039
	Size	,091	,042	,277	2,179	,032	,308	,208	,189	,463	2,160
	X1_Tobins'Q	-,012	,008	-,160	-1,414	,160	,069	-,137	-,122	,584	1,712

– The variables of the Model Y2:

[3] The data collected have been identified over a period of ten years (2009 to 2018) from annual reports of the banks concerned and the statistics published on the web site of the stock markets in question.

[4] Our regression was performed using the GLS. As well, the application has been made to the aid of the program SPSS.

Table 3. Multiple regression results

Model		Unstandardized Coefficients		Standardized Coefficients			Correlations			Collinearity Statistics	
		B	Std. Error	Beta	t	Sig.	Zero-order	Partial	Part	Tolerance	VIF
2	(Constant)	,164	,131		1,247	,215					
	X2_ROAE	,089	,175	,079	,509	,612	,071	,050	,049	,391	2,559
	X3_ROAA	,560	1,524	,051	,368	,714	,090	,036	,036	,490	2,039
	Size	-,011	,020	-,079	-,554	,581	-,043	-,054	-,054	,463	2,160
	X1_Tobins'Q	,000	,004	-,016	-,122	,903	-,032	-,012	-,012	,584	1,712

In the watermark, the regression of Y1 by report to X1, X2, X3 and Size gives us the equation of the following regression:

$$Y_1 = 0,071 - 0,012X_1 + 1.078X_2 - 10,847X_3 + 0,091Size$$

Definitively, the results of the multiple regression show a positive and significant coefficient correlation, associated with the variable X2 if it retains total debt ratio to measure the funding policy of the PBIs in question. Whereas, the results of our study show a very higher negative and significant correlation between Y1 and the financial performance of PBIs (X3).

Moreover, if it retains the long-term debt to measure the funding policy of the banks in question, the results of the linear regression show weak and insignificant relationships between variable Y1 and the explanatory variables (Xi).

Henceforth, we can advance the following hypothesis: the variable Y1 better reflects the financial policy undertaken by the PBIs listed in the stock exchange markets. However, the intervention of the short-term debts, represented by the deposits, in the calculation of the variable Y1 can explain the results found during the empirical validation of our model of research. In this sense, it can be argued that the model which considers the total debts allows us to explain even better the role of financial policy in the control and the discipline of the mangers. Generally, the financial structure constitutes an indirect mechanism of the governance of PBFIs.

Therefore, we can conclude that the assumption of the existence of a negative relationship between the investment opportunities at the level of the banks studied and the increase of debts is confirmed. This confirmation finds its justifications in the literature of the theory of agency which supports that the firms which have more opportunities to maintain profitable investments and are less called to raise the level of their debts to limit the margin of maneuver of managers.

Finally, we note that the variable size is positively correlated with the variable Y1. As well, this coefficient is significant. This result is consistent with the predictions of Rajan and Zingales (1995) and Booth et al. (2001) who thinks that the size can play a role in measurement is the inverse of the probability of bankruptcy.

4 Conclusion

In conclusion, we can say that the governance of PBFIs becomes a very privileged and very debated theme in the academic community and in the world of business. In addition, it provides banking firms a battery of mechanisms that are used to delimit the powers, to influence the decisions and to define the space at the discretion of managers.

The two main goals pursued in this work have been to:

- Introduce the various inputs of theoretical assertions that deal with the means of discipline, control and resolution of conflicts of interest between managers and their shareholders;
- Analyze and empirically test the role of the policy of indebtedness of PBIs listed on the stock markets in the delimitation of the discretionary space of the managers of the banks in question.

In effect, the results of the tests carried out by GLS reveal the existence of a negative correlation between the investment opportunities of banks listed in the stock exchange markets and their financing policies. This result corroborates our hypothesis which is in favor of the existence of a disciplinary role exercised by the policy of indebtedness under a perspective of the theory of the Agency. Furthermore, the work of Jensen (1986), Stulz (1990) and Hart and Moore (1995) confirm that the enterprises that have more opportunities to maintain profitable investment projects have less cash-flows discretionary. Based on this fact, they are relying less on debt as a means to limit the margin of maneuver and the space at the discretion of managers.

However, the present work suffers from certain limits on the theoretical plan and the practical. First, we have focused on the theoretical assertions frequently used by the literature that deals with the means of control and discipline of the managers. As well, we note the existence of a certain imbalance at the level of the importance given to the different theoretical underpinnings chosen. This gap finds its origin in the major interest granted to some theories such as the theory of agency and the theory of transaction costs to the detriment of other theoretical foundations in order to explain the thematic of the governance of PBFIs. Has our sense, this is due to the absence of theoretical bases specific to this problematic.

On the practical level, one can only admit the narrowness of the size of the sample of our study which cannot be considered as representative of all the PBIs. On the other hand, the inability to have the information relating to the long-term debts of all PBIs listed in awards in turn represents one of the shortcomings of this present study.

In addition, it can be argued that the problem of traditional agency between shareholders and managers can result in an agency problem between majority shareholders and minority. In this sense, it would be very useful to follow the vision of Gomes (2000), which presents a model explaining why the minority shareholders are aware that they are not protected in the abuse of power of the majority shareholders, opting frequently for a participation in the capital of companies as minority.

Finally, this work is part of a vision exclusively financial of governance which confers to the policy of debt as main role to align the interests of managers on and those of shareholders. Future research examining the other roles of the financial structure, like that are present in other theories of the governance, would therefore be desirable and may constitute the tracks of important flourishing research.

References

Akhtar, M.F., Ali, K., Sadaqat, S.: Factors influencing the profitability of Islamic Banks of Pakistan. Int. Res. J. Financ. Econ. **1**(66), 125–132 (2011)

Albert, M.: L'irruption du corporate governance. Revue d'économie financière **31**, 11–14 (1994)

Ben Daoud, K.: The financial intermediation participative of Islamic banks. Stud. Islamic Econ. **6**(1 & 2), 29–56 (2013)

Blair, M.M.: Ownership and control: rethinking corporate governance for the twenty-first century. In: Charreaux, G. (ed.) Variation sur le thème: À la recherche de nouvelles fondations pour la finance et la gouvernace. Brookings, Washington (2002)

Boulaksili, A., Hamiche, M., Lahiala, A.: Les spécificités des mécanismes de gouvernance des banques islamiques/The specifications of the mechanisms of governance of Islamic banks. Eur. J. Econ. Financ. Res. **3**(6), 10–35 (2020)

Charreaux, G.: Le statut du dirigeant dans la recherche sur le gouvernement d'entreprise. In: Charreaux, G. (ed.) Le gouvernement des Entreprises, Théories et faits. Economica, Paris (1997)

Charreaux, G.: Variation sur le thème : À la recherche de nouvelles fondations pour la finance et la gouvernance d'entreprise. Finance Contrôle Stratégie **5**(3), 5–68 (2002)

Charreaux, G., Schatt, A.: La recherche française en gouvernance d'entreprise : un panorama. Cahier du FARGO - Centre de recherche en Finance, ARchitecture, pp. 1–38 (2005)

Chatti, M.V.: L'impact de l'application d'une éthique bancaire sur la diversification des banques islamiques. Etudes en Economie Islamique **6**(1 & 2), 1–27 (2013)

Collins, M., Blackwell, D., Sinkey, J.: Financial innovation, investment opportunities and corporate policy choices for large bank holding companies. Financial Rev. **29**(2), 223–247 (1994)

Desbrières, P.: Participation financière, stock-options et rachat de l'entreprise par les salariés. Economica (1991)

Dewatripont, M., Tirole, J.: A theory of debt and equity: diversity of securities and manager-shareholder congruence. Q. J. Econ. **109**, 1027–1054 (1994)

Fama, E.-F., Jensen, M.-C.: Agency problems and residual claims. J. Low Econ. **26**, 327–349 (1983)

Hassan, M.K., Bashir, A.-H.M.: Determinants of Islamic banking profitability. Paper Presented at the Proceedings of the Economic Research Forum (ERF) 10th Annual Conference, Marrakeshe-Morocco, 16, 18 December 2003

Jensen, M.C.: Value maximization, stakeholder theory, and the corporate objective function. J. Appl. Corp. Financ. **14**(3), 8–21 (2001)

Jung, K., et al.: Timing, investment opportunities, managerial discretion, and the security outcome decision. J. Financ. Econ. **42**, 159–185 (1996)

Kaplan, S.: Top executive rewards and firm performance: a comparison of Japan and United State. J. Polit. Econ. **102**(31), 510–546 (1994)

Mallin, C.: Corporate Governance. Oxford University Press, Oxford (2018)

Meckling, M.C.: Theory of the firm: managerial behavior, agency costs and ownership structure. J. Financ. Econ. **3**(4), 305–360 (1976)

Pastré, O.: Le gouvernement d'entreprise, questions de méthodes et enjeux théoriques, dossier «Corporate governance- "Le gouvernement d'entreprise"». Revue d'économie financière **31**, 15–32 (1994)

Rajan, G.R., Zingales, L.: What do we know about capital structure? Some evidence from international data. J. Finance **50**, 1421–1460 (1995)
Rappaport, A.: Expansion management review. Quand l'actionnaire prend le pouvoir (90) (1998)
Smirlock, M.: Evidence on the (Non) relationship between concentration and profitability in banking. J. Money, Credit, Bank. **17**(1), 69–83 (1985)

Toward a New Framework for Process-Aware IoT Discovery

Iman Elkodssi[1], My Driss Laanaoui[2](✉), and Hanae Sbai[1](✉)

[1] Mathematics, Computer Science and Application Laboratory FST Mohammedia, Hassan II University of Casablanca, Casablanca, Morocco
hanae.sbai@univh2c.ma

[2] Computer Systems Engineering Laboratory, Cadi Ayyad University of Marrakech, Marrakesh, Morocco
d.laanaoui@uca.ma

Abstract. People's lives have become more pleasant and uncomplicated in recent years as a result of recent technological advancements. Data sharing and connectivity across systems, devices, and people have been made easier with the introduction of IoT. In reality, the Internet of Things (IoT) provides a platform for ordinary items to become smarter, everyday communication to become more in-formed, and everyday processes to become more intelligent. In this regard, to manage business processes in the new IoT environments, many heterogeneous frameworks have been proposed. Since the number of IoT devices or things is increasing day by day and the number of services is also increasing, while dis-covering these IoT sensitive processes becomes a vital challenge and the traditional strategies of discovery of services are not efficient enough to handle this problem. The problem, approaches of discovering processes from sensor logs do not highlight IOT concepts in the output process models, which make process analysis and decision made more and more difficult. In this paper, we present a new process mining based approach to discover and analyze business process aware IoT.

Keywords: Smart Business Process Management (Smart BPM) · Internet of Things (IOT) · Process-aware IoT

1 Introduction

Business process management is an integral part of a business that automates processes that improve an organization's workflow. While innovation and transformation have become essential in today's environment, Business Process Management (BPM) offers a way to achieve this agility, then the Internet of Things can create information about connecting objects, analyses and make decisions; in other words, it is an approach that connects the physical world with the digital world. It is thought of as a network that connects a physical element using basic IoT components [1]. The integration of IoT into BPM approach introduce new generation of business processes, called "business process aware-Internet of Things". Historically, business processes (BP) are increasingly being

M. Ben Ahmed et al. (Eds.): SCA 2022, LNNS 629, pp. 793–803, 2023.
https://doi.org/10.1007/978-3-031-26852-6_73

adopted to model complex processes occurring in business, operational, and technological contexts. With the advent of the Internet of Things (IoT) and the need to model IoT business activities, many authors have been interested in integrating IoT devices as resources into business processes [2]. The 'bridge' between these processes and IoT concepts, defines how events in the physical world play a role in the BPM. In addition, typical business process-aware IoT, are not linked to a single IoT 'thing', but may require access to many 'things'. So, the role of the business processes is then the synchronization of these 'things' towards the business goal they serve [3]. Consequently, contrary to traditional business process models that focus on modeling human business activities, business process-aware IOT puts much attention on handling business activities that require IoT intervention. Although, the emergence of information technology in general has had a major impact on business processes, and it has become necessary to incorporate their systems and components when modeling business processes to achieve the many benefits offered by the IoT [4]. Therefore, in order, to fill this gap by identifying business process-aware IoT, BPMN 2.0 was evaluated as the most IoT-aware state-of-the-art process modeling approach, [5, 6]. However, the modeling of business process-aware IoT manually is an error prone and consuming task, since the human intervention and the experience of the process designer affect the quality of the process model. Against this background, applying business process mining to discover process-aware IoT seems to be a promising approach, for constructing a process, which represents the real business process aware IoT, and reduce human intervention. In the literature, Process Mining (PM) is approach that helps to automatically discover processes, by analyzing business data, including real-life event logs. It is about three areas:

- Process discovery which require e event log to generate a business process as output
- Process conformance; requires two entries such as an event log and the related business process, to verify if it conforms to the given event log.
- Process enhancement; allows for enriching processes with new information discovered from event logs.

But, the main issue is to apply process mining to sensor logs that record unstructured data generated from sensors. Many approaches have dealt with transforming these sensor logs into an event log. Previews work, we have conducted a comparative study on existing works in process mining sensor log. Our study shows that despite efforts that has been made in this context, existing works on process discovery from sensor logs still insufficient to overcome the main issue for constructing process aware IoT. The challenge is to consider IoT concepts in the process discovered from sensor logs. For this reason, we suggest that the incorporation of IoT concepts in the discovered process, will reduce time and cost of process-aware IoT development and improve ability and quality of discovered models. Hence, in this paper, we propose a new approach for constructing business process-aware IoT models, which combine process mining techniques and new extended process modeling approaches to support IoT concepts. By analogy to approaches that apply process mining to extract traditional business processes. We propose new framework that allow for enriching event logs by IoT concepts as a first step towards discovering business process-aware IoT. The organization of the rest of this paper is as follows. Section 2 is dedicated to the basic concepts. Section 3 deals with the existing works related to the discovery of a process model in an intelligent environment in the context of Process Mining. The comparative study is given in Sect. 4 to developing a new framework in the next section (Sect. 6). Finally, the last part conclusion and perspective summarizes the work and the different points we have to develop in the future.

2 Backround Review

In this section, we present three main concepts related to our work that are smart BPM, Internet of Things and business process-aware IoT.

2.1 Smart Business Process Management (Smart BPM)

Smart Business Process Management is a field that links BPM technology and smart objects. Traditionally BPM systems rely on peoples to observe, control and perform activities. In fact, this dependency between BPM system and peoples influence the capability of BPM systems to realize activities. So the main idea behind smart BPM is to reduce this dependency and improve responsiveness of BPM systems by using smart objects defined by Internet of Things. Uses information technology and analytical techniques in various application areas known as smart home, smart health, smart city, smart energy or smart mobility [7]. All these smart applications integrate sensors, actuators, connectivity and analytic. There are three different levels of smart business process management: Smart multi process management, Smart process model management and smart process instance.

2.2 Internet of Things (IoT)

The Internet of Things (IoT) is a network that links and unites items with the Internet, using standards that guarantee their connection and exchange of information through a range of devices [8, 9]. The representation includes data such as the thing's identity, status, location, and any other corporate, social, or privately important data. Through the use of unique identification, data gathering and transmission, and actuation capacity, the objects provide services with or without human interaction It may alternatively be defined as "the billions of physical devices now linked to the Internet and gathering and exchanging data all over the world." With the introduction of computer chips and the widespread availability of wireless networks, everything from a medication to an airliner may become a part of the youth. In this field, the Internet of Things (IoT) ecosystem has been created by the proliferation of linked devices, where sensors, actuators and beacons are able to communicate fluidly with their environment and with each other to share information or conduct activities. According to [10]. An ecosystem IoT is composed of intelligent agents. Each intelligent agent is considered as an IOT node. An IoT node is made up of a microcontroller, responsible for the calculations, and a set of sensors and doors. By examples of ecosystems: Smart Home, Smart City, Smart Farm, Smart Vehicle, Smart Factory [10] (Fig. 1).

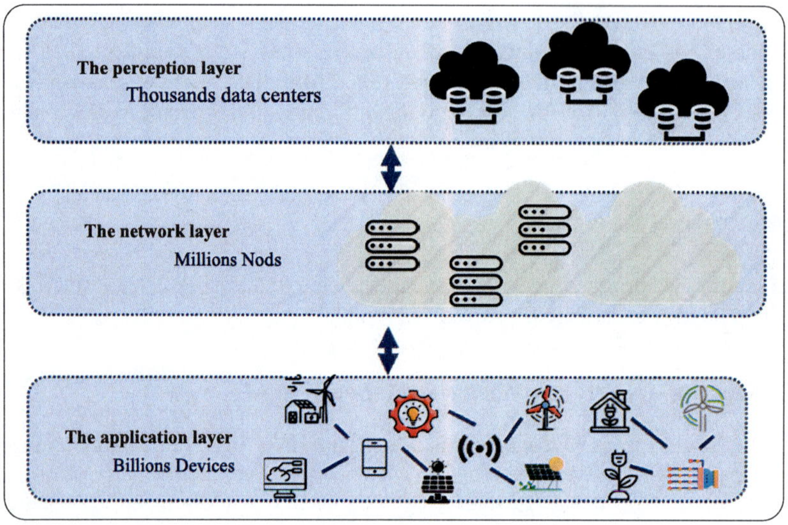

Fig. 1. IoT levels

- The perception layer has sensors and actuators that detect and collect information about the environment.
- The network layer is responsible for the connection, transport and processing of data from sensors and actuators.
- The application layer is responsible for providing the user with specific services and intelligent applications.

Typically, an IoT solution in BPM concepts is made up of the following components: Sensors, Actuators and microcontroller.

- Sensors: It is a device used to detect an event or a physical quantity, such as brightness, temperature, soil moisture, pressure, etc. and that provides a corresponding electrical signal.
- Actuator: a technology complementary to sensors, converts electrical energy into movement or mechanical energy. The actuators allow the energy received to be transformed into a physical phenomenon.
- A microcontroller is an integrated and compact circuit that includes a processor, memory, and input and output devices on a single chip. An MCU is designed to process the raw data captured by the sensors and extract useful information. In the next section, we define the process-aware IoT to understand how processes are linked to smart objects.

2.3 Business Process-Aware IoT

The purpose of an IoT-enabled business process model is to express all relevant IoT information for the process. Therefore, Sensor data are collected, interpreted in order to perform business process activities. These tasks are then triggered in real time to the concerned people who can receive these tasks on graphical interface. For example, the work of [11]. As an example workflow, which is attempted to define and operate these IoT processes. They just added sensors and used process management technologies to introduce smart toilets to pets. The data was obtained on the basis of a gut-related sensor in order to examine the animal's digestive activity, and an alert service was supplied based on the data (Fig. 2).

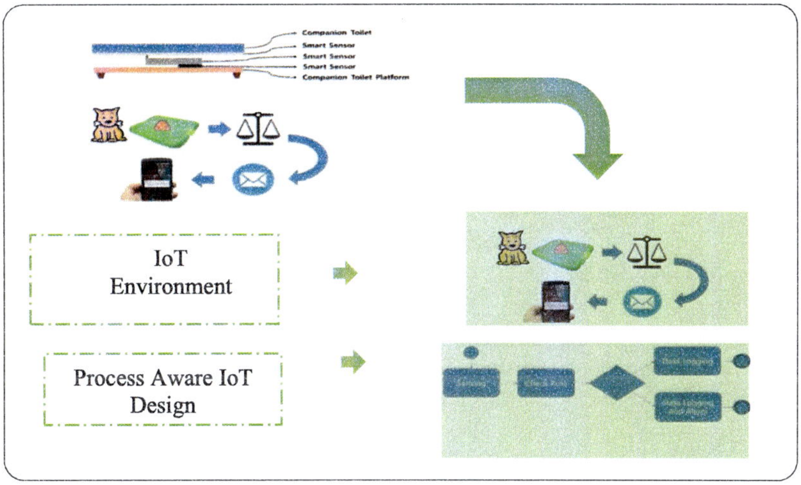

Fig. 2. Example of process aware IoT

The environment is described as a process, and the workflow engine ill automatically execute the environment depending on the given process.

3 Related Work

In the context of process aware IoT, this section addresses the current state of the art. Several studies have begun to employ processes to merge sensors and actuators. BPMN 2.0 has been extended to include an IoT component. The work of [9] expands on the traditional Internet of Things (IoT) conceptual framework to reify an enhanced ubiquitous IoT community collaboration notion known as the process-aware Internet of Things. Despite the fact that many researchers focus on BPMN features when designing BPMN 2.0 for IoT business process modeling, the authors of this paper suggest that BPMN events may better describe IoT characteristics. The BPMN element event for IoT business process modeling is expanded in this study. According to the BPMN 2.0 definition, an event is. BPMN v2.0, according to the authors, is insufficient to define ubiquitous technologies such as sensors and smart readers. As a result, the BPMN model now includes the uBPMN model, which represents ubiquitous technologies. According to the authors, this paradigm is capable of portraying all existing ubiquitous technologies. The model introduces three new tasks: Sensor task, reader task, and collector task. The Task element and the Smart Object in BPMN are supplemented by the BPMN Data Input [12]. Gaaloul's strategy for integrating IoT resources into organizational processes, on the other hand, is based on the formalization of a version of BPMN (widely used for process modeling). The resource element is expanded by an element called resource extension, which integrates both. The process-aware IoT-community framework [11], which is based on the ITU-standardized IoT framework, is described conceptually. The goal of this study is to broaden the standard Internet of Things (IoT) conceptual framework in order to reify the process-aware Internet of Things, which is an advanced ubiquitous IoT-community cooperation idea. The expanded conceptual framework is represented by a reference architecture that may be used as a standardized reference model for the conceptual integration of IoT and process awareness. They also developed some extensions to represent WSN processes in BPMN [8]. They use these qualities to derive the WSN modeling requirements and propose additional BPMN modifications based on them. There is a new task type, a new pool type, and other enhancements.

4 Comparative Study

In this section, we will first introduce the criteria on which our comparison study is based, and then show and discuss the comparative study results. IoT devices are sensors, actuators, gadgets, appliances, or machines that are built for specific purposes and may transmit data over the Internet or other networks. As a result, category refers to the type of sensor, actuator, or reader employed in an IoT context. Extension, to introduce new elements of BPMN The process was modelled with standard BPMN illustrating how IoT devices can be incorporated into BPMN process modeling involves structuring and representing the activities of an organization, usually using graphical notation to visually represent the sequence of activities. Modeling can rely on specialized methods and techniques, and implement process reference frameworks. These process modeling languages are languages dedicated to the execution of processes. They are generally not used directly in the design phases.

Table 1 summarizes the existing literature in terms of the IoT elements that have been proposed for modeling processes aware IoT. Many IoT devices use their Internet connectivity to allow users to access and operate devices, as well as remotely. To consult online databases for information and accept firmware upgrades from the manufacturer. Existing work has focused primarily on the sensor, actuator, and reader. There are many types of IoT devices in the market and these are listed in the table. There are three main categories: on the one hand, the categories of sensors such as (Temperature sensor, Humidity sensor, noise sensor, Air pollution sensors, speed sensors, level sensors, and GPS sensors [12]. On the other hand, there are categories of actuators, namely electrical, hydraulic, mechanical, and pneumatic. There are also categories of readers such as RFID reader and barcode readers [13]. Different types of application scenarios can be constructed using the diversity of resources available on the environment. Are as follows: Temperature control process [14], time banking system process [12], chain management domain [13], military application, health, environmental applications [8], commercial applications and many others. Overall, we discover that the IoT aware of business processes is linked to the extension of BPMN to give a complete and standard modelling. In Concept of IoT-Aware Process Modelling, which is covered by the various works, delivered five concepts for IoT-Aware process modelling. Including "flow object", "Data", "connection object", "swimlanes" and artifacts, they have extended the version of Business Process Modeling Notation 2.0 (BPMN 2.0) [5, 12, 13] based on these six concepts for modeling of the processes associated with IoT. Using these extensions enables the use of business processes to integrate IoT sources. More specifically, [12] have tried to add new task types namely sensing task, actuating task, reading task, collecting task, intermediary task and event stream processing task. And [8] also extended new concepts, at the swim lane and pool level of BPMN 2.0 for IoT integration namely smart object, input event stream, output event stream and physical entity. There is also an article that used BPMN 2.0 to establish processes to model and execute these processes to integrate IoT devices for real application. Like [13], they proposed an approach to integrate IoT resources in the enterprise. The process is based on the extension of the Resource element with an ontology domain and is inspired by the existing semantic model. When it comes to business process modeling languages, it is very important to have process modeling techniques for managing business processes that provide visibility into and document process flows. All works have used Business Process Model and Notation last version 2.0. To define this common framework for the graphic presentation of processes. Process modeling languages is crucial for business process management, as may be observed. Among the modeling languages discussed in prior papers are XML UML, BPEL, SWRL, and OCL [15]. In this context, we can deduce that the extension of BPMN elements has advantages because it allows connect the field of business process intelligence to environment intelligence. In fact, it helps to analyze sensor data from a process point of view. However, existing research focuses only on the extension of the "Activity", "Pool", "Lane" and Artifact" elements.

There are still other elements as IoT elements need also to be considered. So, in the context of business processes, there is a lack of structured and unambiguous descriptions for IoT element, which makes modeling and management more effective.

Table 1. Comparative study.

Article	IOT devices	Categories			BPMN extension	Techniques and methods	Language	Domain application
		Sensor categories	Actuator categories	Reader categories				
[16]	• Sensors • Actuators • Web services	Temperature sensors	Air conditioner actuators	–	Event elements	BPMN	UML XML	Temperature control process
[12]	• AIDC devices • RFIDs • GPS	–	–	• Radio-Frequency RFID • Bar codes • Near Field • Magnetic stripes • Databases • Files • Outcome another process	Smart object Task: • Sensor task • Reader task • Actuator task	• BPMN • Technique based on ubiquitous computing	OCL UML	Time banking system process
[13]	• Sensor • Resource • Actuator • Tag Resource • Workflow resource • Human Resources	–	–	RFID	Ontology BPMN (Devices)	BPMN	EPC UML Activity Semantic Web Rule Language (SWRL)	Chain management domain
[8]	• Wireless • Sensor • Actuator • Network Mobile	• Temperature, humidity, • Vehicular movement, • Lightning condition • Pressure, • Soil makeup • Noise levels	Actuator ventilation	RFID tag	Task: • WSN task Pool: • WSN Pool Grouping structure Annotation	• Model-driven code • Generation techniques BPMN	BPEL UML	Military applications Environmental applications Health applications Commercial applications
[4]	Sensor actuator	• Temperature humidity • Movement • Lightning • Condition pressure • Soil makeup • Noise levels	–	–	Text Annotation Data Object and Participant	BPMN Meta model	–	A process representing physical entity
[15]	Sensor actuator	Temperature sensor	Electronique shelf label	RFID	• Artifact • Annotation • Task: Sensing task	BPMN	UML XML	Retail domain

5 Our Approach

We propose a new framework for the automatic discovery of business processes -aware IoT.

The travelled is at the Cloud level, in which the sensors and actuators are responsible (Fig. 3).

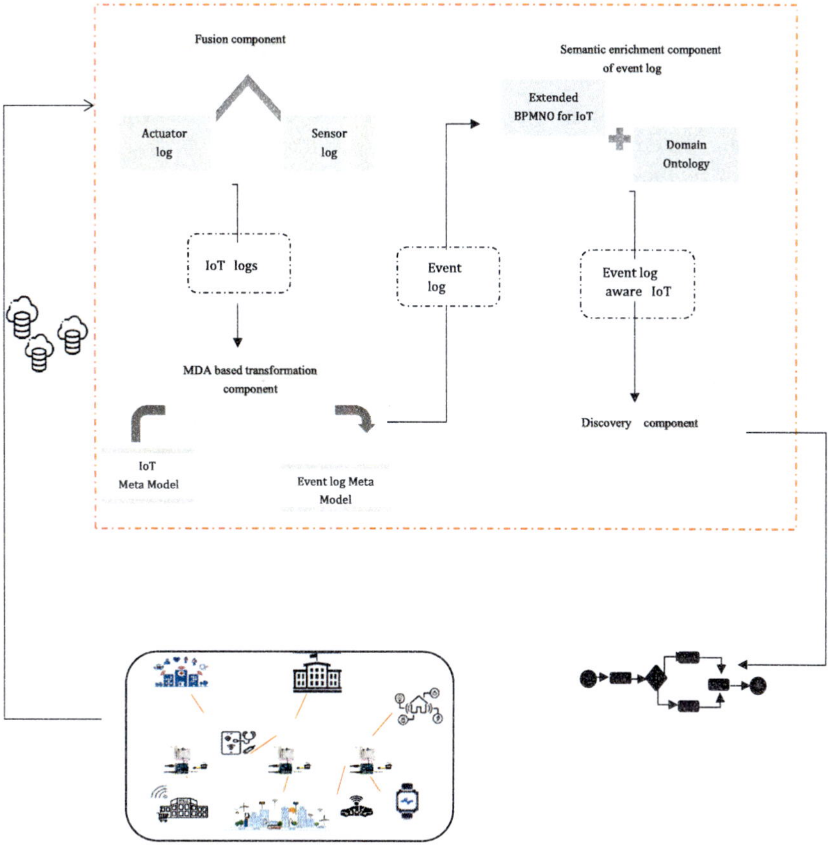

Fig. 3. Our framework

- **Fusion of logs component**: 1st the captured data is recovered, stored in a database and performs processing in order to send commands via the stage in which actuators receive commands to operate. An actuator is said to operate based on inputs from multiple sensors. Our ecosystem is composed of two levels of installation and deployment: Edge and Cloud. Edge level: this is the physical level of the ecosystem, in which the IoT nodes are installed. Cloud this is the level in which the programs deployed, which serves as an intermediary between the sensors and the actuators the sensors capture the physical quantities and send/publish them [10].
- **Model driven Transformation component**: In our approach, we get logs with non-structured also massive data. Considering that the sources of data are more and more varied and we encourage us to transform them into structured event logs, with the proposition of the Model Architecture Drive; amounts to first defining the meta models associated with the merged file and also associating with XES corresponds to know define afterwards how to map a target model from a source model.

- **Semantic enrichment of event log component**: After transforming the data in the event log. We exploit the semantic domain by offering Extended BPMNO for the IoT and Domain Ontology, that we embed in different extensions of process management in an IoT environment in several areas, through a comparative study based on the "Extension" criteria.
- **Discovery component:** Then, the application of Process mining techniques that can automatically discover the Process Aware IOT in which the sensors are highlighted in the output model.

6 Conclusion

The Internet of Things (IoT) includes a vast number of devices that create physics data on a continual basis. While this emitted data can be transferred to the Cloud for business process management, a growing number of scenarios are being examined in which IoT data will be stored, processed, evaluated, and exploited. As a result, IoT data must be disseminated across many settings. In this paper, we examine different extensions of process management in an IoT environment in several areas, through a comparative study. We found that most of the existing works lacks semantic representation at different levels of process elements and the processes are created manually which is a consuming task. So we suggest to apply process mining techniques to discover process aware IoT automatically from sensor logs. The benefits of semantic modeling have already proved itself. Our framework will deal firstly with transformation of sensor logs into a structured event log. Secondly, the event log will be enriched using a new BPMN Ontology for IoT. This allow for generating an event log aware – IoT. Then we will apply process mining to discover business process aware IoT. We are currently working on the first part of our framework which is the transformation of sensor log into an event log.

References

1. Towards a New Metamodel of Internet of Things Ecosystem|SpringerLink. Accessed 18 Apr 2022
2. Yousfi, A., de Freitas A., Dey, A.K., Saidi, R.: The use of ubiquitous computing for business process improvement. IEEE Trans. Serv. Comput. (2016)
3. Meyer, S., Ruppen, A., Hilty, L.: The things of the Internet of Things in BPMN. In: Persson, A., Stirna, J. (eds.) CAiSE 2015. LNBIP, vol. 215, pp. 285–297. Springer, Cham (2015). https://doi.org/10.1007/978-3-319-19243-7_27
4. International Conference on Advanced Communication Technology GIRI, IEEE Communications Society, Institute of Electrical and Electronics Engineers. The 18th International Conference on Advanced Communication Technology: "Information and Communications for Safe and Secure Life" : ICACT 2016 : Phoenix Park, Pyeongchang, Republic of Korea , 31 January–3 February 2016 , Proceeding & Journal (2016). Accessed 18 Apr 2022
5. Semantic Framework for Internet of Things-Aware Business Process Development. IEEE Conference Publication|IEEE Xplore. Accessed 18 Apr 2022
6. Process-Aware Internet of Things: A Conceptual Extension of the Internet of Things Framework and Architecture, KSII TIIS (2016)

7. Brouns, N., Tata, S., Ludwig, H., Asensio, E.S., Grefen, P.: Modeling IoT-aware Business Processes, vol. 42
8. Leek, S., Turnbull, P.W., Naudé, P.: Managing business-to-business relationships: an emerging model. J. Customer Behav. **1**, 357–375 (2002)
9. Ferreira, P., Martinho, R., Domingos, D.: IoT-aware business processes for logistics: limitations of current approaches, 18 April 2022
10. IoT-aware Business Process: comprehensive survey, discussion and challenges. IEEE Conference Publication|IEEE Xplore. Accessed 18 Apr 2022
11. Meyer, S., Ruppen, A., Magerkurth, C. Internet of Things-aware process modeling: integrating IoT devices as business process resources. In: King, R. (ed.) Active Flow and Combustion Control (2018)
12. Chiu, H.H., Wang, M.S.: Extending event elements of business process model for internet of things. In: 15th IEEE International Conference on Computer and Information Technology, CIT 2015 (2015)
13. Atzori, L., Jin, X., Jarvis, S., et al. (eds.): Proceedings - 15th IEEE International Conference on Computer and Information Technology, CIT 2015, 14th IEEE International Conference on Ubiquitous Computing and Communications, IUCC 2015 (2015)
14. Sungur, C.T., Spiess, P., Oertel, N., Kopp, O.: Extending BPMN for wireless sensor networks. In: 2013 IEEE 15th Conference on Business Informatics. IEEE (2013)
15. Context-Aware Business Process Management|SpringerLink. Accessed 18 Apr 2022
16. Challenges of smart business process management: An introduction to the special issue - ScienceDirect. Accessed 18 Apr 2022
17. Saxena, K.B., Vervest, P., Liere, D., Zheng, L.: Business Process Management in a Smart Business Network Environment (2009)
18. Business Process Management (BPM): terminologies and methodologies unified|SpringerLink. Accessed 18 Apr 2022
19. Braun, R., Schlieter, H., Burwitz, M., Esswein, W.: BPMN4CP: Design and Implementation of a BPMNExtension for Clinical Pathways, 12 November 2015
20. Catarinucci, L., de Donno, D., Mainetti, L., et al.: An IoT-aware architecture for smart healthcare systems. IEEE Internet Things J. **2**(6), 515–526 (2015)

An Efficient Framework for the Implementation of Sustainable Industry 4.0

Sara Kachiche[1](✉), Youssef Gahi[1,2], and Jihane Gharib[1]

[1] Laboratory of Engineering Sciences, National School of Applied Sciences, Ibn Tofail University, Kenitra, Morocco
sara.kachiche@gmail.com, {gahi.youssef,jihane.gharib}@uit.ac.ma

[2] School of Electrical Engineering and Computer Science, University of Ottawa, 800 King Edward Ave, Ottawa, ON, Canada

Abstract. The industrial revolution marked a period of development in industrialization which led to economic growth. However, this manufacturing paradigm is not sustainable, and it has emerged as a significant issue in the global market. But with the current transformation of industrial manufacturing through digitalization and the emergence of new technologies known as industry 4.0, this new revolution is set to change and make manufacturing more sustainable while enabling organizations to improve flexibility, efficiency, reduce costs and offer customized products. For this purpose, this paper proposes a framework for the implementation of industry 4.0 by providing an overview of its concepts, drivers, applications, benefits, challenges, and by introducing an efficient approach to ensure the success of its implementation in the context of sustainability.

Keywords: Industry 4.0 · Sustainability · Implementation

1 Introduction

Manufacturing has been through several periods of change, as mentioned in Fig. 1. The first industrial revolution ramped up in the late 18th century in Britain. It was marked by a transition from manual production methods to machines powered by steam or water [1] and led to many scientific discoveries with limited industrialization. For millennia, almost all human power came from muscles. It harnessed some animal power and eventually some wind and water power, but steam power completely revolutionized how much work could be done on behalf of humans. A century later, during the second industrial revolution, mass production and the assembly line were introduced and became widespread [2]. The new power sources, such as oil, gas, and electric power, along with more advanced communication technologies such as telephones and telegraphs, brought mass production and some degree of automation to manufacturing processes. The third industrial revolution, which began in the middle of the 20th century, added computers, advanced telecommunications, and data analysis to manufacturing processes. The digitization of factories began by embedding programmable logic controllers (PLCs) into machinery to help automate some processes and collect and share data [3].

M. Ben Ahmed et al. (Eds.): SCA 2022, LNNS 629, pp. 804–815, 2023.
https://doi.org/10.1007/978-3-031-26852-6_74

We are now in the fourth industrial revolution era, also referred to as Industry 4.0. it is defined as a new level of organization and control over the entire value chain of the life cycle of products. It is geared towards increasingly individualized customer requirements [4]. In the industry 4.0 era, production systems, in the form of Cyber-Physical Production Systems (CPPS), can make intelligent decisions through real-time communication and cooperation between "manufacturing things" [5]. Enabling flexible production of high-quality personalized products at mass efficiency [6].

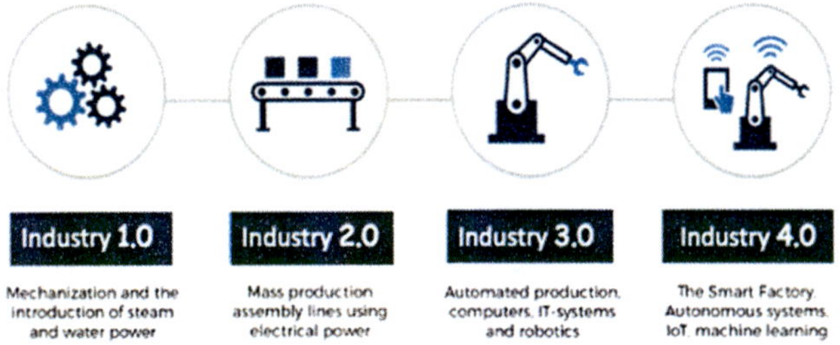

Fig. 1. The four industrial revolutions

The three industrial revolutions have improved the living standards and the quality of life. However, the current production model is not sustainable [7]. On the one hand, industrial production contributes to the degradation of the environment, such as water pollution and climate change. On the other hand, it consumes plenty of nonrenewable resources, such as petroleum and coal [8]. Therefore, the industry needs a radical change, and it is industry 4.0 that addresses this change. The main objective of Industry 4.0 is to implement advancing information technologies so that production will work in an efficient, flexible, and greenway while ensuring higher quality and low cost.

In this paper, after a short introduction, we present a brief literature review of industry 4.0 in Sect. 2, detailing its concept and its core components. In Sect. 3, we pinpoint the various beneficial impacts of its applications for sustainability. In Sect. 4, we identify the different challenges that hinders its implementation and we propose a framework showing a direction for an effective implementation in context of safety and security. Finally, the conclusions and future work are given in Sect. 5.

2 Literature Review

The term industry 4.0 was published for the first time in November 2011 by the German government that resulted from an initiative regarding high-tech strategy for 2020. Since then, this term has been used across Europe [9]. It represents a transformation based on nine foundational technology advances, as illustrated in Fig. 2. Physical and computational systems will be connected across the whole value chain in this revolution. These connected systems can interact, gather and analyze data to predict failure, configure themselves, and adapt to changes [10]. This, in turn, will result in a fully integrated, automated, and optimized value chain, leading to increased efficiency, flexibility, quality, and customization [11].

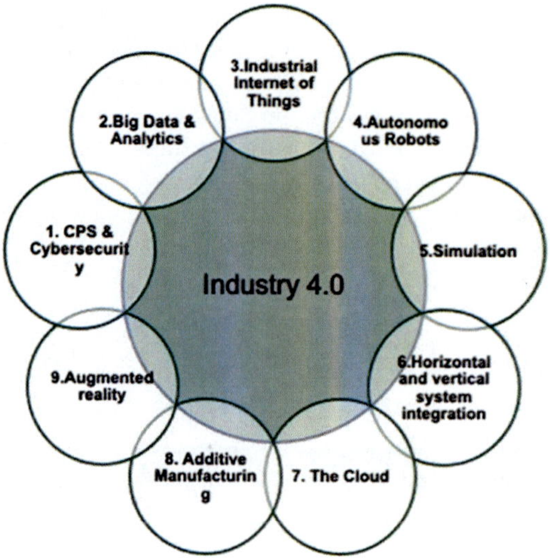

Fig. 2. The nine pillars of technological advancement

2.1 Cyber-Physical Systems (CPS) and Cybersecurity

Cyber-Physical Systems (CPS). The term "cyber-physical systems" emerged in 2006, coined by Helen Gill at the National Science Foundation [12]. CPS are technical systems in which networked computers and robots interact with the physical world [13]. This new emerging technology has many benefits, including increasing efficiency and sustainability. These benefits include: self-driving cars improve safety reduce traffic congestion and air pollution. Mass customization enables the elimination of excessive production and unsold stocks. Thus, it facilitates the reduction of waste during production.

Cybersecurity. Cybercrime has developed remarkably since the first appearance of connected computers. The level of risk faced today is significantly more than it was then. With the increased connectivity and use of standard communications protocols that come with Industry 4.0, the need to protect critical industrial systems from cybersecurity threats increases dramatically [10]. As a result, it is essential to build a secure access management system for machines and users.

Big Data and Analytics. In 2005, Roger Mougalas from O'Reilly Media coined the term Big Data for the first time [14]. It refers to rapidly growing volumes of data and a wide variety of data type (structured data, unstructured data, semi-structured data, streaming data, etc.) that is almost impossible to manage and process using traditional business intelligence tools. The manufacturing world has taken an interest in analytics based on big data only recently. It enables managers to make real-time decisions to improve efficiency, lower cost, and optimize production processes.

Industrial Internet of Things (IIOT). The Industrial Internet of Things (Industrial IoT) is composed of many connected devices. The resulting systems, and even the individual devices that comprise it, can monitor, collect, exchange, analyze, and instantly act on information to intelligently change their behavior or environment, all without human intervention [15].

Autonomous Robots. The first digitally operated and programmable robot was invented by George Devol in 1954 and was ultimately called the Unimate [16]. This later set out the start of the modern robotics industry. Today, Industry 4.0 is on the rise due to automation and the ubiquity of computing power. Autonomous mobile robots are industrial robots that use a decentralized decision-making process for collision-free navigation to provide a platform for material handling, collaborative activities, and complete services within a bounded area [17]. They can complete tasks intelligently and harmonize with minimal human input. They are flexible, autonomous, and cooperative.

Simulation. Simulation has been used in manufacturing since its birth in the 1950s for understanding, improving, and optimizing manufacturing systems [18]. The word "simulation" is defined as "the imitation of the operation of a real-world process or system over time." [19]. This technology enables engineers and experts to study the process or the system in question in a controlled and repeatable environment. Consequently, this helps identify manufacturing bottlenecks and exploit resources to increase throughput and validate the new and existing process.

Horizontal and Vertical System Integration. Nowadays, almost many I.T. systems are not fully connected. Companies, suppliers, and customers are rarely closely linked. Functions from the enterprise to the shop floor level are not fully integrated. But with Industry 4.0, companies, departments, processes, and capabilities will become much more cohesive. Figure 3 shows the different types of integration.

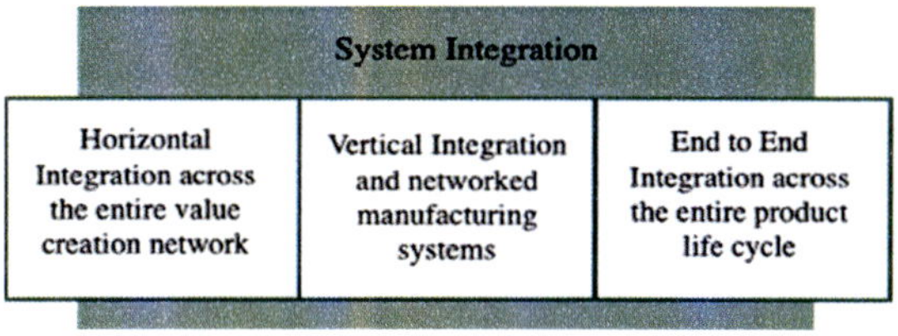

Fig. 3. System Integration

The Cloud. Recently, organizations have been using cloud-based software for data storing, remote access to any related work data, or data analytics. But with Industry 4.0, on the one hand, more data is generated and requires sharing across sites and beyond. On the other hand, the improvement of the performance of cloud technologies results in achieving reaction times of only several milliseconds. As a result, machine data and functionality will increasingly be deployed to the cloud, enabling more data-driven services for production systems.

Additive Manufacturing. Popularly known as 3D printing, is building objects bottom-up, one layer at a time [20]. Because the part is built up instead of being machined from the outside or molded, additive manufacturing makes it possible to produce forms that would be complex to make. It helps to customize the part while changing the details at the last minute because there is no dedicated mold tooling. It eliminates a lot of assembly work. Figure 4 illustrates the difference between traditional subtractive manufacturing and additive manufacturing.

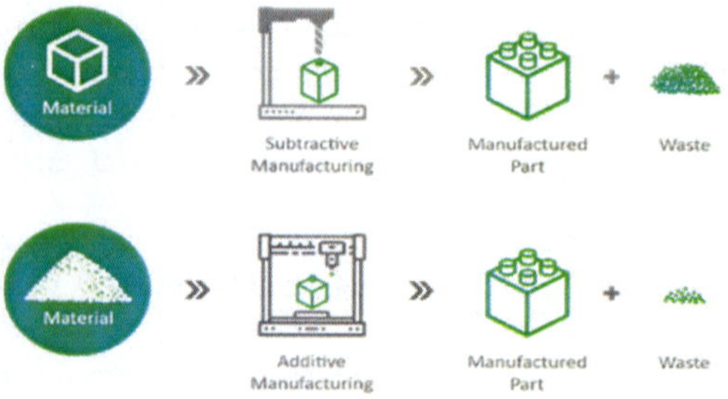

Fig. 4. Additive Manufacturing

Augmented Reality. It is a technology that combines virtual information with the natural world [21]. It uses technologies such as Intelligent Interaction, Multimedia, Sensing, 3D-Modelling, Real-time Tracking and Registration, and more. After simulation, it applies the virtual information generated by computers to the real world. In this way, the two kinds of knowledge complement each other, thus enhancing the real world [22].

3 Sustainability in Industry 4.0

As mentioned before, the emergence of new and novel technologies results in transforming the current value chains. Industries have become more efficient, flexible, and more customer-oriented. Furthermore, industry 4.0 addresses the poor ecological imbalances of the traditional production system and provides a more sustainable model. The implementation of the new technologies contributes to environmental and social sustainability in reducing the environmental impact on production and alleviating the physical workload for workers, and flexibility of work organization. From this new perspective, ample benefits and profits will arise. In the upcoming years, the competitive advantage of manufacturing companies will be driven primarily by the implementation of industry 4.0 technologies. Figure 5 summarizes some of the most prevalent benefits and opportunities.

3.1 Resource and Energy Efficiency

Several scholars agree that industry 4.0 can provide, beyond economic advantages, various contributions for environmental and social sustainability [23]—reduction of

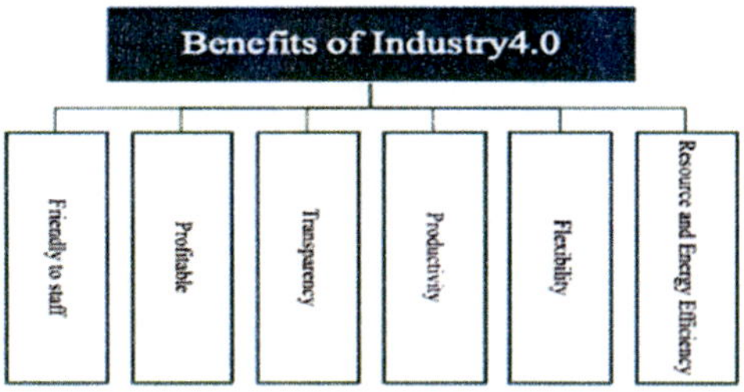

Fig. 5. The benefits of industry 4.0

overproduction and waste reduction of energy consumption. We can establish an accurate knowledge of machine operation based on big data analytics. These intelligent machines operate in a more innovative way that energy consumption can be reduced consequently. The measures include saving energy during breaks and using more new technologies such as operating motors with adjustable speed drives [8]. Furthermore, the use of energy recovery for the whole system provides a significant advantage in reducing waste, especially in the product end-of-life phase [24].

In addition, wrong deliveries, waiting time, and damaged goods can be reduced via industry 4.0 technologies, thereby reducing resource consumption. Further, reducing transportation and travel effort saving of natural resources contributes to the realization of a more favorable production environment for workers in terms of higher physical relief and safety [25].

3.2 Flexibility

The smart, connected, self-aware products can be reconfigured automatically to produce multiple types of products and promote mass customization to meet specific customer requirements. Their capability of sharing data, their status, location, conditions, and more help to improve product quality, customer service, and R&D. The self-organization and dynamic reconfiguration bring robustness in the sense that new machines can join the system in a plug and play-way [8]. As a consequence of the emergence of new innovative business models value, chains are becoming more responsive, increasing competitiveness by eliminating barriers between information and physical structures [26].

3.3 Productivity

The industry 4.0 technologies enable manufacturers to produce more and faster while allocating resources more cost-effectively and efficiently. Compared with the traditional production line, the smart factory can have small lot sizes of different products more efficiently. The production lines will experience less downtime because of enhanced machine monitoring and automated decision-making. The setup time is minimized when switching between different types of products. On the other hand, as the production process is optimized with the help of big data feedback and coordination, the average manufacturing routes are reduced, and the utilization rate of machines and other resources is improved [8]. Once Industry 4.0 is fully implemented, the productivity in the context of the automotive industry alone is expected to increase by 10–20% [10].

3.4 Transparency

The transparency afforded by Industry 4.0 technologies provides stakeholders with vast amounts of useful information needed to make appropriate decisions. Big data analytics allows operators to quantify performance indicators related to machines, products, and systems. Thus, identifying key areas that can benefit from innovation and improvement. This can also facilitate production planning and accelerate response to market demand. When the production ability of the smart factory cannot satisfy a particular product order, the questions about how to improve the system can also be quickly answered [8].

3.5 Profitable

Based on Moore's law, the cost of information technologies will steadily decline while the performance constantly improves. For the future customized demands, with the improved flexibility, resource, and energy efficiency, the operational cost is low compared with the traditional fixed production line. A recent study shows that manufacturers generate 38 percent (average) of product sales from goods with embedded intelligent devices and intelligence. Innovative products create a profit margin of 36 percent (average) vs. 33 percent (average) for other products [27].

3.6 Friendly to Staff

With the assistance of big data analytics, robust software tools, and a more friendly and flexible interface for data acquisition, maintenance and analysis become much more accessible. The machines operate automatically by themselves so that no operators need to perform continuous and unvarying tasks. Even workers all across the value chain can work together to perform remote repair work, as the people and machines can interact with each other through the cloud [8].

4 Proposed Framework for the Implementation of a Sustainable Industry 4.0

As discussed above, the transformation towards industry 4.0 is a noticeable phenomenon in many countries, but its range and intensity vary [28]. To achieve a successful change to Industry 4.0 standard, manufacturing enterprises must implement a clear roadmap [29]. First, we will address the risks and challenges associated with this transformation. In the next step, we propose an initial approach for the implementation of a sustainable industry 4.0.

4.1 Constraints and Challenges of Industry 4.0 in the Context of Sustainability

With the evolving trend of Industry 4.0, driving organization-wide innovation is very difficult. Although the new business model comes with many benefits, it is also rife with challenges. Enterprises face considerable risks upon successful implementation of Industry 4.0 [30]. Numerous authors state that implementing industry 4.0 is a challenging mission, and it is likely to take ten or more years to be realized [9]. Adopting this new manufacturing process involves many aspects and faces many types of difficulties and challenges, including scientific, technological, organizational, economic, and social issues.

Figure 6 provides an overview of the main challenges of implementing Industry 4.0.

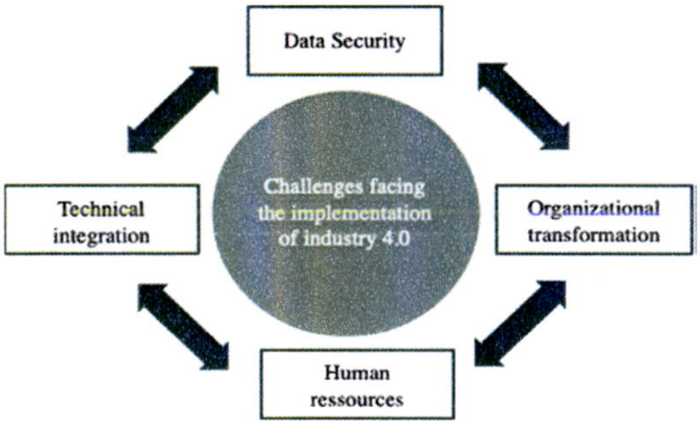

Fig. 6. Challenges of implementation of industry 4.0

Data Security

- Exchanging data bears weak points and the risk of falling victim to cybercrime, unauthorized information access, and industrial spying [31].
- When numerous machines and devices are connected to single or multiple networks in a smart factory, vulnerabilities in any equipment, apparatus, or product could make the system vulnerable to attack. The larger the web and the more interfaces exist, the larger the potential attack surface for cyberattacks [32].
- Security incidents' consequences can be in different areas or dimensions, such as interruption or modification of an operational process or even sabotage to cause harm.
- Manipulating or interrupting such systems could also affect safety, which can have consequences such as environmental damage, injury, or loss of life.

Human Resources

- Creative working processes will have a greater need for the skills required to identify, introduce and implement the new and innovative business opportunities offered by industry 4.0 [33].
- The actual training profile of the industry 4.0 employee has not yet been created. The industry is facing missing talent and capabilities in that regard. [34].

- Factories are often categorized as safety-critical systems as failures of these systems, under certain conditions, can lead to human harm or damage to property or the environment, e.g., due to the use of heavy machinery or hazardous substances [35].

Technical Integration

- The I.T. infrastructure is not entirely ready to support the digital transformation to industry 4.0 [28].
- Diverse systems need to be networked to learn to communicate with each other, and new communications networks need to be developed from scratch.

Organizational Transformation

- The problem of hierarchical organizations hindering the transformation.
- Low innovation culture results in internal resistances are probable, since employees may fear novel technologies and loss of jobs.

From the sustainability perspective, we will focus our discussion on the most challenging aspect for organizations that wish to adopt industry 4.0: safety and security. On the one hand, such systems may not endanger people and the environment – « corporate (internal) safety» and, on the other hand, such devices must be protected for misuse and unauthorized interference - particularly in the area of data misuse, protection against unauthorized interference « Security» [27].

To address the mentioned concerns, we propose an initial approach to implementing sustainable industry 4.0 in the context of safety and security.

4.2 Process of Implementation of a Sustainable Industry 4.0

Within this framework, the objective was to propose an efficient process of implementation of industry4.0 using the PDCA (Plan, Do, Check, Act) approach. This latter stresses the importance of understanding the relevant elements and characteristics of Industry 4.0 technologies that could contribute to attaining the objectives fundamental to environmental sustainability, especially in the context of safety and security. Furthermore, it to ensures the success of a sustainable transformational journey through continual improvement of processes, based on evaluation.

The subsequent steps of implementing a sustainable industry 4.0 are presented in Fig. 7.

Plan Phase: Current state process mapping consists of revising the processes that have the most significant impact on Safety and Security. Subsequently, conducting a risk analysis to figure out where the most operational inefficiencies are occurring.

Do Phase: Future State process mapping: it consists of examining and mapping the impacts of the proposed solutions while mitigating the risks identified.

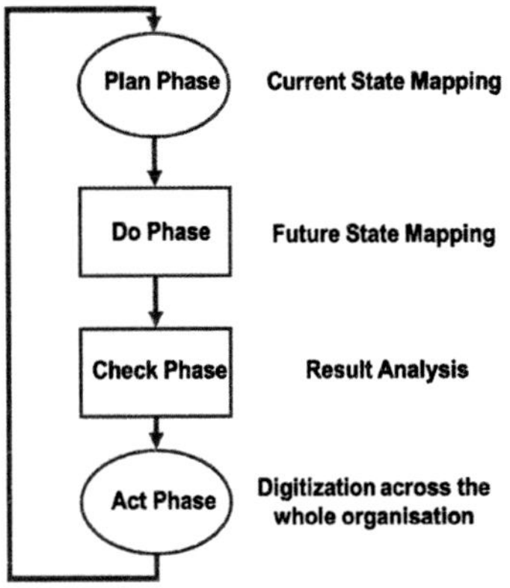

Fig. 7. Process of implementation of a sustainable industry 4.0

Pilot project Management: it consists of creating a pilot project for the implementation. With a smaller scope, especially in the context of safety & security, we ensure an efficient way to properly engage resources, test, discovering discrepancies, refine the solution and demonstrate the business value of the project.

Check Phase: Result Analysis: analyzing information from pilot projects, outline the different issues and possible solutions. Define the needs and requirements for appropriate resources.

Act Phase: Transformation Implementation: implementation of the new solution across the whole organization. Alongside, this transformational process implies the creation of an adaptable corporate culture convinced of the need to pursue the novel industrial paradigm.

Follow up Analysis: Conducting a follow-up analysis to monitor the change and to set an outline for the effective way of implementation.

5 Conclusion and Future Work

In this paper, we have presented a state-of-the-art review of the evolution of industrial revolutions leading to the emergence of industry 4.0. To give an extensive understanding of this development, the different emerging technologies of industry 4.0 have been described based on the current findings in research. Subsequently, other benefits for realizing sustainable manufacturing in Industry 4.0 have been presented as well as the substantial challenges for implementation have been discussed. Finally, an initial approach for sustainable manufacturing performance in industry 4.0 has been defined. Herby, the paper introduces a framework for future research to explore risks that arise during the implementation of industry 4.0 in the context of safety and security. To support the implementation of industry 4.0, the process presented within the framework proposed in this paper must be deepened.

References

1. Xu, X., Lu, Y., Vogel-Heuser, B., Wang, L.: Industry4.0 and Industry 5.0-Inception, conception and perception. J. Manuf. Syst. **61**, 530–535 (2021)
2. Efthymiou, O.K., Ponis, S.T.: Industry 4.0 technologies and their impact in contemporary logistics: a systematic literature review. Sustainability **13**(21), 11643 (2021)
3. How Industry 4.0 technologies are changing manufacturing. IBM research
4. Vaidya, S., Ambad, P., Bhosle, S.: Industry 4.0- a glimpse. Procedia Manuf. **20**, 233–238 (2018)
5. Lu, Y., Xu, X., Wang, L.: Smart manufacturing process and system automation – a critical review of the standards and envisioned scenarios. J. Manuf. Syst. **56**, 312–325 (2020)
6. Wang, S., Wan, J., Zhang, D., Li, D., Zhang, C.: Towards smart factory for Industry 4.0: a self-organized multi-agent system with big data based feedback and coordination. Comput. Netw. **101**, 158–168 (2016)
7. Alkaya, E., Bogurcu, M., Ulutas, F., Demirer, G.N.: Adaptation to climate change in industry: improving resource efficiency through sustainable production applications. Water Environ. Res. **87**(1), 14–25 (2015)
8. Wang, S., Wan, J., Li, D., Zhang, C.: Implementing smart factory of industrie 4.0: an outlook. Int. J. Distrib. Sensor Networks **2016**, Article ID 3159805, 10 pages
9. Mamad, M.: Challenges and benefits of industry 4.0: an overview. IJSOM **5**(3), 256–265 (2018)
10. Rüßmann, M., Lorenz, M., Gerbert, P., Waldner, M.: Industry 4.0: The Future of Productivity and Growth in Manufacturing Industries, April 09, 2015, pp. 1–14
11. Müller, J.M., Kiel, D., Voigt, K.I.: What Drives the Implementation of Industry 4.0? The Role of Opportunities and Challenges in the Context of Sustainability. Sustainability **10**, 247 (2018)
12. Lee, E.A., Seshia, S.A.: Introduction to embedded systems- A cyber-physical system approach", Second Edition, Version 2.2, MIT Press (2017)
13. Scientific Foresight study Ethical Aspects of Cyber-Physical Systems, EPRS, STOA, PE 563.501, June 2016
14. Real Time Innovations Inc, Industrial Internet of Things, RTI FAQ, p. 1 (2015)
15. Mejia, O.C., Marquez, A., Monsreal-Berrera, M.M.: Product Delivery and Simulation for Industry 4.0. Simul. Ind. **4**, 81–95 (2019)
16. Peng, Y.: Simulation and research on some buildings in Ningbo. J. Phys. Conf. Ser. **1549**(3), 032129 (2020)
17. Verhoefa, L.A., Buddeb, B.W., Chockalingamb, C., Nodarb, B.G., van Wijk, A.J.M.: The effect of additive manufacturing on global energy demand: An assessment using a bottom-up approach. Energy Policy **112**, 349–360 (2018)
18. Chen, Y., Wang, Q., Chen, H., Song, X.: An overview of augmented reality technology. J. Phys. Conf. Ser. **1237**(2) (2019)
19. Tianyu, H., et al.: Overview of augmented reality technology. Comput. Knowl. Technol. **34**, 194–196 (2017)
20. Stock, T., Seliger, G.: Opportunities of sustainable manufacturing in industry 4.0. Procedia CIRP **40**, 536–541 (2016)
21. Ender, J., Wagner, J.C., Kunert, G., Guo, F.B., Larek, R., Pawletta, T.: Concept of a self-learning workplace cell for worker assistance while collaboration with a robot within the self-adapting-production-planning-system. IAPGOS **9**, 4–9 (2019)
22. Qiu, X., Luo, H., Xu, G., Zhong, R., Huang, G.Q.: Physical assets and service sharing for IoT-enabled Supply Hub in Industrial Park (SHIP). Int. J. Prod. Econ. **159**, 4–15 (2015)

23. The MPI Group, "Industry 4.0 Drives Productivity- and Profitability- for Manufacturers", Finfings from the MPI 2020 Industry 4.0 Study
24. Pollak, A., Hilarowics, A., Walczak, M., Gąsiorek, D.: A Framework of Action for Implementation of Industry 4.0. an Empirically Based Research. Sustainability, Received: 3 June 2020; Accepted: 14 July 2020; Published: 18 July 2020
25. Duran, J.R., Berges, I., Illarramendi, A.: Towards the implementation of Industry 4.0: A methodology-based approach oriented to the customer life cycle Computers in Industry, vol. 126, April 2021, 103403
26. P.Ghadimi, O. Donnelly, K. Sar, C. Wang and A. H. Azadnia, "The successful implementation of industry 4.0 in manufacturing: An analysis and prioritization of risks in Irish industry ", Technological Forecasting and Social Change, Volume 175, November 2021, 121394
27. Kiel, D., Müller, J.M., Arnold, C., Voigt, K.N.: Sustainable industrial value creation: benefits and challenges of industry 4.0 . Int. J. Innov Manage. **21**(08), 1740015 (2017)
28. Birkel, H.S., Veile, J.W., Müller, J.M., Hartmann, E., Voigt, K.I.: Development of a risk framework for industry 4.0 in the context of sustainability for established manufacturers. Sustainability **11**, 384 (2019)
29. Schmittner, C., Ma, Z., Schoitsch, E.: Combined safety and security development lifecycle. In: 2015 IEEE 13th International Conference on Industrial Informatics (INDIN), pp. 1408–1415. IEEE (2015)
30. Industry 4.0, Challenges and solutions for the digital transformation and use of exponential technologies, Deloitte, Audit. Tax. Consulting. Corporate Finance
31. Dennis, K., Nicolina, P., Yves-Simon, G.: Textile learning factory 4.0 – preparing Germany's textile industry for the digital future. In: 7th Conference on Learning Factories, CLF 2017 Procedia Manufacturing, vol. 9, pp. 214–221 (2017)
32. Jaradat, O., Sljivo, I., Habli, I, Hawkins, R.: Challenges of safety assurance for industry 4.0. In: 2017 13th European Dependable Computing Conference (EDCC)
33. Nagy, O., Papp, I., Szabó, R.Z.: Construction 4.0 Organisational Level Challenges and Solutions. Sustainability **13**, 12321 (2021)
34. Klinc, R., Turk, Ž: Construction 4.0-digital transformation of one of the oldest industries. Econ. Bus. Rev. **21**, 393–410 (2019)
35. Mult, H.C., Sinay, J., Kotianova, Z., Glatz, J.: Risk management in context of industry 4.0. Int. Sci. J. "Industry 4.0", year **iii**(6), 340–342 (2018)

Towards Sustainable Financing Through Local Tax Revenue Forecasting Using Time Series

Nabil Ourdani[1](✉) and Mohamed Chrayah[2]

[1] Faculty of Sciences, Abdelmalek Essaadi University, Tetuan, Morocco
ourdani@gmail.com
[2] National School of Applied Sciences, Abdelmalek Essaadi University, Tetuan, Morocco

Abstract. Talking about the sustainable development of cities and the advent of the smart cities concept goes through the realization of several projects to serve citizen. However, planning these projects, and even their success, is dependent on financing availability. Good revenue forecasting therefore plays a key role for local authorities as it helps in budgeting, planning and financing projects. For this reason, various types of revenue must be forecast for a specific year, using predictive models with a small margin of error. This work aims to use time series processing models, such as auto-regression (AR), auto-regressive moving average (ARIMA) and seasonal ARIMA (SARIMA), for tax revenue prediction of a local authority in Morocco. Data collected concerns monthly tax receipts for the period from January 2000 to December 2020. Conclusions of this study can help experts in the process of preparing local authorities budgets to get more accuracy.

Keywords: Financial sustainability · Local authorities · Prediction · Forecasting · Tax revenue · Auto-Regression · ARIMA · SARIMA · Morocco

1 Introduction

In a process of development and reduction of social and spatial inequalities, citizen must be the main goal. In Morocco, significant achievements have been made in the area of democracy, rights and freedoms as well as in economic, financial and infrastructure terms. This necessarily involves proper implementation of advanced regionalization and local administrative autonomy with a view to better rationalizing territorial organization.

To this end, local finances are increasingly called upon to contribute to economic development and to reduce social and spatial inequalities. This can be done by increasing local authorities' own-source revenues and capital budgets, and reducing the costs of citizen participation [1].

Local authorities play an increasingly important role in the provision of fundamental basic public services. But they also face great challenges. One of the challenges regularly encountered is financial forecasting, planning and budgeting. They are called upon to make sustainable decisions that increasingly aim for financial viability while ensuring the well-being of citizens, sometimes even to the detriment of their financial income [2].

M. Ben Ahmed et al. (Eds.): SCA 2022, LNNS 629, pp. 816–824, 2023.
https://doi.org/10.1007/978-3-031-26852-6_75

Optimal management problem of revenue collection process by local authorities is of paramount importance, not only for benefits it brings, but also as a means of administering an equitable system through establishment of durable budgets. These begin with understanding current income and expenses in the context of a long-term forecast.

Tax revenue is considered one of fundamental sources of budget planning for local authorities in Morocco. Like many cities and towns around the world, local authorities in Morocco used to budget one year at a time assuming that the following year would be very similar to the previous ones 3.

Artificial Intelligence (AI) development and the field of statistics has allowed design of new algorithms so interesting that they are used today in data mining, such as regression analysis 4, neural networks 5 and linear classification models 6, etc. Now, with data analytics, local government managers can understand how changes in revenues, costs, inflation and other exogenous factors will affect council finances - not just next year, but in the future.

Moroccan local authorities collect revenues not only to finance their expenditure but also for the purpose of distribution and allocation in the economy. Collection of revenue and expenditure execution of local authorities is done in by the General Treasury of the Kingdom of Morocco (GTK), a central department of the Ministry of Economy and Finance.

Every month, GTK publish a local finance statistics bulletin. Two fundamental classifications exist in Moroccan local authorities' revenue system: tax revenues and non-tax revenues. The relative importance of these taxes is presented in Table 1 based on data collected from the GTK web portal 7.

Making predictions based on historical data systematically leads us to talk about time series forecasting, an important area of machine learning [8]. Finding suitable model for local tax revenue time series data using appropriate criteria and diagnostic verification is not a trivial task. Time series modeling process goes through stages beginning with model specification, then its adjustment and at the end, its diagnosis to check degree of its accuracy.

There are changing trends in local tax collections over time, and they exhibit high seasonality, making analysis, model selection, and forecasting a complex task. This paper aims to investigate identification of time series pattern from a set of models, perform appropriate diagnostic checks for case study datasets of local tax collections from a city in Morocco and forecast receipts for the next two years.

2 Methodology

Time series is a sequence of values of a variable recorded over time, most often at a regular time interval. It is a composition of four constituent parts: 'level' that every time series has, 'trend' which is an optional increasing or decreasing behavior, 'seasonality' that is an optional repeating patterns and last, the optional 'noise' which represent variability that cannot be explained [8].

Therefore, a time series can be expressed by, an additive model such as

$$X_t = Level + Trend + Seasonality + Noise \quad (1)$$

or by a multiplicative model defined as

$$X_t = Level * Trend * Seasonality * Noise \quad (2)$$

Most often, using the graphical plots of a series helps determine whether a time series is additive or multiplicative.

The search for effective revenue forecasting techniques is deeply rooted in the literature, both for private and public sector. There are several approaches to perform analysis and predictions on time series. Best knowns are: auto-regressive (AR), moving average (MA), auto-regressive moving average (ARMA), auto-regressive integrated moving average (ARIMA) as well as its seasonal variant (SARIMA). This study use AR, ARIMA and SARIMA models.

According to literature [9, 11, 12], AR modeling is a linear regression technique that let predicts future values based on past behavior (each regression term is called the 'lag' which is a previous time point when the measure is taken). MA model is based on lagged values of forecast errors. ARIMA is primarily autoregressive 'AR' and 'I' indicate the differencing of observations to allow for getting series stationarity. And at last, 'MA' indicates the moving average part. ARIMA model is characterized by three terms: p (order of the 'AR'), d (number of differencing required to get stationary) and q (order of the 'MA').

For a time series sequence $\{X_t, t = 1, 2, \ldots, n\}$, ARIMA (p,d,q) model can be written as:

$$\varphi(B)(1 - B)d\,X_t = \theta(B)A_t \quad (3)$$

p is the order of the AR process, d is an order of differences, q is the order of the MA process, At is the white noise sequence, φ is a polynomial of degree p, B is a backshift operator and θ is a polynomial of degree q.

However, an ARIMA model could not analyze time series with seasonal characteristics; therefore, seasonal autoregressive integrated moving average (SARIMA) models have been developed. SARIMA models perform better than the historical average, linear regression and simple ARIMA models for data with seasonal variations. In fact, SARIMA models are capable of taking into account the trend and seasonality. A SARIMA$(p, d, q)(P, D, Q)_s$ model can be written by the following equation:

$$\varphi(B)\Phi(B^s)(1 - B^s)^d X_t = \theta(B)\Theta(B^s)A_t$$

where Φ, Θ, P, D and Q are seasonal counterparts of φ, θ, p, d and q, respectively, and s is the seasonality.

"There are several approaches to fit time series models such as the least square method, method of moment and maximum likelihood method"12. Also, there are several criteria for choosing the most suitable model. Akaike information criterion (AIC), Bayesian information criteria (BIC) and Hannan-Quinn information criterion (HQIC) have been used as an essential tool to choose the best model in this study.

$$\begin{aligned} \text{AIC} &= -2\,\text{LL} + 2\,\text{m} \\ \text{BIC} &= -2\,\text{LL} + \text{m}\log\text{n} \\ \text{HQC} &= -2\,\text{LL} + 2\text{m}\log(\log\text{n}), \end{aligned}$$

m represent the degree of freedom, n represent the sample size and LL is the Log-likelihood function of the parameter of interest.

The model with the lowest (minimum) value of any of the measures is the best.

3 Data

Tax collection data of Moroccan local authorities are kept by the General Treasury of the Kingdom of Morocco (GTK), and "Summary Statistical Bulletins" are published monthly in the GTK web portal.

For this study, data have been collected from GTK after anonymization and a commitment to process it without mentioning the name of the local authority. Data cover all taxes revenues for the period of 1995–2020. After diagnosis of data quality, only the period from January 2000 to December 2020 was retained.

To avoid failures encountered by many machine learning projects, the CRISP-DM (CRoss Industry Standard Process for Data Mining) [13]. Standard process was adopted. This process model is independent of both covered domain and used technology, and has undergone changes to adapt it to the context of machine learning projects. This evolution has preserved the same phases of CRISP-DM by adding a new phase of "Model Management" because it goes through a cycle of recycling and improvement in Machine Learning and needs to be closely monitored14. This allowed us to better plan, organize and implement our project following the six major steps shown below 15 (Fig. 1).

Objective is already known, being long-term forecast of local tax revenues. First step was to understand collected data through exploring and verifying its quality. Then we start to prepare data by selecting only the needed one, cleaning and format it. The AR, ARIMA and SARIMA models were fed by the two variables "Date" and "Amount". Data was split into training and testing where 80% was used for training and 20% was used for testing the accuracy of the models. All analysis is performed using Python as it is a popular language and is becoming first choice for data science projects. Python also provides easy-to-use APIs for most leading machine learning and deep learning libraries to process, model, and visualize time series data16 (Fig. 2).

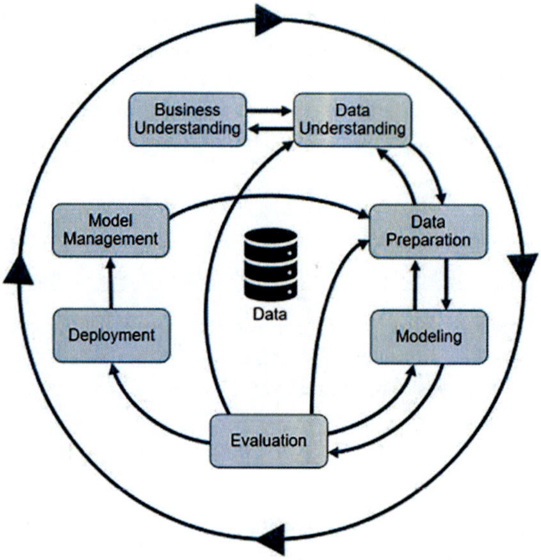

Fig. 1. CRISP-DM Diagram.

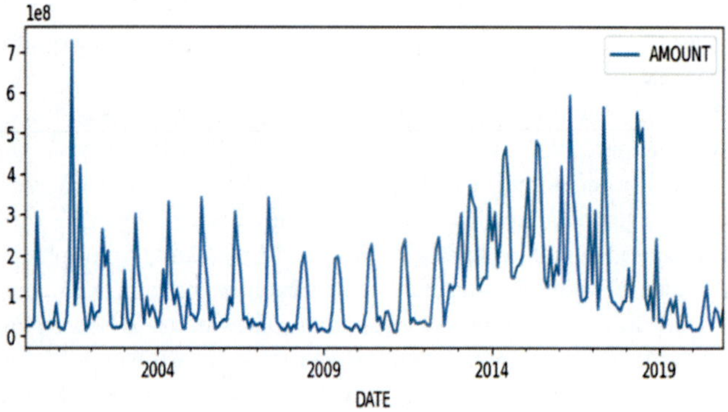

Fig. 2. Ploting clean data

The graph already allows seeing the availability of seasonality in the dataset. This leads us to decompose the time series into multiple components, each representing an underlying pattern category, trend, seasonality, and noise. The 'seasonal_decompose' function from Python's 'statsmodels' library was used for this. 'statsmodels' provides most of the statistical models and tests.

Figure 3 clearly shows the seasonal component of the data. Approach used to make a time series stationary is by differencing it (subtract the previous value from the current value) and verification is done by using Augmented Dickey-Fuller (ADF) 17 which is based on the null hypothesis that the time series is non-stationary. The ADF test gives an indicator called p-value. If p-value is less than the significance level (0.05) then the

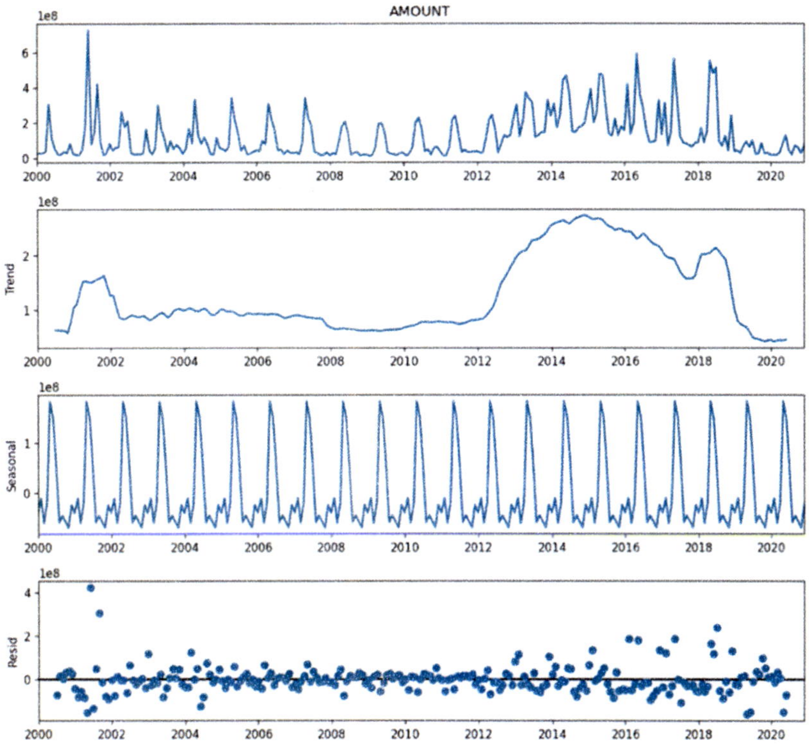

Fig. 3. Time series divided into trend, seasonal and residual components

null hypothesis is rejected and time series is declared stationary. Using Python library 'stattools', the result of p-value for our time series is: 0.346168 which is greater than the significance level (Fig. 4).

```
Results of Dickey-Fuller Test:
p-value = 0.3462. The series is likely non-stationary.
Test Statistic                   -1.870281
p-value                           0.346168
#Lags Used                       14.000000
Number of Observations Used     239.000000
Critical Value (1%)              -3.458011
Critical Value (5%)              -2.873710
Critical Value (10%)             -2.573256
```

Fig. 4. First ADF test's result

The ADF statistic result is greater than the critical values at different levels, and the p-value is also greater than 0.05 and therefore cannot reject the null hypothesis at 90%, 95%, and 99% confidence. Thus, time series data is highly non-stationary. Test was repeated using the normalized log values but it also concludes that our time series is not stationary.

After running the first differentiation on the time series, that's gave the best result with a p-value of 0.000003 and good graphic seasonality (Figs. 5 and 6).

```
Results of Dickey-Fuller Test:
p-value = 0.0000. The series is likely stationary.
Test Statistic                      -5.400865
p-value                              0.000003
#Lags Used                          13.000000
Number of Observations Used        237.000000
Critical Value (1%)                 -3.458247
Critical Value (5%)                 -2.873814
Critical Value (10%)                -2.573311
```

Fig. 5. Second ADF test' result

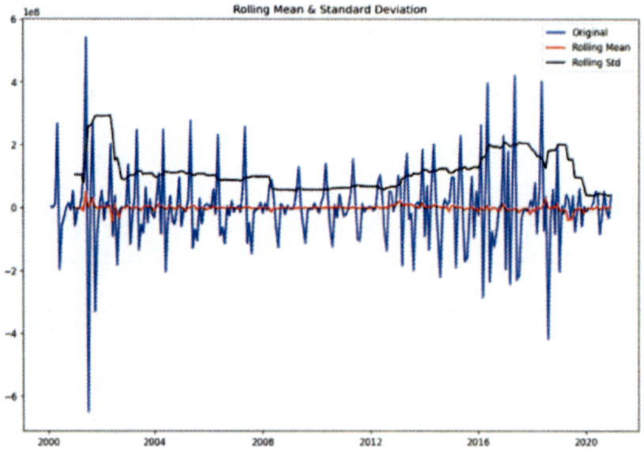

Fig. 6. Stationnarity after differentiation

4 Result and Discussion

Time series data for this study represent monthly tax collected in a Moroccan city from January 2000 to December 2020. Original data consisted of a total of 254 observations giving a date for a month and a total amount of collected tax.

According to data collected, the finding is that over the past two decades, the monthly tax varies upwards during months May, June and July. This was explained by the fact that these are due dates of local tax payments. This explains the finding of seasonality brought up during our analysis. This is clearly visible by looking at the data for each year.

AR, ARIMA and SARIMA models have been processed in Python and compared using the values of AIC, BIC and HQIC. The values for different models for the tax collection are given in Table 1.

Table 1. Assessment of different models.

Model	AIC	BIC	HQIC
AR	8936.404	8953.440	8943.281
ARIMA	9520.129	9544.551	9529.967
SARIMA	8384.124	8394.578	8388.336

While analyzing different used models, SARIMA model has the lowest values of AIC, BIC and HQIC. This allows us to suggest that it is better than the other models for the local tax collection time series of the city concerned.

5 Conclusion, Limitations and Perspective

With the help of Machine Learning, local authorities can achieve more accuracy in forecasting financial revenue. This allows them to project themselves and launch projects that are part of sustainability.

This study has aimed to find a time series model to forecast local tax revenue for a city in Morocco. Efforts to achieve revenue predictions across three models: AR, ARIMA and SARIMA have been possible using Python's statistics libraries which yield great actionable results.

Like any study, ours has certain limitations. First, our dataset concerns only all the tax revenue collected on behalf of a given local authority without taking into account any exogenous factors. That said, the results obtained cannot necessarily be applied to all local authorities in the country. Second, this study has focused on the overall revenue. However, for local authority managers, an analysis by type of tax revenue would be judicious to see the possible points of improvement by nature in order to act on specific points to improve collection. Finally, this study focused on the statistical models AR, ARIMA and SARIMA. Thus, future studies may explore other models.

Considering these limitations, further research should be conducted. For future work, we would like deep into exogenous factors that can relay to explain local revenue collection like tax amnesty periods, inflation or other variables to test the SARIMAX model and other models like LSTM (Long Short-Term Memory) - a Recurrent Neural Network (RNN) - and compare them with the obtained results in this study.

Leaders aim first and foremost to improve tax collection, but also to reduce the cost of this collection. The research will also go in depth towards these aspects to see in what perspective Big Data and Machine learning technologies are likely to provide help in achieving the desired objectives. We also keep in mind the need for taxpayer segmentation to enable local authorities to provide the best services at low cost from a sustainable funding perspective.

Finally, even it's a preliminary case about prediction using model, city's policymakers should keep under consideration the results of this study and specially the concept of using machine learning to get an accurate prediction on the revenue forecast.

References

1. United Nations Human Settlements Programme (UN-HABITAT): The Challenge of Local Government Financing in Developing Countries, https://unhabitat.org/fr/node/109066. Accessed 04 Apr 2022
2. Raich, U.: Fiscal Determinants of Empowerment. Policy Research Working Paper (3705), World Bank, Washington, DC (2005)
3. Maria, E., Babacar, S.: Decentralization, spending efficiency and pro-poor outcomes in Morocco, In: Études et Documents (5), CERDI, France (2018)
4. Jolliffe, I.T.: Principal components in regression analysis. In: Principal Component Analysis. Springer Series in Statistics. Springer, New York (1986)
5. Charu, C.: Aggarwal: Neural Networks and Deep Learning. Springer International Publishing AG, part of Springer Nature(2018)
6. Radhakrishna Rao, C., Toutenburg, H., Shalabh, C.H.: Linear Models and Generalizations, 3rd edn. Springer Series in Statistics (2018)
7. GTK Homepage: Institutional Portal of the General Treasury of the Kingdom of Morocco. https://www.tgr.gov.ma. Accessed 04 Apr 2022
8. Brownlee, J: Introduction to Time Series Forecasting With Python: How to Prepare Data and Develop Models to Predict the Future, Machine Learning Mastery, (2017)
9. Ward Michael, P., Iglesias Rachel, M., Brookes Victoria, J.: Autoregressive Models Applied to Time-Series Data in Veterinary Science (2020)
10. Brockwell, P.J.: Time Series Analysis. International Encyclopedia of Education, Third Edition. Elsevier (2010)
11. Cao, J., Li, Z., Li, J.: Financial time series forecasting model based on CEEMDAN and LSTM, Physica A (2018)
12. Al-Hasani, G., Asaduzzaman, M.: Time Series Modelling Strategies for Road Traffic Accident and Injury Data: A Case Study, Advances in Data Science and Information Engineering, Transactions on Computational Science and Computational Intelligence, Springer Nature Switzerland AG (2021)
13. Wirth, R., Hipp, J.: CRISP-DM: towards a standard process model for data mining. In: Proceedings of the 4th International Conference on the Practical Applications of Knowledge Discovery and Data Mining (2000)
14. Chapman, P., Clinton, J., Kerber, R., Khabaza, T., Reinartz, T.P., Shearer, C., Wirth, R.: CRISP-DM 1.0: step-by-step data mining guide (2000)
15. https://towardsdatascience.com/crisp-dm-ready-for-machine-learning-projects-2aad9172056a
16. Pal, A., Prakash, P.K.S.: Practical Time Series Analysis: Master Time Series Data Processing, Visualization, and Modeling using Python (Packt Publishing) (2017). ISBN 978-1-78829-022-7
17. Dickey, D.A., Fuller, W.A.: Distribution of the estimators for autoregressive time series with a unit root. J. Am. Stat. Assoc. (74) (1979)

A New Hybrid Artificial Intelligence Model for Diseases Identification

Abdallah Maiti[1](✉), Abdallah Abarda[2], and Mohamed Hanini[1]

[1] Computer, Networks, Mobility and Modeling Laboratory FST, Hassan First University of Settat, Settat, Morocco
{abdallah.maiti,mohamed.hanini}@uhp.ac.ma

[2] LM2CE Lab, Faculty of Economics and Management Settat, Hassan First University of Settat, Settat, Morocco
abdallah.abarda@uhp.ac.ma

Abstract. Artificial intelligence is increasingly applied in many fields, specially in medicine to assist patients and physicians. Growing datasets provide a sound basis to adapt machine learning methods to identify and detect some diseases. These later, are often very similar which make difficult their identification by chest X-ray images. In this paper, we introduce a diagnostic AI model that allow to separate, diagnose and classify three various diseases: tuberculosis, covid19 and Pneumonia. The proposed model is based on a combination of Deep Learning using the deep SqueezeNet model and Machine Learning: SVM, KNN, Logistic Regression, decision tree and Naive Bayes. The model is applied to a chest X-ray dataset containing images for each type of disease. To train and test our model, we split the image dataset into two training and test subsets in order to differentiate between different disease types. The accuracy show clearly that our model provides better results of diagnosis and identification.

Keywords: Artificial intelligence · Machine learning · Deep learning

1 Introduction

In recent decades, Artificial intelligence algorithms play an important role in promoting the development of healthcare field. Machine learning overcomes human factors limitations, efficiently processes data through deep learning (DL) neural networks and machine learning (ML), and improves data operations [1]. New artificial intelligence techniques make traditional algorithms unusable or significantly hinder their performance. In this context, the excessive volumes of irrelevant information can considerably complicate the analysis of current scientific trends, especially in a critical and sensitive area such as that related to the health of individuals. False diagnoses of certain diseases - caused by irrelevant information - can lead to major consequences. These erroneous diagnoses are often due to human factors, and to the presence of several complicated data

M. Ben Ahmed et al. (Eds.): SCA 2022, LNNS 629, pp. 825–836, 2023.
https://doi.org/10.1007/978-3-031-26852-6_76

that they had to carefully analyze to make good decisions. Faced with these data analysis complications, the crucial challenge in terms of medical data analysis relies on the fact that the majority of new information collected and stored is unstructured, which makes data mining by conventional computer algorithms excessively difficult and unreliable [2,3].

To overcome this problem, many previous studies use machine learning or deep learning methods -in a separate way-: Several major scientific advancements in terms of the clinical applications of ML in rheumatology were achieved in 2019 [4,5]. Iqbal et al. (2021) focused to present game-changing technology of the future in clinics, by connecting biology with Artificial Intelligence and explain how AI-based assistance help oncologist for precise treatment, they also describe the current trends in the context fields of body oncology, cardiac imaging, and neuroimaging, they focused on the description of possible future developments [7]. Alam et al. (2022) present a machine learning approach for Identification of Malignant mesothelioma etiological factors in an imbalanced Dataset [8]. Ghosh et al. constructed an intelligent diagnostic framework for prediction of heart disease, using a disease dataset. They used three machine learning approaches, Decision Tree (DT), K- Nearest Neighbor (KNN), and Random Forest (RF) in combination with different sets of features [9]. Fusco et al. (2021) present an overview and update on Artificial Intelligence (AI) and COVID-19 using chest Computed Tomography (CT) scan and chest X-ray images (CXR) [10]. Kaplan et al. (2020) studies artificial Intelligence and Machine Learning applied in respiratory medicine and potential role in asthma and COPD Diagnosis [6]. Other articles have compared the performance of machine learning methods and artificial intelligence [11–13].

However, these methods remain very limited for several reasons: first, the use of a only one type of model does not provide a good predictions. Second, some models require very powerful calculation tools. Third, require a very deep architecture and finally, the use of traditional machine learning methods have many difficulties in image recognition. For all these reasons, we propose in our article, a hybrid model that combine traditional machine learning techniques and deep learning methods. Our proposal will be motivated by the relevant results found when fitting these models.

In this paper, we introduce a diagnostic AI model, this model allow to separate, diagnose and classify three diseases: Tuberculosis, covid19 and Pneumonia. The model is based on a combination of Deep Learning using the deep SqueezeNet model and Machine Learning: SVM, KNN, Logistic Regression, Naive Bayes and decision tree. The model is applied to a chest X-ray dataset for each type of disease: Covid19, pneumonia, tuberculosis or normal. To train and test the model, we split the image dataset into two training and test subsets to classify between different disease types. The accuracy of this model show its effectiveness in the identification and diagnosing of diseases.

The proposed model in our paper gives better diagnostic results and better distinguishes between these diseases, using chest X-ray images. We summarize the contributions of our paper as shown below:

- Introducing a new model for the diagnosis and rapid classification between the three lung diseases which have almost the same symptoms, and which doctors find difficulty in diagnosis
- This model is an adaptation between classic machine learning and artificial intelligence in the field of health.
- proposal of an easy-to-implement model for embedded and mobile systems that could help health experts to establish a stable diagnosis of these lung diseases.

This article is organized as follows: Sect. 1 presented a brief reviews and related work on the application of Machine Learning to healthcare Big Data, Sect. 2 describes the adaptation of machine learning and deep learning to healthcare data, end then, describes the model proposed as a solution to the variety and dimension of medical data by using massive unstructured data and machine learning, as well as deep learning (the SqueezeNet architecture). Furthermore, a detailed description of the components of the proposed model and the description of the data set (thorax x-ray images). Section 3 presents the experimental results with evaluation criteria. Finally, some conclusions and perspectives are drown in Sect. 4.

2 Adaptation of Machine Learning and Deep Learning to Healthcare Data

Medical and healthcare data can be categorized as Big Data, not only because of its massive volume and size, but also because of its complexity and timeliness. Health information includes patient data, namely electronic medical reports (EHR), diagnostic reports, physician prescriptions, medical images, pharmaceutical records, and medical journal search data [14]. Today, it is imperative to digitize the data generated by the health industries to improve the quality of treatments, the early diagnosis of diseases to avoid risk factors and for better management of the information system in hospitals. There are tremendous opportunities to impact healthcare quality and productivity using big data analytics. Analytical insights gathered from substantial analysis of medical data can likely improve clinical models and realize efficiencies through thoughtful delivery of care [15].

2.1 Machine Learning

The goal of machine learning is to discover knowledge and make smart decisions. It is used for many real-word applications such as recommendation engines, recognition systems, computing and data mining, and autonomous control systems. Machine learning (ML) is the science that gives computers the ability to learn and predict from data without explicit programming. If a computer program can improve its performance by learning from previous experience, it can be said to have learned. Machine learning uses algorithms that allow computers to iteratively learn from data. The ML process is shown in Fig. 1.

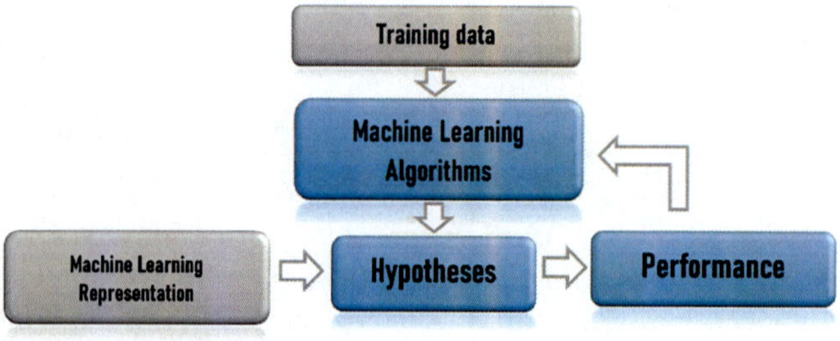

Fig. 1. Machine learning (ML) process

Machine learning in healthcare data can provide several opportunities for the healthcare industry, and experts believe that it can significantly improve healthcare services [16] and contribute to its continuous innovation and Systematic in a Big Data environment.

2.2 Deep Learning in Healthcare Data

Deep Learning algorithms are based on artificial neural networks, and on technologies such as image recognition or robotic vision. Artificial neural networks are inspired by the neurons of the human brain. They consist of several artificial neurons connected to each other. The higher the number of neurons, the deeper the network.

In a deep learning architecture, there is no need for feature engineering anymore. In the context of Big Data, the ability to avoid feature engineering, which is the process of extracting and selecting relevant, informative, and distinctive features from your data, is seen as a great advantage due to the associated challenges to this process. Deep learning uses a hierarchical learning process similar to neural networks to extract data representations from data. It uses multiple hidden layers, and as the data passes through each layer, nonlinear transformations are applied. Since the output of the last layer is simply a transformation of the original input, it can also be used as input for other machine learning algorithms. Deep learning algorithms are therefore an ideal solution to the problem of image classification and recognition. Autoencoders are unsupervised algorithms that can be used for many purposes such as anomaly detection, but in the context of Big Data they usually serve as a preliminary step with neural networks [18]. There are other Deep Learning solutions such as Boltzmann machines [19] which use the stochastic process.

Moreover, the use of an abstract representation also makes these algorithms more flexible and adaptable to the variety of data. Deep learning is an excellent tool for dealing with data heterogeneity. Interestingly, deep learning can be used for both supervised and unsupervised learning [17]. In the context of Big Data, Deep Learning is a great advantage because it makes algorithms less sensitive to veracity challenges such as dirty, noisy and uncertain data [17]. Additionally, its multiple layers of nonlinear transformations address the challenge associated with data nonlinearity. Its depth has been proposed as a way to speed up processing without losing precision [20]. According to the features described, deep learning seems to be well suited to address many of the previously identified challenges such as data feature engineering.

2.3 Evaluation Measure

Table 1. Evaluation measures [22]

N	Metric	Formula
1	Accuracy	$\frac{TP+TN}{TP+FN+TN+FP}$
2	Recall	$\frac{TP}{TP+FN}$
3	F1-Score	$\frac{2*presision*recall}{presision+recall}$
4	Specificity	$\frac{TN}{TN+FP}$
5	Precision	$\frac{TP}{TP+FP}$
6	MCC	$\frac{TP*TN-FP*FN}{\sqrt{(TP+FP)(TP+FN)(TN+FP)(TN+FN)}}$

In order to evaluate the quantitative performance of machine learning and deep learning models, these evaluation measures: Accuracy, Recall, Precision, Specificity and F1 score are calculated statistically from the confusion matrix. Accuracy measures the performance of classification, recall gives the rate of radiological images actually classified among the classes while Precision defines the negative images actually detected. Specificity represents the correctly classified rate of the opposite disease classes. F1-Score is a harmonic mean and gives the combination of recall and precision. The MCC measures the quality of classification performance. According to the confusion matrix, the selected evaluation metrics are defined as follows (Table 1).

The confusion matrix gives the correctly predicted values as well as the wrongly predicted values by the ML models (classifiers) and it is represented as follows (Table 2).

Table 2. The confusion matrix

Confusion matrix	Negative: 0	Positive: 1
Negative: 0	True Negative: TN	False Negative: FN
Positive: 1	False Positive: FP	True Positive: TP

This matrix (is essential to define the different classification metrics such as Accuracy, F1-score or AUC and others measures. TP is true positives representing the number of faulty instances and correctly predicts FP is false positives representing the number of healthy instances but incorrectly predicts. TN is true negatives representing the number of healthy instances and correctly predicts. FN is false negatives representing the number of unhealthy but correctly predicted instances.

3 Experimental Results

3.1 Dataset and Preprocessing

In this work, chest X-ray images were used for the classification of COVID-19, Pneumonia, Tuberculosis and Normal cases. The images are obtained from the Kaggle website which is a web platform that contains the largest Data Science community in the world, this dataset contains 7097 chest X-ray images, divided into four categories of images: 566 images for Covid19, 4265 images for Pneumonia, 691 images for Tuberculosis and 1575 images for normal cases. For the proposed model, we chose a public dataset including chest x-ray images. The aim is to differentiate between similar diseases. The chest X-ray image dataset used in this article is described as follows: For the model training, we used an available chest X-ray dataset. This dataset includes four classes. Figure 2 shows a set of images randomly chosen from the class samples.

We split the dataset into two subsets, a model training set and a test set. The test subset includes 1385 images, or about 20% of the total images available in the main dataset. 5712 images constitute the training subset with a percentage distribution of approximately 80%. Finally, the proposed model was trained and tested to perform the classification between the different types of COVID-19 diseases, pneumonia, tuberculosis or normal cases. All images were visually inspected before using them for training and testing. Low contrast images were identified and preprocessed using a histogram equalization and thresholding operation. It should be noted that image preprocessing techniques have been shown to be effective in the literature for better training machine learning models.

3.2 Proposed Hybrid Model

Machine Learning methods require a structured data or a special pre-processing of data in contrast to deep learning. CNN models achieve high accuracies in

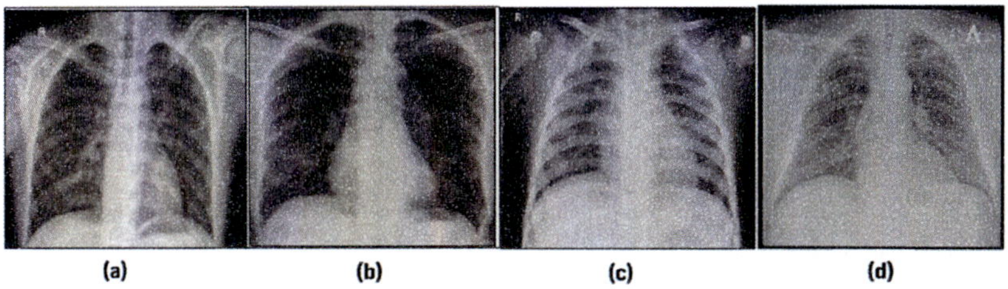

Fig. 2. Chest X-ray: Normal (a), Covid19 (b), Pneumonia (c) and Tuberculosis (d)

image classification problems due to their self-learning capabilities and superior classification results on multi-class problems. However, classic machine learning are not easily adapted to Big Data, especially, in the case of unstructured data and also in the case of the variety of data such as images, texts, audios or videos. This, requires to complete the model with a deep learning approach. The purpose is to extract the images characteristics. Therefore, given these kinds of characteristics, a model based on the deep SqueezeNet architecture and classic machine learning algorithms is proposed in this article for the diagnosis of diseases with similar symptoms Tuberculosis, Pneumonia or COVID-19 from Thorax X-Ray Images The goal is to provide distinctive visual properties as well as a rapid diagnostic system able to classify these properties on new Thorax X-Ray Images and to adapt machine learning algorithms to the variety dimension of Big Data. This method can also be useful for practitioners as they can quickly decide which the treatment plan used in different situations. The proposed model includes six main steps to accomplish the diagnostic procedure of COVID-19, TUBERCULOSIS and PNEUMONIA (Fig. 3)

3.3 Methodology

The proposed framework includes six main steps to accomplish the diagnostic procedure for COVID19, TUBERCULOSIS and PNEUMONIA (Fig. 3).

- Electronic Health Data Repository (a): This is a large collection of quantities of electronic health records (EHRs) of various patients containing data such as patient name, age, gender, images chest X-rays of patients with Pneumonia, Covid-19 or Tuberculosis, etc.
- Data collection (b): In this phase, a subset of data from the electronic repository of health care data is prepared. For example X-ray images of patients with Pneumonia, Covid-19, Tuberculosis or Normal.
- Deep Learning Model (c): After Data Preprocessing, Deep Learning is used in the model to extract features from each image. Indeed, as input the deep learning (DL) models are used to extract the features of the images, by calculating a feature vector for each image. It returns an enhanced data table

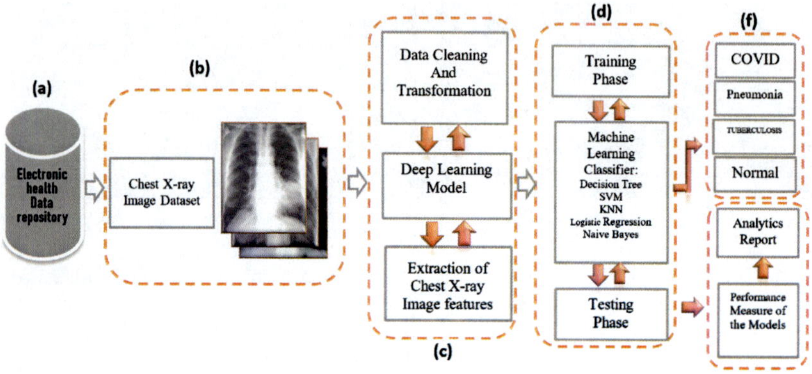

Fig. 3. Proposed hybrid model

with additional columns (image descriptors). In our study, we used a commonly used Deep Learning model, namely SqueezeNet, which is a deep image recognition model. We implemented the SqueezeNet using the weights from the author's pre-trained model.

- Algorithms (Classifiers) of ML (d): Classify diseases with a machine learning classifier. Five traditional machine learning classification methods are tested in the proposed model. In this phase, the models are fed with the features extracted from Deep Learning. Namely that the ML models used are Decision Tree, Support Vector Machine (SVM), K Nearest Neighbor (K-NN), Logistic Regression, decision tree and Naïve Bayes.
- Performance Measurement and Analysis Report (e): Performance analysis of ML models is performed using various metrics such as Accuracy, Precision, Recall and Specificity, F1 Score, etc. extract from the confusion matrix.
- Output (f): Finally, the decision-making of the network with the test phase. The proposed method classifies chest X-ray images into four classes labeled as Pneumonia, Covid-19, Tuberculosis or normal cases (no infection).

The experiment is performed in a Python environment using a workstation with a 6-Core Intel Core i7-8700 processor (3.2 GHz/4.6 GHz), an NVIDIA GeForce RTX 2060 graphics card 6 GB of 32 GB DDR4 RAM. And a 480 GB solid state drive.

As input, deep learning (DL) models are used to extract features from images, calculating a feature vector for each image. In our study, we used the SqueezeNet model [21], which is a deep image recognition model that achieves AlexNet-level accuracy over ImageNet with 50 times fewer parameters. The model is trained on the ImageNet dataset. We have reimplemented the SqueezeNet using the weights from the author's pre-trained model. We use the activations of the flatten 10 pre-softmax layers () as an embedding.

3.4 Results

To assess the efficiency of feature extraction from chest X-ray images. We use the SqueezeNet with the best model parameters for a training process. Classical ML algorithms (LOGISTIC REGRESSION, SVM, KNN, decision tree and NAÏVE BAYES) are trained and tested with training and test subsets of the associated dataset. We use the evaluation measures calculated from the confusion matrix of this classical ML algorithms.

Case 1: Data is Not Augmented (Imbalanced Classes). The Logistic Regression algorithm records a good classification for all classes, the KNN and SVM algorithms exceed more than 90% for the Pneumonia and Tuberculosis classes, and record a classification more than 80% accuracy for both classes (Covid19 and Normal) in the test data. Indeed the normal class distribution, for KNN and SVM algorithms respectively, 12.4% and 24.4% of normal x-ray images are misclassified as pneumonia. This situation shows the negative effect of the unbalanced distribution of the Pneumonia class compared to the Covid19 and Tuberculosis classes.

Table 3. Results of classification of ML algorithms with a test data set (%)

MODELS	AUC (%)	CA (Accuracy) (%)	F1 (%)	Precision (%)	Recall (%)
KNN	97,3	92,8	92,7	92,8	92,8
Tree	81,9	86,1	86	85,9	86,1
SVM	98,4	91,8	91,5	91,7	91,8
NAIVE BAYES	94,7	82,7	83,4	85,9	82,7
LOGISTIC REGRESSION	99	95,3	95,3	95,3	95,3

As shown in Table 3, we can see the detailed classification results of the different classical ML algorithms for the dataset. The logistic regression model with the raw test data set achieves a good prediction rate or good accuracy of 95.30%. While the Accuracy (Precision and Recall) are high (95.30%) which means that the classes are well performed by the model. While the model composed by SqueezeNet and logistic regression shows its robustness and quality of diagnosis of different diseases (Tuberculosis, Pneumonia, Covid19) compared to other models composed with SVM, KNN, Decision Tree and Naïve Bayes.

Case 2: Augmented Data (Classes are Balanced): The second phase of our experiments is to perform the augmented dataset testing process. In this context, we perform an augmentation of the dataset. After preprocessing, the augmented dataset is divided into two sets: training, and test sets. Both sets of data of dataset is 80% training, 20% testing. In this phase, the proposed model is obtained by mixing Deep Learning and Machine Learning (ML) approaches, and it is validated with the separate test dataset.

Table 4. The machine learning classifier ratings measures with an augmented dataset (%)

Models	AUC (%)	Accuracy (%)	F1 (%)	Precision (%)	Recall (%)
KNN	99,5	93,7	93,7	93,7	93,7
Tree	99,4	97,3	97,3	97,3	97,3
SVM	95,5	80,8	80,3	82,5	80,8
NAIVE BAYES	91,7	76	76,6	77,6	76
LOGISTIC REGRESSION	99,6	96	96	96	96

We noticed a huge performance improvement on all classes and for all algorithms except SVM and NAIVE BAYES which experienced a drop in overall accuracy. The Decision Tree and Logistic Regression classifiers recorded better classifications, reaching an overall classification accuracy of 97.30% and 96.00%, respectively. Indeed, there are only 73 chest X-ray images that are misclassified among 2694 images of the test set for the decision tree algorithm, i.e. an error rate of 2.7%. 108 x-ray images from the test set that are misclassified for the Logistic Regression algorithm, i.e. an error rate of 4%. In the proposed diagnostic model, the best model is SqueezeNet from deep learning and Decision Tree from machine learning (ML), the overall accuracy is 97.30%, with an AUC of 99.40%, and an F1 Score, Precision and Recall of 97.30%, which shows that our model is well trained and robust (Table 4).

4 Conclusion and Discussion

In the last decades, artificial intelligence and Big Data are more widely benefiting companies in the healthcare field. Improving the quality of health services helps to provide personalized solutions to particular problems such as diagnosing diseases, predicting epidemics, treating diseases, improving the value of life, increasing avoidable care, etc. This paper concludes that the adaptation of machine learning techniques with big data from the healthcare dataset improves the quality of services. In this study, a hybrid model was developed for the diagnosis of pulmonary inflammation (tuberculosis, Pneumonia, Covid19 and Normal cases). Our contributions are three-fold. First, the adaptation of classic machine learning to Big Data, especially to the dimension of variety of unstructured data, a set of data (thoracic x-ray images), on a large scale to help physicians to diagnose diseases with high similarity. Second, a diagnostic method combining Deep Learning to extract features from chest X-ray images and Classic Machine Learning as a classifier by training the model on a large-scale lung disease dataset. The experimental results show that our method is highly robust and real-time automated based on efficient SqueezeNet architectures in image feature extraction. Third, we study the challenges from classic machine learning to big data in the healthcare field. Therefore, we have demonstrated the utility of SqueezeNet deep networks combined with machine learning to deal with the challenges of the variety of unstructured data (images). The accuracy of our model was 97.30%. The SqueezeNet architecture employed in our model allows feature extraction from images. The experimental results prove the superiority of our proposed model in the healthcare domain. The proposed model shows an excellent classification performances for the diagnosis of Tuberculosis and Pneumonia with chest X-rays.

References

1. Pianykh, O.S., et al.: Improving healthcare operations management with machine learning. Nat. Mach. Intell. **2**(5), 266–273 (2020). https://doi.org/10.1038/s42256-020-0176-3
2. Dash, S., Shakyawar, S. K., Sharma, M., Kaushik, S. : Big data in healthcare: management, analysis and future prospects. J. Big Data **6**(1), 1–25 (2019). https://journalofbigdata.springeropen.com/articles/10.1186/s40537-019-0217-0
3. Hou, R., Kong, Y.Q., Cai, B., Liu, H.: Unstructured big data analysis algorithm and simulation of Internet of Things based on machine learning. Neural Comput. Appl. **32**(10), 5399–5407 (2020). https://doi.org/10.1007/s00521-019-04682-z
4. Guan, Y., et al.: Machine learning to predict anti-tumor necrosis factor drug responses of rheumatoid arthritis patients by integrating clinical and genetic markers. Arthritis Rheumatol **71**(12), 1987–1996 (2019)
5. Kim, K.J., Kim, M., Adamopoulos, I.E., Tagkopoulos, I.: Compendium of synovial signatures identifies pathologic characteristics for predicting treatment response in rheumatoid arthritis patients. Clin. Immunol. **202**, 1–10 (2019)
6. Kaplan, A., Cao, H., FitzGerald, J.M., Iannotti, N., Yang, E., Kocks, J.W.H., et al.: Artificial intelligence/machine learning in respiratory medicine and potential role in asthma and COPD diagnosis. J. Allergy Clin. Immunol. Pract. **9**(6), 2255–2261 (2021). https://doi.org/10.1016/j.jaip.2021.02.014
7. Iqbal, M.J., Javed, Z., Sadia, H., et al.: Clinical applications of artificial intelligence and machine learning in cancer diagnosis: looking into the future. Cancer Cell Int. **21**, 270 (2021). https://doi.org/10.1186/s12935-021-01981-1
8. Alam, T.M., Shaukat, K., Mahboob, H., Sarwar, M.U., Iqbal, F., Nasir, A., et al.: A machine learning approach for identification of malignant Mesothelioma etiological factors in an imbalanced dataset. Comput. J. **65**(7), 1740–1751 (2021). https://doi.org/10.1093/comjnl/bxab015
9. Ghosh, P., Azam, S., Karim, A., Jonkman, M., Zahid, H.M.D.: Use of efficient machine learning techniques in the identification of patients with heart diseases. In: 2021 the 5th International Conference on Information System and Data Mining. Association for Computing Machinery, New York, NY, USA, pp. 14–20 (2021). https://doi.org/10.1145/3471287.3471297
10. Fusco, R., et al.: Artificial intelligence and COVID-19 using chest CT scan and chest X-ray images: machine learning and deep learning approaches for diagnosis and treatment. J. Pers. Med. **11**(10), 993 (2021). https://doi.org/10.3390/jpm11100993
11. Abarda, A., Dakkon, M., Asmi, K., Bentaleb, Y.: Evaluating information criteria in latent class analysis: application to identify classes of breast cancer dataset. Int. J. Data Anal. Tech. Strateg. **13**, 72–87 (2021). https://doi.org/10.1504/IJDATS.2021.114669

12. Abarda, A., Dakkon, M., Azhari, M., Zaaloul, A., Khabouze, M.: Latent transition analysis (LTA): a method for identifying differences in longitudinal change among unobserved groups. Procedia Comput. Sci. **170**, 1116–1121 (2020). https://doi.org/10.1016/j.procs.2020.03.059
13. Abarda, A., Bentaleb, Y., Mharzi, H.: A divided latent class analysis for big data. Procedia Comput. Sci. **110**, 428–433 (2017). https://doi.org/10.1016/j.procs.2017.06.111
14. Shafqat, S., Kishwer, S., Rasool, R.U., Qadir, J., Amjad, T., Ahmad, H.F.: Big data analytics enhanced healthcare systems: a review. J. Supercomput. **76**(3), 1754–1799 (2018). https://doi.org/10.1007/s11227-017-2222-4
15. Kaur, P., Sharma, M., Mittal, M.: Big data and machine learning based secure healthcare framework. Procedia Comput. Sci. **132**, 1049–1059 (2018). https://doi.org/10.3934/bdia.2020005
16. Elayan, H., Aloqaily, M., Guizani, M.: Digital twin for intelligent context-aware IoT healthcare systems. IEEE Internet Things J. **8**(23), 16749–16757 (2021). https://doi.org/10.1109/JIOT.2021.3051158
17. Miotto, R., Wang, F., Wang, S., Jiang, X., Dudley, J.T.: Deep learning for healthcare: status, opportunities and challenges. Bioinform. Briefings **19**(6), 1236–1246 (2018). https://doi.org/10.1093/bib/bbx044
18. L'Heureux, A., Grolinger, K., Elyamany, H.F., Capretz, M.A.: Machine learning with big data: challenges and approaches. IEEE Access **5**, 7776–7797 (2017). https://doi.org/10.1109/ACCESS.2017.2696365
19. Liu, Z., Wang, R., Japkowicz, N., Tang, D., Zhang, W., Zhao, J.: Research on unsupervised feature learning for Android malware detection based on restricted Boltzmann machines. Futur. Gener. Comput. Syst. **120**, 91–108 (2021). https://doi.org/10.1016/j.future.2021.02.015
20. Zhou, L., Pan, S., Wang, J., et al.: Machine learning on big data: opportunities and challenges. Neurocomputing **237**, 350–361 (2017). https://doi.org/10.1016/j.neucom.2017.01.026
21. Lee, H.J., Ullah, I., Wan, W., Gao, Y., Fang, Z.: Real-time vehicle make and model recognition with the residual SqueezeNet architecture. Sensors **19**(5), 982 (2019). https://doi.org/10.3390/s19050982
22. Hemdan, E.E.D., Shouman, M.A., Karar, M.E.: Covidx-Net: a framework of deep learning classifiers to diagnose covid-19 in x-ray images (2020). arXiv preprint arXiv:2003.11055, https://doi.org/10.48550/arXiv.2003.11055

Improving Small Files Processing in HADOOP

Mohamed Eddoujaji[1](✉), Hassan Samadi[1], and Mohammed Bouhorma[2]

[1] Department Doctoral Studies Research Center, National School of Applied Sciences, Agadir, Morocco
m.eddoujaji@gmail.com, hsamadi@uae.ac.ma
[2] Department Doctoral Studies Research Center, Faculty of Sciences and Techniques, Meknes, Morocco
abouhorma@uae.ac.ma

Abstract. The rapid evolution of data has created a quantitative explosion of digital data which has pushed researchers to adopt new ways of analysis suitable for large volumes of data. When we talk about Big data we are talking about a very voluminous data that no traditional information management tool can analyze.

In fact, we create around 2.5 quintillion bytes of data daily [2]. The data to be processed is of multiple nature, database, social media, emails, documents, structured, unstructured data.

As we presented a new approach to the management of small files in a Hadoop platform [1], we will continue to improve the work on this component and share the results of our research and our implementation with you.

Keywords: Data · Big data · Distributed systems · Heterogeneous systems · Hadoop distributed file system · Distributed storage · Distributed computing · SequenceFile · MapFile

1 Introduction

1.1 A Subsection Sample

The port transport industry has stunned world trade by a massive amount which accounts for around 90% of total trade. The shipping industry generates millions of data points daily from seaports, social media feeds, ship and package transfer, logistics, couriers and more [12]. Big Data helps port transportation companies improve their potential by gathering data and analyzing it to find valuable and useful information. Big Data also makes it possible to monitor the movements of ships on the bridge by capturing the data [13]. It also helps port transport companies find the shortest routes to deliver ships. One of the main problems facing the shipping industry is the misplacement of ships on their way from source to destination. This problem has led the industry to turn to smarter technologies like data analytics.

As explained in our first paper, there are many fields that produce tremendous numbers of small files continuously such as analysis for multimedia data mining [3], astronomy [4], meteorology [5], signal recognition [6], climatology [7, 8], energy, and Elearning [9] where numbers of small files are in the ranges of millions to billions. For instance,

M. Ben Ahmed et al. (Eds.): SCA 2022, LNNS 629, pp. 837–852, 2023.
https://doi.org/10.1007/978-3-031-26852-6_77

Facebook has stored more than 260 billion images [10]. In biology, the human genome generates up to 30 million files averaging 190 KB [11] (Fig. 1).

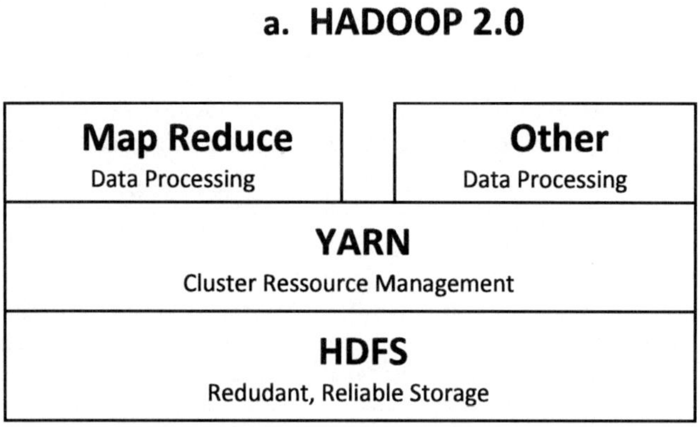

Fig. 1. HADOOP 2.0 main components

1.2 MapReduce

MapReduce is the data processing layer of Hadoop. It deals with large structured and unstructured data stored in HDFS. MapReduce also processes a huge amount of data in parallel. It does this by dividing the work (submitted work) into a set of independent tasks (sub-work). MapReduce works by dividing the processing into phases: Map and Reduce [14, 15].

- Map - This is the first phase of processing, where we specify all the complex logic code. Responsible for the distribution of tasks to the different machines of the cluster [16].
- Reduce - This is the second phase of treatment. Here we are specifying light processing like aggregation/summation. Reduces all machine results into a single summary [17].

It is composed of:

- JobTracker: Node that manages the resources and tasks of a system.
- JobHistoryServer: manages the tasks already accomplished.
- TaskTracker: Agent deployed in each machine, for Map and Reduce tasks (Fig. 2).

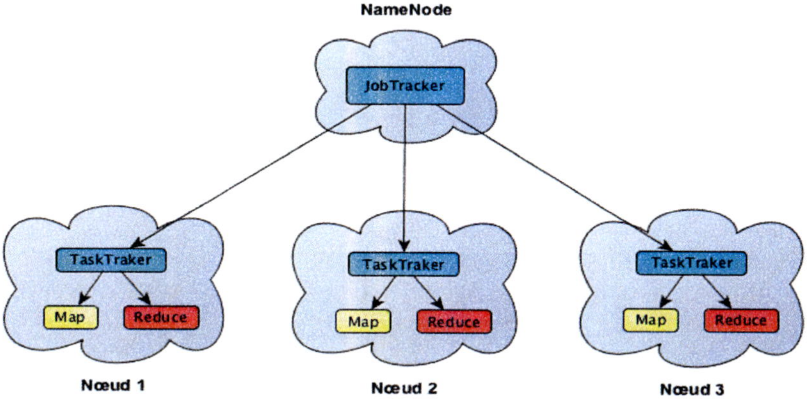

Fig. 2. TaskTracker and JobTracker in a cluster

2 HDFS – Reliable Storage

Due to its massive capacity and reliability, HDFS is a storage system that is very suitable for Big Data. In combination with YARN, this system increases the data management capabilities of the HDFS Hadoop cluster and thus enables the efficient processing of Big Data. Among its main features, there is the possibility of storing terabytes, even petabytes of data [19].

The system is capable of handling thousands of nodes without the intervention of an operator. It allows simultaneous benefits of parallel computing and distributed computing. After a modification, it allows to easily restore the previous version of a data [20].

HDFS can be launched on commodity hardware, which makes it very tolerant to errors. Each piece of data is stored in several places, and can be retrieved under any circumstances. In the same way, this replication makes it possible to fight against the potential corruption of the data [18] (Fig. 3).

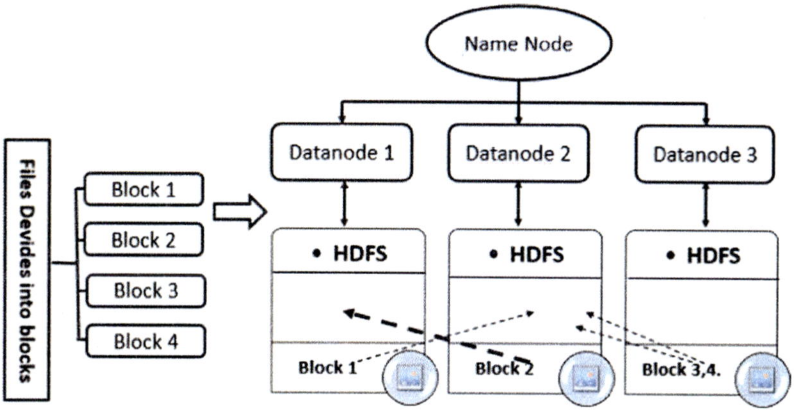

Fig. 3. Storage & Replication of blocks in HDFS

2.1 Kubernetes

Kubernetes was first released in June 2014 and is written in Go. The project was started by Google, which made it open source, and is based on its experience of operating containers at scale, whose objective is to automate development tasks, standardize interfaces based on modular APIs and easy to integrate and operate [21] (Fig. 4).

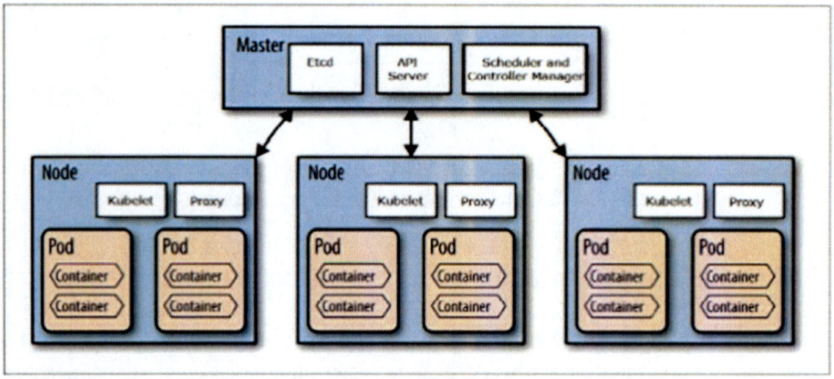

Fig. 4. Hight level architecture

Let's go over the main parts of this high-level architecture [22]:

Kubernetes Master: Master is responsible for maintaining the desired state of the cluster. It manages all cluster nodes. As we can see, the Master is a collection of three processes:

❖ etcd: This component stores the configuration data to which the API server of the master.

Kubernetes can be accessed through a simple HTTP or JSON API.

❖ kube-apiserver: This is the service that manages the entire cluster, including the processing of REST operations, validation and updating of Kubernetes objects, authentication and authorization.

❖ kube-controller-manager: This is the daemon that integrates the basic control loop delivered with Kubernetes, making the necessary changes to match the state.

current to the desired state of the cluster.

❖ kube-scheduler: This service monitors unscheduled pods and links them to working nodes required resources and other constraints.

Kubernetes Nodes: The nodes of a Kubernetes cluster are the machines that run our containers. Each node contains the services needed to run the containers:

❖ kubelet: This is the main agent of the node who ensures that the containers described in the PodSpecs provided by kube-apiserver are running and healthy.

❖ kube-proxy: This is the network proxy executed on each node and which performs a simple transmission of TCP, UDP, SCTP or round-robin transmission over a set of backends.

❖ Runtime container: This is the runtime engine where containers inside pods are executed. There are several container runtime engines for Kubernetes, the most popular of which is the Docker runtime engine.

The following list shows other common terms associated with Kubernetes:

❖ Pods: Kubernetes deploys and schedules containers in groups called pods. Containers in a pod run on the same node and share resources such as file systems, kernel namespaces, and an IP address.

❖ Deployments: These building blocks can be used to create and manage a pod group.

❖ Services: These are endpoints that can be named and can be connected to pods using tag selectors.

❖ Labels: These are key-value pairs attached to objects. They can be used to find and update multiple objects in a single set.

❖ Persistent Volumes: To store data permanently, Kubernetes uses persistent volumes which is a storage element in the cluster.

3 Hadoop YARN vs Kubernetes

The design of Kubernetes allows multiple schedulers to operate in the cluster. Each scheduler manages the resources within its own pods. Kubernetes Scheduler will end low priority and starvation containers to improve resource utilization.

Kubernetes containers can integrate with an external storage system like S3 to provide data resiliency [25].

YARN provides global level resource management such as capacity queues to partition physical resources into logical units. Each logical unit can be assigned a percentage of the cluster resources. The capacity resource sharing system is designed to ensure that resources are available for business priorities instead of using all available physical resources. YARN scores more points when it comes to security. There are more security features in Kerberos, access control for privileged/unprivileged containers, trusted docker images, and placement policy constraints. Most Docker-related securities are closed by default, and the system administrator must manually activate flags to grant more power to the containers [23].

YARN was just a mechanism for sharing resources, while Kubernetes is a complete platform.

For high availability requirements that require more than one node, at least three nodes must be up and running. These nodes are not shared for other workloads, however at the kubernetes level we need a working cluster that can be shared for various workloads.

For YARN It is necessary to install various components on multiple nodes and these components are required for high availability configurations. But for the kubernetes only the Kubernetes cluster must be operational, the code must be part of the Docker image.

Using Kubernetes as an orchestration layer. It hosts various services, including big data tools (Apache Spark or Presto), data science and AI tools (Jupyter, TensorFlow, PyTorch, etc.) and any other application or data microservice (Fig. 5).

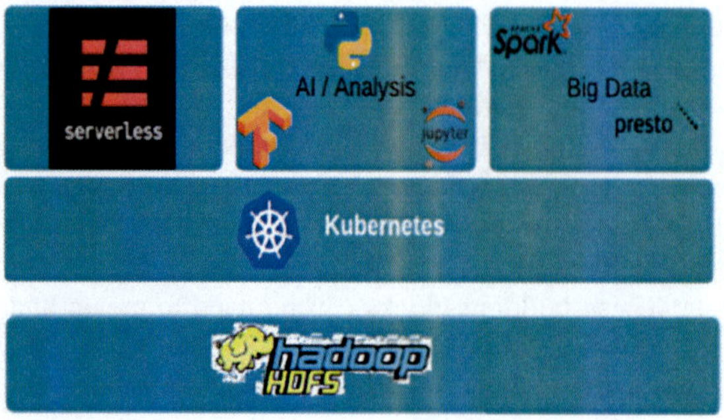

Fig. 5. Kubernetes based on HDFS storage management

4 Final Architecture

4.1 Article Sections

The following parts in this article, will be organized as follows, Sect. 2 lists the existing solutions in the related literature. Section 3 present the proposed approach. Section 4 is dedicated to present our experimental works and results. Finally, and to conclude this research, Sect. 5 for conclusion and expectation.

5 Related Work

Kubernetes containers can integrate with an external storage system like S3 to provide data resiliency.

YARN provides global level resource management such as capacity queues to partition physical resources into logical units. Each logical unit can be assigned a percentage of the cluster resources. The capacity resource sharing system is designed to ensure that resources are available for business priorities instead of using all available physical resources. YARN scores more points when it comes to security. There are more security features in Kerberos, access control for privileged/non-privileged containers, trusted docker images, and placement policy constraints. Most Docker-related securities are closed by default, and the system administrator must manually activate flags to grant more power to the containers.

YARN was just a mechanism for sharing resources, while Kubernetes is a complete platform.

For high availability requirements that require more than one node, at least three nodes must be up and running. These nodes are not shared for The architects below shows the resource manager that we have chosen from where Kubernetes, which is in fact a master-slave type architecture in which certain components (components Master) run the cluster and other components (node components) run the loads of application work (containers) according to the decisions of the Master components.

Master components manage the state of the cluster. This includes accepting customer requests, the scheduling of containers and the execution of control loops to change the current state of the cluster to the desired state, as well as the networking and communications of the entire cluster. These components are the API server (kube-apiserver) which manages requests from Workers nodes, the scheduler (kube-scheduler), which schedules pods on nodes based on of the use of resources as well as we have the controller manager (kube-controller manager) which executes the various controllers of our cluster. And in the end etcd which is a store open-source key-value data that can be accessed by all nodes in the cluster) [27].

Each node (VM) hosts part of our distributed application does so using Docker or one can use a similar container technology, like Rocket from CoreOS. In our we have implemented Apache Spark, HDFS in both nodes, and for NiFi, PostgreSQL, Kafka are deployed at node level 1, and for NiFi zookeeper deployed at node level 2, and for Grafana installed at node level 2. The nodes also run two software additional: kube-proxy, which gives access to our running application, and Kubelet, which receives the commands from the k8s control plane. The nodes run Calico, a network of containers supported by etcd [28].

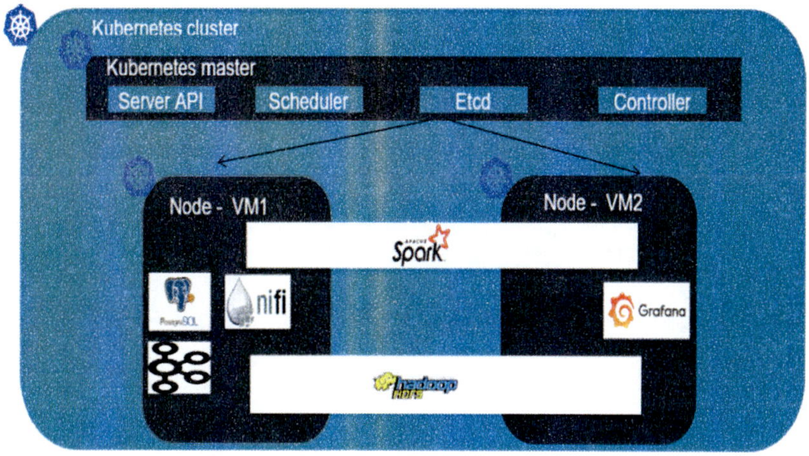

Fig. 6. The operational architecture of the project

5.1 The Architecture of the Software Layers of the Platform

The above figure schematizes the software layers of the platform in the form of phases and level of each phase we will opt for a set of tools.

❖ The data acquisition phase (Data ingestion): This phase is in order to collect data received from multiple sources and automate the flow of data between different systems. For this we have chosen to use Apache NiFi to configure the data movement, and Apache Kafka as a distributed messaging system serves to process and transfer a large volume of data.

❖ The batch processing phase: This phase concerns the sending of the datasets in batch. We chose to perform this batch process with Apache NiFi to collect the source data and store it in HDFS, then process it according to Apache Spark.

❖ The streaming process phase: This phase concerns the sending of recordings of continuous data. We have chosen to carry out this streaming process with Kafka, Apache and Spark Streaming.

❖ The data storage phase: This phase is used to store data in systems database management SGBDRO under PostgreSQL.

❖ The data visualization phase: The last phase of the process. To view the data we opted for the Grafana visualization software (Fig. 7).

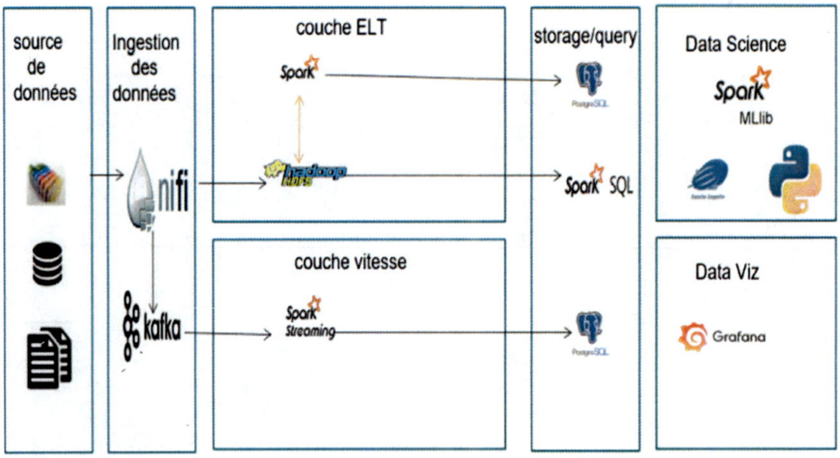

Fig. 7. Platform software layers

6 Proposed Work

6.1 Little Reminder

Hadoop is designed for powerful management of large files, unlike the management of large volumes of small files where the technique always requires improvements, in fact, Hadoop passes each small file to a map () function, which creates a large number of mappers and therefore the solution is not very effective for this kind of traffic. For example, the 100,000 files smaller than 2Mo will need 100 00 mappers, which is very inefficient, which could be a problem throughout the system [14, 27].

The main objective of these improvements and new approaches is on the one hand to solve this problem and on the other hand it is to accelerate the execution of operations (reading and writing) Hadoop. This solution is to combine small files into larger files, this will reduce the number of map () functions that are executed and thus significantly improve performance [24, 28].

7 The Proposed Approach for Small File Management

Currently, it's well known that small files can decrease dramatically the performance of Hadoop clusters. Previous solutions workaround this problem by packaging small files in different formats. Those formats are saved transparently in HDFS as they can be divided into blocks with no specific constraints [19]. Though the performance of MapReduce jobs can be greatly improved based on the way those small files are packaged, none of the adopted mechanisms take in consideration how to organize those small files during the merging phase. The core idea behind our approach is to store files when a client starts a stream that contains small files, combined if relevant, with other client streams into a large file within the same block, and organize them later in an efficient way that we can prefetch the most probable called files first. This was achieved by using a Hadoop File Server Analyzer, see "Fig. 6" (Fig. 8).

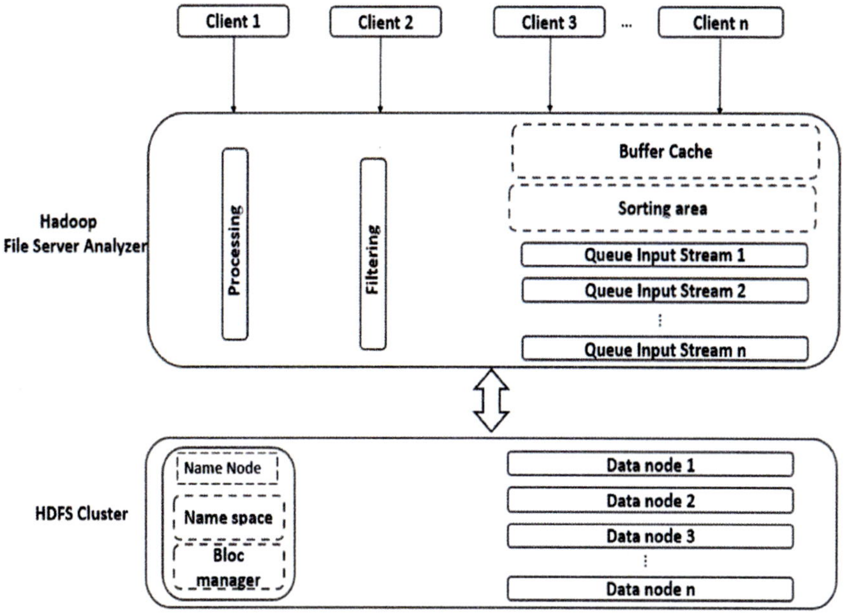

Fig. 8. Hadoop file server analyzer

The first approach proposed in our previous paper, was the installation of the file analyzer server on a hadoop and YARN ecosystem, we noticed a considerable improvement in the performance of the writes of the reads and even in terms of memory consumption.

The next phase is to improve this approach by replacing the YARN component with the Kubernetes container solution to benefit from advantages such as agility, scalability and the use of APIs to communicate with the other components [29].

8 Experimental

8.1 Experimental Environment

The proposed method in this paper is compared with the original HDFS about the usage of NameNode memory, and the performance of MapReduce jobs, during sequential and selective file access. We did a simulation on the Hadoop2.4.0, our cluster consists of:

[1] one NameNode,
[2] 3.10 GHz clock speed,
[3] 8 GB of RAM and
[4] a gigabit Ethernet NIC,
[5] four DataNodes.

All the nodes offer 500 GB Hard Disk, and they are deployed on Ubuntu 14.04. The replication factor is kept as the default value 3 and the block size of HDFS is chosen as 64 Mb. The experimental datasets are basically standard autogenerated, also we can use the public Datasets on [26].

Phases Algorithms

Input:

sFile_List : list of small files to be combined
sFile_Dir: Directory containing small files to be combined
CombinedFile: Name of combined file.
Block_Size: desired block size

Output:

(vfcp
File_Index: index file for small files belonging to CombinedFile.

```
Initialize sFile_Map
if sFile_List is provided as input
for each file fi in File_List
compute code(fi.name)
add fi to sFile_Map with key: code,
values:name(fi), length(fi)
end for
else if sFile_Dir is given as input
for each file f in the directory tree at sFile_Dir
compute code(fi.name)
add f to sF_Map with key: code,
values:name(f), length(fi)
end for
end if
if CombinedFile exists
open CombinedFile for append
LocalIndex= LocalIndex(CombinedFile );
numBlock = length(combinedFile) / Block_Size
cur_offset = length(CombinedFile) % Block_Size
else
create CombinedFile; open CombinedFile for write
Initialize File_Index; set cur_offset= 0, numBlock=0
end if
for each file with code fi in sFile_ap in order
of
increasing length(fi)
name = name(fi), len=length(fi); combinedIn=CombinedFile
 if cur_offset + len > Block_Size
   numBlock++,
   start = numBlock*Block_Size, end= start + len,
  curr_offset = len ++
 else if cur_offset + len = Block_Size
start= numBlock*Block_Size + cur_offset, end= start + len,
curr_offset=0, numBlock++
else
start= numBlock*Block_Size + cur_offset,
end = start + len,
curr_offset=end ++
endif
append f to CombinedFile
Insert key fi into LocalIndex with name, len, start, end, combinedIn as values
end for
close CombinedFile
return File_Index
```

9 Results

We will use the same number and same sizez of files used previously (Fig. 10):

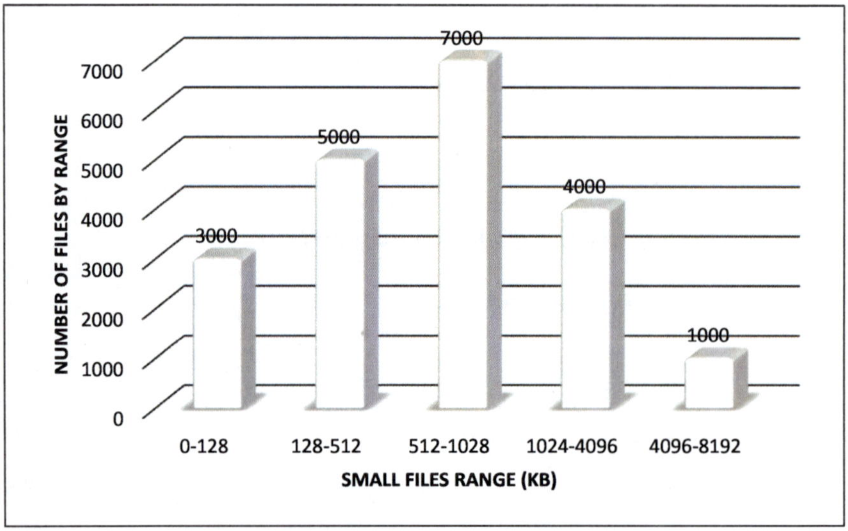

Fig. 9. Distribution of File Sizes in our experiment

9.1 Memory Consumption

Table 1. Comparison of the NameNode memory usage

		Time consumption (s)		
Dataset	Number of small files	Normal HDFS	HFSA algorithm	Enhanced hadoop
1	5000	980	268	300
2	10000	1410	390	540
3	15000	2010	530	600
4	20000	2305	620	710

As illustrated in the figure above, we notice that the memory consumption is too close to the consumption by the previous approach (Table 1).

To confirm if the new architecture is valid to move forward on the subject, we will measure in the following paragraphs the performance and the response times in writing and in reading.

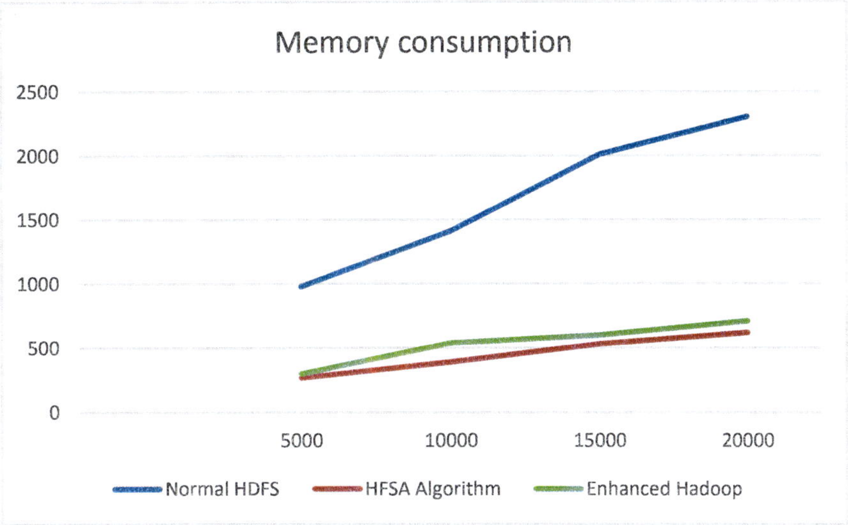

Fig. 10. Figure 9 Memory usage by NodeName

10 Performances Comparison

10.1 Writing Test

Results of writing time are shown in this Fig. 11:

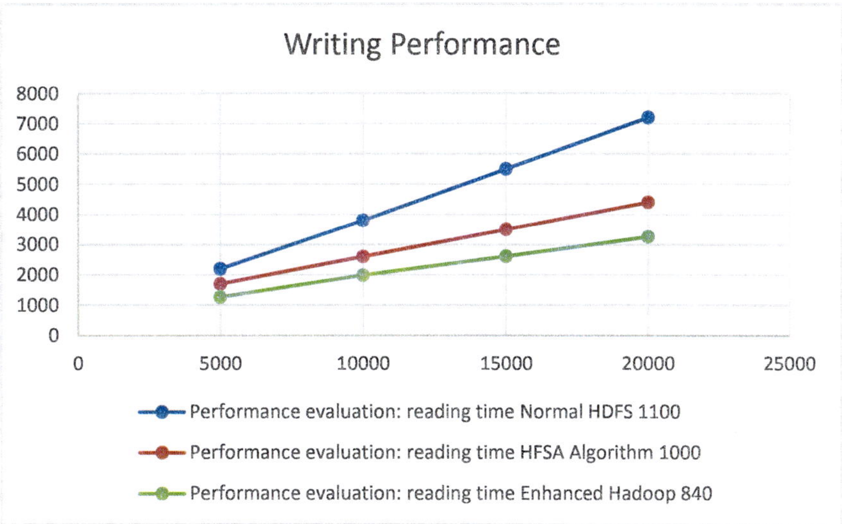

Fig. 11. Performance evaluation: writing time

In our previous approach, we remarked that when the number of files is important the gain in writing performance becomes more important, applying our new improvement, we find that the gain in write performance becomes more and more important as the number of files is large.

For example, we have a performance of 33% for 2500 files beside 16% for 20,000 files.

Through the above comparison, it is proved that the proposed approach can effectively improve the efficiency of file writing.

10.2 Reading Test

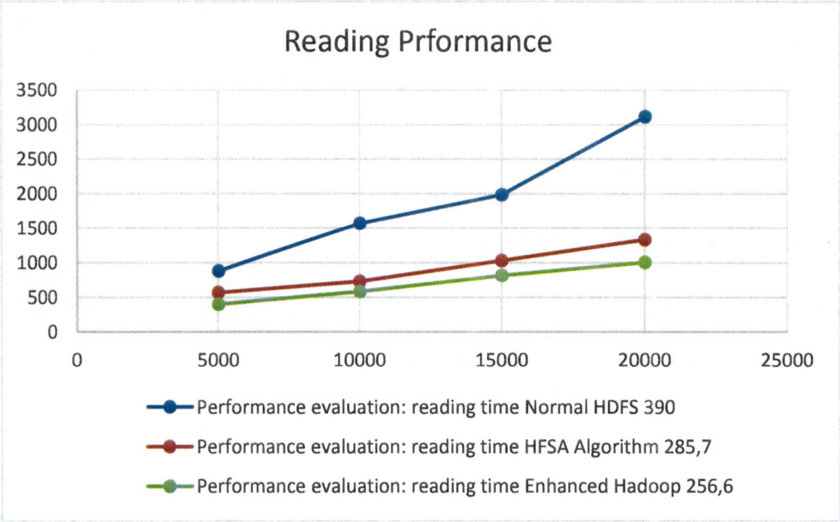

Fig. 12. Performance evaluation: reading time

The average sequence reading time of HFSA is 788,94 s, the average read time of original HDFS is 1586,58 s and the average read of enhanced HDFS using the Kubernetes is 610,9s! (Fig. 12).

Through the above comparison, it is proved that the proposed approach can effectively improve the efficiency of file reading too.

Applying the new enhanced approach, we had significantly enhanced the performance of writing and reading process.

11 Conclusion and Future Works

The work carried out and explained on this paper shows us that we have succeeded in improving the reading and writing performance of the classic HADOOP platform and present a portable and usable model for the port industry where we use small files whose sources are IOTs, PDAs, tablets and mobile phones.

The next step is to continue the improvements using:

Graph theory techniques in the first phase, especially the Dijkstra and Bellman-Moore algorithms, this first phase will be the initial basis that will feed our A * algorithm that we will use in our approach to Artificial Intelligence.

References

1. https://www.forbes.com
2. https://github.com/coreos/flannel
3. Hadoop official site. http://hadoop.apache.org/
4. https://linuxcontainers.org/
5. https://www.docker.com/
6. Eddoujaji, M.: Data processing on distributed systems Storage challenges. NISS21 (2021)
7. Ren, X., Geng, X., Zhu, Y.: An algorithm of merging small files in HDFS. In: 2019 2nd International Conference on Artificial Intelligence and Big Data (ICAIBD)
8. Zhang, Q., Liu, L., Pu, C., Dou, Q., Wu, L., Zhou, W.: A comparative study of containers and virtual machines in big data environment. In: IEEE 11th International Conference on Cloud Computing (CLOUD) (2018)
9. Bhimani, J., Yang, Z., Leeser, M., Mi, N.: Accelerating big data applications using lightweight virtualization framework on enterprise cloud. In: 2017 IEEE High Performance Extreme Computing Conference (HPEC), pp. 1–7. IEEE (2017)
10. Nukarapu, T., Tang, B., Wang, L., Lu, S.: Data replication in data intensive scientific applications with performance guarantee. IEEE Trans. Parallel Distrib. Syst. **22**(8), 1299–1306 (2011)
11. Gao, Y., Zheng, S.: A metadata access strategy of learning resources based on HDFS. In: Proceeding International Conference on Image Analysis and Signal Processing (IASP), pp. 620–622 (2011)
12. Xie, J., Yin, S., et al.: Improving MapReduce performance through data placement in heterogeneous Hadoop clusters. In: 2010 IEEE International Symposium on Parallel & Distributed
13. Mackey, G., Sehrish, S., Wang, J.: Improving metadata management for small files in HDFS. In: IEEE International Conference on Cluster Computing and Workshops (CLUSTR), pp. 1–4 (2009)
14. Vorapongkitipun, C., Nupairoj, N.: Improving performance of smallfile accessing in Hadoop. IEEE International Conference on Computer Science and Software Engineering (JCSSE), pp. 200–205 (2014)
15. Patel, A., Mehta, M.A.: A Novel Approach for Efficient Handling of Small Files in HDFS, pp. 1–14 (2015)
16. Liu, X., Han, J., Zhong, Y., Han, C., He, X.: Implementing WebGIS on Hadoop: a case study of improving small file I/O performance on HDFS. In: IEEE International Conference on Cluster Computing and Workshops, pp. 1–8 (2009)

17. Dong, B., Qiu, J., Zheng, Q., Zhong, X., Li, J., LiA, Y.: Novel approach to improving the efficiency of storing and accessing small files on hadoop: a case study by powerpoint files. In: IEEE International Conference on Services Computing, pp. 1–5 (2010)
18. Carns, P.H., Ligon, III, W.B., Ross, R.B., Thakur, R.: Pvfs: A parallel file system for linux clusters. In: Proceedings of the 4th Annual Linux Showcase and Conference, pp. 317–327. USENIX Association
19. Liu, X., Han, J., Zhong, Y., Han, C.: Implementing WebGIS on Hadoop: A Case Study of Improving Small File I/O Performance on HDFS. IEEE, pp. 1–8 (2009)
20. Yuelong, Z., Xiaoling, X., Yongcai, C.: A strategy of small file storage access with performance optimization. J. Comput. Res. Dev. (2012)
21. W. Cheng, M. Zhou, B. Tong, J. Zhu, "Optimizing Small File Storage Process of the HDFS Which Based on the Indexing Mechanism", 2nd IEEE International Conference on Cloud Computing and Big Data Analysis, 2017
22. Dev, D., Patgiri, R.: HAR+: "Archive and metadata distribution! Why not both? In: IEEE International Conference on Computer Communication and Informatics (ICCCI), p. 16 (2015)
23. Gohil, P., Panchal, B., Dhobi, J.S.: A novel approach to improve the performance of Hadoop in handling of small files. In: International Conference on Electrical, Computer and Communication Technologies (ICECCT), p. 15 (2015)
24. Bende, S., Shedge, R.: Dealing with small files problem in hadoop distributed file system. In: 7th International Conference on Communication, Computing and Virtualization 2016 (2016)
25. Implementing WebGIS on Hadoop: A Case Study of Improving Small File I/O Performance on HDFS
26. https://dataottam.com/2016/09/09/3-solutions-for-big-datas-small-files-problem/
27. Dong, B., et al.: Correlation based file prefetching approach for Hadoop. In: IEEE Second International Conference on Cloud Computing Technology and Science (CloudCom), pp. 41–48 (2010)
28. Alapati, S.R.: Expert Hadoop Administration: Managing, Tuning, and Securing Spark, YARN, and HDFS. Addison-Wesley, America, pp. 300–310 (2016)
29. Luksa, M.: Kubernetes in Action, pp. 16–17. Manning Publications, America (2018)

Hierarchical Classification Method Based on Weighted Barycenter to Resolve the Problem of Group Separation

Sara Jeddin(✉) and Youssef Bentaleb

Engineering Sciences Laboratory, National School of Applied Sciences, Ibn Tofail University, Kenitra, Morocco
{sara.jeddin,youssef.bentaleb}@uit.ac.ma

Abstract. In this paper we propose an improvement of the Barycenter method; this will be done by introducing the notion of weighted center of gravity and by measuring distances and similarities to address the problem of separation of classified groups.

Keywords: Classification · Clusters · Distance · Weighting · Weighted Barycenter

1 Introduction

Classification methods have been used in a wide range in data analysis (Perveen et al., 2016; Umadevi and Marseline, 2017). In response to classification problems, many methods exist, the hierarchical classification known as "hierarchical cluster analysis" is one of the classification methods used to group data on the basis of their similarity (Scheibler & Schneider, 1985). The aim is to identify homogeneous classes within a group of individuals (Lorr, 1983). In practice, hierarchical classification has been used in a number of fields, including the classification of certain mental illnesses (Skinner, 1981).

In data mining there are generally two approaches to performing hierarchical classification: agglomerative hierarchical clustering [8] and divisive hierarchical clustering [8].The method of agglomeration, is that which holds here our attention since it is the most used in practice [10], it consists on starting with n group each containing a single observation and at each step, the two most similar groups are merged and at the end of n steps, a single group containing all n observations is obtained.

In this approach we will choose the aggregation method to calculate the distance between the groups defined by the CAH. Several aggregation methods exist in particular the Single linkage method, complete linkage method [9], ward method [9], Barycenter method [9] and other methods; however, these methods do not permit the separation of individuals grouped by the CHA.

In this article we will refer to the Barycenter method as an aggregation method but with introducing the notion of weighting barycenter in order to solve the problem of group separation.

M. Ben Ahmed et al. (Eds.): SCA 2022, LNNS 629, pp. 853–858, 2023.
https://doi.org/10.1007/978-3-031-26852-6_78

2 Agglomerative Hierarchical Clustering (CHA)

The agglomerative hierarchical clustering starts from singletons and proceeds by successive aggregations, this approach consists of iteratively grouping the closest individuals by gradually building dendrogram which shows the hierarchical relationship between the clusters and groups all individuals into one class.

In each iteration the distance between an individual and a group is calculated, also the distance between two groups.

To do this, it will be necessary to determine the distance between two groups identified by the CAH, for which several aggregation or fusion methods are proposed.

2.1 Position of the Problem

We consider n individuals $\Omega = \{\omega_1, \omega_2, ...\omega_n\}$ randomly extracted from a population.

For each of them, we have p values of p variables $X_1, ..., X_p$.

The characters studied are quantitative.

If we consider a partition in K groups with k center of gravity $(\mu_1, ..., \mu_k)$, we can decompose the total variability of the point cloud noted I_T such as:

$$I_T = \sum_{i=1}^{n} \sum_{i \in k} d^2(x_i, \mu_k) + \sum_{k=1}^{K} n_k d^2(\mu_k, x_G) \tag{1}$$

$$I_T = I_w + I_B; k = \{1, ...K\}$$

where:

- I_T: represents the total variability of the point cloud: it is constant for a fixed dataset.
- I_w: (within variance) represents the dispersion of points around their center.
- I_B: (between variance) represents the separability of groups: to maximize.
- d is a given distance.

Groups A and B are subsets of the data set observations.

The objective is to minimize the increase in intra group variance in each group merge step, the formula is as follows:

$$I_w(A, B) = I_w(A) + I_w(B) \tag{2}$$

$$I_w(A, B) = \sum_{i \in A} d^2(i, \mu_A) + \sum_{i \in B} d^2(i, \mu_B)$$

$$I_w(A \cup B) = \sum_{i \in A \cup B} d^2(i, \mu_{AB}) \tag{3}$$

Then we look for the distance $d^2(A, B)$ between groups A and B, which minimizes the increase in intra-group variance.

However, the hierarchical ascending classification has limits; we note the heaviness of the calculations as soon as we have a large number of data, also several choices of aggregation are proposed and therefore the results differ according to the method of aggregation chosen.

3 Method of Classification by Barycenter

3.1 Barycenter Method (Centroid Method)

The barycenter method is an agglomerative clustering method wish use the barycenter to determine the clusters of a given partition.

We note A and B two classes, or elements of a given partition,

The problem is to define d (A; B), distance between two elements of a partition of Ω.

The CAH algorithm seeks, at each stage, to constitute classes by aggregating the two elements closest to the stage partition previous [5].

At each step of the algorithm, it is necessary to update the table distances (or dissimilarities).

The distance between two classes A and B consists in measuring the distances between clusters $\overline{x}_A$ and $\overline{x}_B$, it is defined as follows:

$$d(A, B) = d(\overline{x}_A, \overline{x}_B) \tag{4}$$

where

$$\overline{x}_A = \frac{1}{n_a} \sum_{i \in A} x_i \tag{5}$$

and

$$\overline{x}_B = \frac{1}{n_b} \sum_{i \in B} x_i \tag{6}$$

- n_a represents the number of individuals in class A
- n_b represents the number of individuals in class B
- Moreover, this method of aggregation does not allow the separation of the obtained classes.

4 Enhanced Barycenter Method

We propose an improvement of the Barycenter method by introducing the concept of weighting.

For all $i \in \{1, \ldots n\}$ the individual ω_i can be represented in R^P by a point m_i of coordinates $(x_{1,i}, \ldots, x_{p,i})$.

We call cloud of points the graphical representation of all these points, it is noted $N = \{m_1, \ldots, m_n\}$.

4.1 Weighted Center of Gravity

We note by $p_{j,k}$ the weighting of the variable X_j relating to an observation k.

The point G_A is the weighted center of gravity of a given class A.

We note by $\overline{x}_j$ the weighted center of gravity of the coordinates $(\overline{x}_1 ... \overline{x}_p)$:

$$\overline{x}_j = \frac{1}{n} \frac{\sum_{k=1}^{p} p_{j,k} x_{j,k}}{\sum_{k=1}^{p} p_{j,k}} \tag{7}$$

With:

$$\sum_{k=1}^{p} p_{j,k} \neq 0$$

4.2 Calculation of Distances Between Classes

The calculation of the distance between two classes A and B is equivalent to calculating the distances between the weighted barycenter of clusters, as follows:

$$d(A, B) = d(G_A, G_B) \tag{8}$$

$$d(A, B) = d(\overline{x}_A, \overline{x}_B)$$

where

$$\overline{x}_A = \frac{1}{n_A} \frac{\sum_{j \in A} p_{j,k} x_j}{\sum_{j \in A} p_{j,k}} \tag{9}$$

And

$$\overline{x}_B = \frac{1}{n_B} \frac{\sum_{j \in B} p_{j,k'} x_j}{\sum_{j \in B} p_{j,k'}} \tag{10}$$

Euclidean Distance Case. We note A and B two classes, or elements, of a given score.

We consider that the data are in the form of a matrix of euclidean distances (n x n) of individuals 2 to 2.

We note:

- $p_{j,k}$ and $p_{j,k'}$ the weights for the two classes A and B.
- $d_{i,j}$ The distance between any two individuals i and j.

The square of the distance between two barycenter is calculated from the matrix of distances between individuals 2 to 2:

$$d^2(g_A, g_B) = \frac{1}{\sum\limits_{i\in A, j\in B} p_{j,k}p_{j,k'}} \sum_{i\in A, j\in B} p_{j,k}p_{j,k'}d_{i,j}^2 \tag{11}$$

The weighted inter-class inertia measures the separation between the weighted classes $p_{j,k}$

$$I_B = \sum_{k=1}^{K} p_{j,k}d^2(g_A, g_B) \tag{12}$$

5 Conclusion

In this paper, we proposed a solution to the problem of separation of classified groups. Knowing that the CAH algorithm seeks, at each stage, to constitute classes by aggregating the two elements closest to the previous stage partition. In each iteration we calculate the distance between an individual and a group, also the distance between two groups.

The Barycenter method consists of calculating the distance between classes by measuring the distances between the cluster's barycenter. This method of aggregation does not allow to separate the obtained classes, the weighted Barycenter method proposed here is inspired by that of the classical Barycenter and aims to improve the existing method by introducing a weighting factor (a weight) for each of the classified individuals and therefore it makes it possible to separate each of the classified elements.

References

1. Forestier, Germain: Connaissances et clustering collaboratif d'objets complexes multisources, Thèse de doctorat de l'Université de Strasbourg, France (2010)
2. Rokach, L., Maimon, O.: Clustering Methods, Data Mining and Knowledge Discovery, Handbook, pp. 321–352. Springer, US (2005)
3. Juditsky, A., Lan, G., Cardot, H., Rivoirard, V., Picard, F.: Classification non supervisée et sélection de variables dans les modèles mixtes fonctionnels. Thèse de doctorat de l'Université de Grenoble, Applications à la biologie moléculaire (2014)
4. Diday, E.: The dynamic clusters method in non-hierarchical clustering. Int. J. Comput. Inform. Sci. **2**(1), 61–88 (1973)
5. Macqueen: Some methods for classification and analysis of multivariate observations. In: 5-th Berkeley Symposium on Mathematical Statistics and Probability, pp. 281–297 (1967)
6. Breiman, L., Friedman, J.H., Olshen, R.A., Stone, C.J.: Classification and Regression Trees. Chapman & Hall (1984)
7. Edwards, A.W.F., Cavalli-Sforza, L.L.: A method for cluster analysis. Biometrics **21**, 362–375 (1965)
8. Matula, D.W.: Divisive versus agglomerative average linkage hierarchical clustering, Classification, a tool of research, éd. Gaul W. et Shader M., Elsevier, pp. 289–301 (1986)

9. Hastie, T., Tibshirani, R., Friedman, J.: The elements of statistical learning, Second edition, Springer Series in Statistics (2001)
10. Balcan, M.F., Liang, Y., Gupta, P.: Robust hierarchical clustering (2014)
11. Kerkeni, N., et al.: Classification des stades de sommeil par des réseaux de neurones artificiels hiérarchiques. IRBM **33**(1), 35–40 (2012)
12. Park, J.-Y., Kim, J.-H.: Online incremental hierarchical classification resonance network. Pattern Recognition (2021)

Detecting Malicious and Clean PowerShell Scripts Among Obfuscated Commands Using Deep Learning Methods and Word Embedding

Seda Kul(✉), Ahmet Manga, Zeki Esenalp, Rıdvan Kaplan, and Ahmet Sayar

Kocaeli University, Baki Komsuoğlu bulvarı No:515, Umuttepe 41001 Kocaeli, Turkey
seda.kul.06@gmail.com

Abstract. Microsoft PowerShell is a new generation command line application that comes installed on Windows machines by default. It provides an environment for programmers, system administrators, and software developers to directly access and use the .NET framework and the features provided by the Windows operating system. It has an important place in the Windows operating system ecosystem in many aspects, such as the ability of system administrators to automate the routine file deletion/creation, start/stop various services, etc., with the help of PowerShell commands. These commands can be designed easily without leaving any traces after running, and various code hiding methods (obfuscation) can circumvent cybersecurity products such as antivirus, endpoint detection, etc. Therefore, it is difficult to obtain information about PowerShell commands in forensic analysis after cyber incidents. This study proposes detecting malicious and clean PowerShell Commands on the Obfuscated Commands dataset, which is open to the public, with deep learning-based classification methods using a word embedding.

Keywords: Cyber security · Information security · PowerShell commands · C# · Natural language processing · Deep neural networks

1 Introduction

Today, Internet, Information, and Communication Technologies (ICTs) fastest growing areas of technical infrastructure development. Along with these developments, the fact that we have continuous internet access causes an increase in cybercrime carried out over the internet. Reports show the growing popularity of using PowerShell [1–3] in cyber attacks. PowerShell has been operating for performing different malicious activities such as gaining control or controlling the server in the system attacked by the person or malware. The difference and diversity of the use of PowerShell in malicious activities is an important consideration that cybersecurity experts must address. Some of the reasons why cybercriminals use PowerShell are as follows [2]: it comes installed on Windows operating systems, can run directly from memory, access is possible with encrypted

M. Ben Ahmed et al. (Eds.): SCA 2022, LNNS 629, pp. 859–868, 2023.
https://doi.org/10.1007/978-3-031-26852-6_79

traffic, works like scripts, can be easily obfuscated and difficult to detect with traditional security tools, many sandbox environments cannot analyze script based malware well.

Recent scientific development in Deep Learning (DL) [4–6] offers many new opportunities for cyber defense. One of the great inventions in DL is contextual embeddings in Natural Language Processing (NLP) processes. Many methods have been proposed in recent years to convert words to vectors [7–10]. Since PowerShell commands are also text data, Natural Language Processing techniques can detect malicious nonPowerShell scripts [12]. However, it is not easy to adapt NLP techniques to detect malicious commands because cyber attackers secretly obfuscate commands to avoid deliberate removal of commands [2].

Hendler et al. [14] proposed to use deep learning models to detect malicious PowerShell commands on their private datasets. Their dataset contains 66,388 data, of which 6,290 are malicious, and 60,098 are clean. Since their dataset set imbalanced, they duplicated every malicious command eight times. They applied different techniques, Natural Language Processing (NLP) based detectors and Convolution Neural Networks (CNN) for text classification. They used different architectures (9-layer CNN, 4-layer CNN, and long short-term memory network), detectors achieved high AUC levels between 0.985–0.990. They reached the best result with the CNN-based classifier with the NLP method. A blog published by FireEye [15] implemented a controlled classifier to detect malicious PowerShell commands that leveraged a tree based detector for PowerShell syntax.

Bohannon and Holmes [16] focused on obfuscated PowerShell scenarios. They offered a technique that uses basic character frequency analysis and the Cosine analogy to detect obfuscated PowerShell scripts. Their analysis was promising about preliminary results, and they indicated a notable difference between obfuscated and non-obfuscated codes. They published the "Invoke-Obfuscation" module created by Daniel Bohannon in 2016 publicly.

Although studies have been carried out on detecting malicious commands written in JavaScript [12, 17], PowerShell has not received much attention in the academic field despite its important position in cyber warfare. Companies such as Symantec [2] and Palo Alto Networks [1] have accomplished most of the work on PowerShell. These publications focus primarily on investigating the PowerShell threat rather than detecting malicious PowerShell activities and developing approaches. These shortcomings in researchers emphasize an urgent need to develop urgent methods to detect such attacks.

In this work, we propose Deep Learning-Based detection models with the NLP method to classify malicious or clean PowerShell commands. PowerShell commands are also handled as text data, and Word2Vec (W2V) [7] is used for this purpose. With our study, we contributed to the literature as follows: We examined and labeled the large realworld dataset published publicly. Using deep learning methods, we were able to detect PowerShell commands as clean or malicious with text-based classification.

The rest of the study is organized as follows. How the dataset is obtained and preprocessed is explained in Sect. 2. Word embeddings of PowerShell tokens given in Sect. 3. Classification method of clean and malicious PowerShell commands with deep learning based detectors using NLP methods is given in Sect. 4. Performance evaluation has been made in Sect. 5. In Sect. 6, conclusion is given.

2 The Dataset

PowerShell is a task automation solution that consists of a command-line shell and a scripting language that works across platforms that compatible with Windows, Linux, and macOS. As stated in the introduction to this paper, attackers can use PowerShell to develop powerful attacks, especially against Windows machines. Attackers have begun to use very simple obfuscation methods to shield the bulk of the command from PowerShell's command line arguments. We describe obfuscation as a set of techniques that modify source codes to make them difficult to understand for human or machine analysts. Studies [1–3] show that obfuscated codes can often be malicious. Malicious codes try to avoid getting caught by getting obfuscated. As commands are complicated, they can be difficult to detect in the AntiMalware Scan Interface (AMSI). Nor can we say malicious for every obfuscated PowerShell command.

For this reason, in our study, the data set created by Daniel Bohannon [16] for "In-voke-Obfuscation" in 2016 was used. There are 365,083 unlabeled scripts in total. Table 1 shows an example of malicious and clean PowerShell commands from our corpus (Table 1).

Table 1. Clean - malicious powershell command

Clean	Malicious
param([string] $infile = $(throw "Please specify a filename.")) $outfile = "$infile.unicode" get-content -Path $infile | out-file $outfile -encoding unicode	powershell.exe -w hidden -nop & "msiexec" url1 = gmail url2 = com /q /i http://office 365id.com/WpnUserService

It is important to reduce the false positives rate of the program and to detect frequently used commands to detect malicious codes. It is important to reduce the false positives rate of the program and to detect frequently used commands to detect malicious codes. Table 2 gives examples and usage of malicious commands. We labeled our dataset ac-cording to these explanations.

For example; The files were labeled malicious by containing suspicious commands such as encrypted command (obfuscated), downloading files, connecting to the web address. In order to label the data set, the commands most commonly used by malicious software were taken. These commands include general commands that communicate with the harmful command control center located outside to achieve persistence in the target system. With these commands, TCP, UDP connection can be opened to an address in the background, or by downloading and running another harmful command on the in-ternet with download string or other methods, it is possible to circumvent security prod-ucts such as antivirus, Endpoint Detection and Response (EDR). Thus, 39,023 commands were reviewed manually, 19,024 commands labeled as malicious, and 19,999 labeled as clean.

Table 2. Usage and examples of malicious commands

<table>
<tr><th>Explanation</th><th>Example</th></tr>
<tr><td>Using encrypted commands</td><td>“C:\Windows\System32\WindowsPowerShell\v1.0\powershell.exe” -Ex BYPasS -noP -W HIddEN -eC IAVwBtAHAAcgBpAHYAZQBsAGUAZwBlAC4AZQBgAZQAdI = =”</td></tr>
<tr><td>In order to circumvent security products; Use of capital, lowercase letters or signs such as “-,.”, then removing these marks with the Replace command</td><td>… $filename.Replace(‘-’,’/’)… $env:temp +’:’ + $name +’.exe…</td></tr>
<tr><td>Downloading files over the Internet</td><td>- Invoke-Expression ((“New-Object Net.WebClient”)).(‘Downloadfile’)…</td></tr>
<tr><td>Switches to listening on a port and waits for a connection when the Powershell command is executed</td><td>New-Object System.Net.HttpListener</td></tr>
<tr><td>Translating and executing the powershell command given as hex as a string</td><td>powershell.exe -e $(sEt-ITem ‘VARIAble:OFS’) + [sTrInG]((36,,61, 32,105,1115, 125) | foREAcH { ([Int] $_-As[char])}) +” $(set-Item ‘VaRIABLe:OFs’ ‘ ‘)”|& ($ENV:CoMSpec[2, 3, 8]-JoIN”)</td></tr>
<tr><td>Using shortened commands</td><td>-nop -w hidden -e < removed ></td></tr>
<tr><td>Execution of malicious commands in ASCII format</td><td>sySTem.TExT.ENcodInG]::asCii</td></tr>
<tr><td>Reading, adding, deleting critical values on the registry</td><td>Get-Item -path HKLM:\SOFTWARE\Microsoft\Windows\CurrentVersion\Run</td></tr>
</table>

2.1 Data Pre-processing

In order to improve detection and evaluation results in the classification model, the data were pre-processed. The steps are as follows:

- PowerShell commands that have been encoded are decoded.
- Fields in the code, such as URLs, are found and extracted using the Named Entity Recognition (NER) model.
- Special characters, punctuation marks (! “# $% & ‘() * +, -. /:; < = > ? @ [\] ^ _‘ {|} ~) are removed.
- Numbers for example as date, IP address are removed.

- Comment lines in the code are removed (lines begins with // are detected)
- Finally, all characters are converted to lowercase characters.

3 Deep Learning Based Classification Using Word Embedding

With a background in architectures and NLP techniques, this section presents the deep learning models we used to detect malicious PowerShell commands. We evaluate and compare the results of Deep Neural Network (DNN), Convolutional Neural Network (CNN), Recurrent Neural Network (RNN), Long Short-Term Memory (LSTM), and Gated Recurrent Unit (GRU) architectures which are given in Sect. 4. Figure 1 gives the flow diagram of the system.

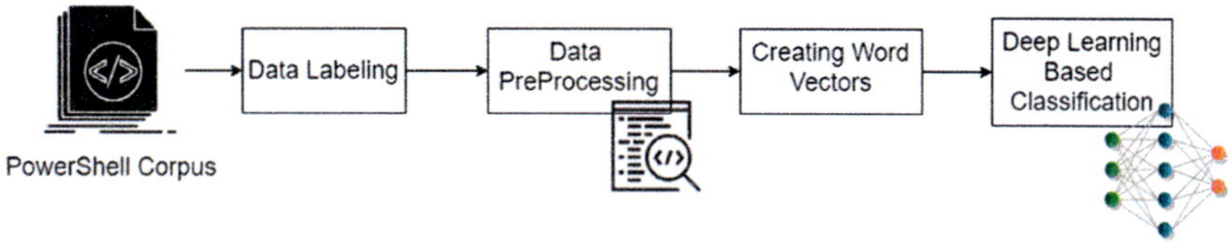

Fig. 1. Deep learning based malicious and clean powershell detection

Since PowerShell commands are also text data, we must first represent words as vectors and present them to the network before classifying them.

3.1 Word Embedding

Input vectors are used in machine learning models. The first step in dealing with text data is to transform it to numerical vectors known as W2V [7–10]. Since PowerShell commands are text data, the same procedure should be followed. One of the most efficient ways to display text data is to use word embedding. It represents not only text data but also contextual information also captures similarity in meaning. Unlike one-hot encoding, which is hardcoded, word embedding represents words as dense word vectors. By using Word Embedding instead of One-Hot encoding, the curse of dimensionality was avoided.

A text corpus is vectorized into a list of integers using the Tokenizer utility. The vocabulary words serve as the dictionary's keys, and each integer corresponds to a value in a dictionary that encodes the entire corpus. With the num words parameter set to 400, which we defined as the vocabulary size.

3.2 Deep Neural Networks

A neural network's main goal is to take a collection of inputs, apply activation func-tions to them, and then output results to solve problems like classification. When com-pared to neural networks, the number of hidden layers in deep learning can be very high.

There are three layers in the Deep Neural Networks model we use, as shown in Table 3; the Embedding, Flatten and Dense layers. There are two services provided by the Embedding Layer. First, they minimize the input's dimensionality. Second, they represent the input in such a way that its meaning is preserved. Each input identifier (usually a word or character, depending on the problem at hand) is converted into a vector representation by the Embedding layer. Starting with the last dimension, the Flatten layer unrolls the values. For example, in our case, when flatten is applied to the Embedding layer with an input shape of [5583, 50], the layer's output shape becomes [279150].

Table 3. Deep neural network architecture

Layer	Shape
Embedding	(5583, 50)
Flatten	(279150)
Dense	1

For calculating the loss function, we use binary cross entropy. Which computes the following average:

$$Loss = -\frac{1}{\substack{output \\ size}} \sum_{i=1}^{\substack{output \\ size}} \left(y_i . \log \hat{y}_i + (1 - yi) \cdot log\left(1 - \hat{y}_i\right)\right) \quad (1)$$

where ŷ is the model output's i-th scalar value, y is the target value.

3.3 Convolution Neural Networks

Instead of concentrating on feature extraction, which is what other neural networks do, CNN takes input as a two-dimensional array and operates directly on the images. CNN's dominant approach provides solutions to identification issues. A one-dimensional convolutional layer is used when dealing with sequential data like text. The rectified linear activation function (ReLU) is used, a linear function that, if the input is positive, outputs the input directly; otherwise, it outputs zero. Binary cross entropy is used for Loss Function. To evaluate the results, we designed two different CNN architectures, with 5 and 7 layers.

The 5-layer CNN architecture consisting of 1 Embedding, 1 Convolutional, 1 Global Max Pooling, and 2 Dense Layers. 7-layer CNN architecture consisting of 1 Embedding, 3 Convolutional, 1 Global Max Pooling, and 2 Dense Layers with adam optimizer as shown in Table 4.

Table 4. CNN architecture

5-layer CNN Architecture	
Layer	Shape
Embedding	(37839, 100)
Conv1D	(37835, 128)
Global_max_pooling1d	128
Dense	10
Dense	1
7-layer CNN Architecture	
Layer	Shape
Embedding	(5583, 50)
Conv1D	(5579, 128)
Conv1D	(5575, 64)
Conv1D	(5571, 64)
Global_max_pooling1d	64
Dense	5
Dense	1

3.4 Recurrent Neural Network, Long Short-Term Memory (LSTM), and Gated Recurrent Unit (GRU)

A recurrent neural network (RNN) is a type of artificial neural network in which nodes are connected in a directed graph that follows a temporal series. This enables it to behave in a temporally complex manner. The recurrent layer of all RNNs has feedback loops which allows them to keep details in memory for a long time. The gradient of the loss function, on the other hand, decays exponentially over time (the vanishing gradient problem). LSTM networks are a form of RNN that employs a combination of special and regular units. A memory cell' is used in LSTM modules, which can store information for long pe-riods of time. When input enters the memory, a series of gates is used to monitor output. They will learn longer-term dependencies according to this architecture. GRUs are similar to LSTMs, but they have a more straightforward structure. They use gates to monitor in-formation flow as well, but they don't use different memory cells and have fewer gates.

To evaluate the results, we designed 3 layer RNN architecture consisting of Embedding, RNN and Dense layers. 5-layer GRU architecture, and 3-layer LSTM architecture shown in Table 5.

Table 5. RNN, GRU and LSTM Architecture

Architecture	Model Layer Information	Additional Information
3-layer RNN	Embedding, RNN, Dense Layers	Shape = 128
5-layer GRU	Embedding, 3 GRU, Dense Layers	Return Sequences = True Shape is 16, 8, and 4 respectively
3-layer LSTM	2 LSTM, 1 Dense	Unit Number = 50, Return Sequences = True

4 Evaluation

In this section, we evaluate how classification algorithms work. Algorithms were evaluated with accuracy and loss metrics given on training and test data. The labeled data set was divided into 75%–25% training-test data. All our experiments were performed on a machine with Intel® Core™ i9-9900KF CPU @ 3.60GHz × 16, GeForce RTX 2080 Ti/PCIe/SSE2. In total, 100 epochs were carried out with 32 batch sizes. Model evaluation of training and test data is given in Table 6. Models' history is given in Fig. 2.

Table 6. Performance results

Deep Learning Architecture	Training Accuracy	Test Accuracy	Average Training Time Per Epoch
3-Layer DNN	0.9991	0.9092	4s 132 us/step
5-Layer CNN	**0.9997**	**0.9667**	12 s 409 us/step
7-Layer CNN	**0.9997**	0.9661	22s 745 us/step
3-Layer RNN	0.9185	0.9029	557s 19 ms/step
5-Layer GRU	0.9982	0.9555	610s 209 ms/step
3-Layer LSTM	0.9597	0.7589	5591s 328ms/step

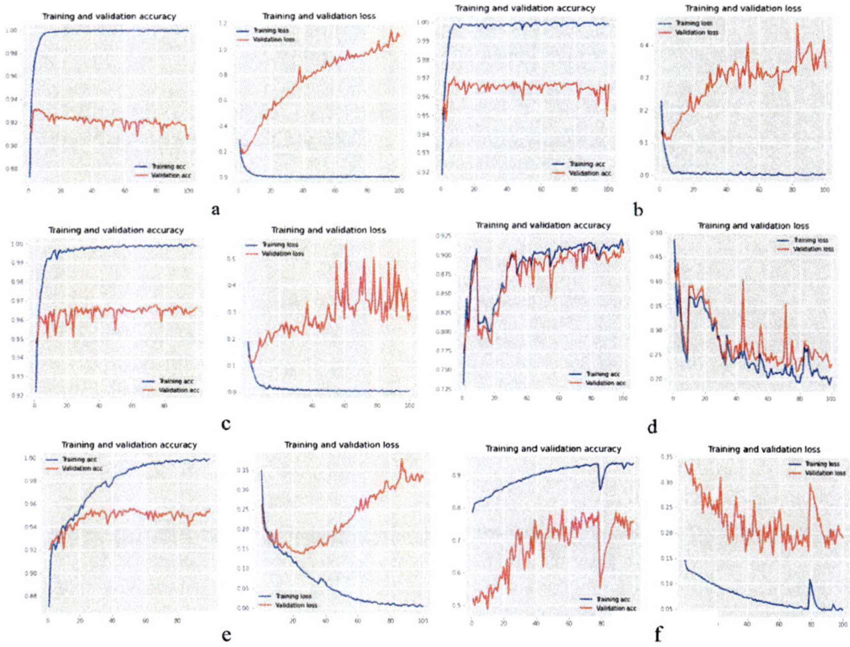

Fig. 2. The history of models shown as follows: (a) 3-Layer Neural Network Deep Learning, (b) 5-Layer CNN, (c) 7-Layer CNN, (d) 3-Layer RNN, (e) 5-Layer GRU, (f) 3-Layer LSTM

5 Conclusions

In this study, the detection of malicious PowerShell commands that target organization such as Security Services, Health Services, Educational Services were examined. Various different deep learning-based classifiers DNN, CNN, RNN, LSTM, GRU, and LSTM, to determine commands were examined and evaluated. Our evaluation results show that deep learning-based classifiers perform high. The best performance was achieved with the 5-Layer CNN classifier. In future studies, the progress of the study will be provided by in-creasing the size of the dataset.

References

1. PaloAlto. Pulling Back the Curtains on EncodedCommand PowerShell Attacks (2017). https://www.researchcenter.paloaltonetworks.com/2017/03/unit42-pulling-back-the-curtains-on-encodedcommand-powershell-attacks/
2. Symantec. The increased use of Powershell in attacks (2016). https://www.symantec.com/content/dam/symantec/docs/security-center/white-papers/increased-use-of-powershell-in-attacks-16-en.pdf
3. FireEye, Malicious PowerShell Detection via Machine Learning (2018)
4. Goodfellow, I.J., Bengio, Y., Courville, A.C.: Deep Learning, Ser. Adaptive Computation and Machine Learning. MIT Press, Cambridge (2016). [Online]. Available: http://www.deeplearningbook.org/

5. LeCun, Y., Bengio, Y., Hinton, G.: Deep learning. Nature **521**(7553), 436–444 (2015)
6. Schmidhuber, J.: Deep learning in neural networks: an overview. Neural Netw. **61**, 85–117 (2015)
7. Mikolov, T., Sutskever, I., Chen, K., Corrado, G.S., Dean, J.: Distributed representations of words and phrases and their compositionality. In: Advances in Neural Information Processing Systems (NIPS), pp. 3111–3119 (2013)
8. Pennington, J., Socher, R., Manning, C.: GloVe: global vectors for word representation. In: Proceedings of the 2014 Conference on Empirical Methods in Natural Language Processing (EMNLP), pp. 1532–1543 (2014)
9. Bojanowski, P., Grave, E., Joulin, A., Mikolov, T.: Enriching word vectors with subword information. Trans. Assoc. Comput. Linguist. **5**, 135–146 (2017)
10. Devlin, J., Chang, M.-W., Lee, K., Toutanova, K.: BERT: pre-training of deep bidirectional transformers for language understanding. arXiv preprint arXiv:1810.04805 (2018)
11. Liu, B., Zhang, L.: A survey of opinion mining and sentiment analysis. In: Aggarwal, C.C., Zhai, ChengXiang (eds.) Mining Text Data, pp. 415–463. Springer US, Boston, MA (2012). https://doi.org/10.1007/978-1-4614-3223-4_13
12. Wang, Y., Cai, W.-D., Wei, P.-C.: A deep learning approach for detecting malicious javascript code. Secur. Commun. Netw. **9**(11), 1520–1534 (2016)
13. Zhang, X., Zhao, J., LeCun, Y.: Character-level convolutional networks for text classification. In: Advances in Neural Information Processing Systems, pp. 649–657. NIPS (2015)
14. Hendler, D., et al.: Detecting malicious powerShell commands using deep neural networks. In: Proceedings of the 2018 on Asia Conference on Computer and Communications Security, pp. 187–197. ACM (2018)
15. Fang, V.: Malicious PowerShell Detection via Machine Learning (2018). https://www.fireeye.com/blog/threat-research/2018/07/malicious-powershell-detection-via-machine-learning.html
16. Bohannon, D., Holmes, L.: Revoke-Obfuscation: PowerShell Obfuscation Detection Using Science (2017). https://www.fireeye.com/blog/threat-research/2017/07/revoke-obfuscation-powershell.html
17. Saxe, J., Berlin, K.: expose: a character-level convolutional neural network with embeddings for detecting malicious urls, file paths and registry keys. arXiv preprint arXiv:1702.08568 (2017)
18. Fukushima, K., Miyake, S.: Neocognitron: a self-organizing neural network model for a mechanism of visual pattern recognition. In: Amari, S.-I., Arbib, M.A. (eds.) Competition and Cooperation in Neural Nets: Proceedings of the U.S.-Japan Joint Seminar held at Kyoto, Japan February 15–19, 1982, pp. 267–285. Springer Berlin Heidelberg, Berlin, Heidelberg (1982). https://doi.org/10.1007/978-3-642-46466-9_18
19. Elman, J.L.: Finding structure in time. Cogn. Sci. **14**(2), 179–211 (1990)

Topological 3D Spatial Interpolation Based on the Interval-Valued Homotopy Continuation

Ali Jamali[1(✉)], Francesc Antón Castro[2], and İsmail Rakıp Karaş[3]

[1] Civil Engineering Department, Faculty of Engineering, University of Karabük, Karabük, Turkey
alijamali@karabuk.edu.tr
[2] School of Mathematics and Information Technology, Yachay Tech University, Urcuqui, Ecuador
fcastro@yachaytech.edu.ec
[3] Department of Computer Engineering, Faculty of Engineering, Karabük University, Karabük, Turkey
ismail.karas@karabuk.edu.tr

Abstract. Estimating unknown values using its surrounding measured values is called spatial interpolation, a vital tool for estimating continuous spatial data such as the earth's surface. Construction of the Digital Elevation Model is one of the most common applications of spatial interpolation methods. There are various global and local interpolation techniques, including Kriging, Inverse Distance Weighted (IDW), Thiessen polygons (TIN), Natural Neighbor (NN), and Spline interpolation. This paper introduces the interval-valued homotopy continuation for 3D spatial data interpolation. Straight lines or algebraic curves can be reconstructed using homotopy continuation between any pairs of 3D data. The novel method of the interval-valued homotopy to restore the topographic surface between spatial data is developed in MATLAB programming language. For a dataset of ASTER GDEM, the presented mathematical algorithm shows better results compared to TIN and IDW methods in terms of Mean Squared Error, Mean Absolute Error, and Root Mean Squared Error with values of 5.2897, 1.53, and 2.299 m, respectively.

Keywords: Homotopy continuation · 3D data interpolation · DEM · Spatial interpolation · 3D GIS

1 Introduction

Surface reconstruction is one of the old issues and yet challenging in many fields, including computer science [1]. The surface of the earth is described mathematically as Digital Elevation Models (DEM) [2]. With field surveys, photogrammetry techniques, and radar, elevation data is collected and interpolated into a 3D surface. There are various global and local interpolation methods for the estimation of spatial data with the concern of the earth's surface. Two well-known spatial data interpolation, including Inverse Distance

M. Ben Ahmed et al. (Eds.): SCA 2022, LNNS 629, pp. 869–879, 2023.
https://doi.org/10.1007/978-3-031-26852-6_80

Weighting (IDW) and Kriging, is commonly used for spatial interpolation. Methods of spatial interpolation are formularized as a weighted mean of input data (see Eq. 1).

$$\hat{z}(x_0) = \sum\nolimits_{(i=1)}^{n} \lambda_i z(x_i) \tag{1}$$

where the number of measured/known points used for the estimation is equal to n, $\hat{z}$ is the approximated value (unknown point) of the point of interest x_0, the measured value at the known point x_i is z, the weight belonging to the known/measured point is defined by λ_i. The spatial interpolation methods' approximation depends on the sampling design, sample size, and data properties. For a given input dataset, a suitable spatial interpolation algorithm is difficult to select. There are no consistent results on the accuracy of the spatial interpolators affected by the mentioned factors [3].

In this study, we are only interested in the topographic surface, as seen as the mathematical hypersurface defining the z coordinate as a function of the x and y coordinates. From this perspective, we are working in a geometric space rather than a geographic space. For this application, we only need the geometric and topological concepts related to the 3D Euclidean affine space, which has its own mathematical richness, different from the richness of the geographic space. Photogrammetry can be seen as a direct application of the mathematical concept of projective space and projective geometry. In this research, a mathematical method called the interval-valued homotopy continuation for data interpolation is presented. The presented method has better results in terms of Root Mean Squared Error (RMSE), Mean Absolute Error (MAE), Mean Squared Error (MSE), and correlation coefficient over two well-known methods, including IDW and TIN.

2 Study Area

Advanced Spaceborne Thermal Emission and Reflection Radiometer Global Digital Elevation Map Version 2 (ASTER GDEM 2) data with the longitude of 645541.86:646535.15 and latitude of 3279321.80:3280229.96 in the WGS84/UTM Zone 39 coordinate reference is used in this study (see Fig. 1). ASTER GDEM is a free spatial dataset that was released in October 2011. ASTER GDEM is a product of The National Aeronautics and Space Administration (NASA) and the Japanese Ministry of Economy, Trade, and Industry (METI). The study area is in Shiraz city, Iran. Shiraz is located in a green plain at the foot of the Zagros Mountains, 4,900 feet (1,500 m) above sea level in the south of Iran.

Dataset has a cell size of 30 m with a minimum elevation of 1568 m and a maximum elevation of 1721 m with a mean value of 1607.38 m.

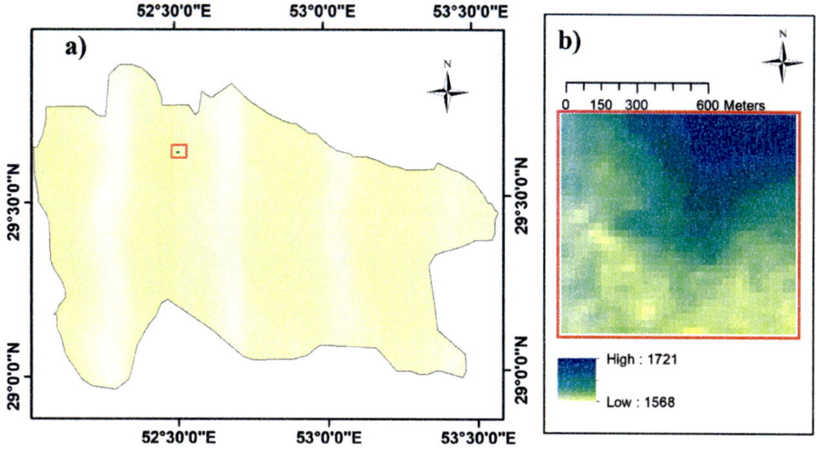

Fig. 1. Study area a) Shiraz city b) Elevation data extracted from ASTER GDEM-2.

3 Interval-Valued Homotopy Continuation

For a dataset of elevations to model the earth's surface, as there is no discontinuity in the n-dimensional space corresponding to the space of real ground, homotopy continuation as a continuous function is used. A homotopy is estimated by $H_{ij}(x, y, \lambda) = (1 - \lambda)f_{ij}(y) + \lambda f_{(i+1,j)}(y)$ where $\lambda = \frac{x - x_i}{x_{i+1} - x_i}$, for two consecutive elevation $z_{ij} = f_{ij}(x, y)$ and $z_{(i+1,j)} = f_{(i+1,j)}(x, y)$. For a set of a combination of linear functions according to parameter λ, H_{ij} is a linear homotopy that transforms $z_{ij} = f_{ij}(x, y)$ into $z_{(i+1,j)} = f_{(i+1,j)}(x, y)$.

For the discussed application of spatial interpolation to model earth's surface, homotopy is employed for the extraction of a topographic surface, which is the graph of a function of the two planimetric variables x and y with the value the z coordinate (elevation). By homotopy, functions whose hypersurface are homeomorphic are considered (using a bijective continuous function, they are continuously deformed, whose inverse is continuous too) to the topographic surface of the study. A function such as a path or a geometric figure is continuously deformed into another path or figure using a homotopy continuation [1, 4]. Having two continuous functions of f_0 and f_1 from a topological space X to a topological space Y, a homotopy is a continuous map $H : X \times [0, 1] \rightarrow Y$ from the topological space X with the unit interval [0, 1] to Y where $H(x, 0) = f_0$, and $H(x, 1) = f_1$, $x \in X$. f_0 is the initial map and f_1 is the terminal map, and λ is the homotopy parameter. If there is a homotopy H taking f_0 to f_1, the two functions f_0 and f_1 are homotopic. Linear and non-linear homotopy are defined by (see Eqs. 2 and 3):

$$H(x, \lambda) = (1 - \lambda)f_0(x) + \lambda f_1(x) \text{ where } \lambda \in [0, 1] \tag{2}$$

$$H(x, \lambda, n) = (1 - \lambda)^n f_0(x) + \lambda^n f_1(x) \text{ where } \lambda \in [0, 1] \tag{3}$$

λ just evolves between 0 and 1, at $\lambda = 0$, the initial map f_0, and at λ=1, the terminal map f_1 are obtained where the equation that presents the homotopy depends on the input of the homotopy, not on the output of the homotopy. The interval-valued homotopy is constructed for a given n + 1 pair of data points (x_i, y_j), i = 0,1,2, …., n-1, j = 0,1,2, …., m-1, as follows (see Eq. 4):

$$H(x, y, \lambda) = \begin{cases} H_{11}(x, y, \lambda), & \text{if } x_0 \le x < x_1 \text{ and } y_0 \le y < y_1 \\ H_{12}(x, y, \lambda), & \text{if } x_0 \le x < x_1 \text{ and } y_1 \le y < y_2 \\ \vdots \\ H_{1,m-1}(x, y, \lambda), & \text{if } x_0 \le x < x_1 \text{ and } y_{m-2} \le y \le y_{m-1} \\ H_{21}(x, y, \lambda), & \text{if } x_1 \le x < x_2 \text{ and } y_0 \le y \le y_1 \\ \vdots \\ H_{2,m-1}(x, y, \lambda), & \text{if } x_1 \le x < x_2 \text{ and } y_{m-2} \le y \le y_{m-1} \\ \vdots \\ H_{n-1,1}(x, y, \lambda), & \text{if } x_{n-2} \le x < x_{n-1} \text{ and } y_0 \le y < y_1 \\ \vdots \\ H_{n-1,m-1}(x, y, \lambda), & \text{if } x_{n-2} \le x \le x_{n-1} \text{ and } y_{m-2} \le y \le y_{m-1} \end{cases} \tag{4}$$

where $H_{ij}(x, y, \lambda)$ can be in the same degree of a polynomial or in different degrees. As far as $H_{ij}(x, y, \lambda)$ is continuous, it can be any function. $H_{ij}(x, y, \lambda)$ are cubic polynomials for this study. Cubic interpolation and linear interpolation are used as two continuous interpolation algorithms. Cubic interpolation is in C^2, and for the interpolation, it requires at least four points whereas linear interpolation requires at least two points where it is in C^0. First, we constructed elevation values (z_{ij}) as a function of y, $z_{ij} = C_{ij}(y)$ using cubic splines (see Eq. 5)

$$C_{ij}(y) = a_{ij} + b_{ij}(y - y_{ij}) + c_{ij}(y - y_{ij})^2 + d_{ij}(y - y_{ij})^3$$
$$, i = 0, 1, \ldots, n-1, \ j = 0, 1, 2, 3, \ldots., m-1.$$

$$C(y) = \begin{cases} C_{11}(x, y), & \text{if } x_0 \le x < x_1 \text{ and } y_0 \le y < y_1 \\ C_{12}(x, y), & \text{if } x_0 \le x < x_1 \text{ and } y_1 \le y < y_2 \\ \vdots \\ C_{1,m-1}(x, y), & \text{if } x_0 \le x < x_1 \text{ and } y_{m-2} \le y \le y_{m-1} \\ C_{21}(x, y), & \text{if } x_1 \le x < x_2 \text{ and } y_0 \le y \le y_1 \\ \vdots \\ C_{2,m-1}(x, y), & \text{if } x_1 \le x < x_2 \text{ and } y_{m-2} \le y \le y_{m-1} \\ \vdots \\ C_{n-1,1}(x, y), & \text{if } x_{n-2} \le x < x_{n-1} \text{ and } y_0 \le y < y_1 \\ \vdots \\ C_{n-1,m-1}(x, y), & \text{if } x_{n-2} \le x \le x_{n-1} \text{ and } y_{m-2} \le y \le y_{m-1} \end{cases} \tag{5}$$

where a, b, c, and d are coefficients. Considering $\lambda = \frac{x - x_i}{x_{i+1} - x_i}$, in the next phase value of each z is estimated as $z_{ij} = H_{ij}(x, y, \lambda) = (1 - \lambda)f_{ij}(y) + \lambda f_{(i+1,j)}(y)$ where

$\lambda \in [0, 1]$, $f_{ij}(y) = a_{ij} + b_{ij}(y - y_{ij}) + c_{ij}(y - y_{ij})^2 + d_{ij}(y - y_{ij})^3$ and $f_{(i+1,j)}(y) = a_{(i+1,j)} + b_{(i+1,j)}(y - y_{(i+1,j)}) + c_{(i+1,j)}(y - y_{(i+1,j)})^2 + d_{(i+1,j)}(y - y_{(i+1,j)})^3$. Results of interpolation using the IDW, TIN, and Ordinary Kriging are presented in Fig. 2. Figure 3 illustrates the results of interval-valued homotopy continuation method.

For the IDW method, distance coefficient p is equal to 2; in Ordinary Kriging, the semi-variogram model is linear, and for the TIN method, the interpolation method is linear.

In summary, the computation procedure pseudo-code of Homotopy continuation is shown in pseudo-code 1.

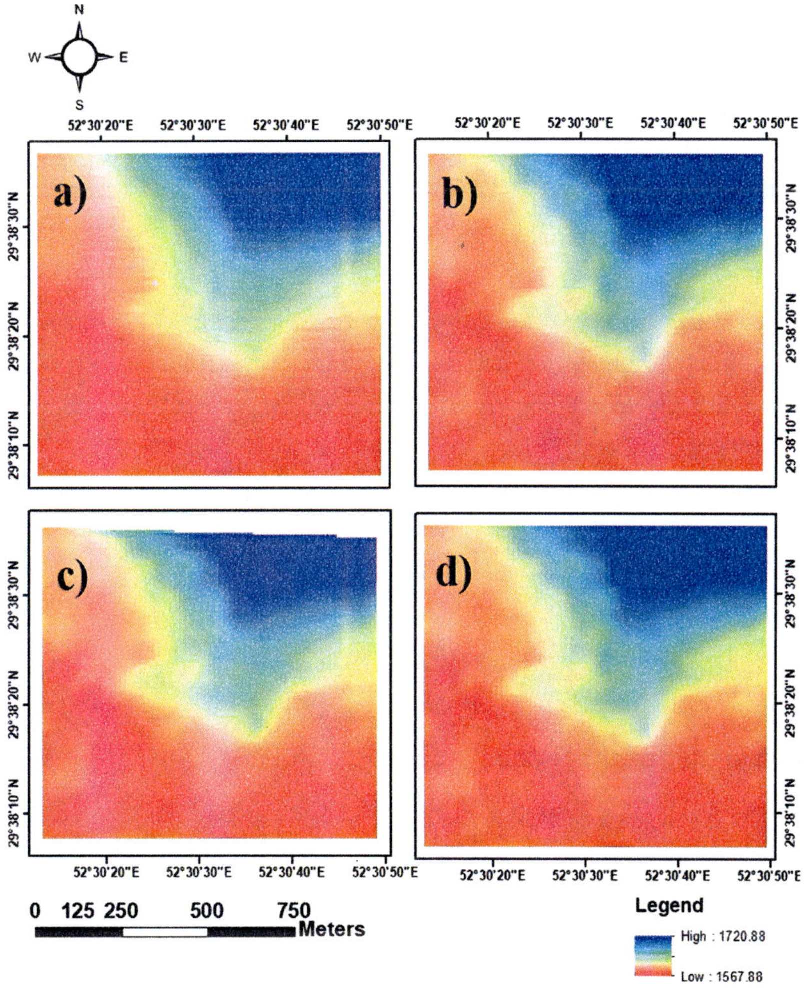

Fig. 2. Interpolation results using a) IDW b) Ordinary Kriging c) TIN d) the interval-valued homotopy continuation algorithms.

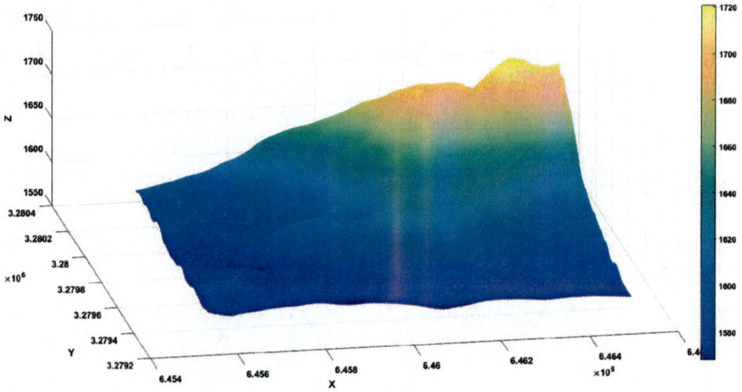

Fig. 3. Results of interpolation using the interval-valued homotopy continuation.

Pseudo-code 1 Computation procedure pseudo-code of Homotopy continuation.

Offline Computation

✓Collect the surface data.

✓Specify the spline and Homotopy directions.

✓Construct the appropriate cubic splines in the 1st direction.

✓Establish the Homotopy relations between the splines along the 2nd direction.

✓Summarize the 2-D piecewise formulas for spanning the whole area.

Employment

✓Get the x, y position.

✓Compute the z value using the piecewise formula.

4 Evaluation of the Results

Based on the validation data consisting of 88 randomly selected points, the accuracy of spatial interpolations methods is evaluated in terms of MAE, RMSE, MSE, and the correlation coefficient (see Eqs. 6–9 and Fig. 4).

$$RMSE = \sqrt{\frac{\sum_{i=1}^{n} (E_{obs,i} - E_{interpolated,i})^2}{n}} \tag{6}$$

$$MSE = \frac{\sum_{i=1}^{n} (E_{obs,i} - E_{interpolated,i})^2}{n} \tag{7}$$

$$MAE = \frac{\sum_{i=1}^{n} |E_{obs,i} - E_{interpolated,i}|}{n} \quad (8)$$

$$R = \frac{\sum_m \sum_n (A_{mn} - \overline{A})(B_{mn} - \overline{B})}{\sqrt{(\sum_m \sum_n (A_{mn} - \overline{A})^2)(\sum_m \sum_n (B_{mn} - \overline{B})^2)}} \quad (9)$$

where E_{obs} is the ground truth elevation data and $E_{interpolated}$ is the estimated elevation from the algorithms used in this research.

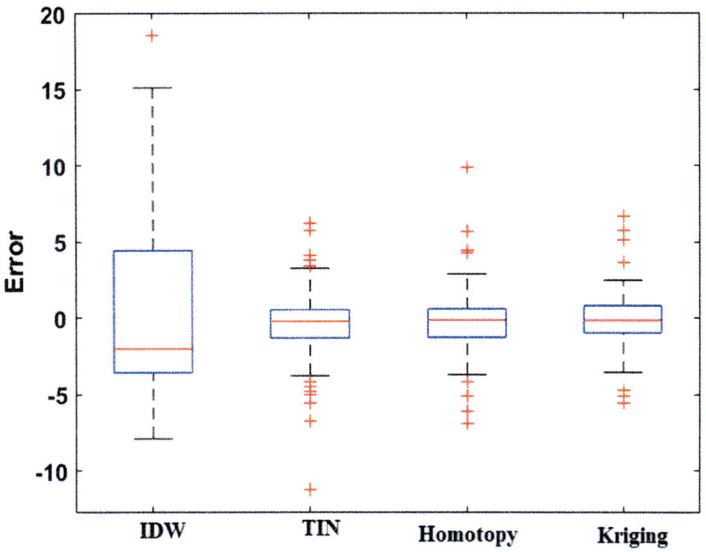

Fig. 4. The error of the interpolation algorithms (in meters).

The interval-valued homotopy continuation with values of 2.299, 1.53, and 5.2897 m in terms of RMSE, MAE, and MSE achieved better results compared to the IDW and TIN techniques. However, in terms of RMSE, MAE, and MSE, the Ordinary Kriging showed the best results with values of 1.9465, 1.35, and 3.7887 m, respectively (see Table 1 and Figs. 5, 6, 7 and 8). Based on the computed correlation coefficient (R), Ordinary Kriging obtained the highest value of 0.9986. On the other hand, the interval-valued homotopy continuation with an R value of 0.9981 achieved better results than that of IDW and TIN methods (with values of 0.9975 and 0.9976).

Table 1. Comparison of the results of elevation interpolations for the validation dataset (in meters).

Interpolation methods	Homotopy	IDW	TIN		Ordinary kriging
RMSE	2.299	5.7391	2.5134		1.9465
MSE	5.2897	32.9369	6.3174		3.7887
MAE	1.53	4.70	1.66		1.35
R	0.9981	0.9975	0.9976		0.9986

The IDW method obtained a mean error value of 0.53 m, a median error value of -2.02 m, a standard deviation value of 5.75 m, a minimum error value of -7.87 m, and a maximum error value of 18.56 m (see Fig. 5).

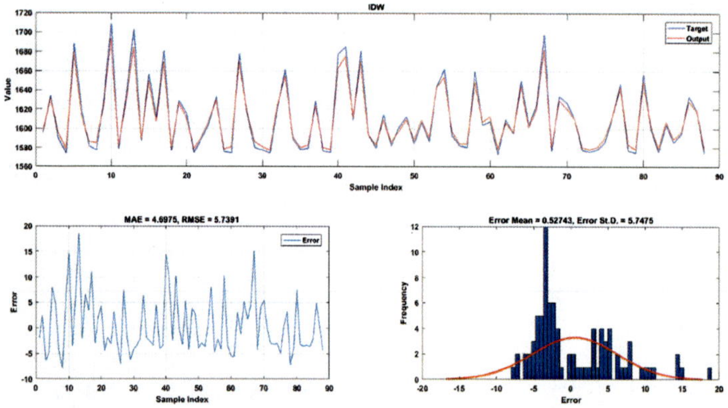

Fig. 5. The error for the estimation of evaluation dataset using IDW (in meters).

The TIN method obtained a mean error value of -0.41 m, a median error value of -0.20 m, a standard deviation value of 2.49 m, a minimum error value of -11.24 m, and a maximum error value of 6.22 m (see Fig. 6).

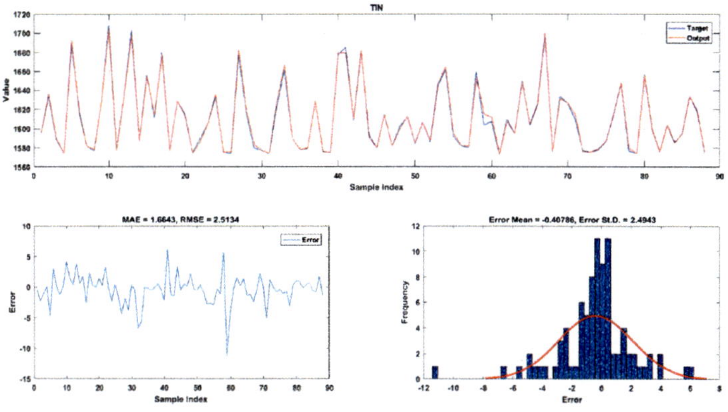

Fig. 6. The error for the estimation of evaluation dataset using TIN (in meters).

For elevation interpolation, the Ordinary Kriging method had a mean error value of -0.18 m, a median error value of -0.16 m, a standard deviation value of 1.95 m, a minimum error value of -5.55 m, and a maximum error value of 6.72 m (see Fig. 7).

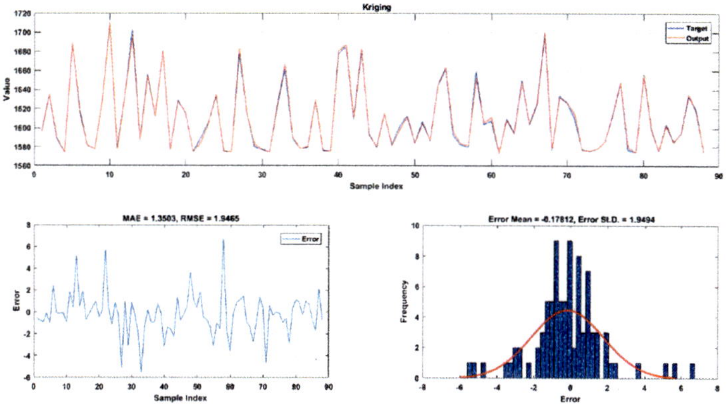

Fig. 7. The error for the estimation of evaluation dataset using Ordinary Kriging (in meters).

The proposed Homotopy continuation method obtained a mean error value of -0.24 m, a median error value of -0.14 m, a standard deviation value of 2.30 m, and a minimum error value of -6.86 m, and a maximum error value of 9.86 m (see Fig. 8).

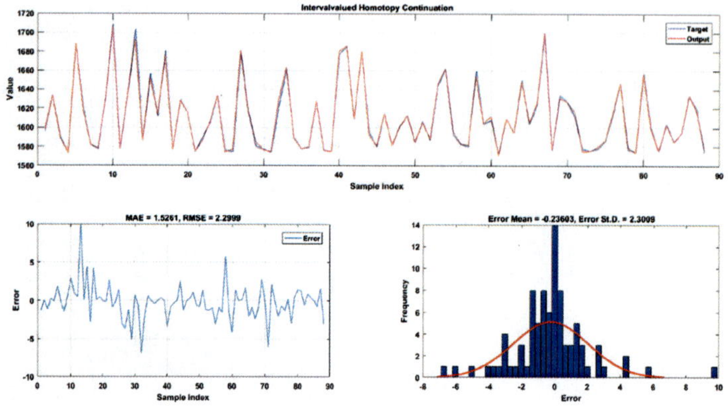

Fig. 8. The error for the estimation of evaluation dataset using the interval-valued homotopy continuation (in meters).

5 Conclusions

In this study, for spatial data interpolation, the interval-valued homotopy is developed in MATLAB, which is evaluated against IDW, Ordinary Kriging, and TIN methods. The interval-valued homotopy showed better results than that of IDW and TIN methods in terms of RMSE, MAE, and MSE. Based on the results for the validation dataset used in this research which consists of randomly selected 88 points of interest, the interval-valued homotopy continuation with the values of 2.299, 1.53, and 5.2897 m for RMSE, MAE, and MSE respectively and value of 0.9981 for the correlation coefficient achieved better results compared to two well-known interpolation algorithms of IDW and TIN. The applications that this method can be beneficial to GIS are as follows:

1. Realistic DTM modeling with cliffs (vertical or slanted), linear or non-linear depression boundaries, thalwegs, and crest lines: they can be achieved through this method of topological DTMs. This is a significant advantage over existing techniques in GIS, which cannot handle non-linear topographic discontinuities and slanted cliffs. This method can also be applied to 3D cityscapes and videogames, offering more realism than existing techniques.

2. Interpolation of other physical variables than elevation (e.g., temperature, atmospheric pressure, salinity, water depth, percentage of vegetal species, percentage of tree species, rock composition percentage, different land usage percentages, etc.) can benefit GIS users to produce specialized maps such as weather maps, bathymetry maps, vegetation maps, geological maps, or land usage maps, etc.

3. The controls on the accuracy provided by the interval-valued homotopy allow GIS users to be able to plot the uncertainty of the DTM models or other thematic maps, which can be of great interest in map fusion. This can also be of benefit in change detection in 3D GIS. The interval-valued homotopy can also allow GIS users to explore the geomorphological changes that affect the 3D DTMs. Homotopy can enable GIS users to visualize such geomorphological changes. Other changes that could benefit

from these techniques are urban developments, city evolution, regional evolution, which are of interest in urbanism and land and urban planning or reorganization.

Conflicts of Interest:. The authors declare no conflict of interest.

References

1. Sharma, O., Anton, F.: Homotopy-based surface reconstruction with application to acoustic signals. Vis. Comput. **27**, 373–386 (2011). https://doi.org/10.1007/s00371-011-0544-4
2. Salekin, S., Burgess, J.H., Morgenroth, J., Mason, E.G., Meason, D.F.: A comparative study of three non-geostatistical methods for optimising digital elevation model interpolation. ISPRS Int. J. Geo-Inf. 7, (2018). https://doi.org/10.3390/ijgi7080300
3. Li, J., Heap, A.D.: A review of comparative studies of spatial interpolation methods in environmental sciences: performance and impact factors. Ecol. Inf. **6**, 228–241 (2011). https://doi.org/10.1016/j.ecoinf.2010.12.003
4. Jamali, A., Antón Castro, F., Mioc, D.: A novel method of combined interval analysis and homotopy continuation in indoor building reconstruction. Null **51**, 520–536 (2019). https://doi.org/10.1080/0305215X.2018.1472253

Author Index

M. Ben Ahmed et al. (Eds.): SCA 2022, LNNS 629, pp. 881–884, 2023.
https://doi.org/10.1007/978-3-031-26852-6